D0075895

PROBABILITY AND STATISTICS FOR MODERN ENGINEERING

THE DUXBURY ADVANCED SERIES IN STATISTICS AND DECISION SCIENCES

Applied Nonparametric Statistics, Second Edition, Daniel
Applied Regression Analysis and Other Multivariable Methods, Second Edition, Kleinbaum, Kupper, and Muller
Classical and Modern Regression with Applications, Second Edition, Myers
Elementary Survey Sampling, Fourth Edition, Scheaffer, Mendenhall, and Ott
Introduction to Contemporary Statistical Methods, Second Edition, Koopmans
Introduction to Probability and Its Applications, Scheaffer
Introduction to Probabilitv and Mathematical Statistics, Bain and Engelhardt
Linear Statistical Models, Bowerman and O'Connell
Probability Modeling and Computer Simulation, Matloff
Quantitative Forecasting Methods, Farnum and Stanton
Time Series Analysis, Cryer
Time Series Forecasting: Unified Concepts and Computer Implementation, Second Edition, Bowerman and O'Connell

THE DUXBURY SERIES IN STATISTICS AND DECISION SCIENCES

Applications, Basics, and Computing of Exploratory Data Analysis, Velleman and Hoaglin
A Course in Business Statistics, Second Edition, Mendenhall
Elementary Statistics, Fifth Edition, Johnson
Elementary Statistics for Business, Second Edition, Johnson and Siskin
Essential Business Statistics: A Minitab Framework, Bond and Scott
Fundamentals of Biostatistics, Third Edition, Rosner
Fundamentals of Statistics in the Biological, Medical, and Health Sciences, Runyon
Fundamental Statistics for the Behavioral Sciences, Second Edition, Howell
Introduction to Probability and Statistics, Seventh Edition, Mendenhall
An Introduction to Statistical Methods and Data Analysis, Third Edition, Ott
Introductory Business Statistics with Microcomputer Applications, Shiffler and Adams
Introductory Statistics for Management and Economics, Third Edition, Kenkel
Mathematical Statistics with Applications, Fourth Edition, Mendenhall, Wackerly, and Scheaffer
Minitab Handbook, Second Edition, Ryan, Joiner, and Ryan
Minitab Handbook for Business and Economics, Miller
Operations Research: Applications and Algorithms, Winston
Probability and Statistics for Engineers, Third Edition, Scheaffer and McClave
Probability and Statistics for Modern Engineering, Second Edition, Lapin
Statistical Experiments Using BASIC, Dowdy
Statistical Methods for Psychology, Second Edition, Howell
Statistical Thinking for Behavioral Scientists, Hildebrand
Statistical Thinking for Managers, Second Edition, Hildebrand and Ott
Statistics: A Tool for the Social Sciences, Fourth Edition, Ott, Larson, and Mendenhall
Statistics for Business and Economics, Bechtold and Johnson
Statistics for Management and Economics, Sixth Edition, Mendenhall, Reinmuth, and Beaver
Understanding Statistics, Fifth Edition, Ott and Mendenhall

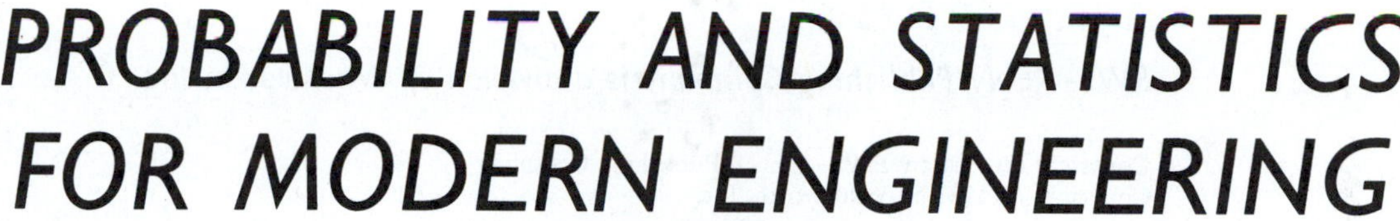

PROBABILITY AND STATISTICS FOR MODERN ENGINEERING

Second Edition

LAWRENCE L. LAPIN

San José State University

PWS–KENT Publishing Company
Boston

PWS-KENT
Publishing Company

20 Park Plaza
Boston, Massachusetts 02116

PWS–KENT Publishing Company is a division of Wadsworth, Inc.

Library of Congress Cataloging-in-Publication Data
Lapin, Lawrence L.
Probability and statistics for modern engineering/Lawrence L. Lapin.—2nd ed.
p. cm.
ISBN 0-534-91654-6
1. Engineering—Statistical methods. 2. Probabilities.
I. Title.
TA340.L36 1990
519.2′02462—dc20 89-36446
CIP

Sponsoring Editor: Michael Payne
Production Coordinator: Wanda Wilking/Robine Andrau
Manufacturing Coordinator: Margaret Sullivan Higgins
Production: Cecile M. Munson/The Cooper Company
Interior Design: Detta Penna
Cover Design: Sally Bindari/Designworks
Cover Photo: Don Landwehrle/Image Bank
Cover Printing: John P. Pow Company, Inc.
Typesetting: Syntax International Pte., Ltd.
Printing and Binding: R. R. Donnelley & Sons

Printed in the United States of America

92 93 94—10 9 8 7 6 5 4 3

Preface

Engineering majors are among the most quantitatively skilled students. But it has been my experience in teaching probability and statistics that engineers experience similar frustrations and anxieties to those commonly associated with less mathematically sophisticated students.

My goal for this book has been to make it easier for engineering students to assimilate probability and statistics. This book should help them discover how these subjects are relevant to their interests and immediate needs because it has been "engineered" especially for them. This book should provide the familiar feel common to popular texts on purely engineering subjects, helping to make a course on probability and statistics a more pleasant and meaningful learning experience.

This thrust distinguishes my book from its competitors. All examples and problems were specifically written with the particular needs of engineering students in mind. The result is a book that should be an order of magnitude more applied than existing statistics texts. Wherever possible, *real data* are used in examples, problems, and illustrations. This orientation favors insight over rigor, practicality over mathematical elegance, and simplicity over formalism. Proofs and derivations have therefore been minimized, appearing only where they reinforce concepts. Although this book emphasizes engineering applications of statistics, the underlying analytical structure is completely presented and difficult topics are still discussed. I have tried to do this by maximizing their intuitive appeal while minimizing the mathematical formalism so abhorred by engineering students.

This *second edition* improves and expands the book in several ways. The new presentation thoroughly integrates the *computer*, and many examples are now illustrated with printouts using SPSS, SAS, and other popular software. A computer thrust has facilitated this edition's broader coverage of descriptive and exploratory statistics; this edition also expands coverage of *graphical displays* and utilizes a greater number of summary statistics.

Another feature of this second edition is expanded topical coverage, including better discussions of *maximum likelihood* and the *method of moments*. Analysis of variance now includes a new chapter, reflecting a more general approach that emphasizes *comparative analysis*, employing Bonferroni, Tukey and Scheffé *collective* confidence intervals. The *Kruskal-Wallis* test is now discussed as an alternative to the F-test. A second new nonparametric method is the *Kolmogorov-Smirnov* procedure for goodness-of-fit testing. Also added to this book are a new chapter on *reliability and life testing* and discussions of the *Weibull* and *negative binomial* distributions. The regression chapters include new sections on model verification and qualification of results.

The second edition provides a smoother, more streamlined presentation. The probability chapters give better support of sampling by previewing key elements of statistical inference. The book better delineates when to use t versus z and when to make approximations, such as the normal for the binomial. (This edition expands on direct use of the binomial in estimation and testing.) The hypothesis testing chapter makes it easier to select an appropriate procedure. Ease of use has prompted resequencing of many discussions in the remaining chapters.

A distinguishing feature of this book is the amount of problem material, which considerably exceeds that of the competition. There are over 800 problems, most with several parts. This second edition contains many new problems that are more challenging and less structured. Many of these new exercises utilize computer printouts and are based on real data. These changes should improve the relevance of homework and provide instructors with a broader choice in making student assignments. Most problems are long enough to impart a flavor unique to engineering, and very detailed solutions are provided in the Instructor's Manual. To give instructors added flexibility in skipping topics, each major section within chapters ends with a separate set of problems. Students will gain added skills by working further problems from the comprehensive problem sets at the end of each chapter. They will also benefit from the abbreviated solutions provided for selected problems.

This book is arranged into four parts. The foundation is laid in Part One, led by a chapter on descriptive statistics and the concept of sampling. Chapter 2 presents basic probability concepts. Three more chapters establish the further probability background essential for understanding the development of statistical methodology. Chapters 3 and 4 cover some of the important probability distributions encountered in key engineering applications, such as reliability and quality control. To provide familiarity with the concepts underlying multivariate analysis, Chapter 5 discusses key properties of bivariate probability distributions. In keeping with the tone of this book, the messy mathematics (Jacobians, integration by parts) of transforming several random variables has been omitted. An optional section presents moment generating functions.

Along with the above chapters, those in Part Two serve to complete the traditional first-term course in statistics. Chapter 6 discusses the role of statistical sampling studies and provides the "nuts and bolts" of how to obtain a random sample. Chapter 7 describes sampling distributions. Chapter 8 is concerned with statistical estimation. Hypothesis testing is discussed next in Chapter 9. Simple regression and correlation are covered in Chapter 10. The remainder of the book consists of eight chapters. Part Three includes a chapter group focusing on multiple variable evaluations. Part Four includes specialized topics. These may be used at the option of the instructor, either to enrich a single-term introduction or to round out a two-term statistics sequence. The following topics were selected because of their special relevance to engineering: multiple regression, including fitting polynomials and use of indicator variables; two-sample procedures, including nonparametric statistics; testing for independence and goodness of fit; analysis of variance and experimental design; statistical quality control; reliability and life testing; and Bayesian statistics.

I wish to thank the many people who have assisted me in writing this book. The following colleagues provided comments and suggestions that were instrumental in the final product: William Astle, Colorado School of Mines; Douglas Bates, University of Wisconsin, Madison; Francis J. Conlan, Santa Clara University; Louis J. Cote, Purdue University; Richard D. DeVeaux, Princeton University; Anant P. Godbole, Michigan Technological University; John E. Hewett, University of Missouri, Columbia; Barry Kurt Moser, Oklahoma State University; Larry J. Ringer, Texas A & M University; José Sepúlveda, University of Central Florida; Dileep R. Sule, Louisiana Tech University; Shie-Shien Yang, Kansas State University; and Farroll Tim Wright, University of Missouri, Rolla.

Special mention goes to Janet Anaya, San José State University, who carefully checked the text and solutions for mathematical and computational accuracy. I am also indebted to my students who helped debug this book.

Lawrence L. Lapin

Contents

PART ONE

FUNDAMENTAL CONCEPTS OF PROBABILITY AND STATISTICS

CHAPTER ONE

Introduction

HARDLY ANYONE CAN get through a day without being exposed to statistics. Although most people envision masses of data when they hear the word, modern statistics is more concerned with interpreting a special kind of data to reach a decision that must otherwise be based on guesswork and hunch.

Today's engineers bear a great responsibility, unique among the professions, for creating works that are safe and reliable. In no other profession are the costs of failure greater, and the use of statistics is crucial to averting failure. Consider a major fault in the design of a large public project such as a dam or a nuclear power plant. Statistical methods are used extensively in evaluating design concepts at all levels and stages of such large projects. The probability of the various hazards that might cause nuclear-reactor accidents, such as tidal waves, earthquakes, or airplane crashes, is routinely evaluated.

Statistical analyses are involved in all the engineering disciplines. In many cases, such analyses are helpful in making choices regarding designs, materials, procedures, technologies, or methods. A mechanical engineer might employ statistics to select materials strong enough to withstand anticipated forces. Electrical engineers need statistics to determine the reliability of subsystems. Civil engineers must design a roof to withstand the weight of snow that would accumulate in the worst imaginable storm. The list of statistical applications is indeed long.

1-1 The Meaning and Role of Statistics

Modern statistics has a very special meaning. The term *statistics* is a broad one and does indeed include the use of masses of numerical data. But the focus of this book is on *using numerical evidence in making decisions.* We especially need such aids when conclusions must be reached in the face of uncertainty. Thus, our notion of statistics must include those methods and procedures that allow us to translate numerical facts into action. And, proper utilization of such tools requires some knowledge of the concepts and theory on which they are based.

A Working Definition of Statistics

To the nonengineer, almost any collection of data constitutes statistics. Indeed, in the general usage of the word, statistics *are* numbers, as in the summaries of baseball results or in "vital statistics." But as an academic discipline, the term has a different, more precise meaning.

As a field of study, modern statistics encompasses a large body of methods and theory that is applied to numerical evidence to help make *decisions* or communicate certain information. (When statistics is referred to as an area of study, the word should be treated as a singular rather than a plural noun.) The solution of a partial differential equation or the inversion of a matrix are, therefore, not statistical exercises. Neither are the readings in a survey, the telemetry recordings from a space probe, or the digitized radar cross sections of ballistic objects. Those data are not statistical in the modern sense unless the numerical information is somehow used in the process of decision making.

Statistical decision making includes another dimension besides choice. This is *uncertainty*. Given a list of weighed items, we can easily pick the heaviest. Perusal of a topographic map can narrow the choice for a dam's location to a single spot. The target acquisition window for satellite-tracking radar can be chosen by extrapolating the satellite's last known trajectory. But none of these actions is a statistical decision unless the numerical evidence is somehow used in resolving uncertainties associated with the choices.

In our working definition of the term, then, **statistics** is the analysis of numerical data for the purpose of reaching a decision or communicating information in the face of uncertainty.

The Role of Modern Statistics

Our working definition of statistics is particularly appropriate to engineers, who, in their quantitative world, must make important decisions in the face of uncertainty. Many new engineers begin their careers as estimators, who establish the prices at which their firms should bid on new projects. From the start, estimators must project future costs, choose vendors for parts and materials, and select appropriate methods and procedures for getting the job done. Wrong choices can lead to improperly low bids, resulting in losses to the firm, or to inordinately high bids and subsequent loss of potential work. Even the smallest projects must be priced under a great deal of uncertainty, and some form of statistics can play a major role in reaching many of the final decisions.

Engineering is a future-oriented profession that requires continuing research. The evaluation of new designs, concepts, procedures, and materials is at the forefront of engineering activity. Which show the most promise and should be implemented immediately? Which require further research? Statistics can be a very valuable tool in such evaluations, since it provides a scientific basis for choice. Through commonly accepted statistical procedures, scientists and engineers can communicate their findings to colleagues throughout the world. The statistical conclusions of one group may save another from pursuing dead-end paths or duplicating work already done.

Descriptive and Inferential Statistics

The emphasis placed on decision making by today's statistics is a recent development. Not many years back, the primary focus of statistics was to summarize or describe numerical data. This area, which is now just one division of the field of modern statistics, is referred to as **descriptive statistics**. Descriptive statistics is mainly concerned with summary calculations and graphical displays, which are still necessary props for statistical evaluations. But the main thrust of modern statistics is making generalizations about the whole (referred to as the *population*) by a thorough examination of the part (called the *sample*). Such conclusions are inferences rather than logical deductions, and this area may be referred to as *inferential statistics.*

Although our major concern is with making inferences, both forms of statistics are included under the umbrella of our definition. Much of this chapter is concerned with descriptive statistics, which provides the mortar and bricks for the more esoteric methods to which the remainder of this book is largely devoted.

Common sense tells us that parts might deviate, perhaps considerably, from the whole. Inferential statistics is therefore largely concerned with the quality of the generalizations made about populations using sample evidence. One need only reflect on the poor performance of many political polls to realize the dangers inherent in drawing conclusions from samples. No sample can be guaranteed to match its target population, but much of the statistical art is concerned with keeping such sampling error within reasonable bounds. The very act of sample selection may incorporate the scientific method so that the potential for error may itself be quantified.

The key to scientific sampling is *random selection* of the units to be observed. Samples that are derived by such a procedure can be evaluated using the concepts of probability theory. Probability and frequency of occurrence provide a foundation for modern statistics. For that reason, Chapters 2 through 5 are devoted to probability theory and applications.

Deductive and Inductive Statistics

A second dichotomy is fundamental to modern statistics. It is based on the concept of logical deduction, whereby the properties of specific cases can be ascribed from the general situation. For example, we can establish .04 as the probability that a randomly chosen student will be an electrical engineer if we know that 4% of all students on campus are taking that major. The application of probability concepts to getting sample results of a particular kind is called **deductive statistics**.

Deductive statistics is where analysis of the sampling process begins; therefore, the early chapters of this book are mainly concerned with that type of statistics. Only by thoroughly understanding how samples are generated from populations with *known* characteristics can we confidently deal with the reverse situation.

It is this second circumstance that is typically encountered in a statistical evaluation. This area of application involves *induction*, or generalizing about the whole from knowledge of just the part. Thus, it is called **inductive statistics** and involves making conclusions about the *unknown* population from the known sample. We have referred to such generalizations as inferences, and the terms *inductive* and **inferential statistics** may be used interchangeably. An exposure to deductive statistics makes it easy to appreciate inductive statistics, the major concern of modern statistics.

Statistical Error

Statistics is partly art and partly science. It is an art because effective application depends so much on judgment and experience in selecting the best methods for a particular evaluation. The choice of procedures is considerable. Nevertheless, the scientific method dominates statistics, as can be seen by the emphasis that modern statistics places on the potential for error. The uncertainties involved in the sampling process justify this emphasis.

To a large extent, sampling error may be quantified. As we shall see, this is achieved mainly through random selection. Also, the number of sample observations must be large enough to keep the chance of error at a controlled level. But scientific sampling includes many additional factors. For example, care must be taken to ensure that observations are not biased. Much of Chapter 6 is concerned with designing a statistical evaluation to minimize the many forms such bias can take. As we shall see, proper statistical procedure minimizes both error and bias.

1-2 Describing Statistical Data

In our daily lives, we have all become familiar with displays and summaries such as pie charts, percentages, and averages computed for every imaginable numerical quantity. These are elements of descriptive statistics, which organizes the data collected in statistical investigations and provides a structural framework for describing and summarizing the results.

The Population and the Sample

The basic statistical element is the **observation**, or single data point. Although we ordinarily associate observed data with numerical values such as volume or mass, they can also take the form of classifications such as "defective" or "nondefective." There are two basic accumulations of observations. The major grouping of observations is given the following definition.

Definition

A **statistical population** is the collection of all possible observations of a specified characteristic of interest.

Under the foregoing definition, we limit the population to the observations. We do *not* include the objects, persons, or things being observed, which are referred to as **elementary units**. A single grouping of elementary units can give rise to any number of populations. Consider, for example, all the students attending a particular university. These persons could serve as the elementary units in populations of grade-point averages, earnings, sexes, majors, heights, ages, and so on. Consider the population of grade-point averages (GPAs): The collection of numerical GPA values, not the students themselves, constitutes the population. Do not confuse a statistical population with a demographic grouping of people (the size of which is often referred to as the "population").

A statistical population has an identity whether all, some, or even no observations are actually made. The population may already exist, as does the collection of starting salaries of all graduating engineers in the past year. Or it can lie in the future, such as the salaries of next year's graduates. Populations can be imaginary or hypothetical and may never actually come into existence. Consider the salaries of students who would have graduated from an experimental program that never got started or the salaries of those who would have worked on a major weapon system that has been cancelled.

For a variety of reasons, the population itself often cannot be totally observed. Usually the number of observations is limited by the availability of money or time. Whenever only a portion of the population is observed, the resulting observations constitute a sample. Theoretical or hypothetical populations that may never come into actual existence are sometimes referred to as **target populations**. They too can be partially observed by means of sampling. We make the following definition.

Definition

A **sample** is a collection of observations representing only a portion of the population.

Most statistical methods focus on the evaluation of samples. Since samples usually form the basis for drawing conclusions about the respective populations, care must be taken in selecting those elementary units to be observed for the sample. As we shall see, the elementary units are ordinarily chosen randomly.

Types of Statistical Data

It is convenient to place populations into two groups, depending on the nature of the observations. When the observations are *numerical values*, the population is referred to as a **quantitative population**. Should the observations instead be *attributes* (such as sex, occupation, or some similar category), the possible observations constitute a **qualitative population**. This dichotomy is an important one, since different procedures are used to evaluate data from the two population types.

There are four types of quantitative data. Most common in engineering evaluations are **ratio data**. These include times and many physical measurements of size, weight, or strength. The arithmetic operations of addition, subtraction, division, and multiplication are all valid with ratio data. This is not the case with **interval data**, with which only addition and subtraction are meaningful. The most common example of this latter type is *temperature*, for which scales are arbitrarily chosen. A 100° *Fahrenheit* day is not twice as hot as a 50° day since 100°/50° = 2 is not a meaningful ratio (and a different ratio applies under the *Celsius* scale).

Moving down the scale spectrum, no arithmetic operations are meaningful with **ordinal data**, which convey ranking in terms of importance, strength, or severity. The *Beaufort* wind scale provides a good example; a value of 3 corresponds to a gentle sea breeze, while a 6 represents a strong breeze and a 9 signifies a strong gale. The change in force between a gentle and strong sea breeze is not equal to that between a strong gale and strong breeze, as the numbers themselves would indicate. At the bottom of the spectrum are **nominal data**, numbers that represent arbitrary codes. An engineering school, for example, might use numbers to denote undergraduate majors: 1 for electrical engineering, 2 for civil engineering, and so on.

The Frequency Distribution

Data organization is a basic need in any kind of statistical evaluation. Consider the values in Table 1-1. These quantities represent the observed times required to complete the calibration inspection of a particular test device. Each number represents a value obtained from a series of stopwatch observations of various laboratory assistants performing the same task. They were recorded in a log and are listed in their original sequence. Since the data points have not been arranged in any meaningful fashion, they are referred to as **raw data**. Descriptive statistics is largely concerned with arranging and manipulating raw data so that they may be interpreted.

One helpful arrangement is achieved by grouping the raw data into categories. Since the observations are numerical values, they may be placed into a series of contiguous blocks called **class intervals**. The number of observations falling into

Table 1-1
Sample Times (in Seconds) to Inspect Test Devices for Calibration

12.8	15.6	13.5	15.7	15.3	15.2	20.1	14.2	12.9	14.0
16.9	14.3	15.5	14.6	13.0	14.7	19.0	13.0	11.3	14.2
14.5	14.8	14.2	13.0	13.1	12.5	16.1	19.1	16.7	13.2
15.0	12.7	13.6	13.3	13.2	14.7	12.9	13.1	17.3	15.4
17.9	13.0	14.3	14.2	15.7	15.6	13.0	13.9	14.2	16.0
12.9	13.1	13.3	12.3	13.1	13.6	13.2	18.5	13.2	13.7
12.6	14.4	14.5	13.9	17.0	13.7	12.7	16.8	13.3	14.7
14.2	13.0	14.6	14.0	12.9	14.7	12.8	12.0	14.2	12.8
13.7	15.2	14.8	13.0	11.7	12.2	13.3	13.8	14.2	14.3
14.7	12.6	18.9	14.3	14.4	15.5	16.8	17.0	13.2	12.9

Time (seconds)	Tally	Number of Inspections
11.0–under 12.0	//	2
12.0–under 13.0	𝍸 𝍸 𝍸 /	16
13.0–under 14.0	𝍸 𝍸 𝍸 𝍸 𝍸 ////	29
14.0–under 15.0	𝍸 𝍸 𝍸 𝍸 𝍸 //	27
15.0–under 16.0	𝍸 𝍸 /	11
16.0–under 17.0	𝍸 /	6
17.0–under 18.0	////	4
18.0–under 19.0	//	2
19.0–under 20.0	//	2
20.0–under 21.0	/	1
	Total	100

Table 1-2
Sample Frequency Distribution of Inspection Times

each interval is then determined. The respective count provides the **class frequency**. Table 1-2 shows the resulting summary for the inspection times, which is referred to as the **sample frequency distribution**. The first class interval is 11.0–under 12.0, the second 12.0–under 13.0, and so forth. Each class interval has a width of 1.0, with the class limits chosen so there is no ambiguity regarding where to place an observation; thus, the upper limit of the first interval is under 12.0, whereas the lower limit of the second is 12.0 exactly. Both the number of classes and the interval widths must be chosen by the investigator.

The frequency distribution for inspection times applies to sample data. Should the data represent the entire population, then the resulting table would provide the population frequency distribution. Ordinarily, only samples are available, and the exact population frequency distribution remains unknown.

The Histogram and Frequency Curve

It is often easier to draw conclusions about the sample data when the frequency distribution has been portrayed graphically. Figure 1-1 shows a **histogram** in which the inspection times are represented by a bar chart having frequency as the vertical axis and the observed values as the horizontal axis. Frequency distributions may be categorized by the basic shapes of their histograms, with various distribution families encountered in certain statistical applications.

An alternative display is the **frequency polygon**, shown for the same data in Figure 1-2. Here a dot is plotted above the midpoint of each interval at a height matching the class frequency. These dots are then connected, with outside line segments touching the horizontal axis one-half of an interval width below and above the lowest and highest intervals, respectively. Notice that the area circumscribed by the frequency polygon is about the same shape and size as that presented by the collective bars of the histogram.

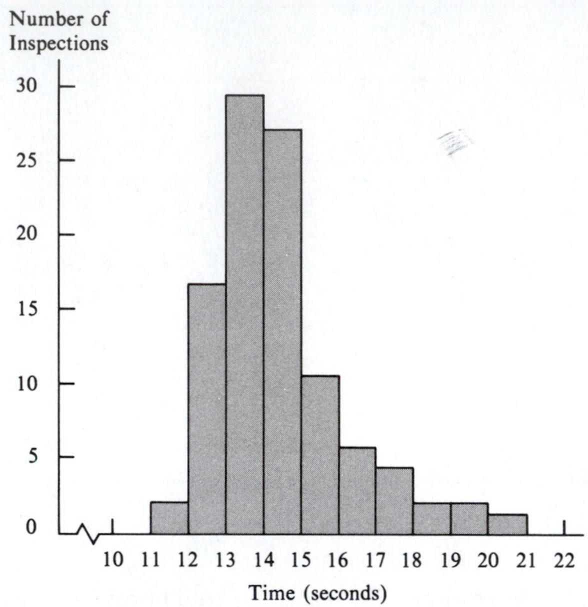

Figure 1-1
Histogram for the frequency distribution of inspection times.

Although the population frequency distribution is not usually available, the sample counterpart suggests the shape it would take when graphed. Figure 1-3 shows the **frequency curve** for calibration inspection times suggested by the sample data, as plotted by either the histogram in Figure 1-1 or the frequency polygon

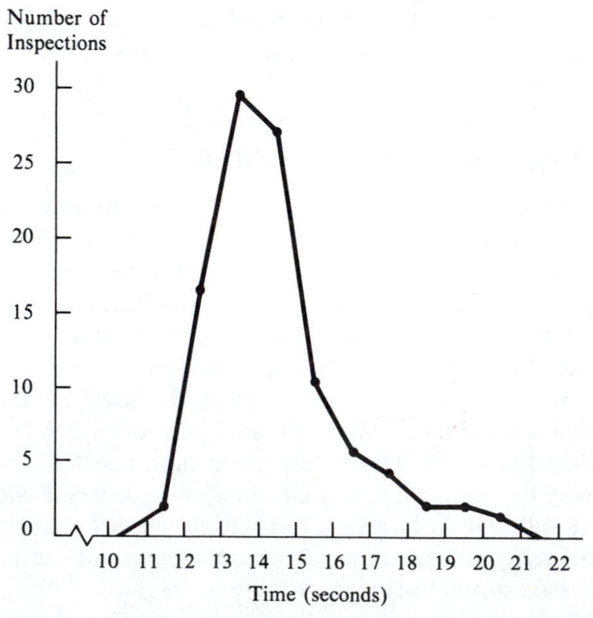

Figure 1-2
Frequency polygon for the frequency distribution of inspection times.

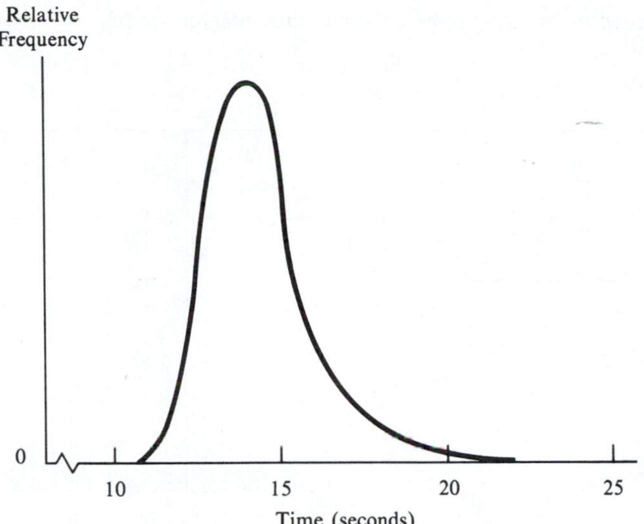

Figure 1-3 *Suggested shape of smoothed frequency curve for the entire population of inspection times.*

in Figure 1-2. Since the size of this particular population is arbitrarily large, the ordinate is given in terms of **relative frequency**. Since the entire population will be very large in relation to its sample, the frequency curve will be smoother than its sample counterparts. (Imagine times measured to the nearest 100 microseconds and a histogram portraying millions of observations graphed with intervals of width .001.)

The Effect of Different Class Interval Widths

The choice of lower limit for the first class interval depends mainly on the smallest observation. The specification of interval width is the primary consideration in specifying the frequency distribution. Wider intervals will result in fewer classes, whereas narrower intervals yield more classes. It is a matter of judgment how these parameters should be set. Too much or too little detail tends to mask the underlying frequency pattern inherent in the raw data.

Consider the histograms in Figure 1-4 constructed from the raw data in Table 1-1. The histogram in (a) has five class intervals of width 2 seconds. It does not capture the full flavor of the frequency pattern nearly as well as the original histogram. Histogram (b) is based on intervals .5 second wide and results in 19 class intervals. Notice that it is quite jagged, providing a sawtooth effect not seen with wider widths. With 27 intervals of width .25 second, histogram (c) does an even poorer job of describing the data.

Unequal Interval Widths

The class intervals need not all be the same width. When raw data have a few extremely large or small values, those may be grouped into a special class.

Figure 1-4 *Histograms for inspection times, using three different class interval widths.*

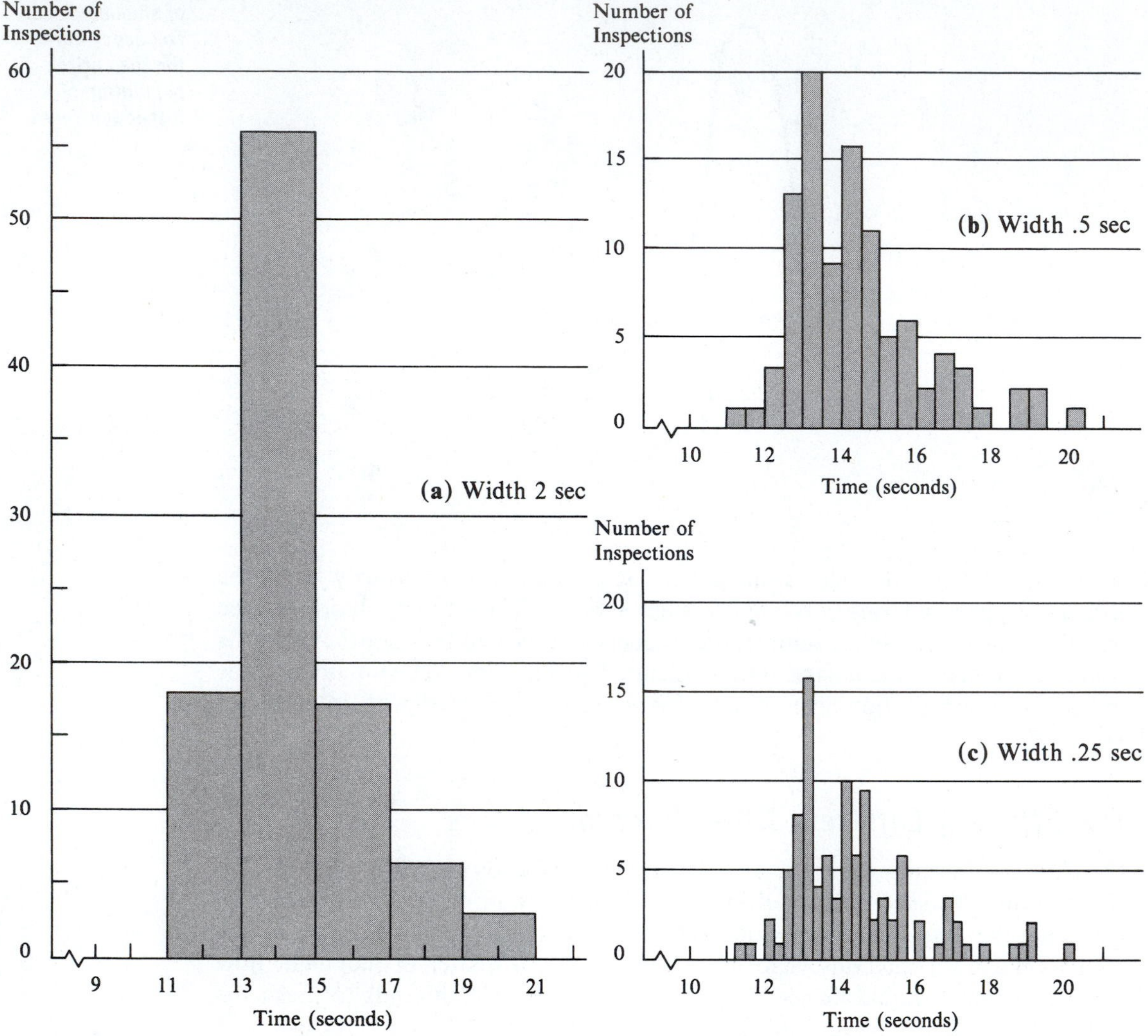

For example, consider the *ages* of a sample of full-time undergraduate students. Raw data might be grouped in class intervals of standard width 2.0 years, and most ages of these predominantly young people would fall below 30. A small group of older undergraduates, however, must also be considered, and their ages could be placed into a single interval, such as 30.0–under 55.0. As a second example, the gasoline mileages for randomly selected cars might cluster around 30 mpg, with standard class interval widths of 2 mpg. The few cars having very low mileages might be lumped into a single class, such as 6.0–under 20 mpg.

Such intervals can be portrayed graphically with no special difficulty. Suppose, for example, that just five student ages fall into the highest interval. The histogram bar for that interval would be 55 − 30 = 25 years wide and would be drawn at the following fractional height:

$$\frac{5}{25} \times 2.0 = .40$$

Thus, each year in the 25-year span represents an average of $\frac{5}{25}$ observation.

Stem-and-Leaf Plots

One summary that is useful for arranging experimental data is the **stem-and-leaf plot**. Here the raw data are arranged tabularly by locating each observation on a "tree." This is done by separating the values into a stem digit and a leaf digit. To illustrate, consider the temperature data in Table 1-3. Scanning these values, we see that the first digits range from a low of 1 to a high of 8. Listing these integers in the first column, the following table is constructed:

```
                                        Leaf
                                     (2nd digit)
             1 | 8
             2 |
             3 | 5 7 8 0 7 8 9
Stem         4 | 7 9 9 6 6 5 7 2 2 3 9 6 9 6 3 9 8 8 3 5 9 7 6 5 6 9
(1st digit)  5 | 1 0 8 5 3 6 6 4 5 0 6 0 2 3 1 2 7 0 5 2 4 3 1 9 2 8 5
             6 | 0 8 8 9 2 0 2 6 2 1 9 0 5 3 9 8 3 6 6 0
             7 | 6 0 2 5
             8 | 0
```

The second digit for each temperature is entered to the right of the line in the row corresponding to its first digit. For example, the first temperature, 47, has a value of 7 (the leaf) in the 4 row (the stem); the second value, 49, has a leaf with a value of 9, also on the row with a stem of 4. The third value, 51, becomes the first leaf, 1, on the row with a stem of 5.

Table 1-3 *April Normal Temperatures (°F) for Selected U.S. Cities*

47	49	51	49	60	46	50	58	46
55	45	47	42	42	68	53	56	56
35	43	54	76	55	50	68	49	46
56	37	38	69	62	60	50	70	72
62	66	49	46	62	52	43	61	53
51	49	30	52	57	69	50	55	52
54	48	60	65	37	53	48	80	
63	51	69	68	63	18	59	38	
43	66	52	39	75	58	45	66	
49	47	46	55	45	60	46	49	

Source: The World Almanac and Book of Facts, 1988, p. 181.

One advantage of the stem-and-leaf plot is that all the original raw data are portrayed. This is useful for later computation of summary statistics. As we shall see, two such statistics—the median and the mode—can be quickly found from the plot itself. The main advantage, however, is that it provides the essential features of the histogram; you can see these easily by rotating the plot 90 degrees. The disadvantage is that the stem-and-leaf plot becomes cumbersome if the number of observations is large or if the data are expressed to more than two significant figures.

Computer-Generated Displays

It can be very time-consuming to generate histograms by hand. Computer assistance can not only eliminate that burden but also avoid potential errors. Several commercial statistical software packages are available. Especially valuable to engineers is SAS, or alternatively, SPSS. Both are illustrated throughout this book. SAS and SPSS are elaborate software packages and are available in either mainframe (on which they were developed) or PC versions. Most universities have one or the other of these programs.

Beginners may be more comfortable with software designed especially for small computers. One popular such package is Minitab. Many special-purpose software packages have also become available for personal computers. One of these, StatGraphics, has superior data display features.

Figure 1-5 shows a histogram for the preceding temperature data that was generated with SPSS.

Figure 1-5 *Computer-generated histogram for April normal temperature data, obtained using SPSS.*

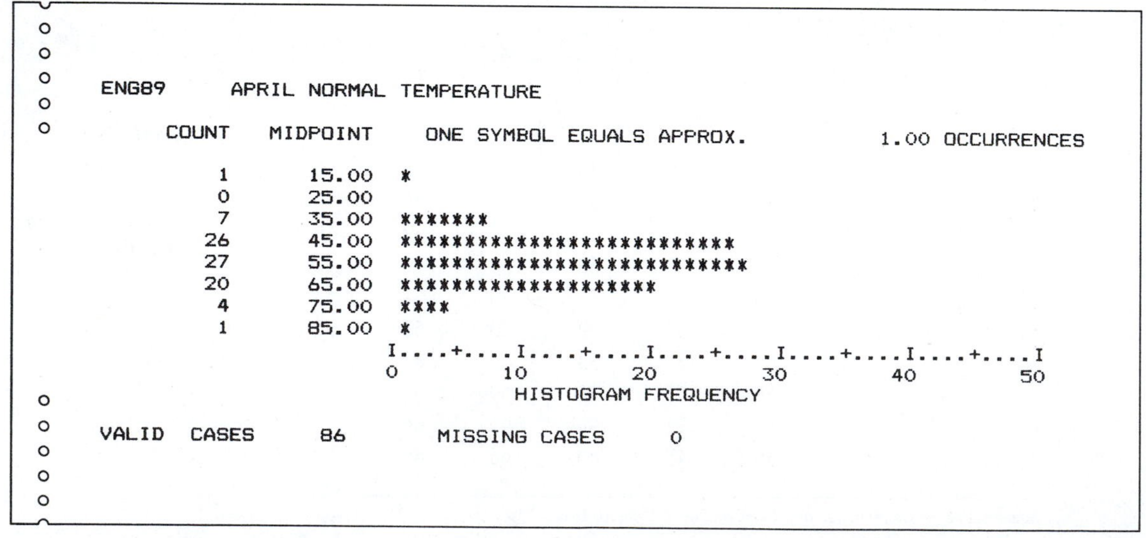

```
ENG89      APRIL NORMAL TEMPERATURE

     COUNT    MIDPOINT    ONE SYMBOL EQUALS APPROX.         1.00 OCCURRENCES

         1       15.00   *
         0       25.00
         7       35.00   *******
        26       45.00   **************************
        27       55.00   ***************************
        20       65.00   ********************
         4       75.00   ****
         1       85.00   *
                         I....+....I....+....I....+....I....+....I....+....I
                         0        10        20        30        40        50
                                      HISTOGRAM FREQUENCY

VALID CASES      86        MISSING CASES      0
```

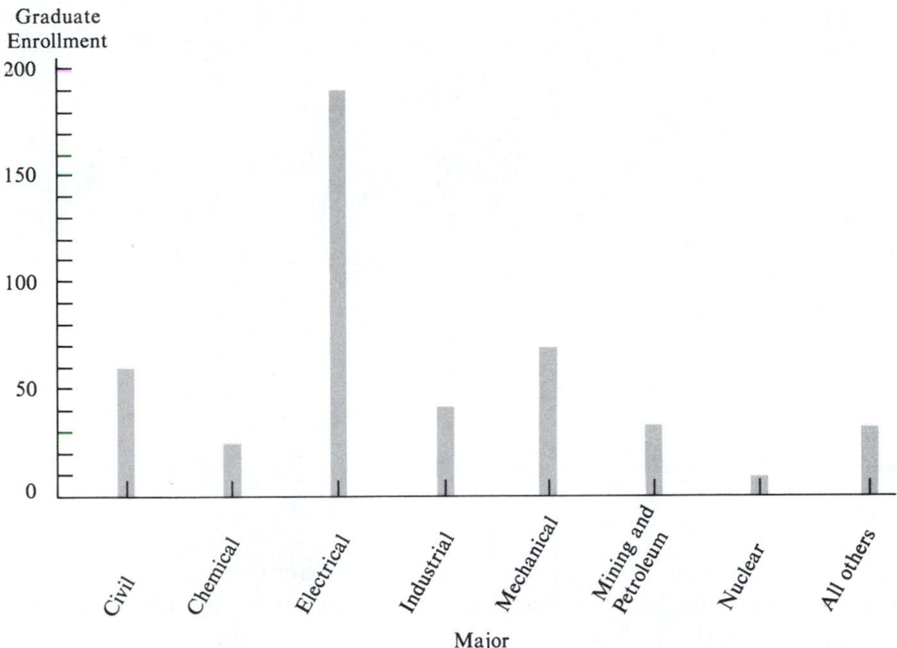

Figure 1-6 *Chart for qualitative frequency distribution of engineering majors.*

Frequency Distributions for Qualitative Variables

There is less choice in portraying qualitative data. Each category or attribute occurs with some frequency, which may be summarized in a table or bar chart. For example, consider Figure 1-6, which represents the frequency distribution for the majors of a randomly selected group of engineering students. Notice that the bars do not touch, which reflects the fact that the chart is a graphical representation of discrete categories. Similar graphs might be appropriate for portraying the frequency distributions of discrete quantitative variables such as number of patents held, number of degrees, or family size.

The **pie chart** is a popular graphical display for qualitative data. Figure 1-7 shows a pie chart generated by a computer using StatGraphics. It represents the frequency distribution for type of accidental death. Since the more frequent observation categories have larger wedges, this type of display quickly conveys the essence of the frequency distribution. The piece of pie is sized according to the category's **relative frequency**, with the angle of the slice corresponding to its proportion of 360 degrees.

Relative and Cumulative Frequency Distributions

Frequency of occurrence is the foundation for much of statistical analysis because it provides a meaningful arrangement of observed data. But there are some drawbacks to basing evaluations directly on the number of data points found in each

Figure 1-7 *Computer-generated pie chart for cause of accidental death (non-vehicular), obtained using StatGraphics.*

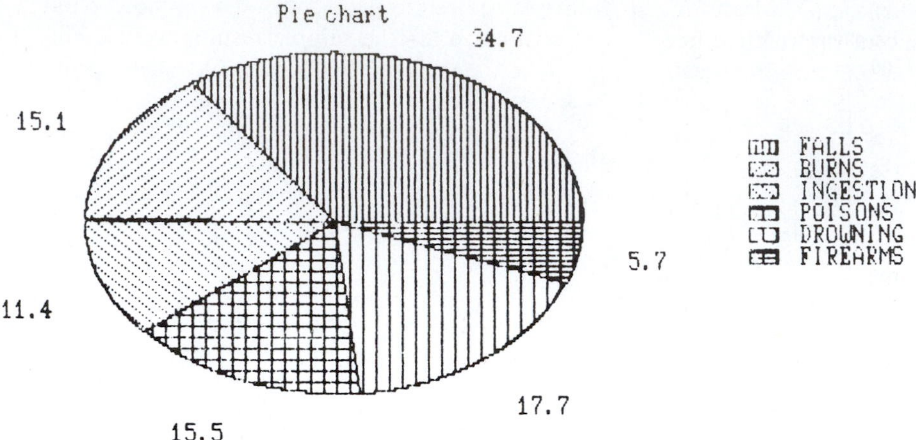

Source: The World Almanac and Book of Facts, 1988, p. 810.

class interval. It is clumsy to compare groups of different size in terms of straight tallies. For example, knowing that there are 254 electrical engineering majors at one university and only 154 at a second says little about the relative importance of that concentration in the two engineering schools unless the respective total enrollments are also included in the comparison. The first school has 1,547 engineering students, so that the proportion, or relative frequency, of electrical engineers is $\frac{254}{1,547} = .164$. The second university has 655 engineering students, and the relative frequency of the electrical concentration is a much higher $\frac{154}{655} = .235$. Thus, electrical engineering is more dominant in the second institution.

The foregoing suggests that it may be helpful to divide each of the original frequencies by the sample size, expressing the distribution in terms of relative fre-

Table 1-4 *Relative Frequency Distribution of Sample Inspection Times*

Time (seconds)	Number of Times (Frequency)	Relative Frequency
11.0–under 12.0	2	$\frac{2}{100}$ = .02
12.0–under 13.0	16	$\frac{16}{100}$ = .16
13.0–under 14.0	29	$\frac{29}{100}$ = .29
14.0–under 15.0	27	$\frac{27}{100}$ = .27
15.0–under 16.0	11	$\frac{11}{100}$ = .11
16.0–under 17.0	6	$\frac{6}{100}$ = .06
17.0–under 18.0	4	$\frac{4}{100}$ = .04
18.0–under 19.0	2	$\frac{2}{100}$ = .02
19.0–under 20.0	2	$\frac{2}{100}$ = .02
20.0–under 21.0	1	$\frac{1}{100}$ = .01
		Total 1.00

quencies. The resulting summary is the **relative frequency distribution**. Table 1-4 shows the relative frequency distribution for the sample results given earlier for the calibration inspection times. After the original frequencies have been converted to relative frequencies, it is easier to compare samples of various sizes taken at different times, in other places, or for other inspection tasks.

It sometimes is advantageous to make another transformation of class frequencies to obtain another form of the frequency distribution. This is the **cumulative frequency distribution**, which is constructed by adding the frequency of each class to the sum of the frequencies for the lower classes. Table 1-5 provides the cumulative frequency distribution for the sample of calibration inspection times. Cumulative frequencies are useful in describing different levels of the variable of interest. We see that 95 out of the 100 times fall below 18.0 seconds, whereas only 2 out of 100 fall below 12.0 seconds.

When plotted on a graph such as Figure 1-8, the cumulative frequency distribution gives another visual summary of the sample. This type of display is sometimes called the **ogive**. Each dot is plotted directly above the upper class limit at a height equal to the cumulative frequency for that interval. These are then connected by line segments, with the lowest line touching the horizontal axis at the lower limit of the smallest class. Such a graph can be useful when the original data are unavailable. The interpolated cumulative frequency of any particular observation value may be read directly from the graph, and the approximate observation level corresponding to any cumulative frequency may be obtained similarly, by reading the graph in reverse. Thus, about 88 out of 100 observations fall below 16.5 seconds, whereas 14.1 seconds is the time below which about half (50) of the observations fall. Often the ordinate is cumulative *relative* frequency, so that the proportion of the observations falling below a particular level may be easily determined from the graph.

Table 1-5 *Cumulative Frequency Distribution of Sample Inspection Times*

Time (seconds)	Number of Times (Frequency)	Cumulative Frequency
11.0–under 12.0	2	2
12.0–under 13.0	16	2 + 16 = 18
13.0–under 14.0	29	18 + 29 = 47
14.0–under 15.0	27	47 + 27 = 74
15.0–under 16.0	11	74 + 11 = 85
16.0–under 17.0	6	85 + 6 = 91
17.0–under 18.0	4	91 + 4 = 95
18.0–under 19.0	2	95 + 2 = 97
19.0–under 20.0	2	97 + 2 = 99
20.0–under 21.0	1	99 + 1 = 100

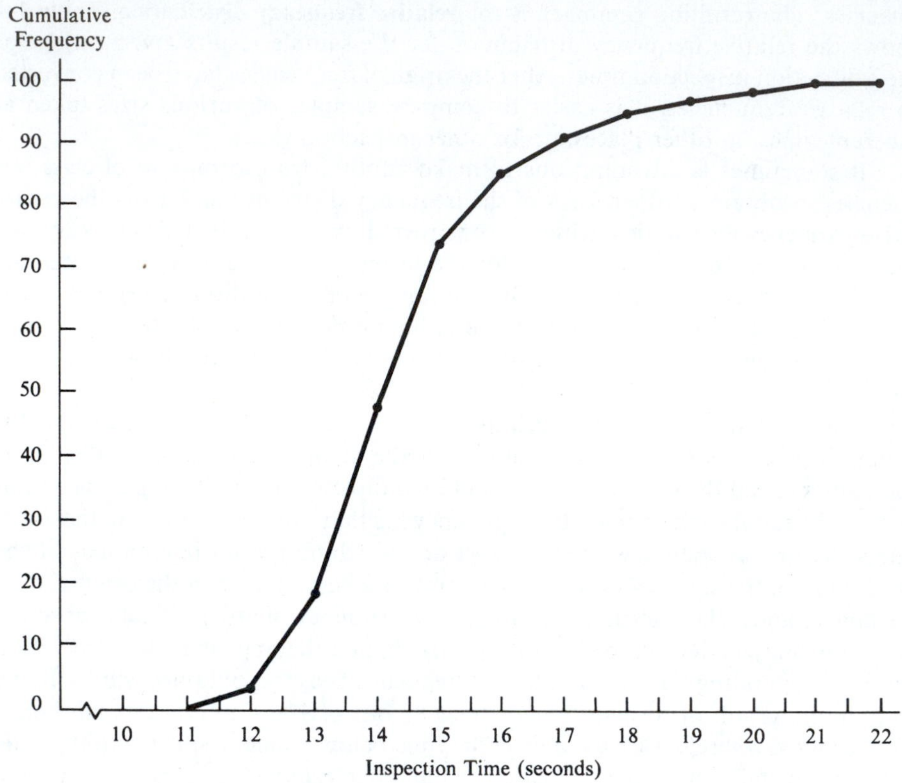

Figure 1-8 *Ogive for cumulative frequency distribution of inspection times.*

Multidimensional Data Displays

Some of the most interesting aspects of statistical investigations arise when there are two or more dimensions to a statistical population, such as chemical yield, processing temperature, and pressure setting; stress and strain; or college major and starting salary. Statistical investigations attempt to uncover relationships between two or more variables, such as how yield responds to different settings of temperature and pressure or how majors differ in salary level. A starting point is often a data display.

Figure 1-9 shows a data display for two quantitative population variables. A computer-generated **scatter diagram** relates concrete strength Y to pulse velocity X. A visual inspection indicates that Y should increase with X. Later chapters show how a line or a curve can be used to summarize that relationship.

A second two-dimensional display arises when population factors are both qualitative. Figure 1-10 shows a computer-generated **crosstabulation** for the active membership of an engineering college alumni association. Each member is counted into the total of a particular cell. There is one row for highest degree received (three levels) and a column for major field (five levels).

Figure 1-9 *SPSS scatter diagram for concrete strength vs. pulse velocity.*

```
DATA LIST /STRENGTH 1-5(2) VELOCITY 7-11(3).
BEGIN DATA.
36.77 4.293
43.33 4.256
44.09 4.337
      :
27.51 4.035
END DATA.
SET PRINTER ON.
PLOT SYMBOLS='*'
/VSIZE=30 /HSIZE=70
/FORMAT=DEFAULT
/TITLE='SCATTER DIAGRAM'
/VERTICAL='COMPRESSIVE STRENGTH (MPa)'
/HORIZONTAL='PULSE VELOCITY (mm/us)'
/PLOT=STRENGTH WITH VELOCITY.
SET PRINTER OFF.
```

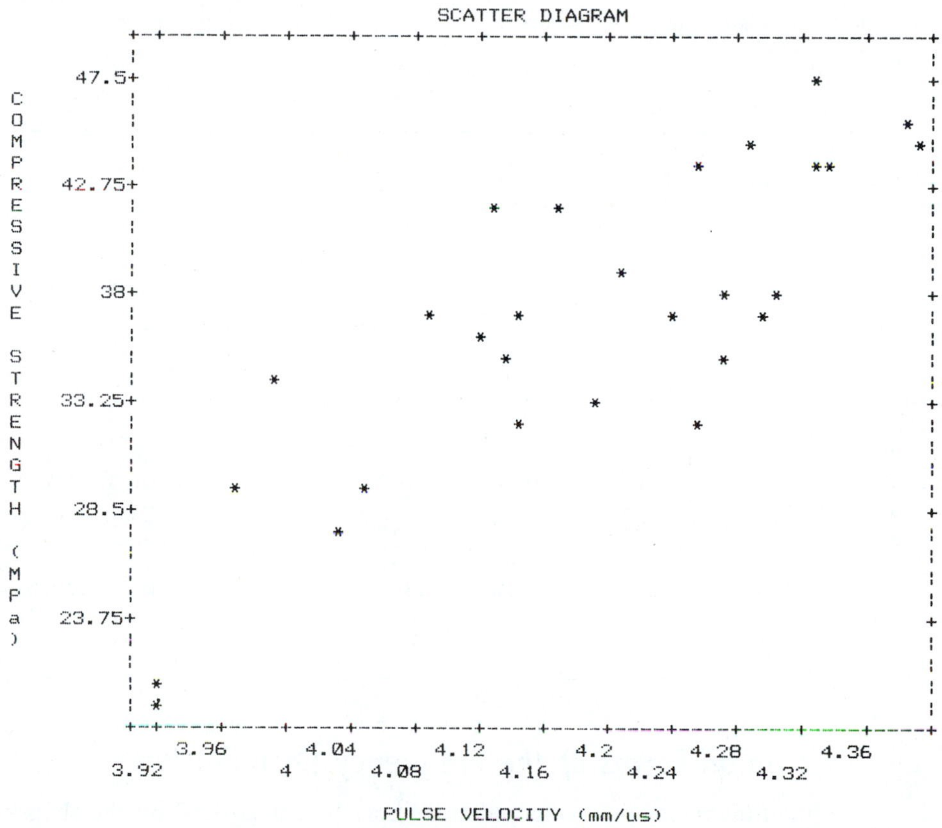

Source: Scanlon and Mikhailovsky, "Strength Evaluation of an Existing Concrete Bridge," *Canadian Journal of Civil Engineering*, Vol. 14 (1987): 150.

Figure 1-10 *SPSS crosstabulation for engineering college active alumni association membership.*

```
DATA LIST FIXED / FREQ 1-5 DEGREE 7 MAJOR 9.
WEIGHT BY FREQ.
VALUE LABELS
  DEGREE  1 'BACHELOR' 2 'MASTER'/
  MAJOR  1 'ELECT' 2 'MECH' 3 'CIVIL' 4 'CHEM' 5 'INDUS'.
BEGIN DATA.
   47 1 1
  113 1 2
   45 1 3
   12 1 4
   24 1 5
   28 2 1
    9 2 2
    3 2 3
    5 2 4
   15 2 5
END DATA.
CROSSTABS  TABLES=DEGREE BY MAJOR
 /OPTIONS=3,4,5
 /STATISTIC=1
FINISH.
```

```
Crosstabulation:      DEGREE
                   By MAJOR

                 Count  :
               Row Pct  :ELECT    :MECH     :CIVIL    :CHEM     :INDUS    :
   MAJOR->     Col Pct  :         :         :         :         :         :  Row
               Tot Pct  :       1 :       2 :       3 :       4 :       5 :  Total
DEGREE         ---------+---------+---------+---------+---------+---------+
                     1  :    47   :   113   :    45   :    12   :    24   :   241
   BACHELOR             :  19.5   :  46.9   :  18.7   :   5.0   :  10.0   :  80.1
                        :  62.7   :  92.6   :  93.8   :  70.6   :  61.5   :
                        :  15.6   :  37.5   :  15.0   :   4.0   :   8.0   :
                        +---------+---------+---------+---------+---------+
                     2  :    28   :     9   :     3   :     5   :    15   :    60
   MASTER               :  46.7   :  15.0   :   5.0   :   8.3   :  25.0   :  19.9
                        :  37.3   :   7.4   :   6.3   :  29.4   :  38.5   :
                        :   9.3   :   3.0   :   1.0   :   1.7   :   5.0   :
                        +---------+---------+---------+---------+---------+
                 Column      75       122        48        17        39      301
                  Total    24.9      40.5      15.9       5.6      13.0    100.0
```

Common Forms of the Frequency Distribution

Quantitative samples or populations may be categorized by the shapes of their relative frequency distributions. Figure 1-11 shows some of the basic forms. The population may be represented either by a relative frequency curve or its cumulative counterpart. In each case, the sample histogram suggests the shape of the underlying population frequency curve. The same mathematical function can be used to describe all curves of a particular shape so that individual members of

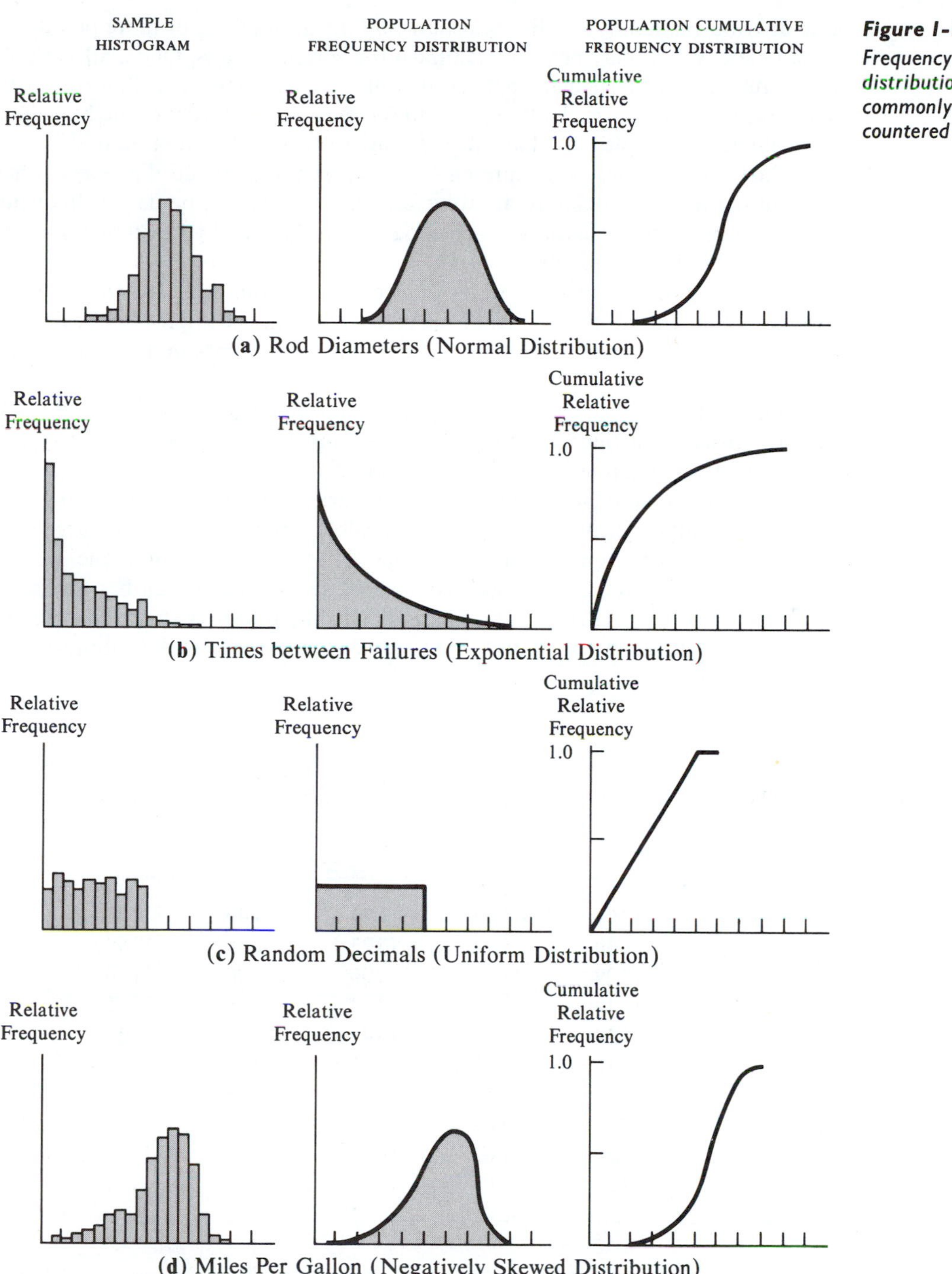

Figure 1-11 *Frequency distributions in commonly encountered forms.*

one distribution family are distinguishable by the level of one or more population parameters, which may be estimated from the sample results. Often, all variables associated with a particular application will have frequency distributions of the same basic form. As we shall see, a common statistical procedure might be used to evaluate all samples that have data fitting a single distribution family.

Part (a) of Figure 1-11 represents the observations of rod diameters, where the sample histogram closely approximates a bell-shaped population frequency curve belonging to the **normal distribution** family. Many physical dimensions are characterized by this distribution.

In (b) the times between failures of electronics components closely fit an **exponential distribution**, the frequency curve of which has a single "tail" extending indefinitely to the right. Exponential distributions are common in reliability and queueing evaluations.

The frequency distribution for the random decimals in (c) is described by the **uniform distribution**. Such a rectangular shape applies to variables that are equally likely to fall anywhere within a fixed interval.

A common characterization for a wide variety of data is an unsymmetrical hill-shaped curve. Such skewed curves are often encountered when there is an upper or lower limit on the possible values. They take two forms. The gasoline mileage data in (d) suggest a **negatively skewed distribution**, which has the longer tail pointing left. This configuration is similar to the **positively skewed distribution** with the longer tail pointing right that was found earlier for the calibration inspection times.

Problems

1-1 The following sample data represent the per pound cost of raw materials used in processing batches of a chemical feedstock:

\$12.01	\$14.90	\$11.24	\$14.40	\$12.87
11.30	16.98	15.23	12.38	12.90
12.29	11.51	12.06	15.06	11.16
10.12	14.82	17.02	12.56	13.95
11.13	12.14	13.41	15.67	12.17
12.15	12.57	14.21	13.21	16.50
13.81	15.58	11.10	12.03	11.20
12.27	17.05	12.57	13.11	13.31
13.84	11.62	12.33	16.31	12.74
14.25	18.63	13.34	13.43	14.78

(a) Using \$10.00 as the lower limit of the first class interval, construct a table summarizing the sample frequency distribution having intervals of width \$1.00.
(b) Plot the above as a histogram.
(c) Plot the above as a frequency polygon.

(d) Of the basic population frequency curve shapes described in the text, which one do the above data most closely match?

1-2 Using the raw materials costs from Problem 1-1, complete the following:
(a) Construct a table for the sample frequency distribution having $2.00 widths for the class intervals, with $10.00 as the first lower limit.
(b) Plot the above as a histogram.
(c) Plot the above as a frequency polygon.

1-3 Using the raw materials costs from Problem 1-1, complete or answer the following:
(a) Construct a table for the sample frequency distribution having $.50 widths for the class intervals, with $10.00 as the first lower limit.
(b) Plot the above as a histogram.
(c) Do you think a more accurate data summary might be obtained with intervals narrower or wider than $.50?

1-4 The following precipitation levels (inches) apply for selected cities during the month of April:

2.9	3.7	3.2	4.0	3.9	2.1	2.9	2.9	1.1
0.4	3.0	3.2	1.0	2.2	5.4	3.5	3.6	4.0
0.7	2.8	1.8	1.5	4.0	4.0	0.3	2.2	3.3
3.8	2.6	2.2	4.2	5.4	4.8	3.3	2.7	1.8
4.4	2.6	2.9	2.0	1.2	3.6	3.9	0.8	3.1
3.1	3.7	0.3	3.7	4.1	4.5	2.3	1.5	3.4
3.4	3.3	1.2	5.9	3.6	3.8	4.0	3.6	
5.0	3.4	2.6	3.3	5.8	0.6	2.9	2.4	
1.6	3.6	0.7	2.9	3.1	2.9	2.0	3.2	
1.2	1.8	3.6	2.7	3.4	2.9	0.5	2.4	

Source: The World Almanac and Book of Facts, 1988, p. 181.

(a) Construct a stem-and-leaf plot.
(b) Using intervals of width 1″ and starting with 0″ as the lower limit of the first, construct a table for the frequency distribution.
(c) Plot the distribution from (b) as (1) a histogram and (2) as a frequency polygon.

1-5 Using standard widths of 5.0 years, plot a histogram for the following age frequency distribution:

Age (years)	Frequency
20–under 25	5
25–under 30	13
30–under 35	17
35–under 40	8
40–under 80	12

1-6 Using standard widths of 5.0 mpg, plot a histogram for the following gasoline mileage frequency distribution:

Mileage (mpg)	Frequency
5–under 20	9
20–under 25	12
25–under 30	14
30–under 35	22
35–under 40	7
40–under 60	4

1-7 A sample of engineering students has been categorized by concentration, level, and financial aid. The number of persons in each category is summarized here:

Concentration	Level	Financial Aid	Number
Electrical	Undergraduate	None	23
Electrical	Undergraduate	Some	6
Electrical	Graduate	None	12
Electrical	Graduate	Some	15
Mechanical	Undergraduate	None	14
Mechanical	Undergraduate	Some	4
Mechanical	Graduate	None	5
Mechanical	Graduate	Some	12
Civil	Undergraduate	None	34
Civil	Undergraduate	Some	10
Civil	Graduate	None	7
Civil	Graduate	Some	24

Construct a table and a bar chart for the sample frequency distribution characterized by (a) concentration, (b) level, and (c) financial aid.

1-8 Referring to the data in Problem 1-7, obtain a crosstabulation of concentration vs. level.

1-9 The following data provide the number of professional U.S. women employed (thousands) in 1986:

A. Engineering/Computer science	347
B. Health care	1937
C. Education	2833
D. Social/Legal	698
E. Arts/Athletics/Entertainment	901
F. All others	355

Source: The World Almanac and Book of Facts, 1988, p. 86.

(a) Construct a bar chart.
(b) Construct a pie chart.

1-10 The following sample data represent the gasoline mileages (in miles per gallon) determined for cars in a particular weight class:

25.1	29.0	34.5	35.7	37.9
34.9	24.3	26.6	27.3	32.0
30.0	34.5	35.3	33.5	36.6
34.8	16.2	13.1	24.5	33.6
28.0	33.9	30.7	32.0	37.7
21.1	31.2	35.6	34.4	25.2
35.9	18.3	29.4	29.5	34.8
29.4	26.4	38.8	36.0	28.7
23.4	35.3	33.7	38.1	28.6
34.2	34.8	39.2	39.9	36.8

(a) Using 10 mpg as the lower limit of the first class interval, construct a table summarizing the sample frequency distribution having intervals of width 5 mpg.
(b) Plot the above as a histogram.
(c) Plot the above as a frequency polygon.
(d) Of the basic population frequency curve shapes described in the text, which one does the above data most closely match?

1-11 Using the gasoline mileages from Problem 1-10, answer the following:
(a) Construct a table for the sample frequency distribution having 2 mpg widths for the class intervals, with 10 mpg as the first lower limit.
(b) Plot the above as a histogram.
(c) Do you think a more accurate data summary might be obtained with intervals narrower or wider than 2 mpg?

1-12 The following data apply to automobiles driven in the United States:

Year	Average Distance (1,000 miles) X	Average Fuel Consumption (gallons) Y
1960	9,450	661
1965	9,390	667
1970	9,980	735
1975	9,630	712
1976	9,760	711
1978	10,050	715
1979	9,480	664
1980	9,140	603
1981	9,000	579
1982	9,530	587
1983	9,650	578
1984	9,790	553
1985	9,830	549

Source: The World Almanac and Book of Facts, 1988, p. 127.

Construct a scatter diagram.

1-13 The frequency distribution of times between arriving teleport messages at the central processing unit of a time-sharing network is as follows:

Time (milliseconds)	Number of Messages
0.0–under 5.0	152
5.0–under 10.0	84
10.0–under 15.0	56
15.0–under 20.0	31
20.0–under 25.0	14
25.0–under 30.0	6
30.0–under 35.0	2

(a) Plot the above as a histogram.
(b) Which one of the basic population shapes described in the text does the above most closely approximate?

1-14 Referring to the interarrival time data in Problem 1-13:
(a) Construct a table showing the relative frequencies. Then find the cumulative relative frequencies for each class interval.
(b) Plot the ogive using cumulative relative frequencies.

(c) Use your graph to determine approximately the durations below which each of the following percentages of interarrival times falls: (1) 90%, (2) 25%, (3) 50%, and (4) 75%.

1-15 The following is the frequency distribution of unscheduled downtime (hours) experienced by a computer mainframe over a sample period:

Downtime	Frequency
0–under 1	35
1–under 2	19
2–under 3	12
3–under 4	7
4–under 5	3
5–under 6	2

(a) Construct a table showing the cumulative number of occurrences for each class interval. From these determine the cumulative relative frequencies.
(b) Plot the ogive using cumulative relative frequencies.
(c) Use your graph to determine approximately the proportion of downtime below (1) 1.5 hours, (2) 3.6 hours, (3) at or above 2.3 hours, and (4) between 1.8 and 4.3 hours.

1-16 The cumulative relative frequency distribution for the depth (in feet) of oil well shafts in a particular region is given here:

Depth of Well	Cumulative Proportion
0–under 1,000	.09
1,000–under 2,000	.35
2,000–under 3,000	.72
3,000–under 4,000	.88
4,000–under 5,000	.95
5,000–under 6,000	.98
6,000–under 7,000	.99
7,000–under 8,000	1.00

Altogether there are 700 wells in the region.
(a) Make a table for the relative frequency distribution.
(b) Using your answer to part (a), determine the original frequency distribution.
(c) Using your answer to part (b), determine the cumulative frequency distribution.

1-17 For each of the following populations, sketch an appropriate shape for the frequency distribution and explain the reasons for your choice:
(a) Content volumes in bottles leaving a filling machine.

(b) Millisecond portion of a computer's clock when processing begins on each new job.
(c) The number of engineers employed in each of several private companies.
(d) Times required to install car bumpers.

1-3 Summary Statistical Measures: Location

Although the frequency distribution arranges the raw data into a meaningful pattern, that summary cannot by itself answer many important statistical questions. For example, an industrial engineer wishing to select the faster of two production methods might obtain sample completion times from pilot runs and then try to reach a decision by comparing the two resulting sample frequency distributions. But unless the sample results are summarized further, beyond frequency tables or histograms, such an evaluation would be cumbersome and might actually cloud the issue. The faster procedure ought to be more clearly indicated by the "average" completion times under the two production methods.

Averages are one class of **statistical measures**. These quantities express various properties of the statistical data. Our discussion begins with measures of **location**. There are two types of location measures. One group expresses **central tendency**. Such quantities indicate the central point around which observations tend to cluster, thus giving a capsule summary of their magnitude. Another group of location measures is concerned with **positions** other than the center and are placed according to frequency of occurrence. Later in this chapter we will describe **variability** or **dispersion**. These quantities express the degree to which observations differ. Even a set of qualitative observations has a summary measure, the **proportion**.

Statistics and Parameters

Summary data measures fall into two major groupings, depending on whether the observations they describe are a population or a sample. When the data constitute a population, each summary measure is referred to as a **population parameter**. But all possible population observations are not ordinarily made, so usually the only available observations are those in a representative sample. A measure that summarizes sample data is called a **sample statistic**. Ordinarily, it is the statistic that is computed from those observations actually made. Important population parameters have counterpart sample statistics that measure the same characteristic, and the latter will often serve to estimate the unknown parameter values.

The Arithmetic Mean

The **arithmetic mean** is the most commonly used and best understood measure of central tendency. Consider the quantitative scholastic aptitude test (SAT) scores achieved by five engineering school applicants: 755, 613, 584, 693, 622. The mean

653.4 is calculated by adding these values and dividing by the number of students (5):

$$\frac{755 + 613 + 584 + 693 + 622}{5} = \frac{3{,}267}{5} = 653.4$$

If the five students represent only a sample from a population of all applicants to a particular engineering school, this calculation is the **sample mean**. If the observations are instead the scores of all the engineering applicants graduating from a particular high school, then they comprise an entire population and 653.4 would then be the **population mean**.

It is convenient to express this procedure symbolically. Consider first the sample mean. Denoting the ith observation in the collection of raw data as X_i and representing the number of observations by n, also referred to as the **sample size**, we denote the sample mean as $\bar{X}$ (X-bar) and compute it as follows:

SAMPLE MEAN

$$\bar{X} = \frac{\sum_{i=1}^{n} X_i}{n}$$

For the engineering applicants we used $n = 5$, with $X_1 = 755$, $X_2 = 613$, $X_3 = 584$, $X_4 = 693$, and $X_5 = 622$ to compute the value of the sample mean, $\bar{X} = 653.4$.

A similar computation provides the **population mean**, which is denoted by the symbol μ (lowercase Greek *mu*). When all of the observations are available, μ is computed in exactly the same way as $\bar{X}$. We reserve the uppercase N to denote the population size. It is rare, however, for μ actually to be computed for most populations encountered in engineering— because of their size, most must be represented by samples. Thus, μ will ordinarily remain an unknown quantity, whereas only $\bar{X}$ actually is computed and serves as an estimator of μ. We may compute the sample mean for the raw calibration and inspection time data from Table 1-1 by first summing all the entries and then dividing by $n = 100$: $\bar{X} = \frac{1{,}434.2}{100} = 14.342$ seconds.

Sometimes, however, the raw data may be unavailable, and the investigator may have access only to the summary provided by the frequency distribution. It is possible in this case to use the grouped data to approximate the value of the sample mean that might be computed were all the raw data accessible. This requires manipulating the quantities describing each class of the frequency distribution. We denote the frequency of the kth class interval as f_k and the **midpoint** as X_k. The following expression may then be used.

SAMPLE MEAN APPROXIMATED FROM GROUPED DATA

$$\bar{X} = \frac{\sum f_k X_k}{n}$$

where $n = \sum f_k$.

The above is equivalent to summing the products of each class midpoint X_k with the relative frequency, f_k/n. Such a calculation is a **weighted average**. Weighted averages play an important role in statistics.

Table 1-6
Calculation of the Mean Calibration Inspection Time Using Grouped Data

Time (seconds)	Number of Times (Frequency) f_k	Class Interval Midpoint X_k	f_kX_k
11.0–under 12.0	2	11.5	23.0
12.0–under 13.0	16	12.5	200.0
13.0–under 14.0	29	13.5	391.5
14.0–under 15.0	27	14.5	391.5
15.0–under 16.0	11	15.5	170.5
16.0–under 17.0	6	16.5	99.0
17.0–under 18.0	4	17.5	70.0
18.0–under 19.0	2	18.5	37.0
19.0–under 20.0	2	19.5	39.0
20.0–under 21.0	1	20.5	20.5
		Total	1,442.0

$$\bar{X} = \frac{\sum f_k X_k}{n} = \frac{1{,}442.0}{100} = 14.42 \text{ seconds}$$

Table 1-6 illustrates the sample mean calculated by using grouped data from the sample frequency distribution found earlier for the calibration inspection times. The value 14.42 computed from the grouped data is only an *approximation* to the sample mean computed directly from the raw data ($\bar{X} = 14.342$). Ordinarily $\bar{X}$ will be computed using just one of the two procedures.

The Median

Another measure of central tendency is the **median**. This quantity often serves as a preferred location for the central value when the observed data involve a skewed frequency distribution. Ordinarily only the **sample median**, denoted by m, is obtained from those observations actually made. It is located by first ordering the data by increasing magnitude and then finding *that value above and below which an equal number of observations lie*. When there is an odd number of observations, the sample median will be the value of middle size.

For example, the SAT scores given earlier may be listed in increasing magnitude:

584 613 622 693 755

The sample median is $m = 622$.

When there is an even number of observations, the sample median is found by averaging the middle two values. As an example, consider the following partial order sequence listing prepared from the original raw data from the calibration inspection times:

47th	48th	49th	50th	51st	52nd	53rd	54th
13.9	14.0	14.0	14.2	14.2	14.2	14.2	14.2

The sample median for the 100 observations is the average of the 50th and 51st values, $m = (14.2 + 14.2)/2 = 14.2$. Since both quantities are the same, the median is the same, too. When there is an even number of observations, the middle two typically will differ. For example, suppose that one more applicant having an SAT score of 648 is added to the original group of engineering applicants, raising n to 6. The middle two values in the order sequence are then 622 and 648, and the sample median is $m = (622 + 648)/2 = 635$.

The procedure for finding m can be very time consuming when n is large, and $\bar{X}$ is ordinarily easier to compute. But the median is an average of **position**, making it often the better representative value. This is especially true with economic data, such as family income. Such data are ordinarily summarized in terms of the median income rather than the mean. The influence of one very rich family could so affect the arithmetic mean that the resulting summary might poorly describe the entire group. But the wealthy family would have "one vote" in placing the median, making it the more democratic measure of central tendency.

The Mode

A third measure of location is the **mode**. Although rarely used alone, this quantity is an important descriptive measure and has some conceptual significance. The mode is *the most frequently occurring value*. For example, suppose that the number of patents held by the members of the mechanical engineering faculty at a certain university are:

0 0 3 7 1 0 2 10 27 15 0 0 1 2

The mode is 0, since that value occurs five times, with 1 and 2 tied for second place with two professors for each.

When the raw data represent values on a continuum, often no single value occurs more than once. In those cases the mode is taken to be *the midpoint of the class interval with the highest frequency*. For example, consider the following grouped sample data:

Class Interval	Frequency
95.0–under 100.0	7
100.0–under 105.0	23
105.0–under 110.0	22
110.0–under 115.0	17
115.0–under 120.0	4
	$n = 73$

The mode is the midpoint of the second class interval, $(100.0 + 105.0)/2 = 102.5$.

When the observations are represented by a frequency curve, the usual descriptive display for entire populations, the mode is the point of highest relative frequency, which is the location of maximum clustering. This feature makes the mode useful in comparing the other measures of central tendency.

Finding the Median and Mode with a Stem-and-Leaf Plot

When the raw data are arranged in a stem-and-leaf plot, the median and mode can be easily found by sorting the leaf digits in ascending value. The following *ordered* stem-and-leaf plot is obtained for the temperature data of Table 1-3:

```
                                         Leaf
                                      (2nd digit)
            1 | 8
            2 |
            3 | 0 5 7 7 8 8 9
   Stem     4 | 2 2 3 3 3 5 5 5 6 6 6 6 6 6 7 7 7 8 8 9 9 9 9 9 9 9
(1st digit) 5 | 0 0 0 0 1 1 1 2 [2 2] 2 3 3 3 4 4 5 5 5 5 6 6 6 7 8 8 9
            6 | 0 0 0 0 1 2 2 2 3 3 5 6 6 6 8 8 8 9 9 9
            7 | 0 2 5 6
            8 | 0
```

The median of the 86 observations is the average of the 43rd and 44th values (the leaf digits for these are shaded):

$$m = \frac{52 + 52}{2} = 52$$

The mode is 49, since the 4-stem has seven 9-leaves, the most frequently occurring of all the leaves on a single stem.

Frequency Distribution Forms and Summary Measures

Figure 1-12 shows three cases ordinarily encountered in statistical investigations that measure physical characteristics or summarize the times for completing an operation. The **symmetrical** frequency curve in (a) represents a population shape often encountered in measurements of dimension. Here the mean, median, and mode all are a common value. In the **skewed** distributions in (b) and (c), the mean is the more extreme population measure, with the median lying between the mode and mean. The direction of skew is expressed by the sign of the difference, mean − median.

The mode concept is used to characterize a population type whose frequency distribution has a double-humped shape, such as those in Figure 1-13. These frequency curves each represent a **bimodal distribution**. A frequency distribution like this is usually obtained when *anthropometric data* (human dimensions) are collected from measurements of men and women. The curve in (a) represents the heights of

Figure 1-12 Positional comparisons of the three measures of central tendency for symmetrical and skewed frequency distributions.

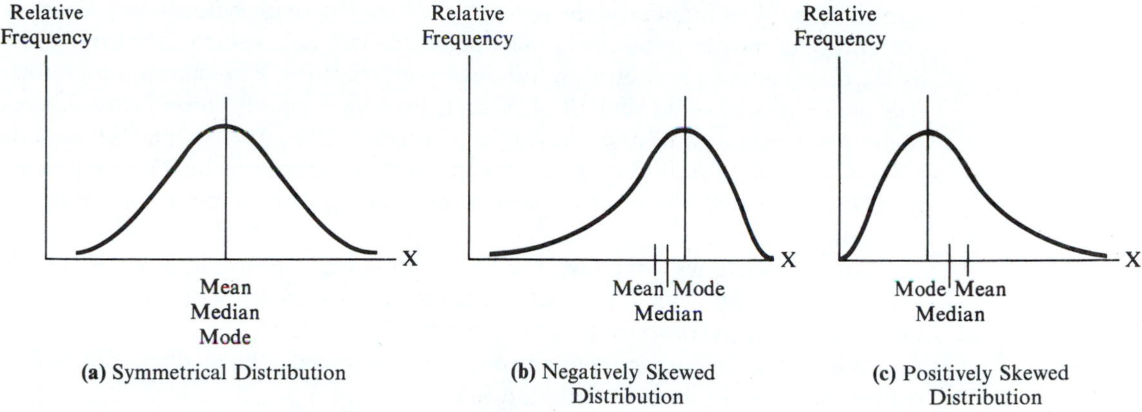

students at a particular university where the sexes are represented equally. It is generally good statistical practice to make separate evaluations of men's and women's heights, treating the populations individually by sex.

Indeed, discovery of a bimodal frequency distribution should signal an investigator that some underlying nonhomogeneity exists in the raw data and that a better description might be obtained by finding the cause and splitting the data into two groups. For example, curve (b) was found for the GPAs of the engineering students in a particular school. Investigation showed that most students whose grades are reflected by the lower hump held jobs, whereas those in the major grouping were largely free of work commitments. Separate evaluations were then made of working and nonworking students. [The lower mode in curve (b) is only a "local" one since it pertains to a small subset of observations. But the frequency curve is still classified as bimodal.]

Figure 1-13 Bimodal frequency distributions.

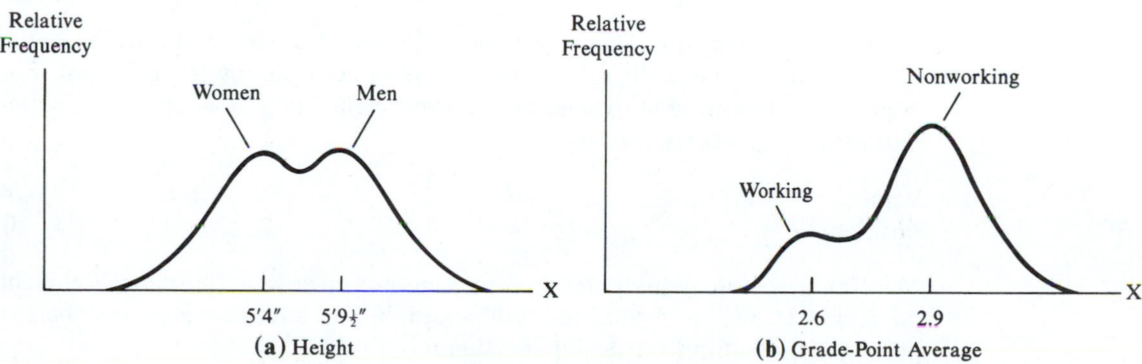

Percentiles, Fractiles, and Quartiles

The final group of location measures includes those concerned with frequency information. Most familiar is the **percentile**, which is a point below which a stated percentage of the observations lie. The percentile will be a value in the same units as the observations themselves, but it need not be unique. For example, an industrial engineer might find that 17 of 20 units involve assembly times below 23 seconds. The percentage of items having such times is $100(\frac{17}{20}) = 85$, and 23 seconds is the 85th percentile. If that engineer finds that 31 seconds is the 95th percentile, then 95% of the 20 test units—19 of them—have assembly times falling below 31 seconds.

An alternative way of conveying this information is to use fractiles. A **fractile** is that point below which a stated fraction of the values lie. Thus, 23 seconds is the .85-fractile and 31 seconds is the .95-fractile.

We have so far encountered one important percentile: the median. *The median is both the 50th percentile and the .50-fractile.*

Frequency information is also expressed in **quartiles**, which divide raw data into four groups of equal frequency. The **first quartile** is the same point as the 25th percentile (.25-fractile), the **second quartile** is the 50th percentile (.50-fractile and also the median), and the **third quartile** is equal to the 75th percentile (.75-fractile).

Example:
Class Standing in Engineering School

The 347 graduates in Bob Blackwell's class were ranked according to their grade-point averages (4.0 being the maximum). Bob's GPA of 3.57 seemed quite high. But that placed him only in the 77th percentile, since 267 (77% of 347) classmates had lower GPAs. Eighty classmates including Bob achieved GPAs of at least 3.57.

It is interesting that the 90th percentile corresponds to a slightly higher GPA of 3.65. Bob is remorseful that just 15 more points on a thermodynamics midterm would have given him an A instead of a B and would have raised his GPA beyond 3.65. That small difference in GPA would have raised him to the 90th percentile and placed him ahead of 312 classmates.

Finding Percentiles from Ungrouped Data When all raw data are available, the percentiles may be established by ranking the values from lowest to highest. For example, the following ten values are sample weights (in grams) of coating materials used in a masking process:

Value	5.3	5.4	5.7	6.0	6.1	6.1	6.2	6.4	6.5	6.6
Position	1	2	3	4	5	6	7	8	9	10

The 10th percentile is any point above 5.3 but not exceeding 5.4 grams. We might use 5.31, 5.33, 5.39, or 5.40 as the 10th percentile. The 20th percentile may be any point greater than 5.4 but less than 5.7 grams.

There is considerable leeway in selecting percentiles directly from raw data. To avoid ambiguity, we adopt the following procedure.

Interpolation Procedure for Locating Percentiles

1. **Sort the raw data in ascending order.** Denote as X_1 the first sorted value, X_2 the second, and so on up to X_n, with n representing the total number of observations. The subscripts represent the *position* of the data value.
2. **Establish the decimal equivalent.** Denote by d the decimal equivalent of the desired percentage point, which must meet the following restriction:

$$\frac{1}{n} \leq d \leq \frac{n-1}{n}$$

Denote by Q_d the corresponding percentile value that is to be found.

3. **Find the relative position of the desired percentile.** This is

$$(n + 1)d$$

Then let k be the largest integer such that

$$k \leq (n + 1)d$$

The desired percentile will lie between X_k and X_{k+1}.

4. **Compute the percentile value.** The following expression applies:

$$Q_d = X_k + [(n + 1)d - k](X_{k+1} - X_k)$$

We illustrate this procedure using the above $n = 10$ sample weights. First we find the 10th percentile ($d = .10$). The relative position of $Q_{.10}$ is $(10 + 1)(.10) = 1.1$. Using $k = 1$, we compute:

$$\begin{aligned} Q_{.10} &= 5.3 + [(10 + 1)(.1) - 1](5.4 - 5.3) \\ &= 5.31 \end{aligned}$$

Next we locate the 25th percentile. The 25th percentile lies between positions 2 and 3, so we have:

$$\begin{aligned} Q_{.25} &= 5.4 + [(10 + 1)(.25) - 2](5.7 - 5.4) \\ &= 5.63 \end{aligned}$$

The 75th percentile lies between positions 8 and 9. We obtain the following:

$$\begin{aligned} Q_{.75} &= 6.4 + [(10 + 1)(.75) - 8](6.5 - 6.4) \\ &= 6.43 \end{aligned}$$

The median is the 50th percentile:

$$\text{Median} = Q_{.50} = 6.1 + [(10 + 1)(.50) - 5](6.1 - 6.1) = 6.1$$

It can be a time-consuming and error-prone task to sort a large set of raw data by hand. It may be easier first to construct an ordered stem-and-leaf plot. Returning to the temperature data given earlier, we can easily count out the positions for any of the $n = 86$ observations.

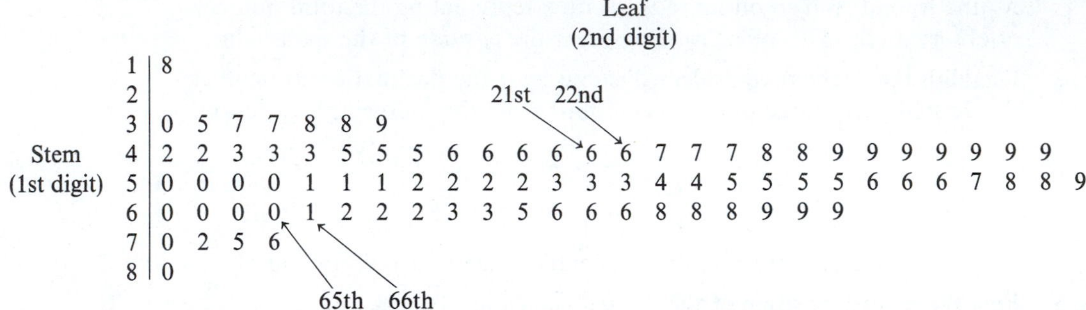

The 25th percentile has relative position $(86 + 1)(.25) = 21.75$, falling above the 21st observation by .75 times the distance between that point and the 22nd observation, both of which have value 46, so that $Q_{.25} = 46$. The 75th percentile is positioned at $(86 + 1)(.75) = 65.25$, so:

$$\begin{aligned} Q_{.75} &= X_{65} + [65.25 - 65](X_{66} - X_{65}) \\ &= 60 + .25(61 - 60) = 60.25 \end{aligned}$$

Finding Percentiles from Grouped Data When the raw data are unavailable, but they have already been grouped into a frequency distribution, the interpolation procedure may be modified slightly to account for the missing details.

Instead of positioning individual observations, we position groups according to the cumulative frequency for class intervals. We denote by k the greatest cumulative class frequency less than or equal to $(n + 1)d$ and represent the *upper limit* of the corresponding interval as X_k. The next higher interval will have cumulative frequency h, with the upper limit designated as X_h. The desired percentile may be computed from

$$Q_d = X_k + [(n + 1)d - k]\left(\frac{X_h - X_k}{h - k}\right)$$

so that Q_d results from a simple linear interpolation between neighboring class limits.

[Should $(n + 1)d$ be smaller than the first cumulative class frequency, the percentile will fall within that first interval. Then $k = 0$, and for $X_k = X_0$ we use the lower limit of the first class interval. The class frequency of that same interval serves as the value for h, with X_h equal to the upper limit of the first interval. Should Q_d fall in the last class interval, h will equal the cumulative frequency for that interval and X_h will equal the upper limit.]

To illustrate, consider the following grouped sample data:

Class Interval	Frequency	Cumulative Frequency
95.0–under 100.0	7	7
100.0–under 105.0	23	30
105.0–under 110.0	22	52
110.0–under 115.0	17	69
115.0–under 120.0	4	73
	$n = 73$	

The 25th percentile is at position (73 + 1)(.25) = 18.5. The greatest cumulative frequency not exceeding 18.5 is 7 for the first interval, so $k = 7$. $Q_{.25}$ will fall in the second interval, and its cumulative frequency is $h = 30$. This percentile lies between the limits $X_7 = 100.0$ and $X_{30} = 105.0$. We compute its value:

$$Q_{.25} = 100.0 + [18.5 - 7]\left(\frac{105.0 - 100.0}{30 - 7}\right)$$
$$= 102.50$$

The 90th percentile lies at position (73 + 1)(.90) = 66.6, so that $k = 52$ and $h = 69$, and

$$Q_{.90} = X_{52} + [66.6 - 52]\left(\frac{X_{69} - X_{52}}{69 - 52}\right)$$
$$= 110.0 + 14.6\left(\frac{115.0 - 110.0}{17}\right)$$
$$= 114.29$$

The median $Q_{.50}$ is located in the middle, at position (73 + 1)(.50) = 37. Using $k = 30$ and $h = 52$, we have

$$\text{Median} = Q_{.50} = 105.0 + [37 - 30]\left(\frac{110.0 - 105.0}{52 - 30}\right)$$
$$= 106.59$$

Finding Percentiles Graphically If the cumulative frequency distribution has already been plotted, percentiles may be read directly from the graph. This is fast and avoids the need for computation.

Figure 1-14 shows the cumulative frequency ogive for the inspection times used earlier. The ordinate now expresses cumulative *relative* frequencies. The 25th percentile is the horizontal coordinate of that point on the curve at height .25:

$$Q_{.25} = 13.25 \text{ seconds}$$

and the 75th percentile is that time where the curve has height .75:

$$Q_{.75} = 15.16 \text{ seconds}$$

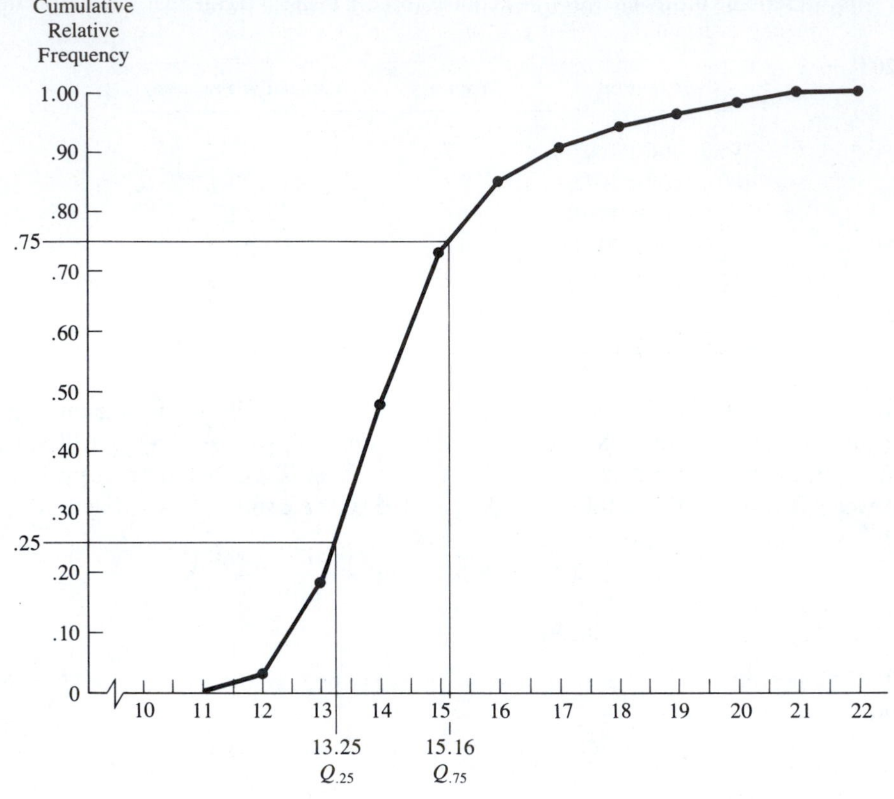

Figure 1-14
Finding percentiles graphically.

The above values should agree closely with those that could be computed from grouped data.

Problems

1-18 The following numbers of positions have been held by a random sample of aerospace engineers during the 10 to 15 years since their graduation:

1	2	3	3	2	5	4	4	3	1
2	1	4	6	5	5	4	2	3	2

Calculate (a) the sample mean, (b) the sample median, (c) the sample mode, (d) the 25th percentile, and (e) the 75th percentile.

1-19 Referring to the materials cost data in Problem 1-1, complete the following:
(a) Calculate the sample mean.
(b) Calculate the sample median.

(c) Determine the difference $\bar{X} - m$. Does this difference indicate that the sample frequency distribution is positively or negatively skewed?

1-20 Referring to the stem-and-leaf plot for the precipitation data in Problem 1-4:
(a) Construct the ordered stem-and-leaf plot.
(b) Find the median.
(c) Find the mode.
(d) Find the following percentiles:
(1) 10th (2) 25th (3) 75th (4) 90th

1-21 Referring to the interarrival time sample frequency distribution in Problem 1-13, calculate (a) the sample mean, (b) the sample median, and (c) the sample mode.

1-22 For each of the following situations, indicate a possible source of nonhomogeneity in the data, and discuss why the population should or should not be split to serve the purposes of the statistical investigation.
(a) A human-factors engineer is establishing new specifications for aircraft hydraulic controls. Actuating one lever requires a considerable amount of strength. The engineer chooses a maximum force that 95% of sample pilots are able to exert while seated in a normal position.
(b) Six months ago a new accident prevention program was initiated by a plant superintendent. For the past year a record has been kept of the number of weekly accidents. An investigation is being conducted to investigate plant safety standards.
(c) A computer software engineer has obtained data on the processing times taken to run payrolls at two identically equipped computer facilities. One center serves New York City clients, the other is located in Nebraska and services clients in neighboring states.

1-23 Referring to the gasoline mileages in Problem 1-10, answer the following:
(a) Calculate the sample mean.
(b) Calculate the sample median.
(c) Determine the difference $\bar{X} - m$. Does this difference indicate that the sample frequency distribution is positively or negatively skewed?

1-24 Referring to the equipment downtime sample frequency distribution in Problem 1-15, calculate (a) the sample mean, (b) the sample median, and (c) the sample mode.

1-25 The following sample data apply to the average yield of a final product (in grams) from each liter of chemical feedstock:

29.84	31.19	31.91	33.78	29.68
32.23	26.23	30.97	31.92	29.19
26.83	31.55	32.00	27.90	27.93
26.59	30.05	26.25	31.23	27.47
31.00	30.02	32.35	29.15	28.82
29.94	30.21	30.62	29.03	31.23
29.81	31.59	30.52	31.01	28.35
27.94	26.68	29.84	30.60	28.10
29.48	27.76	27.59	31.04	30.81
32.22	32.22	27.02	29.03	32.65

Compute (a) the sample mean and (b) the sample median.

1-26 Refer to the data in Problem 1-25. The following frequency distribution applies.

Yield	Number of Batches
25.5–under 26.5	2
26.5–under 27.5	5
27.5–under 28.5	7
28.5–under 29.5	6
29.5–under 30.5	8
30.5–under 31.5	11
31.5–under 32.5	9
32.5–under 33.5	1
33.5–under 34.5	1

(a) Using the grouped data, calculate the sample mean.
(b) Using the grouped data, determine the following percentiles:
 (1) 10th (2) 25th (3) 50th (4) 75th (5) 90th
(c) What is the value of the sample median?
(d) The above statistics are only approximations to the respective values found in Problem 1-25. Determine how much the grouped approximation value lies above or below its counterpart statistic computed from the raw data directly for (1) the sample mean and (2) the sample median.

1-27 The following hourly labor costs were computed for a random sample of small construction projects:

$21.52	$20.76	$21.87	$18.54	$21.36
22.56	22.35	19.81	19.85	24.87
19.83	20.38	21.95	21.33	19.42
23.11	20.73	20.93	20.84	21.23
18.75	23.17	19.05	22.13	19.91
20.50	20.12	21.00	21.92	25.12
22.48	21.75	23.39	20.38	20.13
19.61	19.37	21.05	20.74	20.58
21.11	22.74	20.31	22.05	20.62
20.65	20.84	22.87	22.98	21.75
20.43	24.36	22.17	21.47	21.94
21.61	19.72	21.24	20.26	22.45
19.24	22.25	24.10	23.60	23.05
19.51	20.37	19.73	20.15	19.95
20.48	19.62	20.26	19.84	20.04

Compute (a) the sample mean and (b) the sample median.

1-28 Refer to the data in Problem 1-27. The following frequency distribution applies:

Hourly Cost	Number of Projects
\$18.50–under 19.50	6
\$19.50–under 20.50	23
\$20.50–under 21.50	18
\$21.50–under 22.50	15
\$22.50–under 23.50	8
\$23.50–under 24.50	3
\$24.50–under 25.50	2

Repeat (a)–(d) as in Problem 1-26.

1-4 Summary Statistical Measures: Variability

We have seen several measures of central tendency (location), each of which expresses the center of the observed values in a different way. Often, comparisons of means will be sufficient for evaluating two or more groups of data. However, other summary measures can be used for making comparisons. After location, **measures of variability** provide the next most important descriptive summaries. Such a quantity expresses the degree to which individual observation values differ from each other.

The Importance of Variability

Variability and **dispersion** are the synonymous terms used in statistics to characterize individual differences. The greater the variability between observations, the more they will be spread out. Populations or samples having high variability will have a frequency distribution involving wider class intervals or more classes than a low-variability group measured on the same scale.

The following example will help explain the concept of variability and demonstrates its worth in statistical evaluations.

Example: Comparing Queueing Systems for Computer Workstations

Figure 1-15 shows two frequency polygons constructed from the waiting times experienced by two sample groups of students seeking access to workstations at a university's central equipment room. Two different "queue disciplines" were employed. Under method A, arriving students would stand behind the chair of a student using a terminal, waiting until that or a nearby "unclaimed" workstation was free. Method B is a managed system, whereby arriving students are

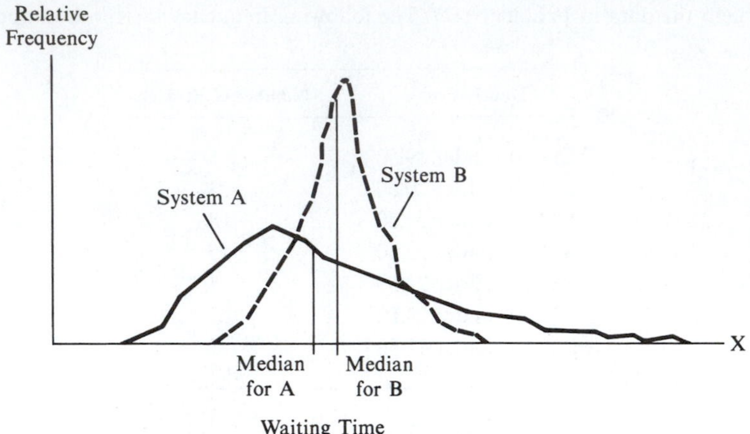

Figure 1-15 *Frequency distributions for waiting time for computer workstations.*

placed in a single queue in order of their arrival. These students are fed one at a time to workstations as they become free.

Notice that sample group B experienced somewhat higher times, as indicated by the level of the sample median. This reflects the "overhead" imposed by the queue manager in determining when workstations are free and in moving the newest student into the work area. But the waiting times under B are less varied and cluster more tightly around the center of the distribution. The anarchy of method A often results in extraordinarily long waiting times. When asked to state their preference for queueing method, the great majority favored method B, which was more predictable, even though it involved longer average waits.

This example shows that variability can provide a second "dimension" for statistical evaluations in which central tendency by itself can often be inadequate for making decisions. There are two main summary measures of variability, one of which takes two forms.

The Range

The simplest measure of variability is the **sample range**, which may be computed by subtracting the smallest observation from the largest. As an example, using the original SAT scores for the five engineering applicants, the highest score was 755 and the lowest was 584. The sample range is

$$\text{Range} = 755 - 584 = 171$$

This value indicates that there is a 171 point spread in SAT scores achieved by the applicants in the sample.

The range is the crudest measure of variability, but it nevertheless can be quite useful. Consider the daily temperature cycle for two locales, both of which experience a mean of 85°F. One cycle might occur in a very pleasant subtropical region where the diurnal temperature ranges from a low of 75° to a high of 95°, with a range of 20°. The other might be in a far less hospitable desert region, where the diurnal temperature can swing from 50° to 120° in the summer and from −20° to 50° in the winter—a 70° range in both cases. An engineer must obviously design different facilities for housing personnel and equipment in the two regions.

One practical difficulty with using the range is that it tends to become larger with increasing sample size. In a group of 100 engineering applicants, the low SAT score might be 485 and the high 783, so that the range would be 783 − 485 = 298. In a group of 10,000 applicants, the extremes might be 350 and 800, giving a range of 450, yet there may be no reason to expect individuals to differ from each other much more in the 10,000-applicant group than in the smaller sample. One way around this distorting effect is to eliminate the **outliers**, or extreme observation values, before computing the range. Doing so can be very subjective, however.

Interquartile Range and Box Plots

A related useful measure of variability, or dispersion, is the *interquartile range*. This represents scatter in the middle 50% of the observations and is the difference between the third quartile ($Q_{.75}$) and the first ($Q_{.25}$).

To illustrate, consider again the ten sample coating material weights (grams):

5.3 5.4 5.7 6.0 6.1 6.1 6.2 6.4 6.5 6.6

Recall that in Section 1-3 we found the following quartile values for these weights:

$$Q_{.25} = 5.63 \qquad Q_{.50} = 6.10 \qquad Q_{.75} = 6.43$$

The interquartile range then is

$$Q_{.75} - Q_{.25} = 6.43 - 5.63 = .80$$

The range and interquartile range may be combined in a **box plot**. Figure 1-16 shows the box plot for the above data. The plot begins with a line segment starting at the minimum observed level and ends with a line segment ending at the maximum value. The box begins at the first quartile, ends at the third, and is divided

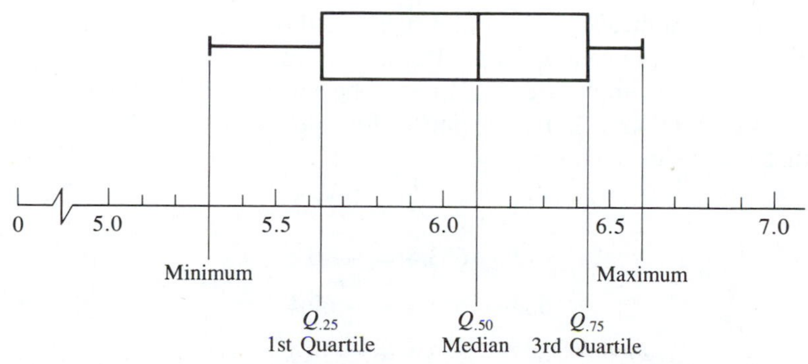

Figure 1-16 *Box plot for sample of coating material weights.*

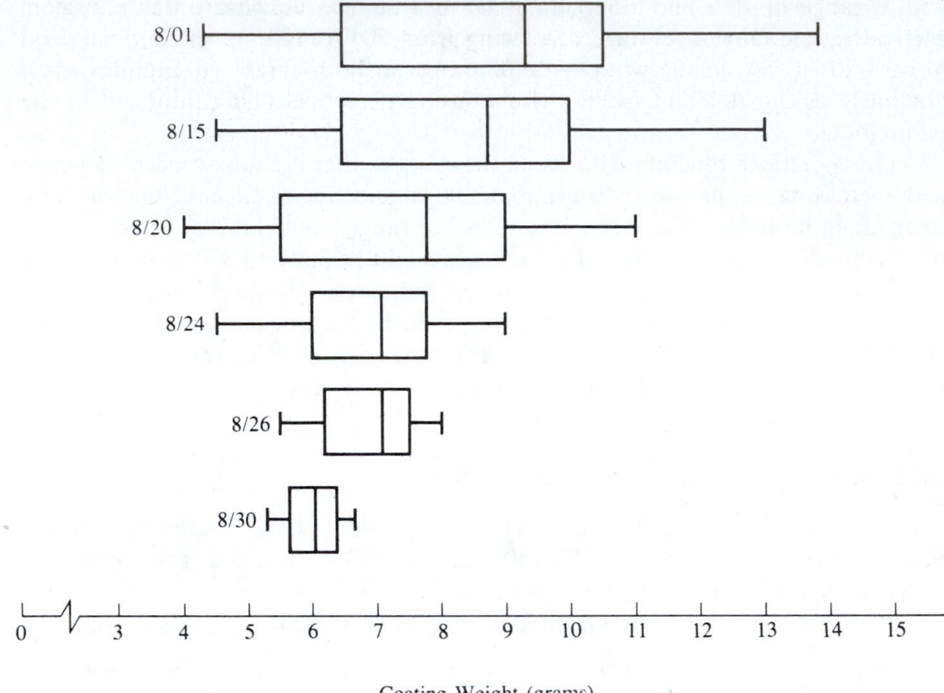

Figure 1-17 Box plots from successive samples of coating weights.

at the median. The overall length of the plot gives the range, while the length of the box provides the interquartile range.

Box plots can be useful when comparing data groups. Figure 1-17 shows box plots from successive samples of coating weights. Notice that as the coating process was refined over time, the sample units received less material, as reflected by progressively lower medians. Concurrently, the weights for individual units became less varied and more uniform, demonstrated by progressively shorter plots and boxes.

The Variance and Standard Deviation

The most important measure of variability is based on the **deviations** of individual observations about the central value. For this purpose, the mean usually serves as the center. For example, we determined the mean of 653.4 for the five SAT scores given in Section 1-3. The deviation for each score may be computed by subtracting that value from each:

$$
\begin{aligned}
755 - 653.4 &= 101.6 \\
613 - 653.4 &= -40.4 \\
584 - 653.4 &= -69.4 \\
693 - 653.4 &= 39.6 \\
622 - 653.4 &= -31.4
\end{aligned}
$$

These deviations may be summarized by a collective measure that considers each deviation. Obviously they cannot be averaged directly, since their mean must be zero (which you can verify as an exercise). One measure of the average deviation is achieved by first taking the absolute values. Unfortunately, absolute values are mathematically cumbersome. Instead, the deviations are first squared. Then the squared deviations are averaged. The resulting summary is the **variance**, which is the mean of the squared deviations about the central value. With the above data, this procedure results in

$$\frac{(101.6)^2 + (-40.4)^2 + (-69.4)^2 + (39.6)^2 + (-31.4)^2}{5} = \frac{19{,}325.2}{5} = 3{,}865.04$$

When all possible observations are taken, the raw data constitute the entire population. The quantity calculated in this manner after a complete enumeration or census is the **population variance**. This quantity is denoted by σ^2 (lowercase Greek sigma, squared) and may be computed from

$$\sigma^2 = \frac{\sum_{i=1}^{N} (X_i - \mu)^2}{N}$$

where X_i denotes the ith observation. Each deviation is found by subtracting the population mean μ from the observation. The divisor is the population size N. In the SAT score illustration we computed the variance of the population of scores for $N = 5$ engineering applicants from a particular high school.

In practice σ^2 cannot be computed directly because the entire population is ordinarily not observed. An analogous measure of variability may be determined with sample data. This is referred to as the **sample variance**, which is represented as s^2. In place of μ, the sample mean $\bar{X}$ is used for computing individual deviations, and the sample size minus 1 replaces the population size. Sample variance is computed as follows:

SAMPLE VARIANCE

$$\mathbf{s^2 = \frac{\sum_{i=1}^{n} (X_i - \bar{X})^2}{n - 1}}$$

We illustrate this equation using the sample times from Table 1-1 for completing the calibration inspections. Subtracting the sample mean $\bar{X} = 14.342$ from each observed value, squaring the differences, and summing, the total of 292.484 is obtained. Thus,

$$s^2 = \frac{292.484}{100 - 1} = 2.9544$$

The expression has a slightly different form than that for σ^2 because its divisor is $n - 1$ rather than n. Reducing the divisor by 1 in computing the sample variance results in a better estimator of σ^2. The theoretical reason for this is described in Chapter 6.

The variance is analogous to the second moment in physics. It not only provides a useful index for gauging individual differences among the observations, but σ is a parameter in certain mathematical functions that are used to represent frequency curves for some populations. The variance also has important properties that are useful to statistical theory and can facilitate statistical evaluations.

As a practical matter, though, the variance is in units different from the observations themselves. In our SAT score illustration the variance is 3865.04 *squared* points. Likewise, the sample variance for the calibration inspection times is in seconds squared. Partly because of the confusion of different units and partly because the variance can be orders of magnitude above (or below) the size of the observed values, statistical evaluations are conducted with the variance's positive square root. The resulting quantity is called the **standard deviation**. The population and sample standard deviations are denoted, respectively, as σ and s.

The standard deviations for our two illustrations are:

$$\sigma = \sqrt{\sigma^2} = \sqrt{3{,}865.04} = 62.17 \qquad \text{(population of SAT scores)}$$

$$s = \sqrt{s^2} = \sqrt{2.9544} = 1.72 \qquad \text{(sample of inspection times)}$$

These values are in the same units as the observations themselves and provide more convenient summaries than the variances. Since one measure can be obtained directly from the other, the variance and standard deviation are used interchangeably.

When computing s^2 (or s) by hand, it is usually convenient to use the following equivalent expression for the sample variance:

$$s^2 = \frac{\sum_{i=1}^{n} X^2 - n\bar{X}^2}{n - 1}$$

One disadvantage of the above is the possibility of roundoff errors. (As an exercise you may verify that this expression is equivalent to the earlier one for s^2.)

As with the sample mean, when the raw data are no longer available the sample variance may be approximated from grouped data provided by the frequency distribution. This is done using the following expression.

SAMPLE VARIANCE APPROXIMATED FROM GROUPED DATA

$$\mathbf{s^2 = \frac{\sum f_k X_k^2 - n\bar{X}^2}{n - 1}}$$

Here the summation is taken over all classes and X_k is the midpoint of the kth class. The sample mean will ordinarily also be approximated from grouped data. As an illustration, the grouped approximation is applied in Table 1-7 with the frequency distribution for calibration inspection times. Notice that the variance computed under this procedure is slightly larger than the value for s^2 computed earlier directly from the raw data. Keep in mind that statistics computed from grouped data are only approximate values.

The Meaning of the Standard Deviation

The population standard deviation is an important index of variability that conveys a great deal of information. It is a fundamental parameter of the mathematical

Table 1-7 *Calculation of the Sample Variance for Calibration Inspection Time Using Grouped Data*

Time (seconds)	Number of Times (Frequency) f_k	Class Interval Midpoint X_k	X_k^2	f_kX_k	$f_kX_k^2$
11.0–under 12.0	2	11.5	132.25	23.0	264.50
12.0–under 13.0	16	12.5	156.25	200.0	2,500.00
13.0–under 14.0	29	13.5	182.25	391.5	5,285.25
14.0–under 15.0	27	14.5	210.25	391.5	5,676.75
15.0–under 16.0	11	15.5	240.25	170.5	2,642.75
16.0–under 17.0	6	16.5	272.25	99.0	1,633.50
17.0–under 18.0	4	17.5	306.25	70.0	1,225.00
18.0–under 19.0	2	18.5	342.25	37.0	684.50
19.0–under 20.0	2	19.5	380.25	39.0	760.50
20.0–under 21.0	1	20.5	420.25	20.5	420.25
			Totals	1,442.0	21,093.00

$$\bar{X} = \frac{\sum f_k X_k}{n} = \frac{1{,}442.0}{100} = 14.42 \text{ seconds}$$

$$s^2 = \frac{\sum f_k X_k^2 - n\bar{X}^2}{n - 1} = \frac{21{,}093.00 - (100)(14.42)^2}{100 - 1} = 3.0238 \text{ seconds}^2$$

function describing the normal curve that fits the frequency distribution of so many populations. But σ is useful in describing many other frequency distributions as well. A theoretical property referred to as **Chebyshev's Theorem** can be used to establish limits within which the proportion of observations in any population must fall. We will save the details of this theorem until Chapter 4. It indicates that at least 75% of all population values will lie within the limits $\mu \pm 2\sigma$, whereas at least 89% will fall inside $\mu \pm 3\sigma$ and 96% within $\mu \pm 5\sigma$. These limits can often be considerably narrowed once the shape of the population frequency curve is identified.

Composite Summary Measures

The various summary statistical measures can be combined to provide meaningful information. Two of these are especially useful in data evaluations.

The **coefficient of variation** relates variability in the sample or population to the mean. It is calculated by dividing the standard deviation by the sample mean:

SAMPLE COEFFICIENT OF VARIATION

$$v = \frac{s}{\bar{X}}$$

The population coefficient of variation is computed by the analogous ratio, σ/μ.

To illustrate, consider again the inspection times from Table 1-1. We have

$$v = \frac{s}{\bar{X}} = \frac{1.72}{14.342} = .120$$

This indicates that the sample standard deviation for the inspection times is only 12% as large as the mean level.

By itself, the standard deviation does not convey the *relative* degree of variability. For example, consider the mean and standard deviation in men's hand length and reach:*

Hand Length	**Reach**
$\mu = 7.44''$	$\mu = 32.33''$
$\sigma = .34''$	$\sigma = 1.63''$
$\sigma/\mu = .045$	$\sigma/\mu = .050$

The respective standard deviations suggest that reach has about five times the variability of hand length. But the relative variability, as expressed by the coefficient of variation, is nearly identical for the two populations.

The coefficient of variation might be used as the basis for sorting statistical distributions into like clusters. Table 1-8 shows the results of an experiment on work-time measurement. Each entry represents sample observations of the time to complete a work element. The coefficients of variation range from .215 to .572. Unfortunately, the researchers found no consistent pattern relating type of work element to coefficient of variation.

Their data illustrate a second composite measure, the **coefficient of skewness**. This quantity expresses the direction of and the degree to which a frequency distribution is skewed; it involves the mean, median, and standard deviation. The following expression is used to compute it:

SAMPLE COEFFICIENT OF SKEWNESS

$$\mathbf{SK} = \frac{3(\bar{X} - m)}{s}$$

To illustrate, consider the inspection times from Table 1-1. We compute:

$$\text{SK} = \frac{3(\bar{X} - m)}{s} = \frac{3(14.342 - 14.2)}{1.72} = 0.25$$

which exceeds zero, indicating a slight *positive skew*. The histograms in Figure 1-4 give visual confirmation of this finding.

The research data in Table 1-8 show coefficients of skewness ranging from 0.298 to 3.224. The researcher concluded that such wide-ranging skewness reflects a lack of homogeneity in the parent work-time distributions.

* *Source:* H. T. E. Hertzberg, G. S. Daniels, and E. Churchill, *Anthropometry of Flying Personnel*, WADC Technical Report 52-321, U.S.A.F. (September 1954).

Table 1-8 *Sample Data from Work-Time Measurement Experiment*

Work Element	Sample Mean (0.01 min) X	Sample Median (0.01 min) m	Sample Standard Deviation (0.01 min) s	Sample Coefficient of Variation v	Sample Coefficient of Skewness SK
1	11.139	9.833	3.338	0.346	1.174
2	5.604	4.613	2.354	0.420	1.263
3	2.540	1.908	0.588	0.232	3.224
4	4.229	3.133	1.068	0.253	3.079
5	9.957	9.081	2.141	0.215	1.227
6	2.913	2.068	1.665	0.572	1.523
7	2.576	1.858	1.451	0.563	1.484
8	5.990	5.070	2.021	0.337	1.366
9	4.467	3.295	2.435	0.545	1.444
10	2.969	2.432	0.881	0.297	1.829
11	3.532	2.790	1.039	0.294	2.142
12	4.465	3.698	1.446	0.324	1.591
13	5.903	4.736	1.832	0.310	1.908
14	3.305	3.197	1.087	0.329	0.298
15	3.210	2.520	1.446	0.452	1.432
16	5.984	5.362	1.789	0.299	1.043
17	4.126	3.378	1.678	0.219	1.337
18	7.270	6.461	2.034	0.263	1.193
19	2.770	2.133	1.063	0.384	1.797
20	6.673	5.914	1.968	0.295	1.157
21	8.530	7.833	1.845	0.216	1.052
22	3.928	3.325	1.688	0.430	1.072
23	5.247	4.604	1.469	0.280	1.313
24	5.094	4.402	1.234	0.242	1.682
25	5.452	4.750	2.079	0.381	1.013
26	3.653	2.958	0.982	0.269	2.123

Source: Knott, Kenneth, and Roy J. Sury, "A Study of Work-Time Distributions on Unpaced Tasks," *IIE Transactions* (March 1987): 50–55.

Problems

1-29 The following sample data represent the observed number of busy teleports in a computer time-sharing network:

10	4	15	17	6	12	9	13	15	5

(a) Calculate the sample range.
(b) Calculate the sample variance and the sample standard deviation.

(c) Find the following fractiles:
(1) .25 (2) .50 (3) .75
(d) Calculate the interquartile range.
(e) Construct the box plot.

1-30 The following grade-point averages apply to a random sample of graduating seniors.

3.65	2.73	2.35	3.09	3.28
3.51	2.86	2.59	3.13	3.24

Repeat (a)–(e) as in Problem 1-29.

1-31 Refer to the data in Problem 1-29.
(a) Compute the sample coefficient of variation.
(b) Find the sample median. Then compute the sample coefficient of skewness.

1-32 Repeat Problem 1-31 using the data from Problem 1-30 instead.

1-33 Refer to the chemical yield data in Problem 1-25. Compute (a) the sample range, (b) the sample variance, and (c) the sample standard deviation.

1-34 Using the grouped data from the sample frequency distribution for interarrival times in Problem 1-13, compute the sample variance and standard deviation.

1-35 For each of the following evaluations discuss why a measure of central tendency may not be a wholly adequate summary by itself.
(a) A human-factors engineer is using a sample of past temperatures experienced in a clean room to determine specifications for a revised air-conditioning system.
(b) An industrial engineer wants to sequence assembly tasks to achieve balance between workstations so that excessive buffer stocks are avoided and worker idle time is minimized. Completion times for each task are used in the evaluation.
(c) In planning for facilities expansion, a construction maintenance superintendent requires data on the repair times of equipment.

1-36 Refer to the hourly labor cost data in Problem 1-27. Compute (a) the sample range, (b) the sample variance, and (c) the sample standard deviation.

1-37 Using the grouped data from the sample frequency distribution for computer downtimes in Problem 1-15, compute the sample variance and standard deviation.

1-5 Summary Statistical Measures: The Proportion

Qualitative observations form an important class of statistical data. The primary summary measure for such data is the **proportion**. When the observations constitute the complete population, this quantity is the **population proportion**, a parameter we denote by the symbol π (lowercase Greek *pi*, which represents the first letter in the word *proportion* and is a quantity between 0 and 1—*not* 3.1416). The analogous measure when the observations are sample data and only representative of some population is the **sample proportion**, which we denote by P. Ordinarily only this sample statistic is computed, so that P serves as an estimator of π. The following expression is used to calculate P.

SAMPLE PROPORTION

$$P = \frac{\textbf{Number of observations in category}}{\textbf{Sample size}}$$

If a sample of $n = 100$ engineers contains 16 civil engineers, then the sample proportion of persons in that discipline is $P = \frac{16}{100} = .16$. A sample of 37 silicon chips might contain three unusable ones, so the sample proportion unusable is $P = \frac{3}{37} = .08$. The manufacturing process for the same chips might consistently produce unusable chips at a lower rate, so the population proportion of unusables might be only $\pi = .05$.

The proportion is an important decision parameter in a wide variety of statistical applications. It is the central focus of many quality-control decisions, where equipment must be adjusted if the proportion of defectives is too high. A major concern in chemical processing is the proportion of volume containing impurities (often expressed as a percentage or in parts per million). The proportion also is an important descriptor of various systems in which the proportion of time in a particular state is crucial.

Example:* *Evaluating Equipment Maintenance Policies

The proportion played a key role in one engineer's evaluation of maintenance policies for papermaking equipment. Under the present policy, random observations provided the following:

Equipment State	Number of Observations	Proportion
Satisfactory	78	$\frac{78}{135} = .578$
Erratic	42	$\frac{42}{135} = .311$
Inoperable	15	$\frac{15}{135} = .111$
	135	1.000

The engineer combined the above proportions with cost information to determine the net cost of the present maintenance schedule, under which a complete adjustment was done whenever the equipment was found erratic. A second maintenance policy of "benign neglect," with repairs made only when the equipment was inoperable, was similarly evaluated. The proportion of time in the latter state was found to be so high that the proposed plan would have been more costly than the present one.

Problems

1-38 The following sample data apply to positions held by recent civil engineering graduates:

	Government (1)	Construction (2)	Design Firm (3)	Graduate School (4)	Other (5)
Men	28	42	17	15	21
Women	17	8	9	17	15

(a) Calculate the sample proportion of women in each position category.
(b) Which category has proportionally the lowest concentration of women?
(c) Which category has proportionally the highest concentration of women?

1-39 Refer to the completion time data in the calibration inspection illustration in Table 1-1. Determine the sample proportion of times (a) below 14.0 seconds, (b) below 16.0 seconds, (c) greater than or equal to 15.0 seconds, and (d) greater than or equal to 18.0 seconds.

1-40 An inspector scraps any production batch yielding a sample proportion of defectives exceeding .10. For each of the following batches (a) determine the sample proportion defective, and (b) indicate whether the batch will be saved or scrapped.

	Batch (1)	(2)	(3)	(4)	(5)
Items inspected	50	75	100	150	200
Items defective	8	3	21	7	15

1-41 Refer to the student data in Problem 1-7. Determine the sample proportion of students who are (a) electrical engineers, (b) graduates, (c) receiving financial aid, (d) undergraduate electrical engineers, and (e) electrical engineers receiving financial aid.

1-42 Refer to the student data in Problem 1-7. Determine the following:
(a) The sample proportion of undergraduate students who receive financial aid.
(b) The sample proportion of graduate students who receive financial aid.
(c) Does the sample evidence indicate that graduates involve a higher or lower proportion of students on financial aid than do undergraduates?

1-43 A plant initiates a policy that the machinery must be adjusted whenever the proportion of broken parts in any hour exceeds 2% of total production during that span. In a particular hour, the following findings were achieved:

	Machine (1)	(2)	(3)	(4)
Number of broken parts	5	23	12	3
Number of satisfactory parts	512	624	387	432

(a) Determine the population proportion of broken parts for each machine.
(b) Which machines will be adjusted?

Comprehensive Problems

1-44 The number of lost days per month of drilling experienced by a sample of offshore platforms during 20 months are given below:

0	5	7	4	4	10	5	2	1	0
2	3	3	6	5	4	4	2	0	1

Determine values for the following sample statistics: (a) mean, (b) median, (c) mode, (d) range, (e) variance, (f) standard deviation, (g) coefficient of variation, and (h) coefficient of skewness.

1-45 The following sample data for completing the framing of a standard tract house are incomplete:

Time (days)	Number of Houses	Relative Frequency	Cumulative Frequency
5–under 6	—	—	—
6–under 7	19	—	27
7–under 8	—	.35	—
8–under 9	15	.15	—
9–under 10	—	—	91
10–under 11	—	—	—
Totals	100	1.00	

(a) Determine the missing values.
(b) Plot the cumulative frequency ogive for these data.

1-46 Suppose that your statistics instructor plans to give five quizzes during the term and that your grade will be based on the computed measure of central tendency, which you get to select now. Would you prefer the mean or the median? Explain.

1-47 The following sample data represent the annual maintenance costs (in thousands of dollars) for heavy earth-moving equipment:

24	21	20	31	25
41	43	32	17	25
32	30	30	26	34
36	39	27	28	31
21	19	25	26	31

(a) Construct a table for the sample frequency distribution using intervals of width 5 thousand dollars, with 15 thousand dollars as the lower limit of the first interval.
(b) Plot a histogram and frequency polygon for these data.

1-48 Refer to the data in Problem 1-47:
(a) Construct a stem-and-leaf plot.
(b) Construct an ordered stem-and-leaf plot.
(c) Using the original ungrouped data, find the following sample statistics: (1) mean, (2) standard deviation, and (3) mode.
(d) Using the original ungrouped data, find the following percentiles: (1) 10th, (2) 25th, (3) 75th.
(e) What is the value of the sample median?
(f) Compute the following sample statistics: (1) range, (2) interquartile range.
(g) Construct the box plot.

1-49 Refer to the data presented in Problem 1-47. Answer the following:
(a) From your grouped data, determine the value of (1) the sample mean, (2) the sample median, (3) the sample mode.
(b) From your grouped data, determine the value of (1) the sample variance and (2) the sample standard deviation.

1-50 Refer to the data presented in Problem 1-47. Answer the following:
(a) Determine the relative frequencies for each class interval.
(b) Determine the cumulative relative frequency for each class interval.
(c) Using your results from (b), plot the ogive. Then determine approximately from your graph the proportion of maintenance costs falling below (1) 23 thousand dollars and (2) 32 thousand dollars.
(d) Determine from your graph the approximate costs below which the proportion of observed values is (1) .25, (2) .50, (3) .75, and (4) .90.

1-51 A sample of projects has been categorized by size and by duration. The following data apply:

Size	Duration	Number of Projects
Small	Short	12
Medium	Short	23
Large	Short	8
Small	Long	18
Medium	Long	15
Large	Long	5

Construct a table for the frequency distribution for the samples characterized by (a) size and (b) duration.

1-52 Refer to the sample data in Problem 1-51. Determine the sample proportion of projects that are (a) small, (b) large, (c) small and short, and (d) large and long.

1-53 A chemical engineer fine-tuned a chemical process by experimenting with control settings. The following completion times (in minutes) were obtained for sample runs, with all settings unchanged during a single day.

Day 1	Day 2	Day 3	Day 4	Day 5
33	34	38	35	38
51	37	30	39	39
22	44	35	37	37
26	28	33	37	38
	27	42		
		32		

The engineer has two objectives: (1) to minimize variability in completion times and (2) to minimize average completion time.

(a) Determine for each day the following sample statistics: (1) mean, (2) standard deviation, (3) first quartile, (4) median, (5) third quartile.

(b) Compute for each day the sample (1) coefficient of variation and (2) coefficient of skewness.

(c) Construct on the same graph a box plot for each day's completion times.

(d) Comment on how near the engineer has come to finding settings that meet his goals.

CHAPTER TWO

Probability

A **PROBABILITY** IS a numerical value expressing the degree of uncertainty regarding the occurrence of an *event*. Probability theory establishes the analytical framework for statistics, allowing us to deduce the composition of samples and to quantify our expectations regarding their properties. These same concepts also lay the foundation for modern decision analysis, which can guide us in picking the best alternative, even when the eventual outcome from that choice is uncertain.

The roots of modern probability theory can be traced back 200 years, when mathematicians entertained themselves by investigating popular gambling games. From such seamy beginnings, probability has evolved into one of the more fascinating branches of mathematics. Even today, devices we ordinarily associate with gambling—cards, coins, and dice—are useful props in setting the stage for the more serious applications of probability.

2-1 Fundamental Concepts of Probability

Probability theory has two basic building blocks. The more fundamental is the generating mechanism, called the **random experiment**, that gives rise to uncertain outcomes. A coin toss may be viewed as a random experiment wherein the eventual outcome "upside face showing" cannot be predicted and is subject to chance. Only slightly more complex is the random experiment of selecting one part for inspection from an incoming shipment; here there are two elements of uncertainty: which particular item gets picked and the quality of that part.

The second structural element of probability theory consists of the random experiment's outcomes, referred to as **events**. With the coin toss, two events, "head" and "tail," apply to the showing face. During a quality-control inspection, the usual events of interest for tested items are "good" and "defective." In either of these illustrations, a great number of other uncertain outcomes might exist. (Consider the angle of Lincoln's nose on the coin relative to the North Pole or the number of spins taken by the coin before landing. A defective part might be too wide, too heavy, or off-color.) However, only those outcomes of primary concern are classified as events.

Elementary Events and the Sample Space

A preliminary step in a probability evaluation is to catalog the events that might arise from the random experiment. Such a listing is made up of **elementary events**, which are the most detailed events of interest. A complete listing provides us with the random experiment's **sample space**.

For example, if we are interested only in the particular *upside showing face* from the coin toss, then we would have

$$\text{Coin toss sample space} = \{\text{head, tail}\} \qquad \text{(showing face)}$$

Alternatively, we might care only about the number of complete *revolutions*, and the same random experiment would yield an entirely different sample space:

$$\text{Coin toss sample space} = \{0, 1, 2, 3, \ldots\} \qquad \text{(revolutions)}$$

A sample space might be portrayed as a list, as above, or in some other convenient form. Sometimes it is conceptually helpful to use a picture, as in Figure 2-1, in which biological symbols are used to represent individual members of a Tau Beta Pi chapter. The random experiment is the selection of one member by lottery. Each person is an elementary event.

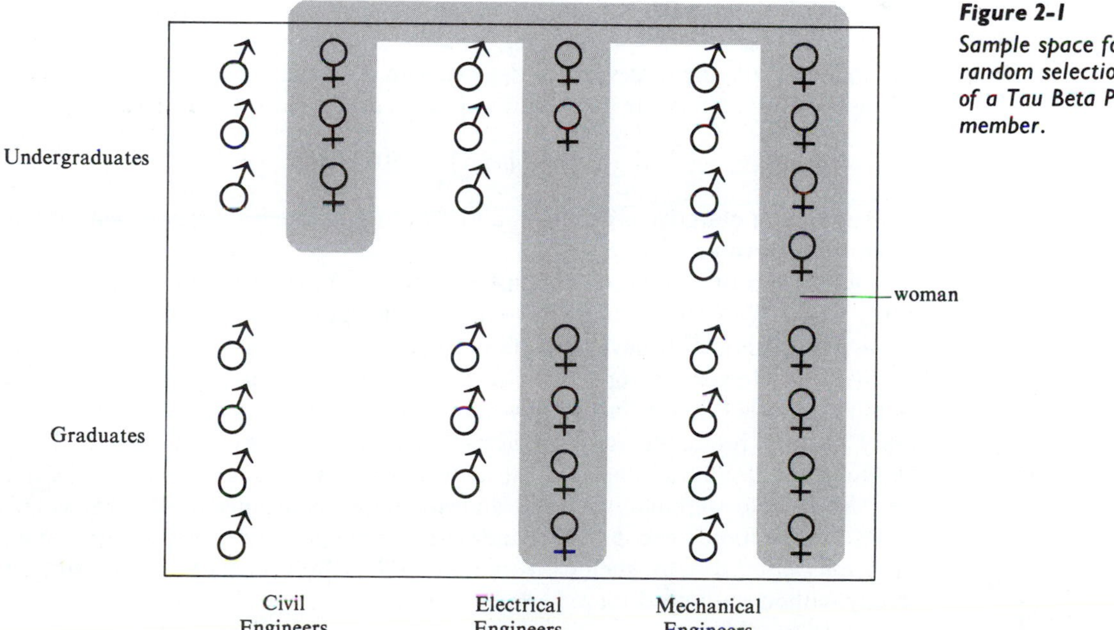

Figure 2-1 *Sample space for random selection of a Tau Beta Pi member.*

Event Sets

The immediate concern when finding a probability is properly identifying the event. This may be achieved by listing all the elementary events that give rise to the desired event. Such a collection is referred to as an **event set**.

Consider again the Tau Beta Pi lottery. A list of the names of female members represents the event set for the event "woman."

$$\text{Women} = \{\text{Ann, Barbara, Betty, Cheryl, Diane, \dots}\}$$

This same event set might instead be pictured. Consider the tinted area in Figure 2-1.

An event set can consist of as few as one element, as in a single coin toss where

$$\text{Head} = \{\text{head}\}$$

It can even be an **empty set**. Referring again to Figure 2-1, there are no female graduate civil engineering majors, so

$$\text{Female graduate civil engineer} = \{\quad\} = \varnothing$$

Later we will see how the fundamental concepts of **set theory** can help explain probability theory.

Basic Definitions of Probability

The most common probability measure expresses the *long-run frequency* for an event occurring in many repeated random experiments. If a perfectly balanced coin is tossed *many times* without bias toward any one side, we should obtain a head in about half the tosses. Thus, the long-run frequency of head is .50. We may then assume that .50 is the probability of head, expressed symbolically as

$$\Pr[\text{head}] = .50$$

This value is an **objective probability**, and there should be no disagreement about how to find its value.

Objective probabilities can often be found, as above, through deductive reasoning alone. We don't actually have to toss the coin to reach the answer, since we have no reason to believe that the head side will show any more or less than half the time. However, suppose our coin has been made lopsided by dropping a few blobs of solder on one face. We would no longer be able to establish the probability value by reasoning alone, and Pr [head] might be a number such as .37 or .59. Only by actually tossing the coin many times, and keeping records of what sides show, could we finally arrive at an estimate of the proper probability value.

Many commonly encountered random experiments involve elementary events that are *equally likely*. In such cases the probability for an event can be deduced directly without any need for experimentation.

PROBABILITY WHEN ELEMENTARY EVENTS ARE EQUALLY LIKELY

$$\mathbf{Pr[event]} = \frac{\textbf{Size of event set}}{\textbf{Size of sample space}}$$

To apply the above "count-and-divide" procedure we need only know (1) how many elementary events constitute the event set in question and (2) the total number of possibilities. For example, we know that "head" is the 1 and only elementary event in the event set for that face, that it is equally likely to occur versus "tail," and that these 2 elementary events constitute the sample space for the showing side from a coin toss. Thus,

$$\Pr[\text{head}] = \frac{1}{2}$$

Consider again the Tau Beta Pi lottery summarized in Figure 2-1. We can apply this procedure to determine the probability that a woman will be selected. There are 17 women (that is, "woman" has an event set containing 17 elementary events, so that its size is 17). The sample space is made up of 38 equally likely elementary events, and

$$\Pr[\text{woman}] = \frac{17}{38} = .447$$

The procedure is valid only when the elementary events are equally likely. Suppose that the Tau Beta Pi lottery is based on selecting one slip from several placed in a hopper. Every time a member attends a chapter event, his or her name gets added to the container of slips. Since individual members might vary in their devotion to chapter affairs, the number of slips bearing a member's name will vary. The elementary events (as defined by a member's name) would no longer be equally likely. The probability of "woman" may then differ—perhaps considerably—from that above.

To find the probability of "woman" in the revised lottery, we would have to *redefine the elementary events* to be slips rather than names. We could then divide the number of slips bearing female names by the total number of slips.

Subjective and Objective Probability

So far we have encountered *repeatable* random experiments. That is, we could generate a whole series of similar events by simply tossing the coin again or repeating the lottery from scratch. Probability values for such experiments—whether obtained by counting and dividing or by experimentation—are long-run frequencies. Since there can be no disagreement about how they are found, we have called them objective probabilities.

Many random experiments are *nonrepeatable*. For example, a wildcat oil well can be drilled just once in a particular location. The long-run frequency for striking oil is a meaningless concept. A new fabrication process for jet engine blades will either be feasible or not feasible; whichever event applies will be found out once and for all. A prototype airfoil either will be acceptable or will have to be redesigned—but any uncertainty will be of a one-shot nature. Again, long-run frequency has no meaning.

Nonrepeatable random experiments must be quantified differently. For this purpose we use **subjective probabilities**, sometimes called *personal probabilities*, which are so identified because two people can legitimately disagree about what the proper value is. Subjective probabilities are most like the betting odds someone might establish for some political or sporting event. Personal probabilities are widely accepted in private-sector managerial analysis, where potential disagreements can ultimately be resolved by a single decision-maker. They are less useful in evaluating public decisions in which societal considerations abound.

Certain and Impossible Events

The basic definitions of probability express the value as a fraction because an objective probability is a long-run frequency. As such, it must be a value between 0 and 1. Events having these extreme probabilities provide two interesting cases.

An outcome having zero probability is called an **impossible event**, since it cannot occur. Returning to the lottery, we found that the empty set

$$\text{Female graduate civil engineer} = \{\ \} = \varnothing$$

applied because there was no student (elementary event) in the above category. The event "female graduate civil engineer" is an impossible lottery event. Applying the basic definition for equally likely elementary events, we have

$$\Pr[\text{female graduate civil engineer}] = \frac{0}{38} = 0$$

All impossible events have zero probability—even when they can be attributed to nonrepeatable random experiments. For example, the U.S. Constitution requires that the president be at least 35 years old, so that

$$\Pr[\text{age of next U.S. president is } 25] = 0$$

An outcome bound to occur is a **certain event** and has a probability of 1. Returning to the lottery, all possible names to be drawn are those of students, who constitute the entire sample space. Hence,

$$\Pr[\text{student is selected}] = \frac{38}{38} = 1$$

In tossing a coin, just two faces, "head" and "tail," are possible. (Should the coin land on its edge or be lost, the random experiment would be incomplete and another toss would have to be made.) Thus,

$$\Pr[\text{head } \textit{or} \text{ tail}] = \frac{2}{2} = 1$$

And according to provisions of the U.S. Constitution,

$$\text{Pr}[\text{age of next U.S. president} \geq 35] = 1$$

Example: Quantum Mechanical Explanation of Photon Filtering

One of the findings of quantum mechanics is that a photon can exist in an ambiguous state until it is measured. Thus, the measured state might be an uncertain event for which a probability might be found. Consider an ideally polarized film and the way in which it filters light that is incident on it at a right angle. Light that is polarized in parallel to the transmission axis is certain to pass through the film, while light that is polarized perpendicular to that axis is certain to be blocked.

Suppose that the light is polarized at some angle θ to the film's transmission axis. The probability that any particular photon will be transmitted is computed from

$$\text{Pr}[\text{transmission}] = \cos\theta$$

Thus, since $\cos(0°) = 1$, parallel light is *certain* to be transmitted, and since $\cos(90°) = 0$, it is *impossible* for perpendicular light to be transmitted. For light polarized at $\theta = 45°$,

$$\text{Pr}[\text{transmission with } \theta = 45°] = \cos(45°) = .70711$$

Source: Abner Shimony, "The Reality of the Quantum World," *Scientific American*, January 1988, pp. 46–53.

Alternative Expressions of Probability

Probabilities are often expressed in forms equivalent to the fraction or decimal.

Odds A popular expression for an event is a specification of odds—either for or against the event. Consider a deck of 52 ordinary playing cards, which contains cards of 13 denominations and 4 suits. For one card drawn at random, we have

$$\text{Pr}[\text{queen}] = \frac{4}{52} = \frac{1}{13}$$

The odds *against* getting a queen are either 48-to-4 or 12-to-1. (The first number is the difference between the denominator and numerator of the probability fraction; the second value is the numerator.) Likewise, the odds *for* getting a queen are 4-to-48 or 1-to-12; that is, the numbers reversed.

Percentages In regular conversation, many find it easier to speak in terms of percentages rather than decimals. Thus, we might express $\text{Pr}[\text{woman}] = .447$ instead as "woman" has a 44.7% chance of occurring.

Chances Another popular probability expression is a statement of chance. For example, instead of saying $\text{Pr}[\text{head}] = .50$, we could state that "head" has a 50-50 chance.

Problems

2-1 Consider the Tau Beta Pi lottery with the sample space shown in Figure 2-1. Find probabilities for the following events:
(a) man (c) civil engineer
(b) graduate (d) mechanical engineer

2-2 Explain how you would go about finding the probabilities for the following events:
(a) A vessel bursts before a pressure of 5,000 psi is applied. It is one of many production items.
(b) One of the defective items from a batch of known size having a known percentage of defectives is selected at random.
(c) A yet-to-be-fabricated prototype will fail.
(d) Some member of your graduating class receives a job offer paying in excess of $50,000 per year.

2-3 Consider the following probabilities:

$$(1)\ \Pr[\text{heart}] = \frac{1}{13} \qquad (3)\ \Pr[\text{two heads}] = \frac{1}{4}$$

$$(2)\ \Pr[\text{failure}] = \frac{65}{100} \qquad (4)\ \Pr[\text{oil}] = \frac{1}{3}$$

(a) Express the probability for each of the above events in the odds-against format.
(b) Express the probability for each of the above events as percentages.
(c) Express the probability for each of the above events in the odds-for format.

2-4 A pilot plant has yielded chemical batches categorized as follows:

	Unusable	Usable
Low in impurities	3	24
High in impurities	21	7

Assuming this experience is representative of full-scale operation, determine estimated probabilities that a production batch will be
(a) low in impurities (e) both low in impurities and unusable
(b) high in impurities (f) both high in impurities and unusable
(c) unusable (g) both low in impurities and usable
(d) usable (h) both high in impurities and usable

2-5 Four students toss a lopsided coin. The following results are obtained:

	Number of Occurrences				Total
	a	*b*	*c*	*d*	*e*
Heads	6	7	24	49	86
Tails	4	8	18	37	67
Total	10	15	42	86	153

Determine each student's estimated probability for "head." Then use their combined experience to estimate the probability of "head."

2-6 In each of the following cases determine a decimal value for the probability of the event:
(a) The odds against "striking oil" are 10-to-1.
(b) There is a 60% chance that we will achieve a "satisfactory" progress report.
(c) There is only a 25-75 chance for vendor "approval."
(d) The odds for "winning" the contract are just 1-to-6.

2-7 Consider a deck of 52 playing cards, from which a card is randomly selected. Sketch the sample space, letting each card be represented by a dot on a grid that has one row for each suit and one column for each denomination. Identify the event sets for the following as areas on your sketch, and calculate the respective probabilities.
(a) ten (b) heart (c) black (d) face card (e) two or three

2-8 Consider the random experiment of tossing a pair of six-sided dice (a red and a white one). The elementary events are the number of dots (1 through 6) appearing on the up sides of each die. Each outcome may be represented as an ordered pair. For example, a 3 for the red die and a 5 for the white die is (3, 5). Give a complete listing of the sample space. Then establish probabilities for each possible *sum* of the dots on the showing faces.

2-9 Consider a photon arriving at a right angle to a filter film. Letting θ represent the polarization angle relative to the transmission axis, find the probability ($\cos\theta$) that the photon will be transmitted when
(a) $\theta = 30°$ (b) $\theta = 60°$ (c) $\theta = 1°$ (d) $\theta = 89°$

2-2 *Compound Events and Event Relationships*

Much of probability theory is concerned with event relationships, which become important when several events are considered as a group. Later in this chapter we will encounter a variety of procedures and shortcuts that are helpful in finding probabilities for various complex outcomes. We will also see how one event's probability might be influenced by the occurrence of some other event.

We begin with **compound events**, the joining of two or more possibilities. We investigate two logical connective operations and review relevant concepts of mathematical **set theory**.

Union of Events

It is important to establish the probability that one or more of a collection of events will occur. The individual events are treated as joined by the logical connective *or*. The resulting compound event is the **union** of its components.*

* We will use the italicized *or* as the symbol for the union operator (rather than the Boolean "bowl").

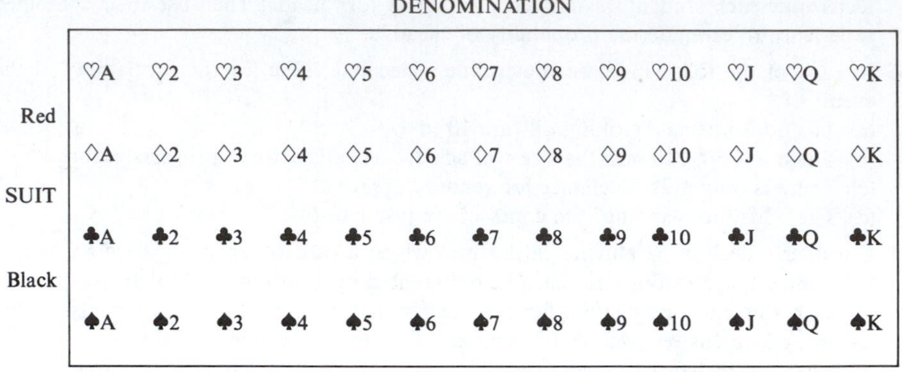

Figure 2-2 *Sample space for drawing one playing card from a shuffled deck.*

We may illustrate this and other important event relationships using a deck of 52 playing cards, each being one of 13 denominations (king, queen, jack, ten, nine, . . . , two, ace) and belonging to one of 4 suits (hearts, diamonds, clubs, spades). The king, queen, and jack are face cards. Figure 2-2 shows the sample space for the random experiment of drawing one card at random. Consider the event that we draw a "face card." This event will occur should any of the events "king," "queen," or "jack" occur. Thus, we can express "face card" as the union of these components:

face card = king *or* queen *or* jack

The union of events may be pictured by grouping together the applicable elementary events of the components. The event set for "face card" is pictured by combining the event sets for "king," "queen," and "jack" (Figure 2-3).

Intersection: Joint Events

The second type of compound event set is created by the **intersection** of two or more components. Consider the sample space in Figure 2-4 for the random experiment of tossing three coins—a nickel, a dime, and a quarter. The elementary events pertain to the showing faces, the letter H denoting head and T tail. The subscripts N, D, or Q indicate which coin. We see that the events "all faces the

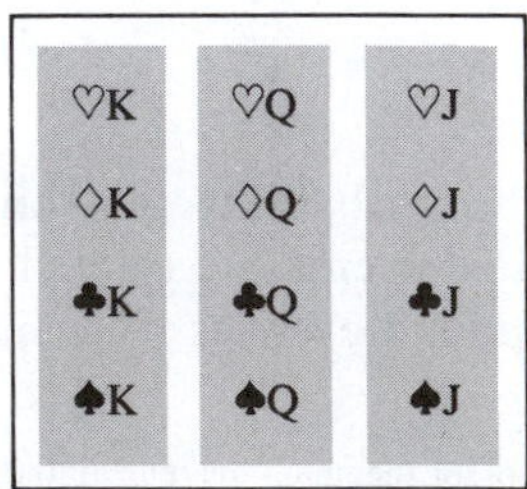

Figure 2-3

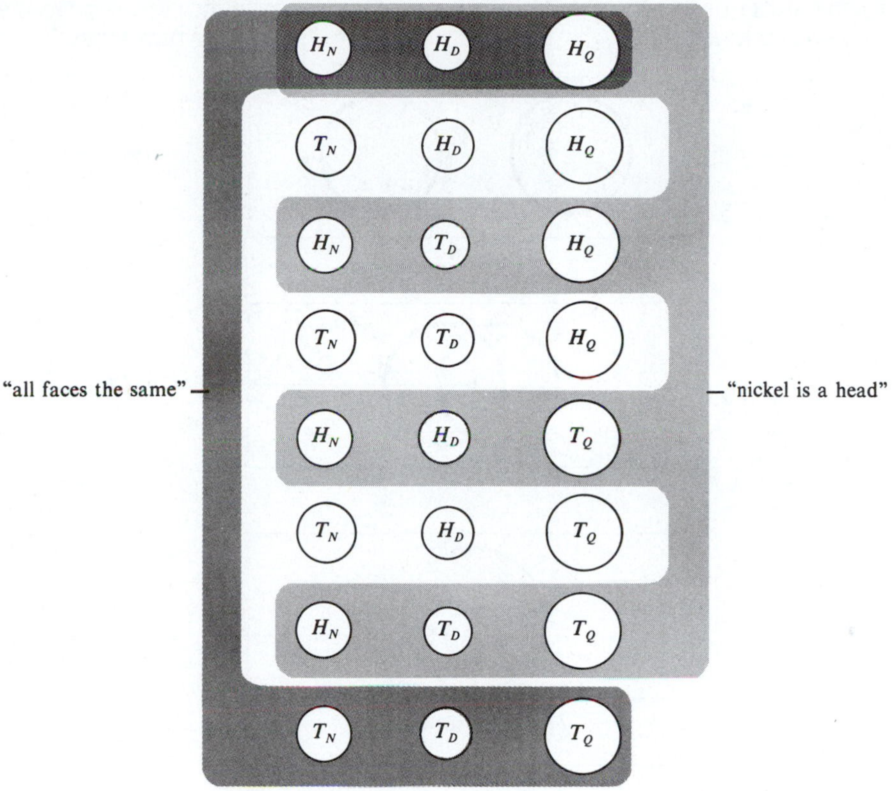

Figure 2-4 *Sample space for tossing three coins.*

same" and "nickel is a head" have the elementary event $H_N H_D H_Q$ in common, or

$$\text{all faces the same } \textit{and} \text{ nickel is a head} = \{H_N H_D H_Q\}$$

The logical operator *and* combines the components into a compound event, often referred to as a **joint event**. The italicized *and* serves as the operator symbol. Thus, *A and B* is a joint event that occurs only when *both* components, event *A* and event *B*, occur. The joint event does not happen if *A* or *B* occur singly.

Figures representing sets are sometimes referred to as **Venn diagrams**. Figure 2-5 shows Venn diagrams of commonly encountered cases for the union and intersection of events.

Next we will consider event **relationships**.

Mutually Exclusive Events

An important probability question is whether or not it is possible for two or more events to occur jointly. If *A and B* is impossible, then *A* and *B* are **mutually exclusive** events. That is, the intersection of the respective sets of mutually exclusive events is the empty set. Should the intersection be nonempty, so that *A* and *B* have elementary events in common, they would not be mutually exclusive.

Figure 2-5
Venn diagrams showing union and intersection possibilities.

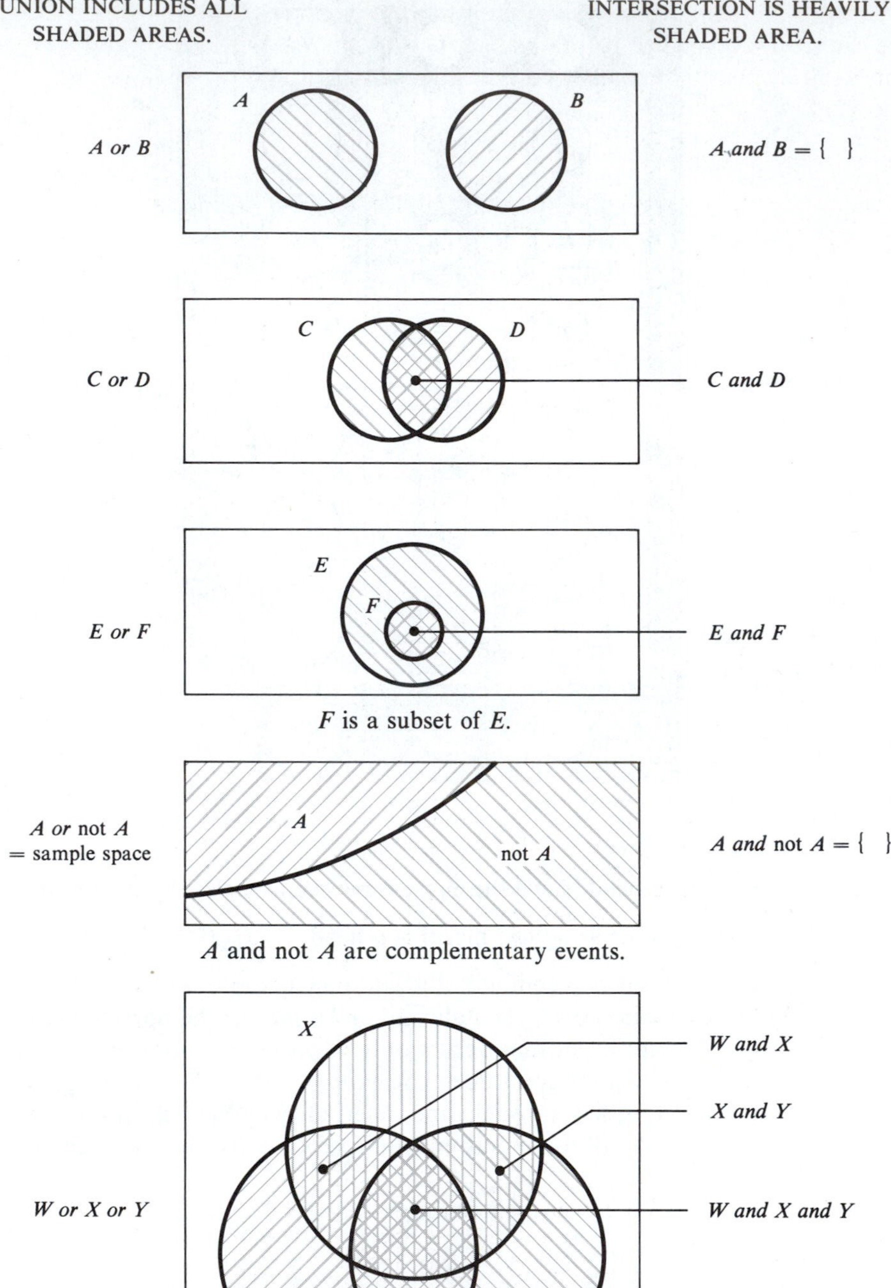

Consider again drawing a playing card at random. Referring to Figure 2-2, we can see that "red card" (either a heart ♡ or diamond ◇) and "spade" (♠) are mutually exclusive events. The respective event sets have no cards in common, so their intersection is empty:

$$\text{red card } \textit{and} \text{ spade} = \{\quad\} = \varnothing$$

Because the intersection of mutually exclusive event sets is empty, the joint event is impossible:

$$\Pr[\text{red card } \textit{and} \text{ spade}] = 0$$

The events "queen" and "heart" are not mutually exclusive because the event sets intersect on a common card, the queen of hearts:

$$\text{queen } \textit{and} \text{ heart} = \{\heartsuit\text{Q}\}$$

We must emphasize that mutual exclusivity is not a trivial concept. As we will see, it can play a significant role in determining how we calculate probabilities.

Collectively Exhaustive Events

For any random experiment, exactly one of the elementary events must occur. A collection of separate events that taken together encompass all elementary events is said to be **collectively exhaustive**. Stated another way, several events are collectively exhaustive if their union includes the entire sample space, so that at least one component event is bound to occur. Thus, events A, B, C, and D are collectively exhaustive if

$$A \textit{ or } B \textit{ or } C \textit{ or } D = \text{Sample space}$$

The union of a collection of collectively exhaustive events is itself a certainty, so that

$$\Pr[A \textit{ or } B \textit{ or } C \textit{ or } D] = 1$$

For drawing a playing card, the following events are collectively exhaustive:

heart, diamond, club, spade

The following are also collectively exhaustive:

face card, red, black

(It does not matter that "face card" and "red" are not mutually exclusive. The collection includes everything, even if some cards are accounted for in more than one way.)

Complementary Events

When properties like those above are combined in a pair of events, the two are **complementary events**. Complementary events are opposites, so that when one occurs the other can't; they are mutually exclusive. But exactly one of them must occur; they are collectively exhaustive. Examples of complementary events are:

some (at least 1) vs. none
head vs. tail
defective vs. satisfactory

Complementary events are very important in calculating probabilities because it is often easier to directly evaluate an event's opposite. As we shall see, the probability value for one member of the pair automatically establishes the probability value of the other.

Example:* *Quality-Control Classifications

Items inspected in the process of quality control may be classified in several ways. Consider a part that is evaluated in three categories: dimension (D), weight (W), and texture (T). In each category a part is classified as satisfactory (S) or unsatisfactory (U). The findings for a particular part are uncertain, and we may represent the possible elementary events symbolically. For example, $D_S W_S T_S$ signifies that the part is satisfactory in all categories.

To be *acceptable*, an item must be satisfactory in at least two categories. A *reworkable* item is an unacceptable item that is satisfactory only in dimension or weight. A *salvageable* item is satisfactory only in texture, and a *reject* is totally unsatisfactory. The sample space and various event sets are pictured in Figure 2-6.

The events listed on each line below are mutually exclusive:

acceptable, reworkable
salvageable, reject
acceptable, salvageable, reject
acceptable, unacceptable

The events in each of the following lists are collectively exhaustive:

acceptable, reworkable, salvageable, reject
acceptable, unacceptable
acceptable, unacceptable, reworkable, salvageable, reject

Since the pair is both mutually exclusive and collectively exhaustive, "acceptable" and "unacceptable" are complementary events. The following groups of events are *not* mutually exclusive:

unacceptable, reworkable (both include $D_S W_U T_U$ and $D_U W_S T_U$)
satisfactory dimension, reworkable (both include $D_S W_U T_U$)
unacceptable, reject (both include $D_U W_U T_U$)

Independent and Dependent Events

A third relationship expresses whether one event does or does not influence the likelihood of another. Two events A and B are **statistically independent** if the probability of A will be the same value when B occurs, when B does not occur, or when nothing is known about the occurrence of B.

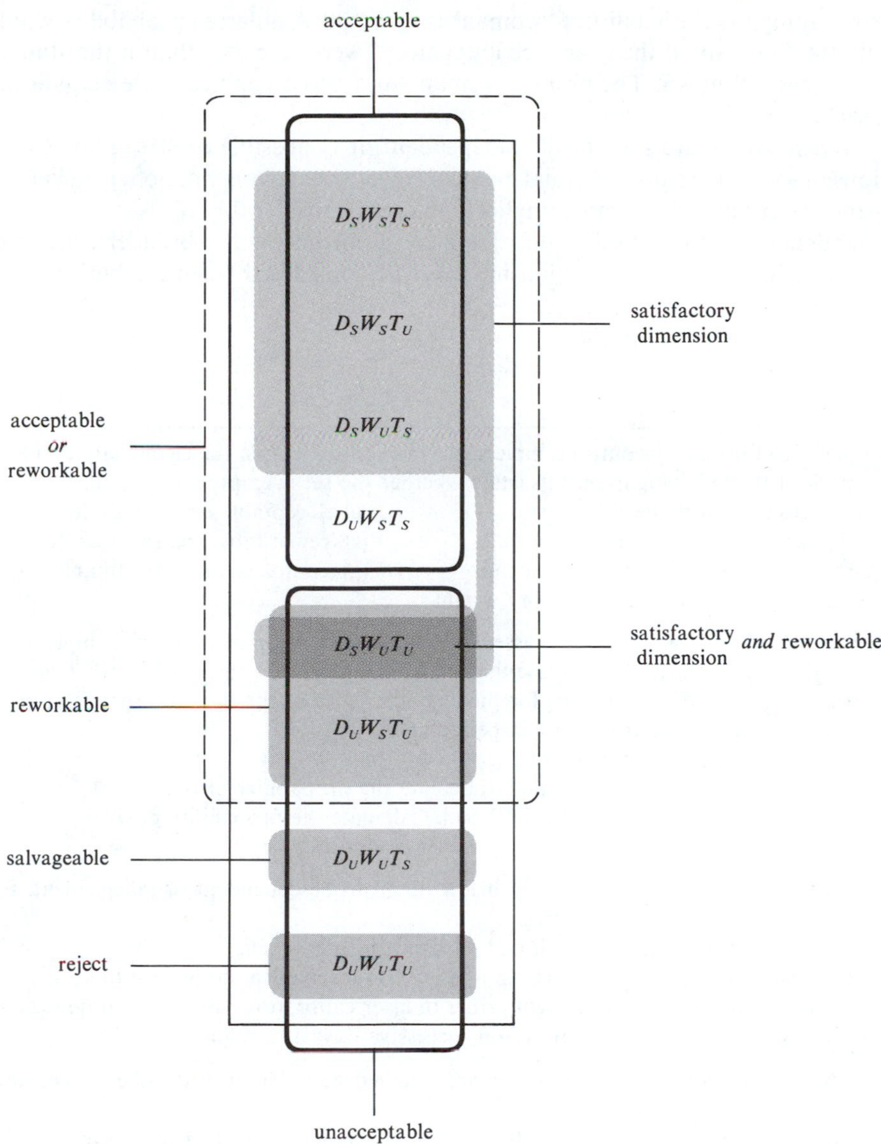

Figure 2-6 *Sample space and event sets for the quality-control inspection of a part.*

As an example, suppose a nickel and quarter are to be tossed separately. The event H_N (head with the nickel) has probability $\frac{1}{2}$, and we have no reason to believe the quarter toss can influence the nickel's, or vice versa. Thus, the probability of H_N will be $\frac{1}{2}$ regardless of whether H_Q or T_Q occurs, or if the quarter isn't even tossed. The events H_N and H_Q are independent.

Next, consider again the student lottery whose sample space is shown in Figure 2-1. For the event "woman" we found the probability value .447 by dividing the number of female students by the total number of students. Now consider just the civil engineering students, which include 3 women out of 10 people; within this grouping the probability of "woman" is $\frac{3}{10} = .30$. A different probability would apply for "woman" if the event "civil engineer" were to occur than if the student major were unknown. The events "woman" and "civil engineer" are **statistically dependent**.

When events are statistically *in*dependent, it is possible to streamline calculations of the probabilities for joint events. Unfortunately, we more often encounter dependent events, which can complicate matters considerably.

Independence (or the lack of it) is also important to the broader aspects of statistical theory, and this relationship must be considered when establishing and selecting analytical procedures.

Problems

2-10 Refer to the quality-control sample space in Figure 2-6. List the elementary events in each of the following event sets (state whether the set is empty):

(a) unsatisfactory weight
(b) satisfactory texture
(c) salvageable *or* satisfactory texture
(d) reworkable *and* unsatisfactory weight
(e) acceptable *or* reworkable
(f) reworkable *and* salvageable
(g) complement of reworkable

2-11 An inspector at a video-game assembly plant is checking out a special shipment of 10 printed circuit chips. Suppose that only 7 are good (G), the rest bad (B). Consider the events G_1, B_1 and G_2, B_2 for the characteristics of the first and second chip inspected. (Testing destroys each inspected chip.)

(a) Calculate the probability of G_1.
(b) Suppose the first chip is good. Calculate the probability of G_2.
(c) Suppose instead the first chip is bad. Calculate the probability of G_2.
(d) Are the events G_1 and G_2 independent or dependent?

2-12 Indicate whether the events listed below ought to be dependent or independent. Explain the reason for your choice.

(a) favorable seismic test result and oil beneath drilling site
(b) score on a spatial relations test and level of mechanical design aptitude
(c) direction of the measurement errors in laser calibrations at two separate sites
(d) noontime retort temperatures on successive days at a refinery

2-13 Three parts from a production line are weighed and classified as either satisfactory (S) or overweight (O).

(a) If we use subscripts 1, 2, and 3 to identify the part, one elementary event will be $O_1S_2O_3$. List the entire sample space. Then list the elementary events in the following event sets:
(b) first part satisfactory
(c) second part overweight
(d) exactly 1 satisfactory
(e) none satisfactory
(f) exactly 2 satisfactory
(g) first part satisfactory *and* second part overweight
(h) first part satisfactory *or* second part overweight
(i) at most 2 satisfactory
(j) at least 2 satisfactory

2-3 Probabilities for Compound Events

We have described two kinds of compound events in which the components are viewed in terms of their union (*or*) or their intersection (*and*). Probability theory allows us to compute probabilities for compound events by arithmetically combining the probabilities of individual components. These procedures are based on either the addition law or the multiplication law.

Applying the Basic Definition

Before we describe these laws, we must emphasize that every compound event is itself an outcome for which the basic definitions of probability apply. When elementary events are equally likely, it is usually possible to compute Pr[*A or B*] or Pr[*C and D*] by simply counting possibilities and dividing by the total number of elementary events.

For example, consider the student lottery discussed earlier and shown again in Figure 2-7. We can use the count-and-divide procedure to find the probability

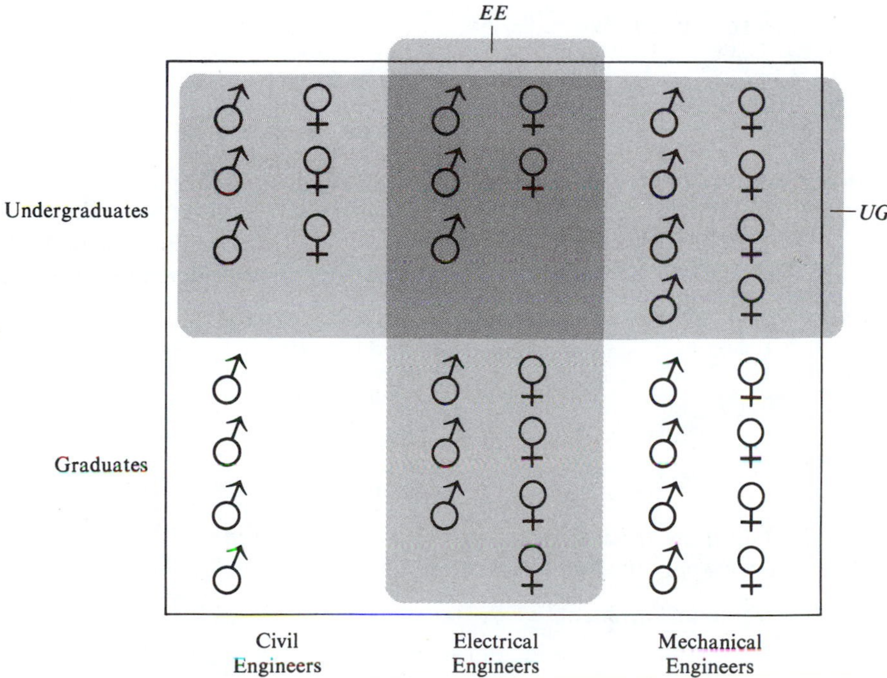

Figure 2-7 *Sample space for random selection of a Tau Beta Pi member.*

that a randomly selected student is either an undergraduate or an electrical engineering major:

$$\Pr[UG \text{ or } EE] = \frac{26}{38}$$

(We make sure to count each undergraduate electrical engineer just once.) Likewise, we may apply the basic definitions to determine the probability for the joint occurrence of the same events:

$$\Pr[UG \text{ and } EE] = \frac{5}{38}$$

Probabilities for joint events are usually referred to as **joint probabilities**.

But in some applications we may not be able to perform these types of calculations—because the elementary events are not equally likely (as with a lopsided coin or a crooked die cube), because we just don't have enough information, or because we cannot catalog the entire sample space. In such cases we might "synthesize" a compound event's probability by arithmetically manipulating the probabilities for its components (when these values are known).

The Addition Law

When a compound event is the union (*or*) of several components, the **addition law** allows us to find its probability by adding together the component event probabilities. There are two versions of this law. We begin with the simpler form.

ADDITION LAW FOR MUTUALLY EXCLUSIVE EVENTS

$$\mathbf{Pr}[A \text{ or } B] = \mathbf{Pr}[A] + \mathbf{Pr}[B]$$

Illustration: *Power Cell Failure Probabilities*

As an illustration, suppose we know that the primary causes for failure of a space power cell are (1) solar radiation, (2) launch vibration, (3) material weakness, or (4) collision. For any particular failure, these causes may be considered as mutually exclusive events, since there can be only one primary reason for failing. Previous testing indicates that

$$\Pr[\text{solar radiation}] = .10$$

$$\Pr[\text{launch vibration}] = .35$$

$$\Pr[\text{material weakness}] = .19$$

$$\Pr[\text{collision}] = .36$$

Causes (2) and (4) are mechanical accidents. To find the probability of this compound event, we apply the addition law:

$$\begin{aligned}\Pr[\text{mechanical accident}] &= \Pr[\text{launch vibration } \textit{or} \text{ collision}]\\ &= \Pr[\text{launch vibration}] + \Pr[\text{collision}]\\ &= .35 + .36 = .71\end{aligned}$$

Table 2-1 *Probabilities for the Number of Power Cell Failures*

Number of Failures	Probability
0	.3679
1	.3679
2	.1839
3	.0613
4	.0153
5	.0031
6	.0005
7 or more	.0001
	1.0000

The addition law for mutually exclusive events extends to any number of components. For instance, consider the probabilities in Table 2-1 for the number of power cell failures in a fixed span of time. The addition law may be used to determine the probability for any particular compound event. For example, "at most 3" is an event occurring whether the number of failures is "exactly 0," "exactly 1," "exactly 2," or "exactly 3." Thus,

$$\begin{aligned}\Pr[\text{at most } 3] &= \Pr[0 \textit{ or } 1 \textit{ or } 2 \textit{ or } 3] \\ &= \Pr[0] + \Pr[1] + \Pr[2] + \Pr[3] \\ &= .3679 + .3679 + .1839 + .0613 \\ &= .9810\end{aligned}$$

Notice that the possible failure events given in Table 2-1 constitute a collectively exhaustive collection (one of these outcomes is bound to occur). The union of all the events is therefore certain and has probability 1. Notice also that the individual event probabilities in Table 2-1 sum to 1. For any collection of mutually exclusive events that are also collectively exhaustive, this must hold.

Application to Complementary Events

We are now ready again to consider complementary events (opposites). The union of any event A and its complement not A is certain, and by the addition law

$$\Pr[A \textit{ or } \text{not } A] = \Pr[A] + \Pr[\text{not } A] = 1$$

From this we get the following:

$$\mathbf{Pr}[A] = \mathbf{1} - \mathbf{Pr}[\textbf{not } A]$$

Thus, if we know the probability of an event's complement, we can subtract that value from 1 to get the probability value for the desired event.

Consider again the power cell failure events in Table 2-1. Suppose we want to find the probability that the number of failures is "at least 1." "At least 1" is the same as "some," which is the opposite of "none." Thus,

$$\begin{aligned}\Pr[\text{at least } 1] &= 1 - \Pr[0] \\ &= 1 - .3679 = .6321\end{aligned}$$

Here it is simpler to subtract a single quantity from 1 than to add together the probabilities of the seven component events for "at least 1." In fact, it is often easier and faster to reach an answer indirectly by evaluating the complementary event.

General Addition Law

The preceding version of the addition law works only for *mutually exclusive* component events. We should keep in mind that mutual exclusivity is the *exception*, so events typically encountered during probability evaluations do not share this property. In the more general case, we use the following.

GENERAL ADDITION LAW

$$\mathbf{Pr}[A \textit{ or } B] = \mathbf{Pr}[A] + \mathbf{Pr}[B] - \mathbf{Pr}[A \textit{ and } B]$$

Notice that the joint probability must be subtracted from the sum of the component event probabilities. This is to avoid double accounting of events.

Consider again the student lottery in Figure 2-7. We may use the general addition law to recompute the probability that the chosen student will be either an undergraduate or an electrical engineer. First we apply the basic definition to find the following probabilities:

$$\Pr[UG] = \frac{19}{38}$$

$$\Pr[EE] = \frac{12}{38}$$

$$\Pr[UG \textit{ and } EE] = \frac{5}{38}$$

Thus, by the general addition law,

$$\begin{aligned}\Pr[UG \textit{ or } EE] &= \Pr[UG] + \Pr[EE] - \Pr[UG \textit{ and } EE] \\ &= \frac{19}{38} + \frac{12}{38} - \frac{5}{38} \\ &= \frac{26}{38}\end{aligned}$$

which is the same probability value found earlier for this compound event.

We see the rationale for subtracting the joint probability for *UG and EE*. Five students are both undergraduates and electrical engineers. They are counted first in establishing $\Pr[UG]$ and a second time in finding $\Pr[EE]$. By subtracting $\Pr[UG \textit{ and } EE]$, these five students are counted just once.

This example explains why the general addition law works, but it provides little motivation for using the law since here the procedure is somewhat roundabout.

After all, we found Pr [*UG or EE*] more simply on page 72 using the basic definition, counting and dividing. In practice, the general addition law is more valuable applied to a situation in which insufficient information exists, a situation in which we can't "count and divide."

The general addition law may be extended to cases involving more than two components.*

Example: Metal Fabrication Faults

Suppose we know that a metal fabrication process yields output with faulty bondings (*FB*) 10% of the time and excessive oxidation (*EO*) in 25% of all segments; 5% of the output has both faults. The following probabilities apply to a sample segment:

$$\Pr[FB] = .10$$

$$\Pr[EO] = .25$$

$$\Pr[FB \textit{ and } EO] = .05$$

The general addition law can be used to find the probability that a sample segment has faulty bonding, excessive oxidation, or possibly both:

$$\begin{aligned}\Pr[FB \textit{ or } EO] &= \Pr[FB] + \Pr[EO] - \Pr[FB \textit{ and } EO] \\ &= .10 + .25 - .05 = .30\end{aligned}$$

Thus, 30% of the output has at least one of the faults.

The Multiplication Law for Independent Events

The second basic law of probability is intended for computing joint probabilities. This is achieved by multiplying together the probabilities of the component events.

MULTIPLICATION LAW FOR INDEPENDENT EVENTS

$$\mathbf{Pr}[A \textit{ and } B] = \mathbf{Pr}[A] \times \mathbf{Pr}[B]$$

To illustrate, suppose a perfectly balanced coin is fairly tossed twice in succession. We denote the showing faces from the first toss by H_1 and T_1 and those from the second by H_2 and T_2. The probability of two heads is the probability of the joint event H_1 *and* H_2. Since the events may be assumed to be independent, the multiplication law provides

$$\begin{aligned}\Pr[H_1 \textit{ and } H_2] &= \Pr[H_1] \times \Pr[H_2] \\ &= \frac{1}{2} \times \frac{1}{2} = \frac{1}{4}\end{aligned}$$

* For three components we have

$$\begin{aligned}\Pr[A \textit{ or } B \textit{ or } C] = {} & \Pr[A] + \Pr[B] + \Pr[C] \\ & - \Pr[A \textit{ and } B] - \Pr[A \textit{ and } C] - \Pr[B \textit{ and } C] \\ & + \Pr[A \textit{ and } B \textit{ and } C]\end{aligned}$$

It is easy to verify this since the random experiment involves exactly four equally likely outcomes:

$$H_1 \text{ and } H_2 \qquad H_1 \text{ and } T_2 \qquad T_1 \text{ and } H_2 \qquad T_1 \text{ and } T_2$$

and $\Pr[H_1 \text{ and } H_2]$ must be equal to 1 divided by 4.

Of course, if we could always count and divide we wouldn't need the multiplication law. Suppose instead that the coin is lopsided, having a 25% chance of "head." Assuming that independence still applies between successive toss outcomes, the multiplication law gives

$$\Pr[H_1 \text{ and } H_2] = \Pr[H_1] \times \Pr[H_2]$$
$$= .25 \times .25 = .0625$$

which is the only way to arrive at the joint probability.

Note that this multiplication law applies only when the component events are *independent*. Consider again the events "undergraduate" and "electrical engineer" for the student lottery in Figure 2-7. These events are *not* independent. If we multiply together $\Pr[UG] = \frac{19}{38}$ and $\Pr[EE] = \frac{12}{38}$ we get an *invalid result*:

$$\frac{19}{38} \times \frac{12}{38} = .158$$

This result does not equal the true joint probability value found earlier, $\Pr[UG \text{ and } EE] = \frac{5}{38} = .132$.

Later in the chapter we will encounter a more general multiplication law that works even when the component events are dependent.

The multiplication law for independent events extends to any number of independent components. It can also be used to evaluate the probability for a joint event that may itself be the elementary event of a complex random experiment.

After using the multiplication law to establish elementary event probabilities, the *addition law* might then be applied to find the probability for a compound event. Consider the following example.

Example: *Noise-Induced Data-Transmission Errors*

Microwave telecommunication systems for data transmission are often subject to short bursts of spurious noise, which can render entire message segments useless. When identified as anomalies by parity-check codes, erroneous segments can sometimes be automatically retransmitted without human intervention. One such system stores the bits from millisecond-long segments in a buffer memory for retransmission should there be a noise interruption.

Suppose that 1% of such data segments suffer noise distortions (N), whereas the remainder are clear (C). Bursts of noise are so short that we may assume independence between C and N events for successive segments. Consider just three segments, identified by 1, 2, and 3. For segment i, we have probability .01 that there is noise (N_i) and .99 that the segment is clear (C_i). The multiplication law for independent events allows us to evaluate the joint probability that all three segments are transmitted clearly:

$$\Pr[C_1 \text{ and } C_2 \text{ and } C_3] = \Pr[C_1] \times \Pr[C_2] \times \Pr[C_3]$$
$$= .99 \times .99 \times .99 = .9703$$

The probability that the second segment is the only one subject to noise is found by applying the multiplication law in evaluating the joint event C_1 *and* N_2 *and* C_3:

$$\Pr[C_1 \textit{ and } N_2 \textit{ and } C_3] = \Pr[C_1] \times \Pr[N_2] \times \Pr[C_3]$$
$$= .99 \times .01 \times .99 = .0098$$

The above joint events are also two of the elementary events in the sample space enumerating all three-segment transmission possibilities. Two more are (N_1 *and* C_2 *and* C_3) and (C_1 *and* C_2 *and* N_3), each having probability .0098. (As an exercise you may find the remaining four elementary events and their probabilities.) To find the probability of exactly 1 noise interruption we may apply the addition law to the respective mutually exclusive component events (the three elementary events involving exactly one N_i):

$$\Pr[\text{exactly 1 noise interruption}]$$
$$= \Pr[C_1 \textit{ and } N_2 \textit{ and } C_3] + \Pr[N_1 \textit{ and } C_2 \textit{ and } C_3] + \Pr[C_1 \textit{ and } C_2 \textit{ and } N_3]$$
$$= .0098 + .0098 + .0098 = .0294$$

Problems

2-14 One Tau Beta Pi member is selected at random, and the sample space in Figure 2-7 applies. Find the following probabilities:
(a) Pr[graduate *or* CE] (c) Pr[ME *or* CE]
(b) Pr[undergraduate *or* woman] (d) Pr[CE *or* EE]

2-15 One Tau Beta Pi member is selected at random, and the sample space in Figure 2-7 applies. Use the count-and-divide method to find the following probabilities:
(a) Pr[graduate *and* woman] (c) Pr[EE *and* man]
(b) Pr[undergraduate *and* man] (d) Pr[CE *and* woman]

2-16 One Tau Beta Pi member is selected at random, and the sample space in Figure 2-7 applies.
(a) Find the probability that the selection will be an *undergraduate* assuming that (1) no other attributes are known and (2) the person's major is known to be mechanical engineering. What relationship applies between the events "undergraduate" and "mechanical engineer"?
(b) Using the count-and-divide method only, find the following:
(1) Pr[undergraduate] (4) Pr[ME *and* undergraduate]
(2) Pr[ME] (5) Pr[CE *and* undergraduate]
(3) Pr[CE]
(c) Use your answers to (b) above. Multiply (1) and (2). Comparing that product to (4), what do you notice? Explain.
(d) Use your answers to (b) above. Multiply (1) and (3). Comparing that product to (5), what do you notice? Explain.

2-17 Consider the data-transmission example on page 76. List all the elementary events for the transmission of three segments. Then use the multiplication law to determine the probability for each elementary event.

2-18 Using your answer to Problem 2-17, complete the following:
(a) Determine the following for the number of noise-interference events:
(1) Pr[exactly 0] (3) Pr[exactly 2]
(2) Pr[exactly 1] (4) Pr[exactly 3]

(b) Apply the addition law for mutually exclusive events to determine the following for the number of noise-interference events:
(1) Pr[exactly 1 *or* exactly 2] (2) Pr[at least 2] (3) Pr[at most 2]

2-19 Consider the quality-control classifications discussed in the example on page 68. Suppose the following percentages of items are satisfactory in the respective categories:

Dimension	80%
Weight	90
Texture	95

Assuming that an item's status is independent from category to category, apply the multiplication law to determine the probability for each elementary event in the sample space shown in Figure 2-6.

2-20 Problem 2-19 *continued.* Use the addition law for mutually exclusive events to evaluate the probabilities for the following events (as defined in Figure 2-6):
(a) acceptable
(b) unacceptable
(c) reworkable
(d) acceptable *or* reworkable
(e) satisfactory dimension
(f) satisfactory dimension *or* salvageable
(g) salvageable *or* reworkable

2-21 A production process yields 90% satisfactory items. The quality events pertaining to successive items are assumed to be independent. Find the probability of finding (1) all satisfactory, (2) none satisfactory, and (3) at least one satisfactory, when the number of selected items is
(a) 2 (b) 3 (c) 4 (d) 5

2-4 Conditional Probability and the Joint Probability Table

We have now seen the important relationships between events, the main forms of compound events, and several ways to compute probabilities for a variety of outcome types. This section consolidates that knowledge and establishes a conceptual framework for further extensions of probability theory. Probabilities themselves can be further categorized, and a convenient tabular array will prove helpful in organizing probability information.

Conditional Probability

Consider how the probability value of one event is affected by the occurrence of another. It is more likely to rain tomorrow if there is solid cloud cover than if there are patches of blue. Electronic equipment is more likely to fail in high temperatures than in moderate ones—and we might be uncertain which operating environment will apply. Items produced in antiquated plants will generally involve more rejects than those created in modern facilities (although we might not know where a particular item is made).

A probability value computed under the assumption that another event is going to occur is a **conditional probability**. For the events "rain" and "solid overcast," the following value might apply:

$$\Pr[\text{rain} \mid \text{solid overcast}] = .90$$

The event "rain" is listed first, and .90 is the probability of that event; the given event, "solid overcast," is listed second, and the vertical bar is a shorthand representation for "given that." A different rain probability applies given some other event:

$$\Pr[\text{rain} \mid \text{blue patches}] = .30$$

Should no stipulations be made regarding sky conditions, we might have

$$\Pr[\text{rain}] = .20$$

which is an **unconditional probability**.

The same basic definitions and laws apply to conditional probabilities as to the unconditional probabilities encountered earlier in the chapter. For example, in drawing a card from a shuffled deck of 52 ordinary playing cards, we determine the conditional probability of "queen" given "face card,"

$$\Pr[\text{queen} \mid \text{face card}] = \frac{4}{12} = \frac{1}{3}$$

by counting the equally likely elementary events in the respective event sets and dividing. The condition "face card" reduces the effective sample space to just the 12 face cards, 4 of which are queens. The task of computing the conditional probability is considerably streamlined through the elimination of extraneous elementary events (here, the nonface cards) prohibited by the condition.

It may instead by possible to compute a conditional probability directly from known probability values using the following:

CONDITIONAL PROBABILITY IDENTITY

$$\mathbf{Pr}[A \mid B] = \frac{\mathbf{Pr}[A \textit{ and } B]}{\mathbf{Pr}[B]}$$

Consider again the events "undergraduate" and "electrical engineer" for the student lottery in Figure 2-7. We have

$$\Pr[UG \mid EE] = \frac{\Pr[UG \textit{ and } EE]}{\Pr[EE]} = \frac{\frac{5}{38}}{\frac{12}{38}} = \frac{5}{12}$$

which expresses the probability that the selected student is an undergraduate given that he or she majors in electrical engineering. Notice that the numerator is the joint probability and the denominator is the given event's probability. A total reversal of the given and uncertain events is achieved by switching divisors:

$$\Pr[EE \mid UG] = \frac{\Pr[UG \textit{ and } EE]}{\Pr[UG]} = \frac{\frac{5}{38}}{\frac{19}{38}} = \frac{5}{19}$$

Of course the same results could have been achieved using the basic definition: With 5 out of 12 electrical engineering majors being undergraduates, we have

$\Pr[UG \mid EE] = \frac{5}{12}$. Likewise, $\Pr[EE \mid UG] = \frac{5}{19}$, since there are 5 electrical engineers out of 19 undergraduate students.

Although the foregoing identity is always true, it may not always be suitable for computing a conditional probability. We must know the values $\Pr[A \textit{ and } B]$ and $\Pr[B]$ in order to use it to evaluate $\Pr[A \mid B]$.

Example: *Crooked and Fair Dice*

A box contains four crooked (C) pairs of dice. One member of each pair has 3 dots on every face and the other has 4 dots on all sides. The box also contains six fair (F) pairs of dice. The faces of each of these dice have a different number of dots from 1 through 6.

One pair is selected at random and rolled. Given that the pair is fair, what is the conditional probability that a 7-sum results when the number of dots on the two up sides are added together?

The answer is

$$\Pr[\text{7-sum} \mid F] = \frac{6}{36} = \frac{1}{6}$$

This is found by *counting and dividing* the respective numbers of possibilities. (You may verify that there are 36 equally likely possibilities, six of which result in a sum of 7.)

The conditional probability identity cannot be used since it requires

$$\Pr[\text{7-sum } \textit{and } F]$$

in the numerator, and the value of this is not presently known.

The Joint Probability Table

Table 2-2 shows the total number of students in the Tau Beta Pi lottery classified in terms of sex (man, woman) and level (undergraduate, graduate). If we divide the number of students in each category by the total, we can determine the respective probability for the corresponding outcome.

These results are shown in Table 2-3, which is referred to as a **joint probability table**. The entries in the individual interior cells are joint probabilities for the re-

Table 2-2 *Two-way Classification of Students in Tau Beta Pi Lottery*

	Level		
Sex	**Undergraduate**	**Graduate**	**Total**
Man	10	11	21
Woman	9	8	17
Total	19	19	38

Table 2-3
Joint Probability Table for Sex and Level Events of Student Lottery

Sex	Level: Undergraduate (U)	Level: Graduate (G)	Marginal Probability
Man (M)	$\frac{10}{38}$	$\frac{11}{38}$	$\frac{21}{38}$
Woman (W)	$\frac{9}{38}$	$\frac{8}{38}$	$\frac{17}{38}$
Marginal Probability	$\frac{19}{38}$	$\frac{19}{38}$	1

spective outcome categories. For example,

$$\Pr[M \textit{ and } U] = \frac{10}{38}$$

$$\Pr[W \textit{ and } G] = \frac{8}{38}$$

Because they appear in the margins, the outside values are called **marginal probabilities**. These represent the probability that the event listed for that row or column occurs. Thus, we have

$$\Pr[M] = \frac{21}{38}$$

$$\Pr[G] = \frac{19}{38}$$

as the marginal probabilities for "man" and "graduate."

All the joint probabilities correspond to mutually exclusive (and collectively exhaustive) outcomes. The addition law for mutually exclusive events provides that the joint probabilities in any row or column sum to the respective marginal probability. We have

$$\begin{aligned}\Pr[M] &= \Pr[(M \textit{ and } U) \textit{ or } (M \textit{ and } G)]\\ &= \Pr[M \textit{ and } U] + \Pr[M \textit{ and } G]\\ &= \frac{10}{38} + \frac{11}{38} = \frac{21}{38}\end{aligned}$$

This reflects that "man" can occur in just one of two ways: jointly with "undergraduate" or with "graduate."

The marginal probabilities in the right-hand margin pertain to collectively exhaustive and mutually exclusive sex events and sum to 1:

$$\begin{aligned}\Pr[M \textit{ or } W] &= \Pr[M] + \Pr[W]\\ &= \frac{21}{38} + \frac{17}{38} = 1\end{aligned}$$

Likewise, from the bottom margin we have, for student level:

$$\Pr[U \textit{ or } G] = \Pr[U] + \Pr[G]$$
$$= \frac{19}{38} + \frac{19}{38} = 1$$

On the basis of the joint probability table, various conditional probabilities can be computed:

$$\Pr[M \mid G] = \frac{\Pr[M \textit{ and } G]}{\Pr[G]} = \frac{\frac{11}{38}}{\frac{19}{38}} = \frac{11}{19}$$

$$\Pr[W \mid G] = \frac{\Pr[W \textit{ and } G]}{\Pr[G]} = \frac{\frac{8}{38}}{\frac{19}{38}} = \frac{8}{19}$$

The two sex events are complementary, so that

$$\Pr[W \mid G] = 1 - \Pr[M \mid G] = 1 - \frac{11}{19} = \frac{8}{19}$$

This calculation, based on the addition law, is allowed because the two probabilities share a *common condition.* (It would not be proper to apply the addition law to conditional probabilities having different conditions.)

Establishing Independence by Comparing Probabilities

Recall that two events are independent if the probability of one is unaffected by the occurrence of the other. We may apply conditional probability concepts to more precisely define independence.

Definition

Two events A and B are statistically **independent** whenever $\Pr[A \mid B] = \Pr[A]$ or $\Pr[B \mid A] = \Pr[B]$.

Stated less formally, A and B are independent if one event has unconditional probability equal to its own conditional probability given the other event.

We may use this fact to establish statistical independence or dependence by comparing probability values. (But keep in mind that independence may be simply assumed to hold between some outcomes—such as the showing faces from two coin tosses—wherein the nature of the random experiments makes the relationship obvious.)

Continuing with the student lottery, we can see that W and G are *dependent* because

$$\Pr[W \mid G] = \frac{8}{19} \quad \text{does not equal} \quad \Pr[W] = \frac{17}{38}$$

This is because the conditional probability of W given G differs from the unconditional probability of W. (You may also establish that $\Pr[G \mid W] \neq \Pr[G]$.) The same relationship applies here for all sex and student-level event pairs.

Consider again the randomly drawn playing card. We have

$$\Pr[\text{queen}\,|\,\text{face}] = \frac{4}{12} \neq \Pr[\text{queen}] = \frac{4}{52}$$

so that "queen" and "face" are dependent. However, "queen" and "heart" are independent, since

$$\Pr[\text{queen}\,|\,\text{heart}] = \frac{1}{13} \text{ equals } \Pr[\text{queen}] = \frac{4}{52} = \frac{1}{13}$$

and the conditional probability of "queen" given "heart" is identical to the unconditional "queen" probability.

Problems

2-22 The architectural and engineering firms in a western region are classified in terms of size and primary type of project. The following incomplete data apply:

	Project Type				
Size	**Commercial (*C*)**	**Industrial (*I*)**	**Public Works (*P*)**	**General (*G*)**	**Total**
Small (*S*)	8				39
Medium (*M*)	6	4	9		
Large (*L*)		7		8	31
Total	29	32		19	100

(a) Determine the missing values and construct a joint probability table for the characteristics of one firm selected at random.

(b) Give the values for the following joint probabilities:
(1) $\Pr[S \text{ and } C]$ (2) $\Pr[M \text{ and } G]$ (3) $\Pr[L \text{ and } I]$ (4) $\Pr[S \text{ and } P]$

(c) Give the values for the following marginal probabilities:
(1) $\Pr[S]$ (2) $\Pr[M]$ (3) $\Pr[P]$ (4) $\Pr[I]$

(d) Compute the following conditional probabilities:
(1) $\Pr[L|C]$ (3) $\Pr[S|G]$ (5) $\Pr[I|M]$
(2) $\Pr[M|I]$ (4) $\Pr[C|L]$ (6) $\Pr[L|P]$

2-23 Consider a random experiment involving three boxes, each containing a mixture of red (*R*) and green (*G*) balls, with the following quantities:

A	B	C
5 *R* 5 *G*	14 *R* 6 *G*	8 *R* 12 *G*

The first ball will be selected at random from box A. If that ball is red, the second ball will be drawn from box B; otherwise, the second ball will be taken from box C.

Let R_1 and G_1 represent the color of the first ball, R_2 and G_2 the color of the second. Determine the following probabilities. (*Hint:* The conditional probability identity will not work.)

(a) $\Pr[R_1]$ (c) $\Pr[R_2 | R_1]$ (e) $\Pr[G_2 | G_1]$
(b) $\Pr[G_1]$ (d) $\Pr[R_2 | G_1]$ (f) $\Pr[G_2 | R_1]$

2-24 Suppose that the following data apply to the architectural and engineering firms in an eastern region:

Size	Project Type				Total
	Commercial (C)	Industrial (I)	Public Works (P)	General (G)	
Small (S)	8	12	14	6	40
Medium (M)	0	6	16	8	30
Large (L)	2	62	40	26	130
Total	10	80	70	40	200

Establish the following pertaining to a randomly selected firm:

(a) S and P are independent. (c) L and G are independent.
(b) M and C are dependent. (d) M and I are dependent.

2-25 Consider the student lottery summarized in Figure 2-7. Construct a joint probability table for major (CE, EE, ME) and level (U, G).

2-26 Refer to Problem 2-25. Find the following conditional probabilities:

(a) $\Pr[EE | G]$ (c) $\Pr[ME | G]$ (e) $\Pr[G | EE]$ (g) $\Pr[G | ME]$
(b) $\Pr[ME | U]$ (d) $\Pr[CE | U]$ (f) $\Pr[U | ME]$ (h) $\Pr[U | CE]$

2-5 The Multiplication Law, Probability Trees, and Sampling

With basic probability concepts and tools firmly in hand, we set the stage for deductive statistics by considering the sampling process. We can now find probabilities for particular sample outcomes. Later in the text, this background will prove useful in developing the concepts of inductive or inferential statistics.

In this section we introduce two new tools while laying the above groundwork. A *general* multiplication law is given that works whether or not component events are independent. To help explain this procedure we introduce a useful display, the probability tree diagram, which conveniently organizes information about random experiments that involve a series of uncertainties. Trees will prove useful later in the book to help explain decision theory.

The General Multiplication Law

The multiplication law introduced earlier requires that the component events be independent. But independence is often the exception. This is especially true in testing and evaluation—in which the desired event and the experimental result should be dependent. For example, a seismic survey result (an uncertainty) should be statistically dependent on the presence of oil (otherwise seismic predictions would be worthless in finding drilling locations).

We have a general version of the multiplication law.

GENERAL MULTIPLICATION LAW

$$\mathbf{Pr}[A \text{ and } B] = \mathbf{Pr}[A] \times \mathbf{Pr}[B|A]$$

and

$$\mathbf{Pr}[A \text{ and } B] = \mathbf{Pr}[B] \times \mathbf{Pr}[A|B]$$

Notice that the probability for the first event in the product is an unconditional one, whereas the second is a conditional probability for the other event, given the first. The multiplication law given earlier is a special case of this because when A and B are independent, $\Pr[A|B] = \Pr[A]$ and $\Pr[B|A] = \Pr[B]$.

Illustration: *Quality-Control Inspector's Probabilities*

To illustrate, suppose that a quality-control inspector has established that 10% of all electrical assemblies fail (F) the circuit test. Twenty percent of the failing items have poor (P) solder joints. This experience provides the following probabilities:

$$\Pr[F] = .10$$

$$\Pr[P|F] = .20$$

The multiplication law gives the joint probability that a particular assembly will fail and have poor solder joints:

$$\begin{aligned}\Pr[F \text{ and } P] &= \Pr[F] \times \Pr[P|F] \\ &= .10 \times .20 \\ &= .02\end{aligned}$$

Altogether, 2% of all assemblies will have both deficiencies.

Illustration: *Oil Wildcatter's Probabilities*

A second illustration is provided by the experience of an oil wildcatter. He has established that there is a 40% chance for oil (O) beneath a particular site. Furthermore, from past experience with seismic testing, the procedure is known to be 80% reliable. That is, for known oil-bearing sites there is an 80% probability of getting a favorable (F) seismic result, and when the site is known to be dry, there is a

90% probability of getting an unfavorable (U) prediction. Thus,

$$\Pr[O] = .4$$
$$\Pr[\text{not } O] = 1 - .4 = .6$$
$$\Pr[F \mid O] = .8$$
$$\Pr[U \mid \text{not } O] = .9$$

The general multiplication law may be used to compute joint probabilities for geological and seismic events:

$$\Pr[O \textit{ and } F] = \Pr[O] \times \Pr[F \mid O] = .4 \times .8 = .32$$

$$\Pr[\text{not } O \textit{ and } U] = \Pr[\text{not } O] \times \Pr[U \mid \text{not } O] = .6 \times .9 = .54$$

Constructing a Joint Probability Table Using the Multiplication Law

In Section 2-4 we saw how to construct a joint probability table for random experiments where there is sufficient information to count and divide. By applying the general multiplication law, in conjunction with the addition law for mutually exclusive events and the properties of complementary events, we can construct a joint probability table even when it is impossible to count and divide.

Consider Table 2-4, the joint probability table for the oil wildcatting illustration. The two joint probabilities computed earlier,

$$\Pr[O \textit{ and } F] = .32 \qquad \text{and} \qquad \Pr[\text{not } O \textit{ and } U] = .54$$

are entered first into the respective cells of a blank table. Also entered are the marginal probabilities found earlier:

$$\Pr[O] = .4 \qquad \text{and} \qquad \Pr[\text{not } O] = .6$$

These initial quantities are shown in boldface in Table 2-4.

Table 2-4
Joint Probability Table for Petroleum Exploration

	Seismic Result		
Geology	**Favorable (F)**	**Unfavorable (U)**	**Marginal Probability**
Oil (O)	**.32**	.08	**.40**
Not Oil (not O)	.06	**.54**	**.60**
Marginal Probability	.38	.62	1.00

The remaining joint probabilities can now be filled in, using the fact that the entries in each row must add up to the marginal totals. After all joint probabilities have been found, the marginal probabilities for the seismic events can be found by summing the respective columns.

The Probability Tree Diagram

An alternative to the joint probability table is another summary display, the **probability tree**. Figure 2-8 shows the probability tree diagram for the oil wildcatting illustration. Each event is represented as a **branch** in one or more **event forks**. A probability tree can be especially convenient for random experiments having events that occur at different times or stages.

In probability trees, time ordinarily moves from left to right. Since the geology events precede the seismic ones, the branches for oil (O) and not oil (not O) appear in the leftmost event fork. Each of these is followed by a separate event fork representing the seismic events, with one branch for favorable (F) and one for unfavorable (U). The complete tree then exhibits each final outcome as a single *path* from beginning to end. The oil wildcatter's tree has four paths: oil-favorable, oil-unfavorable, not-oil-favorable, not-oil-unfavorable. Each of these corresponds to a distinct joint event.

The probabilities for each event are placed alongside its branch. The values listed in the left fork are $\Pr[O] = .4$ and $\Pr[\text{not } O] = .6$. Since that event fork has no predecessor, these probabilities are unconditional ones.

Probabilities for events at later stages will all be conditional probabilities, with the branch or subpath leading to the branching point signifying the given

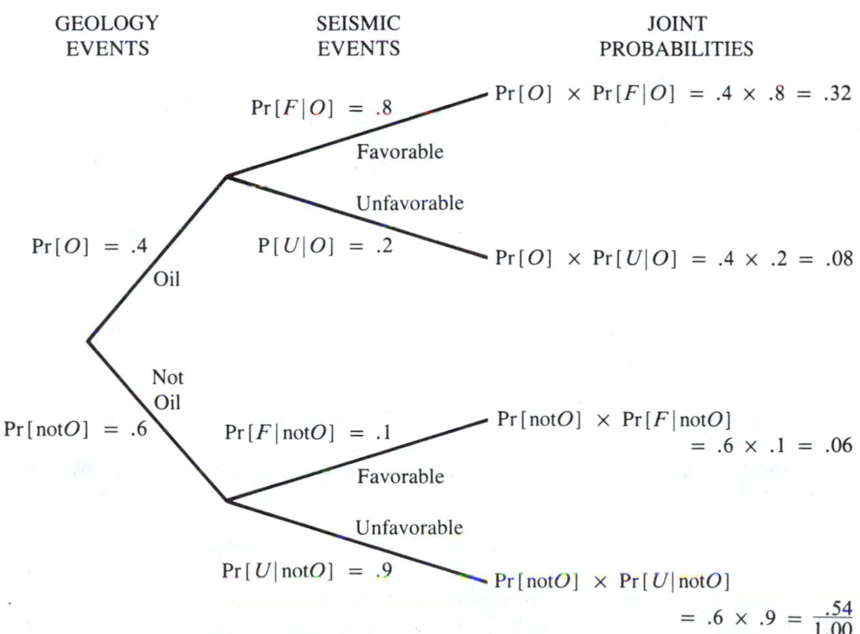

Figure 2-8 *Probability tree diagram for oil wildcatter.*

events. The top seismic event fork lists the probabilities $\Pr[F|O] = .8$ for favorable and $\Pr[U|O] = .2$ for unfavorable; since that fork is preceded by the oil (O) branch, both conditional probabilities involve O as the given event. A completely different set of conditional probabilities, $\Pr[F|\text{not } O] = .1$ and $\Pr[U|\text{not } O] = .9$, apply in the bottom event fork; that branching point is preceded by the not oil (not O) branch, so that not O is the given event.

The events emanating from a single branching point are mutually exclusive and collectively exhaustive, so that exactly one of them must occur. All the probabilities on branches within the same fork must therefore sum to 1.

The probability tree is very convenient for computing joint probabilities. Since each joint event is represented by a path through the tree, its probability is found by multiplying together all the individual branch probabilities along its path. For instance, the topmost path represents the outcome sequence oil-favorable. The corresponding joint probability is

$$\begin{aligned}\Pr[O \text{ and } F] &= \Pr[O] \times \Pr[F|O] \\ &= .4(.8) \\ &= .32\end{aligned}$$

All of the oil wildcatter's joint probabilities are listed in Figure 2-8 at the terminus of the respective event path. Notice again that, because the joint events themselves are mutually exclusive and collectively exhaustive, they sum to 1.

Multiplication Law for Several Events

The general multiplication law can be extended to several components. Consider events $A_1, A_2, \ldots, A_n$. We have

$$\begin{aligned}&\Pr[A_1 \text{ and } A_2 \text{ and } A_3 \ldots \text{ and } A_n] \\ &\quad = \Pr[A_1] \times \Pr[A_2|A_1] \times \Pr[A_3|A_1 \text{ and } A_2] \times \\ &\qquad \cdots \times \Pr[A_n|A_1 \text{ and } A_2 \text{ and } A_3 \ldots \text{ and } A_{n-1}]\end{aligned}$$

Example: *Matching Birthdays*

A surprising result is seen when we determine the probability that people in a group will have at least one matching birthday. It is easiest first to find the probability for the *complementary event*, "no matching birthdays." By considering each person one at a time, we define

$$B_i = i\text{th person does not match birthdays with previous persons}$$

The desired outcome equals the joint occurrence of all the above events for $i = 1, 2, \ldots, n$:

$$\text{No match} = B_1 \text{ and } B_2 \text{ and } B_3 \ldots \text{ and } B_n$$

For simplicity in establishing probabilities for the components it is assumed that all 365 calendar dates are equally likely to be a particular person's birthday, and leap year is ignored. Thus,

$$\Pr[B_1] = \frac{365}{365} \qquad \Pr[B_2 \mid B_1] = \frac{364}{365} \qquad \Pr[B_3 \mid B_1 \text{ and } B_2] = \frac{363}{365} \cdots$$

So that he or she will have no birthday in common with the preceding $i - 1$ persons, the above allows the ith person only $365 - (i - 1) = 365 - i + 1$ possible dates out of the 365 on the calendar. Applying the multiplication law, we have

$$\Pr[\text{no match among } n \text{ persons}] = \frac{365}{365} \times \frac{364}{365} \times \frac{363}{365} \times \cdots \times \frac{365 - n + 1}{365}$$

$$\Pr[\text{at least one match}] = 1 - \Pr[\text{no match}]$$

How large do you think n must be before the odds favor at least one match? (The answer may be found in the footnote below.)

Probability and Sampling

Probability is especially important in summarizing potential sampling results. Because multiple observations are usually involved in statistical evaluations, the probability calculations can be complicated. Probability trees can be helpful in organizing these computations.

Illustration: Quality-Control Sampling with Memory Chips

To illustrate, consider a shipment of 100 memory chips received by a computer manufacturer. Each chip is either good (G) or defective (D). The decision to accept or reject the shipment will be based on a sample of three chips selected at random. Although the inspector cannot know ahead of time how many defectives there are, let's consider a shipment having exactly 90 good chips and 10 defective ones.

Figure 2-9 shows a probability tree that summarizes the essential information. The first observation is represented by the two branches in the leftmost event fork. It is convenient to use the abbreviation G_1 to denote the event that "the first chip is good" and D_1 that "the first chip is bad." The subscripts help distinguish the results from different observations. Each branch for the first observation leads to a separate event fork for the second item, with both forks having a G_2 and a D_2 branch. Together, the initial observations create four distinct circumstances under which the third observation might occur, and each of these is represented by a separate fork having a G_3 and a D_3 branch.

Answer to matching birthday problem. The magical group size is 23. Consider the following.

n	Pr [no match]	Pr [at least one match]
10	.883	.117
23	.493	.507
40	.109	.891
60	.006	.994

Figure 2-9 *Probability tree diagram for a shipment of items sampled without replacement.*

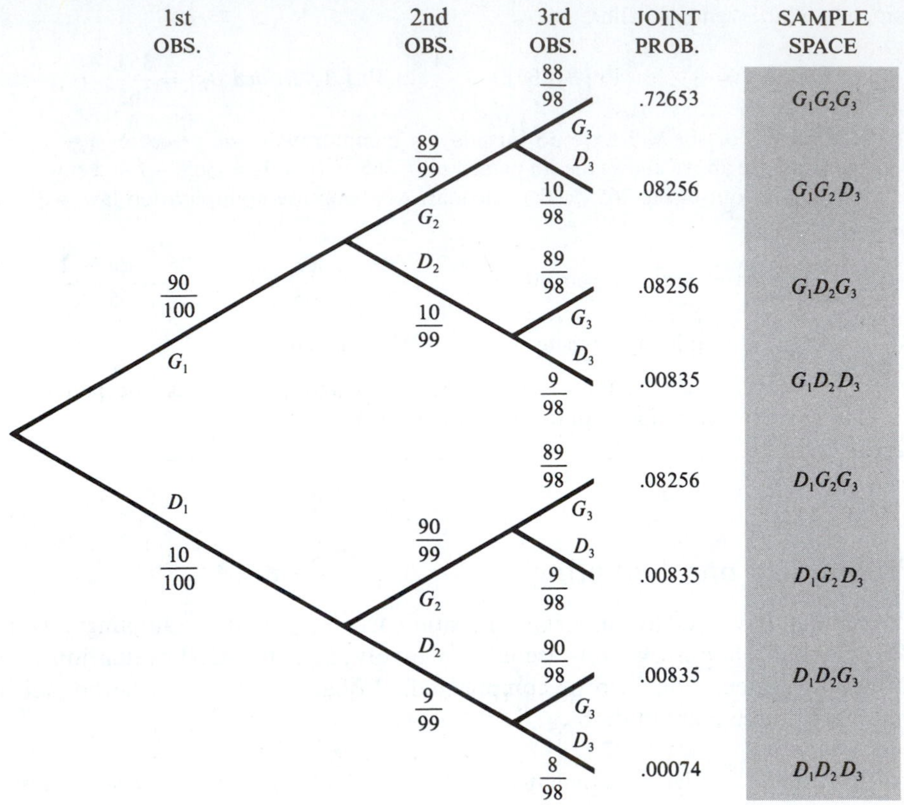

The probabilities for each branch are found using the count-and-divide method, according to the number of remaining good and defective chips up to that point in the tree. The initial fork has probabilities

$$\Pr[G_1] = \frac{90}{100} \qquad \text{and} \qquad \Pr[D_1] = \frac{10}{100}$$

Because the quality mix of the remaining items varies at the second stage, the probabilities for G_2 and D_2 differ in the two event forks for that observation. The shown values are *conditional probabilities.* The top set applies when G_1 has occurred, when there are only 89 remaining good chips and 10 defectives. With 99 chips then available for testing, the count-and-divide procedure provides

$$\Pr[G_2 | G_1] = \frac{89}{99} \qquad \text{and} \qquad \Pr[D_2 | G_1] = \frac{10}{99}$$

Likewise, the lower second-stage fork involves

$$\Pr[G_2 | D_1] = \frac{90}{99} \qquad \text{and} \qquad \Pr[D_2 | D_1] = \frac{9}{99}$$

since at that point—with one defective chip removed—the tree shows that 90 good chips and 9 defectives remain.

The third-stage forks involve probabilities reflecting the respective histories of good and defective items. The fork at the top right is preceded by G_1 and G_2 branches, so the conditional probabilities are

$$\Pr[G_3 | G_1 \text{ and } G_2] = \frac{88}{98} \quad \text{and} \quad \Pr[D_3 | G_1 \text{ and } G_2] = \frac{10}{98}$$

It is easy to verify the remaining third-stage probabilities.

Every path through the tree in Figure 2-9 corresponds to a distinct final outcome, each of which is summarized in the sample space listed in the box beside the tree. The probability for each of these elementary events is found by multiplying together the branch probabilities on its respective path. For instance, to get the second elementary event probability we multiply as follows:

$$\Pr[G_1 G_2 D_3] = \frac{90}{100} \times \frac{89}{99} \times \frac{10}{98} = .08256$$

In doing this, we are actually applying the multiplication law for several joint events:

$$\Pr[G_1 \text{ and } G_2 \text{ and } D_3] = \Pr[G_1] \times \Pr[G_2 | G_1] \times \Pr[D_3 | G_1 \text{ and } G_2]$$

The inspector can summarize the information contained in the probability tree in terms of the number of defectives in the shipment, as shown in Table 2-5. Using the values from the last column, we can determine the probability that a particular shipment would contain more than 1 defective:

$$\begin{aligned} \Pr[\text{more than 1 defective}] &= .02505 + .00074 \\ &= .02579 \end{aligned}$$

Table 2-5 *Probabilities for the Number of Defective Memory Chips When Three Items Are Inspected in a Random Sample Without Replacement*

Number of Defectives	Corresponding Elementary Events	Elementary Event Probability	Defectives Probability
0	$G_1G_2G_3$	.72653	.72653
1	$D_1G_2G_3$	.08256	.24768
	$G_1D_2G_3$	.08256	
	$G_1G_2D_3$	.08256	
		.24768	
2	$D_1D_2G_3$	.00835	.02505
	$D_1G_2D_3$	.00835	
	$G_1D_2D_3$	.00835	
		.02505	
3	$D_1D_2D_3$	.00074	.00074
			1.00000

If she rejects shipments having more than 1 defective item, less than three percent of shipments similar to the above would be unacceptable.

Independent Sample Observations The preceding illustration involves **sampling without replacement**, since inspected sample units are set aside. Although it seems inherently wasteful to do so (and even impossible when testing destroys the items), some inspection schemes would put the inspected items back into the original population, allowing them an equal chance with the rest to be chosen for each subsequent observation. Those plans involve **sampling with replacement**. As we shall see, sampling with replacement simplifies the probability calculations.

Figure 2-10 shows the probability tree diagram that would apply if the inspector sampled with replacement. Notice that all the G branches have identical probabilities of .90, and similarly the D probabilities are all .10. Since sampled items are put back, the probabilities for later quality events are unaffected by what happens earlier. In effect, successive quality events are *statistically independent*. Independence between successive sample observations arises naturally when sampling from a continuing production line, when replacement or nonreplacement would yield identical probabilities due to the theoretically infinite population size.

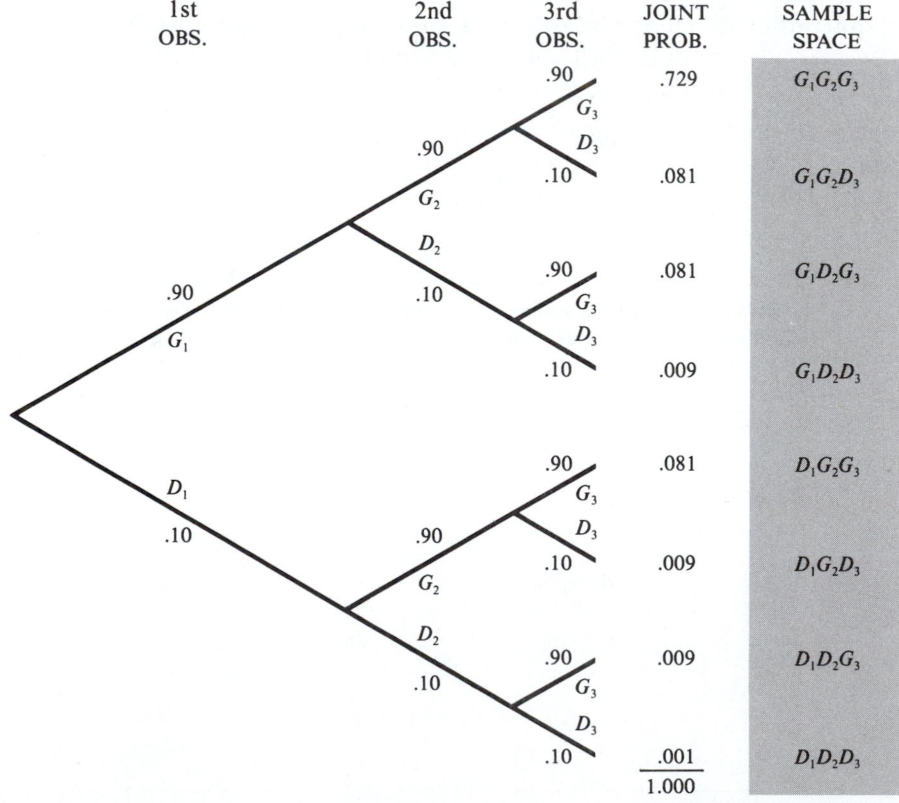

Figure 2-10 *Probability tree diagram for items from a production process sampled with replacement.*

Table 2-6 *Probabilities for the Number of Defective Memory Chips When Three Sample Items Are Inspected*

Number of Defectives	Probabilities for Sampling: *Without Replacement*	*With Replacement*
0	.72653	.729
1	.24768	.243
2	.02505	.027
3	.00074	.001
	1.00000	1.000

The probabilities for the number of defectives found under sampling with replacement are given in Table 2-6. The probabilities are obtained in the same manner as before. To help in comparing the two procedures, the two sets of probabilities are given side by side.

Notice how close the two sets of values are. Since the probabilities computed for sampling without replacement can involve some cumbersome calculations, they are often *approximated* by the more cleanly computed values that strictly apply only when there is replacement.

The sets of probability information in Table 2-6 are examples of *probability distributions,* which are described in detail in Chapter 3. When sampling is done with replacement—or more generally, when successive observations are independent—the *binomial distribution* applies. Under sampling without replacement from finite populations the *hypergeometric distribution* applies.

Problems

2-27 An oil wildcatter has assigned a .30 probability for striking gas (G) under a particular leasehold. He has ordered a seismic survey that has a 90% positive reliability (given gas it confirms gas [C] with probability .90), but only 70% negative reliability (given no gas it denies gas [D] with probability .70).

(a) Establish the following probabilities:
(1) $\Pr[G \text{ and } C]$ (2) $\Pr[\text{no } G \text{ and } D]$

(b) Construct the joint probability table for the geological and seismic events.

(c) Using your answer to (b), find the following probabilities:
(1) $\Pr[G|C]$ (2) $\Pr[\text{no } G|D]$

(d) Construct a probability tree with gas events in the first stage and seismic events in the second.

2-28 A project manager is creating the design for a new engine. He judges that there will be a 50-50 chance that it will have high-energy (H) consumption instead of low (L). Historically, 30% of all high-energy engines have been approved (A) with the rest disapproved (D), while 60% of all low-energy engines have been approved.

(a) Construct a probability tree for the project manager's engine events.

(b) What is the probability that his design will result in an approved engine?

2-29 Repeat Problem 2-27 for a new site where gas has a .20 probability, the positive reliability is 80%, and the negative reliability is 90%.

2-30 Referring to the matching birthday example, determine the probability of (1) no matches and (2) at least one match when the group size is
(a) $n = 5$ (b) $n = 7$ (c) $n = 12$

2-31 Matching birth months problem. Suppose that all months are equally likely to be a random person's month of birth. Determine the probability of (1) no matching birth months and (2) at least 1 matching birth month when the group size is
(a) $n = 2$ (b) $n = 3$ (c) $n = 4$

2-32 A quality-control inspector is testing sample output from a production process for widgets wherein 95% of the items are satisfactory (S) and 5% are unsatisfactory (U). Three widgets are chosen randomly for inspection. The successive quality events may be assumed independent.

(a) Construct a probability tree diagram for this experiment, identify all the sample-outcome elementary events, and compute the respective joint probabilities.

(b) Find the probabilities for the following numbers of unsatisfactory items:
(1) none (3) exactly 2 (5) at least 1
(2) exactly 1 (4) exactly 3 (6) at most 2

2-33 A user of the widgets receives a shipment of 5 dozen containing 95% satisfactory. A sample of 3 widgets is selected *without* replacement. Repeat (a) and (b) from Problem 2-32.

2-34 Two chapters of a civil engineering society each have 60% licensed members (the rest are unlicensed), with 50% on public payrolls (and the rest privately employed). A questionnaire is being sent to a random sample of the members. Consider the characteristics of any one of the members surveyed.

(a) Chapter A has just as many licensed engineers on public payrolls as on private ones. This makes any professional status event *independent* of any employment sector event. Construct a joint probability table by first finding the marginal probabilities and then using the multiplication law to obtain the joint probabilities, which in this case may be expressed as the products of the two respective marginal probabilities.

(b) In Chapter B only 40% of the licensed engineers are public employees. It follows that any professional status event is *dependent* on any employment sector event. This means that no joint probability is equal to the product of the respective marginal probabilities. Construct the joint probability table.

2-6 Counting Methods for Finding Probabilities

In many applications wherein the elementary events are equally likely, probability values may be found by counting the respective elementary events and dividing by the total number of possibilities. This particular approach applies to most sampling situations wherein the units to be observed are taken at random from the underlying population. When the population is tiny, a complete listing of the elementary events is fairly easy to compile. But for even modest sampling experiments the number of possibilities can be so astronomical that a complete cataloging would be impossible.

For example, consider 10 student names. As we shall see, there are over *three million* different sequences in which they can be listed. To list the 26 letters of the alphabet with an entry for each sequence would require more space than is presently contained in all the world's books. There are more possible ways to get heads and tails in 100 coin tosses than the purported number of elementary particles in the known universe! At one picosecond (10^{-12}) per entry, it would take a high-speed computer more years to list each elementary event than are available in the estimated remaining life of our solar system.

And yet we can easily determine the *total number* of possibilities. In this section we describe shortcuts that may be used to count elementary events without actually having to list them all. We do this by treating very large groupings of possibilities as single aggregations that can be totaled directly, once their form and a few key parameter values have been determined.

The Principle of Multiplication

Illustration: *Counting Car Configurations*

Often, possibilities may be counted by multiplying several numbers together. Consider the possible configurations for new cars being fabricated on an assembly line. Considering just the basic car, the following choices might be possible:

5 models	3 rear ends
4 engines	4 body styles
3 fuel systems	3 tire types
2 transmissions	4 tire sizes
3 suspension systems	10 exterior colors

Figure 2-11 shows how we might use a tree to represent the possibilities, with a separate branching point for each choice. The total number of possible distinguishable car types would be

$$5 \times 4 \times 3 \times 2 \times 3 \times 3 \times 4 \times 3 \times 4 \times 10 = 518{,}400$$

which represents more units than one plant can possibly produce in a single year. (A complete version of the tree diagram for this, drawn on the same scale and showing all branches, would be about 5 miles tall.)

The **principle of multiplication** allows us to break the possibility accounting into stages (each having a separate branching point in a tree representation, for each result from earlier stages). In the car illustration, one stage is model specification, another is engine selection, and so on. Each ultimate car configuration outcome or sequence of branches involves one possibility from each stage or branching point. The total number of outcomes from all stages combined is the product of the number of successive possibilities (branches) at each stage.

The available results may vary in type or number from stage to stage, as with the car example, or they may be the same each time. In tossing a coin 5

Figure 2-11 *Tree diagram illustrating the principle of multiplication for counting possible car configurations.*

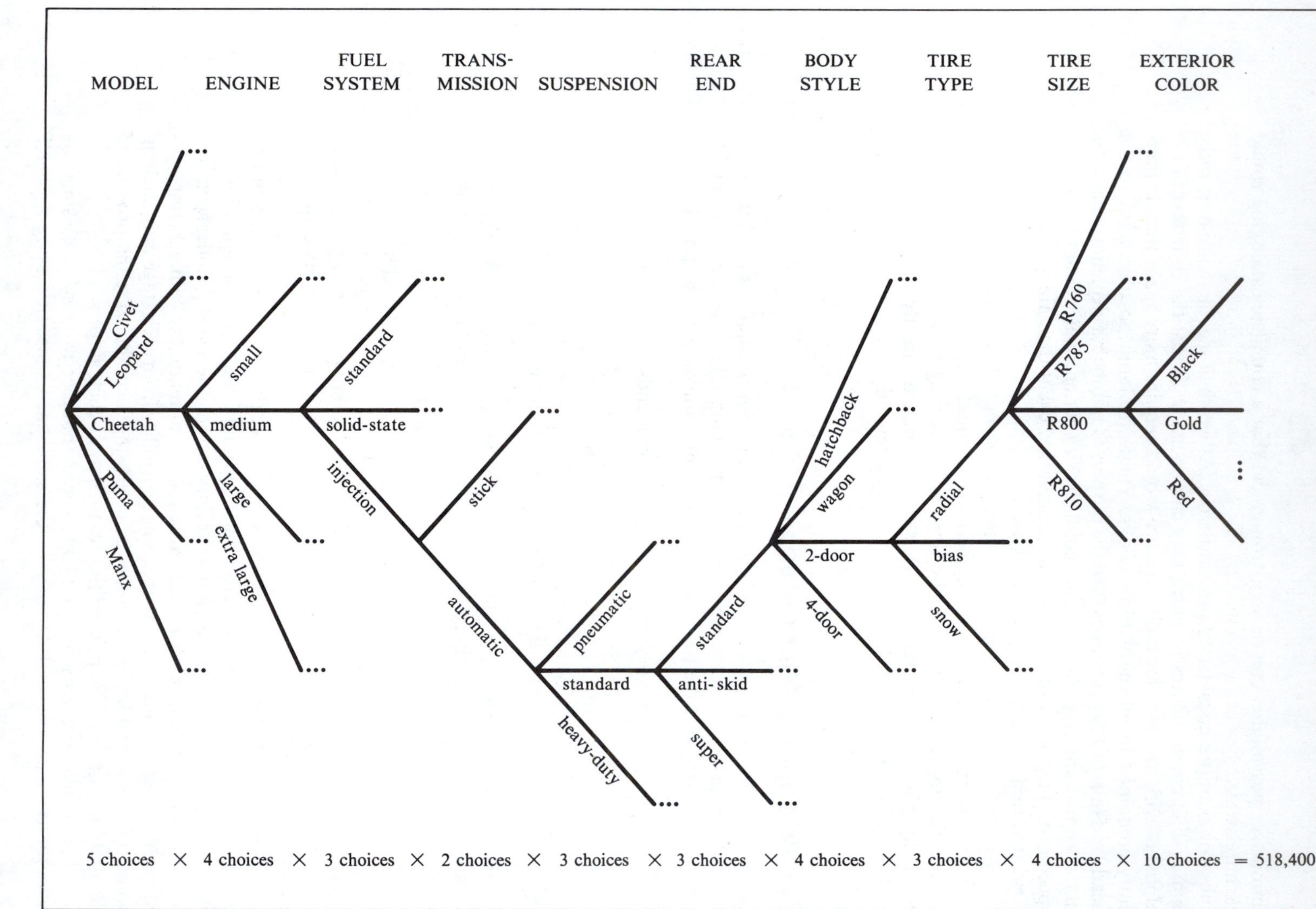

times, the number of showing-side outcomes is $2 \times 2 \times 2 \times 2 \times 2 = 2^5$, with each of the 5 tosses being a stage that involves just 2 possibilities (H or T). A die cube has 6 possible showing sides, so that in 3 tosses the number of possibilities is $6 \times 6 \times 6 = 6^3$. In either case the total number of possibilities is the number of sides raised to the power of the number of tosses.

The possible-stage results might increase or decrease according to some pattern. We next consider a common circumstance encountered in sampling situations wherein there is one less possibility at each stage.

Number of Ways to Sequence Items: The Factorial

We are now ready to lay the foundation for counting possibilities for the random selection of units from a larger group. As a preliminary step, we examine the number of different **sequences** that might be possible when observations are taken one at a time.

Illustration: *The Number of Ways to List Names*

As an illustration, consider the following 5 students: Alan, Belle, Carl, Dave, and Ellen. Including each person just once, how many different ways are there to sequence their names? Although we could make a complete listing of the possibilities, we can extend the principle of multiplication, treating the selection of each successive name as a new stage. All 5 names are possibilities for the first selection, and when the first name is picked, only 4 names remain. Thus, just considering the initial pair of students, the number of possibilities is

$$5 \times 4 = 20$$

We may verify this by lising the initials for each possibility:

AB	*BA*	*CA*	*DA*	*EA*
AC	*BC*	*CB*	*DB*	*EB*
AD	*BD*	*CD*	*DC*	*EC*
AE	*BE*	*CE*	*DE*	*ED*

Suppose *DE* (Dave and Ellen) is the initial pair. Only 3 names remain for the third selection: Alan, Belle, and Carl. Once that choice is made, only 2 names will be left for the fourth pick, and then just 1 name for the final selection. The number of possibilities for these remaining choices is

$$3 \times 2 \times 1 = 6$$

which we again verify by listing:

ABC *ACB* *BAC* *BCA* *CAB* *CBA*

There are this many sequences in initials for every one of the 20 alphabet pairs in the earlier list. It is easy to see that altogether, the total number of 5-name sequences is

$$5 \times 4 \times 3 \times 2 \times 1 = 120$$

Notice that the above product multiplies successive integers and ends with 1. Such products are encountered often when counting possibilities, and they are referred to as **factorials**. For notational convenience, the highest integer followed by an exclamation point is used to represent a factorial. Thus,

$$5! = 5 \times 4 \times 3 \times 2 \times 1 = 120$$

In general, there are $n!$ ways to sequence n objects. Other factorials are

$$3! = 3 \times 2 \times 1 = 6$$

$$6! = 6 \times 5 \times 4 \times 3 \times 2 \times 1 = 6 \times 5! = 720$$

$$7! = 7 \times 6 \times 5 \times 4 \times 3 \times 2 \times 1 = 7 \times 6! = 5{,}040$$

$$10! = 3{,}628{,}800$$

$$20! = 2.4329 \times 10^{18}$$

$$100! = 9.3326 \times 10^{157}$$

Notice that for groups of even modest size n, the total number of sequences $n!$ can be astronomical.

By definition, zero factorial is 1:

$$0! = 1$$

In computing probabilities we aren't usually concerned with sequences involving every item. Instead, the sample units actually selected will be a small subset of the whole. Either of two types of counting approaches may be taken, depending on whether order of selection is an important distinction.

Number of Permutations

A particular sequence involving a subset of items taken from a collection is called a **permutation**. Consider again the 5 students; examples of permutations are

$$ABC \qquad EB \qquad CBEA \qquad DA \qquad EBADC$$

Notice that a permutation might include every possible item, and that these are distinguished not only by *which* items are included but also by the *order* of selection for those items. Thus, BAC, CBA, BCA, ACB, CAB, and ABC are distinct permutations involving the same three initials.

To find the number of permutations of a particular size, we may extend the principle of multiplication. For example, we count the number of permutations of three student names from the original ones by the following product:

5	$\times$	4	$\times$	3	$= 60$
number of choices for first name		number of choices for second name		number of choices for third name	

In general, the number of permutations of r items from a group of size n is the product:

$$n \times (n-1) \times (n-2) \times \cdots \times (n-r+1)$$

For ease of reference we can represent the above product symbolically as P_r^n. For further convenience we can summarize the permutation product in terms of factorials. Multiplying the preceding by $(n-r)!/(n-r)!$ and simplifying we obtain an equivalent expression.

NUMBER OF PERMUTATIONS OF r ITEMS FROM n

$$\boldsymbol{P_r^n = \frac{n!}{(n-r)!}}$$

Thus, considering $n = 5$ names, the number of permutations of size $r = 3$ is

$$P_3^5 = \frac{5!}{2!} = 60$$

When $r = n$, all items are included and $P_n^n = n!$, the number of complete sequences.

As a further illustration, suppose that a random sample of 5 students is taken one at a time without replacement from the 38 members of the Tau Beta Pi chapter described on page 71. The number of sample outcomes—distinguished not only by which particular students may be selected but also by the order or sequence in which they are picked—is equal to the number of permutations of $r = 5$ from $n = 38$:

$$P_5^{38} = \frac{38!}{(38-5)!} = \frac{38!}{33!}$$

To evaluate this equation, it is simplest to factor the numerator down to a factorial matching the denominator and then cancel terms:

$$\begin{aligned}\frac{38!}{33!} &= \frac{38 \times 37 \times 36 \times 35 \times 34 \times 33!}{33!} \\ &= 38 \times 37 \times 36 \times 35 \times 34 \\ &= 60{,}233{,}040\end{aligned}$$

The Number of Combinations

The order of selection—whether we get *ABCDE* or *EBDCA*—is usually of little concern in a probability evaluation of sampling. Instead, the main focus is on *which* particular units get selected.

A subset of a collection of items is referred to as a **combination**. The order in which items are selected does not affect the nature of a combination, which can differ from another only by which particular items are chosen. With three persons, the sequences ABC and CBA represent the same combination of identical initials. But BAC and FED are sequences for two different combinations.

The number of combinations is found by dividing the corresponding number of permutations by the different orders possible for the selected items. In the case of 3 names taken from 5, there are $3! = 6$ possible orders of selection. For example, Alan, Belle, and Carl might arise in any of the following sequences:

$$\begin{matrix} ABC & BAC & CAB \\ ACB & BCA & CBA \end{matrix}$$

These all represent just one combination. Dividing P_3^5 by 6, we have as the number of name combinations:

$$\frac{P_3^5}{3!} = \frac{60}{6} = 10$$

(You may list them all and verify this to your satisfaction.)

We will represent the number of combinations by the symbol C_r^n, and establish the general expression.

NUMBER OF COMBINATIONS OF r ITEMS FROM n

$$\mathbf{C_r^n = \frac{n!}{r!(n-r)!}}$$

This follows directly from the fact that

$$C_r^n = \frac{P_r^n}{r!}$$

We have just computed $C_3^5 = 10$.

Returning to the sample of 5 students from 38, the number of combinations would be

$$\begin{aligned} C_5^{38} &= \frac{38!}{5!33!} = \frac{38 \times 37 \times 36 \times 35 \times 34 \times 33!}{5 \times 4 \times 3 \times 2 \times 1 \times 33!} \\ &= 501{,}942 \end{aligned}$$

This value is not only considerably smaller than the number of permutations found earlier, but it gives us the size for the random experiment's most compact sample space, wherein the elementary events are distinguished only by which particular students are selected (and not by their order of selection, too).

Example:* *Poker-Hand Probabilities

Poker provides a colorful illustration of how we may compute probabilities using the number of hand combinations.* A poker hand consists of 5 cards dealt from a shuffled deck of 52 playing cards. (We will ignore order of draw and consider only one player.) Consider various probabilities for the particular type of hand

* Accounting by permutations would work as well, but the numbers would be considerably larger.

obtained. The number of distinct poker hands is

$$C_5^{52} = \frac{52!}{5!47!} = \frac{52 \times 51 \times 50 \times 49 \times 48 \times 47!}{5 \times 4 \times 3 \times 2 \times 1 \times 47!}$$
$$= 2{,}598{,}960$$

This is the size of the sample space.

Now consider getting a heart flush, which is a hand with all 5 cards belonging to the heart suit. There are 13 hearts altogether, so the number of heart flushes is

$$C_5^{13} = \frac{13!}{5!8!} = \frac{13 \times 12 \times 11 \times 10 \times 9 \times 8!}{5 \times 4 \times 3 \times 2 \times 1 \times 8!}$$
$$= 1{,}287$$

Altogether there are 4 suits that may give rise to a flush, so the number of hands of this type is

$$4(1{,}287) = 5{,}148$$

Thus, the probability of getting a flush may be computed from the basic definition for equally likely events

$$\Pr[\text{flush}] = \frac{5{,}148}{2{,}598{,}960} \approx \frac{1}{505}$$

Now consider the number of straight hands—a run in successive denominations—such as 2-3-4-5-6, 6-7-8-9-10, or 9-10-J-Q-K. Consider the first one. Each successive card may belong to any one of four suits, so that there are 4^5 ways for a 2-through-6 straight to occur. (You might want to try listing different suit arrangements until you are convinced this is so.) Counting ace as high, there are 9 different starting denominations for a straight. Altogether, a complete listing will contain the following number of 5-card straights:

$$9 \times 4^5 = 9{,}216$$

and dividing by the number of possible hands we have

$$\Pr[\text{straight}] = \frac{9{,}216}{2{,}598{,}960} \approx \frac{1}{282}$$

It is evident why flushes beat straights in poker: The latter is almost twice as likely. (Note that 32 straight flushes and 4 royal flushes are included in the counts of both event sets.)

Problems

2-35 Evaluate the following:
(a) 4! (b) 9! (c) 11! (d) 13!

2-36 Evaluate the following:
(a) P_3^7 (b) P_4^8 (c) P_5^{12} (d) P_7^{20}

2-37 Evaluate the following:
(a) C_2^6 (b) C_5^{10} (c) C_8^{13} (d) C_{13}^{52}

2-38 A medium-sized computer facility can be designed by including as modules the following:

CPU—2 speeds
Core—3 sizes
Input/output—1 to 50 channels
Remote memory—1 to 10 units
Teleports—1 to 25

How many different configurations are possible?

2-39 A man's wardrobe has 10 pairs of shoes, 15 pairs of socks, 5 pairs of slacks, 12 shirts, 7 ties, and 4 jackets. How many different complete outfits are available for him to wear?

2-40 Consider a k-sided symmetrical object. Find the number of possible outcomes when the object is
(a) a coin tossed 5 times
(b) a cube rolled 4 times
(c) a tetrahedron (triangular pyramid) flipped 6 times
(d) an octahedron spun 5 times
(e) a hexahedron bounced 4 times

2-41 A restaurant menu lists 2 soups, 4 salads, 7 entrées, 5 beverages, and 4 desserts.
(a) How many complete meals are possible when one item may be chosen from each category?
(b) The short meal involves choice of soup or salad (not both), one entrée, and one beverage. How many meals are possible?
(c) The hearty meal is a complete one plus a second, different entrée and a second, different dessert. How many meals are possible?

2-42 Ten married couples must be seated for dinner at a round table with 20 chairs. How many seating arrangements are possible in each of the following cases?
(a) People are seated randomly without regard to sex or proximity to marital partners.
(b) Couples are seated in adjacent chairs with the woman on the right.
(c) Couples are seated in adjacent chairs, but sexes need not alternate.
(d) The men and women are in separate semicircles.
(e) Sexes are alternated with no acknowledgment of marital partners.

2-43 Refer to the sample space for the Tau Beta Pi lottery in Figure 2-1. Suppose that 5 members are selected randomly without replacement. Determine for each of the following outcomes (1) the number of possible elementary events (ignoring order of selection) and (2) the probability.
(a) all undergraduates
(b) all women
(c) exactly two civil engineers and three electrical engineers
(d) no graduate electrical engineers selected

2-44 Consider a 5-card poker hand dealt from a deck of 52 shuffled playing cards. Find the following probabilities:
(a) royal flush (a 10-through-ace straight flush)
(b) four-of-a-kind (by denomination)
(c) full house (one pair and one triple)

2-45 A pinochle deck is made from two decks of ordinary playing cards by first deleting 2s through 8s and then combining the remaining cards. Suppose a hand of 5 cards is drawn. Determine the following:

(a) number of possible hands (ignoring order of deal)
(b) probability of all aces
(c) probability of a flush

2-7 Revising Probabilities Using Bayes' Theorem

An eighteenth-century mathematician, the Reverend Thomas Bayes, first extended probability concepts so that judgment could be combined with empirical evidence. His procedure starts with an initial measure of chance for an event, called a **prior probability**. This number may be revised upward or downward in accordance with the result of an empirical test, depending on whether that result supports or refutes the event in question. The revised measure is called a **posterior probability**.

Bayes' concepts are justified by the elements of probability theory already described in this chapter. His procedure may be summarized by a single formula referred to as **Bayes' theorem**. What makes Bayes' theorem so important is that it fits in so well with modern decision analysis. The pivotal events in many decisions arise only from nonrepeatable random experiments, so that judgment plays a key role in arriving at the prior probabilities. These therefore must often be **subjective probabilities**. Procedures involving subjective probabilities are sometimes referred to as *Bayesian statistics*.

Bayes' Theorem

Consider a random experiment with two main events, E and not E. Prior probabilities are obtained, usually by judgment:

$$\Pr[E] \qquad \Pr[\text{not } E]$$

An empirical investigation will be made in an attempt to revise the above values. Denoting by R a particular test result, the following *conditional result probabilities* are often available from historical experience with similar random experiments:

$$\Pr[R \mid E] \qquad \Pr[R \mid \text{not } E]$$

The revised probabilities for the main events are predicated on R being given and are themselves conditional probabilities:

$$\Pr[E \mid R] \qquad \Pr[\text{not } E \mid R]$$

The values for the above posterior probabilities may be calculated using Bayes' theorem.

BAYES' THEOREM

$$\mathbf{PR}[E \mid R] = \frac{\mathbf{Pr}[E]\,\mathbf{Pr}[R \mid E]}{\mathbf{Pr}[E]\,\mathbf{Pr}[R \mid E] + \mathbf{Pr}[\text{not } E]\,\mathbf{Pr}[R \mid \text{not } E]}$$

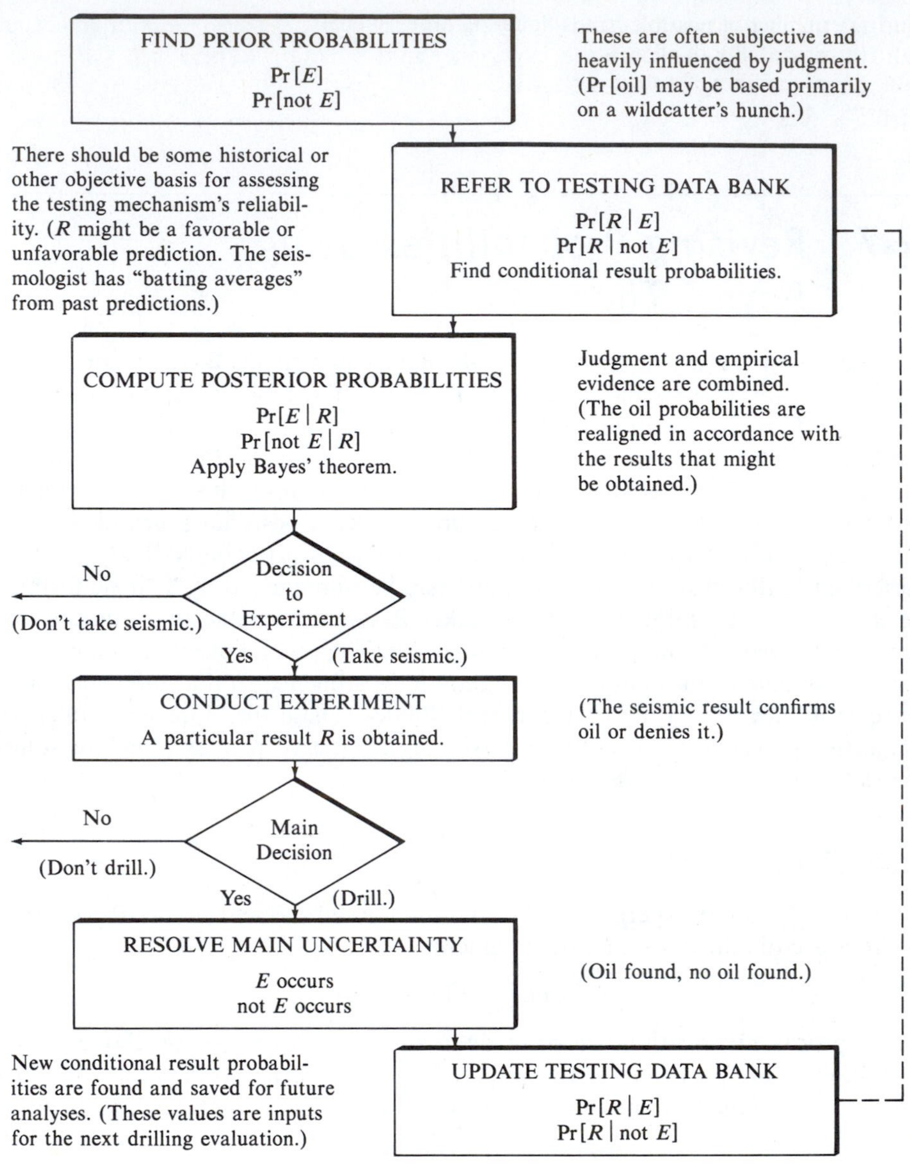

Figure 2-12 *Flow chart showing context of Bayes' theorem.*

This theorem is more than a mere formula. It embodies an elaborate procedure for evaluating decisions, as summarized in Figure 2-12.

Illustration:* *Petroleum Exploration

The use of Bayes' theorem is epitomized by petroleum exploration, where new oil-bearing fields are found by oil wildcatters who must drill risky wells on unproven sites. The main events of interest to a wildcatter are "oil" (O) and "not oil" (not O). Prior probabilities for these events must ordinarily be arrived at through judgment. They are subjective probabilities arrived at largely by hunch and reflect all the wildcatter's exploration experience and his knowledge about the present site. Suppose the following values are reached by a certain wildcatter evaluating a drilling decision at a particular leasehold:

$$\Pr[O] = .35 \qquad \Pr[\text{not } O] = 1 - .35 = .65$$

A panoply of tests are available in petroleum exploration to help predict the uncertain geology. None of these procedures achieves certain forecasts, however. Our wildcatter is considering seismic testing, which has proved to be a reliable predictor. (Decisions regarding what type of test information to get are second in importance only to the main decision, whether or not to drill.) A particular seismological consultant has a pretty good record in making forecasts—80% of the time his procedure has confirmed (C) oil when it was present and 90% of the time it has denied (D) oil when it was known not to be present. These data are based on actual past evaluations and therefore serve to establish the conditional result probabilities:

$$\Pr[C|O] = .80 \qquad \Pr[D|\text{not } O] = .90$$
$$\Pr[D|O] = 1 - .80 = .20 \qquad \Pr[C|\text{not } O] = 1 - .90 = .10$$

To help evaluate whether or not to acquire the seismic test, the wildcatter may apply Bayes' theorem to compute various posterior probabilities. Given that the seismic test confirms oil, so that $R = C$, the posterior probability of oil is

$$\begin{aligned}\Pr[O|C] &= \frac{\Pr[O]\Pr[C|O]}{\Pr[O]\Pr[C|O] + \Pr[\text{not } O]\Pr[C|\text{not } O]} \\ &= \frac{.35(.80)}{.35(.80) + .65(.10)} \\ &= .812\end{aligned}$$

so if this were the actual test result, the probability of "oil" would then be revised upward to .812 from its prior level of .35. And the probability of "not oil" would be $1 - .812 = .188$. Similarly, given that the seismic test denies oil, so that $R = D$, the posterior probability of oil is

$$\begin{aligned}\Pr[O|D] &= \frac{\Pr[O]\Pr[D|O]}{\Pr[O]\Pr[D|O] + \Pr[\text{not } O]\Pr[D|\text{not } O]} \\ &= \frac{.35(.20)}{.35(.20) + .65(.90)} \\ &= .107\end{aligned}$$

And were this to be the actual result, the probability of "oil" would have to be revised downward to .107 from the prior .35. The probability of "not oil" would be $1 - .107 = .893$.

Should the seismic test be used, the wildcatter would use the respective posterior probability for oil that corresponds to the result obtained. But only by drilling will he be able to determine once and for all whether or not there is oil. After drilling, the seismic test will prove to be right or wrong, and the seismologist's "batting averages" would then be updated to provide conditional result probabilities for evaluation in further petroleum exploration.

We have barely scratched the surface of Bayesian statistics. A more complete treatment of these concepts is given in Chapter 17.

Concluding Remarks

Bayes' theorem can be easily proved. The prior probabilities $\Pr[E]$ and $\Pr[\text{not } E]$ are unconditional probabilities for the respective main events. $\Pr[R \mid E]$ and $\Pr[R \mid \text{not } E]$ are given values. The respective probability products provide the joint probabilities for compound events *E and R* and not *E and R*. A posterior probability is a conditional probability for the main event given a particular test result, so it might be computed from the conditional probability identity

$$\Pr[E \mid R] = \frac{\Pr[E \textit{ and } R]}{\Pr[R]}$$

By the multiplication law, the numerator is equal to the product $\Pr[E]\Pr[R \mid E]$. And there are two ways for R to occur, jointly with E or jointly with not E. Thus,

$$\begin{aligned} \Pr[R] &= \Pr[(E \textit{ and } R) \textit{ or } (\text{not } E \textit{ and } R)] \\ &= \Pr[E \textit{ and } R] + \Pr[\text{not } E \textit{ and } R] \end{aligned}$$

which by the multiplication law is equal to the sum

$$\Pr[E]\Pr[R \mid E] + \Pr[\text{not } E]\Pr[R \mid \text{not } E]$$

so that the denominator in the above fraction equals the denominator in the Bayes' theorem expression.

When all else fails, remember that a posterior probability is a conditional probability that can be computed by finding the proper numerator and denominator. Bayes' theorem, then, is nothing more than a restatement of existing concepts.

What makes Bayes' theorem important, however, is how and why it is used. The sources of probabilities employed in making the computation are usually partly judgmental and partly empirical. Bayes' theorem shows how to combine subjective and objective probabilities meaningfully.

Problems

2-46 Suppose a wildcatter assigns a prior probability of .6 for gas (G). A seismic survey might be used that augurs favorably (F) in 85% of applications where gas was known to be present and unfavorably (U) in 90% of sites where no gas was found. Find the posterior probability of gas

(a) given a favorable seismic result

(b) given an unfavorable seismic result

2-47 Refer to Problem 2-46. Suppose a favorable result is obtained. Using your answer from (a) as a revised prior probability for gas, a second magnetic anomaly test is considered. It is known to predict positively (P) 90% of the time when gas is present and to forecast negatively (N) 70% of the time when there is no gas. Find the posterior

probability of gas
(a) given a positive prediction
(b) given a negative prediction

2-48 An aeronautical engineer is evaluating a new airfoil. She presently holds that there is a 75% chance that the wing design can withstand the stresses of normal flight. Wind-tunnel testing on a small-scale model will soon begin; favorable data have been obtained on 90% of all stress-acceptable airfoils tested, whereas unfavorable results were obtained on 95% of all wings that later failed in full-scale testing. The engineer will implement the design if the testing shows at least a 90% chance of acceptable stress-bearing characteristics. She will totally abandon the design concept if that chance is below 20%, and redesign otherwise. Calculate the probability that the wing is acceptable
(a) given a favorable wind tunnel test (What should the engineer do?)
(b) given an unfavorable wind tunnel test (What action should be taken?)

2-49 Consider a card drawn from a shuffled deck of 52 playing cards.
(a) What is the prior probability that it is a queen?
(b) If you notice a reflection indicating that the selected card is a face card, what is the posterior probability that it is a queen?

2-50 A sack contains several boxes, each containing a pair of dice. There are 10 pairs altogether, 3 pairs of which are crooked (one member of the pair has six 3-faces and the other has six 4-faces); the remaining pairs are fair dice. One pair of dice is removed and rolled. You are given only the information regarding the top faces.
(a) What is the prior probability that the selected pair is crooked?
(b) If you know only that a 7-sum shows, what is the posterior probability that the selected pair is crooked?
(c) If you know that a 3-4 combination shows, what is the posterior probability that the selected pair is crooked?

2-51 An admissions committee must select freshmen for its engineering program. Historically, 70% of all admitted students pass (P) their first year. And 70% of the passing students scored at or above (A) 650 on the quantitative section of the SAT, but 50% of the nonpassing students scored below (B) 650. Assume this history applies to next year's class. Consider a particular new freshman.
(a) What is the prior probability that she will pass?
(b) Given that she scores 675 on the quantitative SAT, what is the posterior probability that she will pass?
(c) Given that she scores 580 on the test, what is the posterior probability that she will pass?

Comprehensive Problems

2-52 Consider the faculty at a particular engineering college, categorized in Figure 2-13. One instructor is to be chosen at random. Construct a joint probability table for sex versus employment time.

2-53 Repeat Problem 2-52 for discipline versus employment time.

2-54 One faculty member is chosen from those summarized in Figure 2-13. All persons have the same chance of being selected. Determine the following conditional probabilities:
(a) $\Pr[CE|\text{woman}]$
(b) $\Pr[EE|\text{man}]$
(c) $\Pr[ME|\text{man } \textit{and } FT]$
(d) $\Pr[EE|\text{woman } \textit{and } PT]$

Figure 2-13
Sample space for the characteristics of a randomly selected faculty member.

	Civil Engineering (*CE*)		Electrical Engineering (*EE*)		Mechanical Engineering (*ME*)		Other (*O*)		
Full-Time (*FT*)	♂	♀	♂	♀	♂	♀	♂	♂	♀
	♂	♀	♂	♀	♂	♀	♂	♂	♀
	♂		♂		♂	♀	♂		♀
	♂		♂		♂	♀	♂		♀
	♂		♂		♂		♂		♀
			♂		♂		♂		♀
Part-Time (*PT*)	♂	♀	♂	♀	♂	♀	♂	♂	♀
	♂		♂		♂	♀	♂	♂	♀
	♂		♂		♂		♂		♀
	♂		♂		♂		♂		
	♂		♂		♂		♂		
	♂				♂		♂		

(e) $\Pr[FT \mid \text{woman}]$

(f) $\Pr[FT \textit{ and } EE \mid \text{man}]$

(g) $\Pr[\text{man} \mid CE \textit{ or } EE]$

(h) $\Pr[PT \textit{ and } CE \mid \text{man}]$

2-55 Consider the faculty data in Figure 2-13. A lottery is held to select school membership on the university social committee. Three persons will be selected. Find the following probability that
(a) exactly 2 men will be chosen
(b) at least 1 woman will be chosen
(c) all will be from the electrical engineering department
(d) at least 1 part-timer will be chosen

2-56 Refer to Figure 2-13. Suppose that 5 members are randomly chosen one at a time.
(a) How many combinations are possible?
(b) Ignoring order of selection, find the number of possibilities for each of the following outcomes; then for each outcome compute the probability:
(1) exactly two electrical and three mechanical engineers
(2) all women
(3) all full-timers
(4) no civil engineers selected

2-57 Refer to Figure 2-13. Suppose that 5 members are chosen at random one at a time.
(a) How many permutations are possible?
(b) Including order of selection, find the number of possibilities for each of the following outcomes; then for each outcome compute the probability:
(1) first three are women and last two are men
(2) exactly one of each: part-time civil engineer, full-time electrical engineer, full-time mechanical engineer, part-time other, full-time other
(3) alternating sexes, by order of selection
(4) alternating full- and part-timers, by order of selection

2-58 The following facts are known regarding three events A, B, and C:

$$\Pr[A \text{ or } B] = \frac{3}{4} \qquad \Pr[A \text{ and } B] = \frac{1}{2} \qquad \Pr[B|C] = \frac{1}{2}$$

$$\Pr[B \text{ or } C] = \frac{3}{4} \qquad \Pr[A \text{ and } C] = \frac{1}{4} \qquad \Pr[A|B] = \frac{3}{4}$$

Find the following:
(a) $\Pr[B]$ (c) $\Pr[B|A]$ (e) $\Pr[B \text{ and } C]$
(b) $\Pr[A]$ (d) $\Pr[C]$ (f) $\Pr[C|B]$

2-59 The probability is .95 that a driver in a particular community will survive the year without an automobile accident. Assuming that accident experiences in successive years are independent events, find the probability that a particular driver:
(a) goes 5 straight years with no accident
(b) has at least 1 accident in 5 years

2-60 Consider the elementary events for the sampling experiment in Figure 2-10. Apply the addition law to compute the following compound event probabilities:
(a) exactly 1 defective (b) exactly 2 defectives (c) at least 1 defective

2-61 Repeat Problem 2-60 using the experiment in Figure 2-9.

2-62 The probability of one or more power failures on an automobile assembly line is .10 during any given month. Assuming power events in successive months are independent, find the probability that there will be
(a) no power failures during a 3-month span of time
(b) exactly 1 month involving a power failure during the next 4 months
(c) at least 1 power failure during the next 5 months

2-63 Consider the elementary events for the sampling experiment in Figure 2-10. Determine the following probabilities:
(a) $\Pr[G_1 \text{ and } G_2]$ (c) $\Pr[G_1|G_2]$ (e) $\Pr[G_3]$
(b) $\Pr[G_2]$ (d) $\Pr[G_2 \text{ and } G_3]$ (f) $\Pr[G_2|G_3]$

2-64 Repeat Problem 2-63 using the experiment in Figure 2-9.

2-65 A new-products committee approves funding for 90% of all engineering proposals that later prove to be feasible items to manufacture. The committee turns down 70% of those proposals that later pilot-testing shows to be infeasible. One electrical engineer is promoting her "black box" design. Historically, 20% of similar designs have proved feasible.
(a) What would be an appropriate prior probability that the engineer's design will be feasible to manufacture?
(b) Given that the committee approves the design, what is the posterior probability that it will be feasible?

(c) Given that the committee disapproves the design, what is the posterior probability that it will be feasible?

2-66 Five names will be selected at random without replacement for door prizes from a list containing 20 men and 30 women.
(a) How many different name combinations are possible?
(b) How many combinations of all male names are possible? What is the probability that men will take all the door prizes?
(c) How many combinations are possible when the number of women's names outnumber those of the men? What is the probability that more women than men will take door prizes?

2-67 Three pennies are placed in a box. One is two-headed, one is two-tailed, and one is a fair coin. One coin is selected at random and tossed.
(a) What is the prior probability that it is the two-headed coin?
(b) Given that a head side shows, what is the posterior probability that the two-headed coin was tossed?

2-68 A sample of 4 parts is selected from a production line where 20% are overweight.
(a) Construct a probability tree diagram for this experiment in which each sample observation is treated as a separate stage.
(b) Find the probability that the number of overweights in the sample is:

(1) zero	(3) exactly 2	(5) exactly 4
(2) exactly 1	(4) exactly 3	(6) at least 2

2-69 Suppose that a customer inspects a sample of 4 parts taken without replacement from a shipment of 20 parts from the above plant. Repeat (a) and (b) for Problem 2-68.

2-70 Refer to the crooked and fair dice example on page 80.
(a) Find (1) $\Pr[C]$, (2) $\Pr[F]$, and (3) $\Pr[\text{7-sum} \mid C]$.
(b) Use the multiplication law to find (1) $\Pr[\text{7-sum } \textit{and } F]$ and (2) $\Pr[\text{7-sum } \textit{and } C]$.
(c) Construct the joint probability table using 7-sum and non 7 versus F and C.
(d) Now that $\Pr[\text{7-sum } \textit{and } F]$ and $\Pr[F]$ are known, use the conditional probability identity to verify that $\Pr[\text{7-sum} \mid F] = \frac{1}{6}$.
(e) Given that a 7-sum shows, find the posterior probability that the rolled pair of dice is (1) fair and (2) crooked.

CHAPTER THREE

Random Variables and Discrete Probability Distributions

STATISTICS IS LARGELY concerned with making inferences about unknown population characteristics from known sample data. To lay the groundwork for later discussions of such inductive statistics, we must first understand the process whereby sample outcomes are generated. Doing so requires us to turn the tables temporarily, treating sample outcomes as the unknown; that is, we must use *deductive* statistics. From this reversed vantage point, we view population characteristics as completely specified and the data-generating mechanism as a known entity.

This chapter focuses on the uncertain quantities that may arise from a random experiment. Because the particular level of a variable is uncertain and subject to chance, these quantities are referred to as **random variables**. Our present concern is the *number of events* of a particular type. For instance, we might believe that 99% of all circuit relays are acceptable. If so, how many acceptables will be identified in 100 relays inspected? Or, we might know that once installed, the acceptable relays will fail at an average rate that we might guess to be one per month. How many will fail and need to be replaced in 6 months' time? As we shall see, there is no single answer to these questions. Some quantities are more likely to occur than others, but many results are possible. We will see how to compute probabilities for each outcome. Collectively, the possible values for a random variable and their associated probability values establish a **probability distribution**.

Probability distributions fall into various *classes* or *families*. We will investigate those that have important engineering applications or play an important role in statistical methodology. Within each distribution class, individual probability distributions are distinguished by one or more parameters. For example, statistical quality-control applications frequently involve the **binomial distribution**, which indicates probabilities for getting a particular number of defectives in a sample of fixed size. One binomial distribution may be distinguished from another in terms of the underlying proportion π of defectives or in terms of the number of items inspected n. The **Poisson distribution** provides probabilities for the number of events, such as equipment failures or arrivals at a toll station, occurring during a fixed span of time. Different distributions apply for different average rates λ of failures or arrivals.

The probability distributions described in this chapter are for *discrete* random variables, as we are presently concerned only with how many occurrences there are or in what proportion they happen. Chapter 4 considers *continuous* random variables, which involve such phenomena as time or physical measurements. In either case, it is desirable to have a systematic way to summarize and compare random variables. This is achieved by finding the average outcome or **expected value**. The concept of expected value provides a necessary framework for interpreting and evaluating statistical procedures. It also serves as the central focus of statistical decision theory.

3-1 Random Variables and Probability Distributions

We have seen that all random experiments culminate in an elementary event and that the sample space is the entire collection of these possibilities. Often the elementary events themselves are qualitative outcomes, such as a particular sequence of satisfactories and defectives for n sampled items or a series of successes and failures recorded at a test stand. We usually want to quantify these outcomes in terms of the number of results achieved in a particular category. Before the random experiment, the actual numerical result to be achieved is an unknown and must be treated as a *variable*. This quantity is uncertain, so before the experiment the particular outcome is subject to *random* chance. We use the term **random variable** for such uncertain quantities.

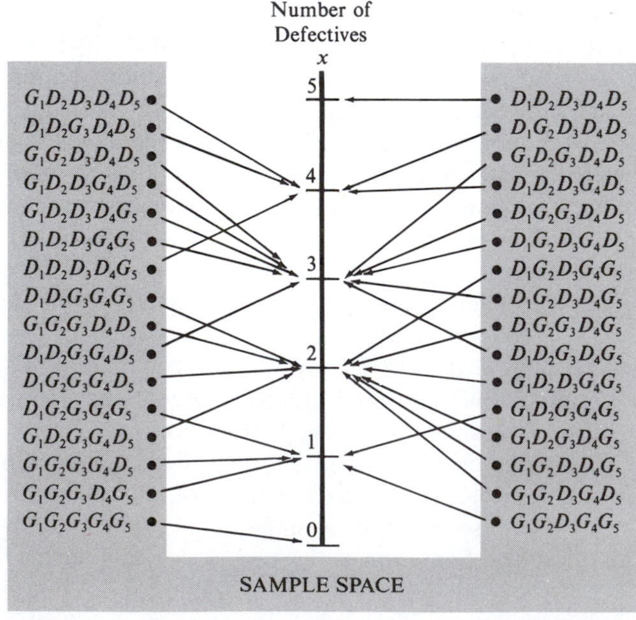

Figure 3-1 *Random variable X (number of defective castings in a sample of size 5) is a function mapping the elements of the sample space onto the real line.*

The Random Variable as a Function

The number of defective castings in a sample and the number of test failures of electronic components are examples of random variables. Mathematically, a random variable is a function that maps every elementary event in the sample space onto the real line. Figure 3-1 illustrates this concept. There X denotes the random variable, the number of defective castings in a sample of $n = 5$. Each observed unit is classified as good (G) or defective (D). Each of the 32 elementary events is represented by a string of G's and D's, with subscripts 1 through 5 designating the sequence of the item. Although there are 32 distinguishable outcomes, there are only 6 levels for X, ranging from 0 (no defectives) to 5 (all defectives).

As a practical matter, statistical applications don't require that we perform any mathematical operations to establish probabilities for X. We can usually match elementary events to a level for X and extend the counting methods of Chapter 2 to find how many outcomes apply to any possible value. This requires a meaningful arrangement of the probability information so that a probability value can be established at each level of X.

The Probability Distribution

The levels for a random variable together with their corresponding probabilities constitute a **probability distribution**. Table 3-1 provides the probability distribution for the number of equipment failures Y that might be encountered over a one-month span of time. There we see that exactly 1 failure has probability $\Pr[Y = 1] = .2707$.

It is standard notation to distinguish the random variable itself by an uppercase letter. Thus, Y is the actual number of failures (whose value is presently unknown). The events from the random experiment are the mutually exclusive and collectively exhaustive outcomes "$Y = 0$," "$Y = 1$," "$Y = 2$," . . . . To avoid

Possible Number of Failures y	Probability $p(y)$
0	.1353
1	.2707
2	.2707
3	.1804
4	.0902
5	.0361
6	.0121
7	.0034
8	.0009
9	.0002
	1.0000

Table 3-1
Probability Distribution for the Number of Equipment Failures over a One-Month Time Span

repetition, they may all be summarized as "$Y = y$" with some specification as to what levels apply for the dummy variable y, which represents any of the possibilities, 0, 1, 2, ...

The probabilities for the outcomes constitute a **probability mass function**, which assigns a number in [0, 1] to each level for the random variable. Denoting this function by p, we have

$$p(y) = \Pr[Y = y] \quad \text{where} \quad y = 0, 1, 2, \ldots$$

For any level of y not listed the probability is $p(y) = 0$, reflecting that those values are impossible. (In the equipment example, only nonnegative integers are listed for y; outcomes such as 1.5 or -5 failures are impossible.) Because $p(y)$ is a probability, for any possible level y

$$0 \leq p(y) \leq 1$$

and because the possible y's correspond to mutually exclusive and collectively exhaustive events, it follows that

$$\sum_{y=0}^{\infty} p(y) = 1$$

A probability distribution may also be summarized by a graph, as in Figure 3-2, which shows the probability mass function for the number of equipment failures Y. There, each possible level y is represented by a spike. The ordinate provides the corresponding probability for each possible number of failures. We see that $p(y)$ is a discrete function that "concentrates its mass" only at those y's that

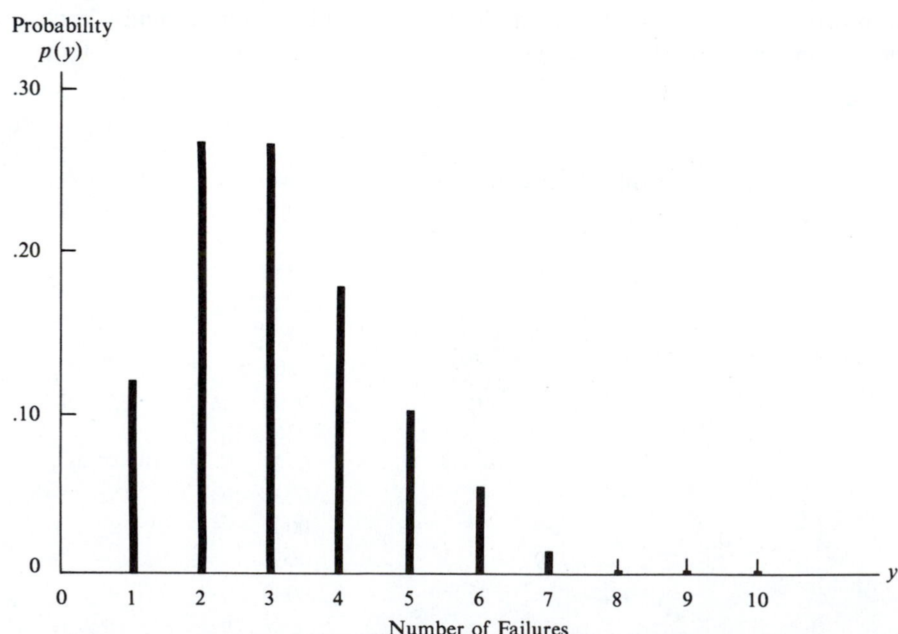

Figure 3-2 *Probability mass function for the number of equipment failures.*

are possible, with the more likely levels being assigned greater probability mass and having taller spikes.

For many classes of probability distributions, the mass function may be characterized by a mathematical expression. Each specific distribution is distinguished by a particular level for one or more parameters. The failure probabilities in Table 3-1 were calculated from

$$p(y) = \frac{e^{-2}(2)^y}{y!} \qquad y = 0, 1, 2, \ldots$$

where the parameter is the mean rate of failures (here set at a value of 2) per month and $e = 2.7182\ldots$ is the base of the natural logarithms. A different set of probabilities will apply for some other rate. This probability mass function is for a Poisson distribution (to be discussed in Section 3-4).

Problems

3-1 Consider the quality-control illustration in Figure 2-10. Determine which elementary events apply to the random variable X, the number of defectives. Then construct a table for the probability distribution for X using the addition law to find the respective probabilities. Sketch this on a graph.

3-2 A computer manufacturer is expecting one of its clients to order a system. The following data apply:

	Probability	Selling Price
CPU (one)		
new model	.50	$250,000
old model	.50	200,000
RAM (one)		
small	.10	50,000
medium	.50	100,000
large	.40	150,000

Assuming that each of the choices is an independent event, construct a table showing the probability distribution for the total selling price of the ordered system.

3-3 Consider an experiment wherein a fair coin is tossed 3 times. Construct a table for the probability distribution for the number of heads obtained.

3-4 An engineering student has been assigned the following probabilities for final grades:

	A	B	C
Thermodynamics	.60	.40	0
Statistics	.50	.50	0
Circuit design	.90	.10	0
Economics	0	.50	.50

(a) Ignoring impossible outcomes, list the sample space for the student's final grades. Assuming independence between course grades, find the probability for each elementary event.

(b) For each elementary event in (a) determine the grade-point average, assuming equal weight for each course. Use A = 4 points, B = 3 points, and C = 2 points. Then construct a table summarizing the probability distribution for the student's GPA X.

3-5 Refer to the dice-tossing random experiment in Problem 2-8. Consider as a random variable the *range* in showing dots (largest value − smallest value).

(a) List the elementary events that correspond to each level of the random variable and determine the corresponding probability value.

(b) Summarize your findings in a table that shows the complete probability distribution.

3-6 A penalty-point system is used by the quality assurance department described on page 68 and summarized in Figure 2-6. For each part unsatisfactory in dimension or weight 1 point is assigned; 2 points are assigned if a part is unsatisfactory in terms of texture. Additional penalties are assigned for overall quality: 3 points for each reworkable; 5 points for salvageables; and 10 points for rejects.

There is a 90% chance that any part will have satisfactory dimensions. But the probabilities are only 80% and 70% that a part will be satisfactory in weight and texture, respectively. The quality events for different characteristics may be assumed to be independent. Let X represent the number of penalty points assessed for any future part. Find the probability distribution for this random variable.

3-2 Expected Value and Variance

Although a detailed description of a random variable is provided by its probability distribution, it is often desirable to summarize that information in a concise form. Such summaries make it easier to compare random variables to determine whether some modification might result in a new, improved random variable. For instance, the failure characteristics of electronic components produced under two different technologies can be compared more easily in terms of their respective failure *rates* (single quantities for each type) than by two columns of time-to-failure probabilities (or *vectors*).

Random variables commonly encountered in engineering have a measure of central tendency or mean, referred to as their **expected values**. Viewing probabilities themselves as long-run frequencies, this number is the long-run average level of the random variable. For example, in a series of 10 coin tosses the average number of heads will be 5. Although many may reach this conclusion by intuition, there is a general procedure for computing expected values directly from the probability distribution.

Another useful summary characterizes how the random variable values tend to differ from each other. A measure of dispersion is provided by the **variance**. Like the expected value, this quantity may be found by an arithmetical operation with the probability distribution.

Expected Value

The expected value of a random variable is simply a weighted average of the possible values, using the respective probabilities as weights. For example, consider the number of dots showing on the top face of a die cube to be rolled. Each face is equally likely, and the probability distribution for the dots random variable is therefore the following:

Dots	Probability
1	$\frac{1}{6}$
2	$\frac{1}{6}$
3	$\frac{1}{6}$
4	$\frac{1}{6}$
5	$\frac{1}{6}$
6	$\frac{1}{6}$
	1

The weighted average of the dots on the showing face is

$$1\left(\frac{1}{6}\right) + 2\left(\frac{1}{6}\right) + 3\left(\frac{1}{6}\right) + 4\left(\frac{1}{6}\right) + 5\left(\frac{1}{6}\right) + 6\left(\frac{1}{6}\right) = \frac{21}{6} = 3.5$$

which is the expected number of dots. Although no single roll can involve 3.5 dots, this value is the average number of dots that will be achieved from many rolls of the die.

In general, the following expression is used to compute the expected value of a discrete random variable.

EXPECTED VALUE OF A DISCRETE RANDOM VARIABLE

$$E(X) = \sum xp(x)$$

In the expression the summation is taken over all possible levels for the random variable. The expected value measures the random variable's **central tendency**, and $E(X)$ is often referred to as the **mean** of X. The foregoing shows that the expected value is analogous to the *first moment* of physics, and this quantity is sometimes identified by that term.

As an illustration of the expected value calculation, consider again the probability distribution for the number of equipment failures, here shown in Table 3-2. The expected value is $E(Y) = 2.0$, indicating that the mean number of failures is 2 per month.

The usual interpretation of an expected value is that it represents the long-run average result from a series of repeated random experiments. Thus, over several years of similar operation the actual tally of month-by-month failures should average to nearly 2.

Table 3-2
Expected Value Calculation for the Number of Equipment Failures in One Month

Possible Number of Failures y	Probability $p(y)$	Weighted Value $yp(y)$
0	.1353	0
1	.2707	.2707
2	.2707	.5414
3	.1804	.5412
4	.0902	.3608
5	.0361	.1805
6	.0121	.0726
7	.0034	.0238
8	.0009	.0072
9	.0002	.0018
		$E(Y) =$ 2.0000

The expected value concept extends to random variables arising from nonrepeatable random experiments. For example, consider the starting salary of one of next year's graduates or the future price of a raw material. Only one value can ever occur for such a random variable, so expected value cannot be thought of as a long-run result. As we have seen, nonrepeatable random experiments must involve subjective probabilities. These measures of chance are best considered as indexes of personal conviction rather than as frequencies. An expected value calculated with subjective probabilities may then be viewed as an "average conviction" as to what number will occur.

Variance and Standard Deviation of a Random Variable

The common measure of dispersion for a random variable is its variance. This measure is also a weighted average, wherein the quantities involved indicate how much individual values differ from the center of the distribution. These quantities are the squared deviations of each possible level from the expected value.

Consider again the dots on a die cube. The deviations are the difference between the respective number of dots and the expected value of 3.5:

$$1 - 3.5 = -2.5$$

$$2 - 3.5 = -1.5$$

$$\cdots$$

The variance is computed by taking a weighted average of the squared deviations:

$$(1 - 3.5)^2\left(\frac{1}{6}\right) + (2 - 3.5)^2\left(\frac{1}{6}\right) + \cdots + (6 - 3.5)^2\left(\frac{1}{6}\right) = 2.917$$

(Unless the deviations are squared first, the positive and negative differences will cancel each other out.) The value 2.917 provides a useful summary of how levels of the dots random variable differ from each other.

The variance is an important statistical concept because it provides a systematic summary of individual differences and because of its convenient mathematical properties. In general, we will use the following expression.

VARIANCE OF A DISCRETE RANDOM VARIABLE

$$\mathbf{Var}(X) = \sum [x - E(X)]^2 p(x)$$

Here, as before, the summation is taken over all possible levels for the random variable. We see that Var(X) is a measure similar to the *second moment* of physics.

Continuing with the example of number of equipment failures Y, Table 3-3 shows the variance computation. The rounded level is Var$(Y) = 2.0$. (Recall that Y has a Poisson distribution. As we shall see, such a random variable will always have a variance equal to its expected value.)

In Chapter 1 we noted that the variance will be in *squared* units. Thus, we have Var$(X) = 2.917$ square dots and Var$(Y) = 2.0$ square failures. As a practical matter, the square root of the variance, or the **standard deviation**, is sometimes used to express dispersion:

$$SD(X) = \sqrt{\text{Var}(X)}$$

We have $SD(X) = \sqrt{2.917} = 1.7$ dots and $SD(Y) = \sqrt{2.0} = 1.41$ failures; these equations convey the same information as the respective variances, but in more convenient units.

Table 3-3 *Variance Computation for the Number of Equipment Failures in One Month*

Possible Number of Failures y	Probability $p(y)$	Squared Deviation $[y - E(Y)]^2$	Weighted Value $[y - E(Y)]^2 p(y)$
0	.1353	$(0 - 2.0)^2 = 4.0$	.5412
1	.2707	$(1 - 2.0)^2 = 1.0$	.2707
2	.2707	$(2 - 2.0)^2 = 0.0$	.0000
3	.1804	$(3 - 2.0)^2 = 1.0$	.1804
4	.0902	$(4 - 2.0)^2 = 4.0$	.3608
5	.0361	$(5 - 2.0)^2 = 9.0$	.3249
6	.0121	$(6 - 2.0)^2 = 16.0$	.1936
7	.0034	$(7 - 2.0)^2 = 25.0$	.0850
8	.0009	$(8 - 2.0)^2 = 36.0$	.0324
9	.0002	$(9 - 2.0)^2 = 49.0$	.0098
		Var(Y) =	1.9988

Problems

3-7 The probability distribution for the number of cars X arriving at a toll booth during any minute is as follows:

x	$p(x)$
0	.37
1	.37
2	.18
3	.06
4	.02

Calculate the expected value, variance, and standard deviation for X.

3-8 The probability distribution for the number of defective items in a random sample is as follows:

x	$p(x)$
0	.35
1	.39
2	.19
3	.06
4	.01

Calculate the expected value, variance, and standard deviation for X.

3-9 The following probability distribution applies to the high temperature X to be experienced in a city when the morning reading is 24°C.

x	$p(x)$
30°C	.05
31	.13
32	.27
33	.36
34	.14
35	.05

Calculate the expected value, variance, and standard deviation for the day's high temperature as expressed in

(a) Celsius degrees

(b) Fahrenheit degrees, using the transformation F° = 32 + 1.8C°

3-10 An environmental test shows that transformer failure characteristics are different at various ambient temperature levels. The following probability distributions apply for the number of failures at low (X) and high (Y) temperature settings.

Number of Failures x, y	Low Setting $p(x)$	High Setting $p(y)$
0	.02	.00
1	.07	.02
2	.15	.04
3	.20	.09
4	.20	.13
5	.16	.16
6	.10	.16
7	.06	.14
8	.03	.10
9	.01	.07
10	.00	.04
11	.00	.02
12	.00	.02
13	.00	.01

Determine the expected number of failures under each setting. On the average, which temperature setting will result in the greatest number of failures?

3-11 Consider the sum of the showing dots X from rolling two die cubes. (You may refer to your answer to Problem 2-8.) Calculate the expected value, variance, and standard deviation.

3-3 The Binomial Distribution

The binomial distribution has wide application in engineering and statistics because it provides a basis for evaluating samples from *qualitative* populations. In such investigations there are ordinarily two complementary categories of major interest—such as defective versus good, satisfactory versus unsatisfactory, operable versus inoperable, or success versus failure. The binomial distribution provides probabilities for the number of observations falling into a particular category.

The binomial distribution is concerned with the events that arise from a *series* of random experiments and is based on several assumptions. These experiments constitute a **Bernoulli process**, named in honor of the pioneer mathematician who first published on the subject.

The Bernoulli Process

A Bernoulli process is epitomized by coin tossing. It is useful to apply the coin analogy in describing the process. The following assumptions can be made:

1. Each random experiment is referred to as a **trial**. The Bernoulli process is a series of trials (such as coin tosses), each having two complementary outcomes. Although designations are arbitrary, one outcome is called a **success** and the other a **failure**. (With coins, either the head or tail side would arbitrarily be designated as a success.)

2. The trial success probability is some constant value for all trials. In coin tossing, Pr[head] is the same for all tosses (although the coin may be lopsided, so that the head probability need not be 1/2).

3. Successive trial outcomes are statistically *independent*. No matter how many successes or failures have been achieved, the probability of success on the next trial cannot vary. (This is an obvious property of coin tosses.)

The Bernoulli process is commonly encountered in engineering applications involving quality assurance. Each new item created in a production process may be considered as a trial resulting in a good or defective unit. Not limited to physical objects, Bernoulli processes have represented voter preferences observed in political polling and product preferences expressed by consumers interviewed during market research investigations.

Binomial Probabilities

The binomial distribution is concerned with how many successes will be achieved in a specified number of trials of a Bernoulli process. This number is a random variable usually denoted by the letter R. Two parameters establish a particular member of this distribution family. One is the trial success probability, denoted by π, and the other parameter is the number of trials (n).

Illustration: *Quality-Control Sampling with Car Parts*

A quality-control illustration will help us develop the procedure used in finding the probabilities. Consider a production line for car parts wherein 10% of the output is defective and must be scrapped, and the rest is good. Treating a defective as a "success," the probability that any particular part is defective is therefore $\pi = .10$. A random sample of $n = 5$ parts is taken.

Figure 3-3 is the probability tree diagram for the car parts investigation, wherein all 5 observation trials have been treated as a combined random experiment. There are $2^5 = 32$ elementary events, distinguishable in terms of good (G) and defective (D) for each successive part. For each part the probabilities are .10 for defective and .90 for good; these values are entered onto the respective branches. The quality events between sample parts are independent, so these same probability values apply for each observed part.

Figure 3-3 *Probability tree diagram for a sample from a car parts production line.*

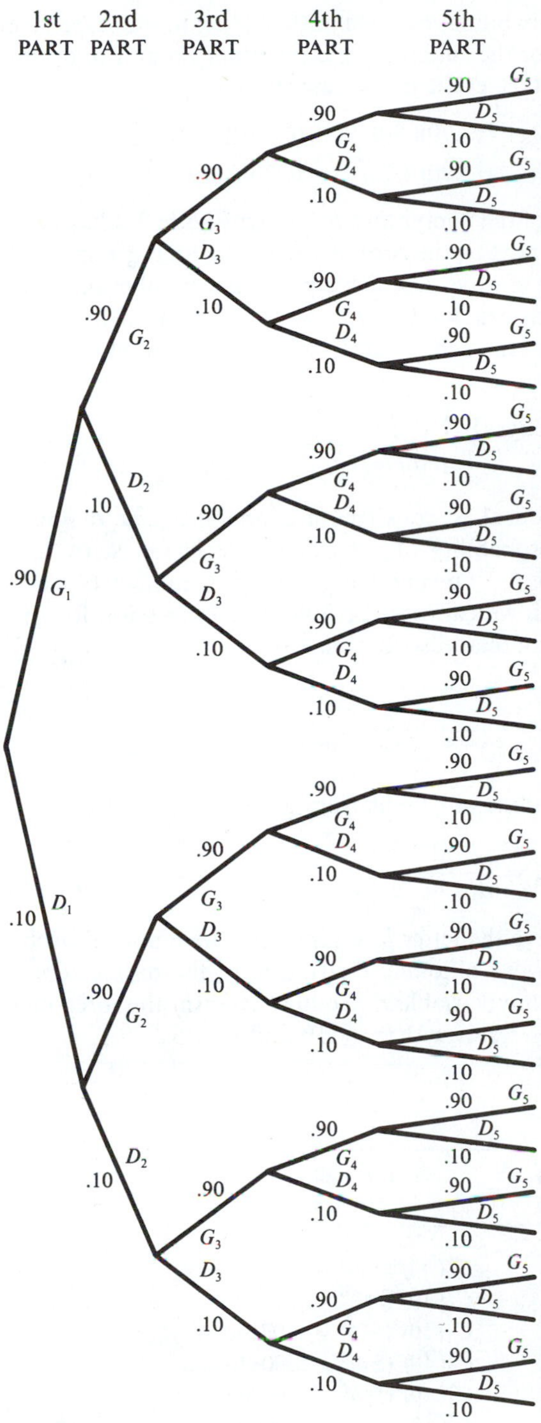

SAMPLE SPACE	PROBABILITY	NUMBER DEFECTIVES	TALLY 0	1	2	3	4	5
$G_1G_2G_3G_4G_5$	.59049	0	•					
$G_1G_2G_3G_4D_5$	.06561	1		•				
$G_1G_2G_3D_4G_5$	.06561	1		•				
$G_1G_2G_3D_4D_5$	.00729	2			•			
$G_1G_2D_3G_4G_5$	.06561	1		•				
$G_1G_2D_3G_4D_5$	.00729	2			•			
$G_1G_2D_3D_4G_5$	.00729	2			•			
$G_1G_2D_3D_4D_5$	.00081	3				•		
$G_1D_2G_3G_4G_5$	.06561	1		•				
$G_1D_2G_3G_4D_5$	.00729	2			•			
$G_1D_2G_3D_4G_5$	.00729	2			•			
$G_1D_2G_3D_4D_5$	.00081	3				•		
$G_1D_2D_3G_4G_5$	.00729	2			•			
$G_1D_2D_3G_4D_5$	.00081	3				•		
$G_1D_2D_3D_4G_5$	.00081	3				•		
$G_1D_2D_3D_4D_5$	.00009	4					•	
$D_1G_2G_3G_4G_5$	.06561	1		•				
$D_1G_2G_3G_4D_5$	.00729	2			•			
$D_1G_2G_3D_4G_5$	.00729	2			•			
$D_1G_2G_3D_4D_5$	.00081	3				•		
$D_1G_2D_3G_4G_5$	.00729	2			•			
$D_1G_2D_3G_4D_5$	.00081	3				•		
$D_1G_2D_3D_4G_5$	.00081	3				•		
$D_1G_2D_3D_4D_5$	.00009	4					•	
$D_1D_2G_3G_4G_5$	.00729	2			•			
$D_1D_2G_3G_4D_5$	.00081	3				•		
$D_1D_2G_3D_4G_5$	.00081	3				•		
$D_1D_2G_3D_4D_5$	.00009	4					•	
$D_1D_2D_3G_4G_5$	.00081	3				•		
$D_1D_2D_3G_4D_5$	.00009	4					•	
$D_1D_2D_3D_4G_5$	.00009	4					•	
$D_1D_2D_3D_4D_5$	.00001	5						•
Totals	1.00000		1	5	10	10	5	1

The product of the branch probabilities on each path leading to the respective end position gives the probability for the culminating elementary event. For example, we have for the fourth elementary event in the sample space

$$\begin{aligned}\Pr[G_1G_2G_3D_4D_5] &= (.90)(.90)(.90)(.10)(.10)\\ &= (.90)^3(.10)^2 = .00729\end{aligned}$$

The number of defectives for each of the compound results in Figure 3-3 has been determined and listed in a column next to the probabilities. The final grouping of columns gives a tally of the number of ways in which a particular number of defectives might occur. There we see that exactly 2 defectives can happen in 10 different ways. Since the elementary event in each of these cases has probability .00729, it follows that

$$\begin{aligned}\Pr[\text{exactly 2 defectives}] &= \Pr[R = 2]\\ &= 10(.00729) = .07290\end{aligned}$$

Of course, it would be impractical to construct a probability tree diagram every time we want to find the probability of a particular level for R. (A tree drawn on the same scale as Figure 3-3 but involving twice the number of trials would be over 20 feet tall.) Instead, we can use the following expression for the probability mass function of the binomial distribution.

BINOMIAL DISTRIBUTION

$$\boldsymbol{b(r; n, \pi) = C_r^n \pi^r (1 - \pi)^{n-r} \qquad r = 0, 1, 2, \ldots, n}$$

Here C_r^n is the number of combinations of r items from a collection of size n,

$$C_r^n = \frac{n!}{r!(n-r)!}$$

Because we use it so often, we reserve the letter b to represent the binomial probabilities. To help distinguish particular binomial distributions, the parameters n and π are often listed after the dummy variable r. We may confirm the foregoing expression by evaluating our earlier result with $\pi = .10$ and $n = 5$:

Table 3-4
Binomial Distribution for the Number of Defective Parts When $n = 5$ and $\pi = .10$

Number of Defectives r	Probability $b(r; n, \pi)$
0	$C_0^5(.10)^0(.90)^5 = .59049$
1	$C_1^5(.10)^1(.90)^4 = .32805$
2	$C_2^5(.10)^2(.90)^3 = .07290$
3	$C_3^5(.10)^3(.90)^2 = .00810$
4	$C_4^5(.10)^4(.90)^1 = .00045$
5	$C_5^5(.10)^5(.90)^0 = .00001$
	1.00000

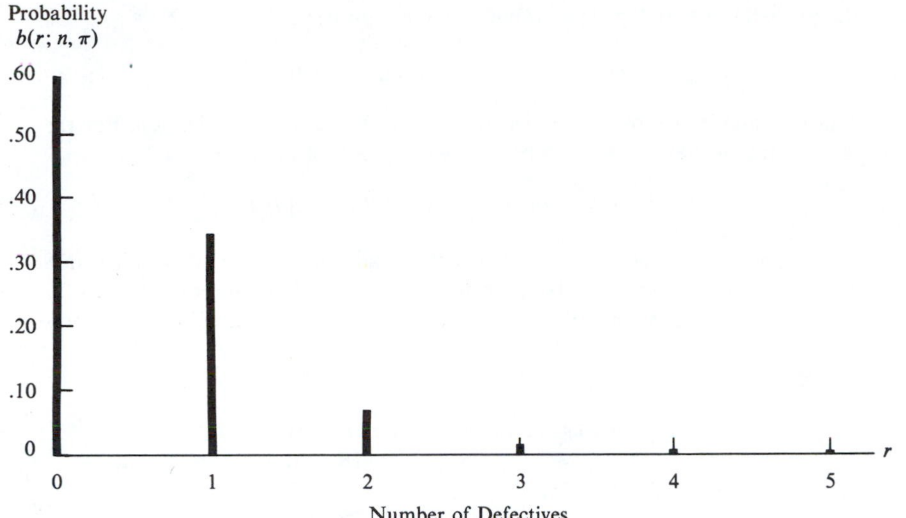

Figure 3-4 *Binomial distribution for the number of defective car parts.*

$$\begin{aligned}\Pr[R = 2] = b(2; 5, .10) &= C_2^5(.10)^2(1 - .10)^{5-2}\\ &= 10(.10)^2(.90)^3\\ &= .07290\end{aligned}$$

The product $(.10)^2(.90)^3$ is the same result achieved by multiplying together all the branch probabilities on any one of the paths having two defectives. Each of those outcomes involves 2 D and 3 G branches. The quantity $C_2^5 = 10$ is the number of such paths and represents the number of ways of getting 2 particular items (as the defective D parts) from 5. In general, there will be C_r^n outcomes involving exactly r successes, each of which has a common probability, $\pi^r(1 - \pi)^{n-r}$.

Table 3-4 gives the complete binomial probability distribution for the number of defective parts. This distribution is graphed in Figure 3-4.

Example: *Opinions Regarding Computer Terminal Display*

A computer manufacturer had to decide the final terminal display configuration for its consumer line of minicomputers. Two possibilities were (1) black-and-white cathode ray tube (CRT) with a hard-copy printer or (2) color CRT with no printer. The final choice depended on the proportion π of potential buyers who would have a definite need for hard copy. A sample of these would-be customers was used to help determine whether that proportion was high or low.

Since the population of potential customers was huge, the opinions of successive persons in the random sample could be assumed independent. This made the binomial distribution appropriate for establishing probabilities for the number of respondents R expressing a need for hard-copy printing.

Using a sample size of $n = 5$, and assuming that

$$\pi = \Pr[\text{a person requires a printer}] = .70$$

the probability that 4 persons would ask for a printer is

$$b(4; 5, .70) = C_4^5(.70)^4(.3)^1 = .3602$$

A larger sample size of $n = 20$ was selected in the actual investigation. For the same π, the probability that 15 persons would require the printer is

$$b(15; 20, .70) = C_{15}^{20}(.70)^{15}(.30)^5 = .1789$$

Of course, π itself was unknown. The actual sample outcome resulted in 8 out of 20 requesting the hard-copy printer. Using the binomial distribution for various levels of π, the probabilities of getting as few or fewer favorable responses are:

Possible Proportion π	**Probability** $\Pr[R \leq 8]$
.30	.8867
.40	.5956
.50	.2517
.60	.0565
.70	.0051
.80	.0001

For high levels of π it is quite improbable that so few favorable responses could have been achieved. The sample evidence tends to favor a lower level for π. Because of this, management decided to drop the printer and use the colored CRT terminal display.

Parameter Levels and Binomial Probabilities

Individual distribution members of the binomial family are distinguished by levels of n and π. Figure 3-5 shows the probability distributions that result when $\pi = .30$ and n is set at different levels. The binomial distribution for the number of successes when $n = 5$ is shown in (a). There are $n + 1$ possible levels for r. When n is raised to 10 in (b), the probability levels are smaller and less concentrated. This spreading and lowering of probabilities is further accentuated when n is raised to 20 in (c). For large n's the binomial distribution tends to have a "bell shape" regardless of π, a feature that will prove useful later when we approximate it by the normal distribution.

Figure 3-6 illustrates how the binomial distribution is influenced by the levels of trial success probability π (holding the number of trials fixed at $n = 5$). Small π's yield distributions with a positive skew that becomes less pronounced as π approaches .5, where a symmetrical distribution applies. As π gets larger, the direction of skew reverses, becoming greater as π gets closer to 1. Notice that complementary π levels .05–.95, .20–.80, and .40–.60 provide probability distributions

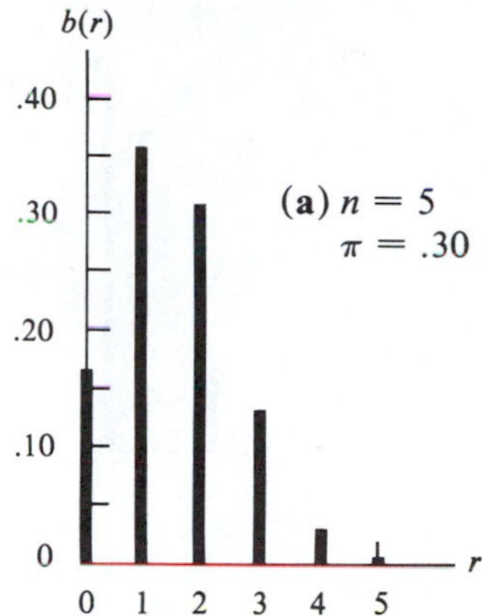

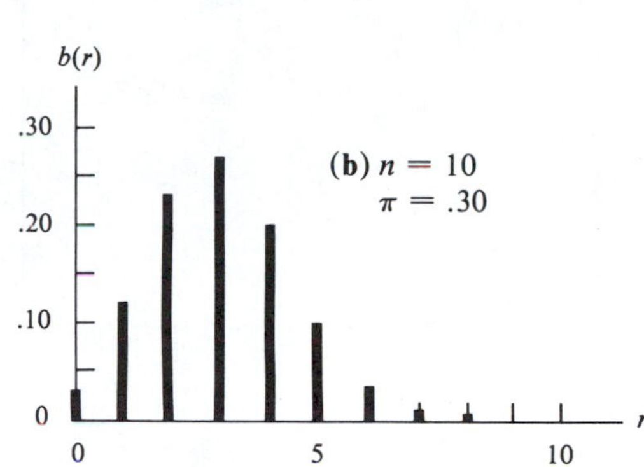

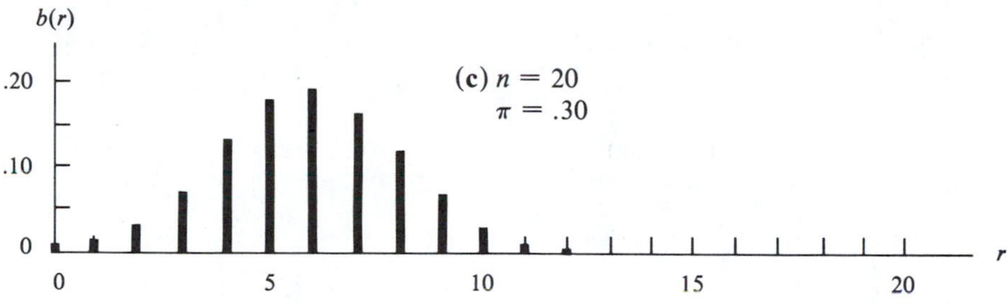

Figure 3-5 *Binomial distributions for three different trial numbers n when the trial success probability π is constant.*

that are mirror images. The probability for r successes is equal to the mirror-image probability for $n - r$ successes.

Expected Value and Variance

The expected value and variance for the number of successes R may be computed using the procedures of Section 3-2. However, these quantities may be more directly computed from the stated parameters using the following expressions.

EXPECTED VALUE AND VARIANCE FOR A BINOMIAL DISTRIBUTION

$$\mathbf{E(R) = n\pi}$$

$$\mathbf{Var(R) = n\pi(1 - \pi)}$$

Thus, the expected number of heads in $n = 100$ tosses of a fair coin, where $\pi = \Pr[\text{head}] = .5$ is

$$E(R) = n\pi = 100(.5) = 50$$

and the variance is

$$\text{Var}(R) = n\pi(1 - \pi) = 100(.50)(.50) = 25$$

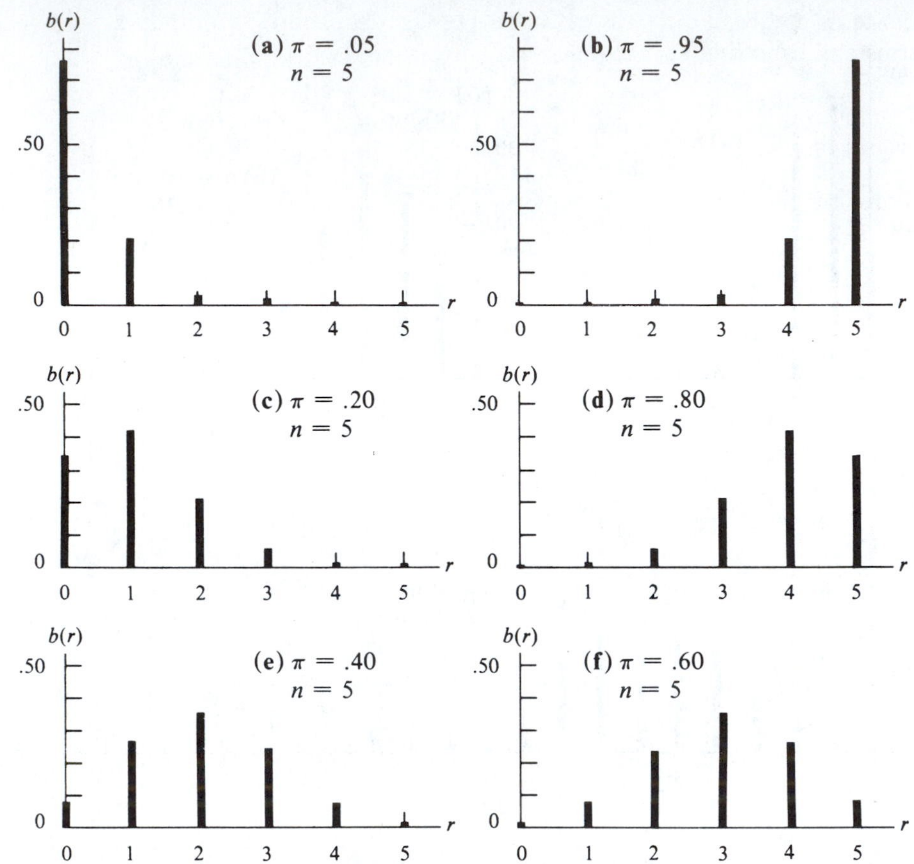

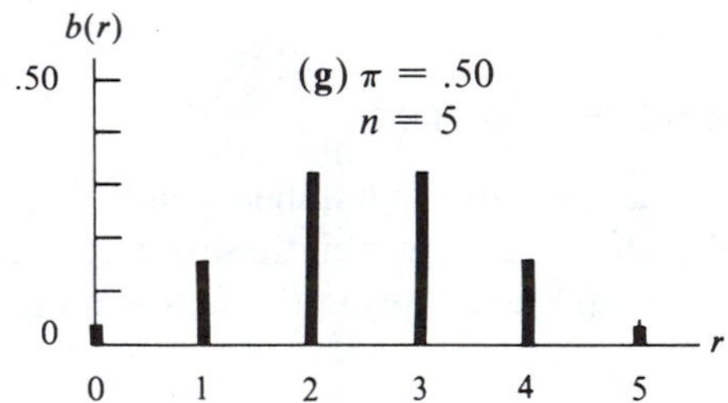

Figure 3-6 *Binomial distributions for different trial success probabilities π when the number of trials is constant.*

It will be helpful to show the derivation for $E(R)$. By the definition of expected value,

$$E(R) = \sum_{r=0}^{n} rb(r; n, \pi) = \sum_{r=0}^{n} rC_r^n \pi^r (1 - \pi)^{n-r}$$

$$= \sum_{r=0}^{n} r \frac{n!}{r!(n - r)!} \pi^r (1 - \pi)^{n-r}$$

The term involving $r = 0$ is zero and may be dropped. Factoring $n\pi$ out of all terms and canceling r's, the preceding is equal to

$$n\pi \sum_{r=1}^{n} \frac{(n-1)!}{(r-1)!(n-r)!} \pi^{r-1}(1-\pi)^{n-r}$$

Representing $r - 1$ by s and $n - 1$ by m, so that $n - r = (n-1) - (r-1) = m - s$, the above is equivalent to

$$n\pi \sum_{s=0}^{m} \frac{m!}{s!(m-s)!} \pi^{s}(1-\pi)^{m-s}$$

The summation includes all binomial probabilities when there are m trials and is therefore equal to 1. Thus,

$$E(R) = n\pi(1) = n\pi$$

It will be left as an exercise for the reader to derive the expression for Var(R).

Notice that $E(R)$ is directly proportional to n and to π, so that greater levels in either parameter will yield an increased level for the expected number of successes. Var(R) also increases in direct proportion to n. However, for a fixed n, Var(R) is greatest when $\pi = .5$, assuming lower levels when π is closer to 0 or to 1.

Cumulative Probability and the Binomial Probability Table

When n is small, it is easy to compute binomial probabilities using a hand-held calculator. But when n is large this becomes an onerous and error-prone chore. To ease the computational burden, the more common binomial probabilities have been tabled.

Such tables are usually constructed in terms of **cumulative probabilities**. A cumulative probability is found by summing the individual probability terms applicable to all the levels of the random variable that fall at or below a specified point. The resulting sums themselves form the **probability distribution function**, defined for discrete distributions in terms of the cumulative sum of the probability mass function. The following expression applies.

BINOMIAL PROBABILITY DISTRIBUTION FUNCTION

$$B(r; n, \pi) = \Pr[R \leq r] = \sum_{x=0}^{r} b(x; n, \pi)$$

We use an uppercase $B(r)$ to distinguish the cumulative binomial distribution function from the binomial probability mass function $b(r)$, which provides individual probability terms.

Table 3-5 shows the distribution function constructed for the number of defective car parts considered earlier.

Table 3-5 *Binomial Probability Distribution Function Constructed for the Number of Defective Car Parts When $n = 5$ and $\pi = .10$*

Number of Defectives r	Probability $\Pr[R = r] = b(r; n, \pi)$	Cumulative Probability $\Pr[R \leq r] = B(r; n, \pi)$
0	.59049	.59049
1	.32805	.91854
2	.07290	.99144
3	.00810	.99954
4	.00045	.99999
5	.00001	1.00000

As an example, the cumulative probability for 2 defectives or less is found by adding together the individual probability terms for 0, 1, and 2 defectives:

$$\begin{aligned}\Pr[R \leq 2] &= \Pr[R = 0] + \Pr[R = 1] + \Pr[R = 2] \\ &= .59049 + .32805 + .07290 \\ &= .99144\end{aligned}$$

Figure 3-7 shows the probability mass function and the probability distribution function graphs of the binomial distribution for the number of successes in $n = 10$ trials when $\pi = .30$. Notice that the probability distribution function plots as a stairway, wherein the cumulative probability at any r is the height of the stairway above ground level at that point. The size of each step equals the probability for getting exactly r successes and is the same size as the respective spike of the probability mass function. For example, the two cumulative probabilities for 2 and 3 successes are $B(2; 10, .30) = .3828$ and $B(3; 10, .30) = .6496$. The step size at 3 is $.6496 - .3828 = .2668$, and is equal to the probability of exactly 3 successes, $b(3; 10, .30)$.

Appendix Table A at the back of the book lists cumulative binomial probabilities for several levels of n and π. This table is a convenient source for binomial probabilities. Cumulative terms are tabled instead of individual terms because most applications consider R events spanning a range of successes.

Illustration: *Preferences for Dash Panel Displays*

As an illustration involving these tables, suppose that a random sample of $n = 100$ persons are asked to state their preferences for two alternative automobile dash panel displays: a sleek, modernistic one and a classical one. The population is sufficiently large for the binomial distribution to represent probabilities for the number of respondents R, favoring the sleek display over the classical one. The assumed proportion favoring the sleek display is 40%, so that $\pi = .40$ serves as the trial success probability.

The probability that 35 or fewer favor the sleek design is

$$\Pr[R \leq 35] = B(35; 100, .40) = .1795$$

whereas the probability that at least 50 like that display better is the complement of 49 or fewer:

$$\begin{aligned}\Pr[R \geq 50] &= 1 - \Pr[R \leq 49] \\ &= 1 - B(49) \\ &= 1 - .9729 = .0271\end{aligned}$$

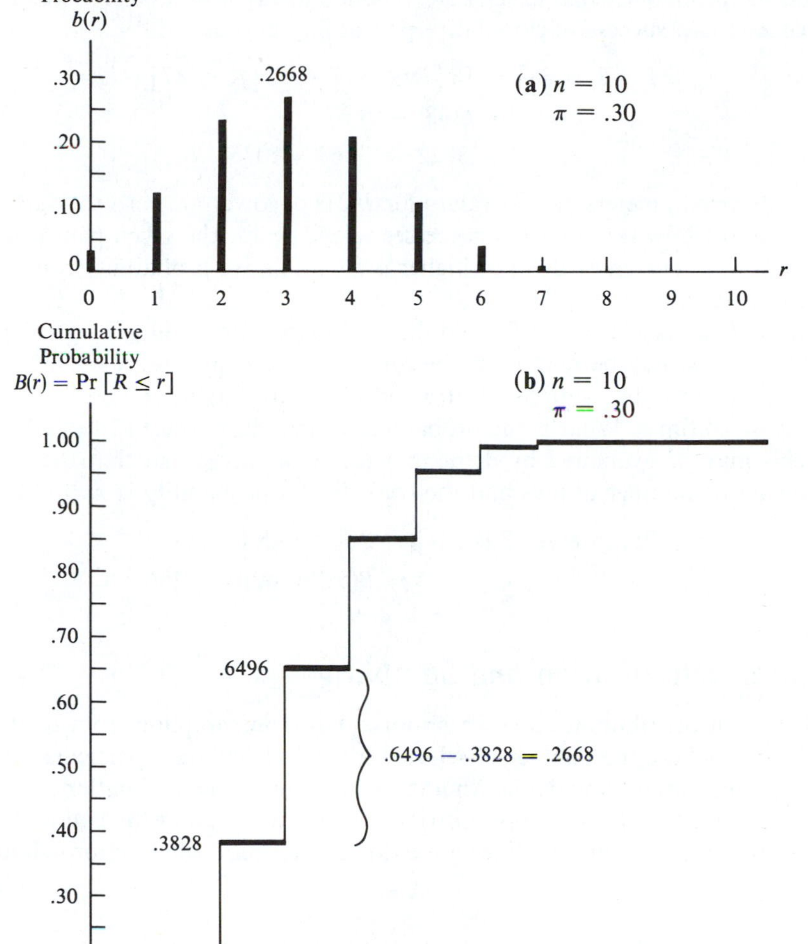

Figure 3-7 *Binomial probability mass function and corresponding probability distribution function.*

The probability that between 40 and 50 inclusively prefer the sleek version is found by subtracting from the cumulative probability for 50 the corresponding value for the unwanted levels 39 or below:

$$\begin{aligned}\Pr[40 \leq R \leq 50] &= \Pr[R \leq 50] - \Pr[R \leq 39] \\ &= B(50) - B(39) \\ &= .9832 - .4621 = .5211\end{aligned}$$

To find the probability that exactly 38 respondents favor that design, we take the difference in two successive cumulative probability entries:

$$\begin{aligned}\Pr[R = 38] &= \Pr[R \leq 38] - \Pr[R \leq 37] \\ &= B(38) - B(37) \\ &= .3822 - .3068 = .0754\end{aligned}$$

At these parameters, tabled values for $B(21)$ or lower are zero to four places, indicating just how rare that few successes would be for the given parameter settings. Likewise, any entry $B(60)$ or higher is a cumulative probability rounding at four places to one.

Table A stops at $\pi = .50$. Should the trial success probability be greater than .50, then values may be read from the mirror-image distribution with the event reversed. For example, suppose that a lopsided coin having .60 as Pr[head] is tossed $n = 20$ times. What is the probability of getting at least 15 heads?

This may be evaluated by defining a *tail* as a success, so that the random variable is the number of tails and the trial success probability is $\pi = .40$. Thus,

$$\begin{aligned}\Pr[\text{at least 15 heads}] &= \Pr[R \leq 5 \text{ tails}] \\ &= B(5; 20, .40) = .1256\end{aligned}$$

Binomial Distribution and Sampling

The binomial distribution plays an important role in sampling from qualitative populations, where the number of selected attributes having a particular characteristic is a key random variable. An important engineering application is quality-control evaluations. There the proportion of defectives defines the overall level of quality, and binomial probabilities are used to determine what actions should be taken.

Example: *Air Force Acceptance Sampling*

The U.S. Air Force contracts for a wide variety of services, such as vehicle maintenance and mess hall operations. In administering these contracts, inspections are required of units randomly selected from the defined *lot* (statistical population). One food service contract stipulates an *acceptable quality level* (AQL) for equipment cleanliness of 6.5% during meal preparations. The allowable proportion of unclean equipment is then $\pi = .065$.

Based on the above AQL, the Air Force requires that a sample size of $n = 13$ be used and that the lot be accepted or rejected according to the following *decision rule* for the observed number of defectives R (mealtimes during which some dirty equipment is found):*

$$\text{Accept lot if } R \leq 2$$

$$\text{Reject lot if } R > 2$$

The value 2 is the *acceptance number*, the maximum number of defectives at which accepting the lot is allowed. If we apply the binomial distribution, we may compute several probabilities.** This decision rule was established so that there will be at least a 95% chance of accepting a lot with the stipulated AQL. We have

$$\Pr[\text{Accept} \mid \text{AQL applies } (\pi = .065)]$$
$$= C_0^{13}(.065)^0(.935)^{13} + C_1^{13}(.065)^1(.935)^{12} + C_2^{13}(.065)^2(.935)^{11}$$
$$= .4174 + .3772 + .1573 = .9519$$

The decision rule is not perfect, since two types of errors can occur: bad lots might be accepted or good ones rejected. For example, consider a poor-quality meal service where the equipment is actually dirty 20% of the time ($\pi = .20$). There is a substantial probability that poor service will be accepted:

$$\Pr[\text{Accept} \mid \text{bad lot } (\pi = .20)]$$
$$= C_0^{13}(.20)^0(.80)^{13} + C_1^{13}(.20)^1(.80)^{12} + C_2^{13}(.20)^2(.80)^{11}$$
$$= .0550 + .1787 + .2680 = .5017$$

And, consider a high-quality meal service where the equipment is only dirty 4% of the time. There is a small probability that the good service will be rejected:

$$\Pr[\text{Reject} \mid \text{high quality } (\pi = .04)] = C_3^{13}(.04)^3(.96)^{10} + \cdots + C_{13}^{13}(.04)^{13}(.96)^0$$
$$= .0135$$

These types of errors are unavoidable in sampling decisions. A good decision rule achieves an acceptable balance between them.

* *Military Standard: Sampling Procedures and Tables for Inspection by Attributes* (MIL-STD-105D), 29 April 1963.

** The Air Force tables are based on the binomial distribution and are only approximately correct. When sampling is done without replacement from a small population, the hypergeometric distribution described in Section 3-5 should be used in computing probabilities for R.

Problems

3-12 A food-packaging apparatus underfills 10% of the containers. Find the probability that for any particular 5 containers the number of underfilled will be
(a) exactly 3 (b) exactly 2 (c) zero (d) at least 1

3-13 Indicate for each of the following situations whether or not the assumptions of a Bernoulli process are met. In all cases, items are produced that are either within or not within tolerance.
(a) To avoid tedium, a lathe operator switches back and forth between "easy" and "hard" types of jobs.
(b) Another machinist works only with a single type of item but gets very sloppy just before breaks, after lunch, and around quitting time.
(c) Each machinist in the plant checks his own work for size. One worker becomes overly careful just after finding himself out of tolerance.
(d) Some equipment has automatic settings that gradually work away from the intended levels.

3-14 Suppose that 20% of the applicants to an engineering school are women. An admissions committee reviews applications in groups of 100. Find the probability that the number of women in the next group is
(a) less than or equal to 15
(b) greater than 25
(c) exactly 27
(d) between 18 and 26, inclusively
(e) less than 10
(f) greater than 40

3-15 Compute the expected number of defectives and the variance when random samples are taken from a production process and the following parameters apply:

(a) $n = 50$ $\pi = .10$ (c) $n = 20$ $\pi = .50$
(b) $n = 40$ $\pi = .20$ (d) $n = 100$ $\pi = .10$

3-16 Construct a table of probabilities for the binomial distribution with $n = 5$ and $\pi = .15$ (rounded to four places). Then compute the cumulative probability for each possible number of successes.

3-17 The proportion π of defective transformer coils wound on automatic spindles is unknown. A quality-control inspector requests tension readjustments whenever more than two coils are found to have broken wires in a sample of $n = 100$. Determine the probability of tension readjustment, assuming

(a) $\pi = .01$ (b) $\pi = .02$ (c) $\pi = .03$ (d) $\pi = .05$ (e) $\pi = .10$

3-18 A student selects his answers on a true/false examination by tossing a coin (so that any particular answer has a .50 probability of being correct). He must answer at least 70% correctly in order to pass. Find his probability of passing when the number of questions is

(a) 10 (b) 20 (c) 50 (d) 100

3-19 A lopsided coin has a 70% chance of head. It is tossed 20 times. Determine the following probabilities for the possible results:

(a) at least 10 heads
(b) exactly 12 heads
(c) fewer than 9 heads
(d) at most 13 heads
(e) between 8 and 14 heads

3-20 A quality-control inspector rejects any shipment of printed circuit boards whenever 3 or more defectives are found in a sample of 20 boards tested. Find the (1) expected number defective and (2) the probability of rejecting the shipment when the proportion of defectives in the entire shipment is

(a) $\pi = .01$ (b) $\pi = .05$ (c) $\pi = .10$ (d) $\pi = .20$

3-21 Consider again the coils in Problem 3-17. The quality-control department is also interested in the uniformity of the wire used in the winding. Only a coil having broken wires is unwound to determine whether it was nonuniform. Suppose that 10% of all coils have spindle-caused broken wires and that 20% of all coils have been wound with nonuniform wire. Altogether 100 coils are selected at random. Determine the probability that the number of coils found to have nonuniform wire is

(a) 0 (b) 1 (c) 2 (d) 3

3-22 An engineer has designed a modified welding robot. The robot will be considered good enough to manufacture if it misses only 1% of its assigned welds. And it will be judged a poor performer if it misses 5% of its welds. (In-between possibilities are not considered.)

A test is performed involving 100 welds. The new design will be accepted if the number of missed welds R is 2 or less and rejected otherwise.

(a) What is the probability that a good design will be rejected?
(b) What is the probability that a poor design will be accepted?

3-23 A sample of $n = 4$ items is selected with replacement from a lot having $\pi = .20$ defective. Construct a probability tree diagram for this experiment. Then, reading the joint probabilities from your tree, compute the values for $b(r; 4, .20)$ using the basic probability concepts of Chapter 2.

3-24 Consider the Air Force acceptance sampling example on pages 132–133. Compute the probability of (1) accepting a bad lot (with $\pi = \pi_b$) and (2) rejecting a good lot (with $\pi = \pi_g$) for each of the following cases:

	(a)	(b)	(c)	(d)
Sample size	$n = 20$	$n = 50$	$n = 125$	$n = 200$
Acceptable quality level	AQL = 4%	AQL = 6.5%	AQL = .4%	AQL = 1%
Acceptance number	2	7	1	3
π_b	.05	.10	.01	.05
π_g	.02	.05	.001	.005

Source: MIL-STD-105D.

3-4 The Poisson Distribution

Next in importance to discrete random variables with a binomial distribution are those belonging to the Poisson distribution family. This distribution is common in waiting-line evaluations, in which it gives probabilities for the number of arrivals at a service facility. It is also frequently used in reliability analysis, in which it can give probabilities for the number of failures.

It will be useful to relate the Poisson distribution to the binomial, which is concerned with the number of successes in a fixed series of trials, each of which can culminate in one of the complementary outcomes. The Poisson distribution concerns only one event (instead of two), such as an arrival at a toll booth or a failure of a satellite power cell. It provides probabilities for *how many events* will occur. The number of Poisson events is uncertain rather than fixed, as with the binomial distribution, where each trial (specified to be n in number) results in one event (a success or a failure). The Poisson events will occur at some average rate over a set time span or within some prescribed space. Thus, the Poisson distribution might provide the probability that 3 accidents will occur during one hour along a stretch of highway, that a data-processing clerk will commit 3 errors while making 1,000 teleprocessing entries, or that a ship will encounter 3 icebergs while traversing the North Atlantic in July.

Like the binomial distribution, the Poisson distribution applies to an underlying stochastic process. Named for the eighteenth-century physicist and mathematician, Simeón Poisson, a **Poisson process** characterizes a random experiment having the special properties described next.

The Poisson Process

A Poisson process gives rise to a series of events occurring at a *mean rate* over time or space. These occur at random, so that no pattern is discernible. Figure 3-8 shows a time record of street lamp burnouts in a city, as recorded by special telemetry equipment. Each lamp failure is an event in an underlying Poisson process wherein lamps have burned out at a historical mean rate of $\lambda = 1$ per day. These occurrences are random over time and without pattern.

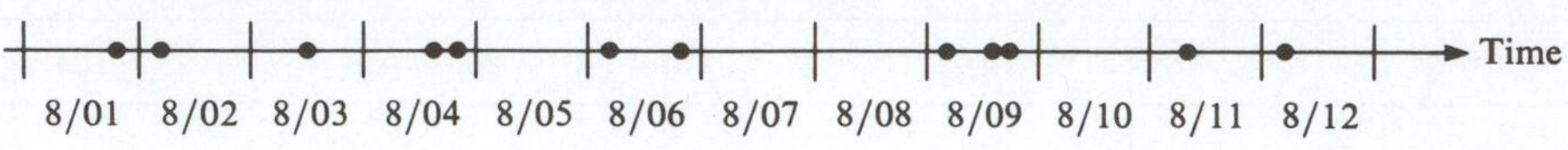

Figure 3-8 *Time record of street lamp burnouts.*

All Poisson processes share the following properties and differ from each other mathematically only by the level for λ.

1. The number of events occurring in one segment of time or space is independent of the number of events in any nonoverlapping segment. A Poisson process has *no memory.*
2. The mean process rate λ *must remain constant* for the entire time span or space considered.
3. The smaller the segment of time or space, the less likely it is for more than one event to occur in that segment. As the segment size tends to 0, the probability of 2 or more occurrences approaches 0.

The Poisson process gives rise to the Poisson distribution discussed here, wherein the concern is how many events will occur in a fixed period of time. It also generates the *continuous exponential distribution* (described in Chapter 4), which has, as its random variable, the amount of time (or space) between events.

Poisson Probabilities

Poisson probabilities may be readily computed for any possible number of events X. These will depend on the specification of the mean process rate λ and the time span t (which may also represent the space expanse). We use the following expression.

PROBABILITY MASS FUNCTION FOR POISSON DISTRIBUTION

$$p(x; \lambda, t) = \Pr[X = x] = \frac{(\lambda t)^x e^{-\lambda t}}{x!} \qquad x = 0, 1, 2, \ldots$$

In keeping with our earlier notation we use p to represent the probability mass function and list the parameters λ and t after the dummy variable x. The Poisson distribution places no upper limit on the number of events possible, so X may turn out to be any nonnegative integer value.

Illustration: *Arrivals at a Toll Booth*

As an illustration, consider the peak morning rush hour at the toll plaza for one of the bridges over San Francisco Bay. Suppose during this period cars arrive at a mean rate of $\lambda = 600$ per hour. We are interested in the number of arrivals X during a 12-second time span ($\frac{1}{5}$ minute), so that $t = (\frac{1}{5})(\frac{1}{60}) = \frac{1}{300}$ hour. (Parameters t and λ must be in compatible units.) We have $\lambda t = 600(\frac{1}{300}) = 2$. Values of $e^{-\lambda t}$ may be read from Appendix Table B at the back of the book. The following probability list is computed:

$$\Pr[X = 0] = p\left(0; 600, \frac{1}{300}\right) = \frac{(2)^0 e^{-2}}{0!} = .1353$$

$$p(1) = \frac{(2)^1 e^{-2}}{1!} = .2707$$

$$p(2) = \frac{(2)^2 e^{-2}}{2!} = .2707$$

$$\vdots$$

$$p(5) = \frac{(2)^5 e^{-2}}{5!} = .0361$$

$$\vdots$$

$$p(10) = \frac{(2)^{10} e^{-2}}{10!} = .0000382$$

$$\vdots$$

This list can never be completed, since X has no upper limit. But Poisson probabilities rapidly converge to 0 as X becomes large.

Parameter Levels and Poisson Probabilities

The probability mass function for the car arrivals is plotted in Figure 3-9. Notice the skew in the pattern of spikes, with probabilities gradually tapering off for larger X's. The degree of skew is most pronounced for low levels of the parameter product λt. For very large levels of λt (such as 10) the Poisson distribution is nearly symmetrical.

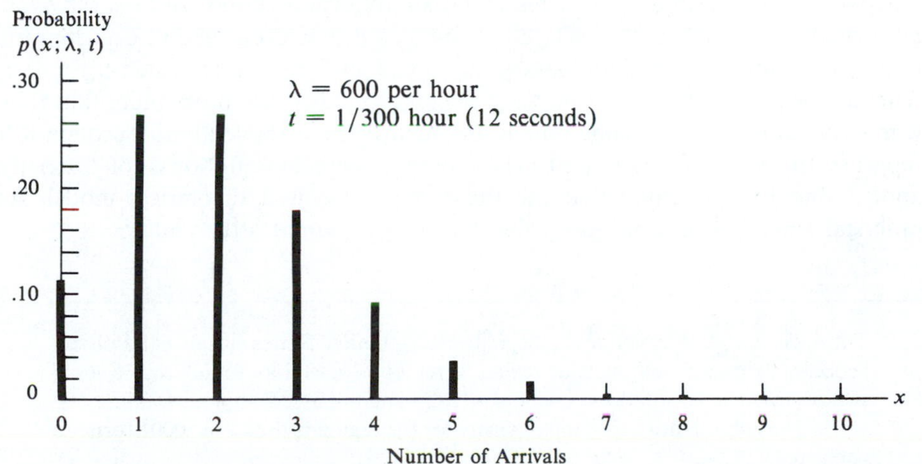

Figure 3-9 Poisson distribution for the number of cars arriving at a toll plaza.

Should the time span be increased, a different parameter applies for t and a new Poisson distribution will apply. For example, raising the time span to 30 seconds, with $t = (\frac{1}{2})(\frac{1}{60}) = \frac{1}{120}$ hour, we have $\lambda t = 600(\frac{1}{120}) = 5$. Thus,

$$\Pr[X = 2] = p\left(2; 600, \frac{1}{120}\right) = \frac{(5)^2 e^{-5}}{2!} = .0843$$

And reducing the time span to 6 seconds makes $t = (\frac{1}{10})(\frac{1}{60}) = \frac{1}{600}$ hour, so that $\lambda t = 600(\frac{1}{600}) = 1$. We have another new probability mass function, and

$$\Pr[X = 2] = p\left(2; 600, \frac{1}{600}\right) = \frac{(1)^2 e^{-1}}{2!} = .1839$$

In these three cases, λ remained the same as we changed t so that the underlying Poisson process was unchanged. A new setting for λ will affect both the probability mass function and the process itself. For instance, the intensity of bridge traffic will fluctuate throughout the day. Thus, some other mean arrival rate will apply during nonrush periods, such as 10 A.M. to 11 A.M. on a weekday. Suppose during this hour the rate is only $\lambda = 150$ cars per hour. Keeping t at $\frac{1}{300}$ hour, $\lambda t = 150(\frac{1}{300}) = .5$, and a completely different probability distribution applies. We have

$$\Pr[X = 2] = p\left(2; 150, \frac{1}{300}\right) = \frac{(.5)^2 e^{-.5}}{2!} = .0758$$

Importance of Poisson Assumptions

Care must be taken that λ not fluctuate over the time span considered, since that would violate one of the Poisson process assumptions. (We could not use a single Poisson distribution to find the probability that in a $t = 8$ hour time span there would be $X = 1{,}000$ arrivals based on an average rate of $\lambda = 200$ cars per hour.) The assumptions of the Poisson process also make the X level in one time period independent of a preceding X. Thus, given an abnormal "flood" of cars, the overall arrival pattern thereafter will be the same as if that event hadn't occurred (or if there had instead been a temporary lull in traffic). There is no "catch-up" effect. Also, although 2 or more cars may arrive close together, the probability that they will arrive at exactly the same time is assumed to be 0. (Several milliseconds will separate the arrivals.) Examples exist wherein these assumptions don't strictly apply—but keep in mind that the Poisson process is a theoretical model. Its practical value lies in how closely the theory approximates the reality.

Example: *Probabilities of Tornado Destruction*

The occurrences of tornadoes in the midwestern United States closely fit a Poisson process. In one 5-million acre region a meteorologist has found that annual swaths traced by tornadoes have historically encompassed an area totaling 500 acres. The annual mean tornado intensity for the region is thus $\lambda = .0001$ tornado per acre.

Consider a city located in this tornado belt having an area of 8,000 acres. What is the probability that it will be hit exactly once during the year?

Setting $t = 8{,}000$ acres, we wish to find the probability that exactly 1 tornado will occur. We have $\lambda t = .0001(8{,}000) = .8$, and the answer is

$$\Pr[\text{damage from exactly 1 tornado}] = p(1; .0001, 8{,}000) = \frac{(.8)^1 e^{-.8}}{1!} = .3595$$

so there is almost a 36% chance that the city will suffer some damage from one tornado.

But the city could be hit by more than one tornado. Being hit at least once is complementary to no hits, so

$$\Pr[\text{no tornado damage}] = \Pr[X = 0] = \frac{(.8)^0 e^{-.8}}{0!} = .4493$$

and

$$\Pr[\text{some tornado damage}] = 1 - .4493 = .5507$$

The probability that no tornado damage occurs within 10 years is (by the multiplication law for independent events)

$$\Pr[\text{no tornado damage for 10 years}] = (.4493)^{10} = .000335$$

so the probability of being hit by at least one tornado sometime within the next 10 years is $1 - .000335 = .999665$, an almost certainty.

Poisson Distribution Function and Probability Table

The following expression gives the Poisson probability distribution function.

POISSON PROBABILITY DISTRIBUTION FUNCTION

$$\mathbf{P(x; \lambda, t) = \Pr[X \leq x] = \sum_{k=0}^{x} p(k; \lambda, t)}$$

In keeping with our notation we use an uppercase P to represent cumulative Poisson probabilities and a lowercase p for the individual terms obtained from the probability mass function.

Like the binomial distribution, Poisson probabilities can be a chore to compute by hand. Appendix Table C provides cumulative Poisson probability values for common levels of the product λt.

As an illustration, suppose that significant noisy spots occur on videotape at a mean rate of $\lambda = .1$ per foot. Consider a segment of tape $t = 200$ feet long. We have $\lambda t = .1(200) = 20$. Using Table C we find various probabilities for the number of noisy spots encountered.

The probability that tape noise is encountered less than 15 times is

$$\Pr[X < 15] = P(14; .1, 200) = .1049$$

whereas the probability that more than 20 of these blemishes will be found is

$$\Pr[X > 20] = 1 - P(20) = 1 - .5591 = .4409$$

and the probability that tape noise will be encountered exactly 20 times is

$$\begin{aligned}\Pr[X = 20] &= P(20) - P(19)\\ &= .5591 - .4703 = .0888\end{aligned}$$

Expected Value and Variance

As with the binomial distribution, the expected value and variance for the number of events may be computed by direct manipulation of the Poisson distribution parameters. We use the following expressions.

EXPECTED VALUE AND VARIANCE FOR A POISSON DISTRIBUTION

$$\mathbf{E(X) = \lambda t}$$

$$\mathbf{Var(X) = \lambda t}$$

The expected number of events is found by multiplying the specified span t times the mean rate λ at which these events occur.

The expected value identity is easily derived. We have

$$E(X) = \sum_{x=0}^{\infty} xp(x; \lambda, t) = \sum_{x=0}^{\infty} x \frac{(\lambda t)^x e^{-\lambda t}}{x!}$$

The first term in this summation is 0 when $x = 0$, so that dropping that term we have

$$E(X) = \sum_{x=1}^{\infty} x \frac{(\lambda t)^x e^{-\lambda t}}{x!}$$

Factoring λt and canceling terms, the above is equal to

$$\lambda t \sum_{x=1}^{\infty} \frac{(\lambda t)^{x-1} e^{-\lambda t}}{(x-1)!}$$

Letting $y = x - 1$, the above equals

$$\lambda t \sum_{y=0}^{\infty} \frac{(\lambda t)^y e^{-\lambda t}}{y!}$$

The summation includes all Poisson probabilities for λ and t and is therefore equal to 1. This establishes that $E(X) = \lambda t$.

It will be left as an exercise for the reader to derive the identity for $\text{Var}(X)$.

Considering again the cars arriving at a toll booth within a specific hour, so that $t = 1$, we see that this is plausible. Twice as many cars are expected during twice that duration, when $t = 2$ hours, and half as many will be expected when the duration is cut by half to $t = .5$ hour. And during commuting times when λ may be triple the off-peak rate, we would expect triple the number of arrivals within equal time spans.

That the variance also equals λt is less obvious, although it is easy to see that long t's will tend to result in greater variation in X than short ones. Likewise, X should vary more with large λ's than with smaller ones.

Poisson Approximation to Binomial Distribution

The probability mass function $p(x; \lambda, t)$ may actually be derived from that for the binomial distribution. This may be accomplished by dividing the span t into many small segments of length Δt and considering each to be a trial having two outcomes: an occurrence or a nonoccurrence of a relevant event. It may be shown that when $n = t/\Delta t$ and $\pi = \lambda \Delta t$,

$$\lim_{\Delta t \to 0} b\left(x; \frac{t}{\Delta t}, \lambda \Delta t\right) = p(x; \lambda, t)$$

Because probability calculations are easier with some distributions than with others, it is a common practice in statistics to use approximations. One popular approximation involves using Poisson probabilities in place of binomial probabilities. When the trial success probability π is small and n is large, it may be established that a binomial probability is close in value to a Poisson probability computed with $\lambda t = n\pi$. In those cases,

$$b(x; n, \pi) \approx \frac{(n\pi)^x e^{-n\pi}}{x!} \qquad x = 0, 1, 2, \ldots, n$$

This approximation steadily improves as n becomes larger. (In the limiting case it is exact.) The following list shows how well this procedure works when the binomial distribution with $n = 100$ and $\pi = .01$ is approximated by the Poisson with $\lambda t = 100(.01) = 1$:

Number of Successes x	Exact Binomial $b(x; 100, .01)$	Approximated by Poisson
0	.3660	.3679
1	.3698	.3679
2	.1848	.1839
3	.0610	.0613
4	.0150	.0153
5	.0029	.0031

Although this illustrates how closely the binomial and Poisson distributions may agree, exact binomial probabilities for most parameter levels to be encountered may be found from the tables at the back of the book. But entries for $n > 100$ are not tabled, nor are probabilities for $\pi < .01$. In those cases $p(x; \lambda, t)$ may still be found from the Poisson probability table. Even if the approximate Poisson probability cannot be found from the table, it may be computed—a far easier task than directly calculating $b(x; n, \pi)$.

Another useful approximation to the binomial distribution is provided in Chapter 4.

Problems

3-25 Transmission line interruptions in a telecommunications network occur at an average rate of 1 per day. Find the probability that the line experiences

(a) no interruptions in 5 days
(b) exactly 2 interruptions in 3 days
(c) at least 1 interruption in 4 days
(d) at least 2 interruptions in 5 days

3-26 Cars arrive at a toll plaza according to a Poisson process.

(a) Find the probability of exactly 5 arrivals in 1 minute when the mean rate of arrivals per minute is
(1) $\lambda = 1$ (2) $\lambda = 2$ (3) $\lambda = 5$ (4) $\lambda = 10$

(b) During the rush period the mean rate is $\lambda = 10$ cars per minute. Find the probability of exactly 5 arrivals in an interval of
(1) $t = .3$ min (2) $t = .5$ min (3) $t = 1$ min (4) $t = 2$ min

3-27 Service calls received by a photocopier maintenance center constitute a Poisson process with a mean rate of 2.5 per hour.

(a) Construct a complete probability distribution for the number X of calls received during 10 to 11 A.M. on a particular Wednesday. Give answers accurate to four places, stopping at the first x whose probability rounds to zero.
(b) Construct the cumulative probability distribution function, using your results from (a).
(c) Compute (1) the expected number of calls and (2) the variance.
(d) Plot on graphs the distributions found in (a) and (b).

3-28 A computer time-sharing system receives teleport inquiries at an average rate of .1 per millisecond. Find the probabilities that the number of inquiries in a particular 50-millisecond stretch will be

(a) less than or equal to 8
(b) greater than 6
(c) between 5 and 12, inclusively
(d) equal to 7
(e) equal to 10

3-29 Use a Poisson distribution to approximate the following binomial probabilities:

(a) $b(5; 100, .05)$
(b) $b(2; 1{,}000, .01)$
(c) $b(9; 500, .04)$
(d) $b(6; 100, .01)$

3-30 An agronomist has found an average of 2 shrubs per acre of jojoba, which serves as the source for transmission oil. Find the following probabilities:

(a) that 5 plants will be found when the number of acres searched is (1) one; (2) two; or (3) three
(b) that no plants will be found in searching the next (1) .5 acre; (2) 1 acre; or (3) 1.5 acres

3-31 A traffic engineer believes that the accidents encountered along a particular stretch of road during rush hours is a Poisson process. Over a duration of 50 weeks, find (1) the expected number of accidents and (2) the probability there will be at least 5 accidents when

(a) there are no roadbed improvements and $\lambda = .1$ rush-hour accident per week
(b) the roadbed is improved, bringing λ down to .04 rush-hour accident per week

3-32 Consider two successive nonoverlapping time segments t_1 and t_2. Show that $p(0; \lambda, t_1 + t_2) = p(0; \lambda, t_1)p(0; \lambda, t_2)$.

3-33 Government approval for a nuclear power plant on the California coast requires a hazard evaluation. Included is a probability analysis of various potentially damaging accidents or natural disasters. Compute the probability of at least one occurrence, (1) in a single year and (2) sometime in the next 100 years, from each of the following potentially damaging events, all of which arise from independent Poisson processes:

(a) impact from an airplane crash, presumed to occur in the vicinity of the generator site at a mean annual rate of 10^{-6}

(b) being hit by a large *tsunami* (tidal wave), known to occur once every 1,000 years with a further chance of $\frac{1}{500}$ of hitting a particular location the width of the generator site

(c) an earthquake causing rupture in the reactor cooling system. This could be only from a Richter-8 or greater shock whose epicenter falls near the generator site. This event is judged to have a mean rate of 10^{-5} per year.

3-34 Cars arriving at a particular highway patrol safety checkpoint form a Poisson process with mean rate $\lambda = 100$ per hour. Each car encountered during a 1-hour period is tested for mechanical defects, and the drivers are informed of needed repairs. Suppose that 10% of all cars in the state need such repairs.

(a) What is the probability that at least 10 cars will be found in need of repairs?

(b) To test the effectiveness of the warnings, the highway patrol will trace the defective cars to determine if the necessary repairs were eventually made. Supposing that only half of the originally warned operators ever fix their cars, what is the probability that at least 6 cars will have been repaired?

3-35 An electronic switching module is expected to malfunction some of the time, misdirecting messages. It must be replaced if the rate of such errors becomes too high. Letting λ represent the mean rate of misdirected messages, the switch is operating according to specifications when $\lambda = .10$ per hour. Should λ reach .50 error per hour, the module ought to be replaced.

There is no way to know λ precisely. A monitoring device may record the number of misdirections in a 10-hour test. Policy is to replace any module that causes more than 2 errors in the test, and otherwise to leave it in place.

(a) What is the probability that a module needing replacement is retained?

(b) What is the probability that a module is replaced even though it is operating according to specifications?

3-5 The Hypergeometric Distribution

We now consider one of the more important discrete probability distributions encountered in statistical sampling applications. The **hypergeometric distribution** provides probabilities for the number of sample observations of a particular category that can be obtained. In this respect it plays the same role as the binomial distribution. But recall that the binomial requires independence between trials. The independence requirement makes that distribution unsuitable for evaluating sampling investigations of small-sized populations, unless sampling is done *with* replacement.

The hypergeometric distribution does not require independence between trials and therefore can be applied when sampling *without* replacement from populations of small size. Thus, it may provide probabilities for the number of defective "black boxes" found in a sample of test items removed from a shipment containing 100, 50, or even 20 items altogether. Since electronic components often must be destroyed during quality assurance testing, sampling must be done without replacement.

Illustration: *Testing a Shipment of Transformers*

To detail the characteristics of the hypergeometric distribution, it will be helpful to look at the sampling process in its basic form. Consider a shipment of 25 transformers from which a sample of 4 will be subjected to a thorough environmental test. (The test items themselves will end up as burned-out hulks with fused coils.) Each transformer in the shipment will ultimately be satisfactory (*S*) or defective (*D*). Figure 3-10 shows the probability tree diagram applicable when the shipment is presumed to have 20% defectives (5 transformers from the entire shipment). Of course, this fact is unknown to the quality assurance department, which must decide to accept or reject the shipment on the basis of the number of defective items R it actually finds in the sample.

Every successive observation is represented in the tree by a fork having one satisfactory and one defective branch. Each path through the tree represents a sequence of observed sample results, which we may view as the elementary events for the complete sampling experiment. Applying basic probability concepts, we can readily find the probability that the sample contains $R = 1$ defective. Referring to the tree, we see that there are four elementary events involving exactly 1 defective. Each of these outcomes is represented on the tree by a heavy-line path.

Consider the topmost of these paths, leading to $S_1S_2S_3D_4$. The branches on this path each have a probability based on what item types were obtained in earlier selections. For the initial observation, 20 out of 25 items are satisfactory, and $\Pr[S_1] = \frac{20}{25}$. Given this event, only 19 satisfactories are left out of 24 remaining items to be selected for the second observation; thus, $\Pr[S_2 | S_1] = \frac{19}{24}$. Similarly, $\Pr[S_3 | S_1 \textit{ and } S_2] = \frac{18}{23}$, since 18 satisfactories will be available for the third observation taken from the 23 remaining items. This leaves 5 defectives in the final 22 items available for the fourth observation, and

$$\Pr[D_4 | S_1 \textit{ and } S_2 \textit{ and } S_3] = \frac{5}{22}$$

The multiplication law allows us to take the product of these branch probabilities in calculating the joint probability for the final outcome:

$$\Pr[S_1S_2S_3D_4] = \frac{20}{25} \times \frac{19}{24} \times \frac{18}{23} \times \frac{5}{22} = \frac{34{,}200}{303{,}600}$$

An identical result is obtained for the three other 1-defective outcomes, so that

$$\begin{aligned}\Pr[R = 1]& \\ &= \Pr[S_1S_2S_3D_4] + \Pr[S_1S_2D_3S_4] + \Pr[S_1D_2S_3S_4] + \Pr[D_1S_2S_3S_4] \\ &= 4\left(\frac{34{,}200}{303{,}600}\right) = \frac{136{,}800}{303{,}600} = .45059\end{aligned}$$

Figure 3-10 *Probability tree diagram for a quality assurance test on a shipment of transformers. Sampling is done without replacement from a small population.*

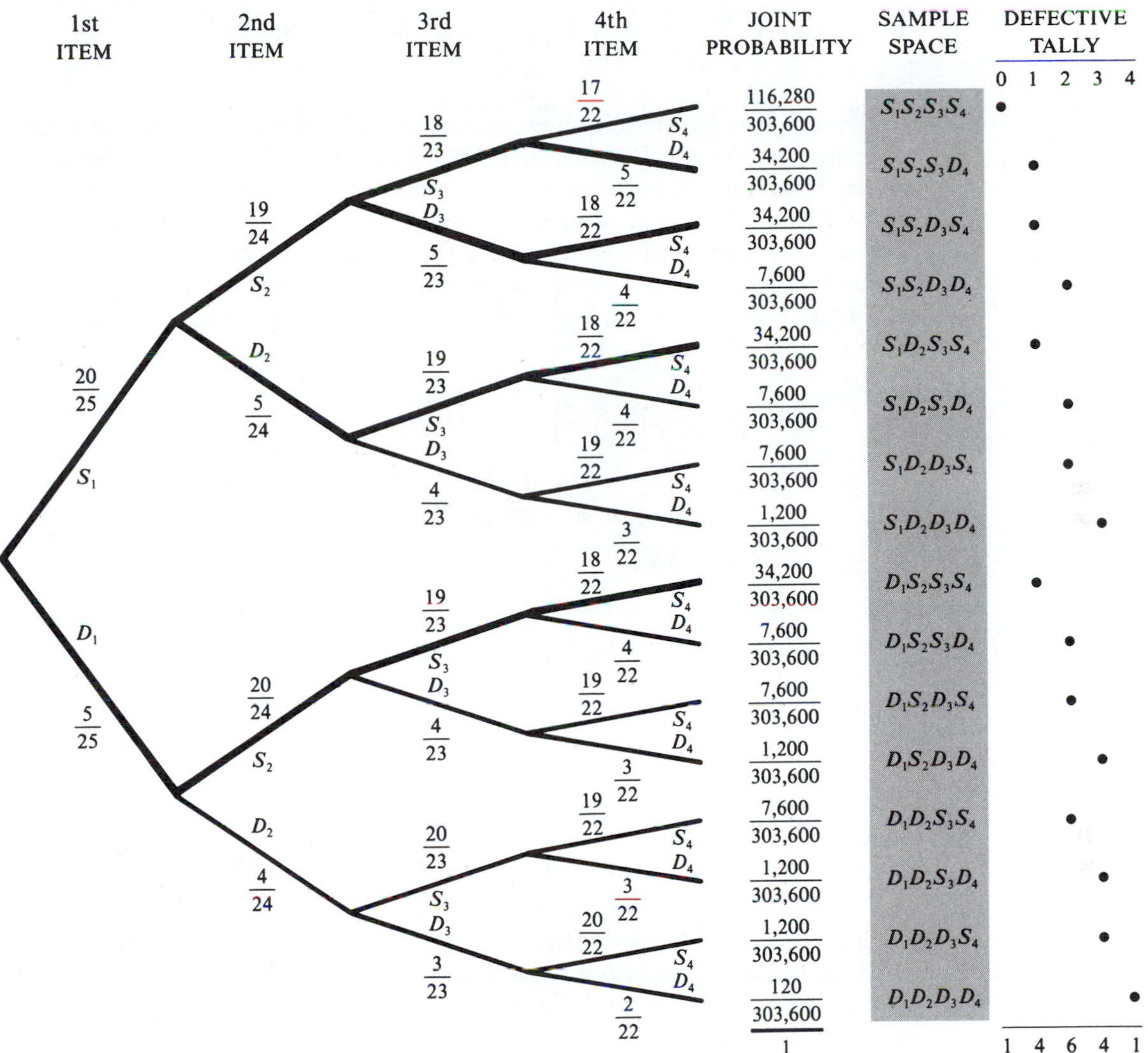

This same probability value may be reached by a somewhat different line of reasoning using event possibility counting methods.

Finding Hypergeometric Probabilities

The sample space in Figure 3-10 does not consider which particular items are selected each time. Envision a more detailed sample space that treats each transformer as a separate entity (identified by its serial number). The number of possible results in that detailed accounting equals the number of *combinations* of 4 specific items from the entire shipment of 25,*

$$C_4^{25} = \frac{25!}{4!21!} = \frac{25 \times 24 \times 23 \times 22 \times 21!}{4! \times 21!} = \frac{303{,}600}{4!} = 12{,}650$$

We can readily determine how many of the above outcomes involve exactly 1 defective transformer. Consider first which one of the five defective transformers will be chosen for the sample; this is the number of ways of picking 1 object from 5, or C_1^5. The other three satisfactory items will be removed from the 20 satisfactories. The number of possibilities for that is C_3^{20}, the total combinations of 3 objects taken from 20. Applying the principle of multiplication, the number of elementary events in the expanded sample space that involve exactly 1 defective is the product of the two combinatorials:

$$\begin{aligned} C_1^5 C_3^{20} &= \left(\frac{5!}{1!4!}\right)\left(\frac{20!}{3!17!}\right) \\ &= \frac{5 \times 4! \times 20 \times 19 \times 18 \times 17!}{1! \times 4! \times 3! \times 17!} = \frac{5 \times 20 \times 19 \times 18}{3!} \\ &= \frac{34{,}200}{3!} = 5{,}700 \end{aligned}$$

Counting and dividing, we have

$$\begin{aligned} \Pr[R = 1] &= \frac{C_1^5 C_3^{20}}{C_4^{25}} = \frac{\dfrac{34{,}200}{3!}}{\dfrac{303{,}600}{4!}} \\ &= 4\left(\frac{34{,}200}{303{,}600}\right) = \frac{136{,}800}{303{,}600} = .45059 \end{aligned}$$

which is the same value found by evaluating the probability tree diagram.

This evaluation suggests a general approach for finding similar probabilities for the number of successes R. Letting N represent the size of the population to be sampled, n the size of the sample, and π the proportion of successes in the population, we use the following expression.

PROBABILITY MASS FUNCTION FOR A HYPERGEOMETRIC DISTRIBUTION

$$h(r; n, \pi, N) = \Pr[R = r] = \frac{C_r^{\pi N} C_{n-r}^{(1-\pi)N}}{C_n^N} \qquad r = 0, 1, \ldots, n \text{ or } \pi N$$

Here the possibilities for the number of successes are all the nonnegative integers up through the smaller of n or πN.

* An even more detailed sample space involves counting the number of *permutations*, which would account for the sequence in which the 4 particular items are chosen. But it is more convenient to deal with combinations instead of permutations. The same probability values will be reached in either case.

Continuing with the transformer illustration, we have $N = 25$ items in the shipment population. A sample of $n = 4$ is selected for testing, and the proportion of defectives in the population is $\pi = .20$. We may use the above expression to evaluate the probability that exactly 2 defectives are found in testing.

$$\Pr[R = 2] = h(2; 4, .2, 25) = \frac{C_2^{.2(25)}C_{4-2}^{.8(25)}}{C_4^{25}}$$

$$= \frac{C_2^5 C_2^{20}}{C_4^{25}} = \frac{\left(\dfrac{5!}{2!3!}\right)\left(\dfrac{20!}{2!18!}\right)}{\dfrac{25!}{4!21!}}$$

$$= \frac{45{,}600}{303{,}600} = .15020$$

(You may verify this by duplicating the earlier probability tree evaluation for the case of $R = 2$.) The remaining probabilities are:

$$h(0) = \frac{C_0^5 C_4^{20}}{C_4^{25}} = \frac{116{,}280}{303{,}600} = .38300$$

$$h(1) = \frac{C_1^5 C_3^{20}}{C_4^{25}} = \frac{136{,}800}{303{,}600} = .45059$$

$$h(3) = \frac{C_3^5 C_1^{20}}{C_4^{25}} = \frac{4{,}800}{303{,}600} = .01581$$

$$h(4) = \frac{C_4^5 C_0^{20}}{C_4^{25}} = \frac{120}{303{,}600} = .00040$$

We may use the following expression for the **hypergeometric cumulative probability distribution function**:

$$F(r) = \Pr[R \le r] = \sum_{x=0}^{r} h(x; n, \pi, N)$$

For example, for the above illustration we compute

$$F(3) = \Pr[R \le 3] = .38300 + .45059 + .15020 + .01581$$
$$= .99960$$

As the following example shows, the hypergeometric distribution has wider application than just for quality assurance.

Example: How Many Users Need Customer Engineering?

Pear Peripherals, a manufacturer of digital computer equipment, provides no routine maintenance for users of its products. The competition, VBM Corporation (Very Big Machines), provides extensive maintenance support through its customer-engineering staff. Pear needs to know if it will be worthwhile to follow VBM's lead.

Pear currently has 70 clients. A random sample of 10 of these will be taken for detailed evaluation to determine how many would be willing to sign a customer-engineering support contract.

Pear does not know what proportion of its clients might be willing to acquire the new service. Nevertheless, the company will provide this service if a majority of sample users express a desire to sign. Consider the number of willing clients R found in a random sample of 10 Pear customers.

Assuming that $\pi = .50$, we may apply the hypergeometric distribution with $n = 10$ and $N = 70$ to calculate the probability that Pear chooses to engage in customer engineering.

$$\Pr[\text{providing customer engineering}] = \Pr[R \geq 6]$$

We find

$$\Pr[R = 0] = \frac{C_0^{.50(70)}C_{10-0}^{.50(70)}}{C_{10}^{70}} = \frac{C_0^{35}C_{10}^{35}}{C_{10}^{70}}$$

$$= \frac{\left(\frac{35!}{0!35!}\right)\left(\frac{35!}{10!25!}\right)}{\frac{70!}{10!60!}} = \frac{35!35!10!60!}{0!35!10!25!70!}$$

Expanding the factorials and canceling terms, this is equivalent to

$$\frac{35 \times 34 \times 33 \times 32 \times 31 \times 30 \times 29 \times 28 \times 27 \times 26}{70 \times 69 \times 68 \times 67 \times 66 \times 65 \times 64 \times 63 \times 62 \times 61}$$

which after further canceling reduces to

$$\frac{29 \times 3}{2 \times 23 \times 67 \times 61} = \frac{87}{188{,}002} = .0005$$

Rather than compute the $R = 1$ probability from scratch, we can use the fact that

$$C_x^k = \left(\frac{k - x + 1}{x}\right)C_{x-1}^k$$

Thus,

$$\Pr[R = 1] = \frac{C_1^{35}C_9^{35}}{C_{10}^{70}} = \frac{\left(\frac{35 - 1 + 1}{1}\right)C_0^{35}\left(\frac{10}{35 - 10 + 1}\right)C_{10}^{35}}{C_{10}^{70}}$$

$$= \left(\frac{35 - 1 + 1}{1}\right)\left(\frac{10}{35 - 10 + 1}\right)\Pr[R = 0]$$

$$= \left(\frac{350}{26}\right)\left(\frac{87}{188{,}002}\right) = \frac{15{,}225}{2{,}444{,}026} = .0062$$

More generally, we may use the fact that

$$\Pr[R = r] = \left(\frac{\pi N - r + 1}{r}\right)\left(\frac{n - r + 1}{(1 - \pi)N - n + r}\right)\Pr[R = r - 1]$$

in computing the remaining probabilities:

$$\Pr[R = 2] = \frac{34}{2}\left(\frac{9}{27}\right)\left(\frac{15{,}225}{2{,}444{,}026}\right) = \frac{86{,}275}{2{,}444{,}026} = .0353$$

$$\Pr[R = 3] = \frac{33}{3}\left(\frac{8}{28}\right)\left(\frac{86{,}275}{2{,}444{,}026}\right) = \frac{271{,}150}{2{,}444{,}026} = .1109$$

$$\Pr[R = 4] = \frac{32}{4}\left(\frac{7}{29}\right)\left(\frac{271{,}150}{2{,}444{,}026}\right) = \frac{523{,}600}{2{,}444{,}026} = .2142$$

$$\Pr[R = 5] = \frac{31}{5}\left(\frac{6}{30}\right)\left(\frac{523{,}600}{2{,}444{,}026}\right) = \frac{649{,}264}{2{,}444{,}026} = .2657$$

Thus, we have

$$\begin{aligned}\Pr[R \le 5] &= .0005 + .0062 + .0353 + .1109 + .2142 + .2657 \\ &= .6328\end{aligned}$$

so that

$$\begin{aligned}\Pr[\text{providing customer engineering}] &= \Pr[R \ge 6] \\ &= 1 - \Pr[R \le 5] \\ &= 1 - .6328 = .3672\end{aligned}$$

Of course, a different result will be reached for some other assumed level of π.

Expected Value and Variance

As with all the probability distributions discussed, the expected value and variance of the number of successes R may be computed directly from the parameter values that apply. We use the following expressions.

EXPECTED VALUE AND VARIANCE FOR A HYPERGEOMETRIC DISTRIBUTION

$$\boldsymbol{E(R) = n\pi}$$

$$\boldsymbol{\mathrm{Var}(R) = n\pi(1 - \pi)\left(\frac{N - n}{N - 1}\right)}$$

Notice that the expected value is identical to that for the binomial distribution. Thus, on the average, successes will occur in the same proportion in the sample as they do in the population. The following results apply to the earlier quality assurance illustration:

$$E(R) = 4(.2) = .8$$

$$\mathrm{Var}(R) = 4(.2)(1 - .2)\left(\frac{25 - 4}{25 - 1}\right) = .56$$

Thus, only .8 defective is expected in the sample, and the variance is .56.

The expression for the variance is also similar to that for the binomial. It will be helpful to compare the variance expressions. Recall that

$$\text{Var}(R) = n\pi(1 - \pi) \qquad \text{(binomial)}$$

which differs from Var(R) for the hypergeometric distribution by the term $(N - n)/(N - 1)$. This quantity is called the **finite population correction factor**. Notice that when the sample size n is near the population size N, the factor may be considerably below 1. It indicates that a smaller variance applies for R when sampling without replacement from a finite population of fixed size N than when sampling from a continuous process wherein the binomial distribution is appropriate and there is no theoretical limit on the population size. Using the parameters from the quality assurance illustration,

$$\frac{N - n}{N - 1} = \frac{25 - 4}{25 - 1} = .875$$

we see that the variance is only 87.5% as large as that for the binomial distribution with the same parameters, $4(.2)(1 - .2) = .64$.

The practical significance of the finite population correction arises when the hypergeometric distribution is being approximated by the normal distribution. We next consider the binomial approximation.

Binomial Approximation to Hypergeometric Distribution

Recall that the binomial distribution also provides probabilities for the number of successes in n trials. That distribution may be used in a probability evaluation for sampling directly from a continuous process, such as an assembly line, when the quality of successive items can be assumed independent. It is also used when the population N is large in relation to the sample size n, although that is not exactly correct unless sampling is done with replacement. And, even when observed items are not replaced, the binomial distribution is still sometimes used. In this case, the binomial is used as an *approximation* to the hypergeometric distribution.

The reason for using the binomial distribution at all is that it is easier to compute values for $b(r; n, \pi)$ than for $h(r; n, \pi, N)$, which involves more factorials and has an additional parameter N. But with modern computational aids (even pocket computers) the advantages from the binomial approximation are not substantial. Users working with probability tables instead of computers will prefer approximating with binomial tables, which take far less space than hypergeometric tables would require.

How good is the binomial approximation? This depends on the settings of the parameters. Generally, the larger the population size N is in relation to the sample size n, the better the approximation becomes. Table 3-6 shows a comparison of the approximate binomial probabilities applicable to the earlier transformer quality assurance illustration. Notice that the binomial probabilities are considerably closer to the true ones when $N = 100$ instead of 25. The overall effect of the larger N is summarized by the finite population correction factor,

Table 3-6
Comparison of Binomial and Hypergeometric Probabilities for the Number of Defective Transformers (when $n = 4$, $\pi = .20$)

Possible Number of Defectives r	Approximate Binomial Probability $b(r; n, \pi)$	Exact Hypergeometric Probability $h(r; n, \pi, N)$ $N = 25$	$N = 100$
0	.4096	.38300	.40333
1	.4096	.45059	.41905
2	.1536	.15020	.15312
3	.0256	.01581	.02326
4	.0016	.00040	.00124

$(N - n)/(N - 1)$, which is .875 when $N = 25$ but .970 when $N = 100$. (When the factor is close to 1, the variance of R is nearly the same level with the binomial and hypergeometric distributions.) The levels for π and n will, to a lesser extent, affect the quality of the approximation.

Problems

3-36 A computer instructor gives his students a 100-line BASIC program. Altogether 10% of the lines have bugs. What is the probability that there will be exactly 5 erroneous lines in a group of 20 lines selected at random?

3-37 The freshman class at an engineering school contains 300 students. Ten percent of them have reserved space in the civil engineering option. What is the probability that there will be 10% or fewer civil engineering optors within the first 20 names from an alphabetical roster?

3-38 Determine for each of the following situations (1) the expected value for the number of successes, (2) the finite population correction factor, and (3) the variance:

	(a)	(b)	(c)	(d)	(e)
N	100	200	50	100	1,000
π	.05	.20	.50	.10	.10
n	10	25	10	5	100

3-39 Consider the Pear Peripherals example when $\pi = .60$. Find the probability that customer engineering will be offered.

3-40 Repeat Problem 3-39 using instead $\pi = .70$.

3-41 A shipment contains 100 printed circuit boards. A sample of 10 boards will be tested. If 2 or fewer defectives are found, the shipment will be accepted. Assuming that 10% of the boards in the shipment are defective, find the probability it will be accepted.

3-42 Find the probability for getting a heart flush in a 5-card poker hand dealt from a complete shuffled deck of ordinary playing cards.

3-43 Repeat Problem 3-42, but consider the probability of exactly 3 queens.

3-44 A quality-control inspector accepts shipments whenever a sample of size 5 contains no defectives, and she rejects otherwise.
(a) Determine the probability that she will accept a poor shipment of 50 items in which 20% are defective.
(b) Determine the probability that she will reject a good shipment of 100 items in which only 2% are defective.

3-45 A professor selects a sample of 5 engineering students from the 25 in her class to state their preferences concerning course content. Each will indicate whether he or she wants more theory or more applications. It is assumed that 40% of the class wants more theory.
Determine the probabilities for all possible levels for the number of students R preferring more theory. (Use five places of accuracy.)

3-46 Determine the probability distribution for the number of defectives in a sample of 5 items taken from a shipment of 500 having 10% defective. (Use five places of accuracy.)
(a) Use exact hypergeometric probabilities.
(b) Use the binomial approximation.

3-6 The Geometric and Negative Binomial Distributions

So far we have discussed the Bernoulli process in conjunction with the binomial distribution, which provides probabilities for the number of successes R in n trials. However, there are other results from a Bernoulli process that may be of interest.

The **geometric distribution** is concerned with the *number of trials* taken in a Bernoulli process before the first success is achieved. Although this is not a common question, the number of trials preceding success can be of paramount concern in applications involving reliability. For example, a satellite power cell may last indefinitely until collision with a micrometeor; each day there is some probability of such a collision. Viewing each day as a trial in a Bernoulli process with a fixed survival probability, we can find the probability that a power cell will survive any specified number of days.

Satellites are ordinarily powered by a group of power cells, so a breakdown of any single unit should not be disabling. The cells are arranged in a parallel configuration. The **negative binomial distribution** gives probabilities for the number of trials such a system will experience before a critical number of successes has been achieved. This is done by treating as the random variable the *number of failures* before the rth success.

The Geometric Distribution

The geometric distribution is itself far less widely encountered than the three distributions discussed previously. But in addition to illustrating a special-purpose probability application in the reliability area, it shows that more than one probability distribution can apply to a particular stochastic process.

Denoting by K the number of Bernoulli trials until the first success, the following expression is used.

PROBABILITY MASS FUNCTION FOR A GEOMETRIC DISTRIBUTION

$$p(k) = \Pr[K = k] = \pi(1 - \pi)^{k-1} \qquad k = 1, 2, 3, \ldots$$

where π is the trial success probability.

The foregoing probability mass function follows directly from basic probability concepts. The probability that the first trial is a success is

$$\Pr[K = 1] = \Pr[\text{success}] = (1 - \pi)^0\pi = (1 - \pi)^{1-1}\pi$$

In order for the first success to occur on the second trial, trial one must be a failure, and the multiplication law provides

$$\begin{aligned}\Pr[K = 2] &= \Pr[\text{failure on first}] \times \Pr[\text{success on second} \mid \text{failure on first}]\\ &= (1 - \pi)\pi\\ &= (1 - \pi)^{2-1}\pi\end{aligned}$$

Similarly, in order for the third trial to be the first success it must be preceded by exactly two failures, so that

$$\begin{aligned}\Pr[K = 3] &= (1 - \pi) \times (1 - \pi) \times \pi\\ &= (1 - \pi)^2\pi\\ &= (1 - \pi)^{3-1}\pi\end{aligned}$$

An analogous result applies to any number of trials.

Illustration:* *Satellite Power Reliability

To illustrate, suppose that each day there is a probability $\pi = .25$ that a satellite power cell will be damaged in a collision. The daily survival probability is therefore $1 - \pi = .75$. (The standard terminology sometimes gets twisted. In this illustration a "success" is a collision and a "failure" is survival.) The probability that the power cell becomes damaged exactly on the 10th day of operation is

$$\Pr[K = 10] = p(10) = .25(.75)^9 = .0188$$

The probability that the power cell lasts exactly until the 20th day (experiencing a collision at that time) is

$$\Pr[K = 20] = p(20) = .25(.75)^{19} = .0011$$

The **geometric probability distribution function** expresses the cumulative probability that the first success occurs on or before the kth trial:

$$F(k) = \Pr[K \leq k] = \sum_{x=1}^{k} p(x)$$

This outcome is complementary to the first success occurring sometime after the kth trial, which can only happen if the first k trials are failures, which has probability $(1 - \pi)^k$. Thus,

$$F(k) = 1 - (1 - \pi)^k$$

For the satellite power cell, we have

$$F(20) = 1 - (.75)^{20} = 1 - .0032 = .9968$$

There is a negligible probability, $1 - .9968 = .0032$, that the power cell will survive past 20 days. There are two ways to improve the power system reliability: (1) reduce the damage probability π or (2) build further redundancy into the system. The first type of improvement might be achieved through shielding. The second can be accomplished by designing a system in which several power cells are arranged in parallel. Later in this section we will investigate how to find probabilities for such a complex system.

Expected Value and Variance We use the following expressions.

EXPECTED VALUE AND VARIANCE FOR A GEOMETRIC DISTRIBUTION

$$\mathbf{E(K) = \frac{1}{\pi}}$$

$$\mathbf{Var(K) = \frac{1 - \pi}{\pi^2}}$$

(A derivation of these results will be left as an exercise for the reader.) The satellite power cell is expected to experience destruction after

$$E(K) = \frac{1}{.25} = 4 \text{ days}$$

of operation. The variance is

$$\text{Var}(K) = \frac{.75}{(.25)^2} = 12$$

Example:* *Busy Signals in Time Sharing

Many time-sharing computer systems operate near their capacity, so users often cannot access the central processor because all teleports are busy. Consider the number of attempts K a client must make to gain a connection. Each call may be viewed as a Bernoulli trial having "busy signal" or "connection" as outcomes. Assuming the system is busy 95% of the time, the probabilities are $\pi = .05$ that any call is completed and $1 - \pi = .95$ that a call gets a busy signal. Thus, the probability that exactly 5 calls are required for a connection is

$$\Pr[K = 5] = p(5) = .05(.95)^4 = .041$$

and a user may expect to place

$$E(K) = \frac{1}{.05} = 20 \text{ calls}$$

until being connected.

How many calls will have to be placed before the odds favor getting a connection? This is the same as finding the smallest k so

$$\Pr[K > k] = .50$$

or

$$(1 - \pi)^k = (.95)^k = .50$$

Taking the natural logarithm of both sides, we seek the smallest k such that

$$k \ln(.95) = \ln(.50)$$

or

$$k = \frac{\ln(.50)}{\ln(.95)} = \frac{-.69315}{-.05129} = 13.51$$

Thus, $k = 14$ calls provides a better than 50-50 chance of getting a connection.

The Bernoulli process has no memory, so regardless of how many attempts have been made so far in trying to gain access, the expected number of calls still needed is going to be the same. Spreading out calls will be no better than placing one right after the other. The quickest way for a high-priority client to gain access is to steadily redial until a connection is made.

The Negative Binomial Distribution

The negative binomial distribution is useful in the same general class of random experiments as the geometric. Denoting by K the number of Bernoulli process *failures* until the rth success, we use the following expression.

PROBABILITY MASS FUNCTION FOR A NEGATIVE BINOMIAL DISTRIBUTION

$$\mathbf{p(k) = C_k^{r+k-1}\pi^r(1 - \pi)^k}$$

with $k = 0, 1, 2, \ldots$ and $r = 1, 2, \ldots$ and where π denotes the trial success probability. When $r = 1$, the geometric distribution arises as a special case.

To see why this expression works, consider a string of Bernoulli trials with outcomes denoted as S or F. In the case where the number of successes is fixed at $r = 2$, then for $K = 0$ there is just one possibility, S_1S_2, and

$$\begin{aligned} p(0) &= \pi\pi = \pi^2 \\ &= C_0^1\pi^2(1 - \pi)^0 = C_0^{2+0-1}\pi^2(1 - \pi)^0 \end{aligned}$$

For $K = 1$ there are two possibilities: $S_1F_2S_3$ and $F_1S_2S_3$. The probability is

$$\begin{aligned} p(1) &= \pi(1 - \pi)\pi + (1 - \pi)\pi\pi \\ &= 2\pi^2(1 - \pi) \\ &= C_1^2\pi^2(1 - \pi) = C_1^{2+1-1}\pi^2(1 - \pi)^1 \end{aligned}$$

Finally, for $K = 2$, the possibilities are $F_1F_2S_3S_4$, $F_1S_2F_3S_4$, and $S_1F_2F_3S_4$, so

$$\begin{aligned} p(2) &= (1 - \pi)(1 - \pi)\pi\pi + (1 - \pi)\pi(1 - \pi)\pi + \pi(1 - \pi)(1 - \pi)\pi \\ &= 3\pi^2(1 - \pi)^2 \\ &= C_2^3\pi^2(1 - \pi)^2 = C_2^{2+2-1}\pi^2(1 - \pi)^2 \end{aligned}$$

In each of these cases, the elementary event string contains $r + k$ letters and ends with an S. For each elementary event there are $r + k - 1$ preceding trials (letters) up to the end, exactly k of which involve failures (F's), so the number of letter combinations is C_k^{r+k-1}. Each elementary event involves r successes (with probability π for each) and k failures (with probability $1 - \pi$ for each). Therefore, every elementary event under a fixed r and k grouping has probability $\pi^r(1 - \pi)^k$ and there are C_k^{r+k-1} of them in each group.

To show how to apply the negative binomial distribution, we will extend the earlier communication satellite illustration. The power *system* involves three identical cells and will survive as long as any one of them remains in working order. Only the active cell is exposed to micrometeor collision, so that on any given day the collision probability $\pi = .25$ applies. The system will survive until the number of collisions reaches three, when all cells will have been wiped out. Using $r = 3$ and letting K represent the number of collision-free days until the third collision, we may apply the negative binomial probabilities.

The probability that the system will have a lifetime of 10 days, so that there will be exactly $K = 7$ collision-free days, is

$$\begin{aligned} p(7) &= C_7^{3+7-1}(.25)^3(.75)^7 \\ &= \frac{9!}{7!2!}(.015625)(.133484) \\ &= .0751 \end{aligned}$$

Table 3-7 gives a partial listing of the probability distribution. Each $p(k)$ term was individually computed, as above. From these the cumulative probabilities for K were computed. The last column shows the power system's probabilities for surviving beyond the indicated lifetime. Notice that power-cell redundancy provides a .0911 probability that the satellite will survive past 20 days. This is much larger than the counterpart probability of .0032 found earlier for a single power cell.

Further improvement in system survivability may be achieved by increasing the level of redundancy. Consider a system with six power cells. Using $r = 6$ and $K = 4$, we find the probability that it will experience a lifetime exactly 10 days long:

$$\begin{aligned} p(4) &= C_4^{6+4-1}(.25)^6(.75)^4 \\ &= \frac{9!}{4!5!}(.00024424)(.31640625) \\ &= \frac{9 \times 8 \times 7 \times 6}{4 \times 3 \times 2 \times 1}(.00007725) \\ &= .0097 \end{aligned}$$

Table 3-7 *Negative Binomial Probability Distribution for Satellite Power System, with Lifetime Survival Probabilities*

Collision-Free Days Until Disablement k	Probability $p(k)$	Cumulative Probability $F(k)$	Days of System Life $r+k$	Probability of Surviving Beyond $r+k$ Days $1-F(k)$
0	.0156	.0156	3	.9844
1	.0352	.0508	4	.9492
2	.0527	.1035	5	.8965
3	.0659	.1694	6	.8306
4	.0742	.2436	7	.7564
5	.0779	.3215	8	.6785
6	.0779	.3994	9	.6006
7	.0751	.4745	10	.5255
8	.0704	.5449	11	.4551
9	.0645	.6094	12	.3906
10	.0581	.6675	13	.3325
11	.0515	.7190	14	.2810
12	.0450	.7640	15	.2360
13	.0390	.8030	16	.1970
14	.0334	.8364	17	.1636
15	.0284	.8648	18	.1352
16	.0240	.8888	19	.1112
17	.0201	.9089	20	.0911

The probability that the six-cell system will survive past 20 days is .6175—much greater than with the smaller r. (As an exercise you may verify this result, repeating with the new r the calculations in Table 3-7.)

Expected Value and Variance We use the following expressions.

EXPECTED VALUE AND VARIANCE FOR A NEGATIVE BINOMIAL DISTRIBUTION

$$E(K) = \frac{r(1-\pi)}{\pi}$$

$$\text{Var}(K) = \frac{r(1-\pi)}{\pi^2}$$

In the preceding satellite illustration, using $\pi = .25$, the expected numbers of collision-free days before the final disabling collision are

$$E(K) = \frac{3(1-.25)}{.25} = 9 \qquad \text{(when } r = 3\text{)}$$

$$E(K) = \frac{6(1-.25)}{.25} = 18 \qquad \text{(when } r = 6\text{)}$$

and the respective variances are:

$$\text{Var}(K) = \frac{3(1 - .25)}{.25^2} = 36 \qquad (\text{when } r = 3)$$

$$\text{Var}(K) = \frac{6(1 - .25)}{.25^2} = 72 \qquad (\text{when } r = 6)$$

Both the expected value and the variance are directly proportional to the level of r.

Reason for the Name "Negative" Binomial Consider the following expansion:

$$C_k^{r+k-1} = \frac{(r + k - 1)!}{k!(r - 1)!} = \frac{(r + k - 1)(r + k - 2) \cdots (r + 1)r}{k!}$$

There are k terms in both the numerator and denominator. If we reverse the signs of the numerator terms, we can rewrite the expansion as

$$(-1)^k \frac{(-r - k + 1)(-r - k + 2) \cdots (-r - 1)(-r)}{k!}$$

Replacing $-r$ by s and listing the numerator terms in reverse sequence, we find that the above is equivalent to

$$(-1)^k \frac{s(s - 1) \cdots (s - k + 1)}{k!} = (-1)^k \frac{s!}{k!(s - k)!} = (-1)^k C_k^s$$

When we make the substitution $-r$ for s, we have established the equivalency:

$$C_k^{r+k-1} = (-1)^k C_k^{-r}$$

Using $p = \pi$ and $q = 1 - \pi$, we can write the probability mass function for the negative binomial distribution as

$$p(k) = C_k^{-r} p^r (-q)^k$$

Problems

3-47 An evenly balanced coin is tossed. Find the probability that the number of tosses made in arriving at the first tail is

(a) exactly 3
(b) greater than 2
(c) at least 1
(d) from 2 to 4, inclusively

3-48 Suppose the time-sharing system described in the text is augmented with new teleports, increasing the connection probability to .20.

(a) Compute the expected value and variance for the number of calls until connection is made.

(b) Determine a listing of probabilities for the necessary number of calls dialed until connection. (Stop at 10 calls.)

(c) From the above, list the cumulative terms of the geometric probability distribution function. How many calls must be placed before the odds favor a connection?

3-49 A food-processing plant uses a servomechanism to control the amount of freeze-dried coffee going into each jar. Whenever the weight of a sample jar's contents falls outside tolerance limits of 16 ± .2 ounces, the filling equipment is shut down and the control is readjusted.

Suppose that 95% of all filled jars fall within tolerance. Find the probability that the number of sample jars weighed before shutting down is

(a) exactly 40 (c) between 30 and 50, inclusively
(b) 20 or less (d) at least 150

3-50 Refer to the coffee-jar filling equipment in Problem 3-49. Suppose that only 90% of the output falls within the specified tolerance. One sample jar is removed every 30 minutes.

(a) How many sample jars can be expected before the equipment is shut down? How long can we expect the equipment to run between shutdowns?
(b) How many jars will have to be sampled before the odds favor shutting down? How many hours of operation does that correspond to?

3-51 Redundancy is designed into the solar-powered communications system of a satellite. On every orbit it transmits a message to the Seychelles ground station. There is a 005 chance that the driving power cell will be inoperable at that time, so that the next backup cell must replace it.

(a) Find the probability that the present cell will continue to power the satellite for (1) at least 50 more orbits, (2) at least 100 more orbits, and (3) at least 200 more orbits.
(b) What is the expected orbit lifetime of a single power cell?
(c) The orbit survival probability would be increased if all power cells were arranged in parallel (instead of as replacements). Then there would be only a .0001 chance that the satellite would become forever inoperable after any given orbit. (1) Find the probability that the new system would survive at least 1,000 orbits. (2) What is the expected orbit lifetime of the proposed communications systems?

3-52 Consider the satellite in Problem 3-51. Suppose that when it is launched it contains two identical power cells and may function as long as one remains operational. Assume further that a substitution of a good for a bad power cell may be made only during passage on the dark side of the orbit.

(a) How many orbits can be expected before one in which the satellite goes permanently silent?
(b) Find the probability that the last orbit is exactly the (1) 100th and (2) 200th.

3-53 At the beginning of the day a machine shop foreman assigns jobs to machines, which may be found suitable or not. If the assigned machine cannot perform the required work, the job is returned to the job queue for reassignment. Assuming that only 5% of the assignments are ever unsuitable, and that a Bernoulli process applies, answer the following.

(a) How many jobs are expected to be assigned before the number needing reassignment reaches (1) 2, (2) 3, (3) 5, (4) 10?
(b) Find the probability that the number of satisfactory assignments before the third reassignment is exactly (1) 5, (2) 10, (3) 20, (4) 50.

Comprehensive Problems

3-54 A data clerk makes entries into a central data bank at a mean rate of 200 per hour. Suppose that 1% of all entries are in error.

(a) What distribution is appropriate for finding the probability of exactly 5 errors in 100 entries? Find that value.
(b) What distribution is appropriate for finding the probability that there will be exactly 5 entries in the next 1 minute? Find that value.
(c) What distribution is appropriate for finding the probability that the clerk will make at least 100 entries before the next error? Find that value.

(d) What probability distribution is appropriate for finding the probability that the clerk will make exactly 100 entries before making the third error? Find that value.

3-55 Message bursts are transmitted through a microwave relay at a mean rate of 10 per second. Two percent of these are garbled by radio interference and must be retransmitted.

(a) What distribution is appropriate for finding the probability that there will be exactly 2 messages transmitted in a particular 1-second span? Find that value.

(b) What distribution is appropriate for finding the probability that there will be exactly 2 garbled messages in the next 50 transmissions? Find that value.

(c) What distribution is appropriate for finding the probability that the microwave relay transmits at least 20 messages before one has to be retransmitted? Find that value.

3-56 Consider the data clerk in Problem 3-54.

(a) Determine the expected number of errors in 100 entries.

(b) Determine the expected number of entries in 1 minute.

(c) Determine the expected number of entries until an error is made. How many entries will be made before the odds favor an error?

3-57 Consider the microwave relay in Problem 3-55.

(a) Determine the expected number of transmissions in a 1-second time span.

(b) Determine the expected number of garbled messages in the next 50 transmissions.

(c) Determine the expected number of messages sent until an error is made. How many messages will be transmitted before the odds favor a garbled message?

3-58 One popular rule of thumb is to use the binomial approximation to the hypergeometric distribution only when the sample size is less than or equal to 10% of the population size. In each of the following situations where this rule allows the approximations, determine (1) the exact hypergeometric probability for $R = 2$ successes and (2) the approximate binomial probability.

(a)	(b)	(c)	(d)
$N = 100$	$N = 50$	$N = 200$	$N = 100$
$n = 10$	$n = 3$	$n = 10$	$n = 5$
$\pi = .50$	$\pi = .20$	$\pi = .05$	$\pi = .10$

3-59 A vial is removed from a chemical process and evaluated at hourly intervals. If the vial contains excessive impurities, the separation tanks are purged. Suppose there is presently a 5% chance that a particular vial will contain excessive impurities.

(a) How many hours are expected before the tanks are purged?

(b) Find the probability that the tanks will be purged by (1) no later than 10 hours, (2) no later than 20 hours, or (3) no later than 50 hours.

3-60 A parallel-component system is to be designed. The system will remain operational as long as there is at least one working component. Assume that each component has a 5% chance of remaining functional at the start of each day, that they cannot be replaced, and that no component is repairable. Assume also that the status of each component is independent of each preceding one and of what happens on the preceding day.

(a) Find the probability that a 2-component system will survive past 2 days.

(b) What daily survival probability must a single component have in order to provide the identical probability to that found in (a)?

3-61 A traffic engineer has monitored the flow of vehicles passing a point in the local "blood alley." The following partial results apply:

	(1) 8–9 A.M. weekday	(2) 8–9 A.M. Sunday	(3) 2–3 P.M. weekday	(4) 10–12 A.M. Saturday
Number of cars	750	40	200	300
Hours observed	5	1	4	3

(a) Determine the mean hourly rate of traffic flow in each of the cases.
(b) Assuming that the underlying pattern of traffic in the respective time period is a Poisson process with that parameter, compute the expected number of passing cars during a 3-minute stretch.
(c) Determine in each case the probability that exactly 5 cars will pass during a 3-minute stretch.

3-62 The number of sample items in a production process has a Poisson distribution with a mean rate of $\lambda = 10$ per hour. A sample is collected hourly. Production itself may be characterized as a Bernoulli process in which 10% of the items produced are defective.
(a) Find the following unconditional probabilities for the number of sample items n obtained: (1) exactly 3, (2) at least 2, (3) no more than 5, and (4) between 8 and 12, inclusively.
(b) Find the following conditional probabilities for the number of defectives given that $n = 10$: (1) exactly 2 defectives, (2) at least 1 defective, (3) at most 5 defectives, and (4) between 1 and 4 defectives, inclusively.

3-63 Consider Problem 3-62 again. Establish the following joint probabilities regarding the next hour's sample:
(a) Ten sample items are obtained *and* there are 2 defectives.
(b) Twenty sample items are obtained *and* the number of defectives is less than or equal to 5.
(c) Between 10 and 12 sample items, inclusively, are obtained *and* the number of defectives is exactly 3.

3-64 Consider again Problem 3-62. Find the unconditional probability that the number of defectives found in the hour is
(a) exactly 2
(b) at least 3
(c) less than or equal to 4
(d) between 1 and 3, inclusively

3-65 Ima Storm is a resident of the tornado belt discussed on page 138. Her home sits on a $\frac{1}{4}$-acre lot.
(a) Determine the probability that Ima's property survives 1 year without being touched by a tornado.
(b) Determine the probability that Ima's property will survive tornado damage until her new 30-year mortgage is paid off.

CHAPTER FOUR

Continuous Probability Distributions

THE PROBABILITY DISTRIBUTIONS encountered in Chapter 3 involve random variables that assume discrete values. We will now consider a wider class of random variables that may assume values anywhere within some specified range of real numbers. These are **continuous random variables**, which are typical of physical measurements or time. The mathematical treatment of probability for the continuous case involves *calculus.*

To introduce continuous probability distributions we will evaluate the random selection of one unit for observation from a quantitative population. The value X to be obtained is a random variable having many possible levels, the probabilities of which may be determined from the underlying population frequency distribution. As we have seen, population data may be summarized by a frequency curve that also represents a **probability density function** for X. This curve is the continuous counterpart to the probability mass function introduced in Chapter 3 for discrete random variables.

4-1 The Underlying Concepts of Continuous Random Variables

To illustrate the concepts of continuous random variables, we consider an application from chemical engineering. Chemical engineers often keep detailed records of the inputs and outputs of various monitored processes. Figure 4-1 shows the frequency curve representing several hundred batches of a technical chemical run at a particular plant. Because of varying characteristics of raw materials and fluctuating operating conditions there was considerable variation in the final product. Each batch resulted in a particular level of active ingredient, expressed in grams per liter.

The detailed data from a randomly chosen batch were to be evaluated with a computer model to determine how closely the model predicts actual yield. Our present concern is the yield X from that selected batch.

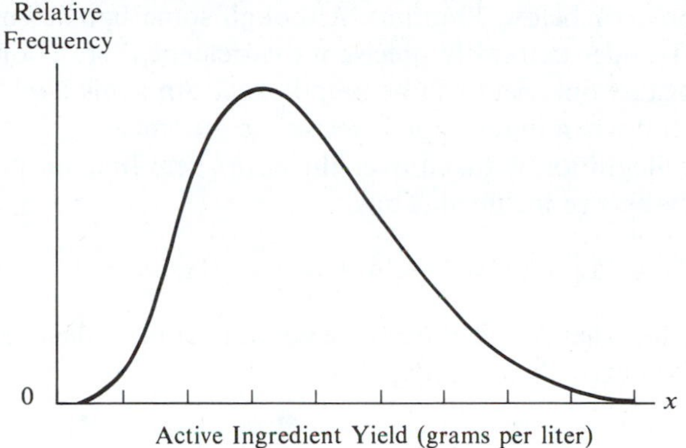

Figure 4-1
Frequency curve for population of active-ingredient yield from batches of raw chemicals processed.

The Probability Density Function

Every continuous random variable has a density function that completely specifies its probability characteristics. Figure 4-2 shows the probability density function for X, which we denote by $f(x)$. Notice that $f(x) \geq 0$ for all levels x; a probability density function never assumes negative values.

The probability that a continuous random variable will fall within an interval is equal to the *area* under the density curve over that range:

$$\Pr[a \leq X \leq b] = \int_a^b f(x)\,dx$$

The desired probability is found by integrating $f(x)$ over the interval of interest.

Because there is no area over a single point (which has no width), the probability is theoretically 0 that a continuous random variable will be precisely equal to some value. For example, the probability that the chosen batch has a yield of exactly 30 grams per liter may be assumed to be 0. This implies that every batch

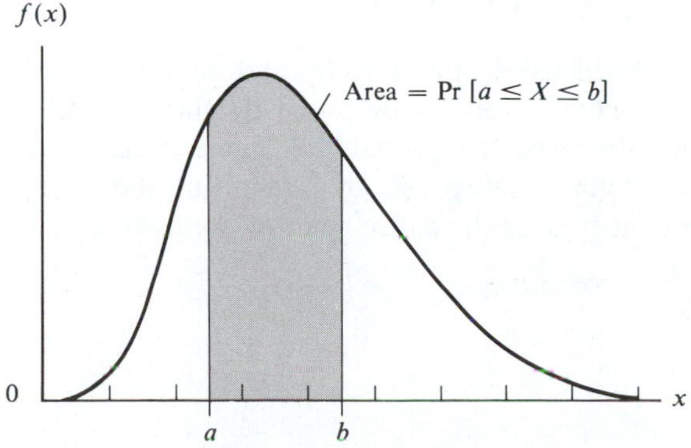

Figure 4-2
Graph of the probability density function for the active-ingredient yield of a randomly selected batch of a chemical.

yield will be above or below 30 grams. Although some batches may have yields very close to 30, under extremely precise measurement there is only an infinitesimal chance that any one yield will be found exact. And this likelihood will more closely approach 0 when under more precise measurement.

When the integration is taken over the entire real line, all possible levels of the random variable are included. Thus,

$$\Pr[-\infty < X < \infty] = \int_{-\infty}^{\infty} f(x)\,dx = 1$$

This reflects the fact that X is certain to be equal to some value, so the area under the entire density curve always equals 1.

Example: Distance between Flaws on Magnetic Tape

A radar telemetry tracking station requires a vast quantity of high-quality magnetic tape. It has been established that the distance X (in inches) between tape-surface flaws has the following probability density function:

$$f(x) = \begin{cases} .01e^{-.01x} & x \geq 0 \\ 0 & \text{otherwise} \end{cases}$$

Suppose one flaw has been identified. The probability that an additional flaw is found within the next 50 inches of tape may be computed by integrating $f(x)$ from 0 to 50:

$$\begin{aligned} \Pr[0 \leq X \leq 50] &= \int_{0}^{50} .01e^{-.01x}\,dx = -e^{-.01x}\Big]_{0}^{50} \\ &= e^{0} - e^{-.50} = 1 - e^{-.50} \\ &= 1 - .6065 = .3935 \end{aligned}$$

The probability that the next flaw will occur after at least 100 inches involves evaluating the integral at limits of 100 and ∞:

$$\begin{aligned} \Pr[100 \leq X < \infty] &= \int_{100}^{\infty} .01e^{-.01x}\,dx = -e^{-.01x}\Big]_{100}^{\infty} \\ &= e^{-1} - e^{-\infty} = e^{-1} - 0 \\ &= .3679 \end{aligned}$$

Expected Value and Variance

Recall that the expected value of a discrete random variable expresses its central tendency. It is a weighted average computed by summing the products of each possible level times the respective probability. The expected value of a continuous random variable is found analogously by integrating the product of the dummy variable and the density over the entire span of possibilities.

EXPECTED VALUE OF A CONTINUOUS RANDOM VARIABLE

$$\mathbf{E(X) = \int_{-\infty}^{\infty} xf(x)\,dx}$$

As an illustration, consider the distance X between magnetic tape flaws in the preceding example. Because $f(x) = 0$ for all $x < 0$, the lower limit of integration will be taken as 0. We have

$$E(X) = \int_0^{\infty} x(.01e^{-.01x})\,dx = (-100 - x)e^{-.01x}\Big]_0^{\infty} = 100$$

This means that on the average there will be $E(X) = 100$ inches of tape between each flaw encountered.

The method of finding the variance of a continuous random variable is also analogous to that of finding its discrete counterpart. Recall that $\text{Var}(X)$ was defined in terms of the weighted average of squared deviations, $[x - E(X)]^2$. For continuous random variables we find the variance by evaluating the integral

$$\text{Var}(X) = \int_{-\infty}^{\infty} [x - E(X)]^2 f(x)\,dx$$

In practice, the foregoing is cumbersome to work with, so we employ the fact that

$$\text{Var}(X) = E(X^2) - E(X)^2$$

and use instead the following expression.

VARIANCE OF A CONTINUOUS RANDOM VARIABLE

$$\mathbf{Var}(X) = \int_{-\infty}^{\infty} x^2 f(x)\,dx - E(X)^2$$

Continuing with the magnetic tape illustration, the variance in distance between flaws is

$$\begin{aligned}\text{Var}(X) &= \int_0^{\infty} x^2(.01)e^{-.01x}\,dx - E(X)^2\\ &= (.01)\int_0^{\infty} x^2 e^{-.01x}\,dx - (100)^2\\ &= (.01)\left(\frac{x^2 e^{-.01x}}{-.01}\Big]_0^{\infty} - \frac{2}{(-.01)}\int_0^{\infty} xe^{-.01x}\,dx\right) - 10{,}000\\ &= 0 + 2\left[\frac{e^{-.01x}}{(-.01)^2}([-.01]x - 1)\right]_0^{\infty} - 10{,}000\\ &= 20{,}000 - 10{,}000 = 10{,}000\end{aligned}$$

This variance is in *square* inches, not a very helpful measure of distance. By taking the square root, we obtain the standard deviation

$$SD(X) = \sqrt{\text{Var}(X)} = \sqrt{10{,}000} = 100 \text{ inches}$$

We will next consider a principle that is helpful in understanding just why the variance and standard deviation are so important in probability and statistics.

Chebyshev's Theorem

An important principle of probability is illustrated by **Chebyshev's theorem**. This result is stated as follows.

> **Chebyshev's Theorem** Let X be any random variable for which the expected value and variance may be found. Then
>
> $$\Pr[E(X) - kSD(X) \leq X \leq E(X) + kSD(X)] \geq 1 - \frac{1}{k^2}$$
>
> That is, the probability is at least $1 - (1/k)^2$ that a value will be obtained lying within k standard deviation units of the expected value or mean of X.

This theorem applies for discrete and continuous random variables. It will be helpful if, before proving Chebyshev's theorem, we discuss its importance.

Chebyshev's theorem indicates that a great deal of information may be gleaned from the variance (or standard deviation). Knowing little else about a probability distribution, we can place a lower bound on the probability that a random variable will fall within $\pm k$ standard deviations of its mean. Figure 4-3 shows how this minimum probability grows as k gets larger, from .75 when $k = 2$, to .96 when $k = 5$ and up, to .99 when $k = 10$. Although we ordinarily have enough information about a random variable's probability distribution to determine this probability more precisely, Chebyshev's theorem suggests that the variance (standard deviation) plays a very important role in describing probability distributions. It is virtually certain that any random variable will lie within a modest number of standard deviations of the mean. Figure 4-3 shows this effect for several common probability distributions to be described in this chapter. In all cases, Chebyshev's theorem provides a lower bound on the exact probabilities.

Figure 4-3 *Graphs of the probability for a random variable falling within $\pm k$ standard deviations of the mean for several important distributions.*

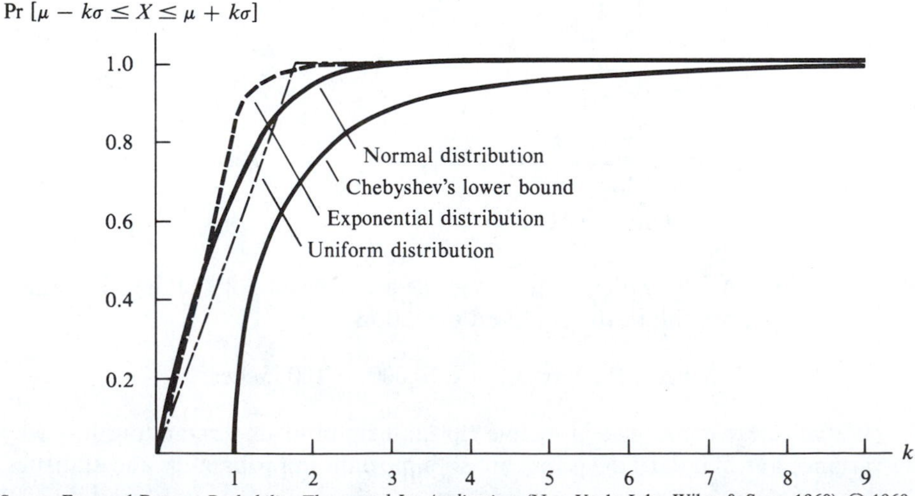

Source: Emanuel Parzen, *Probability Theory and Its Applications* (New York: John Wiley & Sons, 1960). © 1960.

Problems

4-1 Consider the random variable X having probability density function

$$f(x) = \begin{cases} \dfrac{x}{2} & 0 \le x \le 2 \\ 0 & \text{otherwise} \end{cases}$$

(a) Determine the following:
  (1) $\Pr[1 \le X \le 1.5]$ (3) $\Pr[X \le .75]$
  (2) $\Pr[X > .25]$ (4) $\Pr[X < 4]$
(b) Compute the expected value and variance.

4-2 The following cannot be probability density functions. Explain for each why this is so.

(a) $f(x) = \begin{cases} x & 1 \le x \le 4 \\ 0 & \text{otherwise} \end{cases}$ (c) $f(x) = \begin{cases} -1 & -1 \le x \le 2 \\ 0 & \text{otherwise} \end{cases}$

(b) $f(x) = \begin{cases} 1 & x = 10 \\ 0 & \text{otherwise} \end{cases}$ (d) $f(x) = \begin{cases} \dfrac{x^2 - 2}{3} & 0 \le x \le 3 \\ 0 & \text{otherwise} \end{cases}$

4-3 Scrap tubing is left over from fabricating a compressor. Each piece has a length (in inches) represented by the density function

$$f(x) = \begin{cases} \dfrac{1}{10} & 0 \le x \le 10 \\ 0 & \text{otherwise} \end{cases}$$

(a) Determine the probability that a randomly selected tube will be
  (1) less than 4″ (2) greater than 6″ (3) between 5.5″ and 7.5″
(b) Compute the expected value and variance.

4-4 Consider the level of a random decimal, which is a quantity equally likely to fall anywhere between 0 and 1. Give the expression for the probability density function.

4-5 For each random variable having the following probability density functions, determine (1) $E(X)$, (2) $\text{Var}(X)$, (3) $SD(X)$, and (4) $\Pr[.5 \le X \le 1.5]$.

(a) $f(x) = \begin{cases} \dfrac{1}{12} & 0 \le x \le 12 \\ 0 & \text{otherwise} \end{cases}$ (c) $f(x) = \begin{cases} \dfrac{3x^2}{7} & 1 \le x \le 2 \\ 0 & \text{otherwise} \end{cases}$

(b) $f(x) = \begin{cases} 5e^{-5x} & 0 \le x \le \infty \\ 0 & \text{otherwise} \end{cases}$ (d) $f(x) = \begin{cases} x & 0 \le x \le 1 \\ 2 - x & 1 \le x \le 2 \\ 0 & \text{otherwise} \end{cases}$

4-6 Determine levels for the constants that will make $f(x)$ a probability density function.

(a) $f(x) = \begin{cases} \dfrac{x}{a} & 0 \le x \le 4 \\ 0 & \text{otherwise} \end{cases}$ (c) $f(x) = \begin{cases} e^{-cx} & 0 \le x \le \infty \\ 0 & \text{otherwise} \end{cases}$

(b) $f(x) = \begin{cases} bx^2 & 0 \le x \le 1 \\ 0 & \text{otherwise} \end{cases}$ (d) $f(x) = \begin{cases} d(x - .5)^2 & 0 \le x \le 1 \\ 0 & \text{otherwise} \end{cases}$

4-7 Refer to Problem 4-6 and your answers to that problem. Determine for each random variable (1) $E(X)$, (2) $\text{Var}(X)$, (3) $SD(X)$, and (4) $\Pr[.25 \le X \le .75]$.

4-8 Refer to Problem 4-1 and your answers to that problem.
(a) Find the exact probability that each X falls within (1) $\mu \pm 1\ SD(X)$; (2) $\mu \pm 2\ SD(X)$; and (3) $\mu \pm 3\ SD(X)$.
(b) Find the respective limits indicated for these probabilities by Chebyshev's theorem.

4-9 Prove Chebyshev's theorem. In doing this, begin with the integral for $\text{Var}(X)$. Using $\sigma^2 = \text{Var}(X)$ and $\mu = E(X)$, break that integral into three parts, encompassing the ranges $(-\infty, \mu - k\sigma)$, $(\mu - k\sigma, \mu + k\sigma)$, and $(\mu + k\sigma, \infty)$. Then drop one term and use the fact that for the remaining terms, $(x - \mu)^2 \geq k^2\sigma^2$.

4-2 The Normal Distribution

Most frequently encountered in statistics is the **normal distribution**, often referred to as the *Gaussian distribution.* The normal distribution is used so often partly because many physical measurements, such as human height or diameters of ball bearings, provide frequency distributions that closely approximate a **normal curve**. Even examination scores—SAT, GRE, or IQ—share this property.

But the high incidence of its natural occurrence is not the primary reason why the normal distribution is so important in statistics. For a wide range of parameters the normal distribution can be used to closely approximate other probability distributions, such as the binomial and hypergeometric. Such approximations can provide significant computational advantages.

But the normal distribution's importance as a utility distribution is overshadowed by the paramount role it plays in statistical theory. A great many statistical applications involve inferences about population means, and these are usually based on the sample counterparts. As we shall see, the normal curve may be used to provide probabilities for levels of the sample mean $\bar{X}$.

The Normal Distribution and the Population Frequency Curve

To lay the foundation for later statistical analysis, we begin our discussion of the normal distribution by considering the frequency curve for a quantitative population. Envision the measured diameters of a large quantity of precision ball bearings for which the population histogram was constructed and then approximated by a smoothed curve. Figure 4-4 shows the resulting frequency curve. This curve not only describes the underlying population frequency distribution but also takes the shape of a normal curve.

The height of a normal curve above any point x is obtained from the following.

FREQUENCY CURVE AND PROBABILITY DENSITY FUNCTION FOR A NORMAL DISTRIBUTION

$$f(x) = \frac{1}{\sqrt{2\pi}\,\sigma} e^{-\frac{(x-\mu)^2}{2\sigma^2}} \qquad -\infty < x < \infty$$

This expression defines the probability density function for the value X of a random observation from the population.

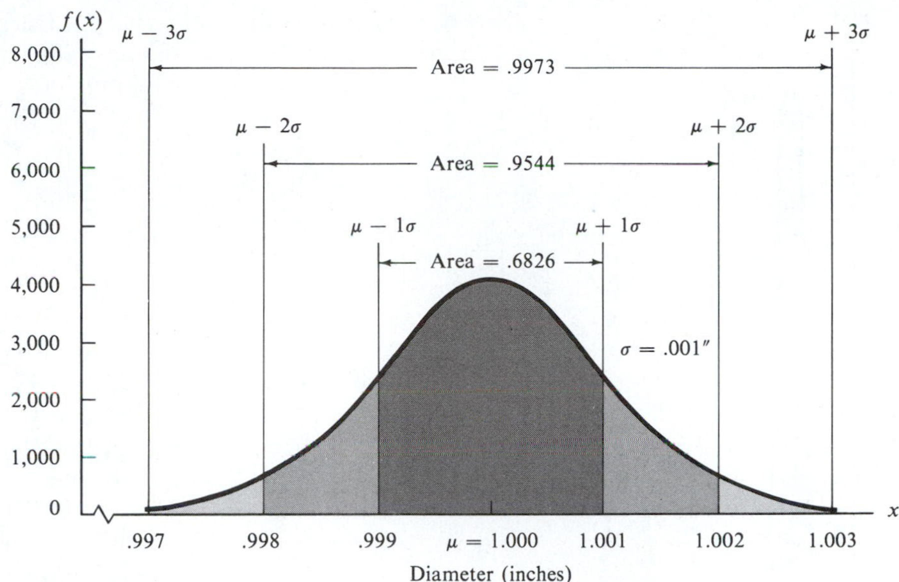

Figure 4-4
Normal curve for diameters of 1″ precision ball bearings.

When plotted, the foregoing function takes the familiar "bell shape." The two parameters μ and σ entirely specify a particular normal curve. The normal curve for diameters of the precision ball bearings in our illustration has parameters $\mu = 1''$ and $\sigma = .001''$, the levels established for the population mean and standard deviation.

Every normal frequency curve is centered on the population mean μ and is symmetrical about this point. Also, this location is both the median and the mode. The tails taper off fairly rapidly for levels of x very far above or below the mean. But it is obvious that the normal curve will never touch the x-axis, since $f(x)$ will be nonzero over the entire real line, from $-\infty$ to $+\infty$.

As we have seen, it is areas under density functions that provide probabilities for continuous random variables. Normal curves have a very convenient property that facilitates finding these: The area under the normal curve between μ and any point depends only on the distance separating that point from the mean, as expressed in units of σ. For instance, the area within the interval $\mu \pm \sigma$ is .6826. This means that about 68% of the precision ball bearing diameters fall within the range $1'' \pm .001''$, or between $.999''$ and $1.001''$. Also, the area spanning $\mu \pm 2\sigma$ is .9544 and that covering $\mu \pm 3\sigma$ is .9973. Over 95% of the precision ball bearings will have diameters of $1'' \pm .002''$, with 5 out of 100 exceeding those limits. Only about 3 out of 1,000 will fall outside of $1'' \pm .003''$.

The two parameters completely specify the location and scale of a normal curve. The mean μ locates the center, whereas the standard deviation σ provides the degree of dispersion. Figure 4-5 shows three normal curves, each representing the frequency distribution of a different population. Each normal curve is centered at the respective population mean. A low level for the standard deviation reflects low dispersion in the underlying population values. This is portrayed by

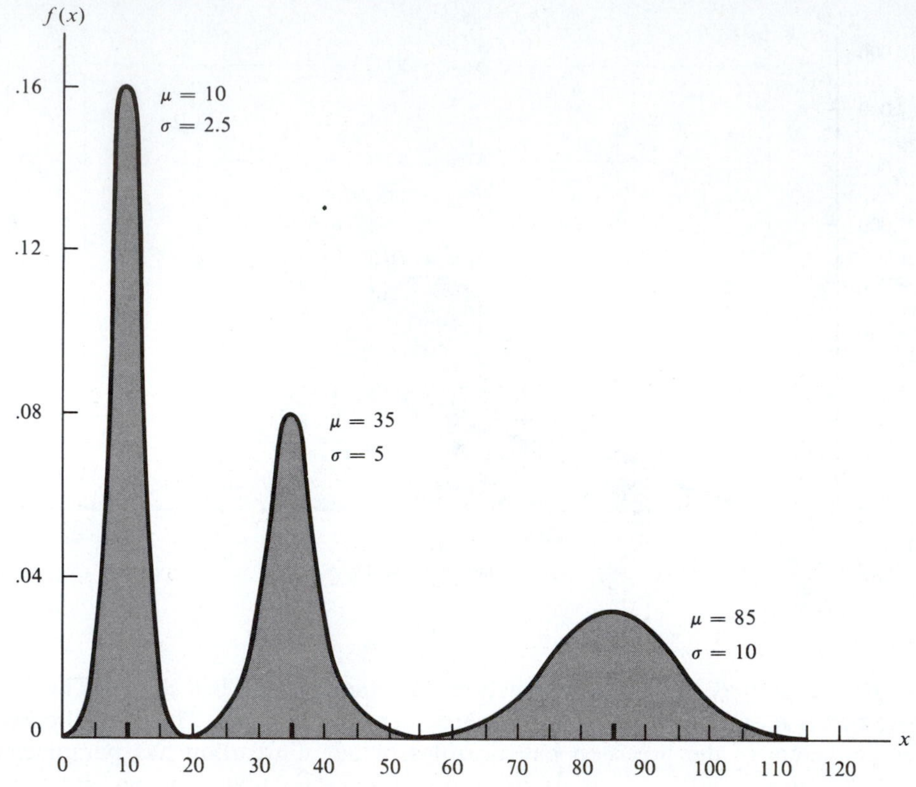

Figure 4-5 *Normal curves for three different populations, each having different mean and standard deviation.*

a normal curve having a tight, peaked "bell." As σ gets larger, reflecting a population with greater dispersion, the corresponding normal curve becomes flatter and more spread out.

The Normally Distributed Random Variable

Consider a sample observation from a population whose frequency distribution may be represented by a normal curve. The value X obtained is referred to as a **normally distributed random variable**. Probabilities may be found for X by determining the respective areas under the normal curve. Since a normally distributed random variable is continuous, nonzero probabilities for X can be found only for possible levels falling within intervals (rather than at single points). It is therefore convenient to work with the following expression.

NORMAL PROBABILITY DISTRIBUTION FUNCTION

$$F(x) = \Pr[X \leq x] = \int_{-\infty}^{x} \frac{1}{\sqrt{2\pi}\,\sigma} e^{-\frac{(y-\mu)^2}{2\sigma^2}} dy$$

Figure 4-6 shows the normal probability density and distribution functions for the diameter X of a randomly selected precision ball bearing. The figure shows that

$$F(1.0005'') = \Pr[X \leq 1.0005''] = .6915$$

and that the area under the normal curve $f(x)$ to the left of x (top, Figure 4-6) is equal to the height of the distribution function $F(x)$ (bottom). As we have seen, the probability value may be obtained by evaluating the integral of $f(x)$ over the appropriate limits. But the normal density function has no closed-form integral, so $f(x)$ must be integrated numerically. That work has already been done, and values for $F(x)$ may be read from a table of normal curve areas.

Before describing that table, we will consider a particular random variable from the normal family.

The Standard Normal Distribution

A very important normal curve characterizes the **standard normal random variable**. Denoted by Z, this random variable is represented by a normal curve with parameters $\mu = 0$ and $\sigma = 1$.

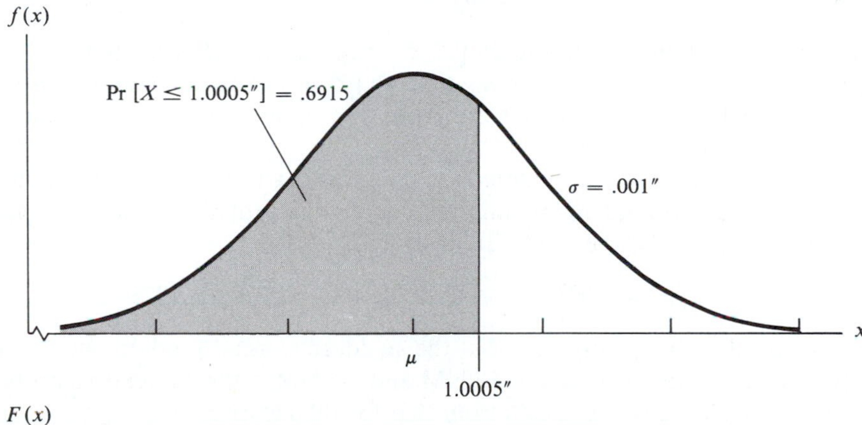

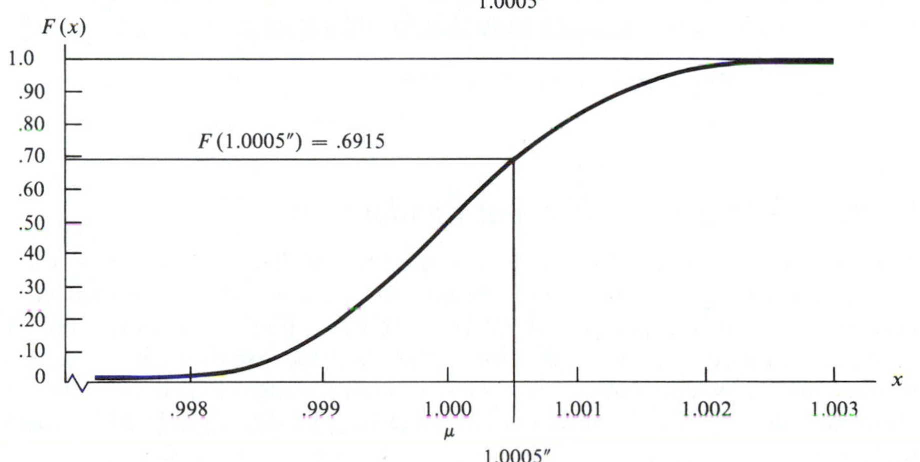

Figure 4-6 *Normal probability density function (top) and normal probability distribution function (bottom) for the diameter of a randomly selected precision ball bearing.*

Illustration:* *Satellite Ranging Errors

As an illustration, suppose that a normal curve applies for the ranging prediction error experienced by a ground station when its tracking radar acquires a satellite target during a particular orbit. The error could be positive (if the predicted range were higher than actual) or negative (prediction lower than actual). The mean error could be 0, and the standard deviation in the population of errors might be 1 nautical mile (NM). The ranging prediction error would then have a *standard* normal distribution. Appendix Table D provides values for this.

STANDARD NORMAL PROBABILITY DISTRIBUTION FUNCTION

$$\Phi(z) = \Pr[Z \leq z] = \int_{-\infty}^{z} \frac{1}{\sqrt{2\pi}} e^{-\frac{y^2}{2}} dy$$

Figure 4-7 shows the areas under the standard normal curve corresponding to several possible cases with the satellite ranging error. The values were obtained from Table D by reading the respective cumulative probability at z, which is also the area to the left of z.

For instance, the probability that the ranging error is less than or equal to 1.50 NM is thus

$$\Pr[Z \leq 1.50] = \Phi(1.50) = .9332$$

The same probability applies to $\Pr[Z < 1.50]$, because Z is continuous and a single point has no probability so that $\Pr[Z = 1.50] = 0$. (With *continuous* random variables, no distinctions are made in evaluating outcomes having "$<$" instead of "$\leq$" or "$>$" instead of "$\geq$".)

The probability that a random variable exceeds a particular amount, such as 2.00 NM, may be found by reading the cumulative probability for the complementary event from Table D and subtracting from 1:

$$\Pr[Z > 2.00] = 1 - \Phi(2.00) = 1 - .9772 = .0228$$

And to calculate the probability that the standard normal random variable falls within a range—say, between −.10 NM and .65 NM—the cumulative probability for the smaller z is subtracted from that for the larger z:

$$\begin{aligned}\Pr[-.10 \leq Z \leq .65] &= \Phi(.65) - \Phi(-.10) \\ &= .7422 - .4602 = .2820\end{aligned}$$

Probabilities for Any Normal Random Variable

The standard normal distribution may be used to obtain probabilities applicable to any normal distribution. This is because the area under any normal curve between two points depends only on how far those points are away from the mean, as measured in standard deviation units. And because the standard normal distribution has a unit parameter $\sigma = 1$, a point's standard deviation distance above or below the center of its normal curve may be directly related to a level of z.

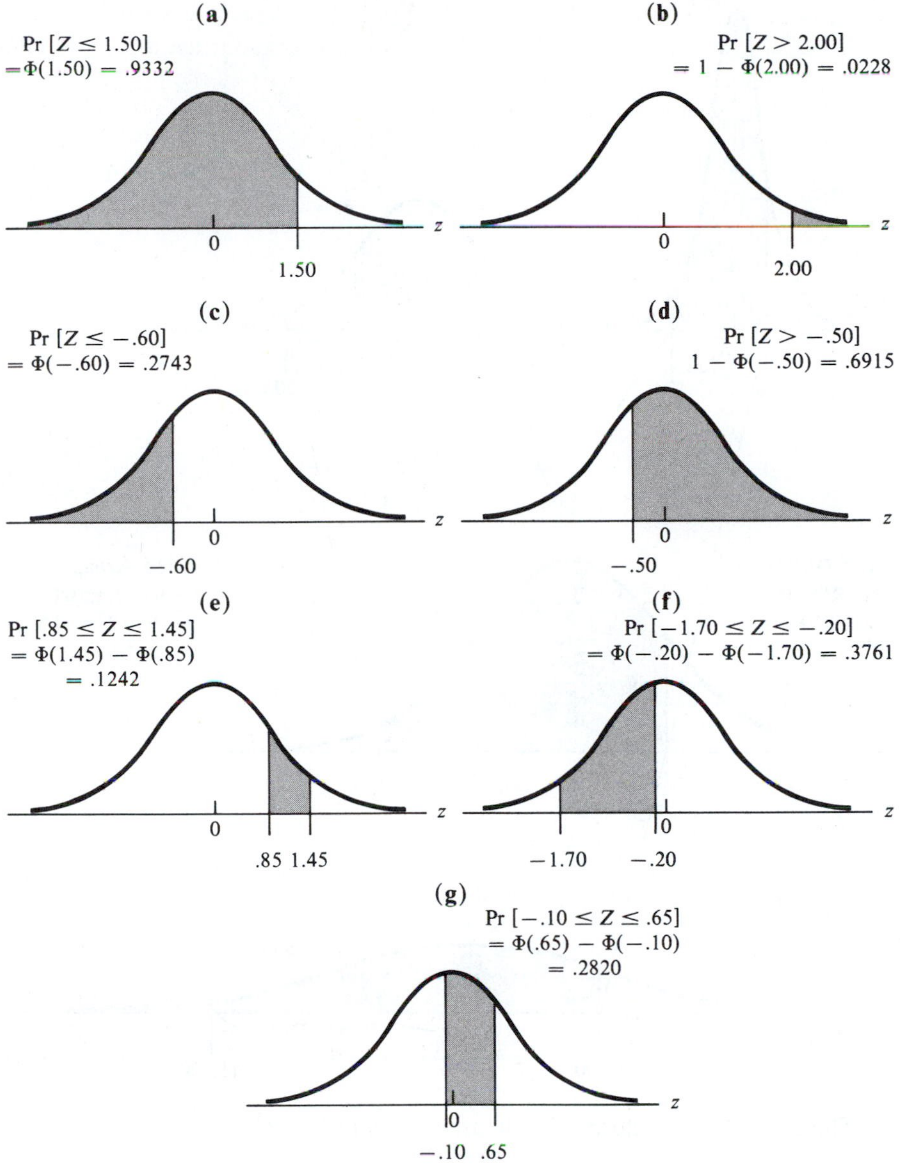

Figure 4-7 *Areas under the normal curve that represent probabilities for the standard normal random variable.*

Figure 4-8 illustrates this concept. Three different normal curves are portrayed there, with the shaded areas of each covering levels of x from a point 1 standard deviation below the mean to 2 standard deviations above it. All these areas are identical, and they match the area under the standard normal curve between $z = -1$ and $z = 2$.

Thus, to find probabilities we need only identify the matching areas under the standard normal curve. This involves finding one or more levels for z, referred to as **normal deviates**.

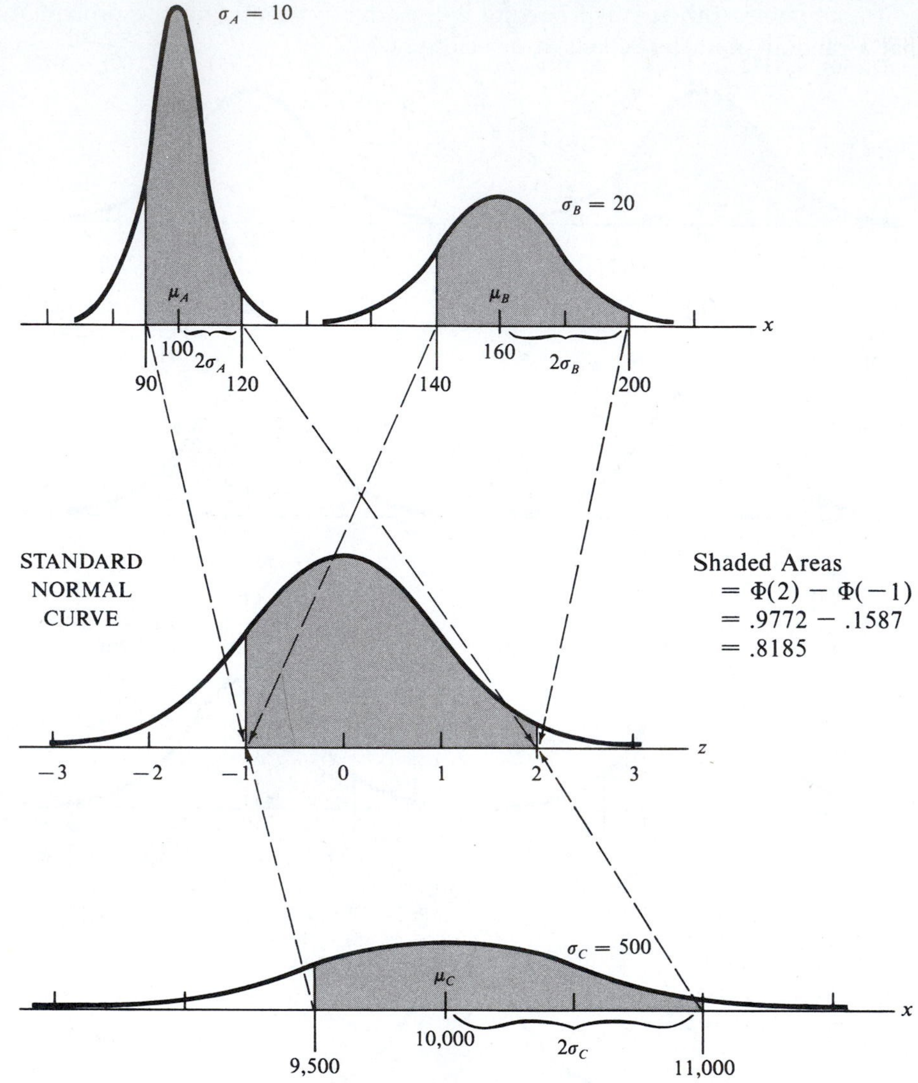

Figure 4-8 *Illustration of how areas for various normal curves match areas under the standard normal curve. In all cases, the shaded areas span a range from 1 standard deviation below the mean to 2 standard deviations above the mean.*

The following transformation is used to identify each.

NORMAL DEVIATE

$$z = \frac{x - \mu}{\sigma}$$

In effect, the probability distribution function for any normal random variable X may be obtained by making the above transformation into the standard normal random variable:

$$F(x) = \Pr[X \le x] = \Pr\left[\frac{X - \mu}{\sigma} \le \frac{x - \mu}{\sigma}\right]$$
$$= \Pr\left[Z \le \frac{x - \mu}{\sigma}\right] = \Phi\left(\frac{x - \mu}{\sigma}\right)$$

Continuing with the ball bearing illustration, we may find the probability that a random diameter X falls at or below 1.0015″:

$$\Pr[X \leq 1.0015''] = \Phi\left(\frac{1.0015'' - 1.000''}{.001''}\right)$$
$$= \Phi(1.5) = .9332$$

that it exceeds .9995″:

$$\Pr[X > .9995''] = 1 - \Phi\left(\frac{.9995'' - 1.000''}{.001''}\right)$$
$$= 1 - \Phi(-.50)$$
$$= 1 - .3085 = .6915$$

and that it lies between .9998″ and 1.0004″:

$$\Pr[.9998'' \leq X \leq 1.0004''] = \Phi\left(\frac{1.0004'' - 1.000''}{.001''}\right) - \Phi\left(\frac{.9998'' - 1.000''}{.001''}\right)$$
$$= \Phi(.40) - \Phi(-.20)$$
$$= .6554 - .4207 = .2347$$

Example: Human Engineering and Nuclear Reactor Controls

The Three-Mile Island nuclear reactor accident highlighted major deficiencies in the state of the art for configuring control panels. Accident investigators reported that operators had far too many dials, lights, and meters for a human to monitor adequately. Instrumentation displays are much more advanced in aircraft controls, in which for many years specialists called *human engineers* have segregated the mundane from the critical in designing cockpit layouts. It was recommended that future reactor control rooms be human-engineered.

One consulting engineer proposed setting up a control room simulator where various locations for key warning signals could be evaluated in detail. For instance, "hot lights" for reactor cooling-system components should be located where they can be easily seen and quickly reacted to. He devised a plan for simulating various hot-light locations by generating hypothetical danger events and timing various operator reactions. Locations could be adjusted until some optimal balance were achieved.

For each panel configuration, the simulation plan called for monitoring the average reaction time between activation of a warning light and the operator's sounding a buzzer alarm. Suppose this random variable X is normally distributed with a mean of $\mu = 45$ seconds and a standard deviation of $\sigma = 8$ seconds. The probability that the alarm is sounded within an average of 1 minute (60 seconds) after a hazardous event is

$$\Pr[X \leq 60] = \Phi\left(\frac{60 - 45}{8}\right) = \Phi(1.88) = .9699$$

whereas the probability that more than 2 minutes (120 seconds) are required is negligible and corresponds to a normal deviate off the table:

$$\Pr[X > 120] = 1 - \Phi\left(\frac{120 - 45}{8}\right) = 1 - \Phi(9.4) \approx 0$$

Expected Value, Variance, and Percentiles

The following facts may be established for a normally distributed random variable:

$$E(X) = \int_{-\infty}^{\infty} xf(x)\,dx = \mu$$

$$\text{Var}(X) = \int_{-\infty}^{\infty} x^2 f(x)\,dx - \mu^2 = \sigma^2$$

where $f(x)$ is the normal probability density function defined earlier. The mean level $E(X)$ equals the center μ of the normal curve and $\text{Var}(X)$ matches the square of the dispersion parameter, σ^2. In effect, specification of the parameters μ and σ completely specifies the particular normal distribution that applies.

The normal curve provides a close fit to the frequency curves for measurements of many dimensions and human characteristics. It may therefore answer important questions pertaining to **percentiles**. A population percentile is *the value at or below which the stated percentage of units lie*. When the population has a normal frequency curve, we may use the following expression.

PERCENTILE VALUE FOR A NORMAL DISTRIBUTION

$$\boldsymbol{x = \mu + z\sigma}$$

Here the normal deviate z corresponds to the stated percentile.

For example, consider one final time the ball bearings, where $\mu = 1.000''$ and $\sigma = .001''$. The 95th percentile is the dimension at or below which 95% of the ball bearing diameters will lie. This is found by reading from Table D the z that corresponds to a cumulative probability of .95. We read the table in reverse, getting*

$$z = 1.64$$

The 95th percentile is thus

$$x = 1.000'' + 1.64(.001'') = 1.00164''$$

This value tells us that 95% of all ball bearings from the underlying population will have diameters $\leq 1.00164''$.

Example:* *Anthropometric Data and Equipment Design

The man–machine interface is an important engineering consideration. Controls should be designed so that it doesn't take an "eight-armed superman" to operate them. (And today, most *women* should be able to operate equipment, too.) Anthropometric data are often used in establishing distances between controls where space is critical, as in aircraft cockpits.

One of the most important variables is *maximum reach*. Assuming that this quantity for adult males is normally distributed, with a mean $\mu = 32.33''$ and a standard deviation of $\sigma = 1.63''$, the following percentiles apply:†

* The cumulative probability entry for the above is .9495. The tabled value for $z = 1.65$ is just as close, but in the case of a tie we will pick the lower z. Although an interpolation procedure will provide more accuracy, to ease our discussions in this book we will just find the closest tabled value.

	1st	5th	50th	95th	99th
z	-2.33	-1.64	0	1.64	2.33
$x = \mu + z\sigma$	28.5″	29.7″	32.3″	35.0″	36.1″

Thus, if design specifications require that a particular control be placed so that 95% of all seated male pilots can reach it without bending, the control must be located no farther than 29.71″ from the seat backrest.

[†] *Source:* H. T. E. Hertzberg, G. S. Daniels, and E. Churchill, *Anthropometry of Flying Personnel*, WADC Technical Report 52-321, USAF, September 1954.

Practical Limitations of the Normal Distribution

The normal distribution has wide application and is often used to represent physical measurements. As a practical matter, it is nearly always an *approximation* to reality rather than a totally realistic representation. For instance, the normal distribution is used to characterize human height data. The height of adult American males might be represented by a normal frequency curve with mean $\mu = 69''$ and standard deviation $\sigma = 2.5''$. Remember that the tails of the normal curve extend indefinitely, so that the height X of a randomly selected male has a finite probability under the normal distribution of exceeding any level—even 10 feet or one mile! Worse yet, there is a probability that X can be negative.

These absurd results apply to levels of the random variable beyond the region of reasonable approximation, ordinarily taken to be $\mu \pm 5\sigma$. Appendix Table D does not provide probabilities for $|z| > 5$. (A mile-tall man would correspond to a normal deviate $z > 25{,}000$.)

The Normal Distribution and Sampling

The normal distribution plays a key role in statistics, due to the following:

1. Many important statistical applications use the computed value of the *sample* mean $\bar{X}$ to make inferences regarding the unknown *population* mean μ.
2. For nearly all populations randomly sampled, *$\bar{X}$ is a random variable that tends to be normally distributed.*
3. When the population standard deviation σ is a known value, then probabilities may be determined for the yet-to-be-computed $\bar{X}$ using a normal curve with mean μ and standard deviation

$$\sigma_{\bar{X}} = \frac{\sigma}{\sqrt{n}}$$

where n denotes the size of the sample.

After laying the conceptual framework in Chapters 5, 7, and 8, we will see why these facts must hold.

Example: *Probabilities for Mean Task Completion Times*

An industrial engineer wishes to estimate the mean time taken to complete an assembly task. She believes that the standard deviation for individual completion times is $\sigma = 25$ seconds, a value obtained from earlier experiments with similar tasks. She plans to make $n = 30$ sample observations. The standard deviation for the random variable $\bar{X}$ is

$$\sigma_{\bar{X}} = \frac{\sigma}{\sqrt{n}} = \frac{25}{\sqrt{30}} = 4.56 \text{ seconds}$$

There is probability $\Phi(z) - \Phi(-z)$ that $\bar{X}$ will fall within the interval $\bar{X} = \mu \pm z\sigma_{\bar{X}} = \mu \pm 4.56z$. Thus, when $z = 2$,

$$\begin{aligned}\Pr[\bar{X} = \mu \pm 4.56(2) = \mu \pm 9.12] &= \Phi(2) - \Phi(-2)\\ &= .9772 - .0228 = .9554\end{aligned}$$

From these calculations, the engineer can see that there is about a 95% chance that her sample result will be within around 9 seconds of the true level for the population mean.

An even larger sample size of $n = 100$ would improve the accuracy. Using the same standard deviation,

$$\sigma_{\bar{X}} = \frac{\sigma}{\sqrt{n}} = \frac{25}{\sqrt{100}} = 2.50 \text{ seconds}$$

There is a .9554 probability that the computed value for $\bar{X}$ will lie within $\mu \pm 2(2.50)$, an accuracy of ± 5 seconds.

Problems

4-10 A satellite range prediction error has the standard normal distribution with mean 0 NM and standard deviation 1 NM. Find the following probabilities for the prediction error:

(a) less than 1.25 NM
(b) greater than $-.75$ NM
(c) less than or equal to -2.50 NM
(d) greater than or equal to 3.5 NM
(e) between 1.5 NM and 1.75 NM
(f) between $-.50$ NM and $-.25$ NM
(g) less than -2.80 NM or greater than .65 NM
(h) less than .65 NM or greater than .75 NM

4-11 Find the following percentiles for the satellite range prediction error described in Problem 4-10:

(a) 10th (b) 25th (c) 85th (d) 95th (e) 99th

4-12 The sitting height of adult males is normally distributed, with mean $\mu = 35.94''$ and standard deviation $\sigma = 1.29''$.

(a) Find the probability that a randomly chosen man's sitting height is (1) less than 33″, (2) between 34″ and 35.5″, (3) between 33″ and 37″, and (4) greater than 40″.

(b) Find the following percentiles:

(1) 5th (2) 10th (3) 75th (4) 95th

4-13 The average active-ingredient yield per liter of raw material for samples of vials may be approximated by a normal distribution with mean $\mu = 30$ grams and standard deviation $\sigma = .2$ gram.

(a) Find the probability that the average yield of a sample is (1) less than 29.55 grams, (2) between 29.5 and 30.25 grams, (3) greater than 30.45 grams, and (4) between 30.15 and 30.35 grams.

(b) Find the following percentiles:

(1) 10th (2) 20th (3) 90th (4) 99th

4-14 The forearm–hand length of adult males is normally distributed, with mean $\mu = 18.86''$ and standard deviation $\sigma = .81''$.

(a) Find the probability that a randomly chosen man's forearm–hand length is (1) less than 19″, (2) between 16.5″ and 17″, (3) between 18″ and 20″, and (4) greater than 21″.

(b) Find the following percentiles:

(1) 1st (2) 5th (3) 50th (4) 60th

4-15 The mean elongation of a steel bar under a particular tensile load has been established to be normally distributed, with parameters $\mu = .06''$ and $\sigma = .008''$. Assuming the same distribution applies to a new bar, find the probability that the mean elongation falls

(a) above .08″
(b) below .055″
(c) somewhere between .05″ and .07″
(d) either below .045″ or above .065″

4-16 A deflection test is performed on several cantilever beams fabricated from a new steel alloy.

(a) Assuming that the mean angle of rotation is normally distributed with parameters $\mu = .035$ and $\sigma = .007$ (both in radians), find the probability that this quantity falls

(1) below .050
(2) above .045
(3) between .03 and .05
(4) between .04 and .055

(b) The mean angle of deflection is also normally distributed with parameters $\mu = .08$ and $\sigma = .02$. Find the probability that this quantity falls

(1) above .11
(2) below .12
(3) between .06 and .075
(4) between .085 and .095

4-17 A ceramics engineer plans to conduct a test on sample scraps of two epoxy substances to determine the average setting time until bonding. Neither the mean nor the standard deviation for this random variable is known although a normal distribution may be assumed to apply. Find the probability that the average bonding time deviates from the unknown μ by no more than 5 hours, assuming that the standard deviation has value

(a) $\sigma = 2$ (b) $\sigma = 4$ (c) $\sigma = 8$ (d) $\sigma = 16$

4-18 A chemical plant superintendent orders a process shutdown and setting readjustment whenever the pH of the final product falls below 6.90 or above 7.10. The sample pH is normally distributed with unknown μ and standard deviation $\sigma = .05$. Determine the probability

(a) of readjusting when the process is operating as intended and $\mu = 7.0$

(b) of readjusting when the process is slightly off target and the mean pH is $\mu = 7.02$

(c) of failing to readjust when the process is too alkaline and the mean pH is $\mu = 7.15$
(d) of failing to readjust when the process is too acidic and the mean pH is $\mu = 6.80$

4-19 Refer to the assembly time example on page 178.
(a) Find the probability when $n = 25$ that the sample mean $\bar{X}$ will fall within the following limits:
(1) $\mu \pm 1.5\sigma_{\bar{X}}$ (2) $\mu \pm 2.5\sigma_{\bar{X}}$ (3) $\mu \pm 3\sigma_{\bar{X}}$ (4) $\mu \pm 4\sigma_{\bar{X}}$
(b) What can you conclude about the accuracy of using $\bar{X}$ to estimate μ and the probability that the resulting estimate will be truthful?

4-20 Refer to the assembly time example on page 178.
(a) Find the level of accuracy (with $z = 2$) obtained when the sample size is: (1) $n = 10$, (2) $n = 50$, (3) $n = 200$, (4) $n = 1{,}000$.
(b) What can you conclude about the accuracy of using $\bar{X}$ to estimate μ and the size of the sample?

4-21 Refer to the assembly time example on page 178.
(a) Find the level of accuracy (with $z = 2$ and $n = 100$) assuming that the standard deviation in assembly times is (1) $\sigma = 5$ seconds, (2) $\sigma = 10$, (3) $\sigma = 15$, (4) $\sigma = 30$.
(b) What can you conclude about the accuracy of using $\bar{X}$ to estimate μ and the level of the population standard deviation?

4-3 The Normal Approximation to the Binomial and Hypergeometric Distributions

The normal distribution plays a major utility role in statistics by approximating other distributions. It generally provides suitable approximations whenever a random variable has a mass or density function assuming the familiar "bell shape." Some distributions even converge to the normal distribution as their parameters approach certain limits. The normal curve is a desirable approximator because it involves only two parameters and the cumulative probabilities can be easily read from a two-page table. Moreover, it is the most familiar distribution because it directly represents many common quantities and because it is encountered so often in statistical inferences.

The normal distribution is particularly advantageous in approximating the binomial distribution. This is because tabled binomial probability values exist only for a few parameter levels. When n or π falls outside the tables, the binomial probabilities can be a challenge to evaluate. The suitability of the normal distribution as an approximation is suggested by the bell-shaped pattern of the spikes for $b(r; n, \pi)$ as n gets large (see Figure 3-5).

It is difficult to relate the *discrete* binomial probability mass function to the *continuous* normal curve directly, but both the binomial and normal involve probability distribution functions that provide cumulative probabilities and are defined over the entire real line. Thus we may work with cumulative probabilities and use the normal version to approximate their binomial counterpart.

The Normal Approximation Procedure

Figure 4-9 shows the binomial distribution function when $n = 10$ and $\pi = .40$. Superimposed on this step function (cumulative probability "stairway") is the cumulative probability S-curve of the approximating normal distribution function. We use the following parameters.

PARAMETERS FOR NORMAL APPROXIMATION TO BINOMIAL

$$\mu = n\pi$$

$$\sigma = \sqrt{n\pi(1 - \pi)}$$

Observe that the right-hand sides of the parameters are the expected value and standard deviation for the binomial random variable R.

Only a few cumulative probability values were needed to sketch the S-curve in Figure 4-9. These were read from Appendix Table D for the normal deviate values calculated from

$$z = \frac{r - \mu}{\sigma} = \frac{r - n\pi}{\sqrt{n\pi(1 - \pi)}}$$

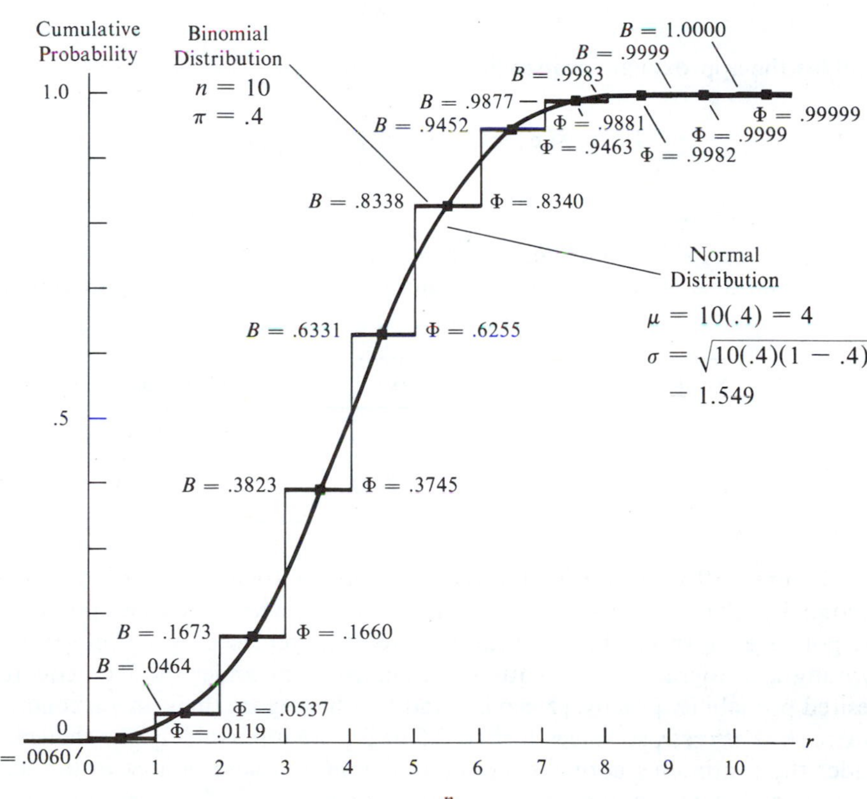

Figure 4-9 Probability distribution functions for binomial and normal distributions.

Recall that the cumulative binomial probability at a particular r is represented graphically by the height of the stairway directly above that point. The cumulative normal curve in Figure 4-9 lies to the right of the stepping points and is consistently about half a step off, a distance of .5 on the r-axis. This suggests that a good approximation to a binomial cumulative probability might be obtained by finding the normal cumulative probability at $r + .5$. This procedure is summarized as follows.

NORMAL APPROXIMATION TO THE BINOMIAL DISTRIBUTION FUNCTION

$$\Pr[R \le r] = B(r) \approx \Phi\left(\frac{r + .5 - n\pi}{\sqrt{n\pi(1-\pi)}}\right)$$

Here Φ denotes the standard normal probability distribution function. Adding .5 in the numerator applies a **continuity correction** to the approximation.

For example, the foregoing procedure provides the following approximation to the cumulative probability that 5 or fewer successes will be obtained:

$$\Pr[R \le 5] = B(5) \approx \Phi\left(\frac{5 + .5 - 10(.4)}{\sqrt{10(.4)(.6)}}\right) = \Phi\left(\frac{5 - 3.5}{1.549}\right)$$
$$= \Phi(.97) = .8340$$

And, as the approximate probability of 3 or fewer successes we have

$$\Pr[R \le 3] = B(3) \approx \Phi\left(\frac{3 - 3.5}{1.549}\right)$$
$$= \Phi(-.32) = .3745$$

Appendix Table A provides the actual cumulative binomial probabilities when $n = 10$ and $\pi = .4$. These actuals are quite close to the above approximations:

Number of Successes	True	Approximation
$r = 3$	$B(3) = .3823$	$\Phi(-.32) = .3745$
$r = 5$	$B(5) = .8338$	$\Phi(.97) = .8340$

Figure 4-10 may be helpful in clarifying the approximation further. The true binomial probability mass function is shown in (a), where the exact probabilities are portrayed as spikes. In (b) those same spikes are replaced by bars of unit width, forming a histogram representation; the shaded area under the bars equals the desired probability exactly. [Keep in mind that histogram (b) is only a conceptual device. R is *discrete*, so values such as 2.5 or 3.5 are impossible.] The shaded area under the continuous normal curve in diagram (c) is close in area to the shaded portion of the histogram. The continuity correction of .5 makes the resulting approximation very close.

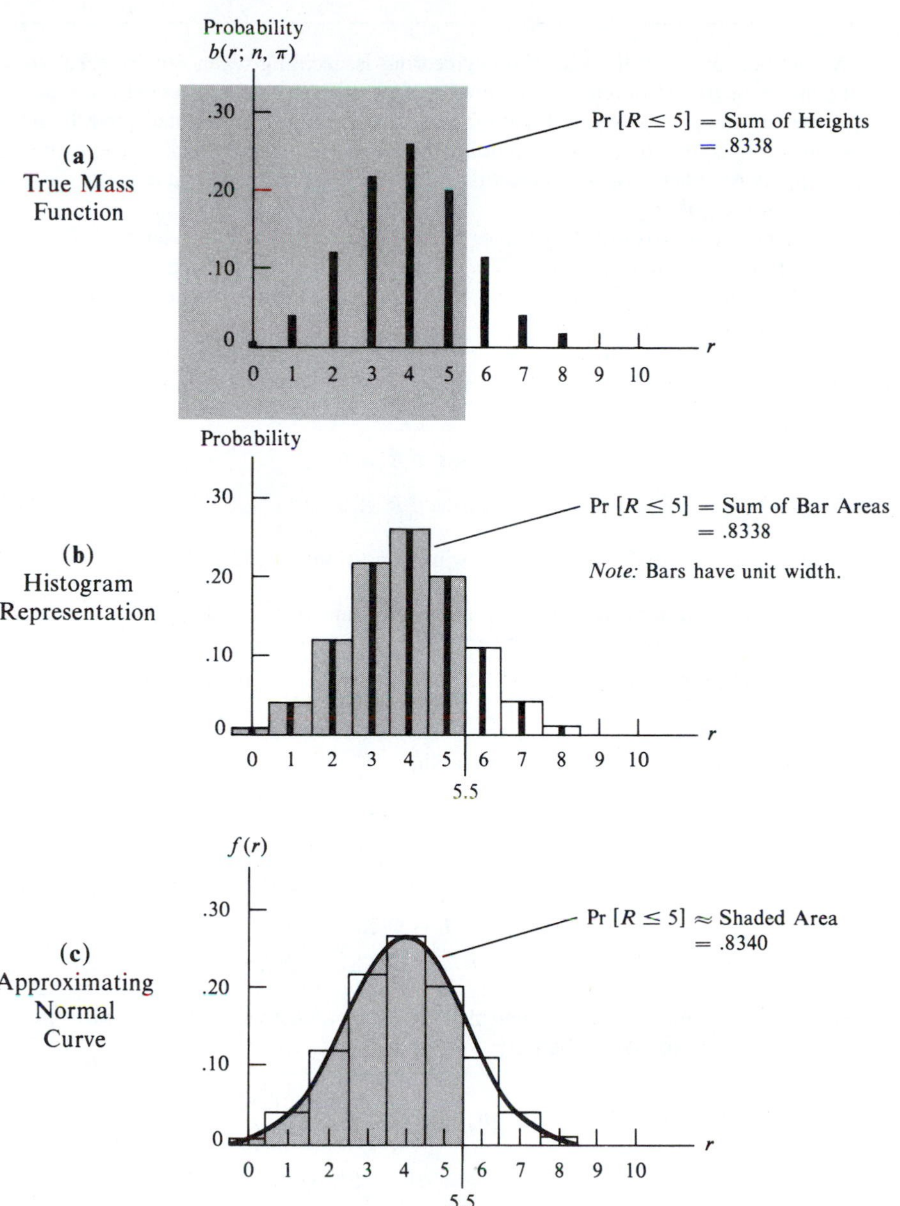

Figure 4-10 *Illustration of normal approximation to binomial distribution.*

Of course, we would never approximate cases when the binomial probabilities are readily available. This procedure becomes valuable when parameters are off the table, such as $n = 350$ with $\pi = .578$ or $n = 1{,}200$ with $\pi = .035$. Even if we have access to a digital computer (which must have the appropriate software) capable of providing exact probabilities, it would be more of a nuisance to find binomial probabilities directly than to use the normal approximation (which involves limited number crunching).

Example:
When to Adjust a Stingy Spigot

A common problem in industrial engineering is deciding when repairs need to be made. Such a requirement may not be obvious when the equipment involved is used in mass production. Erratic operations could go undetected for a long time—leading to an excessive number of faulty items that might have to be scrapped. An efficient way to detect problems early is to sample the items as they come off the line.

One brewery routinely samples 16-ounce cans of beer to determine if the number underfilled is excessive. The proportion π of underfilled cans at any moment is unknown. If π is too great, the filling mechanism must be adjusted—a time-consuming job.

Each hour a random sample of $n = 100$ cans is taken, and their volumes are measured precisely. Letting R represent the number of underfilled cans, brewery policy is to

$$\text{adjust if } R > 6$$
$$\text{leave alone if } R \leq 6$$

This rule was established so the following conditions would be satisfied:

1. There can be at most a 5% chance of unnecessarily adjusting if $\pi = .03$ (when underfilling is not serious).
2. There should be at most a 15% chance of leaving alone if $\pi = .10$ (when underfilling is excessive).

The following approximate probability

$$\begin{aligned}\Pr[R > 6 \mid \pi = .03] &= 1 - \Pr[R \leq 6 \mid \pi = .03] \\ &\approx 1 - \Phi\left(\frac{6 + .5 - 100(.03)}{\sqrt{100(.03)(.97)}}\right) \\ &= 1 - \Phi(2.05) \\ &= 1 - .9798 = .0202\end{aligned}$$

shows that condition (1) is being met. And the adjustment rule provides the following approximate probability:

$$\begin{aligned}\Pr[R \leq 6 \mid \pi = .10] &\approx \Phi\left(\frac{6 + .5 - 100(.10)}{\sqrt{100(.10)(.90)}}\right) \\ &= \Phi(-1.17) = .1210\end{aligned}$$

which meets condition (2).

If the adjustment rule is set more loosely, based on 7 underfilled cans instead of 6, then condition (1) would still be met, but condition (2) would be violated:

$$\begin{aligned}\Pr[R \leq 7 \mid \pi = .10] &\approx \Phi\left(\frac{7 + .5 - 100(.1)}{\sqrt{100(.10)(.90)}}\right) \\ &= \Phi(-.83) = .2033\end{aligned}$$

Likewise, a tighter adjustment rule based on 5 underfilled cans meets condition (2) and violates (1):

$$\Pr[R > 5 \mid \pi = .03] \approx 1 - \Phi\left(\frac{5 + .5 - 100(.03)}{\sqrt{100(.03)(.97)}}\right)$$
$$= 1 - \Phi(1.47)$$
$$= 1 - .9292 = .0708$$

If the original adjustment rule were used with $n = 50$ instead of $n = 100$, condition (2) would be violated. (You may establish this as an exercise.)

Quality of the Approximation

The normal approximation improves as n becomes larger and the closer π becomes to .5. Table 4-1 provides a comparison of the true cumulative binomial probabilities when n is fixed at 10 and π is set at various levels. Notice how close the approximations are when $\pi = .50$. Table 4-2 makes the same comparison holding π fixed at .05 and varying n. A steady improvement in the approximation is achieved as n becomes large (although with such a low π, none of them is very close).

Various rules of thumb have been employed to establish when the normal approximation is valid. Often used is the criterion that both of the following should hold before approximating the binomial distribution:

$$n\pi > 5$$
$$n(1 - \pi) > 5$$

Table 4-1 *True Cumulative Binomial Probabilities for Various Trial Success Probabilities π and Those Obtained by the Normal Approximation When n Is Held at 10*

	$\pi = .01$		$\pi = .05$		$\pi = .10$		$\pi = .50$	
r	B	Φ	B	Φ	B	Φ	B	Φ
0	.9044	.8980	.5987	.5000	.3487	.2981	.0010	.0022
1	.9957	1.0000	.9139	.9265	.7361	.7019	.0107	.0136
2	.9999	1.0000	.9885	.9981	.9298	.9429	.0547	.0571
3	1.0000	1.0000	.9990	1.0000	.9872	.9959	.1719	.1711
4			1.0000	1.0000	.9984	.9999	.3770	.3745
5					1.0000	1.0000	.6230	.6255
6							.8281	.8289
7							.9453	.9429
8							.9893	.9864
9							.9990	.9978
10							1.0000	.9997
Maximum $\lvert B - \Phi \rvert$	.0064		.0987		.0506		.0029	

Table 4-2 *True Cumulative Binomial Probabilities for Various Numbers of Trials n and Those Obtained by the Normal Approximation When π Is Held at .05*

	$n = 10$		$n = 20$		$n = 50$		$n = 100$	
r	B	Φ	B	Φ	B	Φ	B	Φ
0	.5987	.5000	.3585	.3015	.0769	.0968	.0059	.0197
1	.9139	.9265	.7358	.6985	.2794	.2578	.0371	.0537
2	.9885	.9981	.9245	.9382	.5405	.5000	.1183	.1251
3	.9990	1.0000	.9841	.9948	.7604	.7422	.2578	.2451
4	1.0000	1.0000	.9974	.9998	.8964	.9032	.4360	.4090
5			.9997	1.0000	.9622	.9744	.6160	.5910
6			1.0000	1.0000	.9882	.9953	.7660	.7549
7					.9968	.9994	.8720	.8749
8					.9992	.9999	.9369	.9463
9					.9998	1.0000	.9718	.9803
10					1.0000	1.0000	.9885	.9941
11							.9957	.9986
12							.9985	.9997
13							.9995	1.0000
14							.9999	1.0000
15							1.0000	1.0000
Maximum $\lvert B - \Phi \rvert$	.0987		.0570		.0405		.0270	

Approximating the Hypergeometric Distribution

The approximation procedure just described may be adapted in approximating the hypergeometric distribution. Recall that this distribution arises when sampling without replacement from a population of fixed size. As we have seen, hypergeometric probabilities often agree closely with those of the binomial. Recall that both distributions have identical expected values, $E(R) = n\pi$, and that the variance of the hypergeometric equals that of the binomial times the finite population correction factor

$$\text{Var}(R) = n\pi(1 - \pi)\left(\frac{N - n}{N - 1}\right)$$

This suggests that the normal approximation need only be modified by incorporating the correction factor into the normal deviate

$$z = \frac{r + .5 - n\pi}{\sqrt{n\pi(1 - \pi)}\sqrt{\dfrac{N - n}{N - 1}}}$$

and using $\Phi(z)$ to approximate $F(r)$.

Problems

4-22 Compute the approximate binomial probabilities for (1) $R \leq 25$, (2) $R \geq 30$, and (3) $18 \leq R \leq 23$ when
(a) $n = 100$ and $\pi = .27$ (c) $n = 150$ and $\pi = .15$
(b) $n = 300$ and $\pi = .07$

4-23 Suppose that the sample size in the stingy spigot example is reduced to $n = 50$ and the acceptance number is also halved, to 3. Use the normal approximation to determine the probability
(a) of unnecessarily adjusting when $\pi = .03$
(b) of leaving the process alone when $\pi = .10$

4-24 Use the normal approximation to solve Problem 3-22 for the welding robot design.

4-25 Use the normal approximation to solve Problem 3-14 regarding engineering school admissions.

4-26 Use the normal approximation to solve Problem 3-44 regarding acceptance sampling.

4-27 A thermodynamics professor asks a sample of his students whether they would prefer a take-home examination. He is interested in finding probabilities for the number R in the sample who will prefer such a test. Assume that $\pi = .4$, $N = 25$, and $n = 5$.
(a) Calculate the true hypergeometric probabilities $h(r; n, \pi, N)$ to four places.
(b) Determine cumulative probabilities for the true hypergeometric probability distribution function.
(c) Calculate the approximate cumulative probabilities obtained from the normal distribution.
(d) Comparing your answers to (b) and (c), determine the maximum approximation error.

4-28 Repeat Problem 4-23 when the sample size is instead increased to $n = 200$ and the acceptance number is 12.

4-29 Repeat Problem 4-27 using $n = 10$ instead as the sample size.

4-4 The Exponential Distribution

Having wide application in engineering evaluations is the **exponential distribution**. This distribution provides probabilities for the amount of time or space between successive events occurring in a Poisson process—which, as we saw in Chapter 3, characterizes random occurrences under a variety of circumstances. The exponential distribution has been used for the times between arrivals at service facilities, making it a centerpiece in the theory of queues or waiting lines. It plays a similar role in reliability evaluations, in which the exponential distribution gives probabilities for the times to failure. Another major application is in setting production inventories, in which the time between requests for stock is often characterized by the exponential distribution.

All the assumptions of a Poisson process described in Chapter 3 are assumed to hold in those situations where the exponential distribution is applied. Recall that the Poisson process also gives rise to the *discrete* Poisson distribution, which provides probabilities for the number of events in a fixed span of time or space. The exponential distribution treats the span itself as the *continuous* random variable.

Finding Exponential Probabilities

***Illustration:* Times between Arrivals at a Toll Booth**

Figure 4-11 shows a graphical representation of the exponential distribution for the time T between successive arrivals of cars at a toll booth at one of the San Francisco Bay bridges. The random variable applies during a busy time of the day when arrivals occur on the average of 5 per minute. We see from the probability density curve in (a) that the modal value is 0 and that there is a single tail extending indefinitely to the right. As with all continuous probability distributions, probabilities for T are represented by areas under this curve. The area to

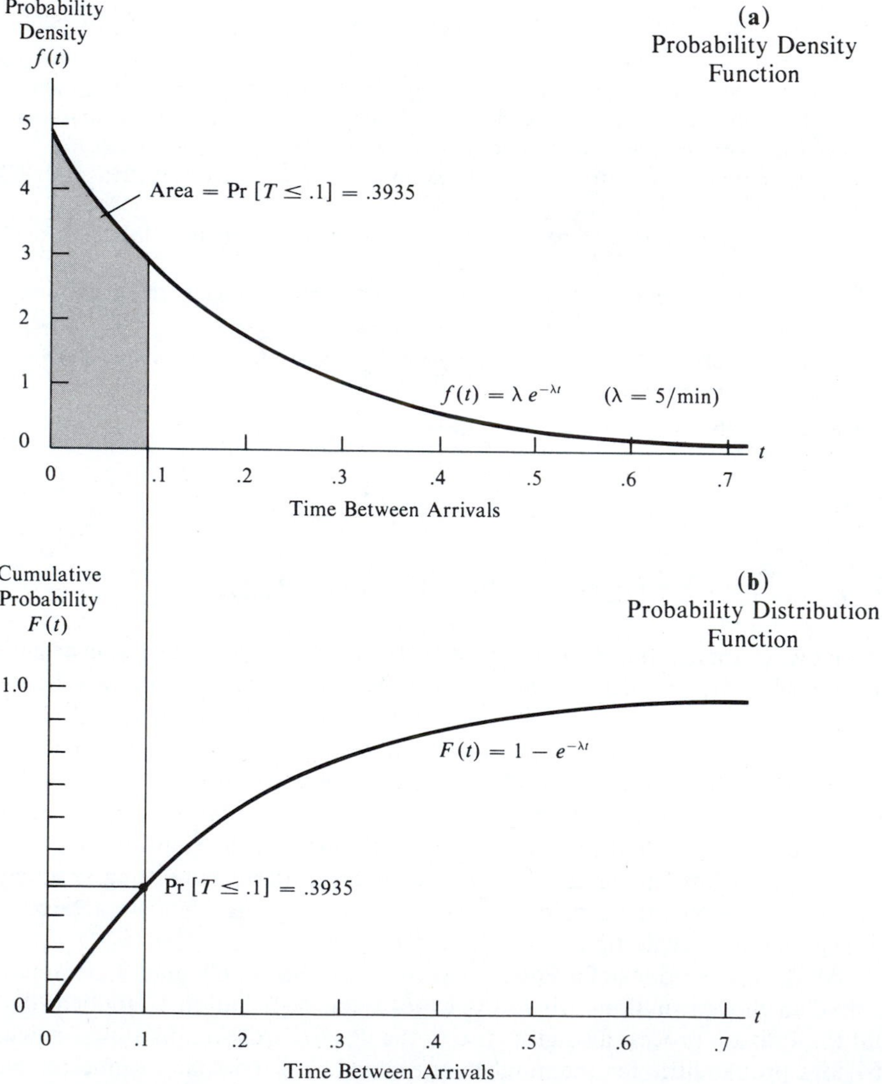

Figure 4-11 *Curves for exponential distribution representing time T between successive arrivals of cars at a toll booth.*

the left of t provides the ordinate for the exponential probability distribution curve at the bottom (b). The latter curve provides cumulative probabilities, which approach 1 as t becomes large.

An exponential distribution has a single parameter. This is the **mean process rate** λ, which expresses on the average how many events occur per unit of time. For the toll booth illustration this level is $\lambda = 5$ per minute. For a specific λ we use the following expression.

PROBABILITY DENSITY FUNCTION FOR
THE EXPONENTIAL DISTRIBUTION

$$\boldsymbol{f(t) = \begin{cases} \lambda e^{-\lambda t} & t \geq 0 \\ 0 & \text{otherwise} \end{cases}}$$

For the toll booth arrivals, $\lambda = 5$ and $f(t) = 5e^{-5t}$ when $t \geq 0$.

The probability that T will fall inside any interval may be found by integrating the function over that interval. We have

$$\Pr[a \leq T \leq b] = \int_a^b f(t)\,dt$$

When the lower limit is 0 and the upper limit is t, we have

$$\begin{aligned} \Pr[0 \leq T \leq t] &= \int_0^t \lambda e^{-\lambda\tau}\,d\tau \\ &= \int_{-\lambda t}^{0} e^x\,dx = e^x\Big]_{-\lambda t}^{0} = 1 - e^{-\lambda t} \end{aligned}$$

This gives the following function.

EXPONENTIAL PROBABILITY
DISTRIBUTION FUNCTION

$$\boldsymbol{F(t) = \Pr[T \leq t] = 1 - e^{-\lambda t}}$$

which applies for all $t \geq 0$. Values for e^{-y} may be read from Appendix Table B.

We may use the foregoing to calculate the probability that the time T between successive car arrivals at the toll plaza is less than or equal to 6 seconds, or .1 minute:

$$\begin{aligned} \Pr[T \leq .1] &= F(.1) = 1 - e^{-5(.1)} = 1 - e^{-.5} \\ &= 1 - .6065 = .3935 \end{aligned}$$

With $\lambda = 5$ cars per minute, it is quite rare for more than a full minute to transpire between arrivals:

$$\Pr[T > 1] = 1 - F(1) = e^{-5(1)} = .0067$$

But the exponential density has no upper limit, so there is a finite probability that several minutes could go by without any arrivals (although this quantity would be tiny).

Expected Value, Variance, and Percentiles

The average time or space between events is equal to the *reciprocal* of the mean rate at which they occur. And the variance is the square of this quantity. We use the following expression.

EXPECTED VALUE AND VARIANCE FOR AN EXPONENTIAL DISTRIBUTION

$$\mathbf{E(T) = \frac{1}{\lambda}}$$

$$\mathbf{Var(T) = \frac{1}{\lambda^2}}$$

For the car illustration, $\lambda = 5$ per minute. Thus, the expected time between arrivals is

$$E(T) = \frac{1}{5} = .2 \text{ minute}$$

and the variance is

$$\text{Var}(T) = \frac{1}{5^2} = .04$$

Taking the square root of the variance, we see that the standard deviation of T is equal to the expected value:

$$SD(T) = \sqrt{\frac{1}{\lambda^2}} = \frac{1}{\lambda} = E(T)$$

Percentiles for the exponential distribution may be found by reading Appendix Table B in reverse and using the fact that for any specified cumulative probability p, there is a corresponding negative power of e:

$$e^{-y} = 1 - p$$

so that

$$t = \frac{y}{\lambda}$$

The level for t that corresponds is the **pth percentile**. In our car arrival illustration we might set $p = .90$. Reading from Table B, the nearest negative power of e to $1 - .90 = .10$ is $y = 2.30$, so that

$$t = \frac{2.30}{\lambda} = \frac{2.30}{5} = .46 \text{ minute}$$

is the 90th percentile for the time between arrivals. This means that the probability is .90 that the time between any two successive arrivals will be .46 minute or less.

Applications of the Exponential Distribution

Recall that one property of a Poisson process is that occurrences in successive time periods are statistically independent events. In effect, *a Poisson process has no memory*. When applied to equipment failures, this requirement stipulates that, given an earlier failure, the chance of failure in the next 100 hours of operations must be the same as it would be if there had been no failure or if the operating history were unknown. It happens that many real-life reliability data demonstrate this characteristic, which is why the exponential distribution is so widely used in reliability evaluations.

Example: Major Electrical Power Disruptions

One electrical utility has infrequently experienced major disruptions to its power grid that cause temporary shutdowns of the entire system. These disturbances are caused by random events such as lightning, transformer failures, forest fires, and airplane crashes. Historically, the grid shutdowns fit the Poisson process, occurring on the average once every 2.5 years. In reliability such a value is often referred to as the **mean time between failures** (MTBF). The mean rate of major power disruption is thus

$$\lambda = \frac{1}{\text{MTBF}} = \frac{1}{2.5} = .40 \text{ per year}$$

What is the probability that there will be at least one disruption within the next year?

The desired event is equivalent to the time of the next disruption being 1 year or less. We have

$$\begin{aligned} \Pr[T \le 1] = F(.40) &= 1 - e^{-.4(1)} = 1 - e^{-.4} \\ &= 1 - .6703 = .3297 \end{aligned}$$

or slightly less than a 33% chance.

About half the disruptions occur because of a cascading effect arising from disturbances originating outside the company network. Special equipment can eliminate these, but the cost would be $1 million. The new investment would reduce λ to .20, so that the annual probability of disruption becomes

$$\begin{aligned} \Pr[T \le 1] = F(.20) &= 1 - e^{-.2(1)} = 1 - e^{-.2} \\ &= 1 - .8187 = .1813 \end{aligned}$$

Company management decided, however, that the benefit in increased reliability did not justify the increased cost.

The Poisson process not only provides a good representation for car arrivals at toll booths but characterizes an entire class of arrival patterns encountered in queueing analysis. In all cases the times between arrivals have an exponential distribution. An interesting special case arises when the time for service is also exponentially distributed.

Example:* *Waiting Times in Telephone Switching

As telephone calls are placed, they must be accepted by the telecommunications network and be routed to their destinations via clear channels. The switching circuitry encountered when a new call is placed may be busy processing earlier calls. While the circuits are busy, all incoming calls must be placed in a holding mode, in effect placed in a queue, until those earlier calls have been processed.

One central telephone exchange receives incoming calls at a rate of $\lambda = 2$ per second. An exponential distribution applies to the time between arriving calls. The equipment has the capacity of switching a call in an average time of .25 second, although with speed-of-light solid-state equipment the most frequent switching time is very near to 0. This feature, plus near adherence to a Poisson process, allows the switching time to be closely approximated by an exponential distribution with a mean rate of $\mu = 4$ calls per second.

Since the interarrival times and service times are both exponentially distributed, it can be established that the waiting time T_W of a particular call for switching to begin is also exponentially distributed, with a mean rate equal to the difference in service and arrival rates, $\mu - \lambda$. Thus, we may evaluate probabilities that an incoming call experiences various waiting times. In the present telephone system, we have $\mu - \lambda = 4 - 2 = 2$ so that the probability that less than 1 second of waiting time will be required is substantial,

$$\begin{aligned} \Pr[T_W \leq 1] &= F(1) = 1 - e^{-2(1)} = 1 - e^{-2} \\ &= 1 - .1353 = .8647 \end{aligned}$$

whereas the probability that there will be more than 3 seconds of waiting time is tiny:

$$\Pr[T_W > 3] = 1 - F(3) = e^{-2(3)} = e^{-6} = .0025$$

There is a danger in using the procedure in the foregoing example when evaluating more general queueing situations. Few service patterns are suitably represented by an exponential distribution. (Imagine a tool cage or barbershop where the most likely service time is zero!)

Problems

4-30 The time between accidents at a congested intersection has an exponential distribution with unknown parameter λ, which may assume values of (1) 1/hour, (2) 2/hour, or (3) 3/hour. Complete the following for each of these levels.

(a) Determine the probability that an accident will happen within 1.5 hours.
(b) Find the expected time between accidents.
(c) Find the variance in time between accidents.
(d) Find the 10th percentile for the time to an accident.
(e) Find the 90th percentile for the time to an accident.

4-31 Cars arrive at a toll plaza at a mean rate of 5 per minute between 7 A.M. and 8 A.M. Find the probability that starting at 7:30 A.M. the toll taker gets no cars until after (a) 7:31, (b) 7:32, (c) 7:30 plus 12 seconds, and (d) 7:30 plus 30 seconds.

4-32 Suppose the telephone switching circuit described in the example is improved, so that service is faster, with an average rate of $\mu = 5$ calls per second. Find the probability that a call's waiting time is

(a) less than 1 second (c) between .2 and .8 second

(b) greater than 2 seconds (d) greater than .5 second

4-33 Referring to the power example in the text, suppose that a disruption causes an average damage loss of \$500,000.

(a) Find the expected loss per year under (1) the present generating system and (2) the proposed improved system.

(b) What are the expected annual savings in damage claims with the improved equipment?

(c) How much time is expected before the proposed equipment improvements are paid for?

4-34 A hospital presently has one back-up emergency electric generator connected to auxiliary circuits that supply power to critical areas. During a blackout, such a unit allows continuation of surgical, emergency, and life-support systems. The mean time between failures ($1/\lambda$) of the generator is 100 hours.

(a) Find the probability that the emergency generator will fail during a 10-hour power blackout.

(b) Suppose a second identical emergency generator is used to provide service in parallel. They operate independently. Both must fail before critical service is halted. Find the probability that this event will occur during the next 10-hour blackout.

4-35 The central processor of a computer may be viewed as a service facility where successive microinstructions are arriving "customers." Suppose these form a Poisson process with mean arrival rate $\lambda = 10^6$ per second. While the processor is handling earlier arrivals, new instructions stack up in a buffer and must wait their turn. The time for completing an instruction (service) may be assumed to be exponentially distributed with a mean rate of 10^7 per second.

Find the probability that the time a particular microinstruction spends in the buffer is

(a) less than 100 nanoseconds (c) more than 1 microsecond

(b) more than 500 nanoseconds (d) between 300 and 400 nanoseconds

4-36 Find the exact probability that any exponentially distributed random variable T will fall within $E(T) \pm kSD(T)$ when (a) $k = 1$, (b) $k = 2$, and (c) $k = 3$.

4-37 Molecules of a toxic chemical eventually decompose into inert substances. Suppose the decomposition time is exponentially distributed with a mean of $1/\lambda$. The half-life of such a persistent poison is that time beyond which the probability is .50 that a particular molecule will remain toxic.

Find the half-life for chemicals whose molecules have an average decomposition time of (a) 1 year, (b) 5 years, (c) 25 years, and (d) 100 years.

4-38 The n power cells in a satellite will be arranged in parallel and will fail at a mean rate of .01 per day. They have independent lifetimes.

(a) Find the probability that any specific cell will fail on or before (1) 100 days, (2) 150 days, and (3) 500 days.

(b) Find the probability that there will be at least 1 cell still working after 600 days, assuming (1) $n = 2$, (2) $n = 5$, and (3) $n = 10$.

(c) What should n be to provide a 90% chance that the satellite's power source will survive at least 600 days?

4-39 Flaws in a reel of high-fidelity radar recording tape occur on the average of once every 10 feet.

(a) Determine the probability that the next recording will begin on a flawless stretch of tape over (1) 5 feet long, (2) 10 feet long, (3) 20 feet long, and (4) 50 feet long.
(b) Over what distance is there a 90% chance that an error will be encountered since the last one?

4-40 Show that for an exponentially distributed variable,

$$\Pr[T > t + h | T > t] = \Pr[T > h]$$

which illustrates that the Poisson process lacks a memory.

4-5 The Uniform Distribution

Although it has much narrower application than those discussed earlier in this chapter, the **uniform distribution** is one of the more important distributions in statistics. It characterizes random variables for which the probability of falling in a particular interval is proportional to that interval's width. The uniform distribution may therefore establish probabilities for the random values generated in computer simulations. Such quantities are similar to the random numbers statisticians use in selecting sample units. This same distribution family has been used in making assessments of damage from natural disasters, accidents, or warfare. For instance, the swath of destruction made by a tornado may have a uniform distribution throughout the area traced by a spawning storm.

Finding Uniform Probabilities

Figure 4-12 shows the probability density and probability distribution functions for the length X of a scrap-lumber segment created during construction of a home. All lengths between 1″ and 11″ are equally likely. (Smaller pieces are considered unmeasurable sawdust, whereas lengths greater than 11″ have further use.)

We use the following expression.

PROBABILITY DENSITY FUNCTION OF A UNIFORM DISTRIBUTION

$$f(x) = \begin{cases} \dfrac{1}{b - a} & a \leq x \leq b \\ 0 & \text{otherwise} \end{cases}$$

which is specified by two parameters, the lower limit a and the upper limit b. For the piece of scrap lumber $a = 1''$ and $b = 11''$. Because all values between a and b are equally likely, the cumulative probability that $X \leq x$ must equal the proportion of points in $[a, b]$ at or below x. Integrating the density, we have

$$\Pr[X \leq x] = \int_a^x \frac{1}{b - a}\,dy = \frac{y}{b - a}\Bigg]_a^x$$

which gives us the following expression.

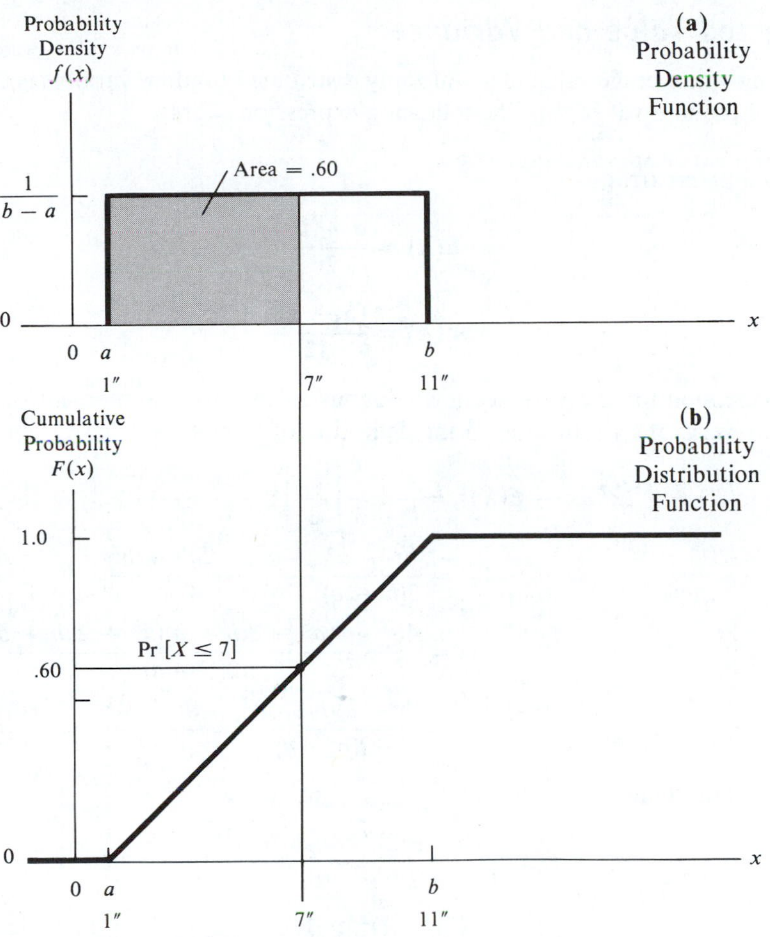

Figure 4-12 *Uniform distribution representing length X of a lumber scrap.*

UNIFORM PROBABILITY DISTRIBUTION FUNCTION

$$\mathbf{F(x) = Pr[X \le x] = \begin{cases} 0 & x < a \\ \dfrac{x-a}{b-a} & a \le x \le b \\ 1 & x > b \end{cases}}$$

The probability that a particular lumber scrap will be at most 7″ long is

$$\Pr[X \le 7] = F(7) = \frac{7-1}{11-1} = .60$$

and the probability that it will lie between 4 and 8 inches is

$$\Pr[4 \le X \le 8] = F(8) - F(4)$$

$$= \frac{(8-1)-(4-1)}{11-1} = .40$$

Expected Value and Variance

The long-run average value of a uniformly distributed random variable is the mid-point of the interval $[a, b]$. The following expressions apply.

EXPECTED VALUE AND VARIANCE FOR A UNIFORM DISTRIBUTION

$$\mathbf{E(X)} = \frac{a + b}{2}$$

$$\mathbf{Var\,(X)} = \frac{(b - a)^2}{12}$$

The expression for the variance is analogous to that for the moment of inertia. We may verify it by returning to the definition of $\text{Var}(X)$. We have

$$\begin{aligned}\text{Var}(X) = \int_a^b x^2\,dx - E(X)^2 &= \frac{1}{b - a}\left[\frac{x^3}{3}\right]_a^b - \left(\frac{a + b}{2}\right)^2 \\ &= \frac{b^3 - a^3}{3(b - a)} - \frac{a^2 + 2ab + b^2}{4} \\ &= \frac{4b^3 - 4a^3 - 3(b - a)(a^2 + 2ab + b^2)}{12(b - a)} \\ &= \frac{(b - a)(b - a)^2}{12(b - a)} = \frac{(b - a)^2}{12}\end{aligned}$$

For the scrap lumber we have

$$E(X) = \frac{1'' + 11''}{2} = 6''$$

$$\text{Var}(X) = \frac{(11 - 1)^2}{12} = \frac{100}{12} = 8.33''^2$$

$$SD(X) = \sqrt{8.33} = 2.89''$$

Example: ***How Much Street Pavement Needs Replacing?***

A paving contractor is paid by the county for jobs that pass a final inspection several weeks after completion. Often the inspector insists on patchwork, and sometimes substantial segments of pavement must be removed and laid again, owing to undermining of the loose fill layer. The length of segments identified to be repaved is a uniformly distributed random variable, the parameters of which depend on the terrain and the ground material.

The probability that a particular segment to be repaved is less than or equal to 20 feet long is shown below for a variety of cases. In each $a = 5$ feet (since small segments are treated as patches rather than as new pavement).

1. Steep hill, granite bedrock: $b = 25$.

$$\Pr[X \leq 20] = \frac{20 - 5}{25 - 5} = .75$$

2. Gradual hill, sandy ground: $b = 50$.

$$\Pr[X \leq 20] = \frac{20 - 5}{50 - 5} = .33$$

3. Flat, clay ground: $b = 10$.

$$\Pr[X \leq 20] = 1, \text{ since } x > b.$$

4. Flat, sandy ground: $b = 30$.

$$\Pr[X \leq 20] = \frac{20 - 5}{30 - 5} = .60$$

Problems

4-41 Suppose the paving contractor in the text example must repave a segment over flat terrain having clay soil. Find the probability that the segment length will be
(a) greater than 7.5 feet
(b) less than 6 feet
(c) between 7 and 9 feet
(d) less than or equal to 6.5 feet

4-42 Scrap wire returned by telephone installers will be recycled. It may be assumed that the reusable scraps are uniformly distributed from 1′ to 15′. (Shorter pieces are set aside for remelting.)
(a) For one piece taken at random find the (1) expected length, (2) variance, and (3) standard deviation.
(b) Determine the probability that the selected wire is (1) longer than 5′, (2) shorter than 10′, (3) between 2′ and 8′ long, and (4) between 5′ and 15′ long.

4-43 A random decimal is a value uniformly distributed over the interval $[0, 1]$.
(a) Find the probability that a random decimal will be (1) greater than .10, (2) less than .23, (3) between .35 and .65, and (4) greater than or equal to .95.
(b) Determine (1) the expected value, (2) the variance, and (3) the standard deviation.

4-44 A harried passenger will miss by several minutes the scheduled 10 A.M. departure time of his flight to London. Nevertheless, he might still make the flight, since boarding is always allowed until 10:10 A.M., and extended boarding is sometime permitted as long as 20 minutes after that time.

Assuming the extended boarding time is uniformly distributed over the above limits, find the probability that the passenger will make his flight, assuming he arrives at the boarding gate (a) at 10:05, (b) at 10:15, (c) at 10:25, and (d) at 10:35.

4-45 Consider again the random decimal in Problem 4-43. Find the probability that a value X will fall within $E(X) \pm kSD(X)$ when (a) $k = 1$, (b) $k = 2$, and (c) $k = 3$.

4-46 A pilot is navigating over the ocean by "dead reckoning" using a malfunctioning compass as the only navigational aid. It is night and overcast. The aircraft is presently 100 NM south of a wide stretch of nearly straight coastline, but there is only enough fuel for 150 NM of flight. The pilot heads "north," following his compass needle.
(a) Find the range of positions (relative to true north) toward which the compass needle must point in order for the aircraft to reach land on or before running out of fuel.
(b) Find the probability that land is reached in time, assuming that the direction of the compass arrow is uniformly distributed (1) over the entire circle, (2) over the entire 180° northern arc.

4-6 The Gamma Distribution

The **gamma distribution** is important because it includes a wide class of specific distributions, some of which underlie fundamental statistical procedures. In addition to serving as a utility distribution, the gamma provides probabilities for yet another random variable associated with Poisson processes. And, as we shall see, the exponential distribution itself is a member of the gamma distribution family.

The Gamma Function

In order to describe the gamma distribution in detail, we must first consider a useful mathematical function, the gamma function.

GAMMA FUNCTION

$$\Gamma(r) = \int_0^\infty x^{r-1}e^{-x}\,dx \qquad r > 0$$

This is used in defining several random variables. The symbol Γ (Greek uppercase *gamma*) is reserved for this function.

We may evaluate the gamma function for various levels of r. When $r = 1$, we have

$$\Gamma(1) = \int_0^\infty e^{-x}\,dx = -e^{-x}\Big]_0^\infty = 1$$

Next, consider $\Gamma(r + 1)$. Integrating by parts, it is easy to establish that

$$\Gamma(r + 1) = r\Gamma(r)$$

and similarly,

$$\Gamma(r) = (r - 1)\Gamma(r - 1)$$

Thus, for any nonnegative integer k, it follows that

$$\Gamma(k + 1) = k!$$

Noninteger values may apply for r. An important class involves values with halves. We have

$$\Gamma\left(\frac{1}{2}\right) = \sqrt{\pi}$$

and for any positive integer k

$$\Gamma\left(k + \frac{1}{2}\right) = \frac{1 \cdot 3 \cdot 5 \cdot \cdots \cdot (2k - 1)}{2^k}\sqrt{\pi}$$

The Probability Density Function

The following expression gives the probability density function for a gamma distribution.

PROBABILITY DENSITY FUNCTION FOR A GAMMA DISTRIBUTION

$$f(x) = \begin{cases} \dfrac{\lambda^r}{\Gamma(r)} x^{r-1} e^{-\lambda x} & x \geq 0 \\ 0 & \text{otherwise} \end{cases}$$

The two parameters λ and r may be any nonnegative values.

A special case of this function occurs when $r = 1$. We have

$$f(x) = \frac{\lambda^1}{\Gamma(1)} x^{1-1} e^{-\lambda x} = \lambda e^{-\lambda x}$$

which is the density function for the exponential distribution. The gamma gives rise to the chi-square distribution (described in Chapter 7) when $\lambda = \frac{1}{2}$ and $r = d/2$, with d being a nonnegative integer parameter.

The expected value and variance may be computed from

$$E(X) = \frac{r}{\lambda}$$

$$\text{Var}(X) = \frac{r}{\lambda^2}$$

Relation to Poisson Process

Recall that a Poisson process characterizes many circumstances wherein events occur over time or space. As we have seen, the time or space encountered between any two events has the exponential distribution. We might consider the segment of time or space occurring until some specified number of events has transpired. The size X of such a segment is a random variable having a gamma distribution with parameters λ representing the mean process rate and r denoting the specific number of events that must transpire as X is reached.

For instance, suppose $r = 2$. We have

$$\begin{aligned} \Pr[X \leq x] &= \int_0^x \frac{\lambda^2}{1!} y^{2-1} e^{-\lambda y}\, dy \\ &= \lambda^2 \int_0^x y e^{-\lambda y}\, dy = \lambda^2 \left[\frac{e^{-\lambda y}}{\lambda^2} (-\lambda y - 1) \right]_0^x \\ &= [-\lambda y e^{-\lambda y} - e^{-\lambda y}]_0^x = -\lambda x e^{-\lambda x} - e^{-\lambda x} - 0 + 1 \\ &= 1 - e^{-\lambda x}(1 + \lambda) \end{aligned}$$

Suppose car arrivals at a toll booth are a Poisson process with mean $\lambda = 5$ cars per minute. The probability that up to 1 minute will elapse until two cars have arrived is

$$\begin{aligned} \Pr[X \leq 1] &= 1 - e^{-5(1)}(1 + 5) \\ &= 1 - 0.006738(6) = .96 \end{aligned}$$

Problems

4-47 Determine the following values:
(a) $\Gamma(5)$ (c) $\Gamma(2)$ (e) $\Gamma(1.5)$ (g) $\Gamma(7.5)$
(b) $\Gamma(6)$ (d) $\Gamma(8)$ (f) $\Gamma(3.5)$ (h) $\Gamma(10.5)$

4-48 Find (1) the expected value, (2) the variance, and (3) the standard deviation for the gamma distributions having the following parameters:
(a) $\lambda = 2, r = 2$ (c) $\lambda = 5, r = 5$
(b) $\lambda = 2, r = 4$ (d) $\lambda = 5, r = 25$

4-49 Find the probability that the time taken for the next two cars to arrive at a toll booth will be 1 minute or less when (a) $\lambda = 1$ per minute, (b) $\lambda = 2$ per minute, and (c) $\lambda = 4$ per minute.

4-50 Consider the gamma distribution with parameters $\lambda = 1$ and $r = 5$. Find the probability that
(a) $X \leq 5$ (b) $X \leq 10$ (c) $5 \leq X \leq 10$

4-51 Consider the gamma distribution with parameters $\lambda = 4$ and $r = 3$. Find the probability that
(a) $X \leq 1$ (b) $X \leq 2$ (c) $1 \leq X \leq 2$

4-7 The Lognormal Distribution

Some variables may give rise to extremely large values. Such data will have positively skewed histograms or frequency curves with long, thin upper tails. When those data are transformed to a logarithmic scale, the resulting display might resemble the bell shape we associate with the normal curve.

The **lognormal distribution** may be useful for representing such variables. We denote the primary random variable by X. The natural logarithm gives a second random variable

$$Y = \ln(X)$$

When Y is normally distributed with mean μ and standard deviation σ, then X has the lognormal distribution with the same parameters. The expected value and variance are different, however, and are given by

$$E(X) = e^{\mu + \frac{\sigma^2}{2}}$$

$$\operatorname{Var}(X) = e^{2\mu + 2\sigma^2} - e^{2\mu + \sigma^2}$$

The lognormal distribution has limited applicability in reliability analysis, where it can be used to characterize maintenance time. (Chapter 18 discusses its shortcomings as a failure-time distribution.) It has natural applications when data are expressed on a logarithmic scale, which is the case in seismology.

Example: *Nuclear Test Ban Treaty Compliance*

To assess Soviet compliance with the Nuclear Test Ban Treaty, the United States employs teleseismic monitoring. The major uncertainty in the situation is the actual yield (in kilotons) of detonations. *Estimated yield* has been characterized as a lognormal random variable. One illustration* represents this in terms of "factor-of-2" uncertainty, with the mode lying at the assumed true yield level. Then, for example, when the given true yield is 150 kt, a specific lognormal density function provides a 95% probability that the estimated yield X will fall between a lower limit of $\frac{150}{2} = 75$ kt and an upper limit of $150 \times 2 = 300$ kt.

The normal curve for $Y = \ln(X)$ will be centered at its mode,

$$\ln(150) = 5.01064$$

which will also be the mean and median, so that $\mu = 5.01064$. The lower limit of 75 kt for X corresponds to a Y of

$$\ln(75) = 4.31749$$

and the upper limit 300 kt corresponds to

$$\ln(300) = 5.70378$$

The distance separating the upper limit from the mean of the normal curve for Y is

$$5.70378 - 5.01064 = .69314$$

The area above the upper limit is

$$\left(\frac{1}{2}\right)(1 - .95) = .025$$

which, from Table D, corresponds to $z = 1.96$. Equating the distance computed above to $z\sigma$,

$$1.96\sigma = .69314$$

it follows that

$$\sigma = \frac{.69314}{1.96} = .3536$$

is the standard deviation of the normal curve for Y.

We may now use $\mu = 5.01064$ and $\sigma = .3536$ to find various probabilities for the estimated yield X. For example, we find the probability that the estimated yield falls at or below 200 kt:

$$\begin{aligned}\Pr[X \leq 200] &= \Pr[Y \leq \ln(200)] \\ &= \Pr[Y \leq 5.29832] \\ &= \Phi\left(\frac{5.29832 - 5.01064}{.3536}\right) \\ &= \Phi(.81) \\ &= .7910\end{aligned}$$

And, the probability that the estimate falls above 250 kt is found as follows:

$$\begin{aligned}\Pr[X > 250] &= \Pr[X > \ln(250)] \\ &= \Pr[Y > 5.52146] \\ &= 1 - \Phi\left(\frac{5.52146 - 5.01064}{.3536}\right) \\ &= 1 - \Phi(1.44) \\ &= 1 - .9251 = .0749\end{aligned}$$

* *Source: Decision Framework for Evaluating Compliance with the Threshold Test Ban Treaty*, Lawrence Livermore National Laboratory, 1988.

Problems

4-52 A random variable Y is normally distributed with mean 4 and standard deviation .5. Find the probability that the lognormally distributed variable $X = e^Y$ falls
(a) below 20 (b) between 30 and 55 (c) above 60

4-53 A random variable X is lognormally distributed with parameters $\mu = 5$ and $\sigma = .4$.
(a) Find (1) $E(X)$ and (2) $\text{Var}(X)$.
(b) Find the following:
(1) $\Pr[X \leq 50]$ (3) $\Pr[X \geq 60]$
(2) $\Pr[X \leq 70]$ (4) $\Pr[X \geq 50]$

4-54 Explain why the natural logarithm of the mode for a lognormal distribution must be equal to the mode of the corresponding normal curve.

4-55 Use the factor-of-2 rule to find the parameters for a lognormally distributed estimate of nuclear yield when the true yield is 200 kt. Then find the probability that the estimated yield will exceed 300 kt.

4-8 Other Common Distributions

The following are several more distributions important to engineering applications of statistics. Detailed discussions of these distributions will be given later in the book, as the topics to which they apply are introduced.

Special-Purpose Distributions: The Weibull

Some applications of statistics involve distributions used primarily for special purposes. Chapter 18 describes reliability analysis. There the **Weibull distribution** is introduced to provide probabilities for failure times of components and systems.

Sampling Distributions

Another class of special distributions includes those that are needed primarily to facilitate a statistical procedure. Such a distribution may generate probabilities for obtaining a particular level for a sample statistic, such as $\bar{X}$ or s^2. Some may be used to approximate the true probabilities that would be generated by more cumbersome distributions. The following three utility distributions are presented in later chapters, each in a variety of applications.

The Student *t* Distribution The Student t distribution is specified by a single parameter and will be described in detail in Chapter 7. Like the normal distribution, the Student t is unimodal and it has a density function that plots as a symmetrical bell-shaped curve.

The Chi-Square Distribution The chi-square distribution is related to the gamma distribution. It is specified by a single parameter, but is generally unimodal and positively skewed. The chi-square distribution will be introduced in Chapter 7.

The *F* Distribution The *F* distribution is important for evaluations involving two or more variables. It is specified by two parameters, and will be introduced in detail in Chapter 7.

Comprehensive Problems

4-56 A new amplifier design is being tested for a radio telescope. Suppose that under a variety of test-signal patterns the average frequency until distortion is normally distributed with mean 35 kilohertz and a standard deviation of 2.5 kilohertz.

(a) Find the 95th percentile for the average distortion frequency.

(b) Determine the probability that the average distortion frequency level falls above that of the present amplifier, in which distortion takes place at an average frequency of 32 kilohertz.

4-57 The launch of a space probe is equally likely to occur at any time between 6 A.M. and 8 P.M. on the same date.

(a) What is the expected launch time?

(b) There is a 90% chance that launch will occur before what time?

(c) What is the probability that the probe will be launched between 9:30 and 10:30 A.M.?

4-58 Several sample observations will be obtained of paint-drying times. The average drying time may be assumed to be normally distributed with a mean of 2 hours and an unknown standard deviation.

(a) Assuming the standard deviation for average drying time is 15 minutes, determine the following percentiles:

(1) 25th (2) 90th (3) 95th (4) 99th

(b) Find the probability that the average drying time will lie between 1.85 and 2.25 hours assuming the standard deviation for that variable is (1) 15 minutes, (2) 10 minutes, and (3) 20 minutes.

4-59 A sample of 150 items is inspected from a population of 1,000. It is believed that 23% of the population is overweight. Assume that the sample items are selected randomly with replacement.

(a) Determine (1) the expected number of overweights to be found, (2) the variance for that quantity, and (3) the standard deviation.

(b) Determine the approximate probability that between 25 and 35 overweight items, inclusively, will be found in the sample.

4-60 Repeat Problem 4-59, assuming that sampling is instead performed without replacement.

4-61 A project engineer is planning to simulate a PERT network for the installation of a data-processing system. She requires a probability distribution for the time taken to complete a government approval activity. This is an uncertain quantity of undetermined probability distribution. Regardless of the distribution form, the expected activity completion time has been established at 20 working hours. She wishes to find the probability that approval will take between 15 and 30 hours.

(a) Approval could be perfunctory—so that it is most likely to require practically no time at all—or it could stretch out considerably for a seemingly indefinite time. What probability distribution would then be appropriate? Use it to find the desired probability.

(b) The pattern prevailing at another government agency might apply. There, approvals take a minimum of 5 hours and never exceed 35 hours—and all values in between those two extremes are equally likely. What probability distribution would then be appropriate? Use it to find the desired probability.

(c) The expected value and modal value might be the same, with values close to that figure more likely than more extreme approval times. Approximately 68% of the times fall within ± 4 hours of the central value. What probability distribution would then be appropriate? Use it to find the desired probability.

4-62 A data clerk makes entries into a central data bank at a mean rate of 120 per hour. Suppose that 2.5% of all entries are in error.

(a) What distribution is appropriate for finding the probability of exactly 30 errors in 750 entries? Find that value. (You may approximate.)

(b) What distribution is appropriate for determining the probability that the clerk will make no entries in the next minute? Find that value.

4-63 The logic wafers in a large computer fail according to a Poisson process with mean rate of 1 every 150 hours.

(a) Determine the probability that the next wafer failure will occur between 90 and 180 hours after the last one.

(b) The average time between events to be experienced in 9 failures may be closely approximated by a normal distribution having the same expected value as the above and a standard deviation of 50 hours. Find the probability that this average falls between the same limits as in (a).

4-64 Power cell failures in a satellite have exponentially distributed times with a mean rate of $\lambda = .01$ per day. Presently only 2 cells remain functioning. They are arranged in parallel and have independent lives, so that the satellite may function as long as at least 1 power cell works.

(a) Find the probability that a particular cell will survive for 200 or more days.

(b) Find the probability that the satellite continues to function for 200 or more days.

4-65 The functional reach of adult men is a normally distributed random variable with mean 32.33″ and standard deviation 1.63″. Consider a randomly chosen man.

(a) Find the probability that his functional reach exceeds 35″.

(b) A second man is chosen. What is the probability that both have reach exceeding 35″?

(c) Find the approximate probability that in a sample of size 100 the number of men having a reach exceeding 35″ is between 4 and 10, inclusively.

4-66 Repeat Problem 4-65, substituting the event "shorter than 30 inches."

4-67 The *measurement signal-to-noise ratio* of the random variable X is defined by

$$\frac{|E(X)|}{SD(X)}$$

Determine this quantity for each of the following:

(a) a normally distributed random variable with parameters $\mu = 10$ and $\sigma = 5$

(b) a uniformly distributed random variable with parameters $a = 5$ and $b = 25$

(c) an exponentially distributed random variable with parameter $\lambda = 2$

(d) a random variable having a gamma distribution with $\lambda = 5$ and $r = 2$

CHAPTER FIVE

Joint Probability Distributions and Expected Value Concepts

WITH THE BASIC concepts of probability firmly in hand, we are now ready to consider in greater detail how to treat several random variables occuring together. To set the stage, this chapter begins with some important properties of expected value. Next, we are introduced to the *joint probability distribution* for two variables, which allows us to extend the concept of statistical independence to random variables. The main portion of this chapter concludes with a discussion of the *sum* and *mean* of several independent random variables. This discussion lays the foundation for certain concepts fundamental to many important statistical procedures. An optional section extends the expected value concept even further through *moment generating functions.*

5-1 Properties of Expected Value and Variance

The concepts of expected value and variance extend to functions of random variables. The expected value of a function $g(X)$ of a random variable may be expressed as

$$E[g(X)] = \sum g(x)p(x)$$

for the discrete case and

$$E[g(X)] = \int_{-\infty}^{\infty} g(x)f(x)\,dx$$

for the continuous case.

The simplest case is when $g(X)$ is a linear function in X. The following is an important property of expected value.

PROPERTY OF EXPECTED VALUE

$$\mathbf{E(a + bX) = a + bE(X)}$$

It is easy to derive the above for the discrete case:

$$\begin{aligned} E(a + bX) &= \sum (a + bx)p(x) \\ &= \sum ap(x) + \sum bxp(x) \\ &= a\sum p(x) + b\sum xp(x) \\ &= a(1) + bE(X) \end{aligned}$$

The derivation is analogous when X is continuous. The case when $b = 0$ establishes that the expected value of a constant is equal to that constant.

This result is particularly important in engineering, where scale transformations often must be made in converting units. For example, if X is expressed in feet, then the expected value in inches is the expected value of $12X$, which is $12E(X)$. It is not necessary to transform each possible level for X to reach that result. Similarly, if a random variable T is in Celsius units, it may be converted to Fahrenheit by the linear function $32 + 1.8T$. The new expected value is obtained from the original by applying this transformation.

The variance is the expected value of the function

$$g(X) = [X - E(X)]^2 = X^2 - 2XE(X) + E(X)^2$$

This leads directly to the following.

EQUIVALENT EXPRESSION FOR THE VARIANCE

$$\mathbf{Var}(X) = E(X^2) - E(X)^2$$

This identity is useful for deriving results for the variance and also provides a shortcut procedure for establishing values.

Another useful result is the following.

PROPERTY OF THE VARIANCE

$$\mathbf{Var}(a + bX) = b^2\mathbf{Var}(X)$$

(You may derive this as an exercise.) This property tells us that the variance of a constant (the case when $b = 0$) is 0. It also tells us that when the units of X are transformed (say, from feet to inches, with $b = 12$), the resulting variance must be the original variance times the square of the scale change (so that when $b = 12$ inches per foot, the variance in the new units is $b^2 = 144$ square inches times the variance in square feet).

An Application to the Proportion of Successes

To further illustrate these properties, consider the proportion of successes achieved in n trials of a Bernoulli process:

$$P = \frac{R}{n}$$

The proportion is a linear transformation of R, the number of successes, with $a = 0$ and $b = 1/n$. In Chapter 3 it was established that

$$E(R) = n\pi \qquad \text{and} \qquad \mathrm{Var}(R) = n\pi(1 - \pi)$$

It follows that

$$E(P) = E\left(\frac{r}{n}\right) = \left(\frac{1}{n}\right)E(R) = \left(\frac{1}{n}\right)(n\pi) = \pi$$

and

$$\text{Var}(P) = \text{Var}\left(\frac{R}{n}\right) = \left(\frac{1}{n}\right)^2 \text{Var}(R)$$

$$= \frac{1}{n^2}[n\pi(1-\pi)] = \frac{\pi(1-\pi)}{n}$$

Thus, the expected proportion of successes is the trial success probability π. (This is not surprising, since the probability itself expresses the long-run frequency of successes.) Notice that $\text{Var}(P)$ is inversely proportional to the number of trials, so that the P's would tend to be more alike for a series of Bernoulli trials when $n = 100$, say, than they would be when $n = 10$.

Problems

5-1 A random variable X has a Poisson distribution with parameters $\lambda = .5$ and $t = 2$. For the transformed random variable $Y = -1 + 2X$ establish (a) $E(Y)$ and (b) $\text{Var}(Y)$.

5-2 For each of the following situations, find (1) $E(Y)$, (2) $\text{Var}(Y)$, and (3) $SD(Y)$:

Units of X	$E(X)$	$\text{Var}(X)$	Units of Y
(a) inches	69	9	feet
(b) feet	100	25	inches
(c) degrees C	100	25	degrees F
(d) degrees F	500	25	degrees C
(e) hours	5	.25	minutes
(f) minutes	30	36	hours

5-3 The mean angle of deflection X in a series of tests on cantilever beams is normally distributed, with parameters $\mu = .08$ and $\sigma = .02$ (radians). Determine (1) the expected value and (2) the variance for each of the following random variables:
(a) $Y = .10 + 5X$
(b) $Z = \dfrac{X - .08}{.02} = -4 + 50X$

5-4 Letting X denote a uniformly distributed random variable over $[10, 20]$, express (1) the expected value and (2) the variance for the following random variables:
(a) $Y = -5 + 2X$ (b) $Z = X/10$

5-5 Letting T denote an exponentially distributed random variable with parameter $\lambda = 10$, express (1) the expected value and (2) the variance for the following random variables:
(a) $Y = 2T$ (b) $Z = 5 + 3T$

5-6 Letting X represent the number of dots showing on the toss of a die cube, determine (1) the expected value and (2) the variance for each of the following random variables:
(a) $Y = X^2$ (b) $Z = 10 + 5X$ (c) $W = (X + 5)^2$

5-7 For discrete random variables, prove that

$$\text{Var}(X) = E(X^2) - E(X)^2$$

5-2 Joint Probability Distributions

Now that we have examined the basic concepts, we are ready to extend our survey of probability by considering two or more random variables together. This background is necessary to explain sampling experiments that involve multiple random observations of a single population. Each observed quantity is a distinct random variable, which we can denote by $X_1, X_2, \ldots, X_n$. Much of statistical inference is based on aggregations of these variables such as the mean or variance.

Direct applications of probability to evaluations in queueing, reliability, communications, and information theory also involve two or more random variables. For instance, a customer's waiting time in a simple queue may be evaluated as a composite of two other uncertain quantities—arrival time and service time. The time to failure of a complex system, such as a communications satellite, is a random variable whose characteristics might be determined from the known failure-time distributions for individual subsystems.

Often a major random variable is described in terms of one or more parameters that are uncertain. In those cases the parameters themselves are additional random variables that further complicate probability evaluations. For example, the number of defective items found by sampling production may have a binomial distribution with parameters π and n. But π itself is usually uncertain and may have a particular distribution, such as the normal. Or, n may be uncertain because of the difficulty in obtaining sample observations.

We begin our discussion of multiple random variables with a presentation of the *joint probability distribution*, often referred to as the *multivariate* probability distribution. This concept is analogous to the joint probability table presented in Chapter 2. Further analogies may be drawn. Conceptually similar to the marginal probabilities identified in the joint probability table, the joint probability distributions give rise to *marginal probability distributions.*

A joint probability distribution may involve any number of component random variables. To make our present discussions easier we investigate in detail the *bivariate* case, which involves just two random variables.

The Bivariate Probability Distribution

Consider two random variables X and Y arising out of a common sample space. To illustrate, we use the temperature X and rainfall Y for a randomly selected city. We may express the probability of any particular result as

$$p_{X,Y}(x, y) = \Pr[X = x \text{ and } Y = y]$$

The above expresses the **joint** or **bivariate probability mass function**, which is summarized for our example by the matrix in Table 5-1 and plotted in Figure 5-1. These data were summarized from rounded values based on the actual 30-year averages for 50 selected major U.S. cities during the month of May. The values $p_{X,Y}$ were found by counting the number of cities falling within the various groupings and dividing by the size of the sample space, 50. The most likely temperature-rainfall levels arise from the 14 cities having rounded temperatures of 65 degrees Fahrenheit and rainfalls of 4 inches:

$$p_{X,Y}(65, 4) = \frac{14}{50} = .28$$

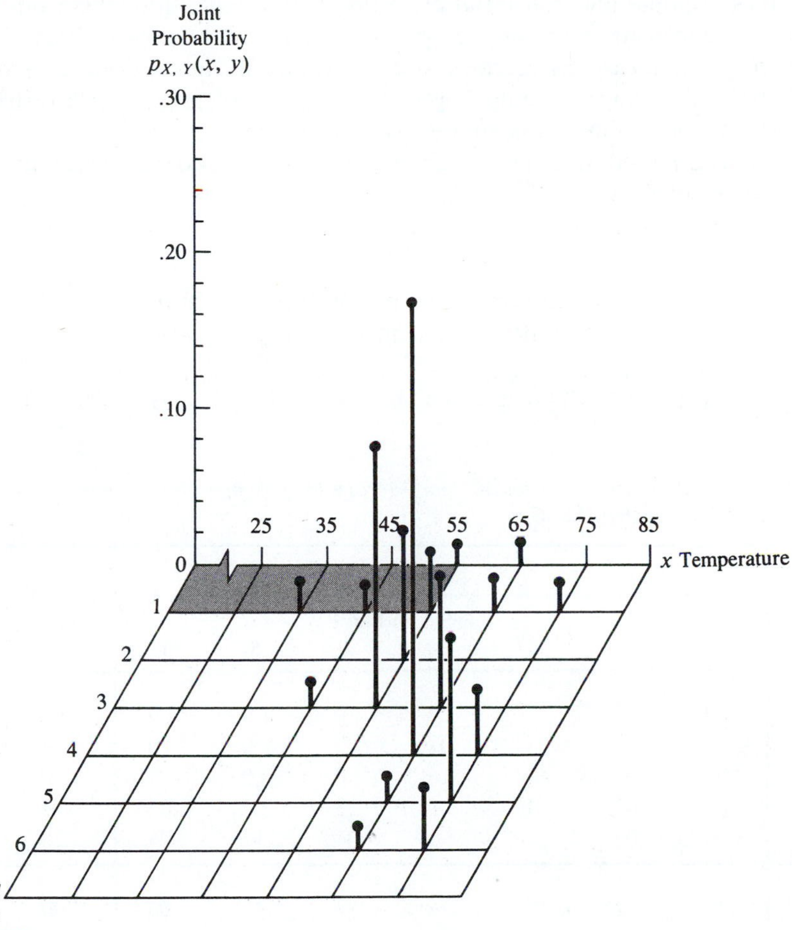

Figure 5-1 Bivariate probability distribution for May temperature X and rainfall Y of a randomly selected U.S. city.

This value is represented on Figure 5-1 by a spike of height .28, rising from the X-Y plane at the coordinates (65, 4).

The **bivariate probability distribution function** is defined in a manner similar to the one-variable case:

$$F_{X,Y}(x, y) = \Pr[X \leq x \text{ and } Y \leq y] = \sum_{x_i \leq x} \sum_{y_j \leq y} p_{X,Y}(x_i, y_j)$$

For our climate example, we have from Table 5-1:

$$\begin{aligned} F_{X,Y}(55, 1) &= \sum_{x_i \leq 55} \sum_{y_j \leq 1} p_{X,Y}(x_i, y_j) \\ &= .02 + .02 + .02 + .04 \\ &= .10 \end{aligned}$$

which represents the probability that the randomly selected city will have temperature and rainfall falling inside the shaded region in Figure 5-1.

This example illustrates a *discrete* bivariate distribution. Were the random variables to instead represent average May rainfall and temperatures for a *future* year, yet to be measured precisely and not rounded, then a *continuous* bivariate probability distribution would apply. The concept of multivariate distributions extends to higher dimensions (more than two variables).

For continuous random variables, the **joint** or **bivariate probability density function**, denoted by

$$f_{X,Y}(x, y)$$

provides the basic description of the probability distribution, and the bivariate probability distribution function is defined in the usual way:

$$F_{X,Y}(x, y) = \Pr[X \leq x \text{ and } Y \leq y] = \int_{-\infty}^{x} \int_{-\infty}^{y} f_{X,Y}(u, v)\, dv\, du$$

Table 5-1 *Joint Probability Distribution for 30-Year Average May Temperature X and Rainfall Y for a Randomly Selected U.S. City*

Temperature °F x	$p_{X,Y}(x, y)$ Rainfall (inches) y: 0	1	2	3	4	5	6	Sum $p_X(x)$
35	0	.02	0	0	0	0	0	.02
45	0	.02	0	.02	0	0	0	.04
55	.02	.04	.08	.16	0	0	0	.30
65	.02	.02	0	.08	.28	.02	.02	.44
75	0	.02	0	0	.04	.10	.04	.20
Sum $p_Y(y)$	.04	.12	.08	.26	.32	.12	.06	1.00

Source: *The World Almanac and Book of Facts*, 1988, p. 181

The density function may be obtained by taking successive partial derivatives of the above with respect to each variable:

$$f_{X,Y}(x, y) = \frac{\partial^2}{\partial x \, \partial y} F_{X,Y}(x, y)$$

Bivariate Normal Distribution

Important to a variety of statistical applications is the **bivariate normal distribution.** Figure 5-2 illustrates one possible frequency graph, wherein the density resembles the cone of a volcano. The probability density function in this case is expressed by

$$f_{X,Y}(x, y) = \frac{1}{2\pi\sigma_X\sigma_Y\sqrt{1-\rho^2}} \times \exp\left[-\frac{1}{2(1-\rho^2)}\left(\frac{(x-\mu_X)^2}{\sigma_X^2} - 2\rho\frac{(x-\mu_X)(y-\mu_Y)}{\sigma_X\sigma_Y} + \frac{(y-\mu_Y)^2}{\sigma_Y^2}\right)\right]$$

Each random variable has its own set of location and scale parameters: μ_X and σ_X, μ_Y and σ_Y. They share an additional parameter, the **correlation coefficient** ρ,

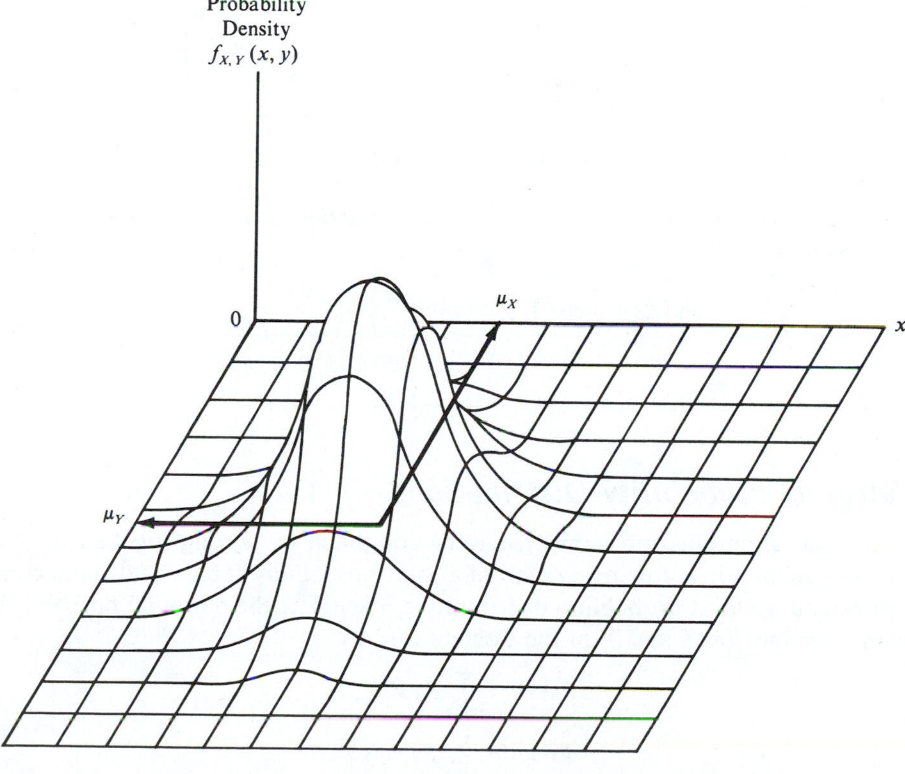

Figure 5-2 Probability density function of a bivariate normal distribution.

which summarizes the degree of *association* between X and Y. (The above requires that $\sigma_X > 0$, $\sigma_Y > 0$, and $-1 < \rho < 1$.)

The multivariate normal distribution has been applied to such phenomena as the velocity vectors of ideal gas particles (in three dimensions) and the heights of fathers and sons. In addition, it is often assumed to apply in a variety of sampling situations where multivariate analysis is employed.

***Example:* Satellite Position Prediction Errors**

Satellite tracking by radar ground stations is a complex task today, with so many objects in the sky. Many of these are continuously monitored by a network of radars. Because of changing conditions and orbit decay, the constants of physical motion must be frequently updated, and the tracking stations must forecast future satellite positions for the next tracking station. The forecasts involve range, azimuth, and elevation as a function of time. The latter two, expressed in degrees (or radians), provide parameters for synchronizing the position of the acquisition radar.

Let X represent the forecast error in azimuth and Y the error in elevation. Suppose that $\mu_X = \mu_Y = 0$, $\sigma_X = 5$ seconds, and $\sigma_Y = 2$ seconds. It is assumed that $\rho = 0$. The bivariate normal density function is

$$f_{X,Y}(x, y) = \frac{1}{20\pi} \exp\left(-\frac{1}{2}\left[\frac{x^2}{5^2} + \frac{y^2}{2^2}\right]\right)$$

Factoring this, the bivariate normal distribution function may be expressed as

$$F_{X,Y}(x, y) = \int_{-\infty}^{x} \frac{1}{\sqrt{2\pi}\,(5)} e^{-\frac{v^2}{2(25)}} \left[\int_{-\infty}^{y} \frac{1}{\sqrt{2\pi}\,(2)} e^{-\frac{u^2}{2(4)}}\, du\right] dv$$

or as

$$F_{X,Y}(x, y) = \Phi\left(\frac{x}{5}\right)\Phi\left(\frac{y}{2}\right)$$

Thus, the probability that the forecast azimuth and elevation errors do not each exceed $+3$ seconds is

$$\begin{aligned}\Pr[X \le 3 \textit{ and } Y \le 3] &= \Phi\left(\frac{3}{5}\right)\Phi\left(\frac{3}{2}\right)\\ &= (.7257)(.9332) = .677\end{aligned}$$

Marginal Probability Distributions

Recall that a marginal probability may be computed by adding together all the cell probabilities in a row or column of a joint probability table. That same concept extends to joint probability distributions. We define the marginal probability mass functions for X and Y in the discrete case by

$$p_X(x) = \sum_{\text{all } y} p_{X,Y}(x, y)$$

$$p_Y(y) = \sum_{\text{all } x} p_{X,Y}(x, y)$$

The marginal probability distributions in the city climate illustration are provided by the margins in the matrix in Table 5-1.

When the random variables are continuous, we define the marginal probability density functions analogously:

$$f_X(x) = \int_{-\infty}^{\infty} f_{X,Y}(x, y)\,dy$$

$$f_Y(y) = \int_{-\infty}^{\infty} f_{X,Y}(x, y)\,dx$$

As an illustration, consider the bivariate probability distribution having density

$$f_{X,Y}(x, y) = \begin{cases} \dfrac{3(x-y)^2}{8} & -1 \le x \le 1; -1 \le y \le 1 \\ 0 & \text{otherwise} \end{cases}$$

We must integrate over y to obtain the marginal probability density function for X:

$$\int_{-\infty}^{\infty} f_{X,Y}(x, y)\,dy = \frac{3}{8}\int_{-1}^{1} (x-y)^2\,dy = \frac{3}{8}\int_{-1}^{1} (x^2 - 2xy + y^2)\,dy$$

$$= \frac{3}{8}\left[x^2y - xy^2 + \frac{y^3}{3}\right]_{-1}^{1}$$

$$= \frac{3}{4}x^2 + \frac{1}{4}$$

Thus, we have

$$f_X(x) = \begin{cases} \dfrac{3x^2 + 1}{4} & -1 \le x \le 1 \\ 0 & \text{otherwise} \end{cases}$$

The marginal probability density for Y is identical to this.

It may be shown that when X and Y have a bivariate normal distribution, the marginal probability distributions for the respective variables are normally distributed, with parameters μ_X, σ_X and μ_Y, σ_Y, respectively.

Problems

5-8 Find the following levels of the joint probability distribution function using the probabilities from Table 5-1:

(a) $F_{X,Y}(55, 3)$ (c) $F_{X,Y}(65, 4)$
(b) $F_{X,Y}(45, 6)$ (d) $F_{X,Y}(75, 6)$

5-9 Find the following levels of the joint probability distribution function for the coordinates of a randomly selected city in Table 5-2:

(a) $F_{X,Y}(95, 30)$ (c) $F_{X,Y}(115, 30)$
(b) $F_{X,Y}(95, 45)$ (d) $F_{X,Y}(85, 40)$

5-10 For the prediction errors of the satellite position example, find the following joint probabilities (values are in seconds):

(a) $\Pr[X > 2 \text{ and } Y > -1]$ (c) $\Pr[X \le .5 \text{ and } Y > 1]$
(b) $\Pr[X \le 1 \text{ and } Y > 2]$ (d) $\Pr[X > 2.5 \text{ and } Y \le -1]$

Table 5-2 *Bivariate Probability Distribution for Longitude and Latitude Coordinates of a Randomly Selected City*

	$p_{X,Y}(x, y)$								
	Latitude y								Sum
Longitude x	20	25	30	35	40	45	50	55	$p_X(x)$
70	0	0	0	0	.02	.01	0	0	.03
75	0	0	0	.02	.10	.02	0	0	.14
80	0	.01	.03	.04	.03	.07	0	0	.18
85	0	0	.01	.05	.09	.02	0	0	.17
90	0	0	.03	.01	.04	.01	0	0	.09
95	0	.01	.02	.04	.02	.01	0	0	.10
100	.03	.01	.01	0	0	0	0	0	.05
105	.01	0	.03	.01	.01	0	0	0	.06
110	0	0	.01	.01	.01	0	0	0	.03
115	0	0	0	.03	0	0	.01	.01	.05
120	0	0	0	.04	.03	0	.01	0	.08
125	0	0	0	0	0	.01	.01	0	.02
Sum $p_Y(y)$	.04	.03	.14	.25	.35	.15	.03	.01	1.00

5-11 A human-factors engineer designing aircraft controls assumes that the arm reach X and sitting height Y for adult men have a bivariate normal distribution with the following parameters:

$$\mu_X = 35'' \qquad \mu_Y = 36''$$
$$\sigma_X = 1.6'' \qquad \sigma_Y = 1.3''$$
$$\rho = .8$$

(a) Express the joint probability density function.
(b) Find the following marginal probabilities for a randomly selected man:
(1) Sitting height $> 38''$ (3) Arm reach $\leq 34''$
(2) Arm reach $> 38''$ (4) Sitting height $\leq 34''$

5-12 The joint probability density function for X and Y is

$$f_{X,Y}(x, y) = \begin{cases} \dfrac{1}{30} & 0 \leq x \leq 5; 0 \leq y \leq 6 \\ 0 & \text{otherwise} \end{cases}$$

(a) Find the following:
(1) $F_{X,Y}(4, 5)$ (2) $F_{X,Y}(2, 3)$ (3) $\Pr[2 \leq X \leq 4 \text{ and } 4 \leq Y \leq 5]$
(b) Determine the marginal probability density function (1) for X and (2) for Y.

5-13 Consider the following joint probability density fuction:

$$f_{X,Y}(x, y) = \begin{cases} x + y & 0 \leq x \leq 1; 0 \leq y \leq 1 \\ 0 & \text{otherwise} \end{cases}$$

(a) Determine the following:
(1) $\Pr[X \leq \frac{1}{2} \text{ and } Y \leq \frac{1}{2}]$ (3) $\Pr[\frac{1}{4} \leq X \leq \frac{1}{2} \text{ and } \frac{1}{4} \leq Y \leq \frac{1}{2}]$
(2) $\Pr[X \leq \frac{1}{2} \text{ and } Y > \frac{1}{4}]$
(b) Find the marginal probability density function (1) for X and (2) for Y.

5-14 Consider a sampling experiment in which the joint probability mass function for the number of sample observations N and the number of sample successes R is provided by

$$p_{R,N}(r, n) = \frac{e^{-2}}{r!(n-r)!} \qquad r = 0, 1, 2, \ldots, n;\ n = 0, 1, 2, \ldots$$

Determine the following:
(a) $\Pr[R = 2 \text{ and } N = 5]$
(b) $\Pr[R = 5 \text{ and } N = 10]$
(c) $\Pr[R \leq 3 \text{ and } N = 5]$

5-3 Conditional Probability Distributions, Multiplication Law, Independence, and Covariance

Now that we have seen how to characterize probability distributions involving more than one random variable, we are ready for the complications arising from the interrelationships between those variables. This may be reflected in the probabilities of one variable, which may be affected by the level of another variable. The resulting distribution is called a *conditional probability distribution*, and it may be obtained from a joint probability distribution in the same way that conditional probabilities for simple events might be computed directly from the joint probability table.

In Chapter 2 we saw how important statistical independence is in determining procedures for computing probabilities. We are now ready to extend the concept of independence to random variables. When it exists, independence may considerably simplify finding a joint probability distribution, allowing us to combine the probability distributions of the component random variables directly. In this respect, statistical independence allows us to do the same thing as in Chapter 2—multiply two marginal probabilities to find a joint probability.

We may express the degree to which dependent random variables are associated. The two measures for doing this will be described. These are fundamental quantities in statistical applications involving more than one variable.

Conditional Probability Distributions

A **conditional probability distribution** characterizes the probabilities of one random variable when the level of the other random variable is fixed at some given value. Recall that for any two events A and B the conditional probability of one event may be found by dividing their joint probability by the probability of the given event:

$$\Pr[A \mid B] = \frac{\Pr[A \text{ and } B]}{\Pr[B]}$$

Extending this concept to random variables, we have the conditional probability mass function for X given $Y = y_j$ obtained from

$$p_{X|Y}(x \mid y_j) = \Pr[X = x \mid Y = y_j] = \frac{p_{X,Y}(x, y_j)}{p_Y(y_j)}$$

which is found by dividing the probability mass function for the joint distribution by that for the marginal distribution for Y when evaluated at $Y = y_j$. In keeping with our previous notation, the subscript $X \mid Y$ denotes a conditional probability distribution.

Using the city climate distribution in Table 5-1, we compute the following conditional probability values:

$$p_{X|Y}(55 \mid 3) = \frac{p_{X,Y}(55, 3)}{p_Y(3)} = \frac{.16}{.26} = .615$$

$$p_{X|Y}(45 \mid 1) = \frac{p_{X,Y}(45, 1)}{p_Y(1)} = \frac{.02}{.12} = .167$$

When X and Y are continuous, we may obtain the conditional probability density function for one random variable given another by dividing the joint density by the marginal density of the given variable:

$$f_{X|Y}(x \mid y) = \frac{f_{X,Y}(x, y)}{f_Y(y)}$$

This is valid whenever $f_Y(y) > 0$.

As an illustration, consider again the densities on page 213:

$$f_{X,Y}(x, y) = \frac{3(x - y)^2}{8}$$

$$f_Y(y) = \frac{3y^2 + 1}{4}$$

These apply for all x and y in $[-1, 1]$ and are zero-valued otherwise. When $f_Y(y) > 0$, the conditional probability density function for X given Y is

$$f_{X|Y}(x \mid y) = \begin{cases} \dfrac{\dfrac{3(x - y)^2}{8}}{\dfrac{3y^2 + 1}{4}} = \dfrac{3(x - y)^2}{6y^2 + 2} & -1 \le x \le 1; -1 \le y \le 1 \\ 0 & \text{otherwise} \end{cases}$$

When X and Y have the bivariate normal distribution with parameters ρ, μ_X, σ_X, μ_Y, and σ_Y, it can be shown that the conditional distribution of Y given

X has the density

$$f_{Y|X}(y|x) = \frac{1}{\sqrt{2\pi}\,\sigma_Y\sqrt{1-\rho^2}} \exp\left[-\frac{1}{2\sigma_Y^2(1-\rho^2)}\left(y - \mu_Y - \rho\frac{\sigma_Y}{\sigma_X}(x-\mu_X)\right)^2\right]$$

For any given level for x, this is the density for a normal distribution with mean $\mu_Y + \rho(\sigma_Y/\sigma_X)(x - \mu_X)$ and standard deviation $\sigma_Y\sqrt{1-\rho^2}$.

Example: Testing Spatial Conceptualization and Manual Dexterity

One large architectural and engineering firm gives a battery of tests to all applicants for entry-level design positions. Two important indicators of success are scores in manual dexterity X and spatial conceptualization Y, which are considered together. Past data suggest that the scores of any particular applicant may be represented by a bivariate normal distribution, with parameters

$$\mu_X = 75 \qquad \mu_Y = 60$$
$$\sigma_X = 10 \qquad \sigma_Y = 15$$
$$\rho = .6$$

The conditional distribution for spatial conceptualization score Y, given a manual dexterity score of $X = x$, is normal, with

$$\mu = 60 + .6(\tfrac{15}{10})(x - 75) = .9x - 7.5$$
$$\sigma = 15\sqrt{1 - (.6)^2} = 12$$

This indicates that manual dexterity has a strong influence on a person's ability to perceive spatial relationships. Consider two applicants, one scoring low on manual dexterity and the other high:

Applicant A	Applicant B
$x = 50$	$x = 90$
$\mu = .9(50) - 7.5 = 37.5$	$\mu = .9(90) - 7.5 = 73.5$
$\sigma = 12$	$\sigma = 12$

Manual dexterity is directly important only in mechanical graphics, which the firm contracts out. But each new hire must score at least 80 in spatial conceptualization to be accepted. We have

Applicant A	Applicant B
$\Pr[Y > 80 \mid X = 50]$	$\Pr[Y > 80 \mid X = 90]$
$= 1 - \Phi\left(\frac{80 - 37.5}{12}\right)$	$= 1 - \Phi\left(\frac{80 - 73.5}{12}\right)$
$= 1 - \Phi(3.54)$	$= 1 - \Phi(.54)$
$= 1 - .99977 = .00023$	$= 1 - .7054 = .2946$

The odds are heavily stacked against "Mr. Five Thumbs."

The influence of the manual dexterity score on the spatial conceptualization probability illustrates the X and Y are *dependent* random variables.

Building Joint Probability Distributions Using the Multiplication Law

Recall that the joint probability for any two events may be found using the multiplication law,

$$\Pr[A \text{ and } B] = \Pr[A] \times \Pr[B|A]$$

This extends to random variables and bivariate probability distributions. In the discrete case we may obtain the bivariate probability mass function from the product

$$p_{X,Y}(x, y) = p_X(x)p_{Y|X}(y|x)$$

The bivariate probability density function is found analogously:

$$f_{X,Y}(x, y) = f_X(x)f_{Y|X}(y|x)$$

Example: *Sample Defectives from an Occasionally Malfunctioning Press*

When operating properly, a press stamps parts with only a 5% defective rate. On occasion, the control mechanism jams and the defective rate doubles. If we assume that jamming occurs 10% of the time, and let X denote the process proportion defective, the following marginal probability distribution applies:

x	$p_X(x)$
.05	.90
.10	.10

If we further assume that the characteristics of successive parts are statistically independent events, one of two binomial probability distributions applies for the number of defectives Y in a sample. Since the error rate X is unknown, these are *conditional* probability distributions:

$$p_{Y|X}(y|x) = b(y; n, x) = C_y^n x^y(1 - x)^{n-y}$$

Finally, we assume that $n = 4$ items are observed. Then, by taking the difference in successive cumulative probabilities, Appendix Table A gives the following:

y	$p_{Y\|X}(y\|.05)$ $b(y; 4, .05)$	$p_{Y\|X}(y\|.10)$ $b(y; 4, .10)$
0	.8145	.6561
1	.1715	.2916
2	.0135	.0486
3	.0005	.0036
4	.0000	.0001
	1.0000	1.0000

Applying the multiplication law, we find the probability that $X = .05$ and $Y = 2$:

$$p_{X,Y}(.05, 2) = p_X(.05)p_{Y|X}(2|.05)$$
$$= .90(.0135) = .01215$$

Applying the multiplication law to all cases, we obtain the following table for completely specifying the joint probability mass function:

x \ y	$p_{X,Y}(x, y)$					
	0	1	2	3	4	$p_X(x)$
.05	.73305	.15435	.01215	.00045	.00000	.90000
.10	.06561	.02916	.00486	.00036	.00001	.10000
$p_Y(y)$	.79866	.18351	.01701	.00081	.00001	1.00000

The marginal probability mass function for the number of defectives in any sample of size $n = 4$ is shown in the bottom margin. We see that there is an overall chance of nearly 80% that no defectives will be found in any particular sample.

The quality assurance engineer needs to find the conditional probability distribution for the error rate X given that three defectives are found. He calculates

$$p_{X|Y}(.05|3) = \frac{p_{X,Y}(.05, 3)}{p_Y(3)} = \frac{.00045}{.00081} = .56$$

It is easy to see that

$$p_{X|Y}(.10|3) = 1 - .56 = .44$$

Given that $Y = 3$ sample defectives are found, there is a .44 probability that the press is malfunctioning. With $Y = 4$, it is 1.

Independent Random Variables

We have defined two events to be **independent** if the probability of either one is unaffected by the occurrence of the other. Independence between A and B may be established by showing that the conditional probability of one event given the other equals that event's unconditional probability,

$$\Pr[A|B] = \Pr[A]$$

If A and B are independent, it follows by the multiplication law that

$$\Pr[A \text{ and } B] = \Pr[A] \times \Pr[B]$$

The same concept extends to random variables. We define two random variables to be **mutually independent** whenever the joint probability distribution function is equal to the product of the marginal probability distribution functions:

$$F_{X,Y}(x, y) = F_X(x)F_Y(y)$$

And whenever this holds for two continuous random variables, so must the following apply to the respective probability density functions:

$$f_{X,Y}(x, y) = f_X(x)f_Y(y)$$

Likewise, if X and Y are independent discrete random variables, the following must hold for their probability mass functions:

$$p_{X,Y}(x, y) = p_X(x)p_Y(y)$$

From this it follows that X and Y are independent whenever

$$f_X(x) = f_{X|Y}(x \mid y) \quad \text{for any } y$$

in the continuous case, and

$$p_X(x) = p_{X|Y}(x \mid y_j) \quad \text{for any } y_j$$

in the discrete case. That is, X and Y are independent whenever the marginal probability distribution of one variable is identical to that variable's conditional probability distribution given any possible level for the other variable.

Should X and Y not satisfy these criteria, they are **dependent** random variables. The property of independence extends to any number of jointly distributed random variables.

Earlier, we considered the densities

$$f_{X,Y}(x, y) = \frac{3(x - y)^2}{8} \qquad -1 \leq x \leq 1; -1 \leq y \leq 1$$

$$f_X(x) = \frac{3x^2 + 1}{4} \qquad -1 \leq x \leq 1$$

$$f_Y(y) = \frac{3y^2 + 1}{4} \qquad -1 \leq y \leq 1$$

which all assume levels of 0 outside the stipulated ranges. The random variables X and Y are dependent, since

$$\frac{3(x - y)^2}{8} = f_{X,Y}(x, y) \neq f_X(x)f_Y(y) = \left(\frac{3x^2 + 1}{4}\right)\left(\frac{3y^2 + 1}{4}\right)$$

This fact is also evident, since

$$\frac{3(x - y)^2}{6y^2 + 2} = f_{X|Y}(x \mid y) \neq f_X(x) = \frac{3x^2 + 1}{4}$$

In the satellite position prediction example in Section 5-2, the forecast errors in azimuth X and in elevation Y were statistically independent. We may verify this by noting that the respective marginal densities are normal and are expressed as

$$f_X(x) = \frac{1}{\sqrt{2\pi}\,5} \exp\left(-\frac{x^2}{2(5)^2}\right) \qquad -\infty < x < \infty$$

$$f_Y(y) = \frac{1}{\sqrt{2\pi}\,2} \exp\left(-\frac{y^2}{2(2)^2}\right) \qquad -\infty < y < \infty$$

The product of these provides the density given for the applicable bivariate normal distribution:

$$\begin{aligned} f_X(x)f_Y(y) &= \left[\frac{1}{\sqrt{2\pi}\,5}\exp\left(-\frac{x^2}{2(5)^2}\right)\right]\left[\frac{1}{\sqrt{2\pi}\,2}\exp\left(-\frac{y^2}{2(2)^2}\right)\right] \\ &= \frac{1}{20\pi}\exp\left(-\frac{1}{2}\left[\frac{x^2}{5^2}+\frac{y^2}{2^2}\right]\right) \\ &= f_{X,Y}(x, y) \end{aligned}$$

As noted, the marginal probability distributions of the individual variables sharing a bivariate normal distribution will themselves be normal. But unless the correlation coefficient has a value of $\rho = 0$ (as in our satellite position prediction example), $f_{X,Y}(x, y)$ will not be equal to the product of the respective marginal densities. Whenever $\rho \neq 0$, X and Y will be statistically dependent (as is the case in the testing example).

The multiplication law for independent random variables extends to higher dimensions. In general, when there are n independent random variables $X_1, X_2, \ldots, X_n$, the **multivariate probability density function** is the product of the respective individual densities:

$$f_{X_1X_2\ldots X_n}(x_1, x_2, \ldots, x_n) = f_{X_1}(x_1)f_{X_2}(x_2)\ldots f_{X_n}(x_n)$$

Expected Value and Covariance

The concept of expected value extends to multiple random variables. Generally any function of two random variables

$$g(X, Y)$$

will have an expected value defined by

$$E[g(X, Y)] = \sum_{\text{all } x}\sum_{\text{all } y} g(x, y)p_{X,Y}(x, y)$$

when the random variables are discrete, and by

$$E[g(X, Y)] = \int_{-\infty}^{\infty}\int_{-\infty}^{\infty} g(x, y)f_{X,Y}(x, y)\,dx\,dy$$

when they are continuous.

The most important bivariate expected value in statistics is the **covariance**, which provides a summary of the degree to which two random variables are associated. This is defined as

$$\text{Cov}(X, Y) = E([X - E(X)][Y - E(Y)])$$

The individual expected values $E(X)$ and $E(Y)$ are defined in the usual manner using the respective *marginal* probability distributions. Those distributions also provide $\text{Var}(X)$ and $\text{Var}(Y)$. It is usually more convenient, however, to use the following equivalent expression.

COVARIANCE OF X AND Y

$$\mathbf{Cov(X, Y) = E(XY) - E(X)E(Y)}$$

To evaluate $E(XY)$ we must use the bivariate probability distribution and obtain the expected value when $g(X, Y) = XY$.

As an illustration, consider again the major city climate example. Using the probability data from before, Table 5-3 has been constructed. In the margins we make the expected value calculations,

$$E(X) = \sum_{\text{all } x} x p_X(x) = 62.60$$

$$E(Y) = \sum_{\text{all } y} y p_Y(y) = 3.30$$

The entries in the body of the table represent the product

$$xyp_{X,Y}(x, y)$$

For instance, when $x = 55$ and $y = 3$, this product is equal to

$$55(3)(.16) = 26.40$$

The sum of the entries in the cells provides $E(XY)$:

$$E(XY) = \sum_{\text{all } x} \sum_{\text{all } y} xyp_{X,Y}(x, y) = 214.70$$

The covariance is thus

$$\begin{aligned}\text{Cov}(X, Y) &= E(XY) - E(X)E(Y) \\ &= 214.70 - (62.60)(3.30) = 8.12\end{aligned}$$

An important fact is that *independent random variables have zero covariance.* In those cases no association or correlation exists between X and Y. Knowledge of the level for one random variable has no effect on probabilities for the other.

Table 5-3 *Covariance Calculations for the City Climate Illustration*

	$xyp_{X,Y}(x, y)$									
Temperature °F x	**Rainfall (inches) y**									
	0	**1**	**2**	**3**	**4**	**5**	**6**	$p_X(x)$	$xp_X(x)$	$x^2p_X(x)$
35	0	.70	0	0	0	0	0	.02	.70	24.50
45	0	.90	0	2.70	0	0	0	.04	1.80	81.00
55	0	2.20	8.80	26.40	0	0	0	.30	16.50	907.50
65	0	1.30	0	15.60	72.80	6.50	7.80	.44	28.60	1,859.00
75	0	1.50	0	0	12.00	37.50	18.00	.20	15.00	1,125.00
$p_Y(y)$	.04	.12	.08	.26	.32	.12	.06	1.00	62.60 $= E(X)$	3,997.00 $= E(X^2)$
$yp_Y(y)$	0	.12	.16	.78	1.28	.60	.36	3.30 $= E(Y)$		
$y^2p_Y(y)$	0	.12	.32	2.34	5.12	3.00	2.16	13.06 $= E(Y^2)$		

The Correlation Coefficient

The covariance may be transformed into a useful index of association. This is the **correlation coefficient**, which expresses the relative strength of the association between X and Y. We use the following expression to calculate it.

CORRELATION COEFFICIENT FOR X AND Y

$$\rho = \frac{\text{Cov}(X, Y)}{\sqrt{\text{Var}(X)\,\text{Var}(Y)}} = \frac{E(XY) - E(X)E(Y)}{SD(X)SD(Y)}$$

Consider again the climate data. Table 5-3 provides the variances for X and Y:

$$E(X^2) = \sum_{\text{all } x} x^2 p_X(x) = 3{,}997.00$$

$$E(Y^2) = \sum_{\text{all } y} y^2 p_Y(y) = 13.06$$

$$\text{Var}(X) = E(X^2) - E(X)^2 = 3{,}997.00 - (62.60)^2 = 78.24$$

$$\text{Var}(Y) = E(Y^2) - E(Y)^2 = 13.06 - (3.30)^2 = 2.17$$

The correlation coefficient is

$$\rho = \frac{8.12}{\sqrt{(78.24)(2.17)}} = .623$$

This value will still apply regardless of the units used to scale X and Y. Had temperature been expressed in degrees Celsius and rainfall in centimeters, ρ would be exactly the same value. This feature makes the correlation coefficient a better measure of association than the covariance, the value of which will fluctuate with the units chosen.

The correlation coefficient computed above is positive, indicating that higher temperatures tend to occur with greater rainfall levels. If the relationship between X and Y were represented on a graph by a straight line, positioned on the X-Y plane at an angle suggested by the spikes in Figure 5-1, then the line would have positive slope. Of course, the correlation coefficient might instead be negative. This is what we would expect for two variables such as remaining aircraft fuel and flight distance. The line for that X, Y pair would have negative slope.

When $|\rho| = 1$, X and Y are *perfectly correlated.* The closer $|\rho|$ is to 1, the stronger is the association between X and Y. With $\rho = .623$, the *May* temperatures and rainfalls (precipitation levels) exhibit a stronger association than might be expected for a *summer* or *winter* month, either of which—because of geographical diversity and climate extremes—might have a considerably smaller ρ. The corresponding bivariate probability mass functions would plot with more scattered spikes, and a linear relationship between X and Y would be harder to see, if one were detectable at all. When $\rho = 0$, no linear relationship exists.

The correlation coefficient will be 0 if X and Y are independent. As we shall see later in this book, ρ is an important measure in statistical inference. The sign of ρ matches the sign of $\text{Cov}(X, Y)$. When $\rho > 0$, variables X and Y are *positively correlated*, which is the case when larger values for X are more likely to occur with larger Y's and smaller X's with smaller Y's. When $\rho < 0$, X and Y are *negatively correlated*, reflecting a tendency for larger X's to occur with the smaller Y's.

In the aptitude-testing example, the correlation coefficient $\rho = .6$ is one of the parameters describing the underlying bivariate normal distribution. In that example, we saw that the manual dexterity score X significantly influences the probabilities for the spatial conceptualization score Y. Had ρ been smaller, the conditional probability distributions for Y would have been more alike, and X would have been a poorer predictor of Y.

Problems

5-15 Determine the values for the following conditional probabilities, using the climate data from Table 5-1:

(a) $\Pr[X = 55 \mid Y = 2]$ (c) $\Pr[Y = 4 \mid X = 65]$
(b) $\Pr[X = 75 \mid Y = 3]$ (d) $\Pr[Y = 1 \mid X = 55]$

5-16 Determine the values for the following conditional probabilities, using the city-coordinate data from Table 5-2:

(a) $\Pr[X = 95 \mid Y = 35]$ (c) $\Pr[Y = 40 \mid X = 85]$
(b) $\Pr[X = 85 \mid Y = 40]$ (d) $\Pr[Y = 45 \mid X = 90]$

5-17 Use the city-coordinate data from Table 5-2 to answer the following:

(a) Prepare a complete table of values for the conditional probability distribution $p_{X|Y}(x \mid 40)$.
(b) Comparing the foregoing to the marginal probability distribution for X, does it appear that knowing the level for Y has a strong effect on probabilities for X? Explain.

5-18 Refer to the design-aptitude testing example in the text. Find the following probabilities:

(a) $\Pr[Y \leq 75 \mid X = 95]$ (c) $\Pr[Y > 90 \mid X = 85]$
(b) $\Pr[Y \leq 60 \mid X = 80]$ (d) $\Pr[Y > 80 \mid X = 70]$

5-19 Referring to Problem 5-9 and your answers to that exercise:

(a) Compute the covariance. (b) Compute the correlation coefficient.

5-20 The following joint probability data apply to the drying time (hours) X and durability (years) Y of paints:

x \ y	5	6	7	8
5	.05	.05	0	0
10	.10	.15	.05	0
15	.05	.10	.20	.05
20	0	.05	.05	.10

(a) Find the marginal probability distributions for X and for Y.
(b) Determine the conditional probability distribution of Y given $X = 10$ hours. Are X and Y independent random variables? Explain.

5-21 Refer to the joint probability data in Problem 5-20 and your answers to that problem.
(a) Compute the covariance. (b) Compute the correlation coefficient.

5-22 A box contains four coins, each having a different probability of head: .3, .4, .5, or .6. One coin is selected at random and tossed three times. Let X represent the head probability for the selected coin and Y the number of heads obtained in the tosses.
(a) Construct a table giving the bivariate probability mass function and the marginal probability distributions.
(b) Given that $Y = 3$, find the conditional probability for (1) $X = .3$, (2) $X = .4$, (3) $X = .5$, and (4) $X = .6$.
(c) Compute (1) $\text{Cov}(X, Y)$ and (2) the correlation coefficient.

5-23 Refer to the human-engineering application in Problem 5-11.
(a) Determine the following conditional probabilities for the sitting height Y given arm reach X (both in inches):
(1) $\Pr[Y > 38 | X = 34]$ (3) $\Pr[Y \leq 35 | X = 35]$
(2) $\Pr[Y > 38 | X = 32]$ (4) $\Pr[Y \leq 35 | X = 36]$
(b) Determine the following conditional probabilities for arm reach given sitting height:
(1) $\Pr[X > 34 | Y = 38]$ (3) $\Pr[X \leq 35 | Y = 35]$
(2) $\Pr[X > 32 | Y = 38]$ (4) $\Pr[X \leq 36 | Y = 35]$

5-24 The following joint probability data apply to fatigue tests to be run on bronze strips. X represents cycles to failure (in 10^5) when alternate strips are bent at a high level of deflection. Y represents the same at a lower deflection level.

x \ y	20	30	40	50
4	.01	.03	.05	.02
5	.03	.10	.08	.04
6	.02	.08	.12	.11
7	.02	.04	.07	.18

(a) Find the marginal probability distributions for X and for Y.
(b) Determine the conditional probability distribution of Y given $X = 5$. Are X and Y independent random variables? Explain.

5-25 Refer to the joint probability data in Problem 5-24 and your answers to that problem.
(a) Compute the covariance.
(b) Compute the correlation coefficient.

5-26 Show that when X and Y are independent continuous random variables, $\text{Cov}(X, Y) = 0$.

5-27 Conveyer system interruptions are a Poisson process, with mean rate $\lambda = 1$ per day. There is a 50% chance that any particular interruption will cause a service failure. The number of service failures Y in any given day is a Bernoulli process, with number of trials equal to the number of interruptions X.
(a) Construct a table for the bivariate probability mass function, including all marginal probabilities. Use just 5 rows, and round the row marginal probabilities to

two decimal places. Express the joint probability values and the column marginal probabilities to three decimal places.

(b) Find the correlation coefficient between number of interruptions and number of failures.

5-4 The Sum and Mean of Random Variables

The sum and mean of several independent random variables are themselves important random variables that play significant roles in statistics. The *sum* has application in a variety of areas where the Poisson process applies, epitomized by queues. It is also useful in applications involving the binomial distribution. But perhaps the most important role is played by the *mean*, a random variable created by averaging several variables. When the variables averaged represent random observations from a population, the *sample mean* is obtained. Much of statistics is concerned with procedures founded on the probabilistic properties of the sample mean.

The Expected Value of the Sum

Consider two independent random variables X and Y. It is easy to see that the expected value of their sum is equal to the sum of their expected values.

If $X + Y$ is treated as a function of the two continuous random variables, its expected value is

$$E(X + Y) = \int_{-\infty}^{\infty} \int_{-\infty}^{\infty} (x + y) f_{X,Y}(x, y)\, dx\, dy$$

When terms are factored, then collected and rearranged, this is equal to the following:

$$\int_{-\infty}^{\infty} x\left[\int_{-\infty}^{\infty} f_{X,Y}(x, y)\, dy\right] dx + \int_{-\infty}^{\infty} y\left[\int_{-\infty}^{\infty} f_{X,Y}(x, y)\, dx\right] dy$$

The second integral in each of these terms gives the respective marginal probability distribution. Thus,

$$\begin{aligned} E(X + Y) &= \int_{-\infty}^{\infty} x f_X(x)\, dx + \int_{-\infty}^{\infty} y f_Y(y)\, dy \\ &= E(X) + E(Y) \end{aligned}$$

and thus,

$$E(X + Y) = E(X) + E(Y)$$

A similar derivation applies to the discrete case. The above property extends to the sum of any number of independent random variables.

EXPECTED VALUE OF THE SUM OF SEVERAL INDEPENDENT RANDOM VARIABLES

$$E(X_1 + X_2 + \cdots + X_n) = E(X_1) + E(X_2) + \cdots + E(X_n)$$

Note that expected values are *additive.* This feature applies whether or not the X's are independent.

An analogous result applies to the variance of the sum of *independent* random variables.

Variance of the Sum

Consider again two random variables X and Y. If we let $Z = X + Y$ and use the fact that $\text{Var}(Z) = E(Z^2) - E(Z)^2$, it follows that

$$\text{Var}(X + Y) = E[(X + Y)^2] - E(X + Y)^2$$

When we apply additivity of expected values, we may reexpress the second term as follows:

$$[E(X) + E(Y)]^2 = E(X)^2 + 2E(X)E(Y) + E(Y)^2$$

The first term may be expanded as

$$E[(X + Y)^2] = E[X^2 + Y^2 + 2XY]$$

which, again using the additivity of expected values, may be expressed as

$$E(X^2) + E(Y^2) + E(2XY) = E(X^2) + E(Y^2) + 2E(XY)$$

Combining the equivalent forms of the two original terms and collecting terms leads to

$$\begin{aligned}\text{Var}(X + Y) &= [E(X^2) - E(X)^2] + [E(Y^2) - E(Y)^2] + 2[E(XY) - E(X)E(Y)] \\ &= \text{Var}(X) + \text{Var}(Y) - 2\,\text{Cov}(X, Y)\end{aligned}$$

When X and Y are *independent*, $\text{Cov}(X, Y) = 0$, and the above reduces to

$$\text{Var}(X + Y) = \text{Var}(X) + \text{Var}(Y)$$

If we extend these arguments to the sum of several independent random variables, the following property applies.

VARIANCE OF THE SUM OF SEVERAL INDEPENDENT RANDOM VARIABLES

$$\mathbf{Var}(X_1 + X_2 + \cdots + X_n) = \mathbf{Var}(X_1) + \mathbf{Var}(X_2) + \cdots + \mathbf{Var}(X_n)$$

A similar derivation applies in the discrete case. The above shows that variability is *additive.* As we include more independent random variables to the sum, the variance of the sum must become greater. There is no canceling effect.

Mean of Several Independent Random Variables

We may apply the properties of expected value and variance to the foregoing results to find the moments for the mean of several independent random variables, which we denote by $\bar{X}$:

$$\bar{X} = \frac{X_1 + X_2 + \cdots + X_n}{n}$$

This is equivalent to a constant $1/n$ times the sum, so that

$$E(\bar{X}) = E\left[\frac{1}{n}(X_1 + X_2 + \cdots + X_n)\right]$$
$$= \frac{1}{n}E(X_1 + X_2 + \cdots + X_n)$$
$$= \frac{1}{n}[E(X_1) + E(X_2) + \cdots + E(X_n)]$$

Similarly,

$$\text{Var}(\bar{X}) = \text{Var}\left[\frac{1}{n}(X_1 + X_2 + \cdots + X_n)\right]$$
$$= \left(\frac{1}{n}\right)^2 \text{Var}(X_1 + X_2 + \cdots + X_n)$$
$$= \frac{1}{n^2}[\text{Var}(X_1) + \text{Var}(X_2) + \cdots + \text{Var}(X_n)]$$

The foregoing results may be summarized by the following property.

EXPECTED VALUE AND VARIANCE OF THE MEAN OF SEVERAL INDEPENDENT RANDOM VARIABLES

$$\mathbf{E(\bar{X}) = \frac{1}{n}[E(X_1) + E(X_2) + \cdots + E(X_n)]}$$
$$\mathbf{Var(\bar{X}) = \frac{1}{n^2}[Var(X_1) + Var(X_2) + \cdots + Var(X_n)]}$$

This result is fundamental to development of procedures used in statistical evaluations involving means.

Common Probability Distributions for the Sum and Mean

It is possible to extend probability theory to derive the probability distribution for the sum (or mean) of several independent random variables. The result obtained will depend on the particular forms of the distributions for the individual variables. The details for doing this are beyond the scope of this book. The following important conclusions are stated without proof.

1. When each X_i is *normally distributed* with parameters μ_i and σ_i, their sum is also normally distributed with parameters $\Sigma\, \mu_i$ and $[\Sigma\, \sigma_i^2]^{1/2}$. Their mean is normally distributed with parameters $\Sigma\, \mu_i/n$ and $[\Sigma\, \sigma_i^2/n^2]^{1/2}$.

   To illustrate, consider the width X_1 and length X_2 of rough planks cut from logs at a sawmill. Suppose that length and width are independent random variables with the following parameters:

Length	Width
$\mu_1 = 12'$	$\mu_2 = .75'$
$\sigma_1 = .5'$	$\sigma_2 = .1'$

The half-perimeter is $X_1 + X_2$, a normally distributed random variable with parameters

$$\mu = \mu_1 + \mu_2 = 12' + .75' = 12.75'$$

$$\sigma = \sqrt{\sigma_1^2 + \sigma_2^2} = \sqrt{(.5)^2 + (.1)^2} = .51'$$

As further illustration, suppose three men's heights are chosen at random from a normal population having mean 70″ and standard deviation 2.5″. Letting X_i denote the height of a randomly chosen man,

$$\mu_1 = \mu_2 = \mu_3 = 70''$$

$$\sigma_1 = \sigma_2 = \sigma_3 = 2.5''$$

The *mean* height of the three men, $(X_1 + X_2 + X_3)/3$, is normally distributed, with parameters

$$\mu = \frac{1}{n}(\mu_1 + \mu_2 + \mu_3) = \frac{1}{3}(70'' + 70'' + 70'') = 70''$$

$$\sigma = \sqrt{\frac{1}{n^2}(\sigma_1^2 + \sigma_2^2 + \sigma_3^2)} = \sqrt{\frac{1}{3^2}[(2.5)^2 + (2.5)^2 + (2.5)^2]} = 1.44''$$

2. When each X_i has a *binomial distribution* with parameters n_i and π, their sum also has a binomial distribution with parameters $\Sigma\, n_i$ and π.

   To illustrate, suppose that two independent random samples taken from a production line are to be pooled. Each observed item will be classified as defective or satisfactory. The assumed proportion of defective items produced is $\pi = .05$. The first sample of $n_1 = 25$ items will contain X_1 defectives, the second will have X_2 defectives from $n_2 = 50$ observations. Each of these random variables has a binomial distribution, with parameters n_1 or n_2 and π. The pooled sample will involve $n_1 + n_2 = 25 + 50$ observations and have $X_1 + X_2$ defectives altogether. The combined number of defectives is also a binomial random variable with parameters $n = 75$ and $\pi = .05$.

3. When each X_i has a *Poisson distribution* with parameter λ_i, their sum over a specified time span t will also have a Poisson distribution with parameter $\Sigma\, \lambda_i$.

Example: Number of Aircraft Repairs

An aircraft maintenance facility performs several different types of repairs. Repair requests in a particular category may be viewed as a Poisson process over time. The following data apply:

Category	Mean Weekly Repair Rate
1. Engine	$\lambda_1 = 2$
2. Avionics	$\lambda_2 = .5$
3. Hydraulics	$\lambda_3 = 1.3$
4. Landing gear	$\lambda_4 = .2$

What is the probability distribution for the total number of repair requests when all these categories are combined?

The overall number of repairs is

$$S_4 = X_1 + X_2 + X_3 + X_4$$

which has the Poisson distribution over duration t with mean rate

$$\lambda = \lambda_1 + \lambda_2 + \lambda_3 + \lambda_4 = 2 + .5 + 1.3 + .2 = 4.0$$

Referring to Appendix Table B, the following probabilities apply for a single week ($t = 1$):

$$\Pr[S_4 \leq 10] = P(10; 4, 1) = .9972$$

$$\Pr[S_4 > 6] = 1 - P(6) = 1 - .8893 = .1107$$

$$\begin{aligned}\Pr[3 \leq S_4 \leq 7] &= P(7) - P(2)\\ &= .9489 - .2381 = .7108\end{aligned}$$

4. When each of r independent T's has an exponential distribution with common parameter λ, their sum will have a gamma distribution with parameters r and λ.

Example: Failure Time of Solar Power System

A space probe's communications system is powered by several solar collectors. It will continue to function as long as at least 1 of the r collectors is functioning. The lifetime of each of these is exponentially distributed with a mean lifetime of 6 months, or a mean failure rate of $\lambda = 2$ per year. The lives of the individual collectors are statistically independent.

How many collectors must be included in the probe so that it may be expected to survive 5 years before all collectors have failed?

The time T to failure has a gamma distribution with parameters $\lambda = 2$ and r. The expected survival time is

$$E(T) = \frac{r}{\lambda} = \frac{r}{2}$$

In order for $E(T) = 5$, there must be $r = 10$ collectors.

Importance of Results to Sampling Applications

Under some conditions, the sample observations obtained from a population are independent random variables. This makes the sample mean $\bar{X}$ a random variable with the properties just described. In such a study, $\bar{X}$ might be used to estimate the population mean μ. In Chapters 7–9 we will see that these results form the basis for statistical procedures involving means.

When the sampled population is very large, each observed value X_i has a common expected value equal to μ and variance equal to the population variance σ^2. For such samples, the results imply that

$$E(\bar{X}) = \mu$$

$$\text{Var}(\bar{X}) = \sigma_{\bar{X}}^{\ 2} = \frac{\sigma^2}{n}$$

As we shall see, when n is large and certain conditions are met, the normal distribution with parameters μ and $\sigma_{\bar{X}}$ applies in finding probabilities for $\bar{X}$.

Problems

5-28 A box containing n dice is shaken and the contents are dropped on a table. Let X_i be the number of dots on the showing face of the ith die cube.
(a) Find (1) $E(X_i)$ and (2) $\text{Var}(X_i)$.
(b) Determine the expected value for the sum of all the showing faces when the number of dice is (1) $n = 5$, (2) $n = 10$, (3) $n = 25$.
(c) Determine the variance for the sum of all the showing faces when the number of dice is (1) $n = 5$, (2) $n = 10$, (3) $n = 25$.

5-29 A random decimal is a uniformly distributed random variable on $[0, 1]$. A series of 12 random decimals will be generated. Letting X_i denote the value obtained for the ith random decimal, complete the following:
(a) Find (1) $E(X_i)$ and (2) $\text{Var}(X_i)$.
(b) Determine (1) the expected value and (2) the variance for the sum of the generated values.
(c) Determine (1) the expected value and (2) the variance for the mean of the generated values.

5-30 A population of women's heights is normally distributed, with mean 64″ and standard deviation 2″. A population of men's heights is also normally distributed, with mean 70″ and standard deviation 2.5″. A man and a woman are chosen at random.
(a) Find the probability that the sum of their heights is (1) $\leq$ 132″, (2) $>$ 140″, and (3) between 138″ and 142″.
(b) Find the probability that the mean of their heights is (1) $\leq$ 65″, (2) $>$ 68″, and (3) between 66″ and 69″.

5-31 The times between arrivals for cars at a toll booth are exponential, with mean rate $\lambda = 5$ per minute.
(a) Determine the time expected to elapse between the 13th car's arrival and the arrival of the 15th car.
(b) Consider the elapsed time between the arrivals of the 8th and the 12th cars and the corresponding duration for the 5th and 7th cars. Determine the expected difference between these two times.

5-32 Suppose the mix of jobs handled by the aircraft repair facility described in the text is changed so that the avionics repair rate doubles and the engine repair rate is halved. Determine the probability that in a particular 2 weeks the total number of repairs is
(a) ≤ 8 (b) > 5 (c) between 2 and 6, inclusively

5-33 A series of observations n is made of the elongation of steel bars under identical tensile loads. Assuming that the resulting quantities may be viewed as independent samples from a normally distributed population with $\mu = .05''$ and $\sigma = .01''$, complete the following:
(a) Specify which probability distribution applies for the sum of 25 observed elongations, and indicate the parameter values.
(b) Determine the probability that this quantity exceeds 1.30″.

5-34 Separate random samples are selected from a production line where 5% of the output is defective. The first two samples comprise 20 items each, the last only 10. Find the probability that the number of defectives in the pooled sample is
(a) ≤ 3 (b) > 5 (c) between 1 and 4, inclusively

Comprehensive Problems

5-35 The following distribution was established by a committee of deans for the rank of 2 leading engineering schools. The rank of the University of California at Berkeley is denoted by X and that for M.I.T. by Y.

X \ Y	1	2	3	4	5
1	0	.15	.07	.06	.03
2	.20	0	.05	.03	.02
3	.10	.06	0	.01	.01
4	.06	.05	.02	0	.01
5	.03	.02	.01	.01	0

Determine the following:
(a) marginal probability distribution for X
(b) marginal probability distribution for Y
(c) conditional probability distribution for X given $Y = 1$
(d) conditional probability distribution for Y given $X = 2$

5-36 Referring to the data in Problem 5-35,
(a) compute (1) $E(X)$, (2) $E(Y)$, (3) $\text{Var}(X)$, (4) $\text{Var}(Y)$, and (5) $E(XY)$
(b) compute (1) $\text{Cov}(X, Y)$ and (2) ρ

5-37 A fair six-sided die cube is tossed. If a "one" or a "two" is obtained for the showing face, a crooked die cube is then tossed that has 3 "three" faces and 3 "four" faces. Otherwise, the original die is tossed a second time. Letting X_1 denote the number of dots obtained on the first toss and X_2 the number on the second, complete the following:
(a) Determine the joint probability distribution for X_1 and X_2.
(b) Find the marginal distribution (1) for X_1 and (2) for X_2.

5-38 Referring to the data in Problem 5-37, find:

(a) the conditional probability distribution for X_2 given (1) $X_1 = 1$ and (2) $X_1 = 5$

(b) the conditional probability distribution for X_1 given (1) $X_2 = 3$ and (2) $X_2 = 5$

5-39 Again referring to the data in Problem 5-37:

(a) compute (1) $E(X_1)$, (2) $E(X_2)$, (3) $\text{Var}(X_1)$, (4) $\text{Var}(X_2)$, and (5) $E(X_1X_2)$.

(b) compute (1) $\text{Cov}(X_1, X_2)$ and (2) ρ.

5-40 A surveyor makes independent tape measurements on three successive segments constituting a baseline. Each reading may be in error by an uncertain amount that is normally distributed with mean 0 and standard deviation .5 millimeters (mm).

(a) Determine the probability distribution for the measurement error of the entire baseline dimension.

(b) Find the probability that the baseline measurement is (1) 1 mm below the true dimension, (2) 2 mm above the true dimension, and (3) within $\pm .5$ mm of the true dimension.

5-41 Refer to the potential nuclear-power-plant accidents in Problem 3-33 on page 143. Find the probability that in any given year at least 1 type of described disaster will occur at the site.

5-42 Consider the *mean* elongation of the 25 bars in Problem 5-33.

(a) What probability distribution applies for this random variable? Specify its parameters.

(b) Determine the probability that the mean elongation falls below .045″.

5-43 A project engineer helping to design a petrochemical refinery is responsible for four nonoverlapping activities that will occur sequentially. She is interested in the amount of time until all of them are finished. Each activity completion time is assumed to be a normally distributed random variable independent of the others. The following data apply:

Activity	**Parameters**	
A	$\mu = 50$	$\sigma =$ 5 days
B	20	3
C	70	10
D	40	4

(a) What distribution applies for the total completion time of these activities? Specify the parameter values.

(b) Determine the probability that the engineer's activities will be completed (1) no later than 200 days, (2) in less than 170 days, (3) in the range from 180 to 210 days.

5-44 An engineer is designing a roof. Because of uncertainties regarding the strengths of materials, the maximum safe load X is uncertain, although it is believed to be normally distributed with mean 100 psi and standard deviation 10 psi. The actual load Y should be considerably less, although after heavy snows Y is assumed to be normally distributed, with mean 80 psi and standard deviation 12 psi. Assuming X and Y are independent, find the probability that at one of these times the actual load will exceed the safe maximum.

5-5 *Optional Section: The Moment Generating Function

The essential properties of a random variable are provided by its distribution, usually summarized in terms of its probability density or mass function. From these, further useful summaries may be obtained. The **moment generating function** allows us to take shortcuts in finding the expected value, variance, and higher-order moments, all of which might be difficult to determine the usual way.

Although we usually think of expected value in terms of the average level of the random variable, we have seen that the concept extends to functions of X. Consider a special function

$$g(X) = e^{uX}$$

where u is any real number. We use this to define the following.

MOMENT GENERATING FUNCTION OF X

$$\boldsymbol{M_X(u) = E(e^{uX})}$$

If X is a discrete random variable, its moment generating function is determined from

$$M_X(u) = E(e^{uX}) = \sum_{\text{all } x} e^{ux} p_X(x)$$

Analogously, for a continuous random variable

$$M_X(u) = E(e^{uX}) = \int_{-\infty}^{\infty} e^{uX} f_X(x)\, dx$$

Properties of Moment Generating Functions

Before we illustrate how to find $M_X(u)$ for a specific distribution, it will be helpful to investigate some of its properties. If we take the first derivative with respect to u, we obtain

$$M'_X(u) = E(Xe^{uX})$$

since the operation affects only the exponential term inside the summation or integral. The second and higher-order derivatives provide

$$M''_X(u) = E(X^2 e^{uX})$$
$$M^{[3]}_X(u) = E(X^3 e^{uX})$$
$$\vdots$$
$$M^{[n]}_X(u) = E(X^n e^{uX})$$

* This section may be skipped with no loss of continuity.

Suppose that after taking the first derivative we set $u = 0$. The exponential term becomes 1, leaving

$$M'_X(0) = E(X)$$

so that the expected value results. We have referred to $E(X)$ as the **first moment** of the random variable, represented symbolically as m_1. Higher-order moments may be found in the same way. In general, we use the following expressions.

MOMENTS OF A RANDOM VARIABLE

$$\boldsymbol{m_k = E(X^k) = M_X^{[k]}(0)} \qquad k = 1, 2, \ldots, n$$

These may be used to obtain expected value and variance:

$$m_1 = E(X) = M'_X(0)$$

$$m_2 = E(X^2) = M''_X(0)$$

$$\text{Var}(X) = E(X^2) - E(X)^2 = m_2 - m_1^{\,2}$$

Moment Generating Functions for Important Distributions

We now derive the moment generating functions for several of the important probability distributions encountered earlier.

Poisson Distribution Denoting by X a Poisson random variable with parameters λ and t, the probability mass function is

$$p_X(x) = \frac{(\lambda t)^x e^{-\lambda t}}{x!} \qquad x = 0, 1, 2, \ldots$$

The moment generating function is defined by

$$\begin{aligned} M_X(u) &= \sum_{x=0}^{\infty} e^{ux} \frac{(\lambda t)^x e^{-\lambda t}}{x!} \\ &= e^{-\lambda t} \sum_{x=0}^{\infty} \frac{(\lambda t e^u)^x}{x!} \end{aligned}$$

We may simplify this using the following fact:

$$e^y = 1 + y + \frac{y^2}{2!} + \cdots + \frac{y^n}{n!} + \cdots$$

Thus,

$$e^{-\lambda t} \sum_{x=0}^{\infty} \frac{(\lambda t e^u)^x}{x!} = e^{-\lambda t} e^{\lambda t e^u}$$

so that after simplification, the following expression applies.

MOMENT GENERATING FUNCTION FOR
A POISSON DISTRIBUTION

$$\mathbf{M_X(u) = \exp[\lambda t(e^u - 1)]}$$

We may use this to verify our earlier expressions for $E(X)$ and $\text{Var}(X)$. We have

$$\begin{aligned} M'_X(u) &= \frac{d}{du}\exp[\lambda t(e^u - 1)] \\ &= \lambda t e^u \exp[\lambda t(e^u - 1)] \end{aligned}$$

and

$$E(X) = m_1 = M'_X(0) = \lambda t$$

Also,

$$\begin{aligned} M''_X(u) &= \frac{d^2}{du^2}\exp[\lambda t(e^u - 1)] \\ &= \lambda t e^u(\lambda t e^u + 1)\exp[\lambda t(e^u - 1)] \end{aligned}$$

and

$$E(X^2) = m_2 = M''_X(0) = \lambda t(\lambda t + 1)$$

so that

$$\text{Var}(X) = m_2 - m_1^2 = \lambda t(\lambda t + 1) - (\lambda t)^2 = \lambda t$$

Binomial Distribution Suppose X has a binomial distribution with parameters n and π. We have

$$p_X(x) = b(x; n, \pi) = C_x^n \pi^x (1 - \pi)^{n-x} \qquad x = 0, 1, 2, \ldots, n$$

The moment generating function for X is defined by

$$\begin{aligned} M_X(u) &= \sum_{x=0}^{n} e^{ux} C_x^n \pi^x (1 - \pi)^{n-x} \\ &= \sum_{x=0}^{n} C_x^n (\pi e^u)^x (1 - \pi)^{n-x} \end{aligned}$$

To simplify this, we use the following fact:

$$\sum_{x=0}^{n} C_x^n p^x q^{n-x} = (p + q)^n$$

so that substituting πe^u for p and $1 - \pi$ for q, the above identity leads to the following expression.

MOMENT GENERATING FUNCTION FOR
A BINOMIAL DISTRIBUTION

$$\mathbf{M_X(u) = [\pi e^u + 1 - \pi]^n}$$

Normal Distribution Using the normal density with parameters μ and σ,

$$f(x) = \frac{1}{\sqrt{2\pi}\,\sigma} e^{-\frac{(x-\mu)^2}{2\sigma^2}} \qquad -\infty < x < \infty$$

we have

$$M_X(u) = \int_{-\infty}^{\infty} e^{ux} \frac{1}{\sqrt{2\pi}\,\sigma} e^{-\frac{(x-\mu)^2}{2\sigma^2}}\, dx$$

Letting $z = (x - \mu)/\sigma$ and then rearranging terms, we find that this is equivalent to

$$\begin{aligned} M_X(u) &= \frac{1}{\sqrt{2\pi}} \int_{-\infty}^{\infty} e^{u(\mu + z\sigma)} e^{-\frac{1}{2}z^2}\, dz \\ &= e^{u\mu}\left[\frac{1}{\sqrt{2\pi}} \int_{-\infty}^{\infty} e^{u\sigma z} e^{-\frac{1}{2}z^2}\, dz\right] \\ &= e^{u\mu + \frac{1}{2}u^2\sigma^2}\left[\frac{1}{\sqrt{2\pi}} \int_{-\infty}^{\infty} e^{-\frac{1}{2}(z - u\sigma)^2}\, dz\right] \end{aligned}$$

The last term in the brackets represents the area under a repositioned standard normal curve and is equal to 1, so the above reduces to the following expression.

MOMENT GENERATING FUNCTION FOR
A NORMAL DISTRIBUTION

$$\mathbf{M_X(u) = \exp\left(u\mu + \frac{1}{2}u^2\sigma^2\right)}$$

Exponential Distribution Using parameter λ, we have the following density for the random variable T:

$$f(t) = \lambda e^{-\lambda t} \qquad 0 \le t < \infty$$

We express the moment generating function by

$$\begin{aligned} M_T(u) &= \int_0^{\infty} e^{ut} \lambda e^{-\lambda t}\, dt = \lambda \int_0^{\infty} e^{(u-\lambda)t}\, dt \\ &= \frac{\lambda}{\lambda - u} \int_0^{\infty} (\lambda - u) e^{(u-\lambda)t}\, dt \end{aligned}$$

This integral equals 1, so we obtain the following expression.

MOMENT GENERATING FUNCTION FOR
AN EXPONENTIAL DISTRIBUTION

$$\mathbf{M_T(u) = \frac{\lambda}{\lambda - u}}$$

Gamma Distribution Using the gamma density with parameters λ and r,

$$f(x) = \frac{\lambda^r}{\Gamma(r)} x^{r-1} e^{-\lambda x} \qquad 0 \le x < \infty$$

we have

$$M_X(u) = \int_0^\infty e^{ux}\left[\frac{\lambda^r}{\Gamma(r)} x^{r-1} e^{-\lambda x}\right] dx$$
$$= \frac{\lambda^r}{(\lambda - u)^r} \int_0^\infty \frac{(\lambda - u)^r}{\Gamma(r)} x^{r-1} e^{-(\lambda - u)x} dx$$

This integral is of the same form as the integral of $f(x)$ with the constant $\lambda - u$ in place of λ, so that it equals 1. Thus we have the following expression.

MOMENT GENERATING FUNCTION
FOR A GAMMA DISTRIBUTION

$$\mathbf{M_X(u) = \left(\frac{\lambda}{\lambda - u}\right)^r}$$

Optional Problems

5-45 Express the moment generating function of a normally distributed random variable with parameters $\mu = 100$ and $\sigma = 10$.

5-46 Express the moment generating function of a Poisson random variable with mean rate $\lambda = 5$ and duration $t = 2$.

5-47 Consider the exponential distribution for the random variable T with parameter λ. Use the moment generating function to show that $E(T) = 1/\lambda$ and $\text{Var}(T) = 1/\lambda^2$.

5-48 Determine for exponential distributions the following moments:
(a) m_3 when $\lambda = 5$ (c) m_5 when $\lambda = 4$
(b) m_4 when $\lambda = 2$ (d) m_6 when $\lambda = 10$

5-49 Suppose X has the binomial distribution with parameters n and π. Use the moment generating function to show that $E(X) = n\pi$ and $\text{Var}(X) = n\pi(1 - \pi)$.

5-50 Determine the expression of the moment generating function for the geometric distribution of a Bernoulli process with parameter π.

5-51 Determine the expression of the moment generating function for the uniform distribution over the interval $[a, b]$.

PART TWO

STATISTICAL ANALYSIS I: BASIC CONCEPTS

CHAPTER SIX

The Statistical Sampling Study

THE MAJOR CONCERN of modern statistics is drawing conclusions about populations. This places the focus on the *sample*, a known and measurable entity representing a population with unknown characteristics. Statistics embodies theory and procedures for using sample information to draw inferences about uncertain populations.

Samples are used for several reasons. They are more economical than a complete enumeration or *census*. They usually allow quick results and are often the most practical source of information, since the very act of observing may destroy the observed units or render them useless. Sampling may even be more accurate than a census.

A sample investigation may be viewed as an experiment similar to those performed by scientists and engineers. Indeed, many physical experiments might properly be considered as sampling experiments because of the uncertainties involved and the inferential nature of the conclusions reached. Of course, all such investigations require careful planning. Planning is especially important in sampling studies because of the potential for *error* when generalizing about that which still remains uncertain. Good sampling design minimizes such error when data collection procedures are carefully thought out and when appropriate methods are then employed to evaluate the eventual sample results.

Some potential for error is unavoidable, however, even in well-designed sampling investigations. By their nature, samples provide an incomplete picture of the population, and there are no guarantees that the partial sample results will be representative of the whole. It is possible, however, to control such errors and limit them to chance influences, usually by choosing observed units at random. Although sample units might be selected primarily through judgment or by convenience, only *random samples* may be systematically evaluated for chance error. Probability may be employed to this end.

The design of the sampling study and the selection of method are fundamental to any statistical investigation. A huge body of knowledge is available to the investigator, and statistics is a discipline as rich and varied as any encountered in engineering. This book can provide only a survey.

6-1 The Need for Samples

There are several reasons why it is more desirable to seek incomplete or partial sample information than the complete data that might be obtained from a census. The need for partial observation rather than complete enumeration is so compelling in most investigations that sampling is taken for granted. Not every production avionics unit will be subjected to a vibration test, only a sample of jet-engine blades will be assessed for strength, and opinions regarding what options should be included in a computer software package are not sought from everybody.

Yet, depending on the purpose of the investigation, sampling might not be satisfactory. We would not want to have our entire course grade determined by the score on a single problem set any more than we would establish our personal bank balances by sampling the month's checks. A program director must have access to the status of all activities under his control, not just the details of some.

Economic Advantages of Samples

A significant advantage of samples is that they are generally more economical than a census. Inspecting a portion of items being produced is obviously cheaper than assessing the entire run in detail. Even in cases where individual physical observations are cheap, we can usually achieve significant savings by relying on samples to trigger quality-control actions when huge quantities of units are involved. Moreover, in some situations the act of observation is very expensive, especially in materials testing. For example, metal-fatigue experiments may involve contorting each test unit thousands of times in special mechanical test stands. Investigations involving human subjects can be elaborate, too. Personal interviews for opinion surveys can cost upwards of $20 for one hour (during which as few as one or two responses might be obtained).

Thus, the economic advantages of sampling are compelling. Scarce research funds must not be squandered on an unnecessarily large number of observations. Not even the most affluent organization can afford to "bust its budget" in overzealous pursuit of total enumeration.

Although sampling can lead to significant savings, it does extract a price in addition to the data collection costs—because no sample is guaranteed to portray the underlying population accurately. For example, production problems will sometimes go undetected because telling evidence is lacking in the particular sample units observed. An entire production run might have to be scrapped later because a sample erroneously indicated that the proportion of defectives was lower than was actually the case. On the other hand, an unrepresentative sample could trigger costly remedial actions. Production might be halted for unneeded repairs. Such errors of omission and commission are especially insidious in basic or applied research, where unrepresentative samples may lead investigators down blind alleys or cause them prematurely to close off avenues that would have led to success.

Balancing the costs and benefits of sampling is an essential element of the statistical art. It is possible to control the incidence of sampling error so that an optimal overall solution is reached. As we shall see, this may be accomplished by judiciously determining the sample size and by establishing an efficient sampling design. The net gain from sampling rather than censusing can be dramatic.

The following example is not a traditional statistical application wherein a sample serves in drawing inferences about an uncertain population. Nevertheless, it graphically illustrates the economic advantages of the sampling concept.

Example:* *Sampling Multiplies Telephone Line Capacity

Voice transmission via telephone is accomplished by converting sound into an electrical analogue. The sound is transformed by a microphone/transmitter into an electronic mode that is relayed over the miles and then converted back into sound by a receiver/speaker. Early telephone equipment was designed entirely to accommodate voice sound, which we know to be quite slow compared with electronic speeds; one conversation required one line.

Essential voice characteristics of the original slow sound wave may be reconstructed from partial readings of the electronic waveform measured at intervals such as once every 100 microseconds. The reconstructed voice is analogous to the "continuous" motion perceived from the discrete frames of a motion picture in that only the signal for every hundredth microsecond of a conversation might be transmitted. In effect, then, a conversation may be transmitted and reconstructed into high-quality sound using only a *sample* of the electronic signal. Each of the 99 microseconds between samplings could be used to obtain a segment from each of 99 other conversations sharing the same line. In effect, sampling would allow for a hundredfold increase in the carrying capacity in a telephone line that otherwise could carry only one conversation at a time.

In this case, transmitting voice samples provides a huge saving over a continuous transmission (really a census). As with any sampling scheme, however, there is a cost. Additional equipment is required to break down, merge, burst, and resynthesize the original voice. Here, however, the savings in transmission costs far outweigh equipment outlays.

Further Reasons for Sampling

There are a variety of reasons other than economic ones for sampling.

Timeliness A census may not be practical because the information thereby gained might take too long to acquire. It could take years to poll all potential users of a product regarding their design preferences, by which time the item will probably have become obsolete. Sample opinion data must be obtained for such an investigation. The exact incidence of defectives cannot be known until the end of a production run, but a plant superintendent cannot wait until that time before ordering necessary equipment adjustments. Instead, he will identify problems early by taking samples. Only a sample of items should be life-tested, for it could take years for an entire population to fail (after which all units would be useless for their intended purpose).

Characteristics of populations might shift over time so that later observations could represent a different entity than earlier ones. This is a common problem in assessing the preferences and attitudes of people, a fickle bunch who are capable of rapid shifts—as political pollsters know too well.

Large Populations As a practical matter, many populations needing measurement are too large for 100% observation to be achieved. Simply counting human beings is such a big job that the United States government conducts a census only once every 10 years, and the poorest nations don't even try. Imagine the problems faced by a food processor if the contents of each of the thousands of cans filled in one day were separately weighed. The census of a continuing process might never be completed, since *all* units will *never* be available for observation; such a population is theoretically infinite.

Destructive Nature of the Observation A large class of statistical investigations involves observations that destroy the units observed. Especially in engineering, test units must be damaged or destroyed to determine strength, durability, or lifetime. In these cases, 100% observation is out of the question: We won't burn out all the light bulbs to find out what proportion were defective! Even surveys involving people might so change the observed individuals that they no longer reflect the population. In-depth interviews seeking attitudes about a product have been known to oversensitize test subjects so much that many of them wish never to encounter the item again—regardless of their initial predilections.

Inaccessible Populations Statisticians often measure populations that may never actually exist. Referred to as **target populations**, these are theoretical constructs that may occur only under laboratory conditions, so the only possible observations are those the investigator creates for the sample. For example, the only thin metal strips ever subjected to 10,000 quick twists on a test bench are those in the investigator's sample. Nevertheless, these strips represent the target population of all such objects that *might be* so twisted. In drug research, only those patients who receive a particular dosage during evaluations might ever receive that therapy—but they are representatives of the target population of all people who would receive that same treatment.

Even when an actual population presently exists, some units may be inaccessible for observation. Consider the automotive engineer wishing to measure carbon deposits on valves of six-cylinder engines. Only engines from cars acquired by the researcher's firm would be available for disassembly.

Accuracy and Sampling

It is commonly believed that data acquired by a census are invariably more reliable than similar information based on sampling. Indeed, if the same data-gathering procedures were used for both the census and the sample, the census results would be superior.

However, the very act of observation can be a demanding one that must be performed under carefully controlled circumstances. The cost of a properly conducted census for even a small population can be so prohibitive that shortcuts might be mandatory. In recent times the observation quality of the U.S. Deciennial Census has been downgraded; before 1970, census takers made personal visits to homes, whereas now the counting is done mainly on the basis of mailed questionnaires. A carefully conducted sampling study can yield higher-quality information than would be obtainable from a sloppy census.

Proper statistical procedure does not end with the making of observations. Data must then be gathered and evaluated. In 1980, a rash of lawsuits charged not only undercounting by the Census but also loss or destruction of data. And even when all census data can be accounted for, its mass can be so overwhelming that terrible mistakes occur.

Example:* *Census Reports Huge Increase in Teenage Widowers

In 1950 the U.S. Census reported that the number of teenage widowers had increased tenfold since 1940. Also during that span of time the number of Indian divorcees grew by a like magnitude.

There was no sociological explanation for these anomalous findings. Dramatic changes in modern lifestyle had been reflected in marital status percentages that shifted by a few percent—not 1,000%, as in these findings. Two statisticians doubted the credibility of the census findings and proposed a scenario to explain what had happened.*

They purport that the oddball findings were the result of a data-processing snafu. In 1950, electronic data processing was used for the first time to tabulate the U.S. Census. Each individual was represented by a punched card. The reporters suggest that the holes in neighboring columns were reversed on a few dozen mispunched cards, so that most of the teenage widowers and Indian divorcees should have been counted as part of much larger groups.

---

* Ansley J. Coale and Frederick F. Stephan, "The Case of the Indians and Teen-age Widows," *Journal of the American Statistical Association*, LVII (June 1962), 338–437.

Problems

6-1 For the following investigations, explain why sample information might be preferred to that from a census:

(a) A personal computer manufacturer must determine the most-preferred position of the CRT display unit.
(b) Metal fatigue characteristics in the wing struts of a particular aircraft are to be evaluated.
(c) The mean lifetime of pressure seals must be established.
(d) Professional engineers' consulting fees are to be summarized.

6-2 Give an example of a type of investigation in which sampling is advantageous primarily for reasons of

(a) economy
(b) timeliness
(c) large population
(d) inaccessibility
(e) destructiveness of observations
(f) greater accuracy

6-3 For each of the following situations, list the important reasons why sampling information would be sought rather than census data:

(a) Materials are being tested for strength.
(b) Customer opinions will be the basis for choosing chassis colors for a line of personal computers.
(c) A determination must be made as to whether the gas buildup inside CRTs is excessive.
(d) Comparisons will be made to determine how long various computers take to complete processing of benchmark programs.

6-2 Designing and Conducting a Sampling Study

As a convenient point of reference, Figure 6-1 shows that a statistical investigation comprises three stages: planning, data collection, and evaluation. You will find it helpful to refer to this flowchart occasionally to find which stage applies to a discussion or problem. Context is especially critical in statistics, because what is a

Figure 6-1 *Stages of a statistical investigation.*

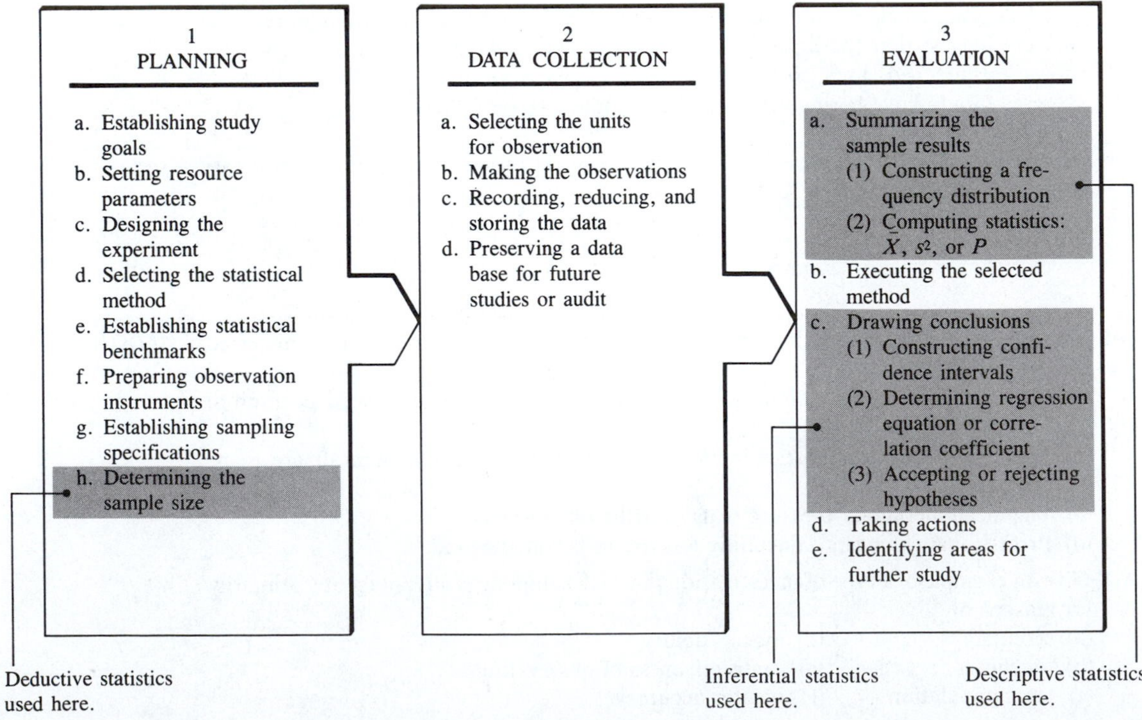

random variable in an earlier stage can be a precise computed value later. Each discussion of method will mainly apply to just one of the three stages.

The Planning Stage

Planning is always important. But good planning is especially important in statistics, and only partly because of the expense involved. Care must be taken that proper procedures are followed because credibility is crucial to investigations involving samples, and statisticians always face the risk that their results will be found invalid.

Several steps are involved in the planning process:

a. *Establishing study goals.* As in any research investigation, goals should be identified by the investigators. The usual goal in quality-assurance testing is to identify problems early so that corrective actions may be taken before excessive damage is done. In product design, more complex goals might apply, and a whole series of decisions regarding alternative configurations will have to be made. Each of these may involve uncertainties, some of which may be largely resolved through sampling. Separate sampling studies might be made in the following areas: assessment of user needs, selection of materials, mode of production, pilot testing, environmental testing, reliability testing, and marketing the product.

Sampling investigations usually terminate with a report or decision. A standards bureau may be concerned only with reporting data for public benefit. A chip manufacturer will be more concerned with a decision, such as picking the most efficient way to layer wafer materials. The study's goals will dictate the kind of information to be provided by the sample.

b. *Setting resource parameters.* Time and funds are the resources universally in short supply. These establish constraints within which the investigation must be conducted. It may not be possible to buy enough sample observations to achieve ideal levels of precision and reliability. A decision may have to be made quickly, so it is not possible to conduct a fully controlled experiment. Laboratory space or personnel may be inadequate for a comprehensive testing program. Part of the statistical art is effective compromise in arriving at a plan that meets the study goals, embodies accepted methods, and is compatible with available resources.

c. *Designing the experiment.* Like a computer software system or a printed circuit, the statistical experiment itself must be carefully designed. The factors to be measured must be identified, the units specified, and the measuring instruments determined. If comparisons are to be made, norms and benchmarks must be established.

Two populations are often compared, one representing a traditional approach and the other an innovative one. A common statistical procedure is to designate the sample from the first population as the **control group** and the second population's sample as the **experimental group**. When several populations are involved, each sample may be considered as a separate **treatment group**.

d. *Selecting the statistical method.* A variety of statistical procedures are described in this book. Conceptually simplest are *estimates* of population parameters, which are generally obtained from counterpart sample statistics. Thus, the population mean distortion in metal rods subjected to a particular force can be estimated by the sample mean. Somewhat more complex is basing a decision on the sample result. For example, do we conclude that a particular metal alloy experiences more or less distortion than another? Such a conclusion may be reached by a *hypothesis testing* procedure.

Statistical testing with two populations provides some interesting design issues. Selecting the sample units for the control and experimental groups independently is simplest conceptually, but significant savings may be achieved by testing with *matched pairs.* In effect, each sample unit in the control group is matched up with a counterpart in the experimental group so that as much as possible of the actual deviation between pairs may be explained by differences in populations. This particular approach is taken in medical research, in which, ideally, identical twins would be placed in opposite groups, so that physiological variations between groups would account for almost none of the differences found in treatments. But identical twins are rare, so individuals are instead paired by sex, age, physical size, lifestyle, smoking habits, and similar factors, and "near twins" are the result.

***Example:** Matched-Pairs Testing of Tire Tread Designs*

A rubber company engineer used sampling in evaluating an improved tread design for a line of radial tires. Having limited test facilities of its own, the company had a policy of field-testing new products. A random group of employees and company associates were given free test tires. In return, they agreed to have tire wear periodically monitored at company stations, where proper pressure was maintained and wheels were kept balanced and aligned. An evaluation based on matched pairs was used to identify any significant differences in tire performance.

Matching was done first by car type, on a model-by-model basis. A heavy-duty Ford pickup was paired with an identical model of similar age and mileage. Likewise, city cars were matched with city cars, single male drivers with single male drivers, and hard drivers with other hard drivers. A wide range of driving environments, vehicle sizes, and operating habits were represented in the test.

For each test pair a coin toss determined which vehicle would be equipped with the experimental tread tires and which would receive the standard ones. After a year of driving, the tire wear rates for all cars were to be determined precisely. Any differences between pairs (other than chance caused) should be largely attributed to the different tire tread designs.

A third major area of statistics is concerned with the *association* between two or more variables. Investigators may be concerned only with the strength of association or *correlation.* Or they might wish to express the level of one variable from that of another directly in terms of a *prediction.* For this purpose, a *regression equation* must be determined.

e. *Establishing statistical benchmarks.* Certain statistical benchmarks must be established *before* the investigation begins in order not to bias the results. The hypotheses being tested must be identified before data are collected. Likewise, the choice of method must include the prior establishment of the particular statistic (mean, median, porportion, and so on) to be used in reaching conclusions.

f. *Preparing observation instruments.* Many statistical investigations involving physical objects are concerned with readily measurable quantities or determinable attributes such as size, weight, or texture. Commonly available equipment would ordinarily be used in making these observations, but special devices, some quite elaborate, may be needed to observe certain characteristics. As noted, strength testing requires special test stands. Traffic engineers often monitor vehicle flow with a pressure-sensitive metering apparatus. Telemetry equipment may be needed to determine times between failures of equipment in remote locations such as orbiting satellites or wilderness pumping stations. Observation instruments can be the major expense of statistical studies, although their economic life may last through dozens or hundreds of separate investigations.

Observations are not limited to physical assessment. Human attitude and opinion may be the fundamental characteristics under investigation. In those cases, an interview or questionnaire serves as the observation instrument.

g. *Establishing sampling specifications.* As we have noted, samples may be unrepresentative—an unavoidable feature of statistics. The *level of precision* in sample results can be controlled, however. For a fixed sample size, this may be achieved by tolerating a less than perfect chance of being correct. This *reliability* probability is also subject to the investigator's control. As we shall see, reliability may be sacrificed for precision, or vice versa. Both may be improved only by paying the price of increasing the sample size.

Precision and reliability may be specified at the outset. Overly ambitious targets can exhaust available research funds, however, so these statistical specifications must be adapted to the study goals and resources. Matters are further complicated in hypothesis-testing investigations, in which a target *significance level* serves to translate sample results into action. This level may also be specified by the investigator, but there are inherent risks—errors of omission and commission—that must be considered and balanced.

h. *Determining the sample size.* As noted, the size of the sample follows from the foregoing specifications for precision and reliability. A probability analysis leads directly to an expression relating sample size to those specifications. In arriving at this expression, the investigator focuses on what types of samples are generated from populations having presumed characteristics; we have referred to such a procedure as *deductive statistics.*

Sample size, then, is not entirely a statistical question; also to be considered are available resources. The final choice is often a compromise between what is desired and what is achievable.

The Data Collection Stage

Once a plan has been established, it must be executed. We refer to this process as the **data collection stage**. Data collection itself requires careful planning, even though there is little need for analysis at this stage. It is a very important part of the sampling investigation, since this is the point where improper technique may lead to those insidious errors that cannot be ascribed to chance alone. One only has to review the blunders of the U.S. Census to appreciate some of the pitfalls associated with data collection.

We may describe this stage by the following four steps:

a. *Selecting the units for observation.* As we shall see, only random samples may be subjected to scientific evaluation. Later in this chapter we will see how *random numbers* may be employed to select the units to be observed.

b. *Making the observations.* Actual observation of physical objects is mainly a mechanical act, ordinarily involving a measurement, a count, or some qualitative reading of one or more attributes. The details for achieving this will have been settled in the planning stage. Special problems may arise in the study of humans because so many subjective elements can influence how a person answers questions. These difficulties and suggested remedies are discussed later in the chapter.

c. *Recording, reducing, and storing the data.* The sample results must be recorded and kept for eventual evaluation. Data may be collected in a raw form so that further processing or refining may be required before analysis. This may be the case with analogue data, such as radar cross sections of moving objects, which may have to be reduced to a digital form. Even in more mundane applications, such as determining tire wear, a series of individual measurements will often be reduced to a single observation quantity.

d. *Preserving a data base for future studies or audit.* Today's sampling study may provide reference points for making comparisons in future studies. All data should be preserved as long as practicable. These could also prove helpful should outside investigation be necessary to verify or audit findings.

The Evaluation Stage

The final phase of a statistical investigation is the **evaluation stage**. This is where most of the statistical methodology is exercised. At this point some action is taken, and the goals of the study are fulfilled. As with the earlier stages, this stage comprises several steps.

a. *Summarizing the sample results.* To derive meaning from raw sample information, those data must ordinarily be summarized. At the very least, those statistics that are necessary to the analytic procedures selected in the planning stage must be computed. Included in such calculations are the sample mean $\bar{X}$ and variance s^2 (for quantitative data) or the sample proportion P (for qualitative data). As we shall see, additional statistics may be required. A more detailed rendering of the sample results can be provided by grouping the data into a frequency distribution, which might even be displayed graphically as a histogram.

This is the step where much of the *descriptive statistics* (discussed in Chapter 1) comes into play. The sample summaries thereby obtained are essential for making conclusions about the still uncertain population.

b. *Executing the selected method.* Most of the creative thinking about procedure and choice of method should have been settled during the planning stage. At this point in the investigation the sample results are known. This knowledge should not be used at the eleventh hour to modify analytic procedures. Doing so could damage the credibility of the entire study. The best policy is to follow the original plan strictly—even if the sample results are disappointing.

c. *Drawing conclusions.* As with executing the method, any conclusions reached should be neither more nor less than what is indicated by the sample. There is no room in statistics for "poetic license." Depending on the original plan, the end results may take the form of estimates for one or more parameters of the underlying population(s). As we shall see, such estimates generally take the form of *confidence intervals* (which reflect the lack of certainty and which acknowledge the inherent imprecision of the sample). A second type of conclusion is to accept or reject a hypothesis established in the planning stage. The third type of conclusion applies when several populations are being investigated and establishes a correlation coefficient or a regression equation.

The conclusions reached are the culmination of the third form of statistics. This category falls under the broad heading of *inferential statistics.* Most methods of modern statistics are directly or indirectly aimed at making inferences; this involves generalizing about the uncertain population from the known sample. Although actual inferences are not made until the evaluation stage, they are contemplated throughout the planning stage.

d. *Taking action.* Three main levels of actions may result from a statistical study. The most passive is simply to report the results, publishing any estimated values for the intended audience. For example, a testing laboratory may publish mean strength statistics for a metal alloy of a particular form. Much scientific evaluation terminates at this level. A more pragmatic level of action is to use the relationship established from the sample to make predictions and forecasts. This will ordinarily involve the regression methods described later in the text. For example, the safe loading for a structure may be predicted from known levels of the supporting components such as weight, strength, or other material properties. The highest level of action is when a hypothesis is accepted or rejected. In quality-control evaluations, the sample may result in the rejection or acceptance of an entire shipment of raw materials.

e. *Identifying areas for further study.* A statistical investigation will often raise more questions than it answers. Many of these issues might be pursued in later studies. It is good statistical technique to inform the client or management of areas where further investigation would be fruitful. No sampling study totally achieves the desired goals (because residual uncertainty always remains). Identification of further areas of study is really a necessary step in completing an evaluation, as the suggestions for improvements help to qualify the present results and compensate for existing inadequacies.

Problems

6-4 Identify the stage of the sampling study during which each of the following actions is most likely to occur:

(a) constructing histograms
(b) deciding the study budget
(c) weighing items
(d) deciding what to study next
(e) making predictions of one variable, using known levels for the other variable
(f) digitizing waveform plots

6-5 In each instance below, indicate whether the primary statistics type is deductive, descriptive, or inferential:

(a) A .50 probability is determined that $\bar{X}$ will fall between 9,500 and 10,500 pounds.
(b) The computed value for $\bar{X}$ is 9,733 pounds.
(c) It is estimated that the population mean is 9,700 $\pm$ 100 pounds.
(d) The computed value of $\bar{X}$ is less than 10,000, so the null hypothesis that $\mu = 10{,}500$ is rejected.
(e) The expression $Y = .005 + .001X$ is found from the sample data.
(f) The computed value for the standard deviation is $s = 25.3$ pounds.

6-6 A quality-control inspector always tests a sample of 10% of every shipment. In view of statistical planning and goals, how might this policy be faulted?

6-7 For each of the following situations, describe how better planning might have improved the statistical investigation:

(a) Because of a shortage of personnel, a receiving department inspects samples of only the more expensive items. In order to reduce the risk of accepting poor shipments, very large sample sizes are used. The types of units inspected are but a fraction of the total value of items ordered.
(b) A testing laboratory has only one vibrating table for durability evaluations. As a consequence, a backlog of jobs has developed, and new-product approvals have been delayed.
(c) In order to obtain very precise and highly reliable estimates consistent with a firm's overall image of quality, large sample sizes were employed. This consumed so many resources and took so long that only half of the required evaluations could be performed.
(d) Because of a company policy of doing all evaluations with in-house personnel, it took 2 years of interviewing to discover that most users of its computers preferred BASIC to FORTRAN.

6-8 You have been charged with estimating the mean starting salary of new graduate engineers in your specialty.

(a) Describe briefly some of the planning you would need to complete before gathering data. Indicate any statistical areas where you feel you would need more background in order to complete your plan.
(b) Describe how you would collect data for your study.
(c) Discuss how you would evaluate your sample data. As in (a), indicate any areas where you feel additional background in statistics would be helpful.

6-3 Bias and Error in Sampling

The greatest problems in any sampling study arise from an unavoidable fact: *The sample may be unrepresentative of the population from which it is drawn.* Statisticians try hard to minimize any deviations between sample results and the true population characteristics and to limit the cause of deviations to chance. We can gain an appreciation of their efforts by becoming familiar with the pitfalls encountered in conducting a sampling study.

Nonsampling Error

Data accumulation, reduction, processing, and storage are themselves actions that may lead to errors and unrepresentative results—even when a census is involved (as we saw with the teenage widowers). The very act of observation may cause distortions that can ruin the credibility of an entire investigation. Errors arising from the manner of data collection or processing fall into the broad category of **nonsampling error**. Avoidance of nonsampling error is largely an art acquired through experience, although valuable lessons can be learned from the mistakes of others.

In engineering, common sources of nonsampling error are measuring instruments that are out of adjustment or in poor calibration. Anybody who has taken a surveying course is familiar with these types of errors, but blunders can occur even in the most sophisticated environments. For example, ground stations in a satellite tracking network had trouble acquiring their targets whenever the orbits had just passed another station where the operators were unaware that their own location parameters had been considerably out of true for several months. Because of this error they had been consistently providing incorrect trajectories to stations farther down the line.

Observations involving people are especially subject to nonsampling error. For a variety of psychological reasons, answers to survey questions may be untruthful. For example, consider the question of age. When asked how old they are, most people will give their age as of their last birthday (and some will lie, out of vanity); a better approach would be to ask for their date of birth.

People may tend to accommodate or to sabotage a survey if they know the purpose of the investigation or the study sponsor. For this reason it might be best not to disclose the purpose of the study. Personal interviews themselves may elicit improper responses. An overenthusiastic interviewer may *induce bias* by unconsciously encouraging the subject to give the "right" answers. People may lie to impress the interviewer; such **prestige bias** has been demonstrated by persons who recognize fictitious names or who have read nonexistent books. Some erroneous answers from human subjects may be inevitable, but these may be minimized by careful wording of questions.

Sampling Error

Any differences between the sample and the population that are due to the particular units selected for observation are **sampling errors**. This type of discrepancy arises whenever the sample is not a perfect image of the underlying population. To some degree, nearly all statistical investigations involve sampling error.

Anthropometric studies of human physical characteristics provide a good example of sampling error. In establishing the frequency distribution for the height of adult males, a sample containing unusually tall men may be quite unrepresentative unless a large number of men have been measured. Although 7-footers are quite rare, having one such person in a sample of 1,000 may not result in a materially distorted portrayal. But it would be ludicrous to use the heights of the basketball team as a sample representing an entire student body.

Sampling error may be controlled, not by excluding tall people, but by allowing all persons an *equal chance* of being selected. This will still allow for disproportionate representation in the sample, but a glaring misrepresentation would be unlikely. The odds against atypical samples increase as the sample size increases.

Ideally, statisticians desire to have sampling error arise only from chance. Too many tall (or short) persons in the sample would then be attributed to bad luck alone. As we shall see, this goal may be achieved if all possible population units have the same chance of being selected for the sample; that is, if the sample is a random one. But this ideal is not always met, and some bias toward certain units might exist.

Sampling Bias

A tendency to favor the observation of some population units over others is called **sampling bias**. Investigators cannot even identify which observations were created from a sampling bias. Nor can probability theory be applied to qualify biased samples. At best, only causes of bias can be identified, not their results. Errors arising from sampling bias are therefore the most insidious ones in statistical applications.

To a large extent, random samples are free from sampling bias—but even they may involve imperfections. Some units may be harder to observe than others, and these might be skipped in favor of the more convenient ones. Skipping observations creates a **bias of nonresponse**; hard-to-get observations may correspond to population units that materially differ from the rest.

As an example of the bias of nonresponse, consider an investigation to determine the durability of a weather coating for a large structure such as a bridge. The least accessible and highest areas may be the most exposed to solar radiation and will therefore probably experience greater deterioration than places easier to inspect.

The effect of the bias of nonresponse is particularly dangerous when investigating people's attitudes. In evaluating a new product design to determine user response, the hard-to-find people (the ones who are very active and never at home) may be the ones who have the most to contribute in terms of constructive criticism. Those nonresponders originally selected for the study should be diligently pursued until it is no longer practicable to observe them.

Example: *Roosevelt Predicted to Lose Presidency*

A classic example of sampling bias took place before the U.S. presidential election in 1936, when Democrat Franklin D. Roosevelt was running against Republican Alfred P. Landon. The then popular *Literary Digest* attempted to forecast the winner on the basis of a sample of voter preferences. Magazine subscription lists and telephone directories were used as the source of names to obtain the attitudes of several million people. A majority of these favored Landon, so the *Digest* predicted he would win by a landslide.

We know the actual results were the reverse. What went wrong? An analysis showed that during the Great Depression economically displaced persons—the ones most likely to vote Democratic—could not afford magazines and telephones. This group was underrepresented in the *Digest* sample, making it strongly biased in favor of the Republican candidate. In spite of the huge sample size, the results were very wrong.

Problems

6-9 Give an example of each of the following types of bias:
(a) prestige bias (b) bias of nonresponse (c) induced bias

6-10 In each of the following cases indicate whether a nonsampling error or a sampling error is involved.
(a) Although room temperatures have varied considerably during measurement, all dimension readings taken of metal objects will be used as sample data.
(b) A random sample of items turns out to contain a higher percentage of unsatisfactory items than are in the population as a whole.
(c) Vehicles on sample stretches of roadway have been counted by a malfunctioning pressure-activated device.
(d) A power company line inspector used sample data to determine the useful remaining life of insulators. These data could be obtained only by climbing poles, and all poles in rough terrain were excluded from the investigation.

6-11 A sample of student opinions about laboratory equipment will be sought. Indicate how nonsampling error might influence the study results in the following cases:
(a) Students are told that their response will be used to justify a larger budget for new equipment.
(b) Students are told that some laboratory sections will have to be curtailed for lack of funding.
(c) Students are told that the faculty have voted for a greater theoretical thrust in the curriculum.
(d) Students are told that employers have complained that recent graduates do not have enough "hands-on" experience.

6-12 In evaluating a new drug, the patients in the control group are provided a *placebo* (a neutral substance), while those in the experimental group are given pills of identical appearance that contain the actual drug. Briefly explain the reason for using a placebo instead of simply not administering any pills to the controls.

6-4 Selecting the Sample

As we have seen, sampling error is a natural consequence in any statistical investigation. To a large extent it may be controlled by randomly selecting the sample units and may be made less significant by increasing the number of observations. Often troublesome is sampling bias, which is hard to predict; it may be minimized mainly by following good procedures in selecting sample units.

We are now ready to describe procedures for selecting the sample. Although the ideal is to minimize sampling error and to avoid sampling bias, some compromise may be desirable or may even be a practical necessity. A variety of methods are available for choosing sample units. Any one of these might be optimal for meeting the study goals within the constraints of time and available resources.

Types of Samples

There are several basic approaches to sample selection. The resulting samples fall into three main categories.

The Convenience Sample Convenience samples result when the only observations made are the convenient ones. Least scientific, a convenience sample is what most people use when they make generalizations from their personal experience. This type of sample may involve substantial sampling bias. Some bias of nonresponse is guaranteed, because it is never convenient to observe the hard-to-find units.

Probability cannot be used to qualify convenience samples in an attempt to measure sampling error. Convenience samples are generally unacceptable in research, and the sample results are not ordinarily suitable for public dissemination. Nevertheless, organizations commonly use this form of sampling internally. Product decisions are often based on tests with employee groups. Brewers enlist office staffs for taste evaluations, food processors use their employees in pilot investigations of new products, and tire manufacturers do the same, providing free tires to personnel in endurance testing experiments. In these cases sampling bias is acknowledged, but its effect is judged unimportant. The cost savings from convenience sampling and its simplicity often outweigh the risk of unrepresentative samples.

The Judgment Sample Judgment samples result whenever some of the observed units are selected by judgment. This approach may be justified in an attempt to ensure sample diversity and to guarantee representation of some groups. Many government indexes, for instance, are based on judgment samples. The Consumer Price Index is established from the quantities and prices of a "market basket" of goods and services that have been chosen by judgment (and it has been criticized as giving too much weight to some items, such as the cost of housing and home mortgages).

Like the convenience sample, probability is of little use in quantifying any sampling error that might result from selecting unrepresentative units. Substantial opportunity for sampling bias exists. Indeed, by its very nature a judgment sample must reflect the biases of the investigator. Nevertheless, such a selection procedure may be justified for its "fairness," for its consistency, and for economic reasons.

The Random Sample When every possible population unit is assigned an equal probability for being selected for observation and inclusion in the sample, the resulting collection is called a **random sample**. An alternative term is **probability sample**. There are several different kinds of random samples.

Only random samples may be qualified by using probability concepts, and nearly all the methods of modern statistics are therefore predicated on random samples. Because so much research is based on statistical investigations employing them, random sampling procedures are often referred to as **scientific sampling**.

The most common form is the simple random sample, to be described next.

Sample Selection Using Random Numbers

We illustrate random sampling with the population of 100 Nobel physics laureates listed in Table 6-1. Ten names will be selected for a **simple random sample**, which gives each population unit (name) an equal chance of being selected.

A straightforward way of selecting our sample would be to write each name on a slip of paper. These could then be placed in a hat and thoroughly mixed, after which 10 slips of paper would be withdrawn, one at a time. Such a physical lottery would be cumbersome. And the randomness of the resulting sample might be doubted by some (as was the randomness of the 1970 U.S. Draft Lottery).

Rather than conducting physical lotteries, it is accepted statistical procedure to select samples using **random numbers**. These values are a succession of digits that were themselves generated by a lottery in such a way that every possible integer is equally likely. The following is a list of 5-digit random numbers taken from the first column of Appendix Table F:

12651	**61**646
81769	**74**436
36737	**98**863
82861	**54**371
21325	**15**732

These random numbers were generated at the RAND Corporation by an electromechanical device. Because any value between 0 and 9 had the same chance of appearing in each digit position, all 5-digit integers had an equal chance of appearing as the successive entries on RAND's list. In addition to equal likelihood, random numbers exhibit no sequencing patterns or serial correlation—each successive value on the list is independent of the preceding ones.

The usual procedure for selecting a sample by random numbers is first to assign each population unit an identification number. The Nobel laureates in Table 6-1 are numbered in chronological sequence. A random number is read from the table, and the population unit with the matching identification number is selected for inclusion in the sample. If the unit has been previously selected or if there is no match, the next random number on the list is used. In our example, the identification numbers have only two digits (except for the number 100, which will be treated as 00); we will achieve our matches by using only the first two digits of the random numbers (shown as boldface in the above list).

Table 6-1 *Winners of Nobel Prize in Physics*

01 Roentgen (1901, Germany)
02 Lorentz (1902, Neth.)
03 Zeeman (1902, Neth.)
04 Curie (1903, France)
05 Becquerel (1903, France)
06 Rayleigh (1904, Gr. Brit.)
07 von Lenard (1905, Germany)
08 Thomson (1906, Gr. Brit.)
09 Michelson (1907, U.S.)
10 Lippmann (1908, France)
11 Braun (1909, Germany)
12 Marconi (1909, Italy)
13 van der Waals (1910, Neth.)
14 Wien (1911, Germany)
15 Dalén (1912, Sweden)
16 Kamerlingh-Onnes (1913, Neth.)
17 von Laue (1915, Germany)
18 Bragg (1915, Gr. Brit.)
19 Barkla (1917, Gr. Brit.)
20 Planck (1918, Germany)
21 Stark (1919, Germany)
22 Guillaume (1920, France)
23 Einstein (1921, Germany)
24 Bohr (1922, Denmark)
25 Millikan (1923, U.S.)
26 Siegbahn (1924, Sweden)
27 Franck (1925, Germany)
28 Hertz (1925, Germany)
29 Perrin (1926, France)
30 Compton (1927, U.S.)
31 Wilson (1927, Gr. Brit.)
32 Richardson (1928, Gr. Brit.)
33 de Broglie (1929, France)
34 Raman (1930, India)
35 Heisenberg (1932, Germany)
36 Dirac (1933, Gr. Brit.)
37 Schrödinger (1933, Austria)
38 Chadwick (1935, Gr. Brit.)
39 Anderson (1936, U.S.)
40 Hess (1936, Austria)
41 Davisson (1937, U.S.)
42 Thomson (1937, Gr. Brit.)
43 Fermi (1938, U.S.)
44 Lawrence (1939, U.S.)
45 Stem (1943, U.S.)
46 Rabi (1944, U.S.)
47 Pauli (1945, U.S.)
48 Bridgman (1946, U.S.)
49 Appleton (1947, Gr. Brit.)
50 Blackett (1948, Gr. Brit.)
51 Yukawa (1949, Japan)
52 Powell (1950, Gr. Brit.)
53 Cockcroft (1951, Gr. Brit.)
54 Walton (1951, Ireland)
55 Bloch (1952, U.S.)
56 Purcell (1952, U.S.)
57 Zemike (1953, Neth.)
58 Born (1954, Gr. Brit.)
59 Bothe (1954, Germany)
60 Kusch (1955, U.S.)
61 Bardeen (1956, U.S.)
62 Brattain (1956, U.S.)
63 Shockley (1956, U.S.)
64 Lee (1957, U.S.)
65 Yang (1957, U.S.)
66 Cherenkov (1958, U.S.S.R.)
67 Frank (1958,U.S.S.R.)
68 Chamberlain (1959, U.S.)
69 Segrè (1959, U.S.)
70 Glaser (1960, U.S.)
71 Hofstadter (1961, U.S.)
72 Landau (1962, U.S.S.R.)
73 Goeppert-Mayer (1963, U.S.)
74 Jensen (1963, Germany)
75 Wigner (1963, U.S.)
76 Basov (1964, U.S.S.R.)
77 Prochorov (1964, U.S.S.R.)
78 Townes (1964, U.S.)
79 Feynman (1965, U.S.)
80 Schwinger (1965, U.S.)
81 Tomonaga (1965, Japan)
82 Kastler (1966, France)
83 Bethe (1967, U.S.)
84 Alvarez (1968, U.S.)
85 Gell-Mann (1969, U.S.)
86 Noel (1970, France)
87 Alfvén (1970, Sweden)
88 Gabor (1971, Gr. Brit.)
89 Bardeen (1972, U.S.)
90 Cooper (1972, U.S.)
91 Schrieffer (1972, U.S.)
92 Giaever (1973, U.S.)
93 Esaki (1973, Japan)
94 Josephson (1973, Gr. Brit.)
95 Ryle (1974, Gr. Brit.)
96 Hewish (1974, Gr. Brit.)
97 Rainwater (1975, U.S.)
98 Mottelson (1975, Denmark)
99 Bohr (1975, Denmark)
100 Richter (1976, U.S.)

The following names constitute the random sample of physics prize winners:

12 Marconi
81 Tomonaga
36 Dirac
82 Kastler
21 Stark
61 Bardeen
74 Jensen
98 Mottelson
54 Walton
15 Dalén

Of course, a different 10 names would be obtained for a different set of random numbers. The starting point on the random number list should itself be selected at random, and it should be decided in advance which digit positions (if any) should be discarded.

Types of Random Samples

The simple random sample procedure just described is the most commonly encountered type, but other types are also important. These are briefly described next.

Sequential Random Sample In the **sequential** variation on the random sample, data collection is made easier by using just one random number as a seed value for selecting all the sample units. For example, suppose our Nobel laureates were selected this way, starting with the random number 03. Every tenth name, starting with 03 Zeeman, would then be selected:

03 Zeeman	53 Cockcroft
13 van der Waals	63 Shockley
23 Einstein	73 Goeppert-Mayer
33 de Broglie	83 Bethe
43 Fermi	93 Esaki

The gap between names is established by the range of identification numbers and sample size. (In a list of a thousand, ranging from 001 to 1,000 (000), a sample of ten would involve every 100th unit, giving 003, 103, 203, . . .) Sequential random sampling has the advantage of being quick and easy to explain. But the first number selected is crucially important, and this approach may (for obscure reasons) be subject to sampling bias.

Stratified Random Sample In stratified random sampling, population units are sorted into categories, each of which must be repeated in the sample; each grouping is called a **stratum**. A separate random sample is then selected from each stratum. Such a procedure is desirable when different groups must be represented, either for the sake of diversity or for comparison with each other.

In our Nobel prize illustration we might stratify the population by nationality of the winners, or by decade of their award. What scheme is used (if any) depends on the purpose of the study.

Example:* *Predicting U.S. Presidential Elections

A popular election-year project of the news media is forecasting the leader in the race for president of the United States. Political pollsters publish current voter preferences. It would be inaccurate to use these numbers to forecast the winner if the election were held at that time—not so much because of omnipresent sampling error as because of the nature of the Electoral College system and the tradition that delegates vote in blocs for the leading popular votegetter in their home states. To complicate the issue further, small states have a disproportionate share of the electoral votes (which are allocated according to the number of a state's seats in Congress). Only a **stratified sample**, involving a separate poll tabulation from each state, can provide a proper basis for predicting the presidential election. (And there will still be sampling error.) As we shall see, this would require about 50 times the budget needed for a single national poll.

Cluster Sample Like the stratified sample, this scheme also divides the population into groups. These are called **clusters**. Random numbers are used, not to select *individual* population units, but to select the clusters by means of simple random sampling. Each unit within a selected cluster is then included in the sample group.

The advantage of cluster sampling is that it can be considerably less expensive than simple random sampling. It is less wasteful to open a few cases of parts and inspect the entire contents of each than to open 24 times as many cases and inspect one or two items from each. It would take far less time for a canvasser to visit every apartment in a few buildings than to visit a few families from nearly every apartment house in the city.

Cluster samples come close to being convenience samples, however, and as such they may involve substantial sampling bias.

Problems

6-13 Using the leading 2 digits in the *second* column of Appendix Table F, select a random sample of 10 Nobel physics laureates from Table 6-1.

6-14 What type of sample does each of the following represent?
(a) the processing times for computer runs done on an investigator's own jobs
(b) circuit schematics used by an electrical engineering professor for a midterm examination
(c) the record of the number of heads obtained from several tosses of a coin
(d) a representative collection of seismic trace plots used by a geologist to explain the characteristics of an oil-bearing stratum

6-15 For each situation below, indicate whether you would recommend a judgment, convenience, or random sampling scheme, or a combination of these. Explain your choice.
(a) An architect is selecting sketches of building types to use in discussing design concepts for the first time with a client.
(b) The engineering dean asks students in his seminar for suggestions about curriculum improvements. Their comments will be used to design a questionnaire on curriculum to be answered by the entire student body.
(c) A professional engineering society wants to compare salaries of engineers in different regions of the country.
(d) An oil company must select a line of minicomputers for its scientific staff. The criteria of selection will be economy and flexibility.

6-16 Determine a sequential random sample of 10 Nobel Prize winners from Table 6-1 using 07 as the seed random number.

6-17 Divide the Nobel laureates from Table 6-1 into nationality strata by decade of award. Then assign a 2-digit identification number, starting at 01, in chronological order, to each name in the respective stratum. Determine a stratified random sample of size 1 from each group. Use the following random numbers for this purpose. (Proceed down the columns, skipping any entry that is too large. Consider no entry more than once.)

11	12	17	06
08	17	07	15
13	09	11	11
08	05	05	01

6-18 Treat each successive 5 names in Table 6-1 as a cluster. Then assign an identification, 01 through 20, to each cluster. Use random numbers 09 and 13 to select the sample. List the names you obtained.

CHAPTER SEVEN

Sampling Distributions

MODERN STATISTICS INVOLVES using known sample information for making *inferences* regarding populations whose true characteristics are unknown. But the procedures of inferential statistics rest on *deductive statistics*, which is concerned with a probabilistic evaluation of the sampling process. Deductive statistics treats the population as the known entity and the sample as the unknown.

Figure 7-1 should help to sort things out. Recall that a statistical sampling study may be viewed in three successive stages: (1) planning, (2) data collection, and (3) evaluation. Deductive statistics applies only in the planning stage—before data are collected. At that point, the population may be *assumed* to have specific characteristics. The main question in stage 1 is: What type of sample will be obtained from a given population? All this takes place under the *pretense* that the population details are *known*. (Of course, no sample would ever be needed to evaluate a completely known population.) Once the data are collected, however, the investigation moves to the evaluation stage. At that time the pretending game is over, and the sample is the known entity that provides the investigator with a basis for characterizing the unknown population. Investigators ordinarily rely on others for the deductive statistics, which has already been incorporated into the theoretical framework of most commonly encountered statistical procedures.

The basic thrust of inferential statistics is drawing conclusions regarding the levels of population parameters, such as the mean μ and standard deviation σ. These conclusions can be based directly on the values of the counterpart sample statistics $\bar{X}$ and s. Before the sample data are in, $\bar{X}$, s, and other sample statistics are uncertain quantities; that is, each is a random variable having its own probability distribution. As a special class, the probability distributions for sample statistics are referred to as *sampling distributions.*

Our discussion of these begins with the sampling distribution for the sample mean $\bar{X}$, the most commonly used sample statistic.

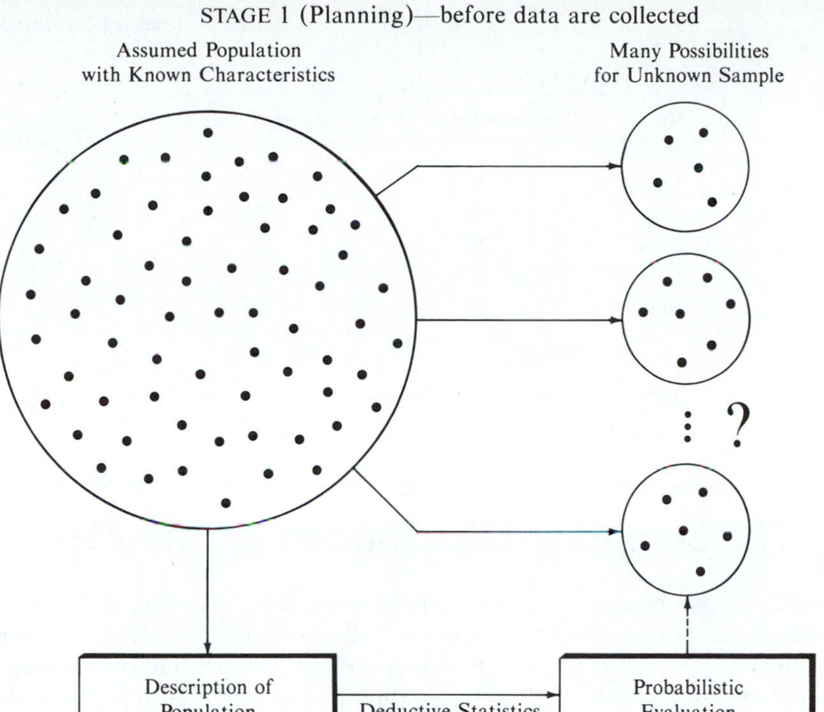

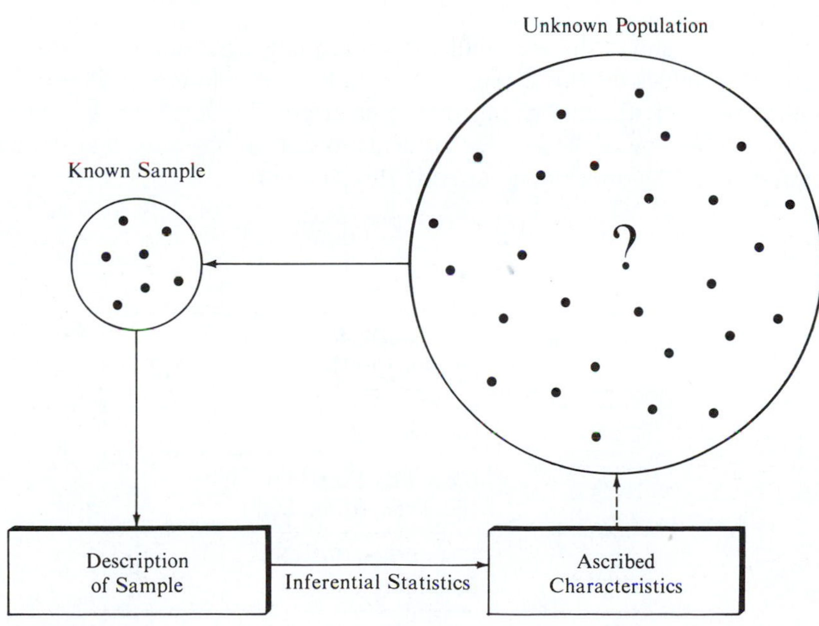

Figure 7-1
Two stages and types of statistics encountered in a sampling study.

Table 7-1
Grade Point Population for Students in Metallurgy Seminar

Name	Examination Grade	Grade Points
Dan	B	3
Ed	C	2
Fran	B	3
Gil	A	4
Ida	C	2

$\mu = 2.8 \qquad \sigma = \sqrt{.56} = .7483$

7-1 The Sampling Distribution of the Mean

We illustrate the sampling distribution concept using the population in Table 7-1. These data represent the midterm examination results for the 5 students attending a senior metallurgy seminar at an engineering college. The population consists of the exam grade points, with an A corresponding to 4 points, a B to 3 points, and a C to 2 points. The mean and standard deviation of this population are $\mu = 2.8$ and $\sigma = .7483$.

A simple random sample of $n = 2$ student grade points will be selected. Although such a tiny sample is unusual, and statistical investigations ordinarily involve much larger populations, the miniature scale of this example will simplify our presentation. We will now establish the sampling distribution of the sample mean grade points.

Table 7-2 shows the probability distribution for the random variable $\bar{X}$. The probability values for this sampling distribution were found by identifying those combinations of students giving rise to each possible level for $\bar{X}$. For instance, $\bar{X} = 3.5$ when one of the selected students has an A (4 points) and the other a B (3 points). The following pairs provide this outcome:

(Dan, Gil) (Fran, Gil)

Table 7-2
Sampling Distribution for Sample Mean in Grade Point Values When Selection Is Done Without Replacement

Possible Mean $\bar{x}$	Applicable Student Combinations	$\Pr[\bar{X} = \bar{x}]$
2.0	(Ed, Ida)	.1
2.5	(Dan, Ed) (Dan, Ida) (Ed, Fran) (Fran, Ida)	.4
3.0	(Dan, Fran) (Ed, Gil) (Gil, Ida)	.3
3.5	(Dan, Gil) (Fran, Gil)	.2

Since there are $C_2^5 = 10$ equally likely combinations of student pairs, each having probability .1, it follows that $\Pr[\bar{X} = 3.5] = .2$.

Expected Value and Variance

Like any random variable, $\bar{X}$ has an expected value and variance. We have

$$E(\bar{X}) = \sum_{\text{all } \bar{x}} \bar{x} \Pr[\bar{X} = \bar{x}]$$
$$= 2.0(.1) + 2.5(.4) + 3.0(.3) + 3.5(.2) = 2.8$$

Notice that this value is equal to the mean of the population, $\mu = 2.8$. The sample mean for any random sample is expected to be equal to the population mean, so we have the following.

PROPERTY OF THE SAMPLE MEAN

$$\mathbf{E(\bar{X}) = \mu}$$

The sample mean has expected value equal to the population mean. Thus, when μ is known, we never actually have to compute $E(\bar{X})$.

The variance of $\bar{X}$ may also be found. Using

$$\text{Var}(\bar{X}) = E(\bar{X}^2) - E(\bar{X})^2$$

we have

$$E(\bar{X}^2) = \sum_{\text{all } \bar{x}} \bar{x}^2 \Pr[\bar{X} = \bar{x}]$$
$$= (2.0)^2(.1) + (2.5)^2(.4) + (3.0)^2(.3) + (3.5)^2(.2)$$
$$= 8.05$$

so that

$$\text{Var}(\bar{X}) = 8.05 - (2.8)^2 = .21$$

and the standard deviation of $\bar{X}$ is the square root of this:

$$SD(\bar{X}) = \sqrt{.21} = .458$$

Standard Error of $\bar{X}$

During the planning stage, a sample statistic must be treated as a random variable. Its standard deviation is usually referred to as the **standard error**. This terminology emphasizes that data have not yet been collected, so the statistic is an uncertain quantity that might assume a variety of values according to its sampling distribution. As we have seen by Chebyshev's theorem, probability limits may be placed on how far a statistic may fall above or below its expected level. These limits are established by a random variable's standard deviation, which thus serves as an index for expressing how much error a statistic may achieve in an attempt to hit its expected value target.

The standard error of $\bar{X}$ is so frequently encountered in statistics that a special symbol is used.

STANDARD ERROR OF THE SAMPLE MEAN

$$\sigma_{\bar{X}} = SD(\bar{X})$$

We have, for the student grade-point illustration,

$$\sigma_{\bar{X}} = .458$$

Because we can find $E(\bar{X})$ from the level of the population mean μ, the value of $\sigma_{\bar{X}}$ may be obtained directly from the population standard deviation (when it is known). This can be a significant computational advantage.

There are two cases, depending on whether or not the sample observations are independent. Independence is guaranteed when sampling is done *with* replacement. When sampling is instead done *without* replacement (the usual case), successive observations will be statistically dependent. That dependence has slight effect on probability values, however, when the population is large in relation to the sample size. As a practical matter, the two cases are (1) *large population* (when as a good approximation to reality, independence is assumed) versus (2) *small population* (when accuracy demands that the sample observations be treated as dependent). The following expression may be used to evaluate the latter case.

STANDARD ERROR OF $\bar{X}$ WHEN A SMALL POPULATION APPLIES

$$\sigma_{\bar{X}} = \frac{\sigma}{\sqrt{n}}\sqrt{\frac{N-n}{N-1}}$$

where σ is the population standard deviation, N is the population size, and n is the sample size. You may recognize the term under the last radical as the finite population correction factor encountered in Chapter 3.

Returning to our grade-point illustration, the population standard deviation is $\sigma = .7483$. With $N = 5$ and $n = 2$, we find

$$\sigma_{\bar{X}} = \frac{.7483}{\sqrt{2}}\sqrt{\frac{5-2}{5-1}} = .458$$

which is identical to the value computed earlier directly from the sampling distribution.

The finite population correction factor, $(N - n)/(N - 1)$, is a fraction that approaches 1 when the population size N is large in relation to the sample size n. In those cases we may use the following expression.

STANDARD ERROR OF $\bar{X}$ WHEN A LARGE POPULATION APPLIES

$$\sigma_{\bar{X}} = \frac{\sigma}{\sqrt{n}}$$

Notice that the standard error of $\bar{X}$ is inversely proportional to the square root of the sample size. This indicates that raising n will lower the value of $\sigma_{\bar{X}}$. The likely deviations in the value of $\bar{X}$ from μ will therefore be smaller. This very important concept is fundamental in statistical planning and is useful in selecting an appropriate sample size.

Of course, $\sigma_{\bar{X}}$ is equal to $\sigma/\sqrt{n}$ when sampling *with replacement*—regardless of the population size. In those cases, the sample observations are *independent* random variables.

Mathematical Justification of Results

In the planning stage of the investigation, the sample observations, $X_1, X_2, \ldots, X_n$, are themselves random variables, and $\bar{X}$ is a function of these:

$$\bar{X} = \frac{X_1 + X_2 + \cdots + X_n}{n} = \frac{S_n}{n}$$

In Chapter 5 we established that when the X_k's are independent random variables, each having expected value μ_k and variance σ_k^2, then it follows that the expected value and variance of their sum S_n are

$$E(S_n) = \sum_{k=1}^{n} \mu_k$$

$$\text{Var}(S_n) = \sum_{k=1}^{n} \sigma_k^2$$

When X_k represents the kth sample observation from a population having mean μ and variance σ^2, it follows that for $k = 1, 2, \ldots, n$ that

$$\mu_k = \mu \qquad \text{and} \qquad \sigma_k^2 = \sigma^2$$

and thus,

$$E(S_n) = n\mu$$

The sample mean is $1/n$ times S_n. It follows that

$$E(\bar{X}) = \left(\frac{1}{n}\right) E(S_n) = \mu$$

$$\text{Var}(\bar{X}) = \left(\frac{1}{n}\right)^2 \text{Var}(S_n) = \frac{\sigma^2}{n}$$

and finally that

$$\sigma_{\bar{X}} = SD(\bar{X}) = \frac{\sigma}{\sqrt{n}}$$

Problems

7-1 Compute the standard error of $\bar{X}$ for the following situations where the population is large:

(a)	(b)	(c)	(d)
$\sigma = 50$ lb $n = 100$	$\sigma = \$16$ $n = 64$	$\sigma = .02$ m $n = 25$	$\sigma = 5$ kHz $n = 36$

7-2 Compute the standard error of $\bar{X}$ for the following situations, where the populations are small and selection is without replacement:

(a)	(b)	(c)	(d)
$\sigma = 5''$	$\sigma = .1'$	$\sigma = \$39$	$\sigma = 3$ grams
$N = 500$	$N = 1{,}000$	$N = 1{,}000$	$N = 200$
$n = 100$	$n = 100$	$n = 169$	$n = 25$

7-3 The midterm examination scores for a population of 6 students are as follows:

Student	Score
T. C.	70
C. E.	80
V. L.	60
M. N.	90
N. S.	80
W. T.	70

(a) Compute μ and σ for the population.
(b) For a sample of size 2 selected without replacement, find $E(\bar{X})$ and $\sigma_{\bar{X}}$.

7-4 Construct a table providing the complete sampling distribution for $\bar{X}$ when $n = 2$ random observations are taken without replacement from the population in Problem 7-3.

7-5 Construct a table providing the complete sampling distribution for $\bar{X}$ when $n = 3$ random observations are taken without replacement from the population in Problem 7-3.

7-6 Consider again the student grade-point population in Table 7-1.
(a) Construct a table providing the complete sampling distribution for $\bar{X}$ when $n = 2$ and the selection is made *with* replacement.
(b) Compute from your table (1) $E(\bar{X})$ and (2) $\sigma_{\bar{X}}$.

7-7 Construct a table providing the complete sampling distribution for $\bar{X}$ when $n = 2$ random observations are taken *with* replacement from the population in Problem 7-3.

7-2 Sampling Distribution of $\bar{X}$ When Population Is Normal

We have seen how the sample mean $\bar{X}$ may be treated as a random variable. The sampling distribution for $\bar{X}$ provides probabilities for the possible levels of this variable. When the characteristics of a population are known, we may find that sampling distribution by exercising basic probability concepts. With a small population such as the student grade points in Section 7-1, this involves counting and grouping possible sample outcomes according to the possible level for $\bar{X}$. A different approach should be taken for large populations, however. In this section we consider sampling from a population whose frequency distribution is characterized by a normal curve. Our evaluation of a normally distributed population will provide the necessary background for finding sampling distributions under more general circumstances.

The Normal Distribution for $\bar{X}$

Property of Sample Mean

When the sample mean is computed for a random sample of size n taken from a *normally distributed population* having parameters μ and σ, the sampling distribution of $\bar{X}$ is *also* normally distributed with mean μ and standard deviation $\sigma_{\bar{X}} = \sigma/\sqrt{n}$.

As we have seen, many physical dimensions for both people and objects are closely represented by the normal distribution. In these cases, knowing μ and σ establishes the sampling distribution for the mean of any sample.

Illustration: *Reach and Hand Length*

As an illustration, consider again the anthropometric data for the reach of adult males first described on page 176. The population may be assumed to be normally distributed, and the following parameters apply:

$$\mu = 32.33''$$

$$\sigma = 1.63''$$

The top normal curve in Figure 7-2 represents the frequency distribution of this population.

Since the above parameters are known, it would be unnecessary for an investigator actually to collect sample data on reach. Samples are necessary only when the population has unknown characteristics. But we may use this population to illustrate the concept of the sampling distribution for $\bar{X}$.

Suppose that a random sample of $n = 100$ men will be selected and the reach measured for each. The sampling distribution for the sample mean is normally distributed, with the same mean as the population's, $\mu = 32.33''$, and standard deviation

$$\sigma_{\bar{X}} = \frac{\sigma}{\sqrt{n}} = \frac{1.63''}{\sqrt{100}} = .163''$$

Notice that this standard deviation is only one-tenth as large as that of the population itself. The sampling distribution for $\bar{X}$ exhibits less dispersion than the population's. This is reflected in the normal curve for $\bar{X}$ at the bottom of Figure 7-2, which has its mass concentrated more tightly about μ.

The shaded areas under each curve further illustrate this concept. Each area represents the frequency of reaches between 32.00″ and 32.50″. The proportion of

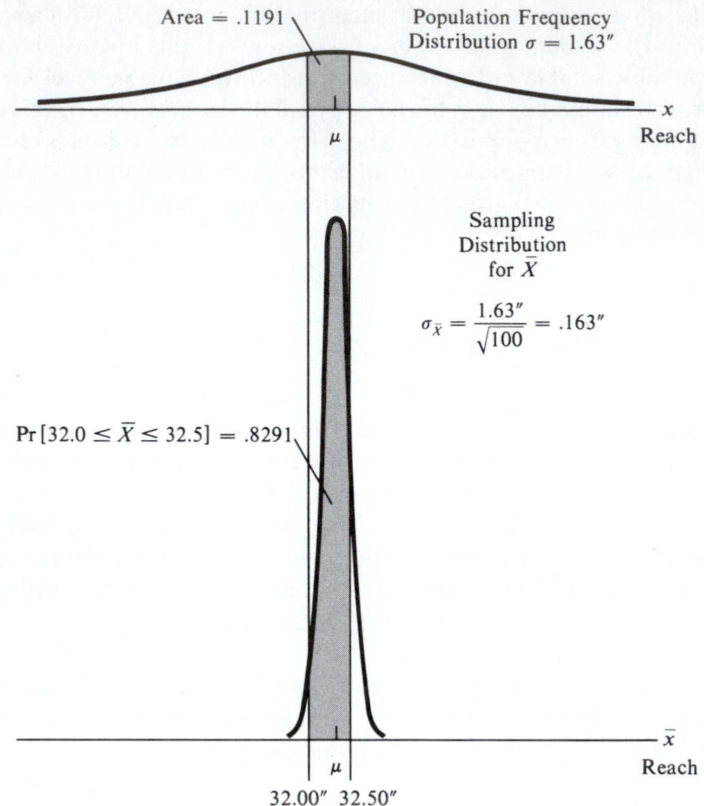

Figure 7-2 *Population frequency curve and sampling distribution for sample mean reach of adult males.*

men in the population whose reaches fall within this range is

$$\begin{aligned}\text{Area} &= \Phi\left(\frac{32.50 - 32.33}{1.63}\right) - \Phi\left(\frac{32.00 - 32.33}{1.63}\right)\\ &= .5398 - .4207 = .1191\end{aligned}$$

But the probability is much higher that the sample mean reach of 100 men will fall inside the same limits:

$$\begin{aligned}\Pr[32.00'' \leq \bar{X} \leq 32.50''] &= \Phi\left(\frac{32.50 - 32.33}{.163}\right) - \Phi\left(\frac{32.00 - 32.33}{.163}\right)\\ &= .8508 - .0217 = .8291\end{aligned}$$

One helpful interpretation of this fact is to imagine several different samples, each of size n, taken from the same population. The respective computed values for $\bar{X}$ will tend to be alike, and their average should be very near to μ. The sample means will be much closer together in value than would a similar collection of individual population values. (We know this because $\sigma_{\bar{X}}$ is smaller than σ.)

The Role of the Standard Error

As we have seen, the standard error of $\bar{X}$ determines how closely to μ the computed sample mean will be likely to fall. A sample involving a small level for $\sigma_{\bar{X}}$ should provide a computed $\bar{X}$ that more faithfully represents the population center than would a sample where $\sigma_{\bar{X}}$ is large. In other words, sampling error decreases as $\sigma_{\bar{X}}$ does.

For normally distributed populations there are two determinants of the level for $\sigma_{\bar{X}}$. These are the population standard deviation σ and the sample size n. (The population size is not a factor, since N is theoretically infinite with normal populations.)

Consider first the effect of σ on the level of the standard error. Since $\sigma_{\bar{X}}$ is equal to $\sigma/\sqrt{n}$, the standard error of $\bar{X}$ is directly proportional to the population standard deviation. In absolute terms, the amount of error in the computed value for $\bar{X}$, relative to the true population mean, will be greater when sampling from a population having a large standard deviation than from one where σ is small.

For example, anthropometric data indicate that adult males have a mean hand length of $\mu = 7.49''$, with a standard deviation of $\sigma = .34''$. The standard deviation for this population is smaller than the earlier one for reach, where $\sigma = 1.63''$. The respective values of $\sigma_{\bar{X}}$ will reflect this difference in σ size. Suppose that a random sample of $n = 25$ men is measured both for hand length and for reach. The respective standard errors are

$$\sigma_{\bar{X}} = \frac{.34''}{\sqrt{25}} = .068'' \qquad \text{(hand length)}$$

$$\sigma_{\bar{X}} = \frac{1.63''}{\sqrt{25}} = .326'' \qquad \text{(reach)}$$

Figure 7-3 shows the respective normal curves. Notice that it is virtually certain that $\bar{X}$ (hand length) will fall within $\mu \pm .25''$, but that there is less than a 56% chance for $\bar{X}$ (reach) falling within .25″ of the mean of the reach population. In absolute terms, the smaller σ for hand length leads to less sampling error, with a computed $\bar{X}$ that is more likely to be close to its μ target (in absolute terms).

***Example:** Distortion Frequencies in Radiotelescopes*

An engineer is evaluating the design for a new radiotelescope amplifier. Prototype testing will determine whether the new device can operate distortion free at higher frequencies than the present unit, where the mean maximum distortion frequency is 50 kHz and the standard deviation is 5 kHz.

The engineer is, of course, uncertain about μ, the mean maximum distortion frequency of the proposed unit. He will recommend his new design only if the sample counterpart figure from $n = 100$ test shots falls above 50 kHz. Assuming that the new amplifier is clearly superior and the true level for μ is 51 kHz, what is the probability that the computed $\bar{X}$ will be so atypically small that the design will fail this test?

Figure 7-3 *Normal curves for adult male anthropometric data.*

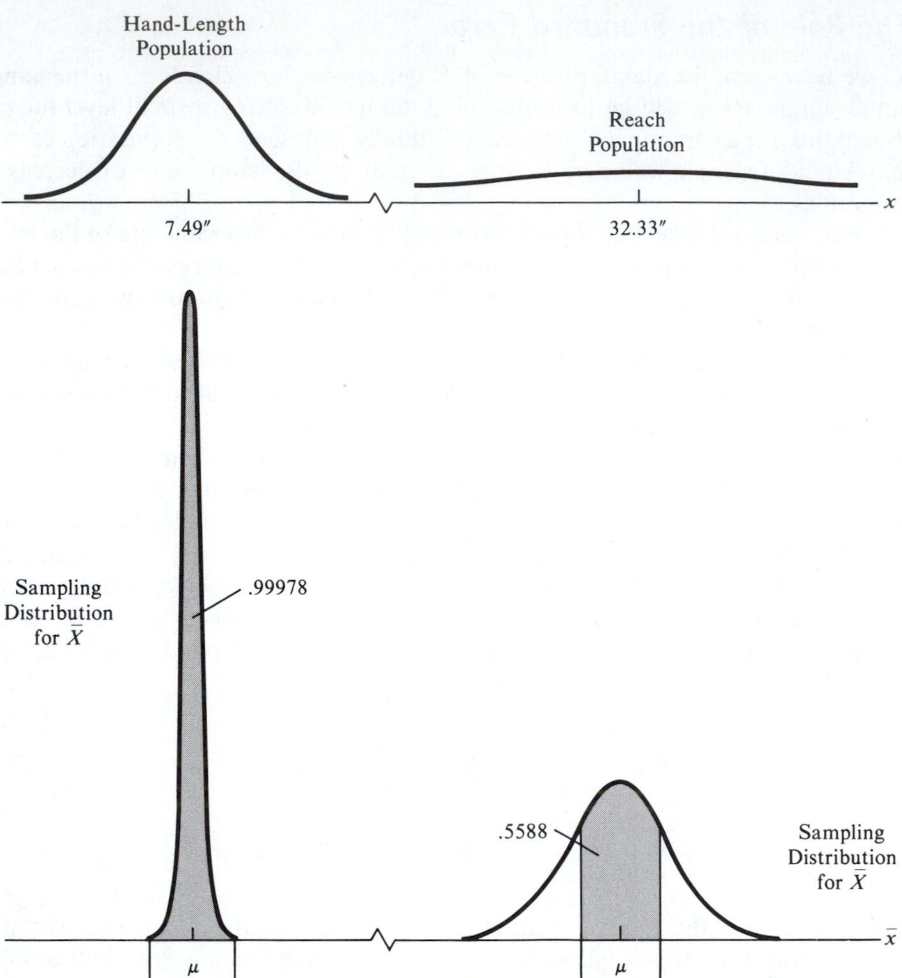

The hand-length population exhibits less variability than the reach population. This is reflected by the smaller standard error of $\bar{X}$ for a sample of 25 men's hand lengths and the tighter clustering of possible values for that mean.

The probability will depend on σ, which, like μ, is unknown. Suppose $\sigma = 5$ kHz (the same as on the present system). Suppose also that the population of maximum distortion frequencies on test shots is normally distributed. Then

$$\sigma_{\bar{X}} = \frac{5}{\sqrt{100}} = .5 \text{ kHz}$$

and

$$\Pr[\bar{X} \leq 50 \mid \mu = 51 \text{ kHz}] = \Phi\left(\frac{50 - 51}{.5}\right) = .0228$$

Should the population standard deviation be greater, say $\sigma = 10$ kHz, then the probability that the amplifier design will fail the test (even though truly better) is substantially greater:

$$\sigma_{\bar{X}} = \frac{10}{\sqrt{100}} = 1.0 \text{ kHz}$$

$$\Pr[\bar{X} \le 50 \mid \mu = 51 \text{ kHz}] = \Phi\left(\frac{50 - 51}{1}\right) = .1587$$

A larger σ gives rise to a larger level for $\sigma_{\bar{X}}$. This increases the chance of obtaining an erroneous sample result.

The second influence on $\sigma_{\bar{X}}$ is the sample size. The standard error of $\bar{X}$ is inversely proportional to the square root of n. Greater fidelity in $\bar{X}$ will therefore be achieved as n becomes larger (which makes $\sigma_{\bar{X}}$ smaller).

Figure 7-4 shows the sampling distributions for mean reach $\bar{X}$ when $n = 25$ and when $n = 100$. Notice that the normal curve for the larger sample size (and smaller standard error) exhibits less dispersion, with possible levels of $\bar{X}$ clustering more tightly about μ. When $n = 100$, the probability that $\bar{X}$ falls within $\mu \pm .25''$ is .8740, compared with only .5588 for the like event when $n = 25$.

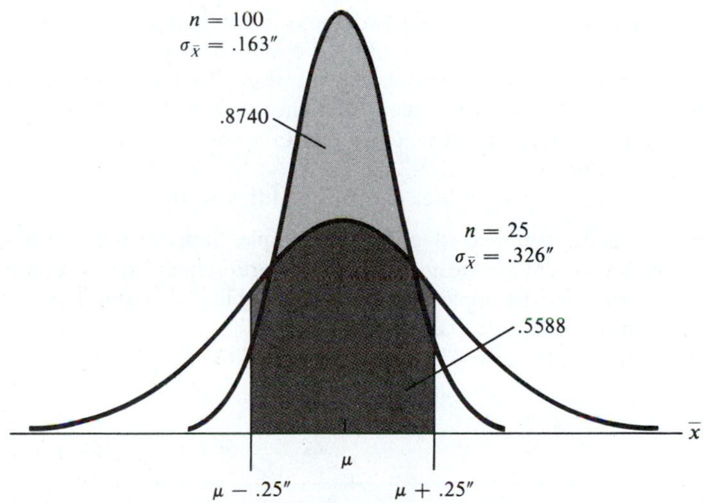

Figure 7-4 *Sampling distributions for mean reach $\bar{X}$ when $n = 25$ and $n = 100$.*

Problems

7-8 The height of adult males in a particular city is normally distributed, with $\mu = 69.5''$ and standard deviation $\sigma = 2.65''$. Find (1) the standard error of $\bar{X}$ and (2) the probability that $\bar{X}$ falls within $\mu \pm .5''$ when a random sample is taken of size
(a) $n = 10$ (b) $n = 25$ (c) $n = 50$ (d) $n = 100$

7-9 The operator reaction time to simulated nuclear-reactor cooling-system failure events is believed to be normally distributed, with a mean $\mu = 45$ seconds and standard deviation $\sigma = 8$ seconds. Find (1) the standard error of $\bar{X}$ and (2) the probability that $\bar{X}$ falls within $\mu \pm 1$ second when a random sample is taken of size
(a) $n = 25$ (b) $n = 100$ (c) $n = 200$ (d) $n = 500$

7-10 The elongation of a steel bar under a particular tensile load may be assumed to be normally distributed, with a mean $\mu = .06''$ and standard deviation $\sigma = .008''$. A sample of $n = 100$ bars is subjected to the test. Find the probability that the sample mean elongation falls
(a) above .062″ (c) between .059″ and .061″
(b) below .0505″ (d) between .0585″ and .0605″

7-11 A sulfonating process yields a final output whose pH fluctuates with each successive barrel. The mean μ for these is unknown, although the standard deviation is assumed to be $\sigma = .25$. This population is assumed to be normally distributed. The chief chemical engineer orders replacement of the control valves whenever the sample mean pH of $n = 25$ barrels falls below 6.90 or above 7.10.
(a) Calculate the standard error of $\bar{X}$.
(b) Determine the probability
 (1) of replacing the valves when the process is operating as intended and $\mu = 7.0$
 (2) of replacing the valves when the process is slightly off target and the mean pH is $\mu = 7.02$
 (3) of failing to replace the valves when the process is too alkaline and the mean pH is $\mu = 7.15$
 (4) of failing to replace the valves when the process is too acidic and the mean pH is $\mu = 6.80$

7-12 A ceramics engineer will be testing two epoxy substances to determine the average setting time until bonding. The population of individual setting times is believed to be normally distributed with some unknown mean μ and standard deviation σ. The test involves $n = 25$ trials. Find the probability that the sample mean setting time deviates from the unknown μ by no more than 1 hour, assuming that the standard deviation has value
(a) $\sigma = 2$ (b) $\sigma = 4$ (c) $\sigma = 8$ (d) $\sigma = 16$

7-13 Consider again the radiotelescope design example. Suppose that the new design is actually inferior, so that the mean maximum distortion frequency is 49 kHz. The same decision criterion will be applied. Find the probability that the new design will be incorrectly not recommended, assuming
(a) $\sigma = 6$ kHz (b) $\sigma = 7$ kHz (c) $\sigma = 10$ kHz

---

7-3 *Sampling Distribution of $\bar{X}$ for a General Population*

One reason the normal distribution is important in statistics is that so many population frequency distributions may be represented by normal curves. But keep in mind that the normal curve is a theoretical construct that provides a convenient approximation to reality. (Remember, it assigns some probability to the mile-tall man!)

But what makes the normal distribution crucial to statistics is the role it plays in representing the sampling distribution for $\bar{X}$. We have just seen that when the population is normally distributed, then the sampling distribution for $\bar{X}$ must *also* be a normal distribution (with the same mean, but with standard deviation $\sigma_{\bar{X}}$). A surprising fact is that this same property can apply approximately even when the population is *not* normally distributed. This fact is established by the **central limit theorem**.

Central Limit Theorem

Because so many statistical procedures are based on the sample mean, the following theorem exhibits one of the most useful properties in statistics.

> Consider a population having mean μ and finite standard deviation σ. Let $\bar{X}$ represent the mean of n independent random observations from this population. Then, as n becomes large, the sampling distribution of $\bar{X}$ tends toward a normal distribution with mean μ and standard deviation $\sigma_{\bar{X}} = \sigma/\sqrt{n}$.

Central Limit Theorem

This theorem permits us to *approximate* the sampling distribution for $\bar{X}$ by an appropriate normal curve—*regardless of the form of the population frequency distribution.*

Figure 7-5 illustrates this concept for 4 underlying populations having frequency curves of radically different shapes. The exact sampling distributions for $\bar{X}$ are pictured in each case for 3 sample sizes. When $n = 2$, the shapes of the respective sampling distributions differ considerably. They become rounder and more symmetrical at $n = 4$, although each case still has a distinctive shape. But from all populations, a smooth, nearly symmetrical bell-shaped curve applies for $\bar{X}$ when $n = 25$. In most statistical applications the sample size will be large enough so that we need not be concerned about the quality of the normal approximation. Indeed, we have already noted that in some sense the normal curve is *always* an approximation anyway.

Finding Probabilities for $\bar{X}$

We may use the normal curve to assign probabilities to possible levels for $\bar{X}$ without worrying too much about the underlying population characteristics.* The central limit theorem requires that the sample observations be independent—a feature guaranteed when the population values are generated from a continuous process (so that the sampled population is theoretically of infinite size) or when sampling with replacement from any population. Although not strictly permitted by the central limit theorem, the normal approximation may be stretched to include sampling without replacement.

* A finite variance is assumed to apply.

Figure 7-5
Illustration of the central limit theorem, showing the tendency toward normality in the sampling distribution of $\bar{X}$ as n increases, for various populations.

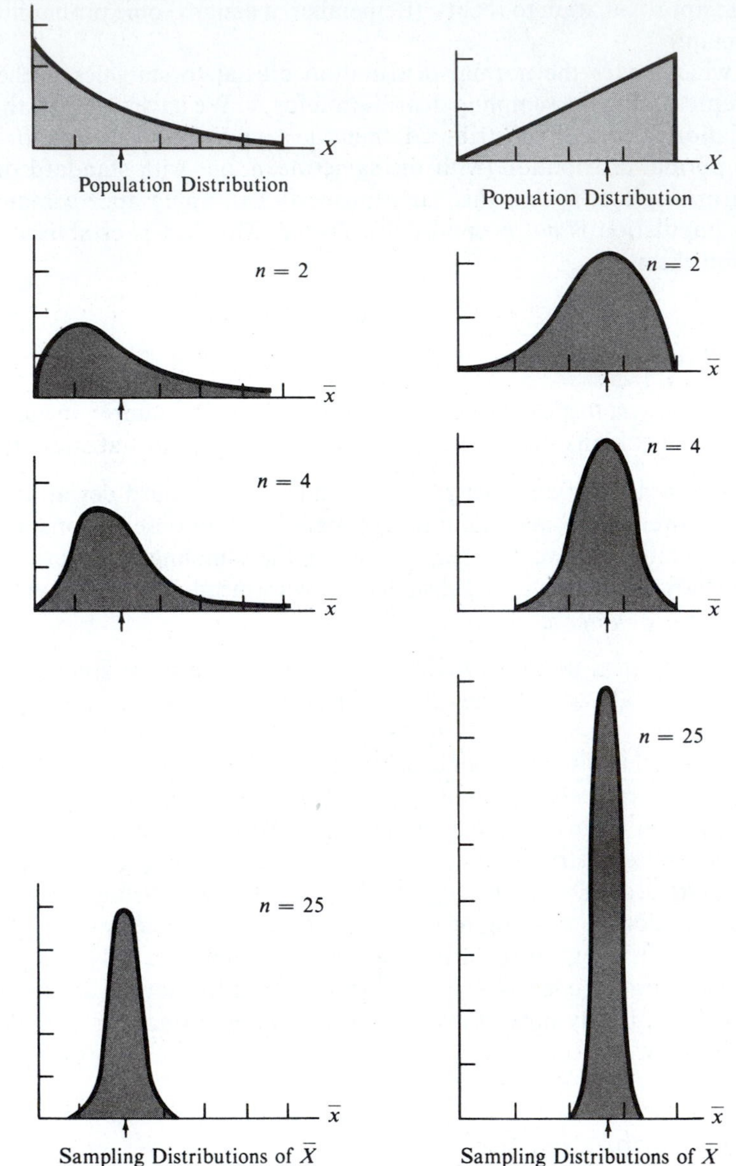

Example: *Mean Time between Failures*

Electronic components often have lifetimes that empirically have been shown to fit an exponential distribution (that is, over time, failures constitute a Poisson process). The underlying population of lifetimes of all components of a particular type being manufactured may then be represented by a frequency curve such as the one at the top left corner in Figure 7-5. Consider one such item for which the mean failure rate λ is unknown. Although it might take on many possible values, the mean rate can be assumed to be a particular value, such as $\lambda = .01$

Figure 7-5
(continued)

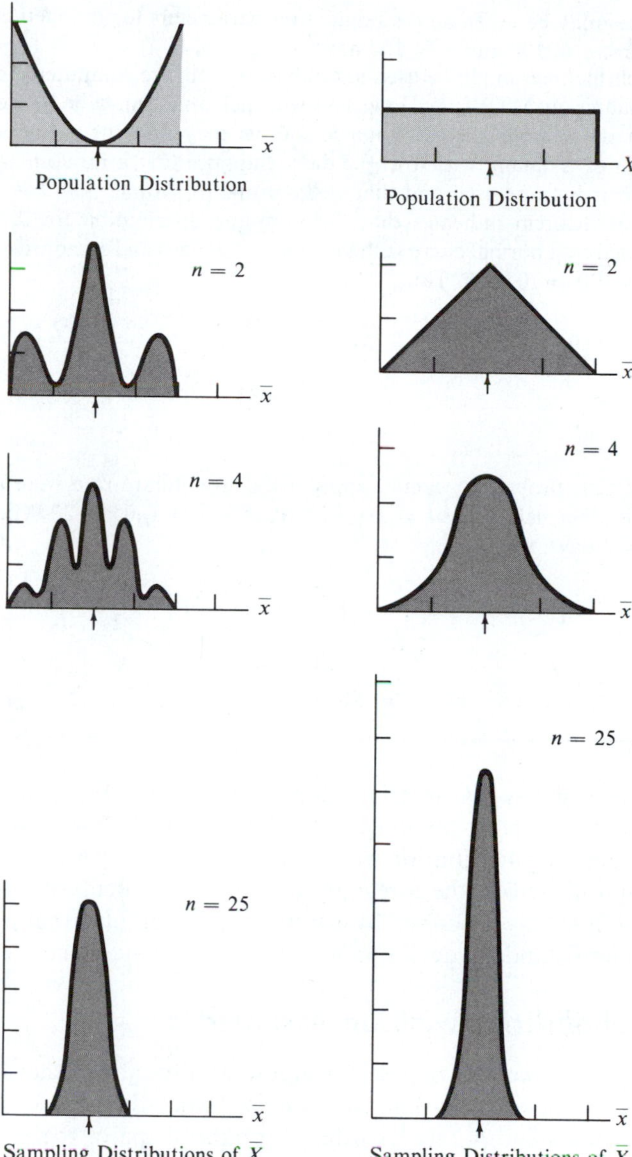

Source: Lawrence L. Lapin, *Statistics for Modern Business Decisions*, 3rd ed. © 1982 Harcourt Brace Jovanovich, Inc., New York.

item per day. In that case the expected time X until an individual item fails is given below, along with the standard deviation:

$$E(X) = \frac{1}{\lambda} = 100 \text{ days}$$

$$SD(X) = \frac{1}{\lambda} = 100 \text{ days}$$

These values may be taken as the population parameters for the lifetimes of all similar items: $\mu = 100$ and $\sigma = 100$ days.

Sample lifetimes might be used to establish μ from the computed mean time between failures, or MTBF, $\bar{X}$. How close will such an estimate lie to the true μ?

From the sampling distribution for $\bar{X}$ we may find the probabilities of events, such as $\bar{X}$ falling within $\mu \pm 5$ days. Suppose that a random sample of $n = 100$ items is to be used and that we continue to assume that $\lambda = .01$. The central limit theorem indicates that the sampling distribution for $\bar{X}$ may be approximated by a normal curve with mean $\mu = 100$ days and standard deviation $\sigma_{\bar{X}} = 100/\sqrt{100} = 10$ days. Thus,

$$\begin{aligned}\Pr[95 \leq \bar{X} \leq 105] &= \Phi\left(\frac{105 - 100}{10}\right) - \Phi\left(\frac{95 - 100}{10}\right) \\ &= .6915 - .3085 \\ &= .3830\end{aligned}$$

A different probability would apply if the true failure rate were assumed to be some other value, say $\lambda = .05$, so that $\mu = \sigma = 1/.05 = 20$. Then $\sigma_{\bar{X}} = 20/\sqrt{100} = 2$ days, and

$$\begin{aligned}\Pr[15 \leq \bar{X} \leq 25] &= \Phi\left(\frac{25 - 20}{2}\right) - \Phi\left(\frac{15 - 20}{2}\right) \\ &= .9938 - .0062 \\ &= .9876\end{aligned}$$

This example shows that to establish probabilities for $\bar{X}$ we must often make detailed assumptions about the underlying population. If those assumptions are incorrect, the sampling distribution will reflect those errors. The exponential population distribution used in the foregoing example is particularly volatile in this respect because it is fully described by a single parameter λ (so that $\mu = \sigma = 1/\lambda$), and the mean and standard deviation are always the same value.

Finding Probabilities with an Assumed σ

It more general cases, we may precisely compute the probability that $\bar{X}$ falls within μ plus or minus some error by knowing σ only. (Thus, there is no need to make possibly erroneous assumptions regarding μ or the shape of the population frequency curve.) Although the true value of σ is also generally unknown when μ is uncertain, a "ballpark" level for σ may be easily found from a related population for which sample data have already been collected.

For example, consider using $\bar{X}$ to estimate the mean instep length of adult males. Suppose that anthropometric data are available only for above-the-waist measurements. Since instep and hand lengths are nearly the same for most people, we might assume that the two populations also have similar variability. During the planning phase for an anthropometric study involving below-the-waist measurements, we might then use the value $\sigma = .34''$, which is known to apply for hand length, to represent the standard deviation for instep length.

Example: *Starting Salaries of New Petroleum Engineers*

The editor of a professional engineering journal wishes to estimate the mean starting salaries of newly graduated engineers in various specialties. Because of that field's growth, petroleum engineers will be categorized separately for the first time. (In previous years, they had been included with chemical engineers or lumped together with mineral engineers.)

Although a large-sample salary data base will be compiled and refined from university and industry sources, that will take a year to complete. A more current (and smaller) sample of starting salaries is needed for immediate publication—and some sampling error must therefore be tolerated. A sample of $n = 25$ petroleum seniors has been selected, and in June a reporter will contact each to determine his or her starting salary. From these responses, the mean starting salary $\bar{X}$ will be obtained. The editor wishes to establish the probability that the sample mean will fall within $\pm\$500$ of the true population mean salary.

Historically, salaries have exhibited a positively *skewed* distribution so that the frequency curves have presented thin upper tails extending tens of thousands of dollars above the mean (and short lower tails touching only thousands to the left of μ). The exact shape of the frequency curve for petroleum engineers' salaries would be similar. The central limit theorem indicates that $\bar{X}$ will have a normal curve regardless, and that the shape of this sampling distribution depends only on the population σ. The standard deviation for petroleum engineers' salaries is of course also unknown, but the editor believes that it will be close to last year's value of \$1,057 for the broad mineral-engineering category. For planning purposes a value of \$1,100 is used.

Under this assumption, the standard error of $\bar{X}$ is $\sigma_{\bar{X}} = \$1{,}100/\sqrt{25} = \$220$, and the desired probability is

$$\begin{aligned}\Pr[\mu - \$500 \le \bar{X} \le \mu + \$500] &= \Phi\left(\frac{\$500}{\$220}\right) - \Phi\left(\frac{\$500}{\$220}\right)\\ &= .9884 - .0116\\ &= .9768\end{aligned}$$

The editor is satisfied with the high reliability that this indicates will be achieved from his sample of petroleum engineers.

Sampling from Small Populations

As noted, the sampling distribution of $\bar{X}$ may be closely approximated by the normal curve even when sampling without replacement from a finite population. In those cases, the following expression applies for the standard error:

$$\sigma_{\bar{X}} = \frac{\sigma}{\sqrt{n}}\sqrt{\frac{N-n}{N-1}}$$

which involves the finite population correction. The latter may usually be ignored when N is large. As a practical rule of thumb, *the finite population correction should be used whenever* n *exceeds 10% of the population size.*

For example, 183 students will be granted bachelor's degrees at a particular engineering school. Of these, only $N = 154$ will enter the job market. A sample of $n = 25$ of these graduates will be polled to determine the starting salaries of the entire class. The standard error of $\bar{X}$ is

$$\sigma_{\bar{X}} = \frac{\$900}{\sqrt{25}}\sqrt{\frac{154 - 25}{154 - 1}} = \$165.28$$

The probability that $\bar{X}$ falls within \$500 of μ is

$$\begin{aligned}\Pr[\mu - \$500 \le \bar{X} \le \mu + \$500] &= \Phi\left(\frac{\$500}{\$165.28}\right) - \Phi\left(\frac{\$500}{\$165.28}\right)\\ &= .9988 - .0012\\ &= .9976\end{aligned}$$

Although the size N of the population and how it relates to the sample size n is important, keep in mind that N is not within the investigator's control. Only the size for n can be chosen. And, since its level will affect the quality of the sample results, n must be chosen carefully. As we shall see, larger n's are more reliable than smaller n's—no matter what N happens to be.

Problems

7-14 Suppose that the failure rate of the electrical components in the example on page 276 is $\lambda = .02$ per day. Find the probability that from a sample size of $n = 25$, the computed $\bar{X}$ will fall
(a) above 55 days (c) between 45 and 60 days
(b) below 40 days (d) either below 35 days or above 65 days

7-15 Suppose that the standard deviation applicable to the petroleum engineers' salaries in the text example should be larger, owing to increasing worldwide oil exploration activity. Using $n = 25$, find the probability that $\bar{X}$ falls within $\mu \pm \$500$ when
(a) $\sigma = \$1{,}500$ (b) $\sigma = \$2{,}000$ (c) $\sigma = \$2{,}500$

7-16 The time that each plane waits for service at an aircraft repair facility may be assumed to be exponentially distributed. A sample of 25 aircraft waiting times will be collected. Assuming that planes are removed from the holding area at a mean rate of 2 per day, determine the probability that the sample mean waiting time
(a) does not exceed .6 day (c) equals $.5 \pm .1$ day
(b) is greater than .7 day (d) is between .3 and .7 day

7-17 A computer generates pseudorandom decimals. Each decimal has an expected value of .5 and a standard deviation of $1/\sqrt{12}$. Consider a particular sequence of 12 numbers.
(a) Determine the standard error for the mean of those 12 random decimals.
(b) Find the probability that the mean of the 12 random decimals is
(1) $\le .4$ (2) $> .65$ (3) $< .35$ (4) $\le .75$

7-18 An engineer wishes to evaluate a high-frequency sonar transducer. The maximum effective echo-ranging distance at a particular frequency should be several miles. The mean effective distance is unknown, although the same standard deviation may be assumed as that of an existing transducer, for which $\sigma = 5$ nautical miles (NM).

An experiment will be conducted with n shots at a series of buoys spaced at various depths and distances throughout a test range. Find the probability that the sample mean maximum effective distance will lie within ± 1 NM of the unknown population mean when the sample size is

(a) 25 (b) 64 (c) 100 (d) 625

7-19 A quality-control inspector accepts shipments of 500 precision $\frac{1}{2}''$ steel rods if the mean diameter of a sample of $n = 100$ falls between .4995″ and .5005″. Previous evaluations have established that the standard deviation for individual rod diameters is .003″.

(a) What is the probability that the inspector will accept an out-of-tolerance shipment having $\mu = .5003''$?

(b) What is the probability that the inspector will reject a near-perfect shipment having $\mu = .4999''$?

7-4 The Student t Distribution

So far our discussion of sampling distributions has focused on $\bar{X}$, which is proper because that statistic is the one most commonly used. When the sample size is large enough, $\bar{X}$ may be represented by a normal curve with mean μ and standard deviation $\sigma_{\bar{X}}$, and this is ordinarily true regardless of the underlying population frequency distribution. Should the population itself be normally distributed, this relation is exact no matter how small n is. In either case, probabilities for $\bar{X}$ may be obtained from areas under the standard normal curve, using the following transformation into the standard normal random variable:

$$Z = \frac{\bar{X} - \mu}{\sigma_{\bar{X}}} = \frac{\bar{X} - \mu}{\sigma/\sqrt{n}}$$

In this section we are still concerned with $\bar{X}$, but now we want to find the sampling distribution for $\bar{X}$ by estimating σ from the sample results (rather than using an assumed value for σ, as we did before). In order to do this, we should use the Student t distribution instead of the normal distribution.

The Student t Statistic

In many statistical applications, the population parameters are estimated from their sample counterparts: $\bar{X}$ for μ and s for σ. This suggests a way around our lack of knowledge about σ—substitute s for σ in the previous expression for Z. Doing this we obtain

$$\frac{\bar{X} - \mu}{s/\sqrt{n}}$$

Before data are collected, $\bar{X}$ and s are random variables, so this expression is a composite random variable that is itself a function of those two quantities.

Can we use this new random variable to assign probabilities in the same fashion that we used Z?

This question was answered by the pioneer statistician W. S. Gosset, who published his results under the pen name "Student." (He used a pseudonym because he worked for a brewery that prohibited employees from publishing their findings.) So that his new random variable would not be confused with Z, Gosset used the letter t, and in his honor, the above quantity is referred to as the **Student t statistic.**

STUDENT t STATISTIC

$$t = \frac{\bar{X} - \mu}{s/\sqrt{n}}$$

The Student t is a continuous random variable whose probability distribution is completely specified by a single parameter, referred to as **the number of degrees of freedom** (df). For the Student t distribution this parameter is determined by the sample size, and here df $= n - 1$. The density curve when df $= 4$ is shown in Figure 7-6.

The degrees of freedom are used by statisticians to indicate the minimal number of free terms in a collection of variables that must sum to a fixed total. The Student t statistic is computed from n sample observations $X_1, X_2, \ldots, X_n$. However, when the level of $\bar{X}$ is fixed, there is freedom to specify the values of $n - 1$ individual X_k's, and the final X_k must balance them all so that $\sum X_k/n = \bar{X}$. In other words, there are only $n - 1$ degrees of freedom for computing $\bar{X}$ (and also s), and hence t.

The probability distribution for t was derived by assuming that each sample observation is selected randomly from a *normally distributed population* having some mean μ and finite standard deviation σ (although the values of these parameters need not be specified). This is a very restrictive condition that narrows the use of the Student t distribution to situations in which the sampled population has a frequency distribution that resembles a bell-shaped normal curve and is nearly symmetrical. Despite this limitation, the Student t distribution may be used in a wide variety of statistical applications.

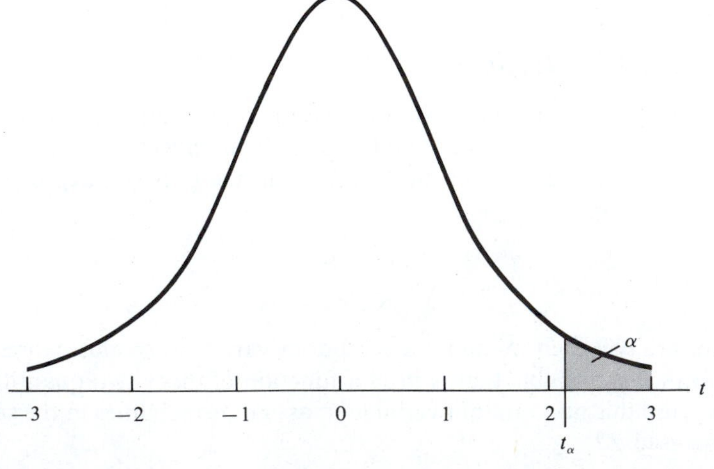

Figure 7-6 *Density curve for the Student t distribution for 4 degrees of freedom.*

The algebraic expression for the Student t density is cumbersome, and areas under the curve must be found through numerical integration. Appendix Table G allows us to evaluate upper-tail probabilities,

$$\alpha = \Pr[t > t_\alpha]$$

For any one of several levels of α, a **critical value t_α** may be read from the table. For example, when df = 4 and the upper-tail probability is $\alpha = .05$, the critical value is read from Table G as $t_{.05} = 2.132$, and thus

$$.05 = \Pr[t > 2.132] \qquad (\text{df} = 4)$$

Similarly, when the upper-tail probability is specified at $\alpha = .01$, the tabled critical value is $t_{.01} = 3.747$, and

$$.01 = \Pr[t > 3.747] \qquad (\text{df} = 4)$$

Appendix Table G has a strange layout compared with the familiar table for areas under the normal curve (Table D). This is partly a space-saving measure, since a separate probability distribution applies for each number of degrees of freedom. Critical values can be listed for only a few levels of α. Also, most statistical applications involve looking up a critical value that matches a particular tail area (rather than matching an α to a particular level of t). There is a separate row of critical values in Table G for each df. For example, when df = 30, then for $\alpha = .01$, we read $t_{.01} = 2.457$ and

$$.01 = \Pr[t > 2.457] \qquad (\text{df} = 30)$$

Like the normal distribution, the Student t is *symmetrical*, so the area above any level t_α must be equal to the area below $-t_\alpha$. For example,

$$.01 = \Pr[t < -2.457] \qquad (\text{df} = 30)$$

Thus, probabilities may be established for any range of values. For instance, there is a .90 probability that the computed t will fall within $\pm t_{.05}$ and a .99 chance that t lies within $\pm t_{.005}$.

The Student t and Normal Curves

Recall that t plays the same role as the standard normal random variable Z, except that Z involves a specified σ whereas t is based on the yet to be computed s. Let's compare these two probability distributions. Figure 7-7 shows the density curve for both distributions. The shapes are similar, although the curve for Student t has thicker tails, so the random variable exhibits greater variability than Z does. This is easy to justify intuitively. Z reflects less uncertainty about where $\bar{X}$ will fall because more population information is available (σ is specified). That same $\bar{X}$ uncertainty is magnified in t because the population variability is also uncertain (σ is unknown).

However, the density curve for t approaches the shape of the standard normal curve as the number of degrees of freedom becomes large. This tends to occur rapidly with large sample sizes.

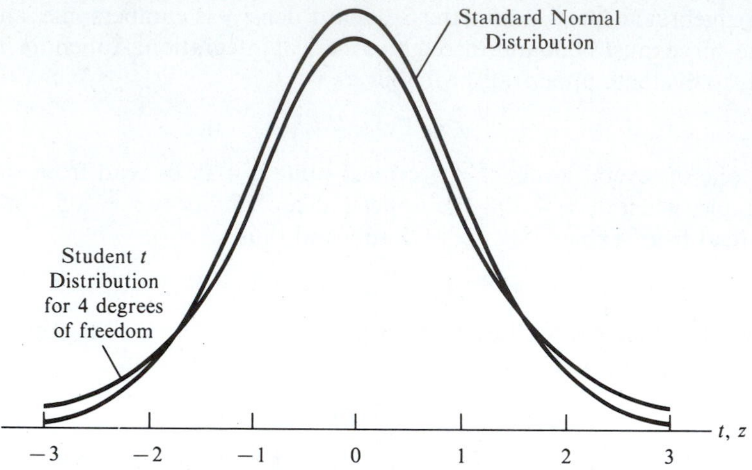

Figure 7-7 *Density curves for standard normal and Student t distributions.*

Example:* *Metal Fatigue and Aircraft Accidents

A Federal Aviation Agency statistician is interested in those aircraft crashes in which metal fatigue in structural members is an indicated cause. He has collected wing segments from 8 different downed aircraft of a particular model and will be performing fatigue tests that measure the elongation of strips under various weight loads. He theorizes that the metal from the downed planes will have weakened with time and stress. Although a larger sample size than 8 would be ideal, the investigation is limited by the actual number of crashes.

Suppose that the mean elongation of new metal of the same type under a particular load is .05″. The statistician will conclude that μ for wing segments in crashed aircraft is larger only if the sample measurements are found to violate that hypothesis significantly. How can he determine whether his test results are significant?

The statistician establishes a rule providing a probability of only 1% of incorrectly concluding that metal strips taken from crash members experience greater distortion than new strips when in fact the two metals have identical characteristics so that $\mu = .05''$. In terms of the Student t statistic, we see that for df $= n - 1 = 8 - 1 = 7$,

$$.01 = \Pr[t > t_{.01}]$$

when $t_{.01} = 2.998$ (from Table G). The following rule therefore meets the stated criterion, as applied to the t computed from the sample data:

Conclude crash and new metals are identical if $t \leq 2.998$.

Conclude crash metals have greater distortion if $t > 2.998$.

The test results provide $\bar{X} = .13''$ with $s = .04''$. Thus,

$$t = \frac{\bar{X} - \mu}{s/\sqrt{n}} = \frac{.13 - .05}{.04/\sqrt{8}} = 5.66$$

Since the computed value exceeds 2.998, the statistician may safely conclude that structural members from crashed aircraft experience significantly greater distortion than new metals.

Problems

7-20 Find the critical values for the following tail areas:

(a)	(b)	(c)	(d)
$\alpha = .05$ df = 11	$\alpha = .10$ df = 6	$\alpha = .005$ df = 25	$\alpha = .0005$ df = 30

7-21 The drying times of 15 sample epoxy substances are assumed to be normally distributed with unknown mean and standard deviation. Find the critical value above which the probability is α that the computed t will fall when
(a) $\alpha = .05$ (b) $\alpha = .01$ (c) $\alpha = .005$ (d) $\alpha = .0005$

7-22 Establish the probability that the computed t from a sample of size $n = 25$ will fall
(a) above 1.711 (d) above 3.091
(b) below .685 (e) between .256 and 3.745
(c) below 2.797

7-23 Suppose that the FAA statistician in the example obtained the following results under different loadings, each with 8 sample metal segments from the crashed aircraft. Compute t and determine for each sample which conclusion should be made, using the rule given in the example.

Loading	Mean Elongation for New Metal	Computed Values	
(a) Light	$\mu = .01''$	$\bar{X} = .008''$	$s = .002''$
(b) Moderate	$\mu = .03''$	$\bar{X} = .07''$	$s = .02''$
(c) Heavy	$\mu = .10''$	$\bar{X} = .18''$	$s = .07''$
(d) Very heavy	$\mu = .20''$	$\bar{X} = .43''$	$s = .12''$

7-24 The texture index for a chemical product has a nominal mean value of $\mu = 100$. On the basis of 25 sample readings, the following decision rule has been established for determining when to replace a kiln thermostat:

Replace whenever $t < -2.492$.
Leave alone otherwise.

For each of the following texture results, indicate whether or not the thermostat should be replaced:

(a)	(b)	(c)	(d)
$\bar{X} = 98$ $s = 4.1$	$\bar{X} = 101$ $s = 5.2$	$\bar{X} = 97$ $s = 3.7$	$\bar{X} = 99$ $s = 4.1$

7-25 A chemical engineer has computed the following results for the active ingredient yields from 20 pilot batches processed under a retorting procedure:

$$\bar{X} = 31.4 \text{ grams/liter}$$

$$s = 2.85$$

Determine the approximate probability for getting a result this rare or rarer if the true mean yield is 29 grams per liter.

7-5 Sampling Distribution of the Proportion

In investigations involving qualitative populations the *proportion* π of units exhibiting a particular attribute or characteristic is the parameter of main interest. As with using $\bar{X}$ to generalize about μ, statisticians usually rely on the computed value for the sample proportion P to draw conclusions regarding the unknown π. The sample proportion for an attribute is found by dividing the number of sample observations R having that attribute by the sample size:

$$P = \frac{R}{n}$$

In Chapter 3 we established that when sample observations are random and independent, R is a random variable with a binomial distribution. That distribution applies as long as the sampled population results from a continuous process or sampling is done with replacement. If sampling is performed without replacement from a finite population, the hypergeometric distribution applies instead. Probabilities for the sample proportion may be obtained from either distribution by matching the possible level of P with the probability for the corresponding level of R.

In either case, in Chapter 4 we saw that it is usually convenient to approximate those probability distributions by an appropriate normal curve, and that such an approximation is usually satisfactory when π is near .5 or when n is large.

Indeed, the central limit theorem may be invoked to establish that a normal curve is appropriate for P when sampling meets the conditions of a Bernoulli process. Then each sample observation, or *trial*, may be represented by a variable having two values: $X = 0$ (if a failure) or $X = 1$ (if a success). The probabilities for the respective trial outcomes are $1 - \pi$ and π, and individual X's have the **Bernoulli distribution**. The binomial random variable R can represent the sum of n Bernoulli X's, and the proportion of successes P is equal to the sample mean $\bar{X}$ of those same observations.

Probabilities for P Using Normal Approximation

In Chapter 3 we saw that the tables ordinarily used in finding exact binomial probabilities are limited to just a few parameter levels. In Chapter 4 we introduced the normal approximation, which works well for a wide variety of levels for n and π and is ordinarily used in statistical applications involving the proportion. From the binomial distribution for R we get the following.

EXPECTED VALUE AND STANDARD DEVIATION FOR P

$$E(P) = \pi$$

$$SD(P) = \sigma_P = \sqrt{\frac{\pi(1-\pi)}{n}}$$

where in keeping with our earlier notation, we use σ_P to denote the standard error of P. When the population size N is small (and sampling is without replacement), σ_P must be modified, as we shall see shortly.

These expressions serve as the parameters for approximating the normal distribution. Recall from Chapter 4 that a continuity correction of .5 was used in finding cumulative probabilities. Applying this, we may use the following expression.

CUMULATIVE PROBABILITY FOR P UNDER NORMAL APPROXIMATION

$$\Pr\left[P \leq \frac{r}{n}\right] = \Phi\left(\frac{(r+.5)/n - \pi}{\sigma_P}\right)$$

As an illustration, suppose that the silicon wafers used in making a particular microcircuit have a final chip yield that contains 9% defectives. A sample of $n = 40$ chips is taken. What is the probability that the sample proportion of defective chips is less than or equal to .15?

Exact binomial probabilities are not tabled for $n = 40$ or $\pi = .09$, so it is convenient to use the normal approximation. The standard error for P is

$$\sigma_P = \sqrt{\frac{.09(.91)}{40}} = .0452$$

The level .15 corresponds to $r = .15(40) = 6$ defective items. Using the normal approximation, the desired probability is found:

$$\Pr[P \leq .15] = \Phi\left(\frac{(6+.5)/40 - .09}{.0452}\right)$$

$$= \Phi(1.60) = .9452$$

Example: When to Replace Circuit Boards

The transmission errors in a teleprocessing system occur randomly over time. The actual proportion is uncertain, but the supervising engineer has established a policy that the transmitter circuit boards must be replaced whenever the sample proportion of erroneous bits in a test transmission is greater than .000025. For this purpose, a test burst of 100,000 bits is used; these are checked by the receiver to see if they match the established test pattern. The supervisor is interested in checking this policy to see if it results in too many unnecessary replacements.

The nominal system specification allows the proportion of erroneous bits to be 2 per 100,000 ($\pi = .00002$). Assuming that this applies for the entire population of transmitted bits, what is the probability that the boards will be replaced unnecessarily after a particular test?

We have

$$\sigma_P = \sqrt{\frac{.00002(.99998)}{100{,}000}}$$
$$= .000014142$$

so that the normal approximation provides

$$\begin{aligned}\Pr[\text{unnecessary replacement}] &= \Pr[P > .000025] = \Pr\left[P > \frac{2.5}{100{,}000}\right] \\ &= 1 - \Pr\left[P \le \frac{2.5}{100{,}000}\right] \\ &= 1 - \Phi\left(\frac{(2.5 + .5)/100{,}000 - .00002}{.000014142}\right) \\ &= 1 - \Phi(.71) \\ &= 1 - .7612 = .2388\end{aligned}$$

This probability was judged by the engineer to be too large. He therefore revised the cutoff level of the sample proportion for replacement from the original .000025 to .00004, for which the following applies:

$$\begin{aligned}\Pr[P > .00004] &= \Pr\left[P > \frac{4}{100{,}000}\right] \\ &= 1 - \Pr\left[P \le \frac{4}{100{,}000}\right] \\ &= 1 - \Phi\left(\frac{(4 + .5)/100{,}000 - .00002}{.000014142}\right) \\ &= 1 - .9616 = .0384\end{aligned}$$

The new cutoff value was accepted by management, although it provides less protection than before against erroneously failing to replace the circuit boards when the true proportion of errors is actually higher than specifications allow.

Sampling from Small Populations

The normal approximation may also be used as the sampling distribution of P when samples are taken from small populations. As noted, the value of σ_P will be computed differently. For this purpose we use the following expression.

STANDARD ERROR FOR P
(SMALL POPULATIONS)

$$\boldsymbol{\sigma_P = \sqrt{\frac{\pi(1 - \pi)}{n}}\sqrt{\frac{N - 1}{N - n}}}$$

As noted, in this case the normal distribution approximates the hypergeometric distribution, and σ_P includes the finite population correction factor.

Problems

7-26 Referring to the circuit board example, suppose that the true proportion of erroneous bits is higher than the specification level. Of course the supervising engineer does not know this and must rely on the $P > .00004$ rule to determine whether the boards should be replaced. Using the same sample size as before, calculate the probability that for a particular test burst the circuit boards will *not* be replaced when
(a) $\pi = .00003$ (b) $\pi = .00004$ (c) $\pi = .00005$

7-27 Repeat Problem 7-26, using a sample size of $n = 200{,}000$ bits instead.

7-28 A sample of 100 parts is to be tested from a shipment containing 500 items altogether. Although the inspector does not know the proportion defective in the shipment, assume that this quantity is .03. Determine the following probabilities for the sample proportion defective:
(a) $P \leq .05$ (b) $P \leq .02$ (c) $P > .04$ (d) $.02 \leq P \leq .03$

7-29 A quality-control inspector accepts shipments whenever a sample of size 20 contains 10% or fewer defectives. She rejects otherwise.
(a) Determine the probability that she will accept a poor shipment of 100 items in which the actual proportion of defectives is 12%.
(b) Determine the probability that she will reject a good shipment containing 150 items, only 8% of which are defective.

7-30 A welding robot needs overhauling if it misses 4.5% of its welds. It is judged to be operating satisfactorily if it misses only .8% of its welds.

A test is performed involving 50 sample welds. If the proportion of missed sample welds is greater than .02, the robot will be overhauled. Otherwise it will continue to operate. Assuming that in-between percentages may be ignored:
(a) What is the probability that a satisfactory robot will be overhauled unnecessarily?
(b) What is the probability that a poorly functioning robot will be left in operation?

7-31 A computer manufacturer rejects any shipment of printed circuit boards in which the proportion of defectives in a sample of 25 boards tested exceeds .16. For each of the following shipments, determine the probability of rejection:

(a)	(b)	(c)	(d)
$N = 200$	$N = 150$	$N = 250$	$N = 75$
$\pi = .10$	$\pi = .12$	$\pi = .15$	$\pi = .18$

7-6 Sampling Distribution of the Variance: The Chi-Square and F Distributions

Although the sample variance s^2 is computed in most sampling investigations involving quantitative data, it is ordinarily an adjunct statistic for procedures that focus on the sample mean. In those cases, direct inferences concern μ only, and no conclusions are made about the population variance (standard deviation). However, there are instances when the major concern is with the value of σ^2 (or

σ) instead of μ. In those cases, *variability* rather than central tendency is the major concern.

For instance, in chemical processing a method having slower reaction times that are consistent may be more desirable than one that is on the average faster but has more varied individual times. Alternative methods would then be compared in terms of the respective reaction-time population standard deviations, with the mean times being of secondary interest. Similarly, although customers are very concerned with their waiting time in a queue (waiting line), they may be more upset about the unpredictability of the amount of wasted time than about its average duration. That is, σ^2 for the waiting time population is often more important than μ. (Have you noticed how people fidget while waiting in line at the post office as earlier patrons mail packages, buy money orders, or send registered letters? These same people might be more tolerant of a similar average delay at a fast-food outlet where the waiting time is more predictable.)

As we have seen with inferences regarding means and proportions, it is appropriate to use the counterpart sample statistic in making conclusions about population variability. Statisticians use the computed value of s^2 to make generalizations about the uncertain level of σ^2. Our present concern is with finding probabilities for the possible levels of s^2.

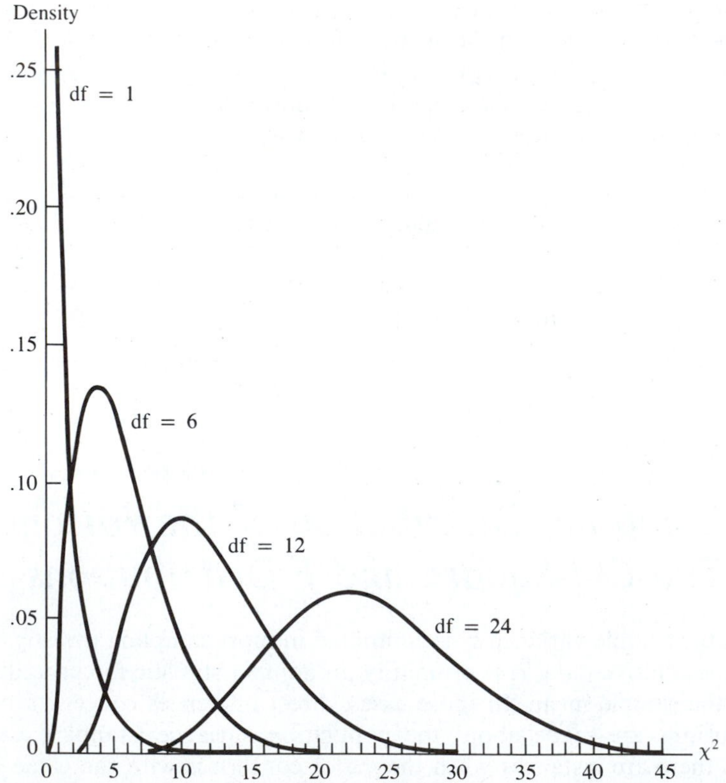

Figure 7-8 *Chi-square density curves for several degrees of freedom.*

The Chi-Square Distribution

The sampling distribution of sample variance s^2 is simplest to obtain by transforming the random variable into the chi-square statistic.

CHI-SQUARE STATISTIC

$$\chi^2 = \frac{(n-1)s^2}{\sigma^2}$$

where the symbol χ is the Greek lowercase *chi* (pronounced "kie") and χ^2 is read "chi-square." This statistic (in the planning stage) is a random variable having the **chi-square distribution**.

Figure 7-8 shows several density curves for the special chi-square random variable used in most statistical applications. Like the Student *t*, the chi-square distribution is specified by a single parameter, the number of degrees of freedom (df). When it is applied to making probability statements about s^2, df $= n - 1$. (There are only $n - 1$ degrees of freedom because in addition to the n observed X's, $\bar{X}$ is used in calculating s^2, and 1 degree of freedom is lost in specifying the level for $\bar{X}$.)

Notice that the chi-square densities are positively skewed, with upper tails extending indefinitely, but the degree of skew becomes less pronounced as the degrees of freedom increase. (As with the Student *t*, the normal curve may be used as an approximation when n is large.) For each upper-tail probability α there corresponds a critical value χ^2_α for the chi-square random variable such that

$$\alpha = \Pr[\chi^2 > \chi^2_\alpha]$$

Figure 7-9 shows the chi-square density curve for 6 degrees of freedom. Two upper-tail areas are given. The respective critical values for these are $\chi^2_{.05} = 12.592$

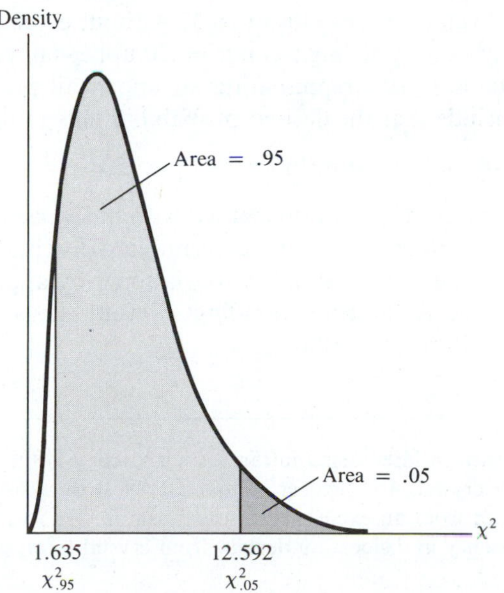

Figure 7-9 *Chi-square density curve for 6 degrees of freedom.*

and $\chi^2_{.95} = 1.635$, so that

$$.05 = \Pr[\chi^2 > 12.592]$$
$$.95 = \Pr[\chi^2 > 1.635]$$

and there is a $.95 - .05 = .90$ probability that χ^2 will lie between 1.635 and 12.592. These values were read from Appendix Table H, which provides critical values of the chi-square statistic for several settings and 30 df levels. Table H resembles the layout for the Student t table. To accommodate the lack of symmetry in chi-square curves, several α levels greater than .5 are listed.

Probabilities for the Sample Variance

The probability of any s^2 event may be found by identifying the corresponding χ^2 event and determining the probability for the latter.

For example, suppose that a human-factors engineer will be collecting a random sample $n = 25$ of the effective reach radii for seated men. The population is assumed to be normally distributed. Although the variance of this population is unknown, anthropometric data for standing reach suggest that an appropriate value would be $\sigma^2 = (1.5'')^2 = 2.25$ square inches. What is the probability that the sample variance will exceed 3 square inches?

This probability is expressed by

$$\alpha = \Pr[s^2 > 3] = \Pr\left[\frac{(n-1)s^2}{\sigma^2} > \frac{(25-1)3}{2.25}\right]$$
$$= \Pr[\chi^2 > 32]$$

To evaluate this probability we need to determine the area under the chi-square curve that lies above 32. For a sample size of $n = 25$, the chi-square density with $\text{df} = 25 - 1 = 24$ applies. Only a few entries are provided in Table H, and we see that the critical value coming closest to 32 without exceeding that quantity is 29.553, a level for χ^2 corresponding to an $\alpha = .20$ upper-tail probability. The next larger critical value is 33.196, representing an upper-tail probability of $\alpha = .10$. Thus, we may conclude that the desired probability falls in the interval,

$$.10 < \Pr[s^2 > 3] < .20$$

Investigators are usually more concerned with the *standard deviation* than the variance. Instead of dealing with a separate sampling distribution, one can obtain the probabilities for the sample standard deviation s by applying the chi-square distribution in evaluating the corresponding s^2 event. Thus, $s > 5$ is equivalent to $s^2 > 25$ and $s = 10$ to $s^2 = 100$.

Example: *Selecting a Laser Crystal for Infrared Spectroscopy*

One criterion used in laser-based infrared spectroscopy is the dislocation density of the laser crystal. An engineer conducted a test to measure this quantity in crystals grown from an experimental composition. He was more concerned with the consistency in dislocation density from crystal to crystal than with the absolute level. Thus, he used the computed displacement statistics from $n = 10$ crystals,

$$\bar{X} = 530/\text{cm}^2$$
$$s = 98$$

to make a comparison to an existing crystal, in which the standard deviation in dislocation density was assumed to be $\sigma = 200/\text{cm}^2$.

Assuming that the population standard deviations (variances) are the same for the existing and experimental crystals, what is the probability that results as extreme as the foregoing, or more so, could have been obtained?

We must find

$$\Pr[s \leq 98 \mid \sigma = 200] = \Pr\left[\frac{(n-1)s^2}{\sigma^2} \leq \frac{(10-1)(98)^2}{(200)^2}\right]$$
$$= \Pr[\chi^2 \leq 2.161]$$

For $10 - 1 = 9$ degrees of freedom, Appendix Table H provides the nearest bracketing critical values of $\chi^2_{.99} = 2.088$ and $\chi^2_{.98} = 2.532$, representing upper-tail probabilities of .99 and .98. The desired event is a lower-tailed one, so that the complementary probabilities apply, and the following conclusion is reached:

$$.01 < \Pr[s \leq 98 \mid \sigma = 200] < .02$$

Assumptions of the Chi-Square Distribution

The chi-square distribution represents a random variable that is the sum of the squares of several independent normal random variables. This means that using the chi-square distribution to represent the sampling distribution for s^2 is theoretically valid only when the individual sample observations are taken from a *normally distributed population.* This assumption of normality is identical to the one underlying the Student t and may often be violated in actual application, so that the chi-square does not technically apply. Nevertheless, the chi-square often serves as a satisfactory approximation to the true sampling distribution even when the population is not normal.

Expected Value, Variance, and the Normal Approximation

It may be established that the expected value and variance of the chi-square distribution summarized in Table H are, for df $= n - 1$:

$$E(\chi^2) = n - 1$$
$$\text{Var}(\chi^2) = 2(n - 1)$$

As noted earlier, the chi-square densities resemble a bell-shaped curve when the sample size (df) is large. As a practical matter, *the normal distribution is used to approximate the chi-square whenever $n > 30$.* The above moments may be used to convert χ^2 into a standard normal random variable by subtracting the expected value and dividing by the standard deviation. The following expression is used.

NORMAL DEVIATE FOR APPROXIMATING CHI-SQUARE

$$z = \frac{\chi^2 - (n - 1)}{\sqrt{2(n - 1)}}$$

Returning to the problem of determining the sitting reach of men, suppose that a sample size of $n = 100$ were used instead. What is the probability that the sample variance will exceed 3 square inches?

Here we use the normal approximation in evaluating the following because $n > 30$ and df $= 100 - 1 = 99$ falls outside of Table H. We have

$$\begin{aligned} \Pr[s^2 > 3] &= \Pr\left[\frac{(n-1)s^2}{\sigma^2} > \frac{(100-1)3}{2.25}\right] \\ &= \Pr[\chi^2 > 132] \\ &\approx 1 - \Phi\left(\frac{132 - (100 - 1)}{\sqrt{2(100 - 1)}}\right) \\ &= 1 - \Phi(2.35) = 1 - .9906 \\ &= .0094 \end{aligned}$$

The F Distribution

A second sampling distribution—**the *F* distribution**—is also associated with the sample variance. This distribution is useful for comparing the variances of *two* samples, which is done by finding their *ratio.*

Consider two normally distributed populations, A and B, each with identical variances. One random sample is selected from each. We let s_A^2 and s_B^2 denote the respective sample variances and n_A and n_B the sample sizes and get the following expression.

F STATISTIC

$$F = \frac{s_A^2}{s_B^2}$$

This random variable has the F distribution, which is completely described by two parameters, $n_A - 1$ and $n_B - 1$. These parameters are referred to as the *degrees of freedom* for the *numerator* and the *denominator*, respectively.

Figure 7-10 illustrates the density curves for the F distribution for three different pairs of degrees of freedom. Notice that the F distribution is positively skewed with possible values ranging upward from zero. As with the Student t distribution, critical values may be determined for a specified upper-tail area α, and these are provided in Appendix Table J. In keeping with the earlier notation, the respective critical values are denoted by F_α, so that

$$\alpha = \Pr[F \geq F_\alpha]$$

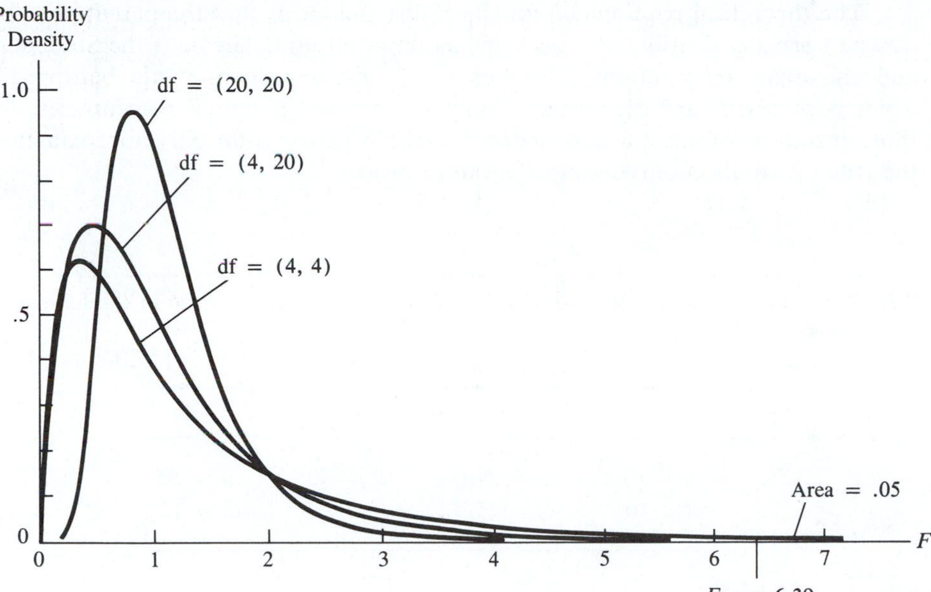

Figure 7-10 *Density curves for F distribution.*

Because there is a different F distribution for each degrees-of-freedom pair, entries are provided for only two levels, $\alpha = .01$ or $\alpha = .05$. Values for $F_{.05}$ are shown in lightface type, while those for $F_{.01}$ are shown in boldface. When the numerator and denominator degrees of freedom are each 4, the critical values are read from row 4 and column 4 of Table J: $F_{.01} = 15.98$ and $F_{.05} = 6.39$. The latter point is the critical value above which the area under the corresponding density curve is .05; that area is shaded in Figure 7-10. The tail area for $F_{.01}$ lies off the page. These values indicate that under that distribution, values taken by the random variable F will be greater than 6.39 only 5% of the time and will exceed 15.98 just 1% of the time.

To illustrate using the F distribution, suppose that a computer science student plans to make two sets of information retrievals from a data base. She will record how long it takes to complete each retrieval. Set A involves $n_A = 5$ retrievals made at lunchtime, between noon and 1 P.M. Set B involves $n_B = 10$ retrievals made in the early morning, between midnight and 1 A.M. She plans to compute the two sample variances in completion time. From Table J, using $5 - 1 = 4$ degrees of freedom for the numerator and $10 - 1 = 9$ for the denominator, we see that there is a .05 probability that s_A^2 will be at least 3.63 times as large as s_B^2:

$$\Pr\left[\frac{s_A^2}{s_B^2} \geq 3.63\right] = .05$$

while there is a .01 probability that s_A^2 will be at least 6.42 times as large as s_B^2:

$$\Pr\left[\frac{s_A^2}{s_B^2} \geq 6.42\right] = .01$$

The theoretical requirement for the F distribution is that the parent *population* be normally distributed. This same assumption underlies both the Student t and chi-square distributions. Although the F distribution may still be applied when populations are nonnormal, doing so involves potential inaccuracies. In those instances we must acknowledge that the F distribution only approximates the true distribution of the sample variance ratio.

Problems

7-32 Determine the upper-tail critical values for the chi-square statistic in the following cases:

(a)	(b)	(c)	(d)
$\alpha = .01$ $\text{df} = 10$	$\alpha = .001$ $\text{df} = 25$	$\alpha = .50$ $\text{df} = 2$	$\alpha = .95$ $\text{df} = 12$

7-33 Compute the chi-square statistic for the following sample results under the assumed σ. Then determine the nearest upper-tail area and the corresponding critical value such that the computed chi-square does not exceed that quantity:

(a)	(b)	(c)	(d)
$s = 10$ $n = 25$ $\sigma = 9$	$s = 8$ $n = 10$ $\sigma = 15$	$s = 215$ $n = 20$ $\sigma = 300$	$s = 53$ $n = 15$ $\sigma = 35$

7-34 The waiting time of time-sharing jobs for access to a central processing unit is believed to be normally distributed with unknown mean and standard deviation. Find the following percentiles for the sample standard deviation s, when a sample of size $n = 25$ is taken and assuming that the true value of the population standard deviation is $\sigma = 1.5$ minutes:

(a) 1st (b) 10th (c) 50th (d) 90th (e) 99th

7-35 Compute the chi-square statistic for the following sample results under the assumed σ. Because of the large sample sizes, the normal approximation must be used. Determine the normal deviate that corresponds and establish the probability for getting a value for s as great or greater:

(a)	(b)	(c)	(d)
$s = 7$ $n = 65$ $\sigma = 5$	$s = 12$ $n = 90$ $\sigma = 15$	$s = 325$ $n = 100$ $\sigma = 300$	$s = 33$ $n = 225$ $\sigma = 30$

7-36 The mean sitting height of adult males may be assumed to be normally distributed, with mean 36″ and standard deviation 1.3″. For a sample size of $n = 100$ men, determine the probabilities for the following possible levels of the sample standard deviation in sitting height:
(a) $s \leq 1.00''$ (b) $s > 1.5''$ (c) $s \leq 1.4''$ (d) $1.25'' \leq s \leq 1.35''$

7-37 The customer waiting times at a certain post office are assumed to be normally distributed, with unknown mean and standard deviation. An industrial engineering student will be monitoring 25 noontime customers, timing their arrivals and service with a watch. As closely as you can, find an interval bracketing the probability that she will observe a standard deviation in waiting time exceeding 5 minutes when the true value of the population standard deviation is:
(a) $\sigma = 4.3$ minutes (c) $\sigma = 6.0$ minutes
(b) $\sigma = 5.4$ minutes (d) $\sigma = 4.7$ minutes

7-38 Consider two random samples of size $n_A = 15$ and $n_B = 20$. Which of the following pairs of variances provides ratios extreme enough to fall into a range that occurs only 5% of the time when the underlying populations have identical variances?

(a)	(b)	(c)	(d)
$s_A^2 = 10.1$	$s_A^2 = 25.4$	$s_A^2 = 50.5$	$s_A^2 = 100.8$
$s_B^2 = 5.3$	$s_B^2 = 9.6$	$s_B^2 = 17.3$	$s_B^2 = 235.6$

7-39 Find the critical values $F_{.05}$ and $F_{.01}$ for the following sampling situations:

(a)	(b)	(c)	(d)
$n_A = 10$	$n_A = 25$	$n_A = 5$	$n_A = 101$
$n_B = 10$	$n_B = 5$	$n_B = 25$	$n_B = 101$

7-40 Denoting by v_A and v_B the numerator and denominator degrees of freedom, let $F_\alpha(v_A, v_B)$ denote the critical value for F such that

$$\Pr\left[\frac{s_A^2}{s_B^2} \geq F_\alpha(v_A, v_B)\right] = \alpha$$

Show that $F_{1-\alpha}(v_A, v_B) = 1/F_\alpha(v_B, v_A)$.

Comprehensive Problems

7-41 The height of adult males in a particular city is normally distributed, with $\mu = 69.5''$ and standard deviation $\sigma = 2.65''$. A sample of $n = 10$ men is to be measured. Find (as closely as possible) the probabilities that (a) $\bar{X}$ falls within $\mu \pm .75''$ and (b) s falls within $\sigma \pm .1''$.

7-42 Refer to the student grade-point population in Table 7-1. Suppose a sample of 2 is taken without replacement. Construct a table showing the complete sampling distribution of
(a) the sample proportion of B's
(b) the sample range (absolute value for difference in observed values)

7-43 Repeat Problem 7-42, but assume instead that the observations are made *with* replacement.

7-44 An engineering society is preparing for a sampling investigation of salaries of ceramics engineers. A random sample of 25 engineers will be selected and their annual incomes determined.
(a) The degree of error in using $\bar{X}$ to estimate μ cannot be known, since the standard error of $\bar{X}$ depends on the unknown population standard deviation. Assuming that this quantity will be the same as that for another income population for which σ was found to be \$1,500, determine the 99th percentile for the sample standard deviation from the present investigation. (There will be a 1% chance that the computed s will exceed this quantity.)
(b) Under the assumption that the true value of the population standard deviation equals the 99th percentile for s found in (a), determine the probability that the sample mean income falls within \$500 of the true population mean.

7-45 An automotive engineer will be conducting crash tests to determine the standard repair costs for cars damaged by hitting a barrier at 5 mph. A sample of $n = 9$ crashes is all her budget allows. She knows neither the mean nor the standard deviation for the population of costs, although she believes the frequency distribution to be nearly symmetrical.

Suppose that the engineer will request a front-end redesign study if the computed sample mean exceeds the historical figure for past models by more than some yet to be determined amount.
(a) On what sample statistic should the engineer base her decision?
(b) What level of the quantity identified in (a) provides a 5% chance of requesting a study when the true mean repair cost is identical to the historical one?
(c) Assuming she uses the critical value found in (b), would she request a redesign study if $\bar{X} = \$242$, $s = \$45$, and the historical mean is $\mu = \$195$?

7-46 Suppose that the engineer in Problem 7-45 will order an extension of crash testing if the sample standard deviation exceeds the historical level by some amount.
(a) On what sample statistic should the engineer base her decision?
(b) What level of the quantity identified in (a) provides a 1% chance of ordering further crashes when the true standard deviation in repair cost is identical to the historical figure?
(c) Assuming she uses the critical value found in (b), would she order an extension if $s = \$45$ and the historical standard deviation is $\sigma = \$30$?

7-47 A plant foreman shuts down a process to replace control valves whenever more than 5% of the items in a sample of 100 are found to be out of tolerance. Use the normal approximation to find the probabilities for the following events:
(a) He fails to replace the control valves when the process is actually yielding 10% off-size items.
(b) He replaces the valves unnecessarily when the process is performing satisfactorily and only 1% of production is out of tolerance.

7-48 Repeat Problem 7-47, but use the binomial distribution to find the exact probabilities.

CHAPTER EIGHT

Statistical Estimation

ONE OF THE basic forms of statistical inference is *estimation.* The news media bombard us with statistical estimates ranging from the latest percentages in presidential popularity to the "average" value of a house. But unless they have had some formal introduction to statistics, few people are aware of its limitations. Too often, sample data are reported as if they were hard facts. Yet we have seen that sampling error is almost inevitable, regardless of how much care went into the design and execution of the statistical investigation.

Science and engineering rely heavily on sample information. Nearly all physical constants, such as $g = 32\ \text{ft/sec}^2$, are based on empirical observations and are really statistical estimates. In large measure the pioneering efforts at evaluating such parameters may be characterized as sampling studies. Practically all modern measurements of physical objects culminate in some sort of estimate—from the level of resistance in a particular substance used in a printed circuit board to the strength reported for structural members of a particular shape, size, and composition.

Mean values and proportions are perhaps the most familiar quantities. The unknown population parameter value μ or π must usually be estimated from sample data. The resulting values are generally based on the computed level for the respective sample statistic, with $\bar{X}$ being used to estimate μ and P serving as the estimator of π. Other statistics might be used in estimating these parameters. For instance, the mean of a symmetrical population might just as well be estimated by the sample median. Although all estimators exhibit the potential for sampling error, some are more accurate than others. Several criteria exist for gauging the suitability of each candidate statistic for estimating a particular population parameter.

Once a particular statistic has been selected, such as $\bar{X}$ for estimating μ, questions remain regarding the form of the estimate itself. Newspaper readers are accustomed to reading reports in which means and percentages are listed as single quantities. By now, we know there are pitfalls in using such statistics for any serious purpose—largely because of the potential for sampling error. That error must be dealt with in a systematic way. The accepted procedure among statisticians is to express estimates in the form of *intervals.* Each result is reported so that explicit consideration is given to the estimate's precision and the likelihood that it will be correct. Satisfactory levels for these end results are often incorporated into the study goals. These objectives can be achieved whenever the planning stage includes a careful choice of procedure and the selection of an adequate sample size.

8-1 Estimators and Estimates

The sample statistic employed in estimating a population parameter is called an **estimator**. An important part of the statistical art is deciding what particular statistic to employ. Several criteria guide in this selection. These help explain why particular statistics are favored in drawing inferences. There are two principal forms a statistical estimate may take.

The Estimation Process

The estimation process begins in the planning stage of the statistical investigation. The population has already been identified and the parameter to be estimated has been determined. An early part of the planning process is deciding which statistic will be the estimator. We shall see how the sampling distribution of the chosen statistic may provide some idea of the precision and reliability of the final estimate. These considerations guide the final selection of the sample size.

The form of the estimate itself will depend largely on the purpose of the study. In a formal report of the study results, the favored form is the **interval estimate**. For example, the mean waiting time μ in a particular queueing situation might be expressed as

$$35 \leq \mu \leq 45 \text{ seconds}$$

The reported value for μ is inexact, and the limits in this inequality provide the end points for the interval estimate of μ, [35, 45]. Interval estimates are usually reported in the inequality form shown. An equivalent expression is

$$\mu = 40 \pm 5 \text{ seconds}$$

The central value (40, here) is often the computed value of the sample statistic serving as the estimator.

The limits of the interval estimate indicate the degree of **precision** involved. A more precise estimate of the mean waiting time would be

$$\mu = 40 \pm 1 \qquad \text{or} \qquad 39 \leq \mu \leq 41 \text{ seconds}$$

The quality of a statistical estimate is measured in two dimensions. Precision or accuracy is one. The other is the **reliability** of the result. The reliability of an estimate is simply the probability that it is correct.

Reliability and precision are competing ends. One can always be improved at the expense of the other. For example, suppose we used sample data to estimate the mean height of male engineering majors. On the basis of a sample measurement of one randomly chosen man we could conclude that the mean height of all men in the population is that person's height plus or minus .01″, say

$$\mu = 5'8.5'' \pm .01'' \qquad \text{or} \qquad 5'8.49'' \leq \mu \leq 5'8.51''$$

a very precise outcome indeed. But how reliable would such a result be? Most would agree on a small probability that the above estimate is valid. With little chance of being correct, an overly precise estimate is useless.

Alternatively, using the same data, we can achieve the following very reliable estimate:

$$\mu = 5'8.5'' \pm 1.00' \qquad \text{or} \qquad 4'8.5'' \leq \mu \leq 6'8.5''$$

an outcome virtually certain to be true (so that the reliability is 100%). But this second estimate is too imprecise to be of any use whatsoever.

Better reliability and satisfactory precision can be achieved by raising the probability that the estimator's value will be close to its target parameter. To achieve this, statisticians seek estimators with sampling distributions that cluster ever more tightly about a central value as n becomes larger. By paying the price of increasing the number of sample observations, an investigator using such an estimator can simultaneously improve precision and reliability.

The second form of estimation involves a **point estimate**. This is a single quantity, such as 5′9.5″ for the mean height of a population of adult males or .64 for the proportion of voters approving of the president. Unlike interval estimates, point estimates give no hint of the sampling error inherent in their generation. They are particularly dangerous when publicly disseminated, because the general public too often assumes such values to be the true parameter levels, without even realizing that sampling was involved.

Point estimates do have their purpose—even in a scientific sampling investigation. Many physical constants are point estimates based on prior sampling experiments. It would be cumbersome to use an interval estimate for g in computing the distance traversed by a falling object in evaluating $d = \frac{1}{2}(g)t^2$. Imagine how much more complicated thermodynamics or classical mechanics would be if all the parameters in any relationship had to be manipulated in interval form. Nevertheless, we must acknowledge that some of the deviation between what an experimenter observes and what a theoretical model predicts is statistical in nature. One component of such error is the sampling error arising from the point estimates originally used to express some of those constants we take for granted.

Point estimates are often employed when the investigation involves two or more unknown parameters, one of which may be the most important. For example, an evaluation of a quantitative population may require both μ and σ, although the latter is usually an ancillary consideration. Ordinarily, an interval estimate would be made of μ, the end points of which are partly determined by σ; because σ is also unknown, the sample standard deviation s often provides a point estimate of that value. In this case, the interval estimate of one parameter embodies a point estimate of another.

Theoretical Rationales for Selecting an Estimator

The estimator used in making statistical inferences from sample data will be a sample statistic, and its value will be computed from the observed data. The following are typical estimators:

$$\hat{\mu} = \bar{X} \qquad \hat{\sigma}^2 = s^2 \qquad \hat{\pi} = P = \frac{R}{n}$$

(Traditional statistical notation denotes the estimator for a population parameter by placing a caret over the symbol for that parameter.) The estimators used in statistical methods are carefully chosen because they exhibit various desirable properties. The following are two common rationales that justify frequently used estimators.

Method of Moments This procedure is based on the moments of a randomly selected observation X. The ***k*th moment** is the expected value

$$m_k = E(X^k)$$

The method of moments involves equating the sample counterparts with the moments for the underlying population distribution and then algebraically solving the resulting system of equations, treating the parameters as the unknowns.

In statistical applications, two moments are especially important:

$$\text{First moment:} \quad m_1 = E(X)$$

$$\text{Second moment:} \quad m_2 = E(X^2)$$

To illustrate the method of moments, suppose that $X_1, X_2, \ldots, X_n$ represent the number of car arrivals at a toll plaza observed in nonoverlapping randomly selected time spans, each of duration t, from a Poisson process having mean rate λ. We have, for any observation,

$$m_1 = E(X) = \lambda t$$

The sample mean is the counterpart to this expression:

$$\bar{X} = \frac{\sum X}{n}$$

Equating the above expressions, we obtain:

$$\bar{X} = \lambda t$$

Solving for λ, we obtain the estimator

$$\hat{\lambda} = \frac{\bar{X}}{t}$$

As a second illustration, suppose that n sample observations are taken from a normally distributed population with parameters μ and σ^2. The estimators for these may be found by first noting the property

$$\text{Var}(X) = E(X^2) - E(X)^2 = m_2 - m_1^2$$

from which it follows that

$$m_2 = \text{Var}(X) + m_1^2$$

The moments for any particular sample observation are

$$m_1 = \mu \qquad \text{and} \qquad m_2 = \sigma^2 + \mu^2$$

The corresponding sample moments are

$$\bar{X} = \frac{\sum X}{n} \qquad \text{and} \qquad \frac{\sum X^2}{n}$$

Equating the respective first moments, we obtain

$$\hat{\mu} = \frac{\sum X}{n} = \bar{X}$$

Equating the respective second moments, replacing μ with the above, we obtain

$$\hat{\sigma}^2 = \frac{\sum X^2}{n} - \bar{X}^2 = \frac{\sum (X - \bar{X})^2}{n}$$

Notice that the sample mean $\bar{X}$ serves as the estimator of the population mean μ. The estimator for the population variance σ^2 resembles the sample variance s^2, except that the divisor is n instead of $n - 1$. (We will see shortly why s^2 was defined with the $n - 1$ divisor.)

The method of moments yields estimators that coincide with or nearly match the population parameter with its counterpart sample statistic. This feature has compelling advantages.

Maximum Likelihood Estimators A second approach for arriving at an estimator involves maximizing a special function. In Chapter 5 it was established that when the sample observations $X_1, X_2, \ldots, X_n$ are treated as *independent* random variables, then the joint probability density or mass function is the product of the respective functions for the individual X_i's.

For example, suppose each observation represents the quality assessment of items from a production line that yields defectives at the rate π. We let $X_i = 0$ if the ith observed item is satisfactory and $X_i = 1$ if it is defective; then each X_i has a Bernoulli distribution with probability mass function

$$p(x_i) = \begin{cases} 1 - \pi & \text{for } x_i = 0 \\ \pi & \text{for } x_i = 1 \end{cases}$$

The joint probability mass function is

$$p(x_1, x_2, \ldots, x_n) = p(x_1)p(x_2) \cdots p(x_n)$$

As a second example, suppose that each X_i is the *time* between successive arrivals of cars at a toll plaza and the process is Poisson with mean rate λ. Then each X_i will have the same exponential distribution, so that the joint probability density function is

$$f(x_1, x_2, \ldots, x_n) = (\lambda e^{-\lambda X_1})(\lambda e^{-\lambda X_2}) \cdots (\lambda e^{-\lambda X_n}) = \lambda^n e^{-\lambda \sum X_i}$$

The **likelihood function** is the joint probability mass or density function evaluated at the actual sample observations. In the first example, the observations would be a string of 0's and 1's, so that the likelihood function would be a product such as

$$p(1, 1, 0, 0, 1, \ldots; \pi) = \pi\pi(1 - \pi)(1 - \pi)\pi \ldots = \pi^k(1 - \pi)^{n-k}$$

involving k defectives and $n - k$ satisfactories. For the car arrivals, the likelihood function would involve the actual values (denoted by capital letters):

$$f(X_1, X_2, \ldots, X_n; \lambda) = \lambda^n e^{-\lambda \sum X_i}$$

The **maximum likelihood estimator** is found by maximizing the likelihood function. This may be found for each of the above by taking the natural logarithm, then taking the first derivative, next setting that result equal to zero, and finally solving for the parameter. The following estimators are obtained:

$$\hat{\pi} = \frac{k}{n} = P \qquad \hat{\lambda} = \frac{n}{\sum X_i} = \frac{1}{\bar{X}}$$

Thus, the maximum likelihood estimator for π is the sample proportion defective. For the mean arrival rate λ it is the sample mean rate (the reciprocal of the sample mean time between arrivals). When the population has two or more parameters (for example, if it has a normal distribution), then partial derivatives must be taken with respect to each, and a system of equations will be solved simultaneously.

The maximum likelihood estimator is that level for the parameter which is most likely to have generated the actual observed results. This makes it especially appealing. Maximum likelihood estimators have other desirable properties to be discussed shortly.

Criteria for Statistics Used as Estimators

Several criteria are employed in assessing the worthiness of a particular statistic as an estimator. We will describe the more common ones.

Unbiased Estimators An estimator is **unbiased** if its expected value equals the parameter being estimated. There are no guarantees that any sample estimate will be correct, but those determined from unbiased estimators will, on the average, hit their targets. In Chapter 7 we noted for the sample mean that

$$E(\bar{X}) = \mu$$

which establishes that $\bar{X}$ is an unbiased estimator of μ. Similarly,

$$E(P) = \pi$$

and P is an unbiased estimator of π.

This property is not a universal one. Consider the sample statistic

$$\frac{\sum_{i=1}^{n} (X_i - \bar{X})^2}{n}$$

which resembles the sample variance s^2. However, the divisor in this expression is n, not $n - 1$ as we have defined that quantity for s^2.

Like σ^2, s^2 expresses the mean of the squared observation deviations (from their own mean). The divisor in computing σ^2 is the population size N. Why didn't we use the entire sample size n as the divisor for s^2, as in the foregoing equation?

The answer is that s^2 as we have been using it ($n - 1$ divisor) is an unbiased estimator of σ^2. That is, $E(s^2) = \sigma^2$, so that on the average, values computed for s^2 for successive samples taken from the same population will be equal to σ^2. This does not apply to the statistic having n as the divisor. Values computed from that biased statistic will tend to undershoot the σ^2 target.

Like $\bar{X}$, the sample median is an unbiased estimator for μ—but only when the population is symmetrical. Unless we know that the population frequency distribution is not skewed, we will consistently overshoot or undershoot μ when using the median as the estimator.

Keep in mind that unbiasedness in an estimator is a theoretical property. This entire chapter is devoted to random samples in which each population unit has the same chance of being selected. Thus, our present discussions assume that there is no sampling bias.

Example: Estimating the Gold Requirements of Logic Circuit Wafers

An industrial engineer for a microcircuit manufacturer is estimating production costs for a new logic chip. Hundreds of chips are produced on a single silicon sheet that is like a wafer cake with as many as 50 layers. The most expensive layers involve gold, which is applied in an aerosol mist. The actual amount of gold consumed cannot be predicted exactly, since the layer consistency will vary from wafer to wafer and the unrecoverable gold loss fluctuates. From a pilot run of 15 batches, the following data were obtained for the gold consumption per wafer:

Amount of Gold (oz) X	$(X - \bar{X})$	$(X - \bar{X})^2$
.15	−.01	.0001
.17	.01	.0001
.13	−.03	.0009
.18	.02	.0004
.15	−.01	.0001
.16	.00	.0000
.17	.01	.0001
.15	−.01	.0001
.17	.01	.0001
.15	−.01	.0001
.19	.03	.0009
.18	.02	.0004
.17	.01	.0001
.15	−.01	.0001
.13	−.03	.0009
$2.40 = \sum X$		$.0044 = \sum(X - \bar{X})^2$
$\bar{X} = 2.40/15 = .16$		$s^2 = .0044/(15 - 1) = .000314$

The engineer uses this sample mean to estimate that the true mean gold consumption per wafer is .16 ounce. If each wafer yields 100 marketable chips, the amount of gold used to make each chip is .16/100 = .0016 ounce. At a current price of \$500 per ounce, the mean value of the gold in each chip is \$500(.0016) = \$.80.

Consistent Estimators An estimator is **consistent** if the precision and reliability of the estimate improve as the sample size is increased. This will be the case if the standard error of the estimating statistic becomes smaller as n is increased. It should be obvious that $\bar{X}$ and P are consistent estimators, since

$$\sigma_{\bar{X}} = \frac{\sigma}{\sqrt{n}} \qquad \text{and} \qquad \sigma_P = \sqrt{\frac{\pi(1-\pi)}{n}}$$

both become smaller as n increases.

Efficient Estimators An estimator is more **efficient** than another if for the same sample size it will provide greater sampling precision and reliability. This will be achieved by the estimator whose sampling distribution clusters more tightly about the parameter being estimated, so the more efficient statistic will be the one having the smaller standard error. The sample mean $\bar{X}$ is more efficient in estimating μ than is the sample median, whose standard error is 1.25 times as large as $\sigma_{\bar{X}}$.

The maximum likelihood estimators discussed earlier share many of the preceding properties. For instance, suppose that the population is normally distributed, with parameters μ and σ^2. It will be left as an exercise to establish that the maximum likelihood estimators (MLE) are

$$\hat{\mu} = \bar{X} \qquad \text{and} \qquad \hat{\sigma}^2 = \frac{\sum (X - \bar{X})^2}{n}$$

Although the MLE for σ^2 is *not unbiased*, it has been established that MLEs *tend* to be unbiased as the sample size is increased and that they *tend* to be the most efficient estimators. For normally distributed populations, however, $\bar{X}$ is the MLE for μ and is (as we have seen) unbiased, making $\bar{X}$ for that distribution the "best" possible estimator for μ.

A substantial portion of statistical theory is devoted to applying these criteria in selecting estimators and in expressing their suitability. However, meeting these criteria does not in itself guarantee that a particular statistic will be a good estimator. As a practical matter, an estimator must be easy to compute and must have a sampling distribution that can be readily determined. A workable procedure for evaluating sample results must then be built around that statistic.

Commonly Used Estimators

The sample mean $\bar{X}$ is a favored estimator of μ because it is unbiased, consistent, and more efficient than many alternative statistics. The sampling distribution of

$\bar{X}$ is particularly convenient. For large sample sizes taken from most commonly encountered populations, the central limit theorem indicates that a normal curve may be assumed for $\bar{X}$. In this chapter we will see how this feature can be used in qualifying estimates based on $\bar{X}$. We will also see how to choose a sample size that guarantees meeting the investigator's precision and reliability goals.

In addition to being unbiased, the sample variance s^2 is a consistent estimator of the population variance σ^2. Its square root may be used in estimating the population standard deviation. In the preceding example, the sample standard deviation for wafer gold content is $s = \sqrt{.000314} = .018$ ounce. This figure may be used to estimate σ, the standard deviation in the population of gold content.

The sample proportion P is an unbiased and consistent estimator of the population proportion π. The following example provides an interesting illustration of using P as an estimator.

Example:* *Work Sampling Eliminates Stopwatch "Tyranny"

Work measurement is a fundamental task of industrial engineers in production management. The information thereby gained may be used to group tasks into comprehensive job packages by finding the mean completion time of each task. Pioneers in the field of work measurement found these durations by timing workers as they performed each step. This was done with a timepiece (stopwatch) and required one watcher for each worker being timed. Not only was this an expensive way to gather data, but it was resented by the workers.

Modern work measurement rarely involves the stopwatch or any kind of continuous monitoring. Rather, the test subject may be observed for a few seconds at random times. A tally is then compiled, showing how often each particular chore was observed. From these data the proportion of time a worker spends on each step or task can be estimated. For example, one industrial engineer's clerk collected the following data for one worker in an 8-hour day:

Task	Number of observations
Mounting	25
Clamping	9
Adjusting	105
Assembling	58
	197

The proportion of observed times during which the worker was adjusting is $P = \frac{105}{197} = .533$. Assuming that the day's work provides a representative sample, for this worker .533 may be used to estimate the proportion of all similar working times spent adjusting items.

This procedure is called **work sampling**, and it can be extended to estimating mean times spent on each task. Continuing with the above, the worker completed 48 parts, and out of the 480-minute shift he was idle 87 minutes. He finished his parts in $480 - 87 = 393$ minutes, averaging $\frac{393}{48} = 8.2$ minutes per part. It is estimated that he spent 53.3% of his productive time adjusting, so that the mean time spent on that task is

$$.533(8.2) = 4.4 \text{ minutes}$$

Work sampling has removed workers from the tyranny of the stopwatch while allowing a many-fold increase in the efficiency of work measurement. (The clerk in this example was able to monitor 17 other workers in the same day. That would have been impossible under continuous time monitoring.)

Problems

8-1 For each of the following situations, indicate whether or not it is possible to take a sample from a population in order to make an estimate within the next few days. Explain your answers.

(a) The mean annual earnings are to be determined for electrical engineers graduating this coming June.

(b) The preferred majors of next year's freshmen engineering students are to be found.

(c) The lifetimes of cathode ray tubes with an experimental phospor coating are to be determined.

(d) A power company must estimate the electricity its customers will require 10 years from now.

8-2 For each of the following parameters, state whether a point or an interval estimate would be more appropriate:

(a) A ballistic coefficient is desired for a recoverable rocket engine to be reused in future satellite launches.

(b) The mean strength of a new structural member is to be reported in an engineering journal.

(c) A forecast value for the cost per unit of a particular ingredient in gasoline must be established for planning purposes.

(d) An engineering journal is publishing mean salaries for engineers in various categories.

8-3 The sample proportion of observations of a particular characteristic is P. Show that NP is an unbiased estimator of the number of items in a population of size N which have that characteristic.

8-4 For a class project, a graduate industrial engineering student observed a coffee shop waitress during 50 random times when she was on station over a 40-hour time span. The following tally summary was obtained:

Task	Number of observations
Taking orders	6
Walking	12
Serving food	8
Cleaning tables	5
Preparing orders	10
Collecting cash	9

An investigation of her checks shows that she served 480 customers during that period. She spent 4.5 of the working hours resting off station.

(a) Estimate the mean time per customer this employee spends serving food.
(b) Estimate the mean time per customer she walked on official business during the study period. If her average pace was 2.5 mph, how far is it estimated she walked during the study period?

8-5 Show that the median M of a random sample of size 3, taken without replacement from the following population, is an unbiased estimator of μ.

105 110 115 120 125 130 135

8-6 Using the information from Problem 8-5, compute $SD(M)$. Then establish the value of $SD(\bar{X})$ (also using $n = 3$). Which estimator, M or $\bar{X}$, is more efficient?

8-7 Find the sampling distribution of the range R (greatest minus smallest values) for a random sample of size 2 taken without replacement from the following student grade-point averages:

2.0 3.0 3.5 4.0 2.5 3.0

Show that R is a biased estimator of the population range.

8-8 Apply the method of moments to show that using the sample observations taken from a population having an exponential distribution with mean rate λ the estimator of this parameter is $\hat{\lambda} = 1/\bar{X}$.

8-9 Suppose that the sample observations $X_1, X_2, \ldots, X_n$ are taken from a population having a normal distribution with mean μ and variance σ^2. Show that the MLEs for μ and σ^2 are $\bar{X}$ and $\sum (X - \bar{X})^2/n$, respectively.

8-2 Interval Estimates of the Mean

As we have seen, the interval estimate is the form favored by statisticians because it provides some acknowledgment of the error inherent in the sampling process. An interval estimate reflects sampling error in two dimensions. First, the end points demonstrate that the sample results do not precisely yield the true mean. Second, because the sampling distribution of $\bar{X}$ may usually be approximated by the normal curve, it is possible to extend probability concepts and quantify the sampling procedure in terms of its reliability.

Precision and reliability goals may be established at the outset, before the data are collected. Our present concern is with the evaluation stage of the sampling study, after the sample results are known. At that time a computed value for $\bar{X}$ is known, and there is no longer any uncertainty about the sample. That computed $\bar{X}$ serves as the estimate of the unknown population mean μ. It defines the **center** of the interval estimate. How do we establish the end points?

Assuming that the population has a finite standard deviation σ, then for a sample of size n, the standard error of $\bar{X}$ is $\sigma_{\bar{X}} = \sigma/\sqrt{n}$. *Before* the data are collected, the normal curve is applicable. At that time, the probability is .95 that $\bar{X}$ will fall within ± 1.96 standard deviations of the mean:

$$\Pr[\mu - 1.96\sigma_{\bar{X}} \leq \bar{X} \leq \mu + 1.96\sigma_{\bar{X}}] = .95$$

This expression can be translated into an equivalent form,

$$\Pr[\bar{X} - 1.96\sigma_{\bar{X}} \leq \mu \leq \bar{X} + 1.96\sigma_{\bar{X}}] = .95$$

so that μ is placed inside the inequalities.

We may generalize the foregoing. Denoting the interval probability for $\bar{X}$ by $1 - \alpha$, the area under that portion of the normal curve over the prescribed interval is also $1 - \alpha$, leaving areas of $\alpha/2$ remaining in each tail. The interval itself may be expressed in terms of the corresponding **critical normal deviate**, which we denote by $z_{\alpha/2}$, above which the upper-tail area under the standard normal curve is $\alpha/2$. With $.95 = 1 - \alpha$, we have $\alpha = .05$ and $\alpha/2 = .025$. Thus,

$$z_{\alpha/2} = z_{.025} = 1.96$$

In general, the $1 - \alpha$ probability interval inside the probability brackets may be expressed as

$$\mu = \bar{X} \pm z_{\alpha/2} \frac{\sigma}{\sqrt{n}} \qquad \text{or} \qquad \bar{X} - z_{\alpha/2} \frac{\sigma}{\sqrt{n}} \leq \mu \leq \bar{X} + z_{\alpha/2} \frac{\sigma}{\sqrt{n}}$$

The critical normal deviate for any specified upper-tail area may be read from Appendix Table E. For instance, $z_{.01} = 2.33$ and $z_{.005} = 2.57$.

The above interval is centered at $\bar{X}$, and its end points depend partly on σ, an often unknown quantity that must be specified before probabilities can be found for $\bar{X}$. They also are affected by the level for $1 - \alpha$ (and hence $z_{\alpha/2}$) and for n, all of which can be established by the investigator. The inequality suggests an appropriate procedure for making an interval estimate of the population mean.

Confidence and Meaning of the Interval Estimate

Again, it is important to establish our perspective. Estimates are made *after* the sample data have been collected. Although there is a probability that any particular sample will place μ inside $\bar{X} \pm z_{\alpha/2}\sigma_{\bar{X}}$, once the actual sample results are available, $\bar{X}$ is a calculated and known quantity.

A probability value is no longer applicable to $\bar{X}$. Once its value has been computed, statisticians employ a related concept to express remaining lack of certainty regarding the still unknown μ. This concept is the **confidence level**.

Definition

Suppose that a sample experiment is repeated many times, each time with a different collection of n observations, and that an interval estimate is obtained from each. The **confidence level** is the percentage of those estimates providing intervals that actually would contain the true value of the population parameter being estimated.

Figure 8-1 illustrates the confidence level concept. A procedure yielding an interval estimate of the mean with a 95% confidence level can be expected to provide correct results only 95% of the time. Were such a procedure repeated 100 times, about 95 of the intervals obtained would contain μ and the remainder—the entire interval lying above or below μ—would be totally off. Of course, only one sample of size n is ordinarily taken, and *there is no way to know whether the estimate obtained is a true one.* Such is the unavoidable price we must pay for

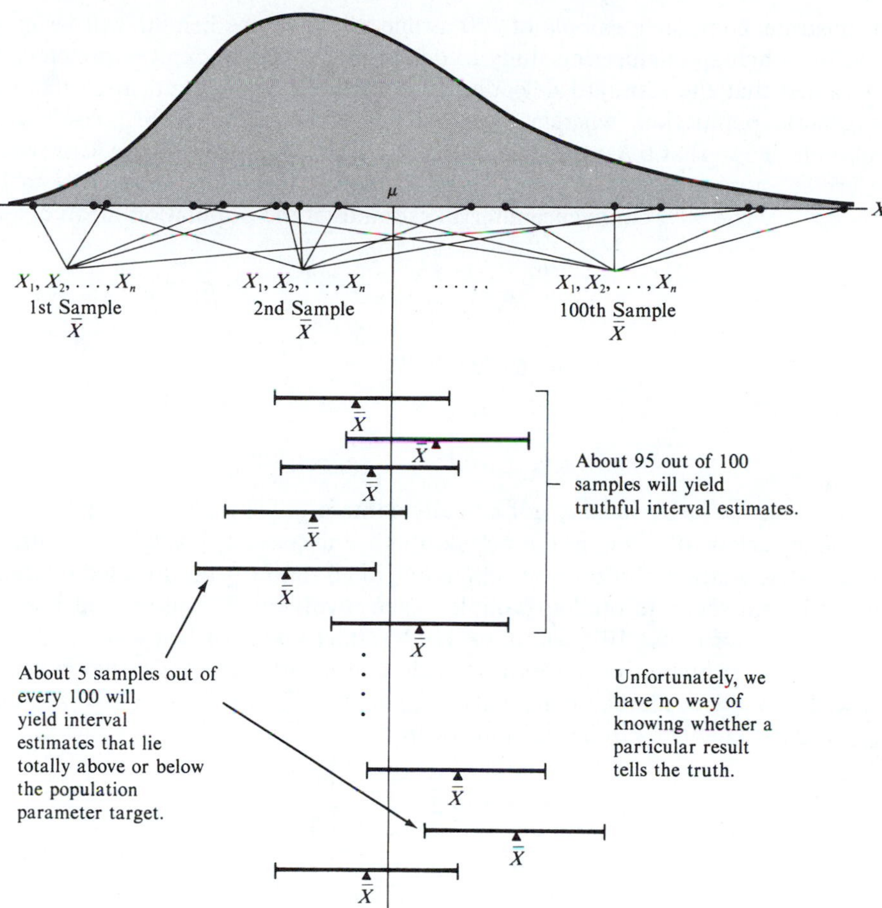

Figure 8-1 *Conceptual illustration of 95% confidence level in making interval estimates of the mean.*

using sample evidence. The only way to improve the situation is to raise the confidence level, using a bigger z (and thus reducing the precision), or to increase the sample size.

Confidence Interval for Mean When σ Is Known

The interval estimate itself is called a **confidence interval**. The simplest confidence interval is when the value of σ is known (say, from an earlier investigation or a related population). When the sample size is large, we use the following expression to compute the $100(1 - \alpha)\%$ confidence interval.

CONFIDENCE INTERVAL ESTIMATE OF THE MEAN WHEN σ IS KNOWN AND n IS LARGE

$$\mu = \bar{X} \pm z_{\alpha/2} \frac{\sigma}{\sqrt{n}} \quad \text{or} \quad \bar{X} - z_{\alpha/2} \frac{\sigma}{\sqrt{n}} \leq \mu \leq \bar{X} + z_{\alpha/2} \frac{\sigma}{\sqrt{n}}$$

Illustration: Heights of Pilots and Submariners

To illustrate, consider a sample of 100 airline pilots whose heights will be measured for a human-engineering study to determine aircraft cockpit parameters. It is assumed that the standard deviation in pilots' heights is identical to that for the general population, wherein $\sigma = 2.44''$. A $100(1-\alpha)\% = 90\%$ confidence level is chosen so that the upper-tail area is $\alpha/2 = .05$. Appendix Table E provides the critical normal deviate $z_{.05} = 1.64$. The sample mean was computed to be $\bar{X} = 69.75''$. The 90% confidence interval estimate of the population mean height is

$$\mu = 69.75'' \pm 1.64 \frac{2.44''}{\sqrt{100}}$$
$$= 69.75'' \pm .40''$$

or

$$69.35'' \leq \mu \leq 70.15'' \qquad \text{(pilots, 90\%)}$$

This interval may or may not actually contain μ. (There is no way to know for certain unless all pilots in the population are measured through a census.) A proper interpretation of the above is that 90% of all similarly constructed intervals obtained from separate random samples, each involving 100 pilots, will bracket the true value of μ, but 10% will not—lying totally above or below μ.

Should the human engineers find such a result too nebulous, they may increase the confidence level, say to $100(1-\alpha) = 99\%$. In that case, $\alpha/2 = .005$ and $z_{.005} = 2.57$. The 99% confidence interval is

$$\mu = 69.75'' \pm 2.57 \frac{2.44''}{\sqrt{100}}$$
$$= 69.75'' \pm .63''$$

or

$$69.12'' \leq \mu \leq 70.38'' \qquad \text{(pilots, 99\%)}$$

The price paid for the increased confidence *level* is a wider, less precise confidence *interval.*

Should the lack of precision in this estimate also be undesirable, the engineers would have to take a larger sample. Suppose that in a related investigation of submariners' heights they used a sample of 500 measurements. Assuming the same standard deviation as before, with a computed sample mean of $\bar{X} = 68.39''$, the following 99% confidence interval is obtained:

$$\mu = 68.39'' \pm 2.57 \frac{2.44''}{\sqrt{500}}$$
$$= 68.39'' \pm .28''$$

or

$$68.11'' \leq \mu \leq 68.67'' \qquad \text{(submariners, 99\%)}$$

Notice that this confidence interval is quite narrow and considerably more precise than the last one found for the pilots.

Both pilots and submariners are selected partly by their physical size. Very tall people or very short people are not allowed into these groups, and so the tails of the general population are underrepresented. That would make it inappropriate for the same standard deviations to be used for pilot and submariner heights as for men in the general population, and the true level for σ in each case may be considerably different than 2.44″. Ordinarily, in a sampling study involving an unknown μ, σ is also unknown and must also be estimated from the sample results.

Confidence Interval Estimate of Mean When σ Is Unknown

The normal curve is inappropriate when the population standard deviation σ is unknown. In those cases, estimating μ involves procedures based on the Student t distribution. This distribution applies to the statistic

$$t = \frac{\bar{X} - \mu}{s/\sqrt{n}}$$

which reflects that σ is estimated by its sample counterpart s.

The following expression is used to calculate the $100(1 - \alpha)\%$ confidence interval.

CONFIDENCE INTERVAL ESTIMATE OF THE MEAN WHEN σ IS UNKNOWN

$$\mu = \bar{X} \pm t_{\alpha/2}\frac{s}{\sqrt{n}} \qquad \text{or} \qquad \bar{X} - t_{\alpha/2}\frac{s}{\sqrt{n}} \leq \mu \leq \bar{X} + t_{\alpha/2}\frac{s}{\sqrt{n}}$$

where, as before, two tail areas are excluded and $t_{\alpha/2}$ is the critical value at $n - 1$ degrees of freedom corresponding to an upper-tail area of $\alpha/2$. For example, when the sample size is $n = 25$ and the confidence level is 95%, the value $t_{.025} = 2.064$ is read from Appendix Table G when df $= 25 - 1$.

We may illustrate this using the example from Section 8-1 for the gold required in the manufacture of a particular logic circuit wafer. The industrial engineer in that example computed the following from a sample of $n = 15$:

$$\bar{X} = .16 \text{ ounce} \qquad s = \sqrt{.000314} = .018 \text{ ounce}$$

A 95% confidence interval is constructed. Using $\alpha/2 = .025$, $t_{.025} = 2.145$ is read from Appendix Table G for df $= 15 - 1$. Thus,

$$\mu = .16 \pm 2.145\frac{.018}{\sqrt{15}}$$
$$= .16 \pm .010 \text{ ounce}$$

or

$$.150 \leq \mu \leq .170 \text{ ounce}$$

This interval does not give a very precise indication of μ. But unless the manager has a budget that can absorb further sampling expense, it will have to suffice.

Example: *Evaluating a Physical Therapy Apparatus*

A mechanical engineer on the staff of a physical therapy research team made a sampling study to evaluate his new design for an exerciser. The device is intended to strengthen the muscles in persons suffering from chronic lower back pain. The availability of only one test device limited the number of test subjects that could be accommodated. A random sample of 12 patients was treated with the exerciser, and the following recovery times (days) were obtained:

15	23	32
18	16	22
41	29	25
27	30	18

The mean recovery time is $\bar{X} = 24.7$ days, and the sample standard deviation is $s = 7.6$ days.

An estimate was made of the mean recovery time for all patients who might receive similar therapy. Using a 95% level of confidence, the engineer used the critical value $t_{.025} = 2.201$ to calculate the following interval estimate:

$$\mu = 24.7 \pm 2.201 \frac{7.6}{\sqrt{12}}$$
$$= 24.7 \pm 4.8 \text{ days}$$

or

$$19.9 \leq \mu \leq 29.5 \text{ days}$$

These results were encouraging, although the device could not be recommended for general use without a controlled experiment that rigorously compared its therapy with more traditional methods. Further sampling investigations were necessary.

Example: *Californium Isotope Production by Irradiating Berkelium*

The data in Table 8-1 show the results of $n = 10$ production runs for producing isotopes. In each run the source material, berkelium (^{249}Bk) is irradiated and chemically treated to provide a californium isotope (^{250}Cf). From the sample results, a 95% confidence interval estimate can be constructed for the mean percentage yield of californium μ. Applying the critical value $t_{.025} = 2.262$ (from Table G) and the sample statistics, $\bar{X} = 16.92\%$ and $s = 8.692\%$, we obtain:

$$\mu = \bar{X} \pm t_{.025} \frac{s}{\sqrt{n}}$$
$$= 16.92 \pm 6.22\%$$

Berkelium Source (mg)	Californium Product (mg)	Isotope Yield (%)
1.5	.160	10.7
1.5	.260	17.3
1.0	.135	13.5
1.5	.204	13.6
1.5	.079	5.3
3.0	1.140	38.0
3.0	.540	18.0
1.2	.235	19.6
3.1	.637	20.5
3.3	.420	12.7
	$\bar{X} = 16.92 \qquad s = 8.692$	

Source: Knauer et al., *Transactions of American Nuclear Society*, Vol. 55, November 15–19, 1987, pp. 239–240.

Table 8-1 *Results of Experimental Production of Californium Isotopes*

or

$$10.70\% \leq \mu \leq 23.14\%$$

This interval is quite wide, reflecting the high degree of variability in isotope yields from run to run.

Large Sample Sizes The critical values for the Student t distribution given in Appendix Table G are available for all degrees of freedom from 1 to 30. Above 30 degrees of freedom, it may be necessary to use linear interpolation. Consider the following example.

Example: *Estimating the Mean Time Between Failures (MTBF) of Floppy Disks*

Floppy disks used in personal computers take a great deal of punishment. A mainframe manufacturer wishes to establish specifications for producers who supply disks to computer users. The desired end is higher standards for the industry. The specifications will depend partly on testing with 3 prototypes, designated as A, B, and C, which differ in composition and density.

Separate random samples of 100 disks of each type have been selected for endurance testing. Entries are made on and read from each disk by the same program on a specially constructed disk drive until disk-caused I/O errors occur. Periodically, a robot removes each test disk to a chamber where temperatures are changed according to a program, fluctuating between 60 and 80 degrees Fahrenheit. Another robot gives the disks an occasional rubdown with a lightly oiled cloth to simulate human handling. Each disk is subjected throughout testing to slight electromagnetic disturbances. The following data were obtained for the observed cumulative drive time until first error (failure):

Disk *A*	Disk *B*	Disk *C*
$\bar{X} = 49$ hours	$\bar{X} = 55$ hours	$\bar{X} = 53$ hours
$s = 8.2$	$s = 10.1$	$s = 7.5$

With $n = 100$, the degrees of freedom for the Student t distribution are $n - 1 = 99$. Using a 95% confidence level, Appendix Table G provides the following critical values:

$$\text{df} = 60 \qquad t_{.025} = 2.000$$

$$\text{df} = 120 \qquad t_{.025} = 1.980$$

Linear interpolation provides the value

$$\begin{aligned} t_{.025} &= 2.000 + \left(\frac{99 - 60}{120 - 60}\right)(1.980 - 2.000) \\ &= 1.987 \end{aligned}$$

The following 95% confidence intervals were obtained:

$$\begin{aligned} \mu &= 49 \pm 1.987 \frac{8.2}{\sqrt{100}} \\ &= 49 \pm 1.63 \text{ hours} \end{aligned}$$

or

$$47.37 \leq \mu \leq 50.63 \text{ hours} \qquad (\text{Disk } A)$$

$$52.99 \leq \mu \leq 57.01 \text{ hours} \qquad (\text{Disk } B)$$

$$51.51 \leq \mu \leq 54.49 \text{ hours} \qquad (\text{Disk } C)$$

Notice that the MTBF confidence interval for Disk *A* lies totally below the intervals for the other two disks. The evidence is strong that *A* exhibits performance inferior to *B* or *C*. The relative performance standings of *B* and *C* are clouded, as reflected by the respective overlapping confidence intervals. Further testing with more floppy disks will be needed to determine which type of disk has the greater true MTBF.

Table G gives critical values when df = 120. The next tabled values are for df = ∞, so that whenever the degrees of freedom fall above 120, the critical values may be taken from the df = ∞ row.

The values from that row are identical to the corresponding normal deviates. This reflects that the density for the Student t approaches the normal curve for large n's. Thus, for large sample sizes the critical normal deviate $z_{\alpha/2}$ can be used in place of $t_{\alpha/2}$.

Theoretical Assumptions Although a theoretical assumption of the Student t is that the *population* be normally distributed, it is usually appropriate for all populations except those having frequency distributions with a pronounced skew.

Confidence Interval When Population Is Small

Once the confidence level is chosen and the corresponding value for $z_{\alpha/2}$ is found, the width of the interval estimate of μ depends on $\sigma_{\bar{X}}$. The standard error of $\bar{X}$ is affected by both n and the population standard deviation σ (which, when unknown, may be estimated by s). When sampling is performed without replacement (the usual case) from a small population, the level for $\sigma_{\bar{X}}$ must reflect the population size N. As we have seen, the finite population correction factor is applied in computing $\sigma_{\bar{X}}$. We may analogously amend the procedure for finding confidence intervals, using the following expressions:

$$\mu = \bar{X} \pm z_{\alpha/2} \frac{\sigma}{\sqrt{n}} \sqrt{\frac{N-n}{N-1}} \qquad (\sigma \text{ known})$$

and

$$\mu = \bar{X} \pm t_{\alpha/2} \frac{s}{\sqrt{n}} \sqrt{\frac{N-n}{N-1}} \qquad (\sigma \text{ unknown})$$

Problems

8-10 Suppose that a fourth type of floppy disk (D) is evaluated in the investigation described in the example. The following results apply: $\bar{X} = 57.4$ hours and $s = 11.1$. Construct a 99% confidence interval estimate for the true MTBF of this disk.

8-11 A sample of $n = 100$ observations is taken from a population of unknown mean wherein the standard deviation is assumed to be $\sigma = 5$ grams. The computed value of the sample mean is $\bar{X} = 29.4$ grams. Construct confidence interval estimates of μ for each of the following levels of confidence:
(a) 90% (b) 95% (c) 99% (d) 99.8%

8-12 A sample investigation of $n = 15$ waiting times at a tool cage provided $\bar{X} = 5.3$ minutes with $s = 1.2$ minutes. Construct confidence interval estimates of μ for each of the following confidence levels:
(a) 90% (b) 95% (c) 99% (d) 99.5%

8-13 The precision of an estimate depends largely on the sample size. In each of the following cases, construct a 95% confidence interval estimate of μ, assuming that the population standard deviation is known to be $\sigma = \$14.50$:

(a)	(b)	(c)	(d)
$n = 36$	$n = 100$	$n = 500$	$n = 1{,}000$
$\bar{X} = \$100.25$	$\bar{X} = \$99.75$	$\bar{X} = \$100.75$	$\bar{X} = \$100.50$

8-14 Construct 95% confidence intervals for each of the following sample results:

(a)	(b)	(c)	(d)
$\bar{X}$ = \$5.55	$\bar{X}$ = 10.1 lb	$\bar{X}$ = 10.8k bits	$\bar{X}$ = 98.1 kHz
s = \$1.29	s = .5	s = 2.3	s = .5
n = 25	n = 10	n = 22	n = 17

8-15 An inspector wishes to estimate the mean weight of the contents in a shipment of 16-ounce cans of corn. The shipment contains 1,000 cans. A sample of 200 cans is selected, and the contents of each are weighed. The sample mean and standard deviation were computed to be $\bar{X} = 15.9$ ounces and $s = .3$ ounce. Construct a 99% confidence interval estimate of the population mean.

8-16 One student measured the heights of 35 male colleagues in her surveying class. It may be assumed that her data constitute a representative random sample of the heights of all men attending the university. She computed $\bar{X} = 70.2''$. Assuming that the population standard deviation is the same as used in the text illustrations, what is her 95% confidence interval estimate of the mean heights of university men?

8-17 The industrial engineer evaluating logic wafers was also interested in the amount of silver required in the photographic stages. For the same 15 batches, the following net usages (ounces per wafer) were determined:

.25	.18	.24	.19	.20
.23	.27	.21	.23	.21
.19	.22	.20	.25	.25

Construct a 95% confidence interval estimate of the mean silver usage per wafer.

8-18 The mechanical engineer who designed the physical therapy device discussed in the example collected the following data for the amount of time (hours) spent by the test patients using his machine:

8	12	26	10	23	21
16	22	18	17	36	9

Construct a 99% confidence interval estimate for the mean time spent until recovery by all back patients who use the same therapy.

8-19 The student in Problem 8-16 computed the sample standard deviation $s = 3.1''$.

(a) Do you think it would be more appropriate for her to use this standard deviation in constructing interval estimates of μ? Explain.

(b) Find her 95% confidence interval estimate of the mean height using $s = 3.1''$ as the standard deviation.

8-20 A car manufacturer obtained the following sample results for the mileage traveled before a transmission overhaul was required in each of $n = 100$ cars tested: $\bar{X} = 76{,}400$ miles and $s = 5{,}250$ miles.

(a) Construct a 95% confidence interval estimate of the mean mileage until a transmission overhaul is needed.
(b) Interpret the estimate obtained in (a).

8-21 A product designer is testing a new consumer battery. A test was conducted in which the new battery was discharged in parallel with the existing model. The following results express the number of hours by which the new battery outlasted the old one:

5	−4	10	15	11	25	−5	17
−5	0	8	12	14	8	1	5

Construct a 95% confidence interval estimate of the mean lifetime advantage of the new battery.

8-22 The data in Table 8-2 provide the exposure rates to various radioactive elements, as measured in randomly selected sites in Riyadh, Saudi Arabia. Construct 95% confidence interval estimates for the mean exposure rates for (a) uranium, (b) thorium, (c) potassium, and (d) cesium.

Table 8-2 *Exposure Rates ($\times 10^{-5}$ Gy/yr) to Radioactive Elements in Riyadh, Saudi Arabia*

Site	Uranium ^{232}U	Thorium ^{232}Th	Potassium ^{40}K	Cesium ^{137}Cs
1	6.2	8.4	7.7	1.1
2	7.2	7.9	10.3	—
3	7.8	10.7	12.0	1.2
4	8.9	11.3	11.8	1.8
5	8.5	8.0	7.1	1.0
6	5.3	7.2	6.6	.4
7	8.7	12.3	7.8	2.1
8	8.3	10.0	9.4	.5
9	5.6	7.2	8.1	.4
10	7.9	10.1	12.0	.3
11	5.7	6.8	7.2	.4
12	7.4	10.1	10.6	3.6
13	8.1	7.2	8.3	.5
14	7.0	9.8	11.6	2.0
15	6.8	6.1	7.6	2.3
16	8.6	6.7	7.6	1.6
17	6.4	7.4	7.0	.8
18	12.2	13.5	9.3	1.2
19	7.0	9.8	11.6	2.0
20	6.0	8.2	8.0	.98
21	7.4	8.2	8.6	8.0

Source: Tawfik et al., *Transactions of American Nuclear Society*, Vol. 55, November 15–19, 1987, pp. 87–88.

8-3 Interval Estimates of the Proportion

Procedures for estimating the population proportion parallel those for the mean. The process is somewhat simplified because only one parameter, π, is involved. Rather than basing procedures on the true sampling distribution for P, which we know is either the binomial (sampling from a large population) or the hypergeometric (small population), when n is large enough we employ the normal approximation.

Estimates When Sampling from Large Populations

For sufficiently large sample sizes, the true binomial sampling distribution for P is closely approximated by the normal distribution with mean π and standard deviation

$$\sigma_P = \sqrt{\frac{\pi(1-\pi)}{n}}$$

In constructing the interval estimate of π, the computed value of P is used in the above to estimate σ_P. The following expression is used to determine the $100(1-\alpha)\%$ confidence interval.

CONFIDENCE INTERVAL ESTIMATE OF THE PROPORTION WHEN THE POPULATION IS LARGE

$$\pi = P \pm z_{\alpha/2}\sqrt{\frac{P(1-P)}{n}}$$

or

$$P - z_{\alpha/2}\sqrt{\frac{P(1-P)}{n}} \leq \pi \leq P + z_{\alpha/2}\sqrt{\frac{P(1-P)}{n}}$$

As an illustration, suppose that an estimate is made of the proportion of operating time spent on unscheduled maintenance for computer mainframes of a particular make and model. A random sample of 100 terminal log entries indicated the computers were down 5 times, so that $P = .05$. A 95% confidence interval is constructed for the population proportion unscheduled downtime, using $z_{.025} = 1.96$ from Appendix Table E:

$$\pi = .05 \pm 1.96\sqrt{\frac{.05(1-.05)}{100}} = .05 \pm .043$$

or

$$.007 \leq \pi \leq .093$$

An interpretation similar to the one used for the mean applies to the above. Were the procedure repeated many times, using a different random sample of 100 log entries each time, about 95% of the interval estimates would correctly bracket the true level of π, but the remaining intervals would lie totally above or below π.

When to Use Normal Approximation Keep in mind that the foregoing procedure is based on the normal approximation to the binomial. That approximation works best for large sample sizes or levels of π that are not extreme. When in doubt, use the following rule.

RULE FOR MAKING NORMAL APPROXIMATION

Use the normal approximation whenever *both* of the following hold:

$$n\pi \geq 5$$

$$n(1 - \pi) \geq 5$$

In applying this rule, the level of π must be estimated from the sample results by the computed level of P. When n does not satisfy these inequalities, confidence intervals may instead be constructed using exact binomial probabilities.

Interval Estimate of the Proportion Using Exact Binomial Probabilities

Using binomial probabilities, an interval of the form

$$\frac{R_1}{n} \leq \pi \leq \frac{R_2}{n}$$

must be found. The end points are determined from the limiting number of successes R_1 and R_2, that set of values satisfying the condition

$$\Pr[R_1 \leq R \leq R_2] \geq 1 - \alpha$$

It will ordinarily be impossible to obtain limits that exactly meet the *targeted* $100(1 - \alpha)\%$ confidence level.

In establishing the interval estimate, a working value of $\pi = P$ (the computed proportion) is assumed.

The first limit R_1 is the level of r that provides the *smallest* cumulative probability such that

$$\Pr[R \leq r \mid \pi = P] > \frac{\alpha}{2}$$

The second limit R_2 is the level of r providing the *smallest* cumulative probability such that

$$\Pr[R \leq r \mid \pi = P] \geq 1 - \frac{\alpha}{2}$$

The *achieved* confidence level is the percentage corresponding to the difference:

$$\Pr[R \leq R_2] - \Pr[R \leq R_1 - 1]$$

(The second term is zero when $R_1 = 0$.)

To illustrate this procedure, consider the preceding example, for which $n = 100$ and $P = .05$. We will use exact binomial probabilities to compute a confidence interval targeted at 95%. From Appendix Table A, using $\pi = .05$ and $\alpha = .05$, the smallest cumulative probability greater than $\alpha/2 = .025$ is

$$\Pr[R \leq 1] = .0371$$

so that $R_1 = 1$. The smallest cumulative probability greater than or equal to $1 - \alpha/2 = .975$ is

$$\Pr[R \leq 10] = .9885$$

so that $R_2 = 10$. The achieved level of confidence is found from the difference

$$\Pr[R \leq 10] - \Pr[R \leq 0] = .9885 - .0059 = .9826$$

to be 98.26%. The corresponding confidence interval is

$$\frac{1}{100} \leq \pi \leq \frac{10}{100}$$

or

$$.01 \leq \pi \leq .10$$

Notice that this interval is wider than the one obtained earlier using the normal approximation. This is partly due to the higher achieved confidence level and partly due to the imprecision in using the normal approximation.

Estimating the Proportion When Sampling from Small Populations

When sampling without replacement from a small population (usually one in which n is more than 10% of the population size), the standard error of P involves the finite population correction factor:

$$\sigma_P = \sqrt{\frac{\pi(1 - \pi)}{n}} \sqrt{\frac{N - n}{N - 1}}$$

Employing this equation, we obtain slightly narrower interval estimates.

The following expression is used to compute the $100(1 - \alpha)\%$ confidence interval.

CONFIDENCE INTERVAL ESTIMATE OF THE PROPORTION WHEN THE POPULATION IS SMALL

$$\pi = P \pm z_{\alpha/2} \sqrt{\frac{P(1 - P)}{n}} \sqrt{\frac{N - n}{N - 1}}$$

As an illustration, suppose an inspector opens a sample of 50 cartons out of a shipment of 300, noting that 6 cartons are underweight. What is the 95% confidence interval estimate for the proportion of underweight cartons in the entire shipment?

The sample proportion of underweight cartons is $P = \frac{6}{50} = .12$. Using $z_{.025} = 1.96$, $n = 50$, and $N = 300$, the following is obtained:

$$\pi = .12 \pm 1.96\sqrt{\frac{.12(1 - .12)}{50}}\sqrt{\frac{300 - 50}{300 - 1}} = .12 \pm .082$$

or

$$.038 \leq \pi \leq .202$$

The inspector is 95% confident that the true proportion of underweight cartons will lie between .038 and .202. Of course π could lie outside these limits, since the sample result might not have been very representative of the whole. There is no way for the inspector to know for sure without opening all cartons in the shipment.

When sampling without replacement from a small population the same restrictions given earlier apply to the normal approximation. When n is too small to justify that approximation, exact *hypergeometric* probabilities may be used to construct confidence limits. (Because those probabilities usually have to be computed from scratch, the hypergeometric is often approximated by the binomial.)

Problems

8-23 Construct 95% confidence intervals for the proportion for each of the following sample results:

(a)	(b)	(c)	(d)
$n = 100$ $P = .23$	$n = 25$ $P = .25$	$n = 500$ $P = .05$	$n = 1{,}000$ $P = .005$

8-24 The proportion of defectives found in a sample of 100 items taken at random from a production line is .10. Construct interval estimates of the true process proportion defective for each of the following levels of confidence:
(a) 90% (b) 95% (c) 99% (d) 99.5%

8-25 For each of the following situations, construct 95% confidence intervals for the population proportion:

(a)	(b)	(c)	(d)
$N = 500$ $n = 100$ $P = .15$	$N = 1{,}000$ $n = 200$ $P = .05$	$N = 5{,}000$ $n = 1{,}000$ $P = .075$	$N = 1{,}000$ $n = 150$ $P = .875$

8-26 An automobile parts distributor found 34 packages containing defective brake linings in a sample of 200 taken at random from a shipment containing 1,000 packages. Construct a 99% confidence interval estimate of the proportion of packages in the entire shipment that contain defective linings.

8-27 The controls in a brewery need adjustment whenever the proportion π of underfilled cans is .015 or greater. There is no way of knowing the true proportion, however. Periodically a sample of 100 cans is selected and the contents are measured.

(a) For one sample, 4 underfilled cans were found. Construct the resulting 95% confidence interval estimate of π.

(b) What is the probability of getting as many or more underfilled cans as in (a) when in fact π is only .01?

8-28 Suppose that the proportion of voters approving the president's performance is to be estimated. Two pollsters found that (1) in May, 75 approved out of 130 persons polled, and (2) in October, 642 approved out of 1,056 persons polled.

(a) For each of these results construct a 95% confidence interval estimate of the true proportion approving.

(b) The difference in confidence limits indicates the amount of precision in an interval estimate. Determine this quantity for each of the intervals you found in (a).

8-29 An industrial engineer's assistant made 50 random observations of the upholstery installer team in an automobile assembly plant. During 12 of the observations the workers were arranging materials beside their work station.

(a) Construct a 95% confidence interval estimate of the proportion of time installers spend arranging materials.

(b) A total of 495 cars passed upholstery installers during the 8-hour shift, during which time the line was in operation 434 minutes. Using your answer to (a), determine an interval estimate for the mean time per car spent by the installation team just arranging materials.

8-30 An industrial engineer finds 10 defectives in a sample of 50 items selected at random from a production line. Use exact binomial probabilities to estimate the true process proportion defective at a targeted 90% significance level.

8-31 A quality-control inspector selects a random sample of 10 items from an incoming shipment of 100 and finds that 3 items are defective.

(a) Ignoring any restrictions on using it, use the normal distribution to construct a 95% confidence interval estimate of the true lot proportion defective.

(b) Using the sample proportion defective as a point estimator of the population proportion defective, construct a table of individual and cumulative hypergeometric probabilities for the number of sample defectives in similar samples.

(c) Using your results from (b), construct an exact interval estimate for the true lot proportion defective with a targeted confidence level of 95%.

(d) Repeat (c), using instead binomial probabilities to approximate the hypergeometric.

(e) Explain which estimate is better. Which procedure do you prefer? Explain.

8-4 *Interval Estimates of the Variance*

Like the mean and proportion, the variance (or standard deviation) is an important parameter often estimated from its sample counterpart. In Chapter 7 we saw that the sampling distribution for s^2 could be represented by a standard version of the *chi-square distribution*.

Making Estimates Using the Chi-Square Distribution

The chi-square distribution also serves as the basis for constructing confidence intervals for σ^2 (or σ). The following expression is used to construct the $100(1 - \alpha)\%$ confidence interval.

CONFIDENCE INTERVAL ESTIMATE OF THE VARIANCE

$$\frac{(n-1)s^2}{\chi^2_{\alpha/2}} \leq \sigma^2 \leq \frac{(n-1)s^2}{\chi^2_{1-\alpha/2}}$$

Notice that owing to the asymmetry of the chi-square distribution there are two critical values, $\chi^2_{\alpha/2}$ and $\chi^2_{1-\alpha/2}$, which must be read from Appendix Table H using $n - 1$ degrees of freedom.

As an illustration, consider the population of waiting times experienced by message segments buffered in a telecommunications satellite until retransmission. Sample results provide a sample standard deviation of $s = 10.4$ microseconds for $n = 25$ messages. A 90% confidence interval estimate of the variance in waiting time for all similar message transmissions is constructed below. The critical values for df $= 25 - 1$ are $\chi^2_{.05} = 36.415$ and $\chi^2_{1-.05} = \chi^2_{.95} = 13.848$. Thus,

$$\frac{(25-1)(10.4)^2}{36.415} \leq \sigma^2 \leq \frac{(25-1)(10.4)^2}{13.848}$$

or

$$71.28 \leq \sigma^2 \leq 187.45$$

Taking the square roots of the end points, we obtain a 90% confidence interval for the population standard deviation:

$$8.4 \leq \sigma \leq 13.7 \quad \text{microseconds}$$

This inequality can be interpreted similarly to the earlier interval estimates. In repeated samples, only about 90% of intervals constructed in this way will actually bracket σ^2 (or σ); the remainder will lie totally above or below the true parameter value.

This procedure assumes that the chi-square distribution is suitable. One of the theoretical assumptions of that distribution is that the *population* be normally distributed. As a practical matter, however, the chi-square works satisfactorily for most populations having frequency distributions not highly skewed.

Making Estimates Using the Normal Approximation

In Chapter 7 it was noted that the chi-square distribution tends toward the normal distribution when n is large. Thus, for large sample sizes (generally those where $n > 30$) that approximation is used in constructing confidence intervals. In those cases the following expression provides the $100(1 - \alpha)\%$ confidence interval.

CONFIDENCE INTERVAL ESTIMATE FOR THE VARIANCE WHEN THE SAMPLE SIZE IS LARGE

$$\frac{(n-1)s^2}{n-1+z_{\alpha/2}\sqrt{2(n-1)}} \leq \sigma^2 \leq \frac{(n-1)s^2}{n-1-z_{\alpha/2}\sqrt{2(n-1)}}$$

In this inequality, the critical values for the chi-square statistic are estimated by the expected value $n - 1 \pm z_{\alpha/2}$ standard errors.

Suppose an engineer wants to estimate the standard deviation for a population of processing times for plotting a waveform with a digital computer. A sample of $n = 100$ plots yields a sample standard deviation of $s = 2.35$ minutes. The following 99% confidence interval was obtained using $z_{.005} = 2.57$:

$$\frac{(100-1)(2.35)^2}{100-1+2.57\sqrt{2(100-1)}} \leq \sigma^2 \leq \frac{(100-1)(2.35)^2}{100-1-2.57\sqrt{2(100-1)}}$$

or

$$4.04 \leq \sigma^2 \leq 8.70$$

which provides

$$2.01 \leq \sigma \leq 2.95 \quad \text{minutes}$$

This estimate of the population standard deviation is fairly precise, but keep in mind that it represents population variability, *not* central tendency.

Problems

8-32 Construct 90% confidence intervals for (1) the population variance and (2) the population standard deviation in each of the following cases:

(a)	(b)	(c)	(d)
$s = .15$	$s = 34.2$	$s = 1.01$	$s = 335.1$
$n = 25$	$n = 15$	$n = 100$	$n = 150$

8-33 The sample standard deviation for $n = 20$ observations was computed to be $s = 12.6$. Construct interval estimates of the population standard deviation for each of the following levels of confidence:
(a) 80% (b) 90% (c) 98%

8-34 Using the silver-content data in Problem 8-17, construct a 90% confidence interval estimate of the standard deviation in wafer silver consumption.

8-35 Using the time data in Problem 8-18, construct a 98% confidence interval estimate of the standard deviation in patient time on the exercise machine.

8-5 Determining the Required Sample Size

At the outset of this chapter we noted that the quality of a statistical estimate has two dimensions, *precision* and *reliability*. Although the desired level of one of these is always attained at the expense of the other, an investigator can achieve high levels of both precision and reliability by increasing the sample size. Should there be constraints on n, perhaps owing to budgetary limitations or a shortage of time, some optimal balance between precision and reliability must be reached.

Let's consider the fundamental relationship among sample size, precision, and reliability. We begin with applications in which the mean is estimated.

Finding Required Sample Size for Estimating the Mean

We begin with the simplest case, in which the investigator has considerable discretion about the sample size and no external restrictions limit the number of observations. The normal distribution may then be assumed to be a satisfactory representation of the sampling distribution of $\bar{X}$. The probability that the sample mean will fall within plus or minus some distance of μ may be expressed by

$$\Pr[\mu - z_{\alpha/2}\sigma_{\bar{X}} \leq \bar{X} \leq \mu + z_{\alpha/2}\sigma_{\bar{X}}]$$

A model for determining n must involve two discretionary parameters:

Desired precision
Specified reliability

The first may be expressed in terms of the allowable deviation d, the maximum distance above or below which the estimate should lie from its target population parameter. This is analogous to a machine tolerance such as $\pm.001''$, which indicates that items must fall within one-thousandth of an inch above or below the specified dimension; in that case, the desired precision is $d = .001''$. Of course, because of sampling error the investigator can never be certain that the maximum deviation won't be exceeded—but the probability of this may be specified. That probability defines the reliability,

$$\text{Reliability} = 1 - \alpha = \Pr[\mu - d \leq \bar{X} \leq \mu + d]$$

The specified reliability equals the area under a unique normal curve for $\bar{X}$. This establishes a critical normal deviate $z_{\alpha/2}$ value, so that this curve must be the one where

$$d = z_{\alpha/2}\sigma_{\bar{X}}$$

The only unknown quantity is the standard error, which for large populations equals $\sigma/\sqrt{n}$. Thus, we have

$$d = z_{\alpha/2}\frac{\sigma}{\sqrt{n}}$$

Solving for n, we obtain the following expression, which may be used to calculate the required sample size for estimating the mean.

REQUIRED SAMPLE SIZE FOR ESTIMATING THE MEAN

$$n = \frac{z_{\alpha/2}^2 \sigma^2}{d^2}$$

where

$$d = \text{desired precision (or maximum error)}$$

$$z_{\alpha/2} = \text{critical normal deviate for specified reliability } 1 - \alpha$$

$$\sigma = \text{assumed population standard deviation}$$

Notice that in addition to precision and reliability, a value must be assumed for the population standard deviation σ.

To illustrate, suppose that an investigator wishing to estimate the mean stretching force required to permanently distort reinforcing bars of a particular type desires precision of $d = 5$ pounds, with a reliability probability of $1 - \alpha = .95$. Previous studies of similar rods indicate that σ should be about 50 pounds. Appendix Table E provides the critical normal deviate $z_{.025} = 1.96$. Thus, the required sample size is

$$n = \frac{(1.96)^2(50)^2}{5^2} = 384 \qquad \text{(rounded)}$$

An expression similar to this applies when the population is small, so that $\sigma_{\bar{X}}$ involves the finite population correction. (You may derive that expression as an exercise.) Another case applies when the sample size itself must be kept arbitrarily small (as is usual in lengthy investigations involving human subjects). We have seen that when n is less than 30, the Student t distribution ordinarily applies. No model for finding n is needed with the Student t, since in such applications n is usually prescribed by circumstance.

Influences on Sample Size

Notice in the foregoing expression that n is directly proportional to the *square* of the reliability normal deviate z and the population standard deviation σ. Also, n is inversely proportional to the *square* of the desired precision d. Thus, doubling $z_{\alpha/2}$ from 2 to 4 (so that reliability is raised from .9544 to .99994) would require a fourfold increase in sample size. That is a stiff price to pay for such a small increment in reliability. However, that same increment in sample size would permit a twofold improvement in precision, allowing d to be cut in half. But improving precision by a factor of 10 raises the required n by a factor of 100. The investigator has no control over σ, but a population having twice the variability of the first will require four times the sample size in order to provide identical precision with the same reliability.

Example: Estimating Height and Hand Length

Imagine a study in which population means for men's heights and hand lengths are to be estimated. For planning purposes, we could use the existing anthropometric tables for Air Force flight crews,* getting

$$\sigma = 2.44'' \text{ (height)} \qquad \text{and} \qquad \sigma = .34'' \text{ (hand length)}$$

Since a different target population applies, a new sampling study will be performed. Although this may be unrealistic, for purposes of illustration we will suppose that heights and hand lengths will be measured independently in different samples. In each case, the desired precision is $d = .10''$ and the required reliability is $1 - \alpha = .99$. What are the respective required sample sizes?

Appendix Table E provides $z_{.005} = 2.57$, and we have

Height	Hand length
$n = \dfrac{(2.57)^2(2.44'')^2}{(.10'')^2}$	$n = \dfrac{(2.57)^2(.34'')^2}{(.10'')^2}$
$= 3{,}932$	$= 76$

Notice that the n needed for estimating mean height is over 49 times as great as the sample size needed to estimate mean hand length. This is because, in absolute terms, heights are considerably more varied than hand lengths, with σ for the former being over 7 times as great. Our model therefore requires that n be over $(7)^2 = 49$ times as great for heights as for hands.

The sample size model does not reflect that the two samples should probably have different d's to account for the differences in the relative sizes of the observed units. For instance, precision of $\pm.10''$ may be adequate for establishing the height of a cockpit passageway (and $\pm1.0''$ or less accuracy might suffice). But that degree of precision would hardly be satisfactory for a glove manufacturer in sizing cutting dies, where $\pm.01''$ might be more appropriate.

* H. T. E. Hertzberg, G. S. Daniels, and E. Churchill, *Anthropometry of Flying Personnel*, WADC Technical Report 52–321, U.S.A.F (September 1954).

This example graphically illustrates the crucial role that σ plays in establishing the required sample size.

Finding a Planning Estimate of σ

Let's place our discussion in context. We are pursuing one aspect of the planning process, before any sample data are collected. At that time the population mean is unknown and must be estimated. But whenever the mean is unknown, the population standard deviation will also ordinarily be unknown. Thus, σ used in the foregoing model is also an unknown quantity. For planning purposes a value must be assumed for σ. If this assumed value deviates from the true population standard deviation by some amount, the sample size computed from the model will over- or understate the actual n requirement by a factor equal to the square of that deviation.

When the orders of magnitude in the X's are presumed to be nearly the same, statisticians often will use a planning σ obtained from an earlier study of a related population. Air Force heights may therefore provide an appropriate ballpark σ for finding what n should be for some other height population.

But in many statistical investigations no closely related data are available. Consider a study measuring processing times for pilot batches to be run under an experimental chemical procedure. Good statistical planning requires an objective determination of the number n of sample batches to be timed. Assuming there is no reference population from which to extrapolate a "ballpark" σ, this quantity must be found in another way. The following procedure has proved to be satisfactory for reaching an approximate value.

We know that, for normally distributed populations, over 99.7% of the values will lie within $\pm 3\sigma$ of the mean. About 1 out of 1,000 will fall below $\mu - 3\sigma$, with a similar fraction falling above $\mu + 3\sigma$. This suggests that two percentiles, the .1 and the 99.9, span a range of 6 standard deviations. Knowing these two quantities, we may compute the assumed population standard deviation from

$$\text{Planning } \sigma = \frac{Q_{.999} - Q_{.001}}{6}$$

Of course, there is no more reason to know the above percentiles than σ itself—but those quantities may be guessed with some confidence. A chemical engineer should have some idea about an absolute minimal processing time (say, 1 hour) and about some maximum amount that it is almost inconceivable to exceed (say, 100 hours). A crude value for σ would then be $(100 - 1)/6 = 16.5$ hours.

Even if the population proves to be nonnormal, we know by Chebyshev's theorem that σ should be close to the above because a substantial percentage of the population will lie within $\mu \pm 3\sigma$, or inside a 6σ central range. Only by actual experimentation could the chemical engineer find a better value for σ, but this procedure at least gives him an n that is of the proper order of magnitude.

Some statisticians employ a two-stage sampling scheme that involves a small pilot sample from which the sample standard deviation s is used as a point estimator of σ. That value in turn is used to establish the n for the main follow-on sample.

Further Sample Size Considerations

It may be impractical or even impossible for an investigator to make as many sample observations as the foregoing procedure would indicate. As noted, the availability of resources or time will ordinarily place an upper limit on n. Some relaxation of precision or reliability goals may therefore be necessary. This is ordinarily done at the expense of the precision at which the estimate will be reported, so that a smaller n will result in wider confidence intervals than originally desired. (Because a minimal level of confidence is usually expected by any statistical audience, there is little room for maneuvering with $z_{\alpha/2}$.) There may be no satisfactory resolution to this dilemma except to live with poorer precision than one would like.

In choosing d, therefore, an investigator must demand no greater precision that what is actually required. A newspaper article on average prices of homes in various regions might require a precision of only $\pm$\$5,000, whereas a report on engineers' salaries might require accuracy of $\pm$\$1,000. An industrial engineer in charge of manufacturing processes for a \$5 item might require precision of $\pm$\$.01 in estimating the mean cost for the required quantity of a particular raw material.

It may help to view the sample size decision itself in economic terms. Figure 8-2 shows that the total cost of sampling might be expressed in terms of two components, *data collection* and *error*. Data collection costs have a fixed component and a variable component, which are represented on the cost line by the y-intercept and the slope, respectively. The error costs are harder to measure, but they encompass the net effect of the wrong choices and actions that result from obtaining samples that seriously over- or undershoot their targets. Generally, error costs are huge when the sample size is small, and they will decrease as the sample size is increased. A hyperbolic shape is a plausible representation, since sampling error should decrease quite rapidly as n rises beyond zero, but the reductions become less dramatic for larger levels of n. Notice that the combined cost of sampling is portrayed as a convex curve achieving minimum cost at the optimal level for n.

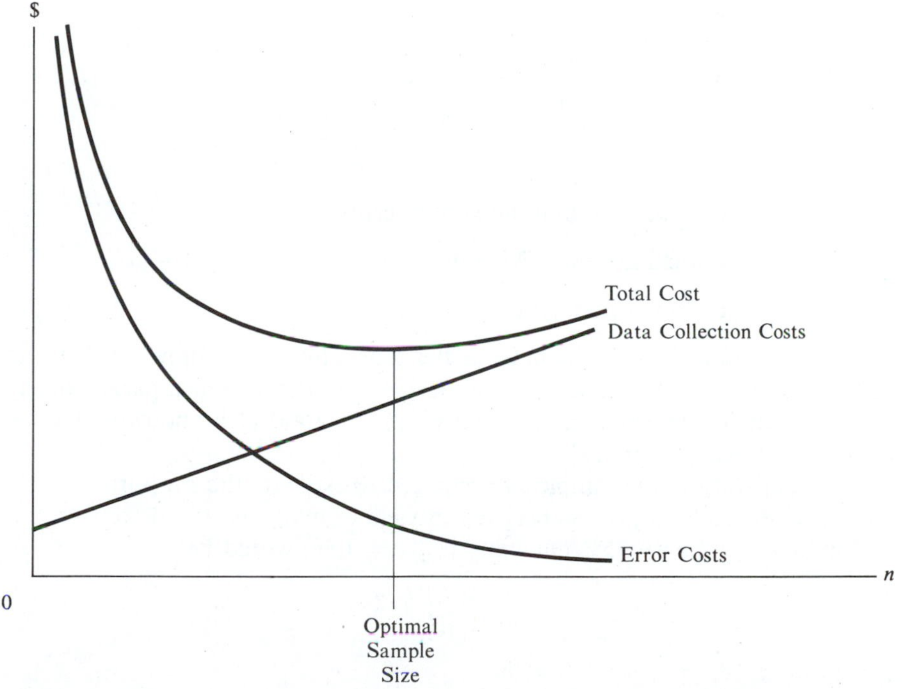

Figure 8-2
The total cost of sampling has two components: data collection costs and error costs.

Example: *The Quality Inspector's Dilemma*

A quality-control inspector for a disk drive manufacturer must pass judgment on incoming shipments. The entire shipment of most parts will be accepted or rejected on the basis of sample evidence. A more flexible arrangement exists with one long-leadtime supplier. Its shipments are always accepted, but a penalty is assessed if the test items in a sample involve mean readings not meeting specifications. Owing to sampling error, an insufficient penalty will sometimes be assessed. On the other hand, a few shipments will be overpenalized. When the completed disk drives are finally tested, the supplier is eventually given credit for the unjustified charges, but there is no way to collect on the undercharges. The extra cost of replacing inferior components and refunding for excessive penalties was judged to be so high that much larger sample sizes were justified on shipments from this supplier than from regular parts vendors. Nevertheless, the total cost of sampling was much lower than it would have been for smaller sample sizes, which would have yielded less precise estimates.

Required Sample Size for Estimating Proportion

A procedure similar to the one described for the mean is used for finding the required sample size for estimating the *population proportion.* Sampling situations in which choice of n is an important issue will ordinarily involve a number of observations large enough so that the normal curve may be assumed as the sampling distribution for P. Duplicating the arguments used earlier for the mean, it is easy to derive the following expression.

REQUIRED SAMPLE SIZE FOR ESTIMATING THE PROPORTION

$$\boldsymbol{n = \frac{z_{\alpha/2}^2 \pi(1 - \pi)}{d^2}}$$

where

$$d = \text{desired precision (maximum error)}$$

$$z_{\alpha/2} = \text{critical normal deviate for specified reliability } 1 - \alpha$$

$$\pi = \text{assumed population proportion}$$

The precision is in the same units as the proportion and will be a value such as .05, .01, or .005. Notice that the formula incorporates π. Thus, paradoxically, to find the proper size for n we must have a pretty good idea about the level of π, the quantity being estimated.

As an illustration, the sample size required to estimate the proportion of defectives π produced by a process believed to yield as many as 10% defectives, to a precision of $d = .02$ with 95% reliability ($z_{.025} = 1.96$) would be:

$$n = \frac{(1.96)^2(.10)(.90)}{(.02)^2}$$

$$= 864$$

In making this determination it was necessary to assume a value for π of .10. Of course, the true value of π could differ, perhaps considerably. Should the true π be greater—say, .15—the required sample size would have to be larger,

$$n = \frac{(1.96)^2(.15)(.85)}{(.02)^2}$$

$$= 1{,}225$$

or else the desired precision or reliability would not be met. On the other hand, the true π might be smaller and more than the desired precision might be obtained.

Over- or undersampling results whenever the assumed π turns out to be smaller or larger than the true level. When estimating the proportion, it is always possible to prevent undersampling, because the product $\pi(1 - \pi)$ achieves a maximum when $\pi = .5$. Using that level in computing the required n guarantees that for the stated reliability the desired precision will be achieved.

Problems

8-36 With a .95 specified reliability, determine the required sample size for estimating the mean in each of the following situations:

(a)	(b)	(c)	(d)
$d = \$100$	$d = .15''$	$d = .40$ pounds	$d = 1{,}000$ kHz
$\sigma = \$1{,}000$	$\sigma = 20''$	$\sigma = 5.6$ pounds	$\sigma = 5{,}000$ kHz

8-37 With a desired precision of .5 gram, determine the required sample size for estimating the mean of a population having an assumed standard deviation of 10 grams for each of the following reliability levels:
(a) .90 (b) .95 (c) .99 (d) .999

8-38 Derive an expression for finding the required size for estimating the mean when sampling from a small population of size N. Use your result to determine n when $N = 1{,}000$ for a population in which the assumed standard deviation is $\sigma = 50$, the specified reliability is .95, and the desired precision is $d = 2.5$.

8-39 Determine the sample size required to estimate the proportion when the specified reliability is .99 and the following apply:

(a)	(b)	(c)	(d)
$\pi = .20$	$\pi = .50$	$\pi = .05$	$\pi = .01$
$d = .02$	$d = .05$	$d = .005$	$d = .001$

8-40 Derive an expression for finding the required sample size for estimating the proportion when sampling is done from a small population of size N. Then for 95% reliability determine the required n when $N = 500$, $\pi = .10$, and $d = .01$.

8-41 A statistical analyst wishes to determine the sample size needed to estimate the mean maximum pressure that can be withstood by an airlock seal. Desired precision is ± 100 psi with reliability .95. She has no feel for an assumed population standard deviation; however, the designer believes that it is almost certain that an individual seal will withstand a test pressure of 5,000 psi, whereas it is extremely unlikely that any seals will survive 6,500 psi.

(a) What value should the analyst use as her assumed standard deviation?

(b) Using your answer to (a), what is the required sample size?

8-42 An industrial engineer must establish the mean time for a worker to complete task A for an individual part.

(a) Determine the required sample size using stopwatch methods when the desired precision is $d = 1$ minute and the specified reliability is .95, assuming that $\sigma = 5$ minutes.

(b) Assuming work sampling is used instead, determine the required n to find the proportion of time for completing task A with a desired precision of $d = .05$, also with .95 reliability. Assume the worst case and use a planning value of $\pi = .5$.

(c) Assuming that the true mean completion time is approximately 20 minutes per part, then the mean task A time obtained in (a) will have precision identical to that obtainable by multiplying the estimate in (b) by 20 minutes. If each stopwatch observation costs \$10 and each work-sampling observation \$.50, which procedure would be less costly?

8-43 The estimated cost for preparing a questionnaire is \$100. It will cost \$10 per observation to have the form filled out by sample subjects. The cost of sampling error is \$1,000 divided by the sample size. What is the optimal sample size?

Comprehensive Problems

8-44 An estimate must be made of the mean quantity of scrap sheetmetal consumed in fabricating home-furnace housings. A random sample of 100 furnaces were monitored throughout the complete production cycle. All raw sheetmetal for individual furnaces was allotted in advance and color-coded. All scraps were set aside and their dimensions noted. The following summarize these amounts: $\bar{X} = 55.3$ square feet and $s = 5.4$ square feet. Construct a 95% confidence interval estimate for the mean amount of scrap per furnace.

8-45 The following costs per 1,000-liter batch were obtained for pilot runs of a technical chemical:

\$105	\$110	\$108	\$97	\$ 99
104	112	103	98	105

(a) Find the value of an efficient, unbiased, and consistent estimator of the mean cost.

(b) Find the value of an unbiased and consistent estimator of the variance in cost.

(c) Find the value of an unbiased estimator of the proportion of costs exceeding \$100.

8-46 Using the data in Problem 8-45, construct 90% confidence interval estimates of the following

(a) the mean cost per batch

(b) the variance in cost per batch

(c) the standard deviation in cost per batch

8-47 The mean drying time of an epoxy coating is to be estimated. Assuming a standard deviation value of 1.5 hours, determine the required sample size under the following conditions:

(a) The reliability for being in error at most .25 hour in either direction is .99.

(b) The reliability for being in error at most .25 hour in either direction is .95.

(c) The reliability for precision of $\pm .5$ hour is .99. How does the resulting sample size requirement compare with your answer to (a)?

8-48 An electrical engineer wishes to estimate the mean signal-to-noise ratio for a radar operating in the Doppler-shift mode while following aircraft at a nominal range and altitude. A minimum value of 10 is anticipated, and the maximum ratio believed possible is 40. How many observations are necessary to estimate the mean signal-to-noise ratio with a precision of ± 2 at a reliability of .95?

8-49 The following sample data were obtained for the time (in minutes) required to hand-assemble a throttle:

11.2	18.5	8.7	12.4	13.5
9.9	12.9	15.4	12.6	16.7
10.2	10.5	14.4	17.7	15.5

(a) Construct a 95% confidence interval estimate of the mean throttle-assembly time.

(b) Construct a 96% confidence interval estimate of the standard deviation in throttle-assembly time.

8-50 For a class project an industrial engineering student used work sampling to estimate the mean times taken by campus bookstore checkers to make change. In 100 random observations, change was being made 23 times. Over the observed span, 154 customers checked out, and 258 employee minutes were logged at the checkstands.

Assuming the total checkout time per customer does not vary from the above, construct a 95% confidence interval estimate of the mean time to make change.

8-51 A civil engineer wishes to find the energy consumption of a proposed sports complex that incorporates new design concepts. The major concern is natural-gas consumption, which will be simulated using Monte Carlo methods.

(a) For planning purposes the engineer must establish a value for the standard deviation in annual gas requirements. It is virtually certain that 10,000 thousand cubic feet (mcf) will be consumed, even in the warmest of years; it is inconceivable that gas requirements in any given year will exceed 100,000 mcf. What value should be used for the standard deviation in annual gas consumption?

(b) Assuming precision requirements for ± 500 mcf for estimating mean annual gas requirements with .99 reliability, how many years should be simulated?

8-52 Using the data in Problem 8-51, consider the following. Because the civil engineer could not afford the computer time needed to simulate the ideal number of years, only 200 years were simulated. The gas consumption summary statistics are $\bar{X} =$ 55,300 mcf and $s = 9{,}250$ mcf.

(a) Construct a 95% confidence interval estimate of the mean annual gas consumption.

(b) The engineer wants to see how much the simulation undersampled. How many more simulated years would be needed to meet precision of ± 500 mcf with .99 reliability, assuming that the standard deviation of the foregoing sample is a satisfactory estimate of the population standard deviation?

CHAPTER NINE

Hypothesis Testing

LIKE ESTIMATION, **hypothesis testing** is one of the basic forms of statistical inference. This broad area of statistics usually culminates in an immediate decision, making it the more dynamic form of statistical inference. Testing is based on two complementary assumptions, or *hypotheses*, regarding unknown populations. Taking various forms, the most common hypotheses involve assumed levels for a population parameter. Later portions of this book describe other types of hypotheses that help to compare two or more populations or are concerned with properties such as the randomness of data, the manner in which two variables are related (if at all), or the shape of the population distribution.

This chapter begins with hypothesis tests for the unknown value of the population mean. These procedures are useful for evaluating decisions involving two complementary actions when the unknown level of μ is the pivotal factor. For example, the decision whether or not to repair equipment might depend on how much the mean reading deviates from some nominal figure. Statistical inferences are needed in such evaluations because μ can be evaluated only indirectly from sample data. For this purpose the sample mean serves as the *test statistic.* Thus, repairs might be requested only if the computed level of $\bar{X}$ exceeds a specified level.

Of course samples may give false readings of reality. The potential for such sampling error can be controlled but never eliminated. Sampling error is managed differently in hypothesis testing than in estimating, where the major concern is with precision and reliability. This control is achieved by establishing a *decision rule* that achieves some optimal balance among the probabilities for taking incorrect actions. For example, the incidence of unnecessary repairs must be kept small, while failure to make needed repairs must be avoided. Hypothesis testing procedures focus on such errors of commission and omission, which may be controlled by judicious determination of the decision rule or by selecting a sample large enough to keep the error probabilities acceptably small.

9-1 Basic Concepts of Hypothesis Testing

To introduce hypothesis testing, we will consider a decision involving the population mean. The sample mean $\bar{X}$ will serve as the test statistic. The following example illustrates the underlying concepts. A general procedure that may be applied in most testing situations is introduced later.

Illustration: *Bacterial Leaching in Ore Processing*

A mining engineer is studying ways to increase the production of metal from a large copper deposit. A substantial amount of copper-bearing ore is presently bypassed because its quality is too low for economical processing at current world prices. The engineer is interested in a new process based on bacterial leaching. This technique isolates key minerals in poor ores through a chemical reaction caused by strains of the bacterium *Thiobacillus*. These microbes convert the iron in the ore from a ferrous form into ferric iron, which then oxidizes the insoluble copper sulfides present in the ore's pyrites. The process culminates in collection of a liquid solution from which copper may be recovered.

The effectiveness of bacterial leaching depends on the particular strain of bacteria used and the composition of the ore. Since there is little history of successful application, a statistical sampling study will be performed to determine whether bacterial leaching should be adopted on a major scale. A sample of 100 loads of ore will be removed from random coordinates throughout the unworkable deposit site and processed by a promising strain of bacteria at a test leaching dump. The final yield, in pounds of copper per ton of ore, will then be determined for each sample load. These data will represent the yields obtainable from bacterial leaching of the more than 100 million tons of target ore.

The success of large-scale bacterial leaching will depend on the level of the mean copper yield μ that will ultimately be achieved from the entire deposit. On the basis of current prices and present processing costs, the break-even level has been established at 36 pounds of recoverable copper per ton of presently unworkable ore. This quantity is the pivotal level for the statistical investigation.

The Structure of a Hypothesis Test

Statistical testing involves two complementary hypotheses. Here these are defined in terms of levels for the unknown population mean μ. The engineer has established the following hypotheses:

Null hypothesis (process is uneconomical)

$$H_0: \mu \leq 36 \text{ lb/ton}$$

Alternative hypothesis (process is economical)

$$H_1: \mu > 36 \text{ lb/ton}$$

Although the adjective "null" has historically been applied to the hypothesis representing no change, this designation is arbitrary. Most often, *the null hypothesis will be the one for which erroneous rejection is the more serious consequence.* For notational convenience, the null hypothesis is represented by H_0, and the opposite or alternative hypothesis by H_1.

Hypothesis testing involves two complementary actions or choices. Our mining engineer must decide whether to adopt full-scale bacterial leaching or to abandon the process for now (and perhaps do further testing). These two actions may be expressed in terms of the foregoing hypotheses as

Accept H_0 (Abandon bacterial leaching for now)

Reject H_0 (Adopt bacterial leaching)

(Rejecting the null hypothesis is the same as accepting the alternative, and accepting H_0 is tantamount to rejecting H_1.) The engineer's choice will be based on the results of the sample investigation.

The decision must be made without knowing which of the two hypotheses is true. A determination must be made whether the sample evidence tends to favor the null hypothesis or refute it. This will be based on the following.

TEST STATISTIC

Sample mean copper yield $\bar{X}$ (lb/ton)

A large level for $\bar{X}$ (say 50 lb/ton) will give support to the efficacy of bacterial leaching, whereas a small value (such as 10) will deny it. Some critical level must serve as the point of demarcation between adopting the process or abandoning it.

After the sample data are obtained, the engineer will compute the test statistic and apply the following.

DECISION RULE

Accept H_0 (abandon bacterial leaching for now) if $\bar{X} \leq 40$ lb/ton

Reject H_0 (adopt bacterial leaching) if $\bar{X} > 40$ lb/ton

The pivotal value of 40 lb/ton serves as the **critical value** of the test statistic. It is sometimes convenient to portray the decision rule in terms of **acceptance** and **rejection regions**, as in the following:

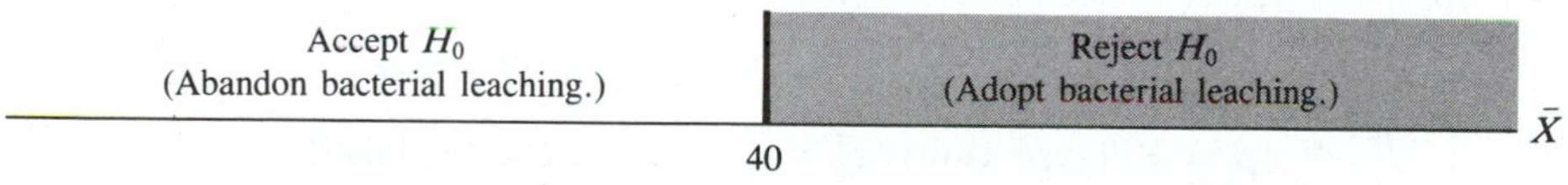

Accept H_0 (Abandon bacterial leaching.)	Reject H_0 (Adopt bacterial leaching.)
← 40	→ $\bar{X}$

The critical value of $\bar{X}$ lies above the 36 lb/ton break-even level for μ that the engineer used to establish the hypotheses. The test statistic's critical value will ordinarily lie outside the range of the null hypothesis. This provides some leeway and reduces the chance that an atypical sample result will cause the wrong decision.

At this point it may be helpful to summarize the structure of the mining engineer's investigation. Table 9-1 arranges the problem in terms of a decision table. There is a column for each of the decision-maker's two choices or acts and

Table 9-1 *Decision Table for Bacterial Leaching Illustration*

	Acts	
Events	**Accept H_0** (abandon process) $\bar{X} \leq 40$	**Reject H_0** (adopt process) $\bar{X} > 40$
H_0 True (process uneconomical) $\mu \leq 36 \quad (= \mu_0)$	**Correct Decision** (abandon uneconomical process)	**Type I Error** (adopt uneconomical process) probability α
H_0 False (process economical) $\mu > 36 \quad (= \mu_0)$	**Type II Error** (abandon economical process) probability β	**Correct Decision** (adopt economical process)

a row for the two uncertain population events, representing the truth or falsity of the null hypothesis. Notice that each act-event combination culminates in a particular outcome. Two of these are correct decisions:

To accept H_0 when H_0 is true
(abandon an uneconomical process)

To reject H_0 when H_0 is false
(adopt an economical process)

Two other outcomes result in **errors**. These are:

Type I error: To reject H_0 when it is true
(adopt an uneconomical process)

Type II error: To accept H_0 when it is false
(abandon an economical process)

Finding the Error Probabilities

A major concern in hypothesis testing is controlling the incidence of the two kinds of errors. The degree of such control may be summarized by the error probabilities:

$$\alpha = \Pr[\text{type I error}] = \Pr[\text{reject } H_0 | H_0 \text{ true}]$$

$$\beta = \Pr[\text{type II error}] = \Pr[\text{accept } H_0 | H_0 \text{ false}]$$

Although both errors are undesirable, rejecting a true null hypothesis is typically the more serious. For this reason, the type I error probability α is also referred to as the **significance level**.

The sampling distribution of $\bar{X}$ may be assumed to be normally distributed with mean μ and standard error $\sigma_{\bar{X}} = \sigma/\sqrt{n}$. Appropriate parameters must be established for the particular normal curve used in finding the error probabilities.

Using the decision rule established earlier for the bacterial leaching study, we may evaluate the error probabilities. Under the null hypothesis $\mu \leq 36$, so a range of values applies. Any value 36 or less is permitted under H_0, such as $\mu = 12.3$, $\mu = 27.5$, or $\mu = 34.99$. Type I error probabilities are usually determined for the worst case, when μ falls exactly at the H_0 limit, which is denoted by μ_0. In the present illustration $\mu_0 = 36$ lb/ton. In computing α it is assumed that $\mu = \mu_0$.

The mining engineer does not know the value for σ any more than he knows μ, but he was able to arrive at a "ballpark" guess. A yield almost certain to be exceeded, 1 pound per ton (which is what would be expected by just washing the ore with water and processing the rinse waste) was subtracted from the present yield of higher-quality, conventionally processed ores—150 lb/ton. Using the procedure of Section 8-6, a range of 6 standard deviations was thereby established. Thus, for planning purposes he found

$$\sigma = \frac{Q_{.999} - Q_{.001}}{6} = \frac{150 - 1}{6} = 24.83$$

which for convenience was rounded to 25 lb/ton. (Keep in mind that this is simply the best guess available for σ, and any probability evaluations based on it may vary considerably from the truth. But *some* value must be used. A better estimate of σ may be obtained from a related set of data or once the sample results are obtained.) Using this quantity, we have $\sigma_{\bar{X}} = 25/\sqrt{100} = 2.5$.

We use the engineer's decision rule to compute the type I error probability

$$\begin{aligned} \alpha &= \Pr[\text{reject } H_0 | H_0 \text{ true}] \\ &= \Pr[\bar{X} > 40 | \mu = 36\,(= \mu_0)] \\ &= 1 - \Phi\left(\frac{40 - 36}{2.5}\right) = 1 - .9452 = .0548 \end{aligned}$$

Thus, there is a better than 5% chance that the engineer will get a sample result so untypically great that he will adopt an uneconomical bacterial leaching program. (This is actually a worst-case figure, since when $\mu = \mu_0 = 36$, the adopted program will exactly break even; for any lower level, such as $\mu = 20$, the chance of incorrect rejection of H_0 will be more remote.)

This appears to be very good. But what about that other error, accepting H_0 when it is false?

Similar computations provide type II error probabilities. There are many ways for the null hypothesis to be false—this will happen in the present illustration whenever the true μ exceeds μ_0. Suppose we assume that $\mu = 42$ lb/ton. We then have

$$\begin{aligned} \beta &= \Pr[\text{accept } H_0 | H_0 \text{ false}] \\ &= \Pr[\bar{X} \leq 40 | \mu = 42] \\ &= \Phi\left(\frac{40 - 42}{2.5}\right) = .2119 \end{aligned}$$

There is a substantial probability that the engineer will get an untypically small level for $\bar{X}$ when indeed the bacterial leaching process would be quite economical.

Figure 9-1 shows the two separate normal curves that apply for the type I and II error probabilities we have just computed. Notice that α is represented by

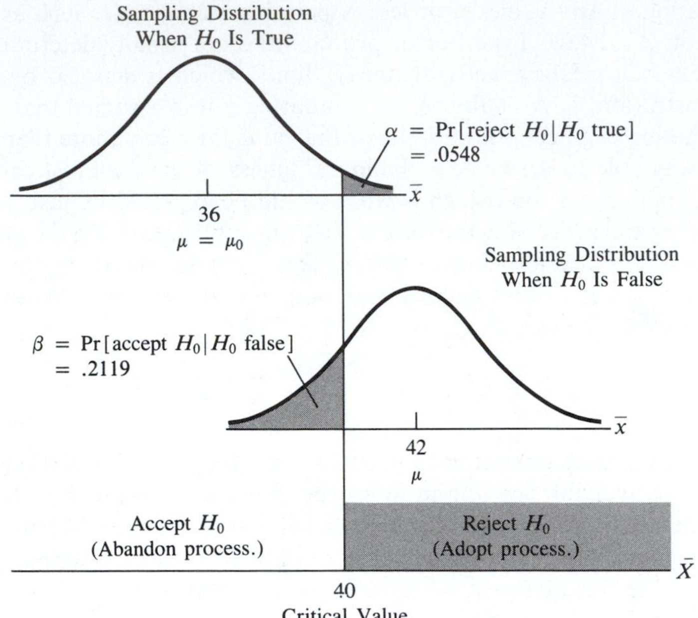

Figure 9-1 *Illustration of hypothesis testing for bacterial leaching decision.*

an upper-tail area in the top curve, whereas β is represented by a lower-tail area in the bottom curve.

Similar type II error probabilities could be obtained for other economical levels for population mean copper yield. Figure 9-2 shows several cases, including $\mu = 50$ (which gives $\beta = .00003$) and $\mu = 39$ (giving $\beta = .6554$).

The dilemma of hypothesis testing—achieving an acceptable balance between α and the spectrum of possible β's—may be resolved by selecting an appropriate decision rule. And, as in statistical estimation, where increasing the sample size will provide greater precision and reliability, a larger n will reduce both the type I and type II error probabilities.

Determining an Appropriate Decision Rule

A new decision rule may be found simply by shifting the critical value for $\bar{X}$ to a new position. Two new decision rules for the bacterial leaching decision are illustrated in Figure 9-3.

In (a) the critical value has been reduced to 38.5 lb/ton. Notice under this rule that it will be more likely to reject a true H_0 (adopt an uneconomical process) with $\alpha = .1587$. But there is less chance of accepting a false H_0 (abandoning an economical process) with $\beta = .0808$. In (b) the critical value for $\bar{X}$ is at 41 lb/ton, with reverse changes providing a lower α (.0228) and a higher β (.3446).

Either of these rules might be an improvement over the original one. That will depend on the attitudes of the engineer and mining company management. The null hypothesis was formulated in such a way that the type I error is the

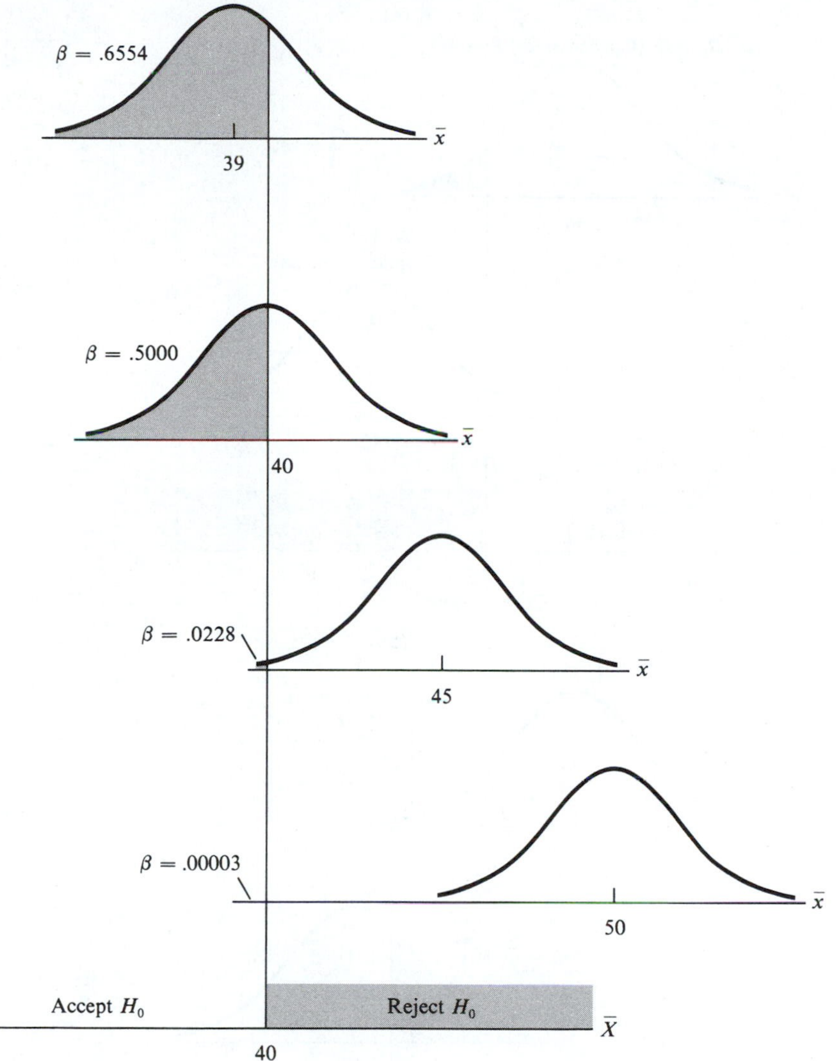

Figure 9-2 *Type II error probabilities for accepting H_0 when it is false, for several different possible means.*

more serious one of adopting an uneconomical bacterial leaching process. This is typical of statistical testing, where a target level (such as 5%) for the probability of this outcome is often prescribed in advance; that in turn establishes a unique critical value and decision rule. Unfortunately, a decision-maker must live with whatever β's the chosen decision rule brings. As we have seen, when the engineer's critical value is 40 there will be a considerable chance that the bacterial leaching process will be abandoned when it might indeed be economical—perhaps considerably so.

Choice of critical value is really a matter of deciding where to position the "fulcrum" so that a proper balance is achieved between α and the β's. If none of the equilibria prove satisfactory, the decision-maker must improve the "leverage," which can be achieved only by increasing the sample size.

Figure 9-3
Error probabilities as they change with a shift in critical value for bacterial leaching decision.

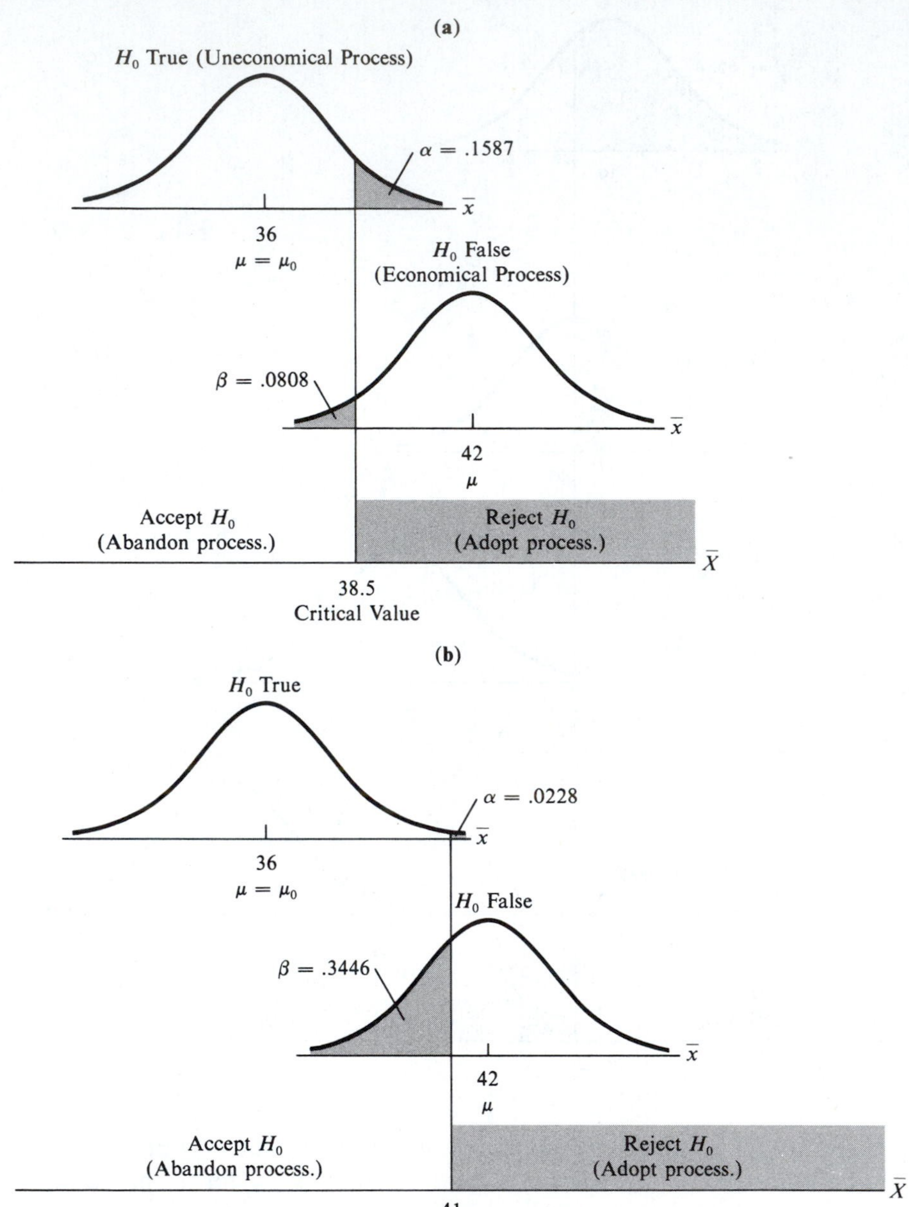

The Effect of Sample Size

Like precision and reliability in estimation, the type I and type II error probabilities of hypothesis testing are competing ends, so that *one* cannot be improved except at the expense of the other. The only way to improve *both* α and β is to reduce potential sampling error by increasing the sample size. Figure 9-4 shows how the

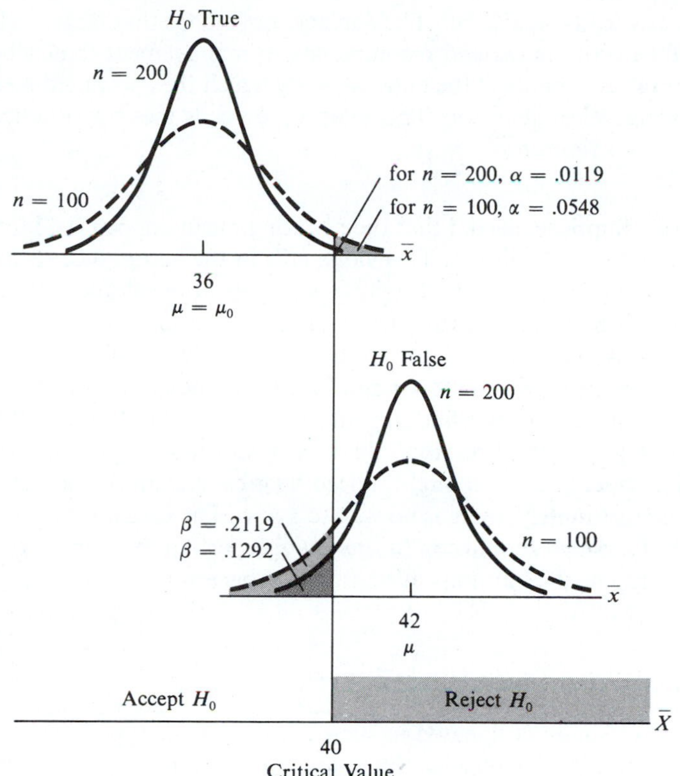

Figure 9-4 *Both error probabilities are reduced when the sample size is increased from 100 to 200.*

mining engineer's situation improves when n is raised from 100 to 200 while the original decision rule is retained. Notice that the normal curves cluster more tightly about the mean, with new levels of $\alpha = .0119$ (when $\mu = \mu_0 = 36$) and $\beta = .1292$ (when $\mu = 42$).

Making the Decision

The preceding discussion applies entirely to the planning stage of a hypothesis testing study. At that time—before data are collected—the hypotheses are formulated, the sample size is established (usually by budgetary considerations), and the decision rule is established (often being dictated by the chosen significance level α). The sample data are then collected. The actual decision is more or less automatic, depending on the computed level of the test statistic.

We will conclude the bacterial leaching illustration by considering two hypothetical scenarios that might result after the data are in.

Scenario One Suppose that the mean yield per ton from the 100 test loads brought to the leaching dump turns out to be $\bar{X} = 47.6$ lb/ton. Since this quantity falls in the rejection region, the null hypothesis that $\mu \leq 36$ must be rejected; the sample results are statistically significant. The engineer concludes that large-scale bacterial leaching will be economical and recommends to management that it be adopted. The engineer takes comfort in the magnitude by which the computed mean exceeds the break-even level of 36 lb/ton. This quantity exceeds that hypothetical level for μ by more than 4 times $\sigma_{\bar{X}}$.

Scenario Two Suppose instead that the resulting mean copper yield from the test batches is only $\bar{X} = 32.1$ lb/ton. This value falls in the acceptance region, and the engineer concludes that bacterial leaching is not economical at this time. (Indeed, under the established decision rule the engineer would take the same action even if the computed $\bar{X}$ were much closer to the critical value of 40 lb/ton—say 38, or even 39.9.) In accepting H_0 the test results are found not to be statistically significant. This does not mean the findings are unimportant, only that H_0 cannot be rejected. The engineer realizes that there is a substantial probability that this decision is incorrect because μ might indeed be greater than the postulated break-even point. Unfortunately, there is no way to know. The case for bacterial leaching can always be reopened should conditions change, perhaps because of rising copper prices or development of an improved strain of bacteria.

Formulating the Hypotheses

As noted, H_0 is usually designated in such a way that the type I error is the most important one to avoid. Consider the following example.

Example: Does a New Type of Battery Last Longer?

A design engineer for consumer products is evaluating a new battery substance based on different chemistry than is presently in use. It gives batteries a mean lifetime of approximately 100 hours. The new battery may last at least as long, on the average, as the present one—or it may die more rapidly. A sampling study will be used, and a decision whether or not to change to the new battery will be based on the observed mean time between failures (MTBF) of test cells composed under the new chemistry. Either of the following assumptions regarding the mean lifetime μ of the new battery might serve as the null hypothesis:

$$\mu \leq 100 \text{ hours} \qquad \text{or} \qquad \mu \geq 100 \text{ hours}$$

Since the company wants to keep its reputation for innovation, the more serious error was judged to be keeping the present battery when the new one lasts longer. (Certainly it would be undesirable to introduce a new battery that doesn't have a longer life, but this was judged far less damaging.) Thus, the following designation was made:

$$H_0: \mu \geq 100 \text{ hours (New battery is at least as good.)}$$

$$H_1: \mu < 100 \text{ hours (New battery has a shorter lifetime.)}$$

A decision rule based on $\bar{X}$ was established so that H_0 would be rejected for shorter computed MTBFs and accepted for longer ones:

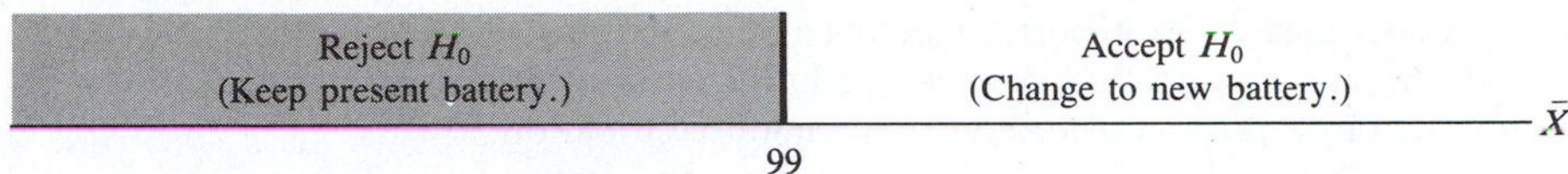

The test described in the battery example is a **lower-tailed test**, since the rejection region coincides with the lower tail of the normal curve for $\bar{X}$ that applies under the null hypothesis. The reversed rejection region in the bacterial leaching illustration indicates that an **upper-tailed test** applies in that application. Some tests involve hypotheses in which the rejection region is split in two, with an intervening acceptance region. The following is an example of such a **two-sided test**.

Example: *Meeting Labeling Requirements*

Packaging items for public consumption can be an exacting task because of regulations or trade practices regarding labeling of the volume or weight of the contents. Too much ingredient is as much to be avoided as too little. Sampling is often used in quality-control investigations concerning weights and measures. Typically the null hypothesis takes the form that the labeled mean is being met exactly, with the opposite alternative. For example,

H_0: $\mu = 32$ grams (Labeling specifications are met.)

H_1: $\mu \neq 32$ grams (Labeling specifications are violated.)

There are two ways for the null hypothesis to be untrue—if there is overfilling ($\mu > 32$ grams) or if there is underfilling ($\mu < 32$ grams). The following form is used for the decision rule:

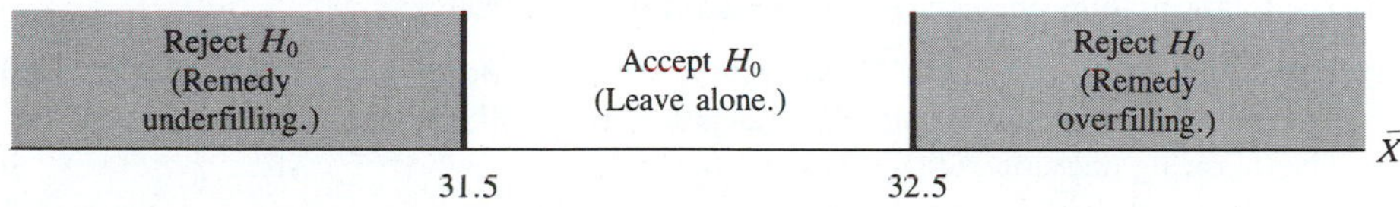

Problems

9-1 Suppose that copper prices have dropped so that the break-even level for the mean yield from bacterial leaching rises to 37 lb/ton. Complete the following:

(a) Reformulate the engineer's hypotheses.

(b) Using the critical value of 40, determine (1) the type I error probability α and (2) the type II error probability β when $\mu = 42$.

9-2 Suppose a mining engineer must decide whether to use a new steel alloy instead of the one presently used to reinforce the mineshaft linings. The null hypothesis is that the present bars are at least as strong as the new ones. He will either accept this assumption, retaining the present reinforcing bars, or reject it and replace them.

State in words (a) the type I error, and (b) the type II error.

9-3 Consider each of the hypothesis testing applications below. Indicate for each outcome whether it is a correct decision, a type I error, or a type II error:

(a) H_0: New power cell's lifetime does not exceed old one's.
  (1) Change to new when old lasts as long or longer.
  (2) Keep old when new lasts longer.
  (3) Keep old when old lasts as long or longer.
  (4) Change to new when new lasts longer.

(b) H_0: Memory chips are satisfactory.
  (1) Reject satisfactory shipment.
  (2) Accept satisfactory shipment.
  (3) Reject unsatisfactory shipment.
  (4) Accept unsatisfactory shipment.

(c) H_0: New design is safe.
  (1) Approve an unsafe design.
  (2) Disapprove an unsafe design.
  (3) Disapprove a safe design.
  (4) Approve a safe design.

9-4 The following hypotheses are to be tested, with $\mu_0 = 100$:

$$H_0: \mu \leq \mu_0 \qquad H_1: \mu > \mu_0$$

Assume that the population standard deviation is $\sigma = 28$ and the sample size is $n = 100$. The following decision rule applies:

Accept H_0 if $\bar{X} \leq 104$

Reject H_0 if $\bar{X} > 104$

Determine the type I error probability α when $\mu = 100$ and the type II error probability β when $\mu = 110$.

9-5 Repeat Problem 9-4 using 106 instead as the critical value. Is the new α larger or smaller? Is the new β larger or smaller?

9-6 Repeat Problem 9-4 using a larger sample size of $n = 150$. Is the new α larger or smaller? Is the new β larger or smaller?

9-7 The following hypotheses are to be tested, with $\mu_0 = 900$:

$$H_0: \mu \geq \mu_0 \qquad H_1: \mu < \mu_0$$

Assume that the population standard deviation is $\sigma = 50$ and the sample size is $n = 100$. The following decision rule applies:

Accept H_0 if $\bar{X} \geq 885$

Reject H_0 if $\bar{X} < 885$

Determine the type I error probability α when $\mu = 900$ and the type II error probability β when $\mu = 875$.

9-2 Testing the Mean

We are now ready to apply the basic concepts of the preceding section to testing the mean. A general procedure has been developed for this application that may be adapted to a wide variety of hypothesis testing situations. Some of these pertain to other parameters than the mean and will be described later in this chapter.

The Hypothesis Testing Steps

All statistical tests involve the same basic steps.

Step 1. *Formulate the hypotheses.* In all cases involving the mean, the decision parameter is the unknown μ. A limiting value μ_0 must be identified for defining H_0. In the bacterial leaching illustration, μ_0 was the break-even point for the procedure. Although the choice of μ_0 is often an economic consideration, other reasons may prevail. The level of μ_0 may be prescribed, as in a labeling specification for package contents. Or, μ_0 might be a benchmark level that corresponds to a present procedure, process, material, or part; we saw such an example in the battery example of Section 9-1.

One-Sided Tests Error considerations usually determine which side of μ_0 to include under H_0. The null hypothesis will be selected from one of the two forms,

$$\mu \leq \mu_0 \qquad \text{or} \qquad \mu \geq \mu_0$$

The two errors are:

1. taking the wrong action when in fact $\mu \leq \mu_0$
2. taking the wrong action when in fact $\mu \geq \mu_0$

The null hypothesis will be H_0: $\mu \leq \mu_0$ if (1) is more serious and H_0: $\mu \geq \mu_0$ if (2) is more serious.

Two-Sided Tests In a two-sided test, the null hypothesis takes the form

$$H_0\colon \mu = \mu_0$$

Since H_0 is untrue both when $\mu < \mu_0$ and when $\mu > \mu_0$, the investigation of the above null hypothesis is a two-sided test.

Step 2. *Select the test procedure and test statistic.* In testing the mean, $\bar{X}$ is the natural test statistic. By the central limit theorem we may ordinarily use the normal curve to represent the sampling distribution of $\bar{X}$. When the population standard deviation is known, $\sigma_{\bar{X}}$ may be computed, and for any given level of μ we may establish probabilities for possible values of $\bar{X}$.

But when σ is unknown, testing procedures employ the following instead.

STUDENT t TEST STATISTIC (σ UNKNOWN)

$$t = \frac{\bar{X} - \mu_0}{s/\sqrt{n}}$$

There are two types of testing procedures, depending on the nature of the decisions to be made. With the **single-decision procedure**, H_0 will be accepted or rejected just once. With a **recurring-decisions procedure**, a series of decisions will be made, and H_0 will be accepted or rejected each time. This dichotomy is useful in selecting the test statistic and determining the form of the decision rule.

Under recurring decisions, one decision rule will be used over and over, possibly being applied by different persons and in several locations. The test statistic should be easy to compute and readily understood by all participants. In a recurring-decisions procedure involving means, the computed mean $\bar{X}$ is ordinarily used and compared to a single critical value $\bar{X}^*$ or a pair of critical values $\bar{X}_1^*$ and $\bar{X}_2^*$. Simplicity of procedure generally requires that the population standard deviation value be stipulated for all test applications, so that σ is treated as a *known* quantity.

Under a single-decision procedure, no stipulation on the level of the population standard deviation is required, and σ may remain *unknown*. The Student t statistic then applies, and the decision rule will be based on it.

Step 3. *Establish the significance level and the acceptance and rejection regions for the decision rule.* As we have seen, the common practice in statistical testing is to base the procedure on avoidance of the more serious type I error. The probability α of this, also referred to as the *significance level*, may then be prescribed in advance. With one-sided applications, the resulting experiment will be a lower- or upper-tailed test, depending on whether small or large values of the test statistic refute H_0. In a two-sided test there are two ways in which H_0 can be rejected. The type I error probability α is split evenly between the two cases. Figure 9-5 summarizes the various situations.

The critical value of the test statistic defines the acceptance and rejection regions. In testing the mean when σ is known, these regions are determined by the **critical normal deviate** z_α (or $z_{\alpha/2}$), for which the normal curve upper-tail area is α (or $\alpha/2$). If σ is unknown, critical values for the Student t statistic t_α (or $t_{\alpha/2}$) are used instead.

When a recurring-decisions procedure applies, so that σ is known and $\bar{X}$ serves directly as the test statistic, the critical normal deviate is used.

CRITICAL VALUES FOR THE SAMPLE MEAN (σ KNOWN)

Lower-Tailed Test (H_0: $\mu \geq \mu_0$):

$$\bar{X}^* = \mu_0 - z_\alpha \sigma_{\bar{X}}$$

Upper-Tailed Test (H_0: $\mu \leq \mu_0$):

$$\bar{X}^* = \mu_0 + z_\alpha \sigma_{\bar{X}}$$

Two-Sided Test (H_0: $\mu = \mu_0$):

$$\bar{X}_1^* = \mu_0 - z_{\alpha/2} \sigma_{\bar{X}}$$

$$\bar{X}_2^* = \mu_0 + z_{\alpha/2} \sigma_{\bar{X}}$$

Note: In computing the above,

$$\sigma_{\bar{X}} = \frac{\sigma}{\sqrt{n}} \qquad \text{for large populations } (n \leq .10N)$$

$$\sigma_{\bar{X}} = \frac{\sigma}{\sqrt{n}} \sqrt{\frac{N-n}{N-1}} \qquad \text{for small populations } (n > .10N)$$

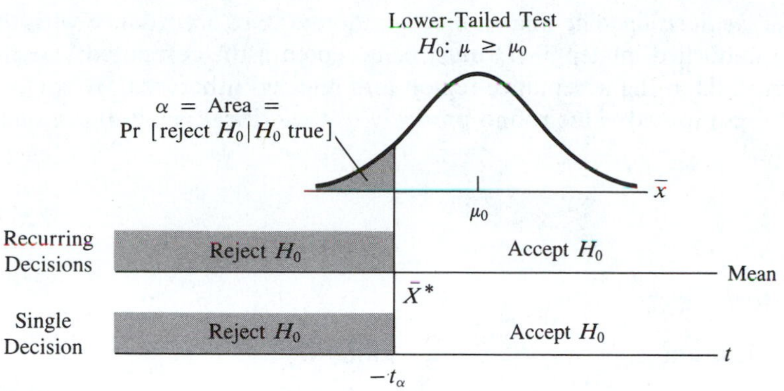

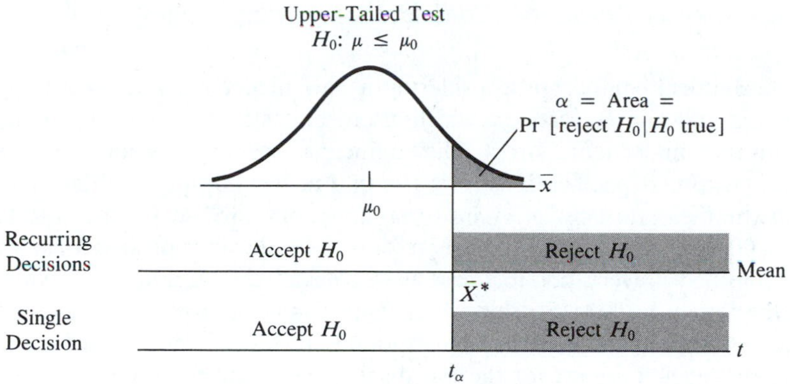

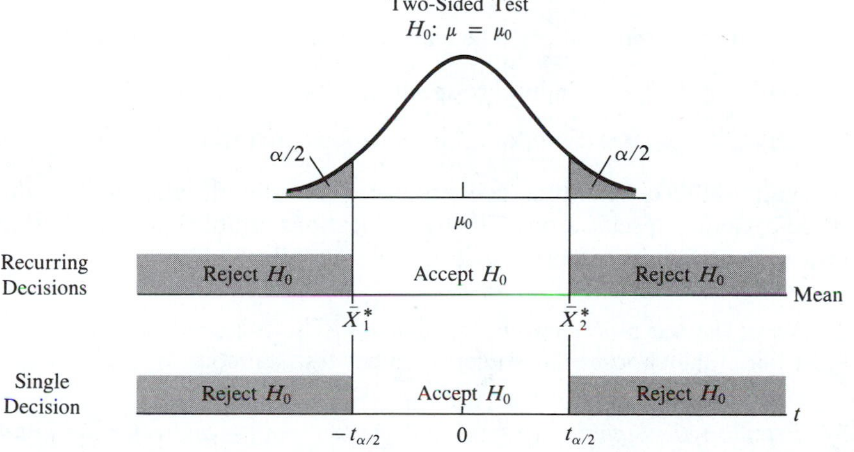

Figure 9-5 *Three forms of hypothesis tests for mean.*

Step 4. *Collect the sample data and compute the value of the test statistic.* This step incorporates the time-consuming and expensive second stage of the statistical sampling study. No revisions to the earlier steps should be based on the particular results obtained here.

Step 5. *Make the decision.* The final choice is automatic, in accordance with the decision rule established in step 3. H_0 must be accepted if the computed value of the test statistic falls in the acceptance region and rejected otherwise. By convention, should the computed value round precisely to the critical value, H_0 must be accepted.

Upper-Tailed Tests

Two detailed illustrations of upper-tailed tests follow.

Single-Decision Procedure Consider the following application of stress testing.

Illustration: *Axial Stress Testing of a New Alloy*

A mechanical engineer is considering a new nickel–chrome–iron alloy. She has ordered 100 sample castings, which are to be tested at a materials laboratory for endurance under axial stress. The engineer is seeking a metal strong enough to meet customer specifications for parts in a new stamping machine. These require that the mean number of cycles to failure μ obtained in vibration testing exceed 500,000. The target population represents the endurance measurements that might be obtained if every possible casting—not just the sample 100—were subjected to the test procedure. So that quick results may be obtained, a lower endurance figure was specified than is ordinarily found in stress testing. Accordingly, unusually extreme levels were set for the test displacement and force parameters.

The following hypothesis testing steps were taken.

Step 1. *Formulate the hypotheses.* The engineer selects as her hypotheses:

$$H_0: \mu \leq 500{,}000 \text{ (Endurance specifications are not exceeded.)}$$

$$H_1: \mu > 500{,}000 \text{ (Endurance specifications are exceeded.)}$$

The designations reflect her strong desire to avoid choosing the alloy when it does not exceed customer specifications. These specifications establish $\mu_0 = 500{,}000$ as the limiting value for the mean number of cycles to failure.

Step 2. *Select the test procedure and test statistic.* The engineer's decision will be made just once. She chooses the Student t as her test statistic.

Step 3. *Establish the significance level and the acceptance and rejection regions for the decision rule.* Because she regards so seriously the type I error of rejecting H_0 when it is true, the engineer chooses a 1% significance level, so that $\alpha = .01$. With df $= 100 - 1 = 99$, Appendix Table G provides the critical value $t_{.01} = 2.369$ (using linear interpolation). Large positive levels for t will refute H_0, and the test is upper-tailed. The decision rule is shown in Figure 9-6.

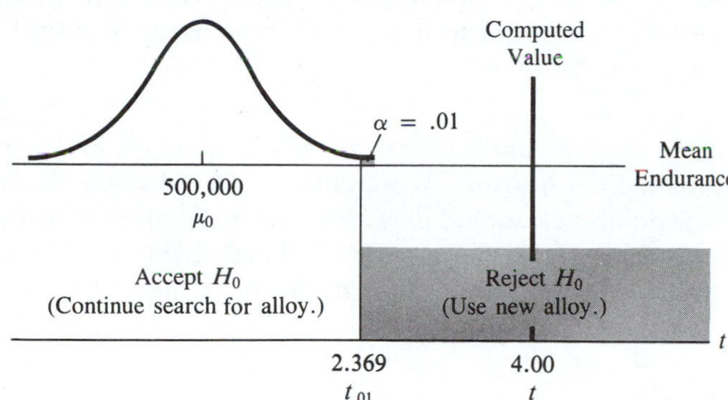

Figure 9-6

Step 4. *Collect the sample data and compute the value of the test statistic.* The test results show a computed sample mean of $\bar{X} = 519{,}500$ cycles to failure, with a sample standard deviation of $s = 48{,}732$. The computed value of the test statistic is

$$t = \frac{519{,}500 - 500{,}000}{48{,}732/\sqrt{100}} = 4.00$$

Step 5. *Make the decision.* The computed mean exceeds the critical value and falls in the rejection region. The engineer must therefore reject H_0 and adopt the new alloy in the stamping machine.

Recurring-Decisions Procedure The following illustration details how hypothesis testing can be employed to establish a permanent testing policy.

Illustration:* *Deciding When to Calibrate Chemical Controls

A chemical engineer must set operational policies for the manufacture of a chemical base. At one stage, sample vials will be selected at random time intervals, and the level of impurities measured for each. A tolerable mean level of impurities μ for all potential vials is 150 parts per million. Anything lower would be satisfactory. The engineer needs to find a decision rule that technicians may use to decide whether or not to recalibrate control settings, thereby reducing the incidence of impurities. A new sample of vials will be taken every eight hours, so the calibration decision must be made three times a day. The following hypothesis testing steps apply.

Step 1. *Formulate the hypotheses.* The engineer selects as his hypotheses:

$$H_0\colon \mu \leq 150 \text{ ppm}$$

$$H_1\colon \mu > 150 \text{ ppm}$$

The pivotal level for the mean is thus $\mu_0 = 150$. With these hypotheses the type I error is unnecessarily recalibrating when the impurity levels are actually tolerable. The type II error is failing to recalibrate when impurities are intolerably high.

Since impurities can be reduced—at a modest cost—in a later stage of processing, and since recalibration is very expensive, the former error is judged to be more serious.

Step 2. *Select the test procedure and test statistic.* The sample mean impurity level $\bar{X}$ (ppm) serves as the test statistic. To establish his decision rule, the engineer uses a value for the population standard deviation of $\sigma = 20$ ppm. This figure was obtained historically from similar processing of related chemicals. The standard deviation is assumed to be stable even though the mean fluctuates and is therefore treated as a known quantity.

Step 3. *Establish the significance level and the acceptance and rejection regions for the decision rule.* The engineer decides that a 5% significance level will adequately protect against the type I error. To keep the incidence of the type II error low, he relies on a fairly substantial sample size of $n = 25$ vials.

Because a continuous process is involved, the population is theoretically infinite, and the following applies for the standard error of $\bar{X}$:

$$\sigma_{\bar{X}} = \frac{20}{\sqrt{25}} = 4.0 \text{ ppm}$$

Since *large* $\bar{X}$'s will refute the null hypothesis of tolerable impurity levels, the test is *upper-tailed.* Using $\alpha = .05$, the critical normal deviate is found from Table E to be $z_{.05} = 1.64$, and the critical value for the sample mean is

$$\bar{X}^* = 150 + 1.64(4.0) = 156.6 \text{ ppm}$$

The decision rule in Figure 9-7 applies.

Figure 9-7

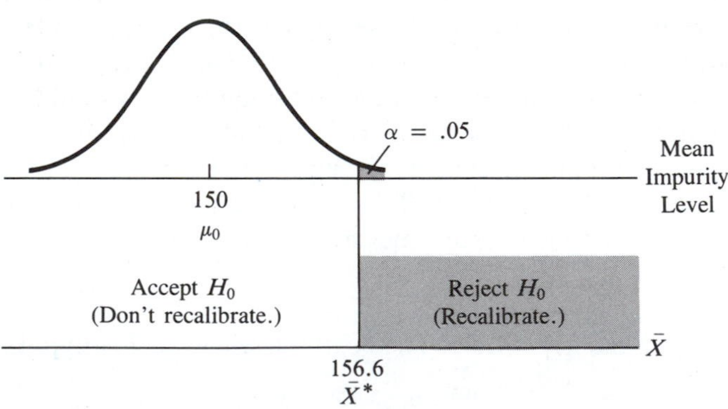

Step 4. *Collect the sample data and compute the value of the test statistic.* This step is conducted every eight hours, with the mean impurity level computed for every 25-vial sample.

Step 5. *Make the decision.* As with step 4, a decision is made every eight hours, in accordance with the above decision rule.

Lower-Tailed Tests

Both procedures with a lower-tailed test are described using the following detailed examples.

Recurring-Decisions Procedure The chemical engineer in the preceding example must set another testing policy.

***Illustration:** Deciding When to Purge Separation Tanks*

The chemical engineer finds a decision rule for a later stage of processing the same chemical base. At that stage, the major concern is with the final yield of an active ingredient. As before, a sample of vials will be selected at random time intervals, with the grams per liter of that ingredient determined for each. Company standards are that the mean yield μ should be at least 35 grams per liter. A decision rule will be used by technicians to translate sample results into action. For yields smaller than the critical level, the separation tanks will have to be purged, thereby increasing the efficiency of the reactions. This decision will be faced only once every 24 hours. The following hypothesis testing steps apply.

Step 1. *Formulate the hypotheses.* The engineer selects as his hypotheses:

$$H_0\colon \mu \geq 35 \text{ grams/liter}$$

$$H_1\colon \mu < 35 \text{ grams/liter}$$

The pivotal level for the mean is thus $\mu_0 = 35$. With these hypotheses, the type I error is unnecessarily purging the tanks when the minimum yield requirements are being met. The type II error is failing to purge the tanks when it is really necessary to do so. The first error is judged more serious since that would involve unnecessary expense. The type II error is less expensive, since low-yield chemicals can be enriched as needed during the final processing stage.

Step 2. *Select the test procedure and test statistic.* The sample mean yield $\bar{X}$ (grams/liter) serves as the test statistic. To establish his decision rule, the engineer uses a value for the population standard deviation of $\sigma = 1$ gram/liter, established in a similar way to that in the preceding test.

Step 3. *Establish the significance level and the acceptance and rejection regions for the decision rule.* The engineer decides that a 1% significance level will adequately protect against the type I error. A sample size of $n = 100$ vials has been determined to be most cost-effective.

Figure 9-8

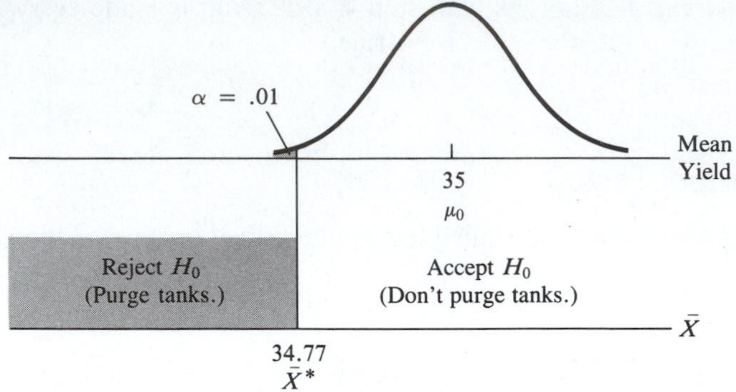

As with the preceding test, the population is theoretically infinite, and the following applies for the standard error of $\bar{X}$:

$$\sigma_{\bar{X}} = \frac{1}{\sqrt{100}} = 0.1 \text{ gram/liter}$$

Since *small* $\bar{X}$'s will refute the null hypothesis of acceptable yield level, the test is *lower-tailed*. Using $\alpha = .01$, the critical normal deviate is found from Table E to be $z_{.01} = 2.33$, and the critical value for the sample mean is

$$\bar{X}^* = 35 - 2.33(0.1) = 34.77 \text{ grams/liter}$$

The decision rule in Figure 9-8 applies.

As before, the test is applied over and over again with a new sample each time. A fresh $\bar{X}$ is computed every day, and the tanks are purged whenever $\bar{X}$ falls into the above rejection region.

Single-Decision Procedure The following illustration considers another chemical engineering application where the hypothesis testing is done only once.

Illustration: *Selecting the Faster Fermentation Temperature*

A chemical engineer is pilot-testing a fermentation process for the manufacture of a pharmaceutical product. All design parameters have been selected except for the temperature at which fermentation itself will take place. Current temperature control hardware is designed to operate at a low nominal level, although theoretically an environment 10°C higher should provide a faster fermentation rate. Since the higher-temperature operation will require specially designed control devices, that environment will be incorporated into the final design only if sample batches ferment significantly faster than the nominal mean time of 30 hours. A series of 50 pilot batches are to be run at the hotter temperature.

The following hypothesis testing steps apply.

Step 1. *Formulate the hypotheses.* Expressing by μ the mean fermentation time of all future batches that might be processed under high temperatures, the engineer establishes:

H_0: $\mu \geq 30$ hours (Higher-temperature fermentation is not faster.)

H_1: $\mu < 30$ hours (Higher-temperature fermentation is faster.)

The limiting case is the nominal processing time $\mu_0 = 30$ hours. The $\geq$ orientation was selected for H_0 because the more serious (type I) error is to choose the higher-temperature process when it is not actually the faster one.

Step 2. *Select the test procedure and test statistic.* Because the population standard deviation is not known, the test statistic is the Student t.

Step 3. *Establish the significance level and the acceptance and rejection regions for the decision rule.* The engineer establishes a significance level of $\alpha = .05$. This level reflects his conservatism regarding the possible type II error (accepting H_0 when H_0 is false). The probabilities (β's) of this second error (not designing the higher-temperature process when in fact it is faster) would be uncomfortably large with a smaller α. The critical value for t is found from Table G to be $t_{.05} = 1.678$ (using linear interpolation). The test is lower-tailed, since a fast mean processing time $\bar{X}$ under the higher-temperature environment will refute H_0. Thus, negative values for t lying below -1.678 will lead to rejection. Figure 9-9 applies.

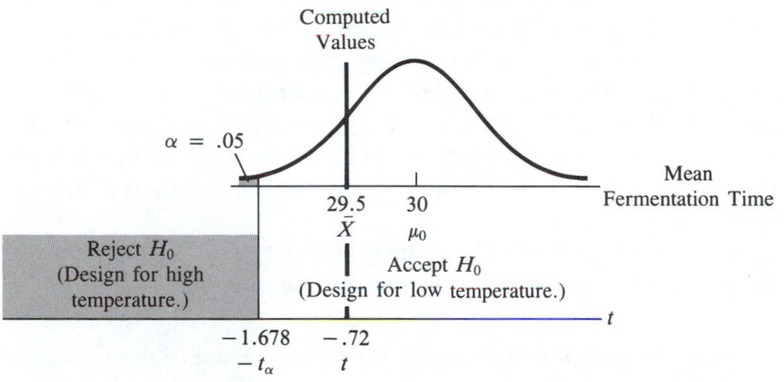

Figure 9-9

Step 4. *Collect the sample data and compute the value of the test statistic.* The following test results were obtained:

$$\bar{X} = 29.5 \text{ hours}$$

$$s = 4.91$$

This provides the computed Student t statistic

$$t = \frac{29.5 - 30}{4.91/\sqrt{50}} = -.72$$

Step 5. *Make the decision.* This quantity falls inside the acceptance region, so the engineer must accept H_0 and design the process for low fermentation temperatures.

In accepting the null hypothesis that the low-temperature environment results in faster fermentation, there is the danger that the type II error has been committed. It is good statistical procedure to consider the potential for such error. The engineer might be incorrectly designing for low fermentation temperatures when the actual value of μ is faster than nominal. Of course that could not be established without further testing. As we have seen, the incidence of such type II errors may be controlled by using a larger α or by increasing the sample size. We will discuss this point further in Section 9-5.

Two-Sided Tests

Two-sided tests also involve either single-decision or recurring-decisions procedures.

Recurring-Decisions Procedure In the following illustration, a routine maintenance policy is established.

Illustration: Replacing a pH Regulator

The superintendent of a new chemical plant is establishing operational policies, such as when to perform critical maintenance. When practical, those actions will be based on sample data. This is the case for a process in which the pH of an intermediate stage must be maintained at a neutral level. A control valve keeps the mean batch pH at 7.0. Owing to the caustic nature of the fluids flowing through it, the valve must occasionally be replaced. This action will be based on the mean pH of 100 sample vials taken at random times from the separation tank containing the output from that stage.

Hypothesis steps 1–3 show how the policy for valve replacement was established. Steps 4 and 5 apply to a particular sample outcome.

Step 2. *Select the test procedure and test statistic.* Since a permanent policy is being established, a *recurring-decisions procedure* applies. The computed mean pH $\bar{X}$ for the sample vials serves as the test statistic. The standard deviation for individual test vials has long been established at .50 on the pH scale. This establishes the standard error for $\bar{X}$: $\sigma_{\bar{X}} = .50/\sqrt{100} = .05$.

Step 1. *Formulate the hypotheses.* The following hypotheses apply:

$$H_0: \mu = 7.0 \qquad \text{(Process is neutral.)}$$

$$H_1: \mu \neq 7.0 \qquad \text{(Process is not neutral.)}$$

Step 3. *Establish the significance level and the acceptance and rejection regions for the decision rule.* It is very expensive to shut down the process to replace the control valve, so plant policy permits just a 1% chance of doing this unnecessarily. The significance level for each sample test is therefore $\alpha = .01$. The critical normal deviate is $z_{.005} = 2.57$, and the critical values for the sample mean are

$$\bar{X}_1^* = 7.0 - 2.57(.05) = 6.87 \text{ pH}$$

$$\bar{X}_2^* = 7.0 + 2.57(.05) = 7.13 \text{ pH}$$

The acceptance and rejection regions are shown in Figure 9-10.

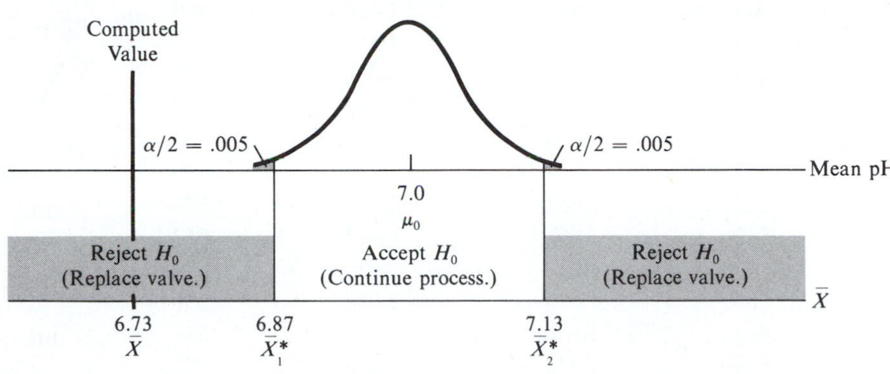

Figure 9-10

Step 4. *Collect the sample data and compute the value of the test statistic.* The computed mean pH for a particular test turned out to be $\bar{X} = 6.73$.

Step 5. *Make the decision.* Since this value falls in one of the two rejection regions, H_0 must be rejected. At the present time it must be concluded that the process mean pH is not neutral (and is too acidic). The process will be shut down to replace the control valve.

Single-Decision Procedure The following illustration shows how a two-sided hypothesis test can be used in conjunction with a pure research experiment.

Illustration:* *Fuel-Element Powers—Comparing Diffusion-Theory Calculated and Measured Values

A researcher performed an experiment with reactor fuel elements that involved gamma rays emitted during decay. For particular reactor core locations (presumed here to be selected randomly), he computed the power produced by the element, using diffusion theory. The theoretical values were divided by their measured powers to give the ratio of calculated-to-experimental (C/E) values shown in Table 9-2.

We will use the sample data to establish whether the gamma-scan measuring system coincides with diffusion theory. If it does, then the C/E values should average to 1. The following steps apply.

Step 1. *Formulate the hypotheses.* We let μ denote the mean calculated-to-experimental ratio. The following apply:

H_0: $\mu = 1$ (Measures coincide with diffusion theory.)

H_1: $\mu \neq 1$ (Measures deviate from diffusion theory.)

Table 9-2
Comparison of Diffusion-Theory Calculated and Measured Fuel-Element Powers

(1) Core Position	(2) Calculated Diffusion Theory C	(3) Measured Gamma Scan E	(4) C/E Ratio X
A-2	.641	.675	.950
A-3	.767	.802	.956
A-4	.924	1.016	.909
A-5	1.074	1.126	.954
A-6	.928	1.034	.897
A-7	.794	.792	1.002
A-8	.605	.634	.954
B-3	.925	1.016	.910
B-5	.987	1.003	.984
B-7	.867	.911	.952
C-2	1.227	1.253	.979
C-4	1.096	1.058	1.036
C-5	1.109	1.178	.941
C-6	1.372	1.378	.996
C-8	1.179	1.245	.947
D-2	.792	.840	.943
D-3	1.318	1.231	1.071
D-5	1.064	1.023	1.040
D-7	1.233	1.134	1.087
D-8	.790	.803	.984
E-2	1.035	.974	1.063
E-4	1.241	1.146	1.083
E-6	1.234	1.130	1.092
E-8	1.067	1.026	1.040
F-3	.758	.717	1.057
F-5	.867	.788	1.100
F-7	.766	.721	1.062

Source: Hobbs, *Transactions of American Nuclear Society* Vol. 55, November 15–19, 1987, pp. 195–197.

This is a *two-sided* test, since ratios either below 1 or above 1 refute the null hypothesis.

Step 2. *Select the test procedure and test statistic.* This is a *single-decision procedure.* Since the population standard deviation σ is unknown, the Student t statistic is used.

Step 3. *Establish the significance level and the acceptance and rejection regions for the decision rule.* A level of $\alpha = .05$ is used here as the significance level. This gives the two-sided test portrayed in Figure 9-11. In published research, it is common to use traditional significance levels, such as .05 or .01.

Figure 9-11

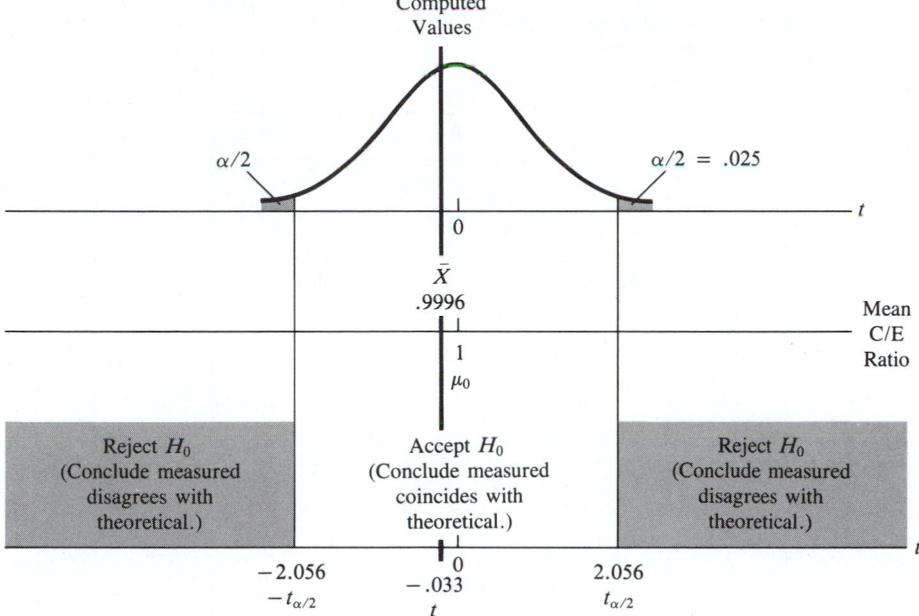

Step 4. *Collect the sample data and compute the value of the test statistic.* The following values are computed from the results in Table 9-2:

$$\bar{X} = .9996 \qquad s = .0626$$

From these, the computed value of the test statistic is:

$$t = \frac{.9996 - 1}{.0626/\sqrt{27}} = -.033$$

Step 5. *Make the decision.* The computed t falls inside the acceptance region. We must therefore accept the null hypothesis and conclude that the gamma-scan measured values coincide with those calculated by diffusion theory.

Hypothesis Testing and Confidence Intervals A two-sided hypothesis test may be explained in terms of statistical estimation. An equivalent procedure for conducting such a test would be to construct a $100(1 - \alpha)\%$ confidence interval from the sample results. Then if μ_0 falls inside the interval, H_0 must be accepted; otherwise it must be rejected. α is the proportion of such intervals that will fail to bracket μ_0 when it is the true mean. In other words, when this procedure is used, the type I error of rejecting H_0 when it is true will occur at frequency α.

For the fuel-element powers experiment, $1 - \alpha = .95$. Using the sample results from before, we construct a 95% confidence interval estimate for μ:

$$\mu = \bar{X} \pm t_{\alpha/2}\frac{s}{\sqrt{n}} = .9996 \pm 2.056\frac{.0626}{\sqrt{27}} = .9996 \pm .0248$$

or,

$$.9748 \leq \mu \leq 1.0244.$$

Figure 9-12
Summary of hypothesis testing procedures for mean.

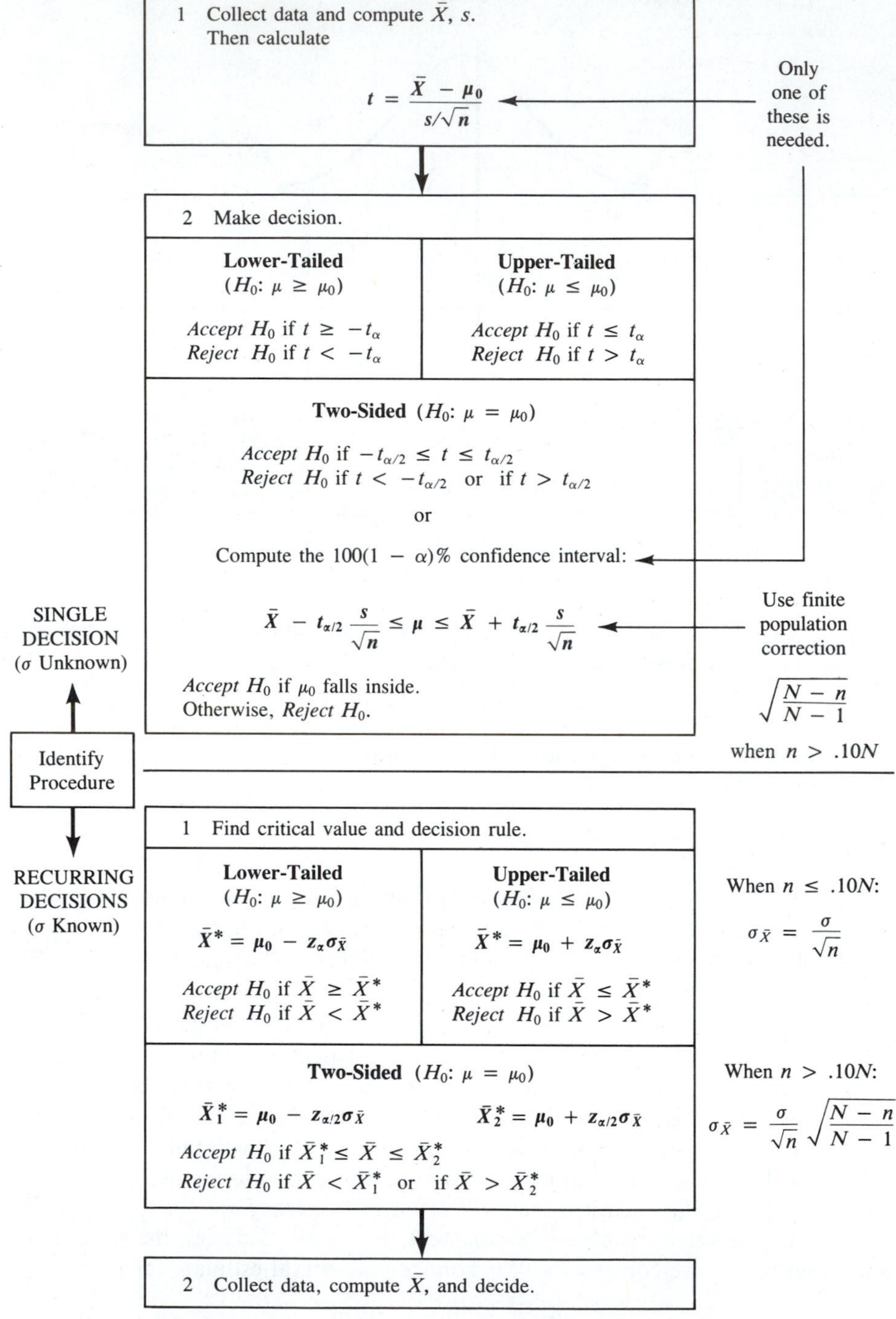

Again, since this overlaps the hypothesized value of $\mu_0 = 1$, we must *accept* the null hypothesis and reach the same conclusion as above.

We may interpret using the confidence interval as follows. If the experiment were duplicated several times, and if a similar confidence interval were constructed for each new set of sample values, then we would expect 95% of those intervals to contain μ_0 within their respective limits. But about 5% of them would have limits that lie totally above or below μ_0; for these, H_0 would be rejected. Thus, 5% of the time the type I error would be committed.

Summary of Testing the Mean

Figure 9-12 summarizes hypothesis testing with the mean. When σ is unknown, the Student t distribution applies, and critical values t_α or $t_{\alpha/2}$ may be found from Appendix Table G. For large sample sizes, when interpolation from that table is impossible, critical values may be read from the df = ∞ row. (These correspond to the normal deviates z_α and $z_{\alpha/2}$.)

Problems

9-8 The mechanical engineer from the stress-testing illustration performs another stress test on a new material, where σ = 600,000 cycles to failure and H_0: $\mu \leq$ 10,000,000. Suppose a sample size of $n = 50$ castings is tested at a significance level of $\alpha = .05$.
(a) Assuming that the same choices must be made, identify the acceptance and rejection regions.
(b) The computed mean is 10,115,000. What action should the engineer take?

9-9 Suppose that the chemical engineer in the fermentation-temperature illustration had reported identical sample results, but for a larger sample size of $n = 100$ instead. Perform hypothesis testing steps 3–5.

9-10 The plant superintendent in the pH regulator illustration found that too much supplemental processing was required because faulty control valves were not replaced often enough. He raised the significance level to $\alpha = .05$.
(a) Perform hypothesis testing step 3.
(b) What action should be taken for each of the following sample results?
(1) $\bar{X} = 6.85$ (2) $\bar{X} = 7.24$ (3) $\bar{X} = 6.59$ (4) $\bar{X} = 7.07$

9-11 Refer to the data in the logic circuit wafer example on page 305. Should the industrial engineer accept or reject the null hypothesis that $\mu \geq .20$ ounce? Use $\alpha = .01$.

9-12 Consider a sample of size $n = 100$ taken at random from a small population of size $N = 500$. The null hypothesis is that $\mu \leq$ 5,000. Suppose the population standard deviation is known to be $\sigma = 150$ and that $\bar{X}$ will serve as the test statistic.
(a) Assuming that a significance level of $\alpha = .01$ is desired, find the critical value for the sample mean and determine the decision rule.
(b) Should H_0 be accepted or rejected if the computed sample mean turns out to be $\bar{X}$ = 5,060?

9-13 A plant produces rods with a specified 1 cm mean diameter. The standard deviation is .01 cm. A decision rule must be established for determining when to correct for over- or undersized output. That choice will be based on a sample of 100 rods, under the stipulation that there be only a 1% chance of taking corrective action when the mean diameter is exactly on target.

(a) Perform hypothesis testing steps 1–3.
(b) What act should be taken if $\bar{X} = .993$ cm?
(c) Suppose that $\bar{X} = 1.0023$ cm. Construct a 99% confidence interval estimate for the mean diameter. Does this contain μ_0? What action should be taken?

9-14 Refer to the data in the physical therapy apparatus example on page 314. Should the mechanical engineer accept or reject the null hypothesis that $\mu \leq 20$ days? Use $\alpha = .05$.

9-15 The plant in Problem 9-13 will soon be producing a .5 cm rod. The standard deviation is unknown. Again, corrective action will be based on a sample of 100 rods, with the same error goal.

(a) Perform hypothesis testing steps 1–3.
(b) What act should be taken if $\bar{X} = .497$ cm and $s = .0075$ cm?
(c) Construct a 99% confidence interval for the mean diameter when $\bar{X} = .508$ cm and $s = .013$ cm. Does this contain μ_0? What action should be taken?

9-16 An engineering department manager wants to decide whether or not to include a computer-aided design (CAD) software package in his budget. He is skeptical about the vendor's time-savings claims. A test has been conducted using pairs of nearly equally skillful engineers who independently design the same item—one using CAD and the other unassisted. For $n = 8$ parts the following percentage time-savings by CAD over the unassisted design were obtained:

80	10	37	26
45	29	44	5

At the 5% significance level, can the manager conclude that CAD will indeed yield savings in time over present design methods?

9-17 The following sample-failure data (thousands of miles) were obtained for a type of catalytic converter:

62.3	44.4	49.2	63.3	47.6	60.1
37.4	55.8	57.5	58.3	56.2	54.3

(a) Construct a 95% confidence interval estimate for the mean distance to failure.
(b) At the 5% significance level, must you accept or reject the null hypothesis that the catalytic converter will last, on the average, 50 thousand miles?

9-18 A biological reduction process is being evaluated as an alternative to a conventional technique for which the mean yield of active ingredients is $\mu = 10$ grams/liter with a standard deviation of $\sigma = .5$ gram/liter. The biological method will be an effective replacement for the present procedure if it can be assumed to provide a yield of more than 12 grams/liter. A total of 100 batches will be pilot-tested under bacterial reduction.

(a) If test results indicate that significant improvement in yield might be made, the biological procedure will be adopted. There should be only a 5% chance that this will be done when the new technique's yield is not high enough to be effective as a replacement. Assuming the same standard deviation in yield applies whichever procedure is used, perform hypothesis testing steps 1–3.
(b) The test results provide a yield of 13.3 grams/liter. What action should be taken?

9-19 A quality-control inspector for a microwave transmitter manufacturer is assessing a shipment of 400 crystal controls. The actual broadcast frequency will depend on the resonant frequency, which will vary slightly from crystal to crystal, but the mean level should achieve the rated target of .55 mHz. The exact standard deviation for the individual crystal frequencies is unknown. A random sample of 50 crystals from the shipment will be tested and the resonant frequency determined for each. The entire shipment will be rejected if the observed mean is significantly above or below the rated level and accepted otherwise. The inspector wants just a 1% chance of rejecting a shipment in which the mean frequency exactly matches the rated level.

(a) Formulate the inspector's hypotheses.

(b) Sample results provide $\bar{X} = .5503$ mHz and $s = 485$ Hz. Construct a 99% confidence interval estimate of the mean resonant frequency for the entire shipment. What action should the inspector take?

9-20 A new test stand might be acquired by the inspection department of GizMo, an instrumentation fabricator, if the mean time to check out a standard "black box" is significantly less than that for the present unit. The new test stand will save enough operating costs to pay for itself only if it can "burn in" all new instrument boxes in 15 hours or less, on the average. The equipment vendor allows GizMo to time-test a sample of 50 boxes. No educated guesses have been made regarding the target population's characteristics.

(a) Assuming that GizMo wants no more than a 1% chance of buying the new test stand when it will not save enough to justify its cost, perform hypothesis testing steps 1–3.

(b) The test results provide a mean burn-in time of 13.5 hours with a standard deviation of 3.4 hours. What action should be taken?

9-21 The final stage of a chemical process is sampled and the level of impurities determined. The final product is recycled if there are too many impurities and the controls are readjusted if there are too few (which is an indication that too much catalyst is being added). If it is concluded that the mean impurity level is the nominal level of $\mu = .01$ gram/liter, the process is continued without interruption. The standard deviation of impurities has been established to be $\sigma = .005$ gram/liter. A sample of $n = 100$ specimens is measured every 24 hours and the level of impurities is determined for each.

(a) Perform hypothesis testing steps 1–3, assuming a significance level of 5%.

(b) What action should be taken for each of the following sample results for the mean level of impurities?

(1) $\bar{X} = .0095$ (2) $\bar{X} = .0088$ (3) $\bar{X} = .0112$ (4) $\bar{X} = .0104$

(c) Find the type II error probability when the true mean impurity level is:

(1) $\mu = .009$ (2) $\mu = .012$ (3) $\mu = .008$ (4) $\mu = .014$

9-22 The world average exposure rate to ^{238}U is 12.0 ($\times 10^{-5}$ Gy/yr).* Using the sample results from Table 8-2, should you conclude the mean exposure rate to this radioactive element in Riyadh, Saudi Arabia, is identical to the worldwide figure? Use $\alpha = .05$.

9-23 The following world exposure rates ($\times 10^{-5}$ Gy/yr) apply to two radioactive elements:*

(a)	(b)
^{232}Th	^{40}K
14.0	9.0

* A. Towfik et al., *Transactions of American Nuclear Society*, Vol. 55, November 15–19, 1987, p. 89.

Using the sample results from Table 8-2, test the respective null hypotheses with $\alpha = .05$ that Riyadh, Saudi Arabia, has an identical exposure rate to the worldwide figure.

9-3 Testing the Proportion

Tests of the proportion are next in importance to those for the mean. The proportion is especially important in statistical quality control, where a major concern is whether the level of defectives in a shipment is high or low; similarly, the overall quality of a production process might be measured by the proportion of defective output. Proportions are also fundamental in work measurement, where work sampling is often used instead of stopwatch procedures. The proportion is also a key parameter in specifying a Bernoulli process, and a considerable amount of probability analysis might be based upon an assumed level.

The testing procedures established for the mean are easily extended to the proportion. The sample proportion P is a natural test statistic for experiments regarding π, but sometimes the number of successes R is employed equivalently. The appropriate sampling distribution for P or R is either the binomial or the hypergeometric—the latter applying when sampling without replacement from small populations. A normal curve closely approximates these underlying distributions over a wide range of π's and n's, and tests of the proportion are most often based on that approximation.

Testing the Proportion Using the Normal Approximation

The sample proportion P is a convenient test statistic for testing the value of the population proportion. As with testing the mean, we divide applications into single-decision and recurring-decisions procedures. Figure 9-13 summarizes these. (A continuity correction of $.5/n$ is used in computing critical values for P.)

The normal approximation may be applied whenever the sample size is sufficiently large. We will use it whenever both of the following are true:

$$n\pi \geq 5$$

$$n(1 - \pi) \geq 5$$

A Lower-Tailed Test

A lower-tailed test is described in the illustration at the top of page 368.

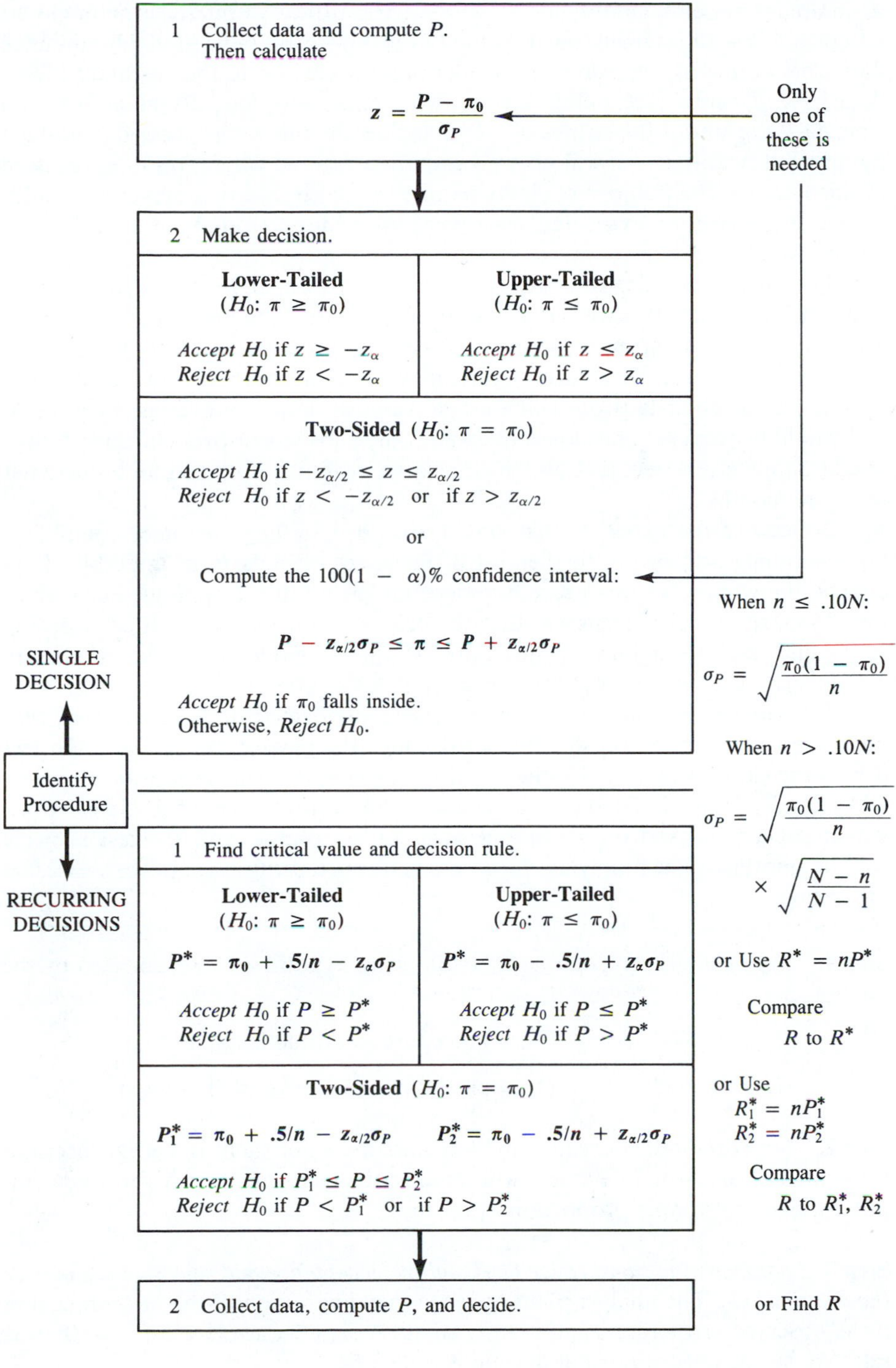

Figure 9-13 *Summary of hypothesis testing procedures for proportion.*

***Illustration:* Is Double Data Entry Necessary?**

A computer systems analyst for a large electric utility wondered if it might be practical to cut the tremendous expense of the data processing staff. Hundreds of data clerks presently encode meter-reader tags for entry into the customer-billing data base. To minimize billing errors, two clerks independently enter the data from each tag, and if the entries do not coincide, the tag is reprocessed. Although the possibility that errors will occur is extremely remote when tags are processed in this manner, the number of clerks required is twice that of a single data entry system. Furthermore, extra data processing is required to make the verification comparisons.

The analyst wonders if significant cost savings might be achieved by abandoning double-entry processing in favor of entering the data just once—obviously a much more error-prone procedure. Under the proposed scheme, large errors would be filtered out by comparing usage with the average of previous months; only those customer tags on which usage is above or below average by a factor of 3 would be recycled once for verification. Any subsequent over- or underbilling would ultimately be rectified by the cumulative readings obtained for a customer in future months.

Because of the expenses that would arise in handling customer complaints for overbilling and owing to the cost of uncollected funds from underbilled accounts, the efficacy of this procedure depends on a fairly low single-entry error rate. Management consensus was that any new-procedure error rate not exceeding .5% would result in savings for the company. The pivotal value for the proportion π of all erroneous entries was therefore set at $\pi_0 = .005$.

A sampling experiment was initiated and performed under conditions that duplicated very closely those anticipated under the new procedure. For the test period the clerks were asked to be especially careful, as their work was not going to be totally verified. To preclude lackadaisical performance, each clerk's work would still be checked on a random basis for accuracy—and the test subjects were so informed. The following steps were performed using a sample of $n = 2{,}000$ meter tags:

Step 1. *Formulate the hypotheses.* The following hypotheses were selected by the analyst:

H_0: $\pi \geq .005$ (π_0) (New procedure is impracticable.)

H_1: $\pi < .005$ (New procedure may be an improvement.)

Step 2. *Select the test procedure and test statistic.* Since there is a single decision to be made, the normal deviate z will serve as the test statistic. This will be computed using the sample proportion of errors P.

Step 3. *Establish the significance level and the acceptance and rejection regions for the decision rule.* The analyst plans to recommend adoption of the new procedure if the observed error rate is sufficiently small. A significance level of $\alpha = .05$ was selected, so the critical normal deviate is $z_\alpha = 1.64$.

The decision rule is summarized in Figure 9-14.

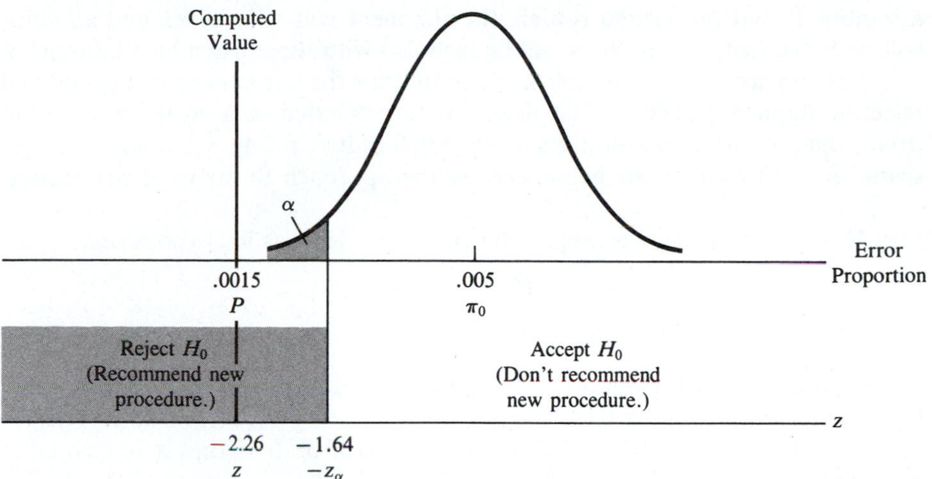

Figure 9-14

Step 4. *Collect the sample data and compute the value of the test statistic.* Under H_0, the standard error of P is

$$\sigma_P = \sqrt{\frac{.005(.995)}{2{,}000}} = .00155$$

Only 3 of the sample tags were in error, so the computed error proportion was only $P = \frac{3}{2{,}000} = .0015$. The computed value for the normal deviate is

$$z = \frac{.0015 - .005}{.00158} = -2.26$$

Step 5. *Make the decision.* The computed value falls below the critical value, and the analyst must reject the null hypothesis and conclude that the new procedure might provide savings in operational costs. The analyst therefore recommends the new procedure.

An Upper-Tailed Test

To illustrate an upper-tailed test of the proportion, consider the following quality control illustration.

***Illustration:** Establishing a Policy for Receiving Microcircuit Chips*

An electrical engineer for a computer manufacturer wishes to establish the policy for inspecting and receiving shipments of microcircuit chips for installation onto printed circuit boards (pcb's). Presently, all chips are installed without any testing. Defective chips are removed during later-stage testing of the pcb's. He believes that it will be cheaper to first test a sample of chips. Then, if the sample proportion defective is small, the shipment will be *accepted* and be released directly for assembly. If that proportion is high, the shipment will be *rejected*, and all chips will be tested first; no defectives will be included with those assembled onto pcb's.

The engineer needs to establish where to draw the line between accepting and rejecting shipments of chips. This determination is based on a break-even level of .10 for the proportion of defectives in any 1,000-unit shipment. Choosing a sample size of $n = 50$, he uses the hypothesis testing approach to arrive at the answer.

Step 1. *Formulate the hypotheses.* The engineer selects as his hypotheses:

$$H_0: \pi \leq .10$$

$$H_1: \pi > .10$$

The pivotal level for the population (shipment) proportion defective is the break-even level $\pi_0 = .10$. With these hypotheses, the type I error will be to reject a good-quality shipment, while the type II error will be to accept a poor-quality shipment.

Step 2. *Select the test procedure and test statistic.* The sample proportion defective P serves as the engineer's test statistic for this recurring-decisions procedure.

Step 3. *Establish the significance level and the acceptance and rejection regions for the decision rule.* The engineer decides that a 1% significance level will adequately protect against the type I error. Since neither $n\pi_0 = 50(.10)$ nor $n(1 - \pi_0) = 50(.90)$ fall below five, the sample size is large enough for him to make the normal approximation. The critical normal deviate is $z_{.01} = 2.33$. In finding the critical value P^*, he first computes the standard error of P that applies under H_0:

$$\sigma_P = \sqrt{\frac{.10(1 - .10)}{50}} = .0424$$

(The population size of $N = 1{,}000$ is great enough that the finite population correction may be ignored.) For this *upper-tailed* test, the critical value is computed

Figure 9-15

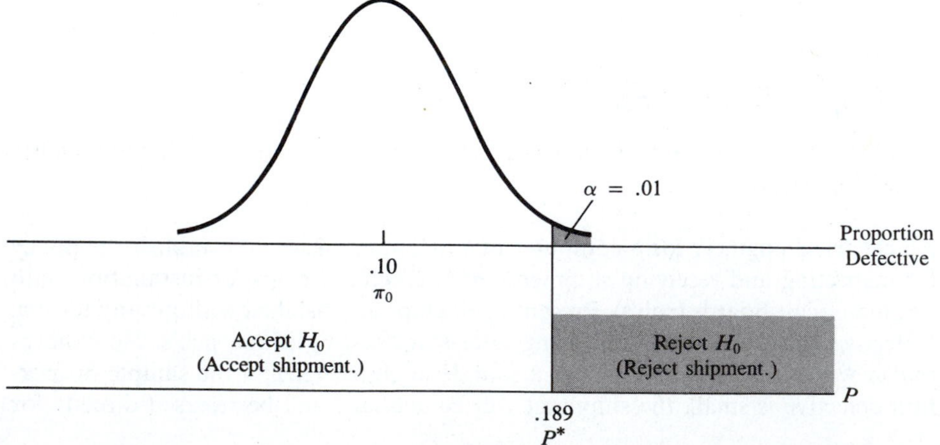

as follows:

$$\begin{aligned} P^* &= \pi_0 - \frac{.5}{n} + z_{.01}\sigma_P \\ &= .10 - \frac{.5}{50} + 2.33(.0424) \\ &= .189 \end{aligned}$$

The decision rule in Figure 9-15 applies.

Hypothesis testing steps 4 and 5 will be performed for each shipment of chips received. The following results would be achieved for five future shipments:

Number of Defectives R	Proportion Defective $P = R/n$	Action
3	.06	Accept
9	.18	Accept
10	.20	Reject
7	.14	Accept
12	.24	Reject

After reviewing this decision rule, the engineer chooses to simplify the procedure by not requiring the inspectors to compute the sample proportion.

Alternative Procedure for Testing Proportion

When several persons may be involved in repeated applications of the test, it may be easier to streamline the procedure by basing the decision rule on the number of observed successes R as the test statistic. The critical value is found by translating P^* into its number counterpart:

$$R^* = nP^* \qquad \text{(integer value)}$$

where R^* is always the nearest *integer* $\geq nP^*$ for a lower-tailed test and $\leq nP^*$ for an upper-tailed test. The form of the decision rule is analogous. For two-sided tests, R_1^* is the nearest integer $\geq nP_1^*$ and R_2^* the nearest integer $\leq nP_2^*$.

Using this alternative procedure, the engineer in the preceding illustration computes

$$nP^* = 50(.189) = 9.45$$

so that rounding down he gets the critical number of sample defectives $R^* = 9$ and arrives at the following decision rule:

Accept H_0 (Accept shipment.)	Reject H_0 (Reject shipment.)
R $(\leq)$ 9 $= R^*$	$(>)$

Regardless of whether P or R serves as the test statistic, identical results will be obtained.

Testing with Binomial Probabilities

When n is too small to make the normal approximation, the decision rule may instead be based directly on binomial or hypergeometric probabilities. (The latter always apply when sampling without replacement from a finite population, but can ordinarily be approximated by the binomial when N is large.) Once π_0 and H_0 have been set and n chosen, the decision rule follows directly from the *targeted* significance level α. It will ordinarily take the R form.

Lower-Tailed Tests The critical value R^* is that level of r providing the *smallest* cumulative binomial probability such that

$$P[R \leq r \mid \pi = \pi_0] > \alpha$$

This gives a type I error probability

$$\Pr[\text{reject } H_0 \mid \pi = \pi_0] = \Pr[R < R^* \mid \pi = \pi_0] \leq \alpha$$

which cannot exceed the targeted significance level.

Upper-Tailed Tests The critical value R^* is that level of r providing the *smallest* cumulative binomial probability such that

$$\Pr[R \leq r \mid \pi = \pi_0] \geq 1 - \alpha$$

Like the above, this guarantees that the type I error probability,

$$\Pr[\text{reject} \mid \pi = \pi_0] = \Pr[R > R^* \mid \pi = \pi_0] \leq \alpha$$

will not exceed the targeted significance level.

Two-Sided Tests The lower critical value R_1^* is the level of r providing the *smallest* cumulative binomial probability such that

$$\Pr[R \leq r \mid \pi = \pi_0] > \frac{\alpha}{2}$$

The upper critical value R_2^* is that level of r providing the *smallest* cumulative binomial probability such that

$$\Pr[R \leq r \mid \pi = \pi_0] \geq 1 - \frac{\alpha}{2}$$

To illustrate, suppose the same engineer in the preceding illustration wishes to establish a decision rule for a relay module. These come in shipments of $N =$ 5,000 and he wishes to use $\pi_0 = .05$ to decide how to dispose of these items.

Because testing is so expensive, he can only afford to use $n = 20$. Since $n\pi_0 = 20(.05) = 1$ is less than 5, the normal approximation should not be used, and his decision rule will be based on binomial probabilities.

Using a significance level of $\alpha = .05$ for this upper-tailed test, the critical number of sample defectives has the smallest cumulative binomial probability greater than or equal to $1 - .05 = .95$. From Appendix Table A, we read

$$\Pr[R \leq 3] = .9841$$

as the required value, so that $R^* = 3$. The following decision rule applies to relays:

$n = 20$:

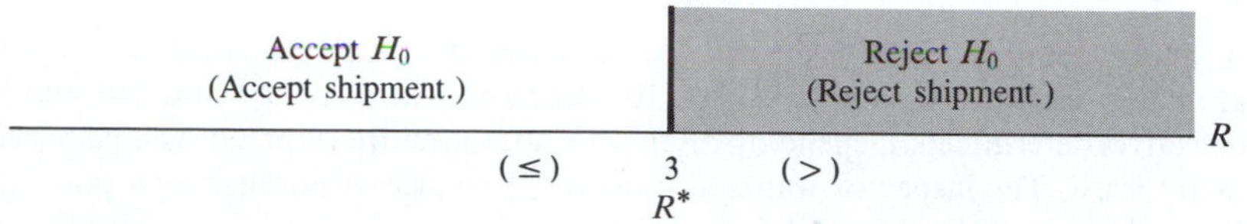

Accept H_0 (Accept shipment.)	Reject H_0 (Reject shipment.)
$(\leq)$	$(>)$

3
R^* R

The *achieved* significance level is lower than the *targeted* .05 level and is equal to

$$\Pr[R > 3 \mid \pi = .05] = 1 - .9841 = .0159$$

The engineer was disappointed to find that using this rule results in very high probabilities for accepting shipments that are really bad (committing a type II error). For instance, when the true proportion defective is $\pi = .20$,

$$\Pr[\text{accept}] = \Pr[R \leq 3 \mid \pi = .20] = .4114 \qquad (n = 20)$$

Not wanting to change the targeted significance level, he chooses instead to increase the sample size to $n = 50$. Returning to Appendix Table A, for that sample size, a new level for r and smallest cumulative probability is found

$$\Pr[R \leq 5 \mid \pi = .05] = .9622 \qquad (n = 50)$$

so that $R^* = 5$. The new decision rule is as follows:

$n = 50$:

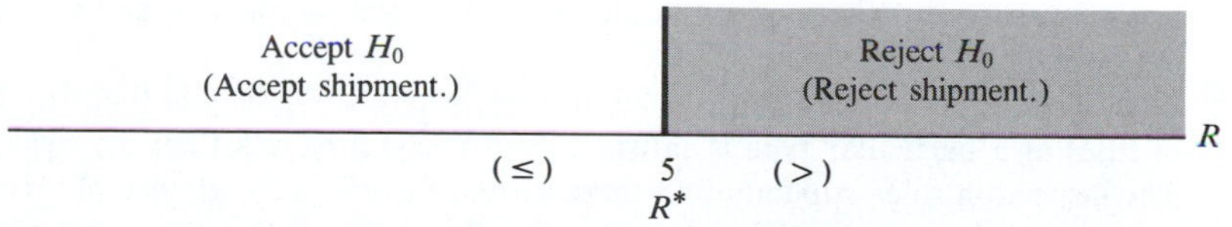

Accept H_0 (Accept shipment.)	Reject H_0 (Reject shipment.)
$(\leq)$	$(>)$

5
R^* R

Although this rule raises the achieved significance level slightly to $1 - .9622 = .0378$, there is now better protection against the type II error of accepting bad shipments. In the case of $\pi = .20$, the new sample size and decision rule provide

$$\Pr[\text{accept}] = \Pr[R \leq 5 \mid \pi = .20] = .0480 \qquad (n = 50)$$

Problems

9-24 Suppose that the systems analyst described in the data-entry illustration used $\pi_0 = .0035$ and a 1% significance level instead. Assuming the same sample results, what action should the analyst take?

9-25 For the data-entry illustration in the text, suppose that $\pi_0 = .007$ had been used with a 10% significance level.

(a) Determine the critical value and resulting decision rule.

(b) What decision will be reached if the number of sample defectives is

(1) 25? (2) 6? (3) 10? (4) 31?

9-26 Consider again the microcircuit illustration on page 369. Assume that the given decision rule with $n = 50$ will be used. Using the normal approximation, determine the type II error probability for incorrectly accepting a shipment when the time proportion of defectives is

(a) .12 (b) .15 (c) .17 (d) .19

9-27 A quality-control inspector must decide about a very large shipment of 10-ohm amplifier resistors. A sample of $n = 100$ resistors will be tested and the number of defectives determined. Depending on that result, the entire shipment will be purchased or returned. The inspector wants at most a 5% chance of purchasing a poor-quality batch (having 10% or more defectives).

(a) Using the number of defectives in the sample as the test statistic, perform hypothesis testing steps 1 and 3. Use the binomial distribution directly instead of the normal approximation.

(b) What action should be taken when the number of sample defectives is

(1) 4? (2) 6? (3) 5? (4) 7?

9-28 A statistician is testing the null hypothesis that exactly half of all engineers will still be in the profession 10 years after receiving their bachelor's. He took a random sample of 200 graduates from the class of 1979 and determined their occupations in 1989. He found that 111 persons were still employed primarily as engineers.

(a) Construct a 95% confidence interval estimate for the proportion of engineers remaining in the profession. (Ignore the continuity correction.)

(b) At the 5% significance level, should the statistician accept or reject his hypothesis?

9-29 Consider fuses of a particular type that are inspected by both the producer and the consumer.

(a) The manufacturer defines as marketable any batch of fuses in which the proportion of defectives does not exceed .07. The inspection rules are based on the requirement that there should be no more than a 1% chance of scrapping any marketable batch. This outcome is referred to as the **producer's risk**. A sample of 500 fuses will be tested from a production run of 4,000. Perform hypothesis testing steps 1–3. What action should be taken if the sample is found to contain 8% defectives?

(b) One of the users of the fuses is a laboratory, where the policy is that any batch of fuses of a particular type is satisfactory if it contains less than 5% defectives. The inspection rules stipulate that there should be only a 5% chance of accepting an unsatisfactory batch. This outcome is referred to as the **consumer's risk**. A sample of 100 fuses is to be tested from a shipment of 500. Perform hypothesis testing steps 1–3. What action should be taken if the sample is found to contain 2 defectives?

9-30 The number of switches placed on line in a communications system depends on the level of traffic. At 20 random times over a brief interval it is determined what proportion of the time all switches are busy. An additional switch is then placed on line if this quantity exceeds a critical level. A similar rule is applied to determine when to remove one of the switches, which is done whenever the sample proportion falls below a smaller critical value.

(a) Assume that there should be only a 5% chance of unnecessarily adding a switch when all the switches are busy no more than 50% of the time.
  (1) Perform hypothesis testing steps 1–3.
  (2) What action should be taken if all switches are busy in 65% of the observed sample times?

(b) Assume that there should be only a 10% chance of incorrectly removing a switch when all the switches are busy at least 30% of the time.
  (1) Perform hypothesis testing steps 1–3.
  (2) What action should be taken if all switches are busy in 25% of the observed sample times?

9-31 You want to establish a policy for accepting or rejecting shipments of parts. Each lot contains 1,000 units. Using a pivotal level of .01 for the lot proportion defective, answer the following:

(a) Is the normal approximation appropriate?

(b) Using $n = 20$ and an α target of .05, determine the critical value for the number of sample defectives. Express the corresponding decision rule.

(c) Find the type II error probability given that the true lot proportion defective is .10. Do you think that the rule found in (b) is satisfactory? Comment on how you might improve the test.

(d) Increase the sample size to 100.
  (1) Find the new critical value.
  (2) Express the decision rule.
  (3) Determine the type II error probability when the true lot proportion defective is .10.

9-32 Consider the dilemma faced by a workstation manufacturer receiving shipments of keyboards from a vendor. These arrive in lots of 50 each. Based on the number of sample defectives, a decision rule for accepting or rejecting shipments needs to be established so that there is at most a 5% chance of rejecting a good shipment having a maximum of 10% defective keyboards.

(a) For a sample of size 5, construct a table of cumulative hypergeometric probabilities.

(b) Using the above, find the critical value and express the decision rule.

(c) Find the critical value when binomial probabilities are used to approximate the hypergeometric. Is there any change from (b)?

9-33 An engineer wishes to establish a policy for deciding when to accept or reject shipments. He wants to have just a 5% chance of rejecting a good shipment having only .06 defective and just a 10% chance of accepting a really bad shipment having .25 defective. Answer the following assuming that the normal approximation is appropriate.

(a) Find an expression when H_0 is true that treats n and P^* as unknown quantities and z_α and π_0 as givens.

(b) Let z_β represent the normal deviate that corresponds to the accept probability with P^* when $\pi = \pi_1$. Find an expression when H_1 is true that treats n and P^* as unknown quantities and z_β and π_1 as givens.

(c) Assuming that the normal approximation is appropriate, find the sample size n that will achieve the engineer's objective. (Ignore the continuity correction.)

(d) Using that sample size, find P^* and express the engineer's decision rule.

9-4 Testing the Variance

Although far less frequently encountered than tests of the mean or proportion, similar procedures may be applied to the variance. Recall that this parameter is important to situations where consistency, or the lack of it, is crucial. For example, although the mean queueing time is important to waiting persons, they feel more comfortable when they can closely predict how much time must be sacrificed. People are more tolerant of long lines that move at a fairly predictable pace—lines in which the variability in waiting time is small and the length of wait may be gauged by the number of persons in line.

Testing with the Chi-Square Statistic

Testing procedures for the variance are analogous to those for the parameters already discussed. As in those tests, the analogous sample statistic—in this case the sample variance s^2—will be the basis for accepting or rejecting a null hypothesis regarding σ^2. As we have seen, the chi-square distribution may be used to find probabilities for the level of s. But it is more convenient instead to make the following transformation.

CHI-SQUARE TEST STATISTIC FOR THE VARIANCE

$$\chi^2 = \frac{(n-1)s^2}{\sigma_0^2}$$

This random variable has the standard chi-square sampling distribution.

Hypotheses regarding σ^2 take the same form as those for μ. Figure 9-16 summarizes the single-decision procedure for testing the variance.

Recall that the chi-square density is asymmetrical, and that the table entries all involve the $>$ orientation. For this reason the one-sided critical values are designated by the subscript $1-\alpha$ in the lower tail and α in the upper tail, with analogous values for two-sided tests. The quantities may be read from Appendix Table H using $n-1$ for the number of degrees of freedom.

Illustration: *Consistency in Chemical Reaction Times*

A chemical engineer wants to determine if a low-pressure purification procedure for a synthetic fiber will result in an improvement over the present high-pressure process. Although the latter is believed considerably faster and has a low mean processing time, the time needed to complete a batch varies considerably—creating difficulties in scheduling plant facilities. The high-pressure process has taken as long as 112 hours to complete but has terminated as quickly as 33 hours, and the standard deviation has been established at $\sigma = 15$ hours. The chi-square procedure is based on the variance, which for the present process is $\sigma^2 = (15)^2 = 225$. It will be worthwhile to change to low-pressure processing if the new procedure provides a σ^2 half as large.

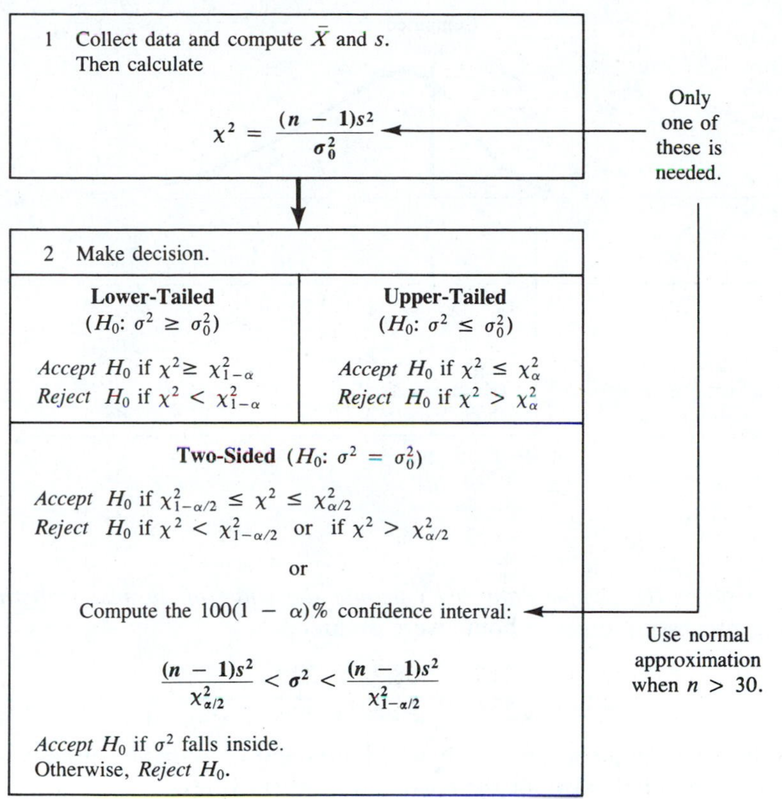

Figure 9-16 *Summary of single-decision procedure for hypothesis test for variance.*

Using $\sigma_0^2 = 112.5$, the following hypothesis testing steps were taken.

Step 1. *Formulate the hypotheses.* Wanting to protect most against the error of selecting the low-pressure process when in fact its processing times are not sufficiently consistent, the following apply:

H_0: $\sigma^2 \geq 112.5$ (The low-pressure process is too varied to justify a switch.)

H_1: $\sigma^2 < 112.5$ (The low-pressure process times are consistent enough to warrant switching.)

These indicate that the test will be a lower-tailed one.

Step 2. *Select the test procedure and test statistic.* The chi-square statistic is used.

Step 3. *Establish the significance level and the acceptance and rejection regions for the decision rule.* The engineer chooses a significance level of .01. A small value for s^2, and hence χ^2, will tend to refute the null hypothesis. The decision rule in Figure 9-17 was determined on the basis of a sample of 10 pilot batches processed at the lower pressure.

Figure 9-17

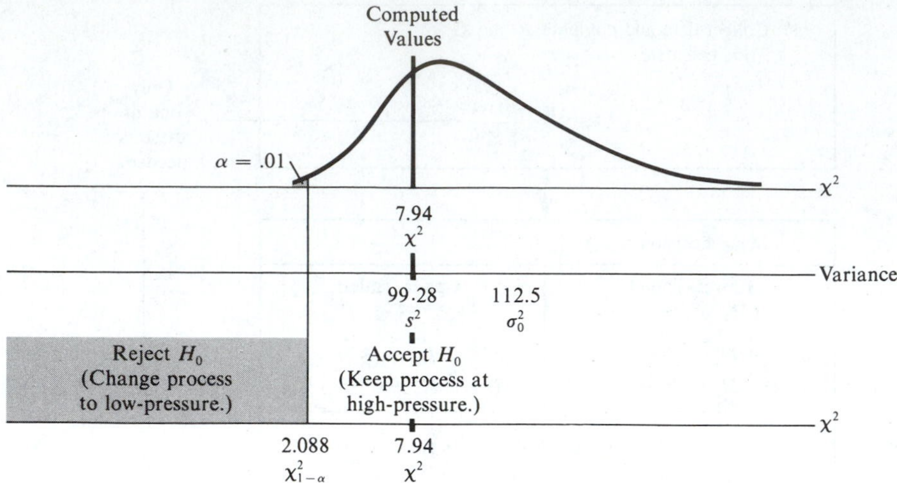

Step 4. *Collect the sample data and compute the value of the test statistic.* The following processing times in hours were obtained:

35.4	44.7	58.9	28.6	37.5
49.1	52.6	55.3	36.2	50.0

From these, the sample mean is $\bar{X} = 44.8$ hours and the sample variance is $s^2 = 99.28$. The computed value for the chi-square test statistic is

$$\chi^2 = \frac{(10 - 1)(99.28)}{112.5} = 7.94$$

Step 5. *Make the decision.* The computed test statistic falls in the acceptance region. Although the computed variance is smaller than σ_0^2, it is not significantly low enough to reject the null hypothesis. The engineer must therefore accept H_0 that the low-pressure process is not sufficiently consistent to warrant changes. The high-pressure procedure will be retained.

Testing with the Normal Approximation

Recall that the chi-square density becomes symmetrical for large sample sizes, approaching the shape of the normal curve. A rule of thumb is that the normal approximation may be used whenever $n > 30$. (Table H stops at 30 degrees of freedom.) In those cases the chi-square test statistic is transformed.

NORMAL DEVIATE TEST STATISTIC FOR VARIANCE (LARGE n)

$$z = \frac{\chi^2 - (n - 1)}{\sqrt{2(n - 1)}} = \frac{(n - 1)(s^2 - \sigma_0^2)}{\sigma_0^2\sqrt{2(n - 1)}}$$

Figure 9-18

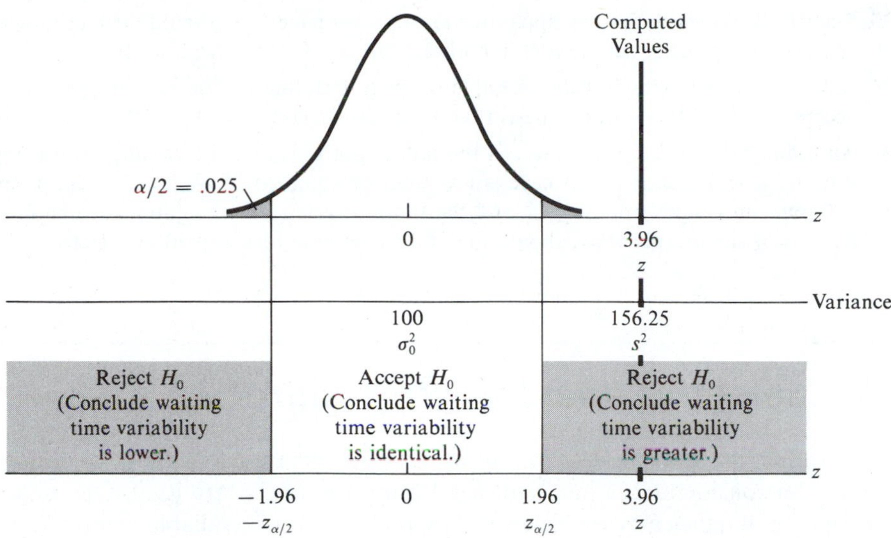

The critical normal deviate z_α applies in one-sided tests, and $z_{\alpha/2}$ is used in two-sided tests.

As an illustration, suppose that an industrial engineer wishes to test the null hypothesis that the standard deviation or variance in waiting times by partially completed assemblies at a particular station is the same as at another one where the standard deviation is known to be 10 minutes (σ_0). His null hypothesis is that $\sigma^2 = (10)^2 = 100$. For this two-sided test he chooses a significance level of $\alpha = .05$. A random sample of 100 assemblies are tagged and timed as they flow through the workstation. The decision rule in Figure 9-18 applies.

The computed value of the sample standard deviation is $s = 12.5$, so that $s^2 = (12.5)^2 = 156.25$, and

$$z = \frac{99(156.25 - 100)}{100\sqrt{198}} = 3.96$$

which falls in the upper rejection region shown in Figure 9-18. The engineer concludes that waiting times at the test station involve greater variability than at the reference location.

Problems

9-34 An educational researcher believes that even though foreign engineering students on the average score lower than all students on a particular graduate achievement examination, their drastically different backgrounds should account for greater variability. As her null hypothesis she assumes that this is not so, and that the variance in foreign students' scores is $\leq 2{,}000$ (σ^2), the variance of all test scores. With a random sample of 25 foreign student scores, the computed variance is $s^2 = 2{,}749$. At a 5% significance level, can she conclude that foreign students' test scores are more varied than those of all students?

9-35 Refer to the physical therapy apparatus example on page 314. Should the mechanical engineer accept or reject the null hypothesis that $\sigma^2 = 100$? Use $\alpha = .10$.

9-36 Refer to the logic circuit wafer example on page 305. Should the industrial engineer accept or reject the null hypothesis that $\sigma^2 \leq .0015$? Use $\alpha = .01$.

9-37 An industrial engineer wishes to test the null hypothesis that the variance in waiting time by machinists at a tool cage is less than or equal to 25. A random sample of 100 machinist times were logged, and the sample variance was computed to be 41.4. At the 5% significance level, should the null hypothesis be accepted or rejected?

9-5 Selecting the Test Procedure

Our introduction to hypothesis testing would not be complete without discussing some of the considerations involved in selecting the procedure itself. One major concern is how efficiently the selected test makes use of available sample data. This is partly determined by the form of the hypotheses and the particular test statistic employed. In comparing procedures, the better test is the one providing lower probabilities for the two types of errors. Ordinarily, this aspect may be improved by increasing the sample size, but investigators usually are required to keep sampling expenses within budgetary limitations.

Some Important Questions

Several fundamental questions must be resolved in planning a statistical testing experiment.

What particular test should be used? The choice of test depends partly on the kind of decision to be made. Often this involves comparing a new procedure with an existing method, and the objective is to select the better way. Often quality may be expressed in terms of a mean μ, the better method involving the larger or smaller level. But there may be other ways to measure the same things. For example, instead of using the mean copper yield as the basis for evaluating the bacterial leaching decision at the outset of the chapter, the mining engineer could conceivably use the metabolizing rate or some other biological yardstick. A queueing evaluation might be based on the mean waiting time, but the determining factor might instead be the proportion of time the service facility is busy. In many applications—for instance, work measurement—either the proportion or the mean might serve as the measure of effectiveness; the test statistic $\bar{X}$ or P coincides. The better test—the more efficient one—will be the one that provides lower error probabilities for the same cost. Often, two or more populations must be compared when the parameters of each are unknown. More complex controlled experiments are then necessary; we will encounter procedures for evaluating some of these in later chapters. In some cases the populations are better represented by **nonparametric tests**.

How do we evaluate a test in terms of its protection against erroneous decisions? Both the type I error (rejecting H_0 when it is true) and the type II error (accepting

H_0 when it is false) are undesirable. But we must tolerate some probability for each as the unavoidable price of basing our choice on a sample. Ordinarily the probability (α) of the more serious type I error is specified in advance, and the decision rule is selected to match. But if the resulting rule provides unacceptably large chances (β's) for the type II error, an investigator may choose to revise it. One useful tool employed by statisticians for this purpose is the operating characteristic curve.

Are there any underlying assumptions that might limit the scope of a testing procedure? A test statistic may have limited validity arising from the probability theory and assumptions used to derive its sampling distribution. For instance, we have been using the normal curve to find error probabilities for $\bar{X}$ and to set decision rules. But doing so is nearly always an approximation. The quality of that approximation may be very poor for small sample sizes (and it may be totally inappropriate for certain unusual populations). When n is small and σ is unknown, more accurate procedures might be based on the Student t distribution. But even then, the population itself is theoretically assumed to be normal (which is usually not the case). Similar problems arise with the chi-square distribution. When the theoretical underpinnings of a traditional procedure are seriously violated by circumstances, statisticians may substitute a nonparametric test that presupposes very little about the sampled population. Fortunately, such violations often are not serious; a procedure that is workable anyway is **robust** with respect to a violation of that assumption.

The Operating Characteristic Curve

In all the testing situations encountered so far, there is a range of possible values for the decision parameter. When sufficient information is known about the underlying population, it is possible to compute the conditional probability for accepting H_0 (or rejecting H_0) for any given value in that range. It may be useful to plot several of these probabilities on a graph, as in Figure 9-19, which provides the **operating characteristic curve** (or OC curve) for the bacterial leaching decision from the beginning of the chapter.

The ordinate provides $\Pr[\text{accept } H_0 \mid \mu]$ for levels of the population mean copper yield μ on the abscissa. For any population mean >36 $(=\mu_0)$ pounds per ton, the null hypothesis is false and the process is economical. The corresponding points on the curve are the type II error probabilities β. When $\mu = \mu_0$, the probability of accepting H_0 is exactly $1 - \alpha$. Below 36, the probability of accepting H_0 (then a correct action) approaches 1. Notice that the β's are quite large, even when μ considerably exceeds the break-even level of 36. These indicate that the mining engineer faces a high probability of incorrectly accepting a false null hypothesis—not recommending an economical bacterial leaching procedure. This stems from the choice of decision rule, with the critical value of $\bar{X}^* = 40$.

If the mining engineer finds these error probabilities too high, he has two choices. (1) He may drop the critical value to a lower level. The effect of this would be to shift the OC curve to the left, but that results in a lower "accept" probability when $\mu = 36$, reducing $1 - \alpha$ and raising α. This increase in the type I

Figure 9-19 *Operating characteristic curve for bacterial leaching decision.*

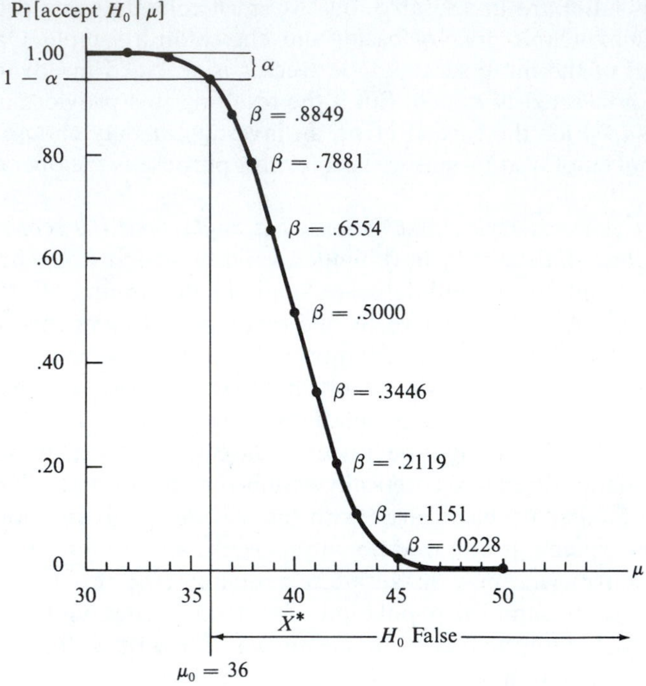

Figure 9-20 *Bacterial leaching OC curves when n = 100 and n = 200. Also shown is the limiting case, when there is no uncertainty.*

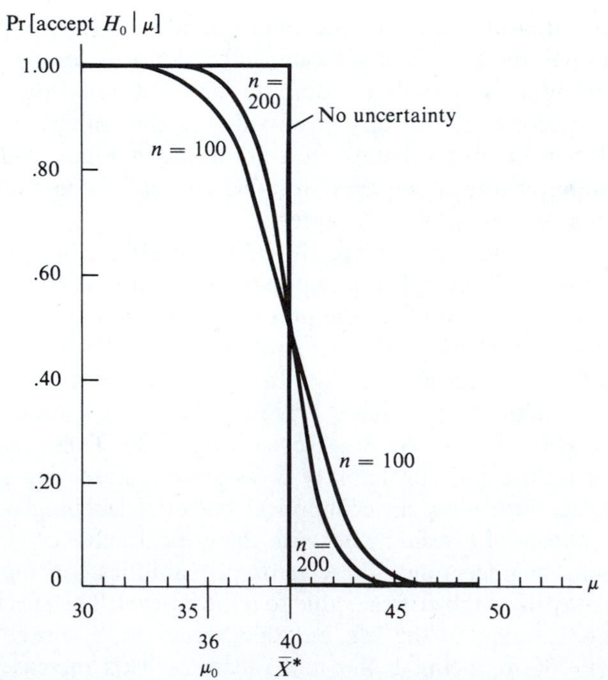

error probability may be even more unappealing than the original β's. (2) A second remedy would be to increase the sample size. Although this increases the cost of the experiment, the reduction in error probabilities may be worthwhile.

Figure 9-20 shows the original OC curve for $n = 100$ and the OC curve when the sample size is raised to $n = 200$. Notice that except where the curves cross, $n = 200$ provides higher probabilities for accepting H_0 for low levels of μ and lower probabilities for high μ's. Both curves achieve the same .50 accept probability when $\mu = \bar{X}^*$. The limiting case is provided by the rectangular portion, where n is infinite and there is no uncertainty. (When the decision rule is shifted and $\bar{X}^* = \mu_0$, no error can occur in that limiting case.)

Figure 9-21 shows the shape of OC curves for the three types of test orientations encountered in this chapter. Some statisticians prefer to place the probability of rejecting on the ordinate, which provides the **power curve**. That display provides exactly the same information as the OC curve.

The OC curve may prove to be a valuable tool for evaluating a proposed test and decision rule. But as a practical matter, it is not always possible to construct

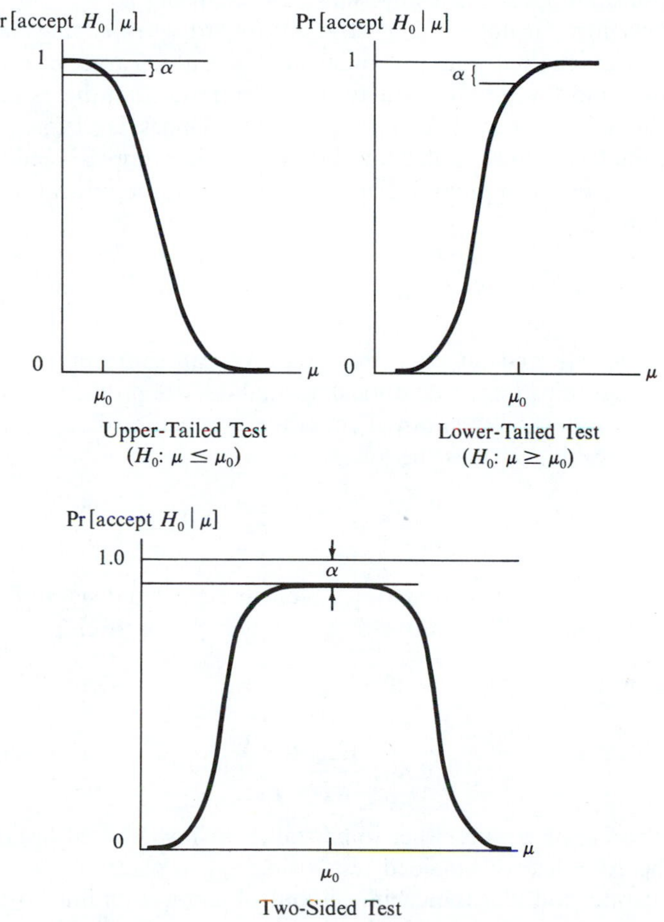

Figure 9-21 *Basic OC curve shapes.*

such a curve. There may be many unspecified ways for the type II error to occur, and it may be difficult or impossible to compute β for each. For instance, unless σ is known, the β probabilities in testing μ are indeterminate. Some procedures, such as the nonparametric ones, do not even have abscissa units for constructing a graph.

Finding the Required Sample Size for Testing the Mean

As we have seen, larger sample sizes improve the overall discrimination of the testing procedure. This is reflected in lower probabilities for both the type I and type II errors. But just how large should the sample size be?

The answer lies with the error probabilities α and β. Earlier in the chapter we saw that as the position of the critical value $\bar{X}^*$ shifts in one direction, α will rise and β (with μ fixed) will fall; and, an opposite placement of $\bar{X}^*$ lowers α and raises β. The optimal position of the critical value is where α and β are in balance. The choice of sample size n accomplishes this balancing act.

The procedure for doing this is straightforward when testing the mean with a known level for the population standard deviation σ. Target levels must be established for α and β, with the stipulation for the latter that the assumed population mean lie at a level μ_1. Like μ_0, μ_1 is a benchmark level for μ that will lie above μ_0 if the test is lower-tailed and below μ_0 if it is upper-tailed.

In the case of an upper-tailed test (H_0: $\mu \leq \mu_0$), the critical value for $\bar{X}$ is obtained from

$$\bar{X}^* = \mu_0 + z_\alpha \sigma_{\bar{X}} = \mu_0 + \frac{z_\alpha \sigma}{\sqrt{n}}$$

Since μ_0, σ, and α—and thus z_α—are fixed, $\bar{X}^*$ can shift only by increasing or decreasing n. Denoting by z_β that critical normal deviate providing a lower (upper) tail area equal to β when the normal curve for $\bar{X}$ is centered at μ_1, we will adjust n so that the resulting $\bar{X}^*$ has the following property:

$$\bar{X}^* = \mu_1 - \frac{z_\beta \sigma}{\sqrt{n}}$$

The required n is found by setting equal the right-hand sides of these expressions and solving for n. This provides the following expression.

REQUIRED SAMPLE SIZE IN TESTING THE MEAN

$$n = \left[\frac{\sigma(z_\alpha + z_\beta)}{\mu_1 - \mu_0}\right]^2$$

This expression applies with either upper-tailed or lower-tailed tests. (The procedure may be extended to two-sided tests using $z_{\alpha/2}$ in place of z_α.)

To illustrate, consider the chemical control calibration illustration on pages 353–355, where the mean impurity level μ is tested using $\bar{X}$. This involves μ_0 =

150 ppm and $\sigma = 20$ ppm. The targeted significance level is $\alpha = .05$, so that $z_\alpha = z_{.05} = 1.64$. Suppose now that the chemical engineer wants to choose n so that he may achieve a type II error probability of $\beta = .10$ for not recalibrating when the impurity level is an intolerably high $\mu_1 = 160$ ppm. The corresponding normal deviate is $z_\beta = z_{.10} = 1.28$. He needs a sample size of

$$n = \left[\frac{150(1.64 + 1.28)}{160 - 150}\right]^2 = 34.11$$

Limitations of Hypothesis Testing

Hypothesis testing as described in this chapter was originally designed for applications in biology and medicine. Within budgetary limitations, avoidance of error is an overriding consideration in those fields. Thus, choosing a procedure that limits the incidence of the more serious error is perhaps the best decision-making criterion. But that criterion may not be best for decision making in other areas.

Later in this book we will encounter Bayesian decision-making procedures where a wider focus is made. Error is just one aspect of Bayesian evaluations, which are based on a broader **statistical decision theory**. These more general procedures allow us to establish probabilities for the level of μ (or π). It is possible to expand the dimensions of the statistical evaluation to consider a payoff measure that quantifies each outcome. Within that analytical framework it is even possible to consider such fundamental questions as whether or not to sample at all. (It may be better to decide without the bother and benefit of sampling!)

Problems

9-38 Suppose that the mining engineer in the bacterial leaching illustration changes his significance level to $\alpha = .01$ and uses a sample size of 150 batches.

(a) Determine his new decision rule when $\bar{X}$ is the test statistic.

(b) Construct an operating characteristic curve using the decision rule from (a).

9-39 Construct the OC curve for the biological reduction process decision in Problem 9-18.

9-40 Construct the OC curve for the rod decision in Problem 9-13.

9-41 Suppose that the mining engineer in the bacterial leaching illustration uses as targets $\alpha = .05$ with $\mu_0 = 36$ and $\beta = .05$ with $\mu_1 = 44$. Assuming that $\sigma = 25$, find the required sample size.

9-42 Suppose that the mining engineer in the bacterial leaching illustration uses as targets $\alpha = .01$ with $\mu_0 = 36$ and $\beta = .10$ with $\mu_1 = 42$. Assuming that $\sigma = 25$, find the required sample size.

Comprehensive Problems

9-43 For each of the following situations, (1) specify the test statistic, (2) give its critical value, and (3) indicate whether the test is lower-tailed, upper-tailed, or two-sided. In each case use $\alpha = .05$ and assume a large population unless otherwise indicated.

(a) H_0: $\mu \leq 50$, $\sigma = 10$, $n = 100$ (b) H_0: $\mu \geq 100$, $\sigma = ?$, $n = 100$

(c) $H_0: \mu = 75, \sigma = ?, n = 20$
(d) $H_0: \pi \leq .20, n = 100$
(e) $H_0: \pi \geq .40, n = 50, N = 200$
(f) $H_0: \mu \leq .50, \sigma = .15, n = 25, N = 150$
(g) $H_0: \mu \geq 1.46, \sigma = ?, n = 25$
(h) $H_0: \mu = 25, \sigma = 2, n = 200, N = 1{,}000$

9-44 The following data (minutes) apply for a random sample of processing times for a chemical reaction:

2.3	6.7	3.8	5.0	4.9	6.1
4.4	5.2	3.9	4.8	4.6	5.7
5.3	4.7	4.2	4.7	5.7	4.8

Can you conclude that the mean of all processing times exceeds 4 minutes? Use $\alpha = .05$.

9-45 The following data were obtained for the amount of time (seconds) taken by a proposed computer system to compile a sample of short FORTRAN programs:

.6	1.0	1.8	4.8	4.2
4.4	2.4	3.5	3.9	3.9
4.7	3.0	2.3	2.9	1.0

(a) At the 5% significance level, test the null hypothesis that the mean compilation time for all short FORTRAN programs run on the system is at least 4 seconds. Should it be accepted or rejected?
(b) At the 5% significance level, test the null hypothesis that the standard deviation in compilation time is less than or equal to .9 second ($\sigma^2 \leq .81$). Should it be accepted or rejected?

9-46 A pharmaceutical loader has been set to insert exactly 5 milligrams into each capsule of a particular drug. Periodically a sample of 100 capsules is taken and the contents of each are measured. The standard deviation has been established at .2 milligram per capsule. Corrective action will be taken whenever the computed mean is significantly great or small.
(a) Determine a decision rule when a 1% chance is permitted for taking unnecessary action.
(b) Apply the above to find the probability of failing to take needed correction when the true mean is (1) 5.05 milligrams? (2) 4.97 milligrams? (3) 5.07 milligrams? (4) 4.98 milligrams?

9-47 A program director for a large computer maker is evaluating a circuit-designing software package. Not only is the program purported to ease the engineers' tasks, but its author also claims that the finished products will be more economical because fewer logic elements will be needed. To test this theory, the software was borrowed and used to redesign 20 prototype circuits already completed in the traditional way. The resulting computer-generated designs all worked and involved a mean percentage reduction in components of 3.4% ($s = 4.2\%$).
(a) At the 1% significance level, can the project manager reject the null hypothesis that the software designs involve at least as many elements as do traditional ones?
(b) What is the smallest significance level at which the null hypothesis can be rejected?

9-48 Refer to the data in Problem 9-44. What is the lowest significance level at which you can conclude that the variance in processing time exceeds .25 minute^2?

9-49 Refer to the data in Problem 9-45.
(a) Construct a 99% confidence interval estimate for the mean. Should the null hypothesis that $\mu = 3.50$ minutes be accepted or rejected at the 1% significance level?

(b) Construct a 98% confidence interval estimate for the variance. Should the null hypothesis that $\sigma^2 = 1.0$ be accepted or rejected at the 2% significance level?

9-50 Only 18 persons in a sample of 100 students indicated a desire for a full-scale summer program. Can it be concluded that the majority of students don't want such a program? Use $\alpha = .01$ for incorrectly making that conclusion. Assume a large population and apply the normal approximation.

9-51 Specifications for a component require that the MTBF exceed 100 hours. The actual MTBF of any shipment is unknown, although it is assumed for each that the lifetime standard deviation is 10 hours. There should be only a 5% chance of concluding that a shipment meets specifications when it does not. A statistician takes a random sample of 50 items from a particular shipment of 300 and computes the MTBF to be 102 hours. What conclusion does he reach?

9-52 The failure rate of jet engine fan blades subjected to a particular test condition has been specified to be less than .05. An entire production batch found to violate that requirement will be scrapped; otherwise they will be assembled into engines. Assuming just a 10% chance of erroneously assembling with a poor-quality batch of fan blades, determine the maximum number of failed blades in a sample of 100 that would still allow the batch to be used in assembly. (Use the binomial distribution directly, assuming the batch size is large.)

9-53 Consider a one-sided hypothesis test of the mean where σ is known and N is large. Find the level for n that should be used in each of the following:

(a) $\mu_0 = 90$, $\mu_1 = 100$, $\sigma = 25$, $\alpha = .05$, $\beta = .10$
(b) $\mu_0 = 50$, $\mu_1 = 30$, $\sigma = 15$, $\alpha = .01$, $\beta = .05$
(c) $\mu_0 = 1.52$, $\mu_1 = 1.55$, $\sigma = .5$, $\alpha = .05$, $\beta = .10$
(d) $\mu_0 = 1{,}100$, $\mu_1 = 1{,}000$, $\sigma = 100$, $\alpha = .10$, $\beta = .20$

CHAPTER TEN

Regression and Correlation Analysis

SO FAR OUR sampling applications have focused on a single population, with observed values denoted by X. In this chapter we extend sampling to incorporate a second population, represented by Y. Our present concern is predicting values for one population on the basis of observations taken from the other. Such predictions are important in engineering when Y cannot be observed directly; in such cases its value must be forecast from a known value for X. A statistical procedure provides a function, $Y = f(X)$, that converts values for the **independent variable** X into levels for the **dependent variable** Y. For example, the unknown time Y to process a raw chemical might be expressed in terms of the known density X of the starting solution. By measuring and timing sample batches of the chemical, a relationship may be established between these two quantities, such as $Y = 9.5 + .45X$.

The process is called regression analysis. *Regression* is the term originally given in the pioneering work of Sir Francis Galton, who used the heights of fathers to predict how tall their sons would become. Galton found that children of tall persons on the average grew to be shorter than their parents, whereas the male offspring of short people were on the average taller than their fathers. He reported that the stature of the second generation "regressed" toward the general population mean. (But the overall frequency distribution of heights remained stable from one generation to the next.) The term *regresssion* has since been adopted by statisticians to represent the entire class of applications in which one or more variables are used to predict another.

Engineers encounter some of the elements of regression analysis whenever they fit a curve to experimental data. But regression analysis has a wider scope than "curve-fitting," primarily because it is a rigorous procedure that controls prediction errors. A considerable refinement to the freehand fairing of a curve, the *method of least squares* is the central regression technique that minimizes the collective deviations of the individual observations from the resulting regression surface. An accompanying mathematical function, or *regression equation*, completely specifies the resulting line or curve. This feature ensures that any two analysts can achieve identical results from the same given data and assumed shape—permitting greater objectivity and making the curve-fitting more of a science than an art.

Regression analysis calls upon the entire panoply of probability theory and extends those procedures of statistical inference encountered so far in this book. Thus, the regression surface is just a starting point. The predictions it provides may be qualified by means of confidence intervals, and hypothesis testing procedures can verify the nature of the corresponding regression equation. The probability concept of covariance extends to regression's companion procedure, *correlation analysis*, which is concerned with the strength of association between X and Y.

10-1 Linear Regression Using Least Squares

The basic application is **linear regression**, wherein Y is related to X by a straight line, $Y = a + bX$. After collecting a set of sample data points (X, Y), an investigator arithmetically combines those observations using the method of least squares, which finds the slope b and Y-intercept a providing the best fit. Because it is just a sample summary, the resulting expression is referred to as the **estimated regression equation**. The regression line may then be used to predict the level for the dependent variable Y that best corresponds to a particular setting of the independent variable X. The computed Y-intercept and slope are only estimates of the true line parameters. Since a and b will vary from one sample to the next, they are referred to as **estimated regression coefficients**.

***Illustration:** Video-Display Terminal Transactions*

Figure 10-1 shows how a line might provide a satisfactory basis for predicting the unknown video-display-terminal time Y required to complete a transaction involving a known number of inputs X. The plotted points are **raw data** obtained from a random sample of 25 transactions made at computer terminals by materials-control clerks in an electronics assembly facility. Such a graphical display of sample data is sometimes called a **scatter diagram**. The method of least squares was used to determine the regression line fitting the observed data.

The Method of Least Squares

The **method of least squares** finds that particular line where the aggregate deviation of the data points above or below it is minimized. Rather than measuring each point's separation in terms of the physical distance, the procedure is instead based upon *vertical* deviations, which are then squared. This not only eliminates difficulties in measuring perpendicular line segments, but it provides summary statistics having desirable properties.

To distinguish the actual and predicted values of the dependent variable we use Y_i to denote the ith observation and $\hat{Y}(X)$ to denote a predicted value for a given level of X. For the ith observation the independent variable value is X_i, and

$$\hat{Y}(X_i) = a + bX_i$$

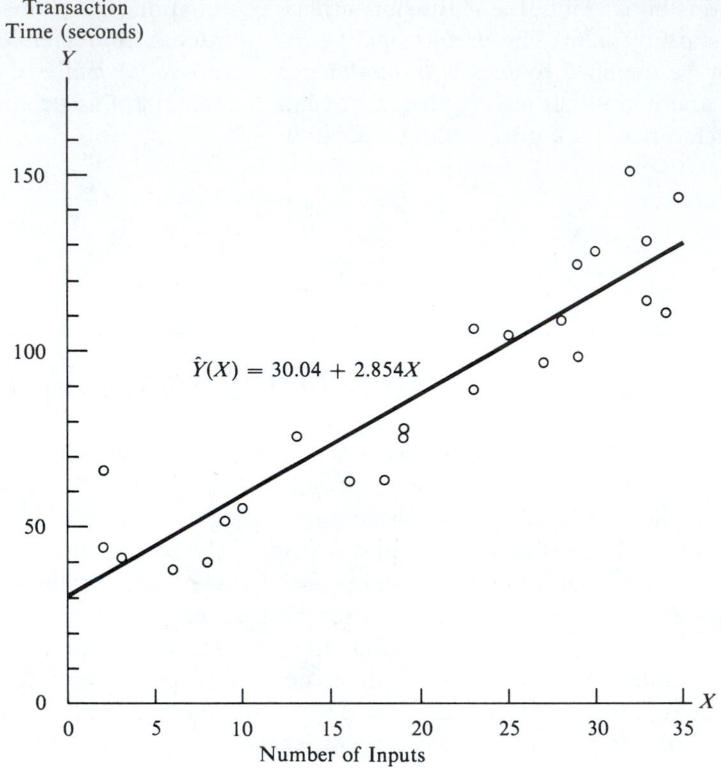

Figure 10-1 *Scatter diagram with regression line for terminal transactions.*

The goal of least-squares regression is to find the values of a and b that minimize the following:

$$\sum_{i=1}^{n} [Y_i - \hat{Y}(X_i)]^2 = \sum_{i=1}^{n} [Y_i - (a + bX_i)]^2$$

Taking the partial derivatives of this expression with respect to a and to b, and then setting the resulting terms equal to 0 and simplifying, we obtain the following.

NORMAL EQUATIONS FOR LINEAR REGRESSION

$$\sum_{i=1}^{n} Y_i = na + b \sum_{i=1}^{n} X_i$$

$$\sum_{i=1}^{n} X_i Y_i = a \sum_{i=1}^{n} X_i + b \sum_{i=1}^{n} X_i^2$$

(The term *normal* in this context refers to a property of linear algebra, rather than to the normal distribution.)

A simultaneous solution of the normal equations provides us with the following expression for the slope b. (For simplicity the subscripts are dropped.)

$$b = \frac{n\sum XY - \sum X \sum Y}{n\sum X^2 - (\sum X)^2}$$

It is easiest to express a in terms of the value found for b by transforming the first normal equation, getting

$$a = \frac{1}{n}\left(\sum Y - b\sum X\right)$$

We may simplify this by using the fact that $\bar{X} = (\sum X)/n$ and $\bar{Y} = (\sum Y)/n$ and use the resulting expressions to compute the following.

ESTIMATED LINEAR REGRESSION COEFFICIENTS

$$\mathbf{b} = \frac{n\sum XY - \sum X \sum Y}{n\sum X^2 - (\sum X)^2}$$

$$\mathbf{a} = \bar{Y} - b\bar{X}$$

To illustrate, consider the raw data for the terminal transactions in Table 10-1. Substituting the intermediate calculations into the above, we find the following estimated regression coefficients:

$$b = \frac{25(52{,}670) - (506)(2{,}195)}{25(13{,}130) - (506)^2} = 2.854$$

$$a = 87.80 - 2.854(20.24) = 30.04$$

The estimated linear regression equation is thus

$$\hat{Y}(X) = 30.04 + 2.854X$$

This regression line may be used to predict the terminal time requirements for any particular transaction for which the number of inputs X is known. For instance, should $X = 32$ entries be necessary for controlling a particular material, the predicted terminal time would be

$$\hat{Y}(32) = 30.04 + 2.854(32) = 121.4 \text{ seconds}$$

Values obtained in this manner would be useful for planning purposes, such as forecasting the data processing resources necessary for handling a new product line or an increase in orders.

Rationale and Meaning of Least Squares

Figure 10-2 will help us understand the underlying rationale for the method of least squares. Each observed data point lies above or below the regression line. The vertical deviation for the ith observation is denoted by the difference $Y_i - \hat{Y}(X_i)$. For example, the 22nd observation has coordinates (32, 151) and lies

Table 10-1 *Raw Data for a Sample of Terminal Transactions and Intermediate Calculations for Finding Estimated Regression Coefficients*

Observation i	**Number of Entries** X	**Transaction Time (seconds)** Y	XY	X^2	Y^2
1	2	66	132	14	4,356
2	19	77	1,463	361	5,929
3	6	37	222	36	1,369
4	23	106	2,438	529	11,236
5	10	55	550	100	3,025
6	23	89	2,047	529	7,921
7	9	52	468	81	2,704
8	30	128	3,840	900	16,384
9	18	63	1,134	324	3,969
10	25	104	2,600	625	10,816
11	19	76	1,444	361	5,776
12	2	44	88	4	1,936
13	27	97	2,619	729	9,409
14	28	109	3,052	784	11,881
15	8	40	320	64	1,600
16	29	124	3,596	841	15,376
17	29	98	2,842	841	9,604
18	16	63	1,008	256	3,969
19	33	131	4,323	1,089	17,161
20	3	41	123	9	1,681
21	34	111	3,774	1,156	12,321
22	32	151	4,832	1,024	22,801
23	13	76	988	169	5,776
24	33	114	3,762	1,089	12,996
25	35	143	5,005	1,225	20,449
	$506 = \sum X$	$2{,}195 = \sum Y$	$52{,}670 = \sum XY$	$13{,}130 = \sum X^2$	$220{,}445 = \sum Y^2$

$$\bar{X} = \frac{506}{25} = 20.24 \qquad \bar{Y} = \frac{2{,}195}{25} = 87.80$$

above the estimated regression line by

$$Y_{22} - \hat{Y}(32) = 151 - 121.4 = 29.6 \text{ seconds}$$

This deviation is portrayed by the heavy line segment connecting the observed point to the regression line. Each vertical deviation represents the amount of *error* associated with using the regression line to predict a new transaction's terminal time. The values $a = 30.04$ and $b = 2.854$ minimize this error.

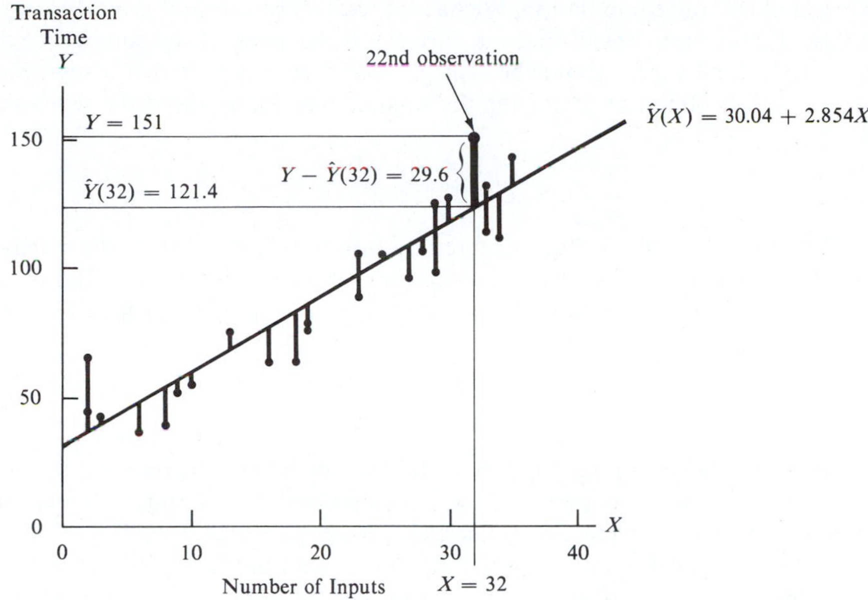

Figure 10-2 *Fitting a regression line to the terminal transaction data.*

Some of the observation points fall below the regression line and provide negative deviations. It is easy to establish that

$$\sum [Y - \hat{Y}(X)] = 0$$

so the positive and negative deviations cancel. This is one reason why the procedure is based on *squared* deviations. The estimated regression line obtained for the transaction sample minimizes the sum of the squared deviations for all the data points.

The value of the slope, $b = 2.854$, indicates that the terminal time needed for any particular transaction is estimated to increase by 2.854 seconds for every additional data input required. The Y-intercept, $a = 30.04$, tells us that regardless of the number of inputs, any transaction is estimated to require 30.04 seconds of setup time—including accessing the operating system, logging in and out, and other chores that do not depend on the complexity of the transaction.

Knowing that a particular transaction involves $X = 30$ inputs, $\hat{Y}(30) = 115.7$ seconds may be used to estimate the terminal time requirements. But the proper interpretation of that time is that *on the average* all transactions involving the same number of inputs will take 115.7 seconds. This quantity is a point estimate, and individual terminal times will vary from transaction to transaction, even when they appear to have identical input requirements.

Measuring the Variability of Results

The fundamental measure of variability in regression analysis is the degree of scatter exhibited by the data points about the regression surface. In seeking a summary value for this scatter, or dispersion, we may take guidance from our treatment of the single population. We have seen that the variance is a satisfactory measure of dispersion. Recall that the variance is the mean of the squared deviations of individual values about the central value. This suggests that a regression analogue may be found by averaging the squared deviations about the regression *line*:

$$\frac{\sum [Y - \hat{Y}(X)]^2}{n}$$

The square root of the mean squared deviations is referred to as the **standard error of the estimate** about the regression line. We will slightly modify this expression before taking the square root, obtaining as the standard error of the estimate

$$s_{Y \cdot X} = \sqrt{\frac{\sum [Y - \hat{Y}(X)]^2}{n - 2}}$$

For consistency with our past notation, we use the letter s to indicate that the calculations are based on sample data. The subscript $Y \cdot X$ indicates that the deviations are about the regression line that provides predicted levels for Y given X. The sum is divided by $n - 2$, which makes $s_{Y \cdot X}$ an unbiased estimator of the true variance in Y values about the regression line. By subtracting 2, we acknowledge the loss of 2 degrees of freedom owing to the fact that $\hat{Y}(X)$ is defined in terms of two statistics, a and b, calculated from the same data.

Notice the resemblance of the foregoing expression to that for the sample standard deviation of the individual Y observations, which we designate by s_Y,

$$s_Y = \sqrt{\frac{\sum (Y - \bar{Y})^2}{n - 1}} = \sqrt{\frac{\sum Y^2 - \frac{1}{n}(\sum Y)^2}{n - 1}}$$

This is calculated without reference to the independent variable X. The sample standard deviation is the mean of the deviations about Y, which is the center of the observed data. We may view s_Y as a measure of the **total variability** in Y. Ordinarily, deviations about Y will be greater than those about the regression line, so the level of s_Y will be greater than that for $s_{Y \cdot X}$. This feature is illustrated in Figure 10-3, in which two frequency curves are sketched. The curve on the left represents the distribution of individual Y's and has variability estimated by s_Y. On the right is a more compact frequency curve representing the vertical deviations about the regression line, with variability represented by $s_{Y \cdot X}$.

Ordinarily, it is easier to use the following expression.

STANDARD ERROR OF THE ESTIMATE

$$\mathbf{s_{Y \cdot X} = \sqrt{\frac{\sum Y^2 - a \sum Y - b \sum XY}{n - 2}}}$$

Returning to our terminal transactions, we calculate the standard error of the estimate using the values for a and b found earlier:

$$s_{Y \cdot X} = \sqrt{\frac{220{,}445 - 30.04(2{,}195) - 2.854(52{,}670)}{25 - 2}} = 13.49$$

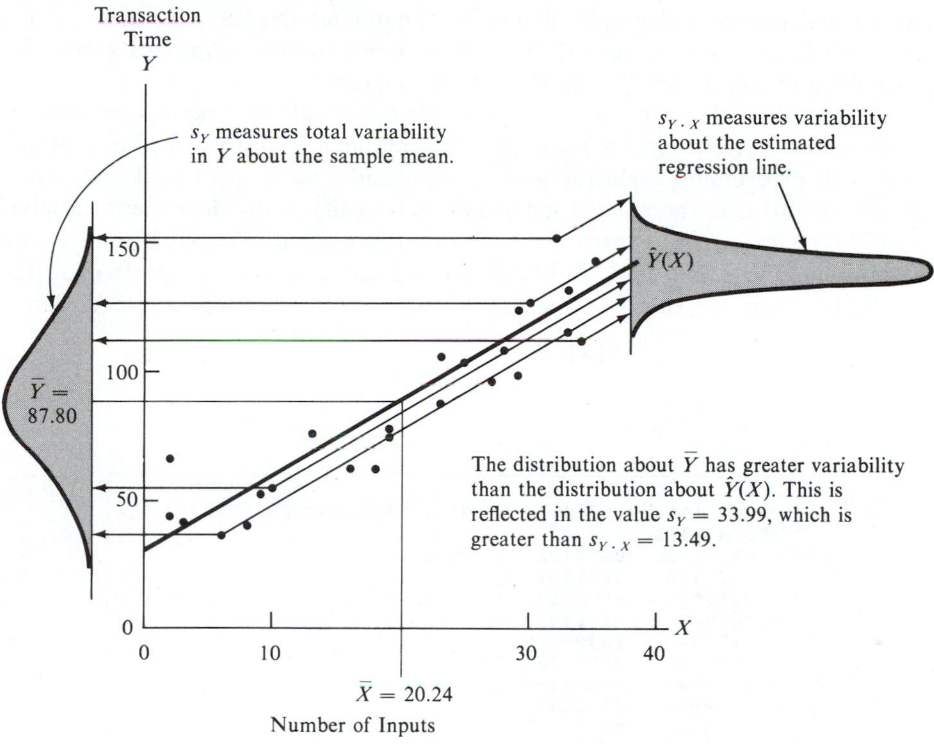

Figure 10-3 *Illustration of total variability and variability about the estimated regression line.*

When knowledge of X is ignored so that no regression line is employed, the following measure of variability applies to the same sample results:

$$s_Y = \sqrt{\frac{220{,}445 - \frac{1}{25}(2{,}195)^2}{25 - 1}} = 33.99$$

The value of s_Y is considerably larger than the computed level of $s_{Y \cdot X}$, which indicates that the variability about the regression line is smaller than the total variation. This reflects the general fact that the regression line will provide a more reliable prediction of the mean level for Y based on a known X than what would be achieved using Y alone as the estimator for the population mean.

Computer-Assisted Regression Analysis

Hand computation of regression coefficients can be a very time-consuming chore, with a great potential for error. Plotting a scatter diagram by hand can also be tedious. Regression evaluations are therefore usually performed with computer assistance.

Most statistical software packages (and even many general-purpose programs) will compute the coefficients of the estimated regression line. The better packages will even plot scatter diagrams. Older software, originally designed for mainframe computers, has limited graphical display capabilities. Newer packages, written primarily for personal computers, will often provide complete graphics, some even displaying the regression line on the scatter diagram.

Figure 10-4 shows a scatter diagram for the results of a regression analysis made with the SPSS software package. The experiment was performed in conjunction with experiments concerning the design and operation of coal conversion plants. In this run, the natural logarithm of viscosity is the dependent variable, with the reciprocal of temperature X as the independent variable. The raw data are provided in Table 10-2. Although the regression line is not plotted on the graph, the computer run provided the following estimated regression equation:

$$\hat{Y}(X) = -4.275 + 1{,}282.8X$$

Figure 10-4
SPSS scatter diagram for coal liquid viscosity data.

```
DATA LIST /VISCOSITY 1-10(7) TEMPERATURE 12-22(7).
BEGIN DATA.
-.0382212  .0032982
-.1775728  .0031939
-.3202052  .0030912
-.3043536  .0030905
-.4218994  .0030021
-.5093268  .0029430
-.5318791  .0029146
-.6474093  .0028273
-.6348783  .0028321
-.7582192  .0027427
END DATA.
PLOT SYMBOLS='*'
/VSIZE=30 /HSIZE=70
/FORMAT=DEFAULT
/TITLE='SCATTER DIAGRAM'
/VERTICAL='VIS TRANSFORM'
/HORIZONTAL='TEMP RECIP'
/PLOT=VISCOSITY WITH TEMPERATURE.
```

```
                        SCATTER DIAGRAM
          ++-----+-----+-----+-----+-----+-----+-----+--+
        0+                                              +
V        :                                            * R
I        :                                              :
S        :                                              :
         :                                        *     :
T   -.25+                                               +
R        :                                  *           :
A        :                                              :
N        :                            *                 :
S        :                                              :
F    -.5+                        *                      +
O        :                     *                        :
R        :                                              :
M        :              *                               :
         :                                              :
   -.75+ *                                              +
          ++-R--+-----+-----+-----+-----+-----+-----+--+
            .0028      .00296      .00312      .00328
     .00272      .00288      .00304      .0032

                          TEMP RECIP
```

Table 10-2 *Viscosity Data for a Coal Liquid*

Average Temperature (K)	Viscosity (mPa·s)	Temperature Reciprocal X (1/K)	Viscosity Transform Y(ln [mPa·s])
303.2	.9625	.0032982	−.0382212
313.1	.8373	.0031939	−.1775728
323.5	.7260	.0030912	−.3202052
323.1	.7376	.0030905	−.3043536
333.1	.6558	.0030021	−.4218994
340.1	.6009	.0029430	−.5093268
343.1	.5875	.0029146	−.5318791
353.7	.5234	.0028273	−.6474093
353.1	.5300	.0028321	−.6348783
364.6	.4685	.0027427	−.7582192

Source: Byers and Williams, *Journal of Chem. Eng. Data.* Vol. 32, 1987, pp. 349–354.

Problems

10-1 Consider the following sample results, where the number of data points X is used to predict computer processing time Y (in seconds):

X	Y	X	Y	X	Y
105	44	211	112	55	34
511	214	332	155	128	73
401	193	322	131	97	52
622	299	435	208	187	103
330	143	275	138	266	110

(a) Plot the above on a scatter diagram.
(b) Use the method of least squares to determine the expression for the estimated regression line. Then plot the line on your scatter diagram.
(c) Compute the sample standard deviation for computer processing time.
(d) Compute the standard error of the estimate for computer processing time.
(e) Determine the predicted processing time when the number of data points is
(1) 200 (2) 300 (3) 400 (4) 500

10-2 The following sample observations have been obtained by a chemical engineer investigating the relationship between weight of final product Y (in pounds) and volume of raw materials X (in gallons):

X	Y	X	Y
14	68	22	95
23	105	5	31
9	40	12	72
17	79	6	45
10	81	16	93

(a) Plot the given data on a scatter diagram.
(b) Use the method of least squares to determine the expression for the estimated regression line. Then plot the line on your scatter diagram.
(c) Compute the sample standard deviation for final product weight.
(d) Compute the standard error of the estimate for final product weight.
(e) Determine the predicted final product weight when the volume of raw material in gallons is (1) 5, (2) 10, (3) 15, and (4) 20.

10-3 The following data were obtained from a stress test on rods fabricated from an experimental alloy:

Lateral Strain ϵ_Y	Longitudinal Strain ϵ_X	Lateral Strain ϵ_Y	Longitudinal Strain ϵ_X
.0006	.002	.0035	.010
.0011	.003	.0046	.015
.0014	.004	.0069	.020
.0016	.005	.0086	.025
.0022	.006	.0102	.030

(a) Use the method of least squares to determine the estimated regression equation for predicting lateral strain from longitudinal strain.
(b) **Poisson's ratio** is the lateral strain divided by the longitudinal strain. If the true Y-intercept is zero, what is the estimated value of this quantity provided by your regression results?

10-4 An industrial engineer wishes to establish a relationship between cost per production batch of laminated wafers and size of run. The following data have been obtained from previous runs:

Size X	Cost Y	Size X	Cost Y
1,550	$17,224	786	$10,536
2,175	24,095	1,234	14,444
852	11,314	1,505	15,888
1,213	13,474	1,616	18,949
2,120	22,186	1,264	13,055
3,050	29,349	3,089	31,237
1,128	15,982	1,963	22,215
1,215	14,459	2,033	21,384
1,518	16,497	1,414	17,510
2,207	23,483	1,467	18,012

(a) Use the method of least squares to find the estimated regression equation.
(b) The *fixed cost* of a production run is the component of total expense that does not vary with quantity produced, whereas the *variable cost* is the added cost per

incremental unit. Use your answer from (a) to find for any production run (1) the fixed cost and (2) the variable cost.

(c) For a run of 2,000 units, use your regression line to predict (1) the total production cost and (2) the average cost per unit.

10-5 The following data apply for a sample of engineering students:

Grade-Point Average (GPA)	Graduate Achievement Examination Score	Grade-Point Average (GPA)	Graduate Achievement Examination Score
3.3	550	3.7	650
3.9	670	2.8	450
2.7	510	2.9	520
3.1	480	3.9	710
3.2	450	2.6	420
3.8	710	3.2	570

(a) Using a student's GPA as the independent variable, find the estimated regression equation for predicting that person's achievement score.

(b) Using a student's achievement score as the independent variable, find the estimated regression equation for predicting that person's GPA.

(c) Plot the equations from (a) and (b) on a graph. Should the results be collinear? Explain.

10-6 A stress-strain relationship is to be established from the following sample data collected for steel rods of a particular composition:

Load P (pounds)	Diameter D (inches)	Elongation δ (inches)	Length l (inches)
5,000	.50	.06	22
1,000	.25	.07	31
1,500	.25	.13	32
2,000	.25	.11	30
10,000	.50	.12	24
15,000	.50	.14	18
10,000	1.00	.02	15
20,000	1.00	.05	15
30,000	1.00	.07	12
12,000	.50	.13	24
4,000	.25	.20	36
3,000	.25	.25	36
14,000	.50	.17	24
16,000	.50	.18	25
25,000	1.00	.06	13

(a) Compute the stress (σ) and strain (ε) for each sample point using the following relationships:

$$\sigma = \frac{P}{A}\text{(psi)} \qquad \varepsilon = \frac{\delta}{1}$$

where A is the *area* of the right section. Then plot the resulting points on a scatter diagram with σ as the ordinate and ε the abscissa.

(b) Treating stress as the dependent variable and strain as the independent variable, find the estimated regression line and plot it on your graph.

(c) The **modulus of elasticity** E may be estimated by the slope of the regression line. What is the point estimate for this quantity?

10-7 The following viscosity data were obtained from an experiment involving a toluene and tetralin mixture:

Average Temperature (K)	Viscosity (mPa · s)	Average Temperature (K)	Viscosity (mPa · s)
298.2	.7036	368.2	.3383
308.2	.6177	308.2	.6155
318.2	.5530	328.2	.4902
328.2	.4909	348.2	.4031
338.2	.4404	358.2	.3676
348.2	.4004	368.2	.3369
358.2	.3655		

Source: Byers and Williams, *Journal of Chem. Eng. Data*, Vol. 32, 1987, pp. 349–354.

(a) Using the reciprocal of temperature as the independent variable X, find the equation for the estimated regression line for predicting the natural logarithm Y of viscosity.

(b) Using the fact that viscosity $= e^Y$, find the estimated viscosity when the temperature is (1) 300 K, (2) 320 K, and (3) 350 K.

10-8 The following data were obtained from an experiment involving measurements of infrared energy levels for a chemical substance at various temperatures.

Temperature (K)	Temperature Reciprocal X (1/K)	Libration Bandwidth Y(cm $^{-1}$)
17.9	.056	43
16.9	.059	43
16.1	.062	34
14.9	.067	35
14.5	.069	38
14.1	.071	31
13.0	.077	29

Source: Baciocco et al., *Journal of Chem. Physics*, Vol. 87, 15 August 1987, pp. 1913–1916.

(a) Find the estimated regression equation using the temperature reciprocal as the independent variable and libration bandwidth as the dependent variable.
(b) Determine the estimated libration bandwidth when the temperature is (1) 14 K, (2) 15 K, and (3) 16 K.

10-2 Assumptions and Properties of Linear Regression Analysis

It is very important that investigators be aware of the assumptions and properties of linear regression analysis. These provide the basis for extending the concepts and procedures of statistical inference to making predictions about the dependent variable and drawing conclusions regarding the nature of the regression line itself.

Assumptions of Linear Regression Analysis

To help explain the assumptions of linear regression, we will expand our video-display-terminal illustration. Consider each possible required terminal time Y for *all* transactions that are to be completed by materials-handling clerks for a specified number of inputs X. For any fixed X the values of Y constitute a population having some mean, which we denote by $\mu_{Y \cdot X}$. A separate population applies to each level for X.

Figure 10-5 shows how several Y populations may be characterized in the regression context. Notice that there are three dimensions. In addition to the axes

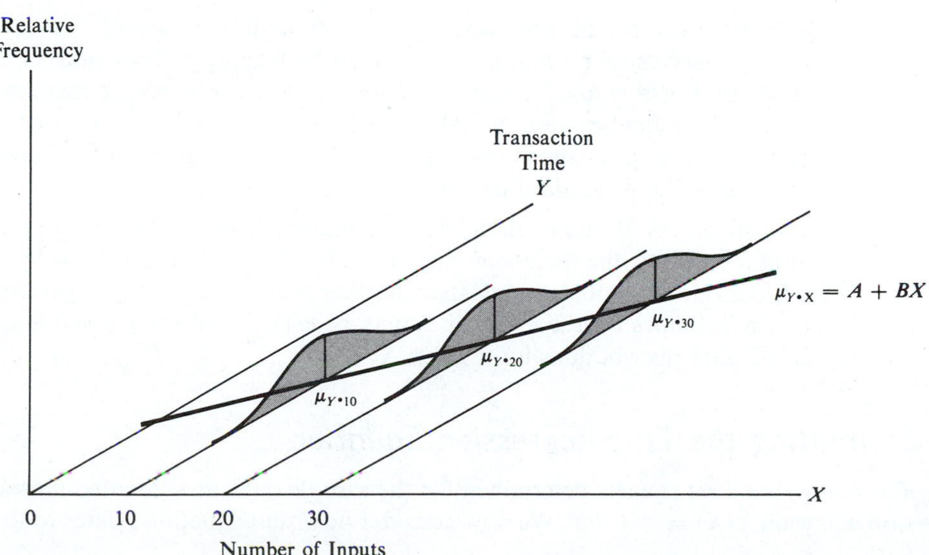

Figure 10-5 *Conditional population distributions with conditional means on the true regression line.*

for the independent and dependent variables, we include a third vertical axis for relative frequency. Each population of Y's may be represented by a frequency curve having center at some specified distance $\mu_{Y \cdot X}$ from the X-axis. We therefore refer to $\mu_{Y \cdot X}$ as the **conditional mean** of Y given X. There will be a different frequency curve for each X. Illustrated are the populations when the number of inputs is $X = 10$, $X = 20$, and $X = 30$. The respective conditional means are denoted by $\mu_{Y \cdot 10}$, $\mu_{Y \cdot 20}$, and $\mu_{Y \cdot 30}$. We can make the following four assumptions about the populations for Y:

1. The conditional means $\mu_{Y \cdot X}$ all lie on the same straight line, given by the equation

   $$\mu_{Y \cdot X} = A + BX$$

   which is the expression for the **true regression line**. This result follows from an assumption about individual observations. For the ith sample pair the level of the dependent variable may be expressed in terms of these parameters by

   $$Y_i = A + BX_i + \epsilon_i$$

   where ϵ_i is a **random error term**. For all i, the error term is assumed to be a random variable having expected value 0 and constant variance. It follows that

   $$E(Y_i) = A + BX_i$$

   so that for any given level X the corresponding observed Y has expected value equal to $\mu_{Y \cdot X}$.
2. All populations have the same standard deviation, denoted by $\sigma_{Y \cdot X}$, no matter what the value of X is. This follows from the assumption that $\text{Var}(\epsilon_i)$ is the same fixed value for all observations. Thus, treating $A + BX_i$ as a constant,

   $$\text{Var}(Y_i) = \text{Var}(A + BX_i + \epsilon_i) = \text{Var}(\epsilon_i)$$
3. Successive sample observations are independent. This means that the value of successive Y's will not be affected by whatever other values are obtained. For this to be the case, the successive error terms are assumed to be independent random variables, and $\text{Cov}(\epsilon_i, \epsilon_j) = 0$ for all possible pairs. It is further assumed that the error terms are normally distributed with mean 0 and standard deviation $\sigma_{Y \cdot X}$.
4. The value of X is known in advance. Although X might itself be an uncertain quantity, the regression model treats this independent variable as a fixed value. The only thing that matters is the uncertainty regarding Y. Probabilities for the levels of Y may be established using *conditional* probability distributions for a given X.

Estimating the True Regression Equation

The method of least squares determines, for the sample data, an estimated regression equation $\hat{Y}(X) = a + bX$. We now consider how that equation relates to the true regression equation $\mu_{Y \cdot X} = A + BX$. The **true regression coefficients** A and B are usually unknown and are ordinarily estimated from the sample counterparts, a and b.

Of course the computed values for a and b will depend on the particular sample results obtained. The equation $\hat{Y}(X) = 30.04 + 2.854X$, relating the terminal time for completing a transaction having X inputs, was the result of the particular 25 sample transactions actually observed. The regression equation computed for a different set of 25 transactions would probably differ considerably from the one we found.

The estimated regression coefficients a and b found by the method of least squares are *unbiased estimators* of the true coefficients A and B. Thus, if the same regression procedure were applied to a series of random samples, then the average of the a's would tend toward A and the b's would cluster about B. As we have seen, unbiasedness is a very desirable property for an estimator. Another feature of estimators derived from least squares is that they are the *most efficient* of all possible unbiased estimators of the linear regression coefficients. Thus, a and b minimize the chance sampling error involved in establishing a relationship between X and Y. This makes predictions determined from $\hat{Y}(X) = a + bX$ the most reliable ones obtainable for a fixed sample size. The a and b computed by the least-squares procedure are also *consistent* estimators of A and B; this means that, on the average, they become progressively closer to their targets as the sample size is increased. (The standard error for both a and b decreases as n increases.)

These properties place the least-squares estimators in a special class. When the Y's are normally distributed, a and b are *maximum-likelihood estimators.*

10-3 Statistical Inferences Using the Regression Line

Inferences in regression analysis take two forms. One category involves predictions of the dependent variable. The second area concerns inferences regarding the regression coefficients A and B.

Prediction and Confidence Intervals in Regression Analysis

The main purpose of regression analysis is predicting the level of the dependent variable. Usually such a prediction will involve finding the level for the conditional mean $\mu_{Y \cdot X}$. For this purpose, the estimated regression line provides the point estimate $\hat{Y}(X)$. Continuing with our terminal transaction illustration, we may predict the mean terminal time required for all transactions involving 25 inputs. In that case $X = 25$, and the best point estimate of $\mu_{Y \cdot X}$ will be the fitted Y value from the regression line. We have:

$$\begin{aligned}\hat{Y}(25) &= a + b(25)\\ &= 30.04 + 2.854(25)\\ &= 101.39\end{aligned}$$

This value can also be used to estimate the terminal time for a particular transaction. To distinguish a mean value from an *individual value*, either of which is estimated for a given X by $\hat{Y}(X)$, we will use the special symbol Y_I.

As we have seen with estimates in general, the point estimate is deficient in acknowledging the presence of imprecision and uncertainty due to sampling error. In making predictions from the regression line, it is therefore desirable to extend the concept of the interval estimate. For this purpose, estimates of individual values of the dependent variable are usually expressed in the form of **prediction intervals**.

Confidence Intervals for the Conditional Mean

We establish prediction intervals just as we did the confidence intervals encountered earlier. For the mean these take the form

$$\mu = \bar{X} \pm t_{\alpha/2} \frac{s}{\sqrt{n}}$$

where $s/\sqrt{n}$ is the estimator of $\sigma_{\bar{X}}$.

In like fashion, a prediction interval for the conditional mean for Y may be established. We use

$$\mu_{Y \cdot X} = \hat{Y}(X) \pm t_{\alpha/2} \text{ estimated } \sigma_{\hat{Y}(X)}$$

The standard error of $\hat{Y}(X)$ is denoted by $\sigma_{\hat{Y}(X)}$ and represents the amount of variability in possible Y's for the specified X. In the terminal transaction illustration, a somewhat different regression line from the one found earlier, such as $\hat{Y}(X) = 28 + 3.1X$ or $\hat{Y}(X) = 32 + 2.7X$, might be obtained after applying the least-squares procedure to the data obtained from another random sample collected in a similar fashion. Thus, a value for $\hat{Y}(25)$ that is slightly different from 101.39 might be computed from the regression line from another sample. For each setting of X, the potential set of $\hat{Y}(X)$ values will have a standard deviation of $\sigma_{\hat{Y}(X)}$.

Variability in $\hat{Y}(X)$ has two components. These are additive, so we have

$$\sigma^2_{\hat{Y}(X)} = \begin{matrix}\text{Variability in the}\\ \text{mean of } Y\text{'s}\end{matrix} + \begin{matrix}\text{Variability caused by the}\\ \text{distance of } X \text{ from } \bar{X}\end{matrix}$$

The first source of variability is analogous to that for the sample mean, which, as we have seen, depends on population standard deviation and the sample size. A second source is the distance separating X from $\bar{X}$. In Figure 10-6 the estimated regression lines have been plotted for different samples taken from the same transaction population. Notice that although the X's are of the same magnitude in each case and $\bar{X}$ is identical for each sample, the levels of a, b, and $\bar{Y}$ differ some-

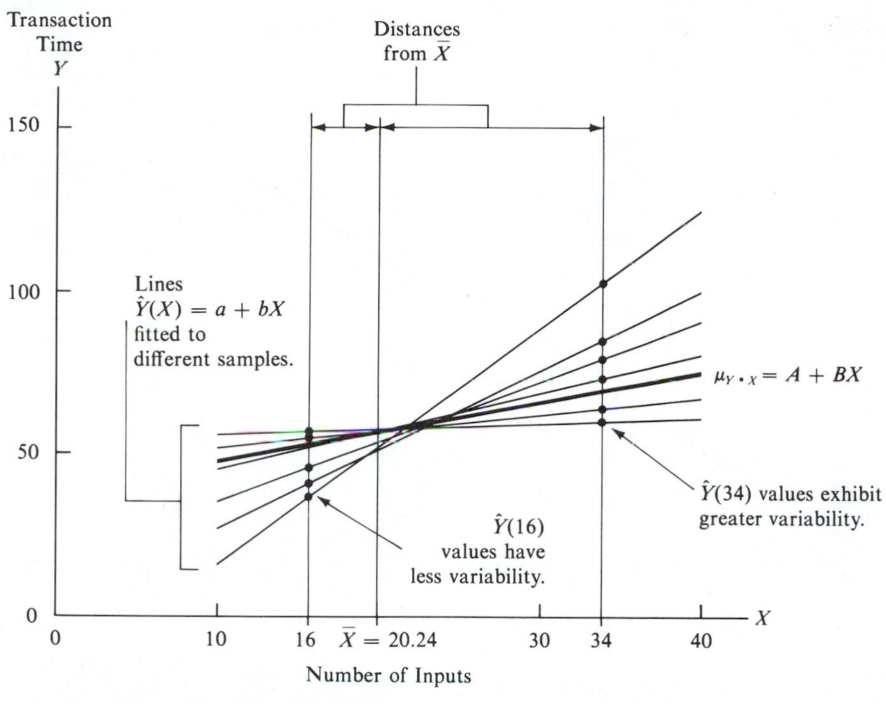

Figure 10-6 *Illustration of variability in predicted values for Y that is due to the particular estimated regression line obtained from the sample. Notice that this variability increases with the distance from the mean for the independent variable.*

what from sample to sample. The lines tend to diverge, with the total separation increasing as X becomes farther from $\bar{X}$. Thus, the values for $\hat{Y}(X)$ become more varied the farther X is from $\bar{X}$.

Although we won't derive it here, it may be established that for small sample sizes (generally those involving $n < 30$), the following expression provides the $100(1 - \alpha)\%$ confidence interval.

CONFIDENCE INTERVAL FOR THE CONDITIONAL MEAN USING SMALL SAMPLES

$$\boldsymbol{\mu_{Y \cdot X} = \hat{Y}(X) \pm t_{\alpha/2} s_{Y \cdot X} \sqrt{\frac{1}{n} + \frac{(X - \bar{X})^2}{\sum X^2 - (1/n)(\sum X)^2}}}$$

The $1/n$ under the radical provides the first component of variability. The term involving $(X - \bar{X})^2$ establishes the magnitude of the second source of variability. Appendix Table G provides $t_{\alpha/2}$ for $n - 2$ degrees of freedom. When transactions involve $X = 25$ inputs, the mean terminal time $\mu_{Y \cdot X}$ may be predicted with 95% confidence using the above expression. For $25 - 2 = 23$ degrees of freedom, using $\alpha/2 = .025$, we have $t_{.025} = 2.069$, and we obtain

$$\begin{aligned} \mu_{Y \cdot 25} &= \hat{Y}(25) \pm t_{.025} s_{Y \cdot X} \sqrt{\frac{1}{n} + \frac{(25 - \bar{X})^2}{\sum X^2 - (1/n)(\sum X)^2}} \\ &= 101.39 \pm 2.069(13.49) \sqrt{\frac{1}{25} + \frac{(25 - 20.24)^2}{13{,}130 - (\frac{1}{25})(506)^2}} \\ &= 101.39 \pm 6.10 \end{aligned}$$

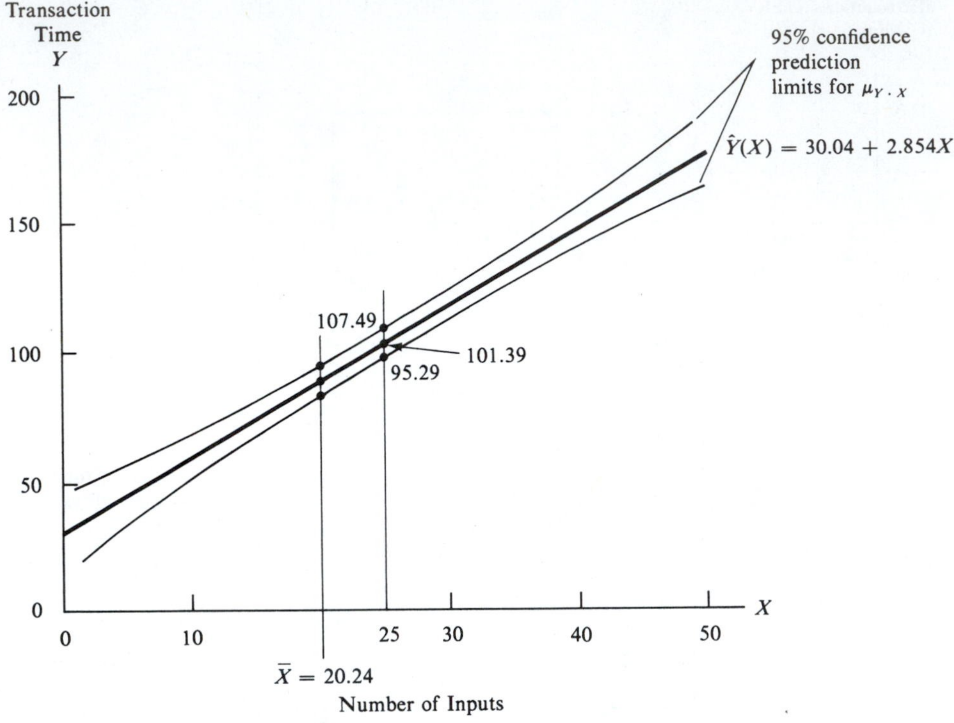

Figure 10-7 *Confidence limits for predictions of the mean terminal transaction time.*

or

$$95.29 \leq \mu_{Y \cdot 25} \leq 107.49$$

From this, we may conclude that the mean terminal time to complete a transaction involving 25 inputs is between 95.29 and 107.49 seconds. We are 95% confident this is so. (Which really means that if the regression procedure were repeated several times, and in each case $\hat{Y}(X)$ fitted to the data and an interval constructed similar to the above, then 95% of such intervals would contain the true value of $\mu_{Y \cdot X}$.)

Analogous confidence intervals may be computed for other numbers X of transaction inputs. Confidence limits may thereby be obtained over the entire range of X. This has been done for the terminal transaction data. Figure 10-7 shows the resulting graphical display. Notice that the width of the confidence *band* increases with the distance X falls farther above or below the mean.

Prediction Intervals for an Individual Y Given X

Predicting an individual value of Y given X is similar to estimating the conditional mean. Consider a prediction of the next 25-input transaction. The same point estimator applies as before, 101.39 seconds. As with the conditional mean, we account for sampling error by reporting our prediction in terms of a confidence interval. The following expression provides the $100(1 - \alpha)\%$ prediction interval.

PREDICTION INTERVAL FOR AN INDIVIDUAL Y USING SMALL SAMPLES

$$Y_I(X) = \hat{Y}(X) \pm t_{\alpha/2} s_{Y \cdot X} \sqrt{\frac{1}{n} + \frac{(X - \bar{X})^2}{\sum X^2 - (1/n)(\sum X)^2} + 1}$$

This expression is identical to the counterpart for $\mu_{Y \cdot X}$, except for the addition of $+1$ under the radical. That term reflects a third source of variability not encountered earlier. The component reflects the fact that *even if we knew the true regression line, individual Y's will still exhibit variability about their conditional mean.* Thus, given X, individual Y's will still have standard deviation $\sigma_{Y \cdot X}$, which is estimated by $s_{Y \cdot X}$.

We construct a 95% prediction interval for $Y_I(25)$, again using the terminal transaction data. Substituting the appropriate values, we have

$$\begin{aligned} Y_I(25) &= \hat{Y}(25) \pm t_{.025} s_{Y \cdot X} \sqrt{\frac{1}{n} + \frac{(25 - \bar{X})^2}{\sum X^2 - (1/n)(\sum X)^2} + 1} \\ &= 101.39 \pm 2.069(13.49) \sqrt{\frac{1}{25} + \frac{(25 - 20.24)^2}{13{,}130 - (\frac{1}{25})(506)^2} + 1} \\ &= 101.39 \pm 28.57 \end{aligned}$$

or

$$72.82 \leq Y_I(25) \leq 129.96$$

This interval is considerably wider than the one computed for the conditional mean. This should be no surprise, since it reflects the individual variability in Y's from transaction to transaction. This prediction interval expresses the single terminal time for a *particular transaction* involving 25 inputs, rather than the mean of them all. Its added width would be present even if the true regression line were known.

Dangers of Extrapolation

In making predictions from the regression line it is important that investigators be cognizant of the dangers of extrapolation. This occurs whenever a value of Y is found for a level of X falling outside the observed range for that independent variable. In the terminal transaction illustration, no sample observation involved more than 35 required inputs. Suppose that a forecast is desired for the mean terminal time for very long transactions involving $X = 100$ inputs. Would the estimated regression line established earlier be suitable?

There is no sure answer to this question. It could be argued that the data clerks might process such a transaction considerably faster, on the average, than our regression line would predict. This conclusion could be justified because the repetitive nature of such long input runs permits an "economy of scale," reducing overall time requirements. But the reverse might just as well be argued: Longer transactions will be more complex, and the data clerks will become so bogged down in detail that each transaction will take longer than extrapolating the line would indicate; also, long input strings will be monotonous and therefore slower. Of course either conclusion is pure conjecture—and that is precisely the point! The only way to know the response of Y to an unobserved X is to collect sample data that correspond. Extrapolation assigns values using a relationship that has been measured for different circumstances, so that any conclusions thereby reached are totally without supporting evidence.

Inferences Regarding the Regression Coefficients

The second class of inferences encountered in regression analysis involves the nature of the regression line. These generalizations apply to either the Y-intercept A or the slope B. The latter is often the greater concern in engineering, since it indicates the response Y to a one-unit change in X. Conclusions about A are usually of secondary importance. For example, a metallurgist might use regression analysis to develop a mathematical relationship between alloy concentrations and strength properties. The investigator's major concern may be determining the extra shearing force needed to permanently distort a rod for a unit change in alloy material, rather than predicting a particular force. Thus, B may be more important than $\mu_{Y \cdot X}$.

The Slope The level of B may be estimated from its sample counterpart, b. The standard error of b may be obtained from

$$s_b = \frac{s_{Y \cdot X}}{\sqrt{\sum X^2 - (1/n)(\sum X)^2}}$$

Using this expression, procedures have been established for making confidence interval estimates of B and testing its value.

The following expression is used to construct a $100(1 - \alpha)\%$ confidence interval.

CONFIDENCE INTERVAL ESTIMATE
OF TRUE SLOPE B

$$\boldsymbol{B = b \pm t_{\alpha/2} \frac{s_{Y \cdot X}}{\sqrt{\sum X^2 - (1/n)(\sum X)^2}}}$$

where b is the computed slope of the estimated regression line and $t_{\alpha/2}$ involves $n - 2$ degrees of freedom.

Again referring to the terminal transaction data, we may construct a 95% confidence interval estimate of the slope B of the true regression line; this quantity represents the increase in terminal time per additional input. Using $\alpha/2 = .025$ and $25 - 2 = 23$ degrees of freedom, Appendix Table G provides $t_{.025} = 2.069$. We have

$$B = 2.854 \pm 2.069\left(\frac{13.49}{\sqrt{13{,}130 - (\frac{1}{25})(506)^2}}\right)$$
$$= 2.854 \pm .519$$

or

$$2.335 \le B \le 3.373$$

We therefore estimate with 95% confidence that there will be between 2.335 and 3.373 seconds of extra terminal time for each additional data input.

It is also possible to test hypotheses regarding B. Ordinarily, such a test is used to establish whether *any* significant relationship exists between X and Y. A testing procedure will find whether or not the two variables are correlated, and if so, whether the slope of the true regression line is positive or negative. Such a test is useful when the only concern is the existence of a relationship or its direction.

To test the null hypothesis that $B = 0$, a two-sided procedure applies, and it is necessary only to construct the confidence interval estimate that matches the desired significance level. If the interval spans zero, H_0 must be accepted, and it must be rejected otherwise. A one-sided test will involve the following.

TEST STATISTIC FOR SLOPE B
OF TRUE REGRESSION LINE

$$t = \frac{b}{\dfrac{s_{Y \cdot X}}{\sqrt{\sum X^2 - (1/n)(\sum X)^2}}}$$

The Intercept Inferences regarding the intercept are not as common as for the slope. The standard error of a is

$$s_a = s_{Y \cdot X}\sqrt{\frac{1}{n} + \frac{\bar{X}^2}{\sum (X - \bar{X})^2}}$$

From this, we get the following expression.

CONFIDENCE INTERVAL ESTIMATE
OF THE TRUE INTERCEPT

$$A = a \pm t_{\alpha/2} s_{Y \cdot X}\sqrt{\frac{1}{n} + \frac{\bar{X}^2}{(X - \bar{X})^2}}$$

where $t_{\alpha/2}$ corresponds to $n - 2$ degrees of freedom.

Continuing with the terminal transaction data, we find the 95% confidence interval estimate for the intercept of the true regression line. We first compute

$$\sum (X - \bar{X})^2 = \sum X^2 - n\bar{X}^2 = 13{,}130 - 25(20.24)^2 = 2{,}888.56$$

Then, for $25 - 2 = 23$ degrees of freedom, Appendix Table G provides $t_{.025} = 2.069$, and we compute:

$$A = 30.04 \pm 2.069(13.49)\sqrt{\frac{1}{25} + \frac{(20.24)^2}{2{,}888.56}} = 30.04 \pm 11.90$$

or

$$18.14 \le A \le 41.94$$

Problems

10-9 Refer to the data in Problem 10-1 and your answers to that exercise.

(a) Construct a 95% confidence interval estimate of the conditional mean computer processing time when the number of given data points is 250.

(b) Construct a 95% prediction interval estimate of the computer processing time for an individual job having 320 data points.

10-10 Refer to the data in Problem 10-2 and your answers to that exercise.

(a) Construct a 99% confidence interval estimate of the conditional mean final product weight when the raw-materials volume is 20 gallons.

(b) Construct a 99% prediction interval estimate of the final product weight in an individual batch having 25 gallons of raw material.

10-11 Refer to Problem 10-1 and your answers to that exercise.

(a) Construct a 90% confidence interval estimate of the slope of the true regression line.

(b) At the 10% significance level, must you accept or reject the null hypothesis that processing time and the number of data points are uncorrelated variables?

10-12 Refer to Problem 10-2 and your answers to that exercise.

(a) Construct a 95% confidence interval estimate of the slope of the true regression line.

(b) At the 5% significance level, must you accept or reject the null hypothesis that final product weight and raw-materials volume are uncorrelated variables?

10-13 Using your intermediate calculations and answers to Problem 10-3, construct a 95% confidence interval estimate for Poisson's ratio for the experimental alloy represented in that exercise.

10-14 Using your intermediate calculations and answers to Problem 10-4, answer the following:

(a) Construct a 95% confidence interval estimate for the mean cost of batches involving 1,000 items.

(b) Construct a 95% confidence interval estimate for the variable cost of items produced.

10-15 Refer to the student data in Problem 10-5.

(a) Construct a 95% confidence interval estimate of the "unconditional" mean examination score for the students in the sampled population.

(b) Using the sample mean GPA as the given level for X, construct a 95% confidence interval estimate for the conditional mean for examination score (Y).

10-16 Refer to Problem 10-6 and your intermediate calculations and answers to that exercise. Construct a 95% confidence interval estimate of the modulus of elasticity for the steel represented by the test data.

10-17 Researchers performed an experiment to evaluate an electromagnetic bearing for the vibrational control mechanism for supercritical shafts. The following data were obtained in tests where the shaft was displaced:

Programmed Damping Rate X	Damping Coefficient Y
0	0.013
199	0.123
272	0.178
397	0.279
496	0.357
595	0.407
744	0.501

Source: Bradfield et al., *Proceedings of the Institution of Mechanical Engineers*, Vol. 201, No. 3, March 1987, pp. 201–212.

(a) Find the estimated regression equation.
(b) For a programmed damping rate of 400, construct a 95% confidence interval estimate of the *mean* level of the damping coefficient.
(c) Construct a 95% prediction interval estimate for the level of the damping coefficient for a particular *individual* trial displacement when the programmed damping rate is 500.

10-4 Correlation Analysis

Regression analysis is largely concerned with predicting the level of the dependent variable Y for a known level of the independent variable X. It therefore focuses on finding an appropriate function expressing the relationship between X and Y. Correlation analysis is more concerned with the nature of the relationship between the two variables, with a central focus on the *strength* of that relationship.

Recall that the regression model assumes that X is fixed at some value, so that Y is the only random variable explicitly considered. In the more general case, X may be an uncertain quantity as well. Correlation analysis does not single out one variable to be the dependent one. Both X and Y are treated equally and are considered to be random variables having a joint probability distribution of an assumed form.

Correlation analysis gives a valuable perspective of bivariate relationships, and thus explains many of the concepts that underlie the regression model, making correlation analysis a helpful complement to regression analysis.

The Bivariate Normal Distribution

The most useful characterization of the X-Y relationship is provided by the **bivariate normal distribution**, which was introduced in Chapter 5. It will be helpful

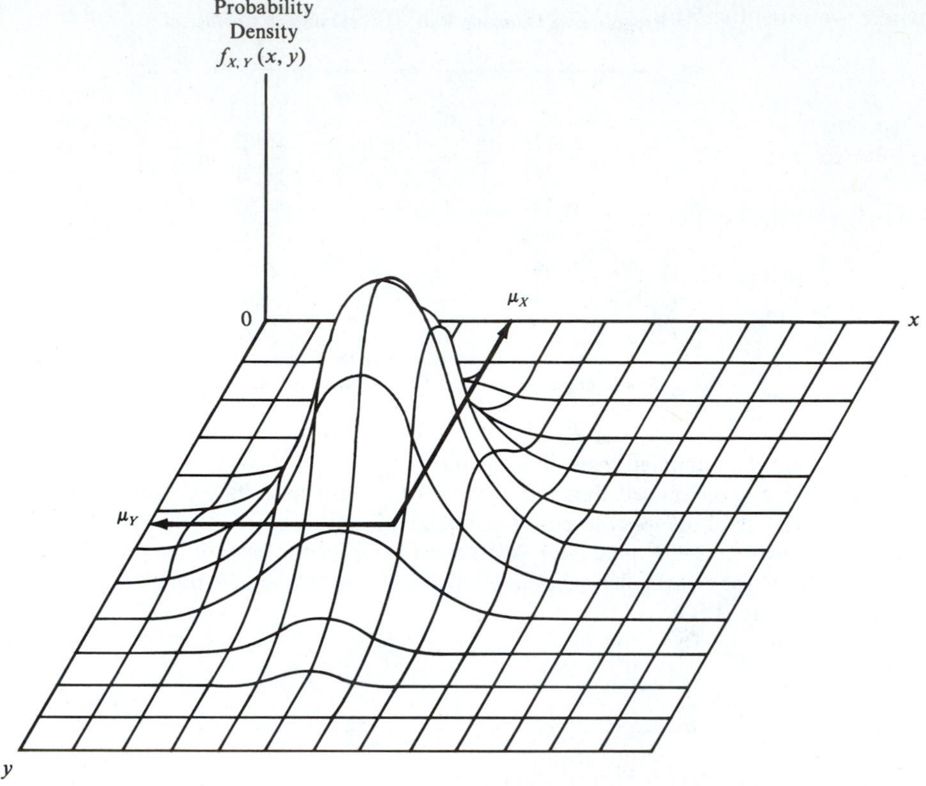

Figure 10-8 Probability density function for bivariate normal distribution.

to look again at the density function for this distribution,

$$f_{X,Y}(x, y) = \frac{1}{2\pi\sigma_X\sigma_Y\sqrt{1-\rho^2}}$$
$$\times \exp\left[-\frac{1}{2(1-\rho^2)}\left(\frac{(x-\mu_X)^2}{\sigma_X^2} - 2\rho\frac{(x-\mu_X)(y-\mu_Y)}{\sigma_X\sigma_Y} + \frac{(y-\mu_Y)^2}{\sigma_Y^2}\right)\right]$$

This function expresses the theoretical population frequency surface illustrated in Figure 10-8. The marginal probability distributions of X and Y and the populations these variables represent are each themselves normally distributed, with respective parameters μ_X, σ_X and μ_Y, σ_Y. The joint density has an additional parameter ρ, which is the **population correlation coefficient**. In Chapter 5 we saw that this quantity is related to the covariance, so that we may use the following.

POPULATION CORRELATION COEFFICIENT

$$\rho = \frac{\sigma_{XY}}{\sigma_X\sigma_Y}$$

where we introduce the double-subscripted sigma to denote the covariance:

$$\sigma_{XY} = \text{Cov}(X, Y) = E(XY) - E(X)E(Y)$$

The values of the population parameters are ordinarily unknown and must be estimated from their sample counterparts.

The bivariate normal density geometry is useful for explaining underlying relationships. Imagine a horizontal slice through the density surface. When this is projected onto the X-Y plane, the resulting contour will be an ellipse. The center of the projected ellipse will fall at (μ_X, μ_Y) and its major axis will indicate the direction of relationship between X and Y. That axis will have positive slope for $\rho > 0$ and negative slope when $\rho < 0$. The shape of the ellipse (and the density surface itself) is determined by σ_X, σ_Y, and ρ.

In Chapter 5 we saw that the density function for the conditional probability distribution of one random variable may be found by dividing the joint density by the marginal density for the given variable. Thus, we obtain

$$f_{Y|X}(y|x) = \frac{f_{X,Y}(x, y)}{f_X(x)}$$

When the bivariate normal distribution applies, for a given level X, the above density is also normally distributed with conditional mean and variance:

$$\mu_{Y \cdot X} = A + BX$$

$$\sigma^2_{Y \cdot X} = \sigma^2_Y(1 - \rho^2)$$

A and B are the true regression parameters and can be expressed in terms of the bivariate parameters by

$$A = \mu_Y - \mu_X \rho \frac{\sigma_Y}{\sigma_X}$$

$$B = \rho \frac{\sigma_Y}{\sigma_X}$$

Thus, as we saw earlier, the conditional means for Y fall on the true regression line, which in turn depends on the nature of the relationship between X and Y.

Sample Correlation Coefficient

The population correlation coefficient may be estimated from the sample results. It may be calculated in an analogous fashion from the sample counterparts to the population parameters. For this purpose we introduce the **sample covariance**,

$$s_{XY} = \frac{\sum(X - \bar{X})(Y - \bar{Y})}{n - 1} = \frac{\sum XY - \left(\frac{1}{n}\right)\sum X \sum Y}{n - 1}$$

which serves as the point estimator of the population covariance σ_{XY}. The following expression may be used.

SAMPLE CORRELATION COEFFICIENT

$$r = \frac{s_{XY}}{s_X s_Y} = \frac{\sum XY - \left(\frac{1}{n}\right)\sum X \sum Y}{\sqrt{\left[\sum X^2 - \left(\frac{1}{n}\right)(\sum X)^2\right]\left[\sum Y^2 - \left(\frac{1}{n}\right)(\sum Y)^2\right]}}$$

where, in keeping with the present notation, s_X denotes the sample standard deviation computed from the observed levels for X. The value of r computed from this is ordinarily the only known measure of correlation available to the investigator.

The correlation coefficient expresses the strength of the *linear* relationship between X and Y. Figure 10-9 illustrates several cases. The scatter diagrams in (a) and (b) portray sample data where all the observed points fall on a single line. These ideal situations indicate that X and Y are *perfectly correlated*, so the computed values for the sample correlation coefficient are precisely *one*, in absolute value. The slope of the lines indicates the direction of the relationship. In (a), where $r = +1$, X and Y are *positively correlated*. The data in (b) indicate that the variables are *negatively correlated*.

More typical sample results are shown on the scatter diagrams in (c) and (d), where the observed data points cluster about the respective fitting lines. Notice that the data cluster more tightly about the line in (c), where $r = +.97$, than they do in (d), where $r = -.61$. The variables are more highly correlated in (c), indicating a stronger relationship between X and Y. The closer r is in absolute value to 1, the greater the degree of correlation, and the data will be less scattered about the fitting line. The diagrams in (e) and (f) exhibit *zero correlation* between X and Y. In both cases, $r = 0$, and a horizontal line provides the best fit to the data. Although the data in diagram (f) indicate a pronounced nonlinear relationship, that is not reflected in r, which expresses the strength and direction of *linear* relationships only.

As an illustration, the sample correlation coefficient is calculated using the terminal transaction data from Table 10-1.

$$r = \frac{52{,}670 - (\frac{1}{25})(506)(2{,}195)}{\sqrt{[13{,}130 - (\frac{1}{25})(506)^2][220{,}445 - (\frac{1}{25})(2{,}195)^2]}} = .92$$

Not surprisingly, this indicates a strong positive correlation between a transaction's number of inputs X and the required terminal time Y.

Inferences Regarding the Correlation Coefficient

Of course, the population correlation coefficient ρ is ordinarily unknown. Inferences may be made regarding this quantity. The same types of generalizations may be made with ρ that we have encountered for other population parameters.

The usual hypothesis test of primary interest is one that establishes whether or not any correlation exists at all between X and Y. This involves the following:

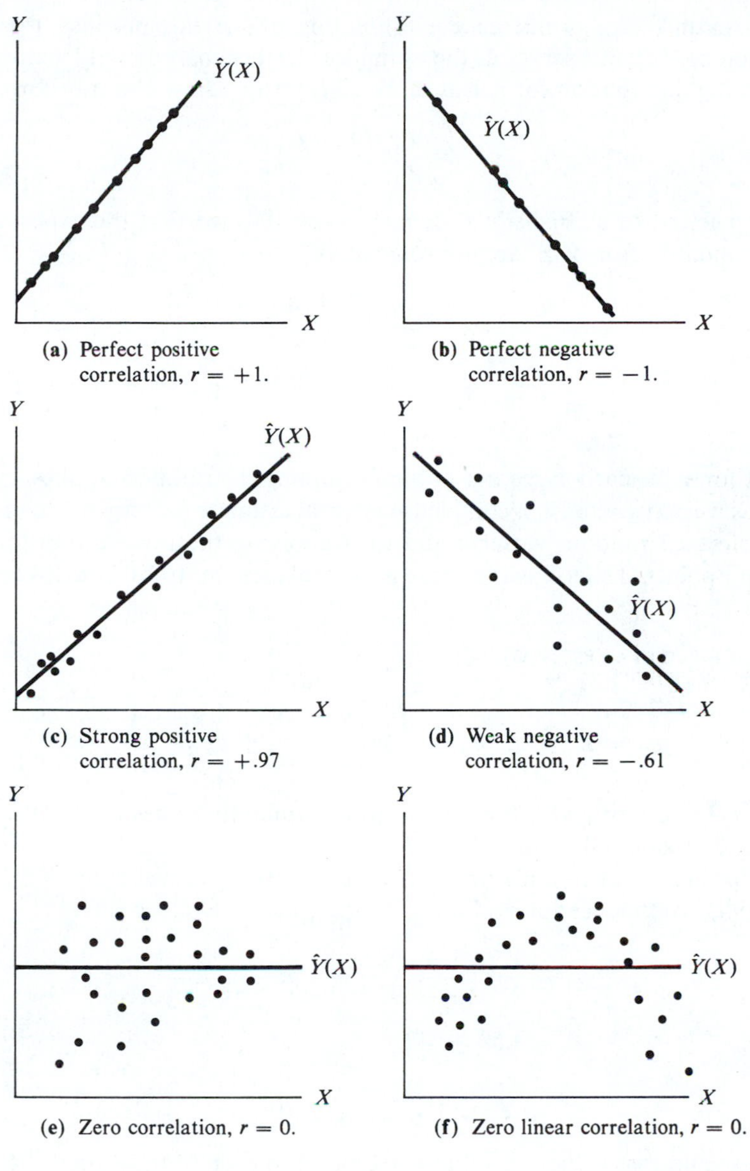

Figure 10-9 *Scatter diagrams for various degrees of correlation.*

$$H_0: \rho = 0$$

$$H_1: \rho \neq 0$$

Testing this is identical to testing the following hypotheses regarding the slope B of the true regression line:

$$H_0: B = 0$$

$$H_1: B \neq 0$$

The procedure for the latter test was described in Section 10-3.

The second form of inference is estimating ρ. For this purpose, the sample correlation coefficient r serves as the estimator. Rather than directly manipulating the sampling distribution for r, it is more convenient to use the transformation

$$Z' = \frac{1}{2}\ln\left(\frac{1+r}{1-r}\right)$$

which is referred to as **Fisher's Z**. It may be established that the expected value of and standard error of Z' are approximately

$$\mu_{Z'} = \frac{1}{2}\ln\left(\frac{1+\rho}{1-\rho}\right)$$

$$\sigma_{Z'} = \frac{1}{\sqrt{n-3}}$$

and that for sufficiently large n a normal sampling distribution applies.

Thus we may establish a confidence interval estimate for ρ by first considering the transformed random variable and then applying the inverse transformation to the end points. The following expression provides the $100(1-\alpha)\%$ confidence interval.

CONFIDENCE INTERVAL ESTIMATE FOR THE MEAN OF FISHER'S Z'

$$\mu_{Z'} = Z' \pm z_{\alpha/2}\sigma_{Z'} = \frac{1}{2}\ln\left(\frac{1+r}{1-r}\right) \pm z_{\alpha/2}\frac{1}{\sqrt{n-3}}$$

Appendix Table I provides the inverse transformation values for converting the estimate to an interval for ρ.

Returning to our terminal transaction illustration, we will use $r = .92$ to construct a 95% confidence interval for ρ. Using $z_{.025} = 1.96$, we have

$$\mu_{Z'} = \frac{1}{2}\ln\left(\frac{1+.92}{1-.92}\right) \pm 1.96\frac{1}{\sqrt{25-3}}$$

$$= 1.59 \pm .42$$

or

$$1.17 \le \mu_{Z'} \le 2.01$$

Reading the nearest values of ρ that correspond to each of these limits, we obtain the following 95% confidence interval estimate for ρ:

$$.82 \le \rho \le .96$$

Problems

10-18 Refer to the data in Problem 10-1 for the number of data points X and processing time Y. Calculate the sample correlation coefficient.

10-19 Refer to Problem 10-18 and your answer to that exercise:

(a) Construct a 95% confidence interval estimate for the population correlation coefficient.

(b) Does the above interval contain zero? At the 5% significance level should the null hypothesis that $\rho = 0$ be accepted or rejected?

10-20 Refer to the data in Problem 10-2 for the weight of the final product Y and volume of raw materials X. Calculate the sample correlation coefficient.

10-21 Refer to Problem 10-20 and your answer to that exercise:

(a) Construct a 99% confidence interval estimate for the population correlation coefficient.

(b) Does the above interval contain zero? At the 1% significance level should the null hypothesis that $\rho = 0$ be accepted or rejected?

10-22 The following data pertain to drying time and solvent content of sample batches of an experimental coating substance:

Solvent Content X	Drying Time Y	Solvent Content X	Drying Time Y
2.5%	2.3 hours	3.4%	1.9 hours
2.7	2.1	3.5	1.9
3.0	2.2	4.0	1.6
3.1	2.1	4.2	1.6
3.2	2.0	4.5	1.5

Calculate the sample correlation coefficient.

10-23 Refer to Problem 10-22 and your answer to that exercise:

(a) Construct a 90% confidence interval estimate for the population correlation coefficient.

(b) Does the above interval contain zero? At the 10% significance level should the null hypothesis that $\rho = 0$ be accepted or rejected?

10-24 The following data pertain to seasonal rainfall and augmented reservoir storage:

Seasonal Rainfall (inches) X	Reservoir Storage (acre-feet) Y	Seasonal Rainfall (inches) X	Reservoir Storage (acre-feet) Y
25.3	45,200	25.1	46,700
17.2	37,750	28.3	52,450
19.4	39,480	20.5	39,700
22.1	41,150	15.2	32,250
23.4	42,650	18.6	37,420

Calculate the sample correlation coefficient.

10-25 Refer to Problem 10-24 and your answer to that exercise:

(a) Construct a 95% confidence interval estimate for the population correlation coefficient.

(b) Does the above interval contain zero? At the 5% significance level should the null hypothesis that $\rho = 0$ be accepted or rejected?

10-26 Artificial intelligence researchers have studied algorithms for solving constraint-satisfaction problems. Using the same problem on each, they obtained the following approximate numbers of consistency checks:

RBT X	**ABT** Y
850	550
1,200	1,950
1,275	700
1,600	2,250
1,800	700
1,900	1,950
2,150	3,600
2,200	2,250
2,400	700
2,450	1,750
2,600	1,750
2,800	700
2,900	1,750
5,000	1,750
5,250	3,575

Source: Dichter and Pearl, *Artificial Intelligence*, Vol. 34, No. 1, December 1987, p. 31.

(a) Plot these data on a scatter diagram.

(b) The data suggest that there is little correlation between X and Y. Test the null hypothesis of zero correlation. What must you conclude at the $\alpha = .05$ significance level?

10-5 Assessing the Quality of the Regression

An important issue in regression analysis is assessing its overall quality. It is valuable to measure how well the regression explains the relationship between the dependent and independent variables. An index that quantifies the strength of that relationship is the *coefficient of determination.* Since the regression model rests on theoretical assumptions, an *analysis of residuals* may pinpoint potential troublesome areas that could invalidate certain conclusions.

The Coefficient of Determination

Although correlation analysis may be performed apart from regression analysis, it may be extended to strengthen regression evaluations. Recall that the estimated regression equation $\hat{Y}(X)$ is found by minimizing the sum of the squared deviations about the regression line. For any particular observation there are *three* deviations, as indicated by the terms in the following expression:

$$(Y - \bar{Y}) = (\hat{Y}(X) - \bar{Y}) + (Y - \hat{Y}(X))$$

The term on the left expresses the *total deviation* in the observed Y about the central value for the entire sample. Consider again the regression line in Figure 10-10 found earlier for the terminal transaction data. For the 22nd sample observation we have $(Y - \bar{Y}) = 63.2$. This quantity may be expressed as the sum of two component deviations. The first of these indicates how the Y predicted from regression deviates from the center of the sample. From Figure 10-10 we find $(\hat{Y}(X) - \bar{Y}) = 33.6$, a difference that is explained by the regression line. That term is referred to as the **explained deviation**. The second term indicates how much the individual Y deviates from the line itself and is referred to as the **unexplained deviation**. For the 22nd transaction this quantity is $(Y - \hat{Y}(X)) = 29.6$.

The unexplained deviations are denoted by e_i and each is referred to as a residual. These are defined by the difference

$$e_i = Y_i - \hat{Y}(X_i)$$

The e_i's may be thought of as the *observed* errors. Although similar, these differ from the *actual* errors ϵ_i of the theoretical model, the values of which remain unknown.

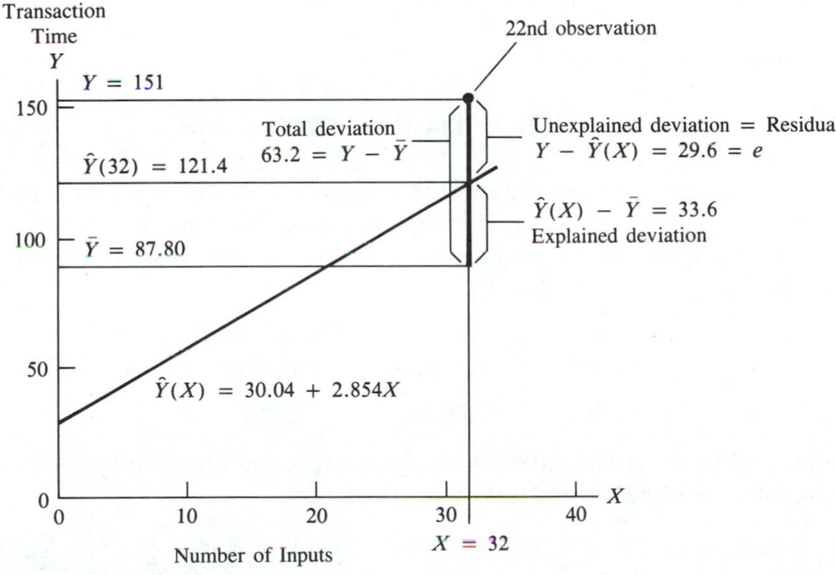

Figure 10-10 *Illustration of explained, unexplained, and total variation in regression analysis.*

Squaring these deviations and summing, we extend the deviation relationship to the entire collection of observations:

$$\begin{array}{ccccc} \text{Total variation} & = & \text{Explained variation} & + & \text{Unexplained variation} \\ \sum_{i=1}^{n} [Y_i - \bar{Y}]^2 & = & \sum_{i=1}^{n} [\hat{Y}(X_i) - \bar{Y}]^2 & + & \sum_{i=1}^{n} [Y_i - \hat{Y}(X_i)]^2 \\ SSTO & = & SSR & + & SSE \end{array}$$

The total variation expresses to what degree individual Y's deviate from their mean Y without regard to the regression relationship. This is summarized by the following.

TOTAL SUM OF SQUARES

$$\boldsymbol{SSTO = \sum_{i=1}^{n} [Y_i - \bar{Y}]^2 = \sum_{i=1}^{n} Y_i^2 - n\bar{Y}^2}$$

As Figure 10-10 shows, the distance between Y and the regression line at X_i explains a portion of the deviation in the observed Y_i from its mean Y. The explained variation summarizes the collective squared distances between the regression line $\hat{Y}(X)$ and the sample mean $\bar{Y}$. This is summarized by the following.

REGRESSION SUM OF SQUARES

$$\boldsymbol{SSR = \sum_{i=1}^{n} [\hat{Y}(X_i) - \bar{Y}]^2 = b\left[\sum_{i=1}^{n} (X_i - \bar{X})(Y_i - \bar{Y})\right]}$$

The final component in total variation involves the residuals and gives a measure of the overall observed error in the sample, the error sum of squares.

ERROR SUM OF SQUARES

$$\boldsymbol{SSE = \sum_{i=1}^{n} [Y_i - \hat{Y}(X_i)]^2 = SSTO - SSR}$$

The method of least squares chooses the regression coefficients a and b so that the above quantity is minimized. This term expresses the collective dispersion in Y about the regression line. The regression line leaves those deviations *unexplained*.

Rearranging terms, the identity

$$\begin{array}{ccccc} \text{Explained variation} & = & \text{Total variation} & - & \text{Unexplained variation} \\ SSR & = & SSTO & - & SSE \end{array}$$

is used to construct a useful index. Dividing the explained variation by total variation provides the **sample coefficient of determination**:

$$r^2 = \frac{\text{Explained variation}}{\text{Total variation}} = \frac{SSTO - SSE}{SSTO} = \frac{\sum [Y - \bar{Y}]^2 - \sum [Y - \hat{Y}(X)]^2}{\sum [Y - \bar{Y}]^2}$$

An equivalent expression is

$$r^2 = \frac{\text{Explained variation}}{\text{Total variation}} = 1 - \frac{\sum (Y - \hat{Y}(X))^2}{\sum (Y - \bar{Y})^2}$$

The sample coefficient of determination expresses the proportion of the total variation in Y explained by the regression line.

The population coefficient of determination is denoted by ρ^2, and r^2 serves as an estimator of that quantity. During the regression analysis, r^2 may be calculated from the sample data, and the following mathematically equivalent expression is ordinarily used.

SAMPLE COEFFICIENT OF DETERMINATION

$$\mathbf{r^2 = \frac{a \sum Y + b \sum XY - \frac{1}{n}(\sum Y)^2}{\sum Y^2 - \frac{1}{n}(\sum Y)^2}}$$

For the terminal transaction data, the sample coefficient of determination is

$$r^2 = \frac{30.04(2{,}195) + 2.854(52{,}670) - (\frac{1}{25})(2{,}195)^2}{220{,}445 - (\frac{1}{25})(2{,}195)^2} = .849$$

This indicates that 84.9% of the total variation in terminal time Y may be explained by the regression line relating this variable to the number of inputs X. This leaves 15.1% of the variation unexplained by regression.

It is sometimes convenient to express the sample coefficient of determination equivalently as

$$r^2 = 1 - \frac{s_{Y \cdot X}^2}{s_Y^2}\left(\frac{n-2}{n-1}\right)$$

For the terminal transaction data this gives

$$r^2 = 1 - \frac{(13.49)^2}{(33.99)^2}\left(\frac{25-2}{25-1}\right) = .849$$

Relation to Correlation Coefficient The choice of r^2 to denote the sample coefficient of determination is no coincidence. The following relationship applies:

$$\text{Coefficient of determination} = (\text{Correlation coefficient})^2$$

Thus, the absolute value of the correlation coefficient is the square root of the coefficient of determination:

$$|r| = \sqrt{r^2} \qquad \text{and} \qquad |\rho| = \sqrt{\rho^2}$$

The sign of the r or ρ will match the sign of the slope of the corresponding regression line. Thus, for the terminal transaction illustration, we see that the sample correlation coefficient is

$$r = \sqrt{.849} = +.92$$

the value found in Section 10-4.

This relationship provides another perspective on why the coefficient of determination gauges the strength of the X-Y relationship. In the special case when all the data points fall directly on the regression line, so there is a perfect corrrelation and $r = +1$ or $r = -1$, all of the variation in Y is explained and r^2 must then be equal to 1. At the other extreme, a horizontal regression line is obtained when X and Y exhibit zero correlation, so all of the variation in Y is unexplained and r^2 must then be equal to 0.

Appropriateness of Model: Analysis of Residuals

The residuals for the terminal transaction data are computed in Table 10-3. A beginning point in analyzing these is a residual scatter plot, as shown in Figure 10-11.

All inferences made from the regression relationship are based on the theoretical properties discussed in Section 10-2. If the underlying population and

Table 10-3
Residuals for Terminal Transaction Data

Observation i	Transaction Time Y_i	Number of Entries X_i	Predicted Time from Regression Line $\hat{Y}(X_i) = 30.04 + 2.854X_i$	Residual $e_i = Y_i - \hat{Y}(X_i)$
1	66	2	35.748	30.252
2	77	19	84.266	−7.266
3	37	6	47.164	−10.164
4	106	23	95.682	10.318
5	55	10	58.580	−3.580
6	89	23	95.682	−6.682
7	52	9	55.726	−3.726
8	128	30	115.660	12.340
9	63	18	81.412	−18.412
10	104	25	101.390	2.610
11	76	19	84.266	−8.266
12	44	2	35.748	8.252
13	97	27	107.098	−10.098
14	109	28	109.952	−0.952
15	40	8	52.872	−12.872
16	124	29	112.806	11.194
17	98	29	112.806	−14.806
18	63	16	75.704	−12.704
19	131	33	124.222	6.778
20	41	3	38.602	2.398
21	111	34	127.076	−16.076
22	151	32	121.368	29.632
23	76	13	67.142	8.858
24	114	33	124.222	−10.222
25	143	35	129.93	13.070

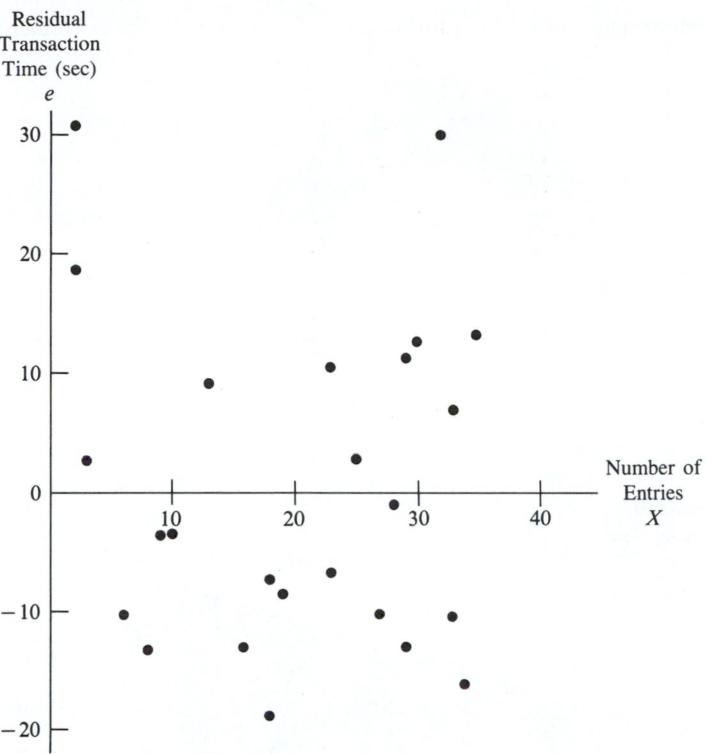

Figure 10-11
Residual scatter plot for terminal transactions.

sampling process deviate from the model described there, the regression analysis will be based on false assumptions. An analysis of the residuals can be helpful in uncovering model violations.

The terminal transaction data residuals reflect the underlying linear relationship found earlier, with the individual *e*'s falling within a narrow horizontal band. Such a plot has the ideal appearance for a regression analysis. Figure 10-12 shows three plots where key model assumptions are violated.

The scatter plot in Figure 10-12(a) illustrates a *nonlinear* relationship, with the *e*'s exhibiting a pronounced curve. As will be seen in Section 10-6, one way to rectify this would be to transform Y or X. In Chapter 11 we will see how the regression procedure may be extended for fitting a polynomial to the raw data.

Figure 10-12(b) shows a plot representing the regression analysis for a hypothetical experiment where the computer processing time Y for a network analysis is related to the number of nodes X. Notice that the band of *e*'s gets wider as X increases. This indicates a *nonconstant variance* in the *e*'s as the level for X becomes greater, and violates the theoretical assumption that errors have constant variance $\sigma^2_{Y \cdot X}$. That difficulty is sometimes rectified by transformation of the variables. For instance, when there is increasing variance in the *e*'s, a regression with $Y' = \sqrt{Y}$ might provide a constant variance.

Figure 10-12(c) shows a residual plot of a hypothetical regression where Y is the time needed to adjust temperature settings on batches run in a chemical process

Figure 10-12
Residual scatter plots reflecting violations of the underlying regression model.

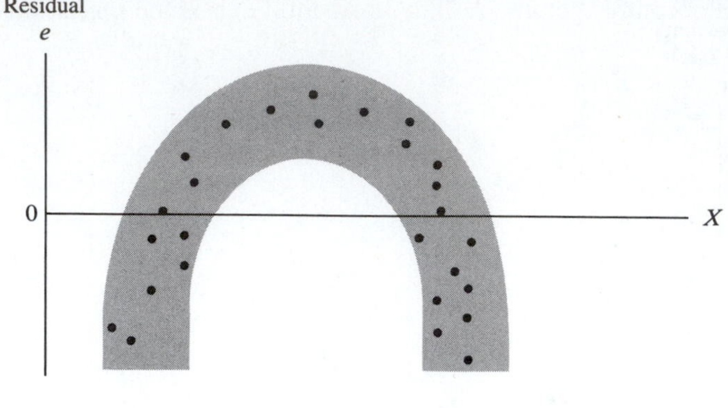

(**a**) Nonlinear Relation

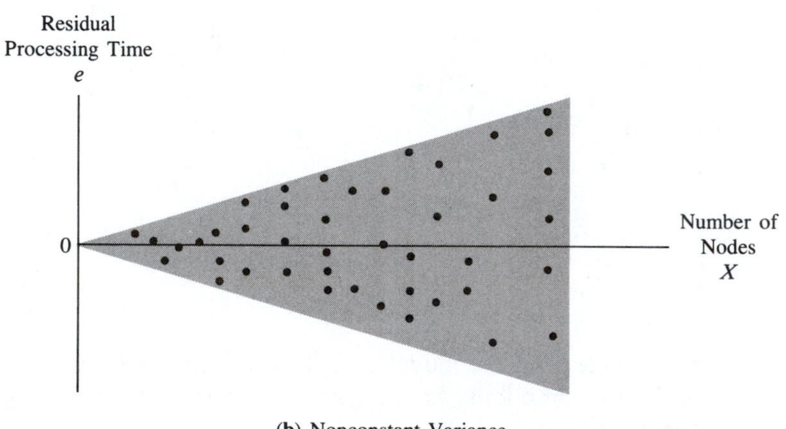

(**b**) Nonconstant Variance

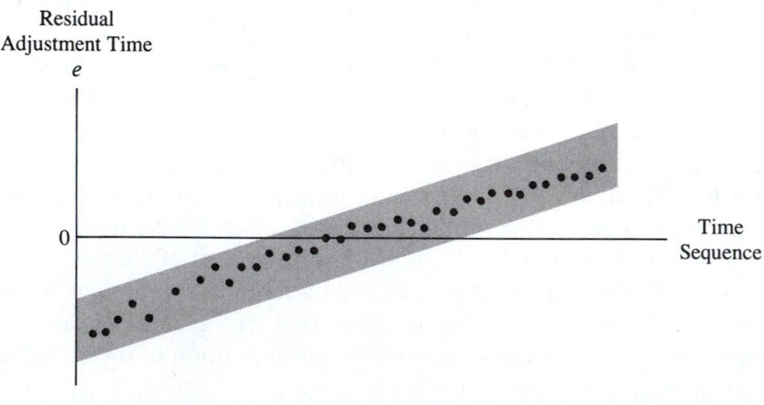

(**c**) Nonindependent Error Terms

involving final output of volume X. The horizontal axis is the *time sequence* of the observation (rather than X). Notice that the e's are increasing over time, indicating a *lack of independence*. This could be due to a variety of explained causes, such as operator fatigue. (That difficulty might be avoided by collecting all data from batches run on different days, at about the same time of day.)

Residual analysis extends to two other types of model violations where a conventional scatter plot is less useful. One involves **outliers**, isolated observations that are so extreme that they don't appear to belong with the rest. If an identifiable cause exists (such as a power outage or an absent operator), the data point might be safely discarded.

Another violation not readily identified on a standard residual plot is non-normality in the e's. Chapter 15 describes *goodness-of-fit tests* that can establish how to conclude whether or not the e's are normally distributed.

Problems

10-27 Refer to the data in Problem 10-1 and your answers to that exercise. Calculate the sample coefficient of determination. What percentage of the variation in processing time Y is explained by the regression line using the number of data points X as the independent variable?

10-28 Refer to the data in Problem 10-2 and your answers to that exercise. Calculate the sample coefficient of determination. What percentage of the variation in pounds of final product Y is explained by the regression line using gallons of raw materials X as the independent variable?

10-29 Refer to the data in Problem 10-24 and your answer to that exercise. What percentage of the variation in reservoir storage Y could be explained by the estimated regression line in which seasonal rainfall is the independent variable?

10-30 Refer to the artificial intelligence experiment data in Table 10-4, page 426.

(a) Using the original X-Y units, determine the sample correlation coefficient. From this, compute the sample coefficient of determination. What percentage of the variation in skill index might be explained by the estimated regression *line*?

(b) Duplicate (a) using instead X with $\ln Y$. What percentage of the variation in skill index is explained by the estimated regression *curve*?

10-31 Refer to the data in Problem 10-4 and your answers to that exercise. Compute for each observation the residual values. Plotting these, what do you conclude regarding consistency with the theoretical assumptions of the regression model?

10-32 Repeat Problem 10-31 using the data in Problem 10-7, as transformed, and your answers to that exercise.

10-6 Curvilinear Regression

Regression analysis also extends to relationships other than straight lines. We now consider **curvilinear regression**, where the dependent variable Y may best be described in terms of X by a nonlinear function. A broad class of nonlinear

Table 10-4 Computer-Chess Results for Artificial Intelligence Study

Number of Games X	Skill Index Y	$\ln Y$	$X \ln Y$	X^2
3	16	2.773	8.318	9
7	34	3.526	24.682	49
8	40	3.689	29.512	64
12	93	4.533	54.396	144
14	125	4.828	67.592	196
9	50	3.912	35.208	81
9	48	3.871	34.839	81
4	18	2.890	11.560	16
11	70	4.248	46.728	121
13	110	4.700	61.100	169
12	85	4.443	53.316	144
8	45	3.807	30.456	64
7	40	3.689	25.823	49
6	33	3.497	20.982	36
3	20	2.996	8.988	9
$126 = \sum X$		$57.402 = \sum \ln Y$	$513.500 = \sum X \ln Y$	$1{,}232 = \sum X^2$

$$\bar{X} = \frac{126}{15} = 8.40$$

regression functions may be established using the procedures described so far. This is accomplished by transforming X or Y so that the underlying curve relating Y to X may be expressed as a line involving the transformed variables.

Although a variety of two-variable relationships may be treated as essentially linear, an alternative approach is to fit a polynomial to the raw data directly. That more general procedure, **polynomial regression**, will be described in Chapter 11.

Logarithmic Transformations

In some applications the relationship between X and Y might best be described by the following multiplicative model:

$$\hat{Y}(X) = ab^X$$

Taking the natural logarithm of both sides, we may express this equivalently as

$$\ln \hat{Y}(X) = \ln a + X \ln b$$

which is the expression for a **logarithmic regression line** relating the transformed dependent variable $\ln Y$ to X. The logarithmic regression line is found by using the method of least squares exactly as before.

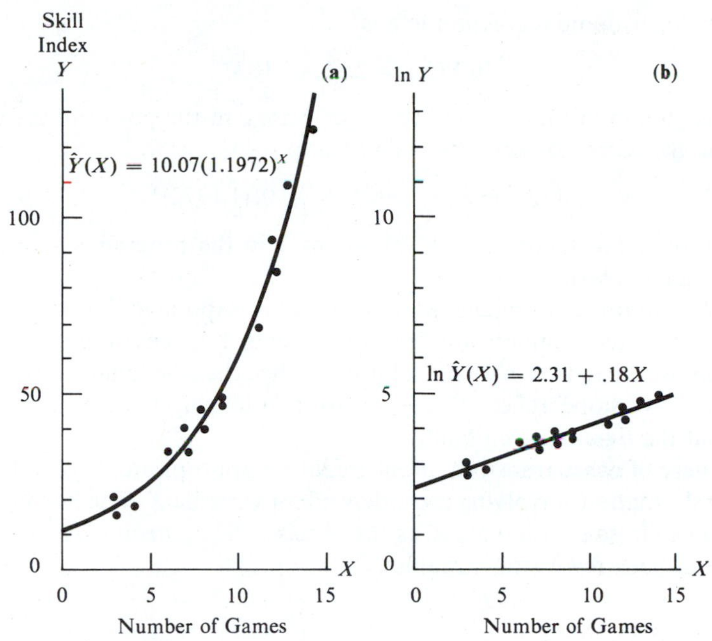

Figure 10-13 *Using least squares to find estimated regression curve (a) by first finding logarithmic regression line (b).*

ESTIMATED REGRESSION COEFFICIENTS FOR THE LOGARITHMIC REGRESSION LINE

$$\ln b = \frac{n \sum X \ln Y - \sum X \sum \ln Y}{n \sum X^2 - (\sum X)^2}$$

$$\ln a = \frac{\sum \ln Y}{n} - \bar{X} \ln b$$

(These computations may be performed using logarithms to the base 10. Natural logarithms are computationally more convenient, however.)

Illustration: *Artificial Intelligence and Computer Chess*

Consider the data in Table 10-4 resulting from an artificial intelligence investigation involving computer chess games. The skill index levels Y have been established for a particular heuristic chess algorithm after several successive games X played against an established program. Each data point resulted from a reinitialized computer run with new program parameters; computer time limitations determined the number of games played in each run. In relating Y to X, it is plausible to assume exponential improvement in chess-playing skill as the number of games is increased. This is demonstrated by the scatter diagram in Figure 10-13 (a), where the data closely fit a regression curve having appropriate form.

The regression curve was found using the computed values from Table 10-4:

$$\ln b = \frac{15(513.50) - 126(57.402)}{15(1{,}232) - (126)^2} = .18$$

$$\ln a = \frac{57.402}{15} - 8.40(.18) = 2.31$$

so that the logarithmic regression line is

$$\ln \hat{Y}(X) = 2.31 + .18X$$

This line is plotted in Figure 10-13 (b). If we raise e to the power of the preceding expression, the corresponding regression curve is

$$\hat{Y}(X) = e^{2.31 + .18X} = 10.07(1.1972)^X$$

which indicates that there is a 19.72% increase in the program's skill level with each chess game played.

The regression curve might assume a *negative* exponential form, as in Figure 10-14 (a), where the number of failed components Y in environmental testing is plotted against the time X between failures. Notice that the logarithmic regression line has negative slope, reflecting the inverse relationship between the number of failures and the time between failures.

A variety of other transformations might be appropriate. Figure 10-15 illustrates transformations involving the independent variable X. The line in (b) results from a change in scale using $\log X$ as the abscissa. The line in (d) reflects a transformation of both regression variables.

Reciprocal Transformations

Another family of curvilinear transformations involves reciprocals. These may be useful when there is an asymptote for one of the variables. One of the forms shown in the table on page 429 might be appropriate.

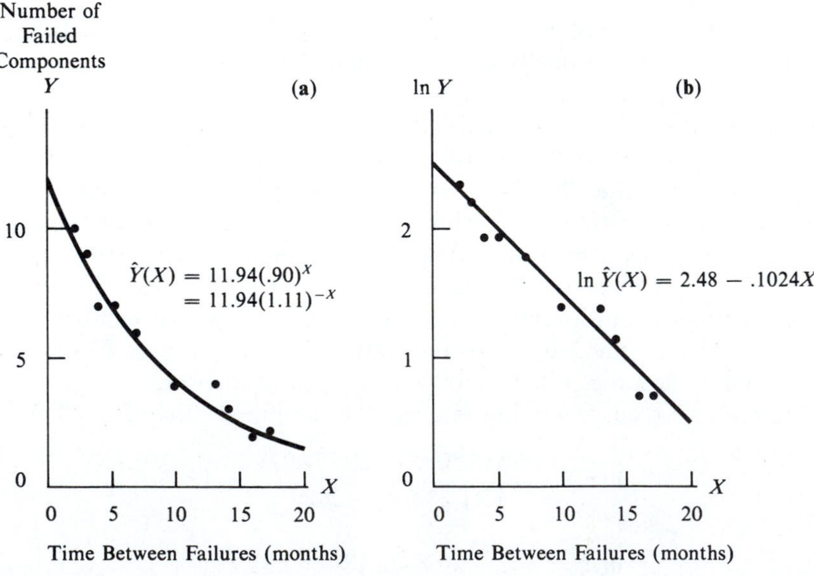

Figure 10-14 *Estimated regression curve (a) and estimated logarithmic regression line (b) for components in environmental test.*

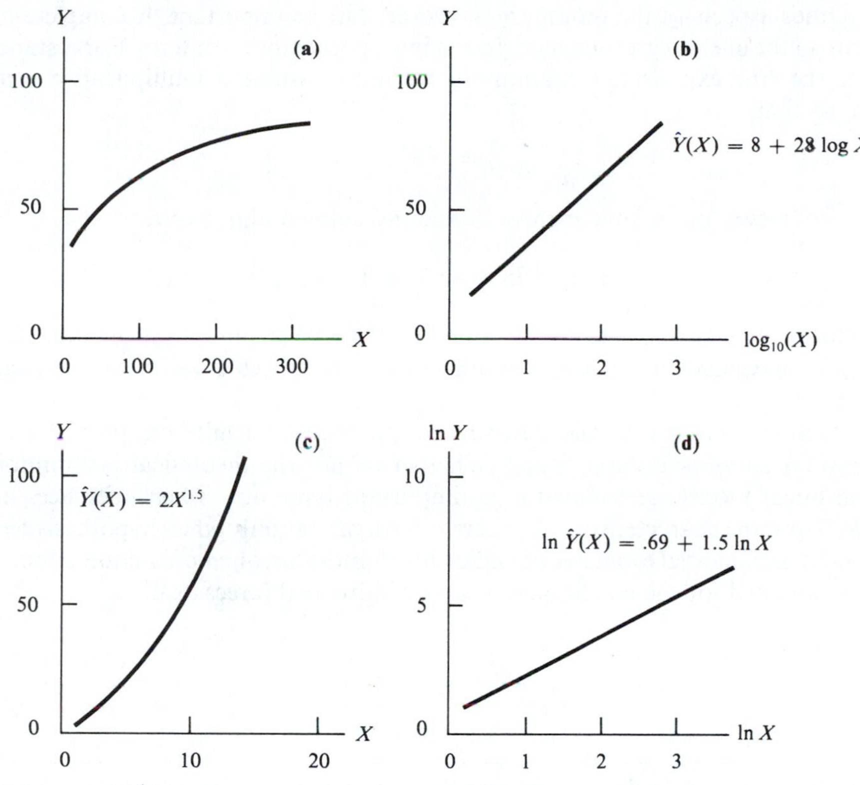

Figure 10-15 *Further logarithmic transformations for obtaining linear regression relationships.*

Original Relationship	Transformation	Linear Relationship
$Y = a + \frac{b}{X}$	$X' = \frac{1}{X}$	$\hat{Y}(X') = a + bX'$
$\frac{1}{Y} = a + bX$	$Y' = \frac{1}{Y}$	$\hat{Y}'(X) = a + bX$
$\frac{1}{Y} = a + \frac{b}{X}$	$X' = \frac{1}{X}, \quad Y' = \frac{1}{Y}$	$\hat{Y}'(X') = a + bX'$

Effect of Transformation on Linear Model

Recall that the estimated regression coefficients minimize the squared vertical deviations by the observed points from the regression surface. Under a logarithmic transformation, this minimization is based on deviations in transformed units rather than the original ones. It is ordinarily of little consequence.

Other aspects of the procedure, however, can be important. A complete extension of the linear regression model must incorporate the error term. For instance, under the true exponential relationship we must assume a multiplicative error term, so that

$$\hat{Y}_i = AB^X \epsilon_i$$

This provides a logarithmic regression line involving additive error:

$$\ln Y_i = \ln A + X \ln B + \ln \epsilon_i$$

The linear regression model described in Section 10-3 requires that the $\ln \epsilon_i$ error terms be independent random variables having 0 expected value and constant variance.

As a practical matter, the least-squares procedure for lines has proved satisfactory for curvilinear curve-fitting. Whether or not the theoretical assumptions of the linear model are violated is an important issue only when inferences are made. Thus, the characteristics of the error term matter only when hypothesis tests or confidence interval estimates are made for A and B or when prediction intervals are constructed for the conditional mean or individual forecasts.

Problems

10-33 Consider the following data relating the area of aquifer contamination Y (acres) to the time X (years) after release of a toxic chemical:

X	Y	X	Y
.5	1.5	5.1	35.8
1.3	4.8	5.6	44.5
2.4	5.3	6.2	68.7
3.6	10.1	7.3	165.6
4.4	19.7	8.1	253.4

(a) Plot these data on a scatter diagram.
(b) Use curvilinear regression in fitting a curve of form $\hat{Y} = ab^X$. Plot your curve on the diagram found in (a).

10-34 Using the artificial intelligence sample data in Table 10-4, determine the coefficients for a regression equation of the form $\hat{Y} = aX^b$.

10-35 Using the data in Problem 10-33, determine the coefficients for a regression curve of the form $\hat{Y} = aX^b$.

10-36 An FAA engineer has collected the following aircraft landing accident data:

Number of Crashes	Nearest Distance from Runway
32	100 ft
15	200
6	1,000
2	2,000
1	5,000

Determine the logarithmic regression line using distance as the independent variable.

10-37 The following data apply to a test performed on a sample of bitumen removed from a tar sand deposit:

Temperature X (°F)	Viscosity Y (poise)	Temperature X (°F)	Viscosity Y (poise)
750	50	620	820
800	15	650	400
700	100	680	150
850	9	710	110
590	950	550	1,200

(a) Determine the coefficients to the following regression equation:

$$\hat{Y} = ab^{(1,000/X)}$$

(b) Plot the corresponding curvilinear regression line and the transformed data points, using

$$X' = \frac{1,000}{X}$$

$$Y' = \ln Y$$

10-38 An evaluation was made of a method for decontaminating stainless steel cannisters used in disposing of radioactive waste. A corrosive concentration of varying strength was applied to sample cannisters and the weight loss measured. The following data were obtained:

Cerium Concentration M	Cannister Weight Loss (mg)	
	One Hour	*Three Hours*
.0025	2.9	4.0
.0100	7.1	9.4
.0280	15.4	44.0
.0570	41.0	113.0
.1150	101.5	139.8

Source: Bray and Thomas, *Transactions of American Nuclear Society*, Vol. 55, November 15–19, 1987, pp. 230–231.

(a) For the one-hour results, find the estimated curvilinear regression equation for estimating cannister weight loss from cerium concentration. Use a natural logarithmic transformation on both the independent and dependent variables.
(b) Plot the scatter points and sketch the regression surface using (1) arithmetic scales and (2) log-log scales.

10-39 Repeat Problem 10-38 using instead the three-hour results and using logarithmic transformations to the base 10.

Comprehensive Problems

10-40 The following sample data provide the reaction time and retort chamber temperature for refining a chemical feedstock:

Temperature X (°F)	Time Y (hours)	Temperature X (°F)	Time Y (hours)
200	3.3	240	2.8
205	3.3	245	2.7
210	3.0	250	2.8
215	3.1	255	2.8
220	3.0	260	2.6
225	2.9	265	2.6
230	2.9	270	2.5
235	2.8		

(a) Plot these data on a scatter diagram. Find the estimated regression equation and plot this on your graph.
(b) Compute (1) the sample standard deviation in reaction time and (2) the standard error of the estimate from regression. Can a significant amount of the variation in temperature be explained by the regression analysis? Explain.

10-41 Refer to the data in Problem 10-40:
(a) Determine the predicted reaction time under the following retort temperatures: (1) 220°F, (2) 245°F, (3) 250°F.
(b) Construct a 95% confidence interval for the mean reaction time under each of the temperatures in (a).
(c) Construct a 95% prediction interval for an individual batch reaction time under each of the temperatures in (a).

10-42 Refer to the data in Problem 10-40:
(a) Compute the sample coefficient of determination. From this, determine the value for the sample correlation coefficient. What proportion of the variation in Y is explained by the estimated regression line?
(b) Construct a 95% confidence interval estimate for the slope of the true regression line. At the 5% significance level, must the null hypothesis of zero slope be accepted or rejected?
(c) In light of your answer to (b), should the null hypothesis of zero correlation between reaction time and retort temperature be accepted or rejected? Use $\alpha = .05$.

(d) Construct a 95% confidence interval estimate for the population correlation coefficient.

10-43 A government transportation engineer has collected the following data for an experimental synthetic fuel used in test track runs with a benchmark automobile:

Speed (miles/hour) X	Mileage (miles/gallon) Y	Speed (miles/hour) X	Mileage (miles/gallon) Y
40	30.0	60	28.3
42	29.7	62	27.9
45	29.3	63	28.1
48	29.1	65	27.7
50	29.2	68	27.3
53	28.7	70	27.0
55	28.6	72	26.9
57	28.7		

(a) Plot the data in the table on a scatter diagram. Find the estimated regression equation and plot this on your graph.
(b) Compute (1) the sample standard deviation in mileage and (2) the standard error of the estimate from regression. Can a significant amount of the variation in mileage be explained by the regression analysis? Explain.

10-44 Refer to the data in Problem 10-43:
(a) Determine the predicted mileages at the following speeds: (1) 50 mph, (2) 55 mph, (3) 60 mph.
(b) Construct a 95% confidence interval for the mean mileage at each of the speeds in (a).
(c) Construct a 95% prediction interval for an individual track-run mileage at each of the speeds in (a).

10-45 Refer to the data in Problem 10-43:
(a) Compute the sample coefficient of determination. From this determine the value for the sample correlation coefficient. What proportion of the variation in Y is explained by the estimated regression line?
(b) Construct a 99% confidence interval estimate for the slope of the true regression line. At the 1% significance level, must the null hypothesis of zero slope be accepted or rejected?
(c) In light of your answer to (b), should the null hypothesis of zero correlation between mileage and speed be accepted or rejected? Use $\alpha = .01$.
(d) Construct a 99% confidence interval estimate for the population correlation coefficient.

10-46 Consider the chemical feedstock refining data in Problem 10-40.
(a) Transforming reaction time, fit a logarithmic regression line to the data and determine the estimated regression curve.
(b) Plot the above curve on a scatter diagram representing the original data. Over the ranges of observation, does the curvilinear model seem to provide a good fit to the data?

10-47 Consider the following data:

X	Y	X	Y
1.0	10	3.6	21
1.3	16	4.1	23
2.1	11	4.5	35
2.4	13	4.7	44
2.9	16	5.0	59

(a) Calculate the sample correlation coefficient.
(b) Consider the transformation $Y' = \ln Y$. Calculate the sample correlation coefficient under the transformation.
(c) Which linear regression would explain the greatest percentage of variation in the dependent variable, the one with the original units or the one with transformed units? Explain.

10-48 The following disabling injury data apply to California workers:

	Injuries per Million Hours	
Year	*Burns*	*Amputations*
1953	0.6165	0.2330
1954	0.5537	0.2045
1955	0.6091	0.1934
1956	0.5551	0.1871
1957	0.5184	0.1740
1958	0.5290	0.1680
1959	0.5684	0.1529
1960	0.5123	0.1315
1961	0.5032	0.1382
1962	0.4915	0.1273
1963	0.5005	0.1321
1964	0.5226	0.1498
1965	0.5503	0.1553
1966	0.5098	0.1783
1967	0.5178	0.1483
1968	0.4999	0.1397
1969	0.5037	0.1252
1970	0.4726	0.1151
1971	0.4426	0.0961
1972	0.4522	0.0905
1973	0.5171	0.0853
1974	0.5084	0.0734

Year	Injuries per Million Hours	
	Burns	*Amputations*
1975	0.4652	0.0494
1976	0.5249	0.0462
1977	0.5840	0.0584
1978	0.5939	0.0394
1979	0.5914	0.0340
1980	0.5492	0.0366
1981	0.5009	0.0486
1982	0.4608	0.0408
1983	0.4527	0.0427
1984	0.4691	0.0452
1985	0.4483	0.0474

Source: Robinson, *American Journal of Public Health*, Vol. 78, No. 3, March 1986, p. 279.

(a) Treating year as the independent variable and burns as the dependent variable, plot a scatter diagram.

(b) Making any suitable transformations of variables, find the estimated regression equation.

(c) *Extrapolating*, predict the burn rate for 1990. Discuss why you cannot attach any statistical confidence to such a number.

10-49 Repeat Problem 10-48 using amputations as the dependent variable.

PART THREE

STATISTICAL ANALYSIS II: EVALUATIONS WITH SEVERAL VARIABLES

CHAPTER ELEVEN

Multiple Regression and Correlation

IN CHAPTER 10 we saw how the value of one variable may be used in predicting another variable. We can do this by using sample data to fit an equation expressing the dependent variable Y in terms of an independent variable X. Since there is a single independent variable, such a procedure is often called *simple regression.* The underlying concepts and procedures extend to more than one predictor, so *multiple regression* incorporates several independent variables in making predictions of Y. Multiple regression ordinarily explains more of the total variation in Y, so that it often provides more precise estimates than simple regression. The advantage of using several variables extends also to *multiple correlation analysis,* which considers how the several variables relate to each other.

By using several independent variables in multiple regression, an investigator may make greater use of sample information. For example, an industrial engineer can better predict the cost of making a particular production run if, in addition to the quantity of raw materials, he knows how much labor will be available. Similarly, a chemical engineer ought to make a better yield prediction if she knows not only the temperature and pressure settings but also impurity levels for the raw feedstocks. A software engineer can more finely gauge a job's processing time if, besides knowing the number of input data points, he also knows memory requirements and the amount of output.

11-1 Linear Multiple Regression Involving Three Variables

Linear multiple regression analysis extends simple linear regression by considering two or more independent variables. Extending our earlier notation, the independent variables will be denoted as $X_1, X_2, X_3, \ldots$. In the case of two independent variables we use the following.

ESTIMATED MULTIPLE REGRESSION EQUATION

$$\hat{Y} = a + b_1X_1 + b_2X_2$$

As in simple regression, $\hat{Y}$ represents the value for Y computed from the estimated regression equation. With the dependent variable Y and the two independent variables X_1 and X_2, a total of three variables is considered. A graphical representation of the scatter diagram will be three dimensional.

Regression in Three Dimensions

Figure 11-1 represents a three-dimensional scatter diagram. The estimated regression equation is represented by a **regression plane**, which, like the regression line, is positioned to fit the observed data points. As in simple regression, the method of least squares establishes the particular plane providing the best fit.

The Y-intercept a is the level of Y when both X_1 and X_2 are zero. The values of b_1 and b_2 are referred to as the **estimated partial regression coefficients**. These provide the slope of the respective lines obtained by slicing the regression plane with cuts parallel to the respective axes. The constant b_1 expresses the net change in Y for a unit increase in X_1 when X_2 is held at a constant value. Likewise, b_2 represents the response of Y to a unit increase in X_2 when X_1 is fixed at some level.

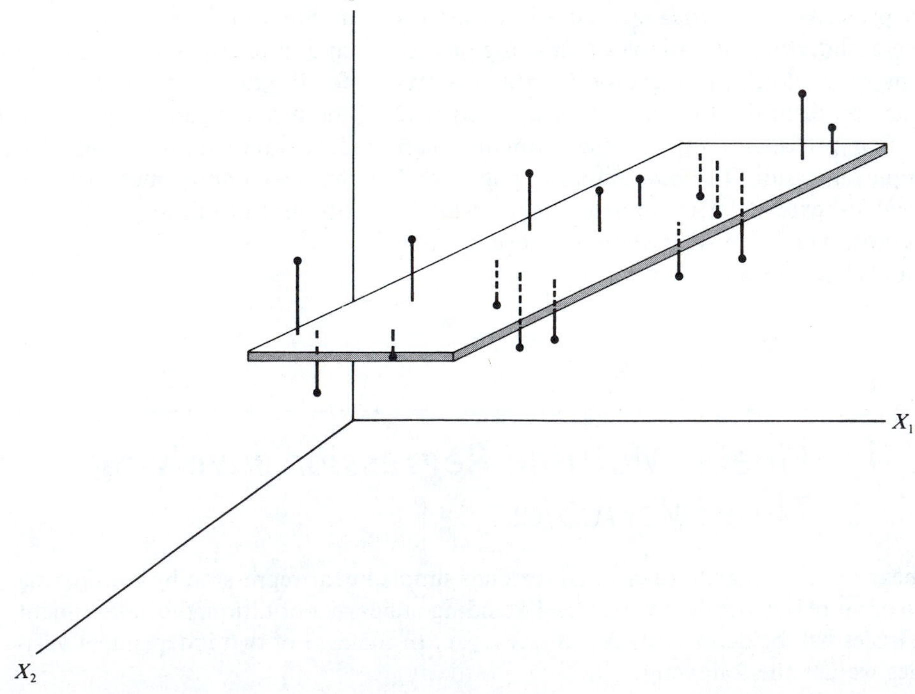

Figure 11-1 *Scatter diagram and regression plane for multiple regression with three variables.*

Example: Hydrological Survey: Predicting River Flow

Multiple regression has been used to assist in a hydrological survey for selecting a dam site. The seasonal mean river flow Y (in cubic feet per second, or cfps) past a particular point is estimated by the equation

$$\hat{Y} = 2{,}500 + 4{,}900X_1 + .23X_2$$

where X_1 represents the annual rainfall (inches) and X_2 the average release (also in cubic feet per second) from upstream dams.

Here $a = 2{,}500$ cfps represents the predicted flow that might be achieved if there were no rainfall and no releases (so that all river flow resulted from aquifer seepage). The partial regression coefficient $b_1 = 4{,}900$ indicates that the mean river flow would increase by 4,900 cfps for each additional average inch of rainfall in the drainage basin, regardless of releases from upstream dams. The constant $b_2 = .23$ signifies that only 23% of the water volume released from dams ever reaches the downstream observation location, with the rest lost to evaporation, seepage, and earlier withdrawal; this quantity applies regardless of what the rainfall happens to be. Thus, we predict the river's flow in a year having a mean rainfall of $X_1 = 18.4$ inches and when the average dam release is $X_2 = 150{,}000$ cfps:

$$\hat{Y} = 2{,}500 + 4{,}900(18.4) + .23(150{,}000) = 127{,}160 \text{ cfps}$$

Of course, the actual river flow under the stipulated conditions might differ from this amount, perhaps considerably.

A regression equation ordinarily does not explain all the variation in Y. Some of it might be explained in terms of factors not included. In the hydrological survey example, mean temperature might serve as an additional independent variable, and its inclusion in the regression evaluation might further explain variation in river flow and thereby improve predictions of Y. Furthermore, a regression evaluation is ordinarily based on sample data only, which makes it subject to sampling error. Finally, even if the true regression equation were known, individual Y's involving exactly the same X's would still exhibit some variability.

The following theoretical assumptions lay the necessary groundwork for extending regression analysis to qualifying such predictions.

Assumptions of Multiple Regression

The theoretical assumptions of multiple regression analysis are straightforward extensions of the properties described in Chapter 10 for simple regression. As before, there is a true regression plane

$$\mu_{Y\cdot 12} = A + B_1X_1 + B_2X_2$$

for which the method of least squares provides estimates. The conditional mean of Y given X_1 and X_2 is denoted by $\mu_{Y\cdot 12}$. The values of the true regression coefficients, A, B_1, and B_2, are unknown quantities that will be estimated by their sample counterparts a, b_1, and b_2. For any setting of the two independent variables X_1 and X_2 there is a corresponding point on the true regression plane, the height of which is $\mu_{Y \cdot 12}$. This point serves as the center of the corresponding Y population. Like the simple linear model, each such population has a common variance, denoted by $\sigma^2_{Y \cdot 12}$.

As in simple regression, each sample observation may be expressed as follows:

$$Y_i = A + B_1 X_{i1} + B_2 X_{i2} + \epsilon_i$$

where the error terms ϵ_i are assumed to be independent and normally distributed, with mean 0 and standard deviation $\sigma_{Y \cdot 12}$.

11-2 Multiple Regression Using the Method of Least Squares

As in simple regression analysis, the regression coefficients are found by fitting a plane through the observed data. Here too, the vertical deviations $Y - \hat{Y}$ are the basis for the evaluation. Again, the method of least squares minimizes

$$\sum (Y - \hat{Y})^2 = \sum [Y - (a + b_1 X_1 + b_2 X_2)]^2$$

Mathematically, the regression parameters are found by taking the respective partial derivatives of the above and setting each resulting expression equal to zero. That provides three equations in three unknowns. As in simple regression, these are referred to as **normal equations**.

NORMAL EQUATIONS FOR MULTIPLE REGRESSION

$$\sum Y = na + b_1 \sum X_1 + b_2 \sum X_2$$
$$\sum X_1 Y = a \sum X_1 + b_1 \sum X_1^2 + b_2 \sum X_1 X_2$$
$$\sum X_2 Y = a \sum X_2 + b_1 \sum X_1 X_2 + b_2 \sum X_2^2$$

In finding the estimated regression coefficients the simplest procedure is to directly solve these equations simultaneously rather than to derive further expressions.

Illustration:* *Predicting Gasoline Mileage

To illustrate finding the multiple regression equation we consider the problem of predicting gasoline mileage Y (in miles per gallon). The independent variables are fuel octane rating X_1 and average speed X_2 (miles per hour). The sample data in Table 11-1 were obtained from 20 test runs with cars of the same make driven with different fuels and at various speeds.

Table 11-1 *Mileages, Octanes, and Speeds for 20 Automobile Test Runs*

Run	Gasoline Mileage (mpg) Y	Octane X_1	Average Speed (mph) X_2
1	24.8	88	52
2	30.6	93	60
3	31.1	91	58
4	28.2	90	52
5	31.6	90	55
6	29.9	89	46
7	31.5	92	58
8	27.2	87	46
9	33.3	94	55
10	32.6	95	62
11	30.6	88	47
12	28.1	89	58
13	25.2	90	63
14	35.0	93	54
15	29.2	91	53
16	31.9	92	52
17	27.7	89	52
18	31.7	94	53
19	34.2	93	54
20	30.1	91	58

The intermediate calculations that provide the inputs for expressing the normal equations are shown in Table 11-2. Substituting the appropriate column totals, the following normal equations are obtained:

$$604.5 = 20a + 1{,}819b_1 + 1{,}088b_2$$

$$55{,}066.5 = 1{,}819a + 165{,}535b_1 + 99{,}063b_2$$

$$32{,}905.2 = 1{,}088a + 99{,}063b_1 + 59{,}626b_2$$

Solving these equations simultaneously, the following estimated regression coefficients are determined:

$$a = -63.535$$

$$b_1 = 1.1789$$

$$b_2 = -.24743$$

Ordinarily, multiple regression results will be obtained with the assistance of a digital computer. (A detailed discussion of computer applications is provided in Section 11-3.) It would be quite a chore to obtain accurate intermediate results using hand computations—even with the assistance of a calculator. Solution of the normal equations with such large coefficients by hand would also be a challenging task.

Table 11-2 *Intermediate Calculations for Obtaining Regression Coefficients*

Y	X_1	X_2	X_1Y	X_2Y	X_1X_2	X_1^2	X_2^2	Y^2
24.8	88	52	2,182.4	1,289.6	4,576	7,744	2,704	615.04
30.6	93	60	2,845.8	1,836.0	5,580	8,649	3,600	936.36
31.1	91	58	2,830.1	1,803.8	5,278	8,281	3,364	967.21
28.2	90	52	2,538.0	1,466.4	4,680	8,100	2,704	795.24
31.6	90	55	2,844.0	1,738.0	4,950	8,100	3,025	998.56
29.9	89	46	2,661.1	1,375.4	4,094	7,921	2,116	894.01
31.5	92	58	2,898.0	1,827.0	5,336	8,464	3,364	992.25
27.2	87	46	2,366.4	1,251.2	4,002	7,569	2,116	739.84
33.3	94	55	3,130.2	1,831.5	5,170	8,836	3,025	1,108.89
32.6	95	62	3,097.0	2,021.2	5,890	9,025	3,844	1,062.76
30.6	88	47	2,692.8	1,438.2	4,136	7,744	2,209	936.36
28.1	89	58	2,500.9	1,629.8	5,162	7,921	3,364	789.61
25.2	90	63	2,268.0	1,587.6	5,670	8,100	3,969	635.04
35.0	93	54	3,255.0	1,890.0	5,022	8,649	2,916	1,225.00
29.2	91	53	2,657.2	1,547.6	4,823	8,281	2,809	852.64
31.9	92	52	2,934.8	1,658.8	4,784	8,464	2,704	1,017.61
27.7	89	52	2,465.3	1,440.4	4,628	7,921	2,704	767.29
31.7	94	53	2,979.8	1,680.1	4,982	8,836	2,809	1,004.89
34.2	93	54	3,180.6	1,846.8	5,022	8,649	2,916	1,169.64
30.1	91	58	2,739.1	1,745.8	5,278	8,281	3,364	906.01
604.5 $=\sum Y$	1,819 $=\sum X_1$	1,088 $=\sum X_2$	55,066.5 $=\sum X_1Y$	32,905.2 $=\sum X_2Y$	99,063 $=\sum X_1X_2$	165,535 $=\sum X_1^2$	59,626 $=\sum X_2^2$	18,414.25 $=\sum Y^2$

The computed values for a, b_1, and b_2 provide the estimated multiple regression equation

$$\hat{Y} = -63.535 + 1.1789X_1 - .24743X_2$$

We may use this equation to predict the gasoline mileage from any particular test run. Suppose that the fuel is rated at 90 octane and that the test car will be driven at an average speed of 60 miles per hour. With $X_1 = 90$ and $X_2 = 60$, the predicted mileage is

$$\hat{Y} = -63.535 + 1.1789(90) - .24743(60) = 27.72 \text{ mpg}$$

The proper interpretation of $b_1 = 1.1789$ is that each unit increase in fuel octane rating brings about an estimated increase in gasoline efficiency of 1.1789 miles per gallon. Furthermore, this mileage increase is estimated to be the same no matter what the average speed happens to be. Likewise, $b_2 = -.24743$ signifies that gasoline consumption increases with speed, so that mileage is estimated to decline by .24743 mpg for each additional mile per hour—regardless of the fuel octane rating.

The Y-intercept $a = -63.535$ by itself has little meaning since the origin represents zero octane and zero speed. It will help if we shift the origin to minimum practical levels for the independent variables. The lowest fuel octane rating in the sample was 87, and an optimal mileage figure for the test car is assumed to be 45 mpg. Considering the transformations

$$X_1' = X_1 - 87$$
$$X_2' = X_2 - 45$$

the estimated regression equation may be expressed as

$$\begin{aligned}\hat{Y} &= -63.535 + 1.1789(X_1' + 87) - .24743(X_2' + 45) \\ &= 27.895 + 1.1789X_1' - .24743X_2'\end{aligned}$$

Under the shift in origin, the Y-intercept is 27.895 mpg, which may be interpreted as the estimated mileage when octane and speed are at $(X_1', X_2') = (0, 0)$, or their minimal practical levels $(X_1, X_2) = (87, 45)$.

Advantages of Multiple Regression

Let's compare the multiple regression results to what would be achieved with separate simple regression evaluations. This will highlight the advantages of simultaneous consideration of the two independent variables.

Figure 11-2 shows three two-dimensional scatter diagrams for the gasoline mileage data from Table 11-1. In Figure 11-2(a) mileage is plotted against octane. Notice the fairly high sample correlation coefficient value of .74 for these two variables, indicating that octane rating X_1 by itself might provide adequate predictions of gasoline mileage Y. It will be convenient if we use double subscripts to denote which variables are involved, so $Y1$ indicates that the association between Y and X_1 is being represented by $r_{Y1} = .74$.

A similar figure applies to each variable pair. In relating gasoline mileage Y to average speed X_2, the sample correlation coefficient is $r_{Y2} = .081$. This second relationship is not a very strong one, as reflected by the scatter diagram in (b), which exhibits a pattern of almost zero correlation. This seems to indicate that average speed will have little effect on gasoline mileage, so by itself X_2 will provide poor predictions of Y. Furthermore, this appears to contradict our earlier multiple regression finding that Y is estimated to decline by nearly .25 mpg for each additional mph of average speed. How can we resolve this paradox?

Our answer lies in the scatter diagram in Figure 11-2(c), where octane rating X_1 is plotted against average speed X_2. There is a fairly strong positive relationship between these variables, confirmed by the sample correlation coefficient value of $r_{12} = .53$. For the particular sample test runs in the investigation, the higher speeds tended to occur when higher-octane fuel was used, while low octanes prevailed for the lower speeds. The effect of speed on gasoline mileage is camouflaged by this strong interaction between speed and octane.

Our example illustrates a general inadequacy in using separate two-variable evaluations to establish how the dependent variable relates to several independent variables. Although the sample correlation between mileage Y and speed X_2 is only $r_{Y2} = .081$, it would be a blunder to discard X_2 from the analysis because of that small correlation alone. This is because X_2 exhibits a significant correlation with X_1, $r_{12} = .53$, so that the influence of average speed X_2 on mileage Y can

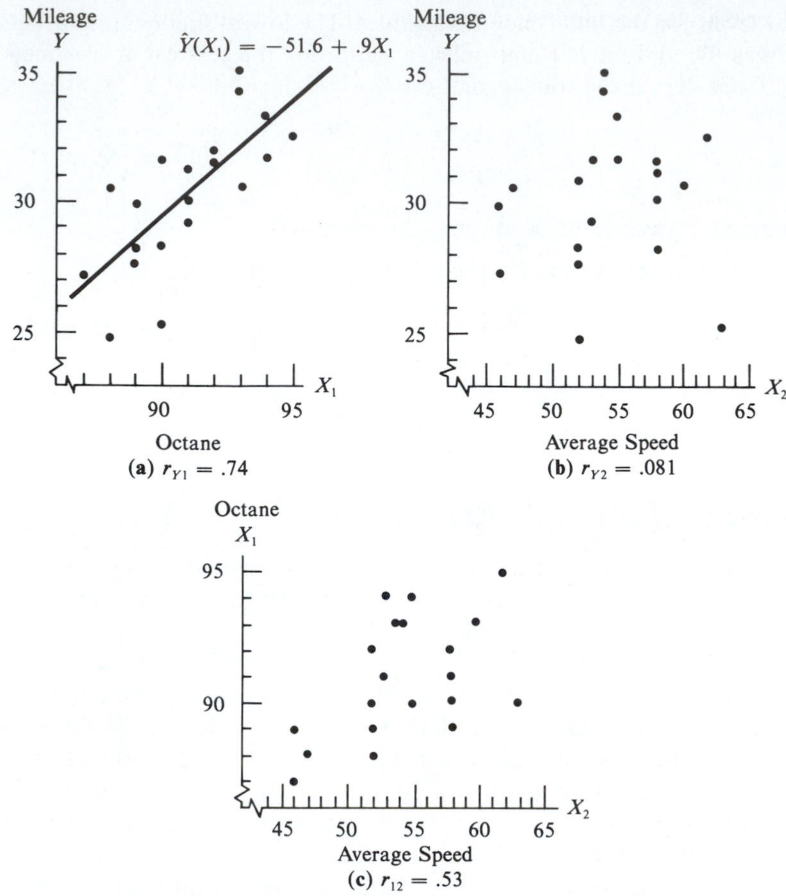

Figure 11-2 *Scatter diagrams for variable pairs in gasoline mileage illustration.*

be explained only by considering how Y relates to *both* X_1 and X_2 through multiple regression. A predictor variable can be discarded only after its interactions with the other independent variables have been assessed. Later in the chapter we will present the method for finding the necessary information to evaluate the benefits of including each possible predictor in the evaluation.

Residuals and the Standard Error of the Estimate

The least squares procedure fits a regression surface Y through the observed data points so that the sum of squared vertical deviations, $Y - \hat{Y}$, is minimized. In Chapter 10 we used the square root of the average of the squared deviations about the regression line, $s_{Y \cdot X}$, to define the standard error of the estimate for values of Y about the regression line. In the same way we will define for the regression plane the standard error of the estimate for values of Y:

$$S_{Y \cdot 12} = \sqrt{\frac{\sum (Y - \hat{Y})^2}{n - 3}}$$

which summarizes variability in Y about the regression plane. The subscript $Y \cdot 12$ shows that two independent variables X_1 and X_2 are being used to predict Y. The divisor $n - 3$ reflects that three degrees of freedom are lost in estimating the regression coefficients. This makes $S^2_{Y \cdot 12}$ an unbiased estimator of the variance of Y about the true regression plane.

The vertical deviations, $Y - \hat{Y}$, are referred to as *residuals*. This terminology is often used because $\sum (Y - \hat{Y})^2$ represents the variation in Y left unexplained by the regression analysis. This unexplained variation is what is left over, or the "residual," from the evaluation. Including a further independent variable in the regression analysis might bring about a reduction in unexplained variation. The residuals resulting from an expanded regression would then be smaller than the current ones.

The predicted values and residuals computed for the gasoline mileage investigation are shown in Table 11-3. Notice that, except for rounding errors, the residuals (vertical deviations) sum to zero—a consequence of the least squares procedure. The standard error of the estimate for the gasoline mileage multiple regression is

$$S_{Y \cdot 12} = \sqrt{\frac{45.4751}{20 - 3}} = 1.64$$

Table 11-3 *Actual, Predicted, and Residual Values from Three-Variable Regression Analysis in Gasoline Mileage Investigation*

Actual Mileage Y	Predicted Value $\hat{Y}$	Residual Value $Y - \hat{Y}$	Squared Deviation $(Y - \hat{Y})^2$
24.8	27.34	−2.54	6.4516
30.6	31.26	−.66	.4356
31.1	29.39	1.71	2.9241
28.2	29.70	−1.50	2.2500
31.6	28.96	2.64	6.9696
29.9	30.00	−.10	.0100
31.5	30.57	.93	.8649
27.2	27.65	−.45	.2025
33.3	33.67	−.37	.1369
32.6	33.12	−.52	.2704
30.6	28.58	2.02	4.0804
28.1	27.04	1.06	1.1236
25.2	26.98	−1.78	3.1684
35.0	32.74	2.26	5.1076
29.2	30.63	−1.43	2.0449
31.9	32.06	−.16	.0256
27.7	28.52	−.82	.6724
31.7	34.17	−2.47	6.1009
34.2	32.74	1.46	2.1316
30.1	29.39	.71	.5041
604.5	604.51	−0.01	45.4751

This expression requires that all the residuals be computed first. For this reason, it is ordinarily easier to use instead the following equivalent expression.

STANDARD ERROR OF THE ESTIMATE

$$S_{Y \cdot 12} = \sqrt{\frac{\sum Y^2 - a\sum Y - b_1 \sum X_1 Y - b_2 \sum X_2 Y}{n - 3}}$$

Using the data from Table 11-2 and the regression coefficients found earlier for the gasoline mileage investigation, we have

$$S_{Y \cdot 12} = \sqrt{\frac{18{,}414.25 - (-63.535)(604.5) - 1.1789(55{,}066.5) - (-.24743)(32{,}905.2)}{20 - 3}}$$

$$= 1.6352 \text{ mpg}$$

This summarizes the degree to which the gasoline mileage data points are scattered about the regression *plane*.

One demonstration of multiple regression's power is provided by comparing the standard error to its counterpart in a simple regression analysis involving mileage and octane only. Using the data from Table 11-2, the following estimated regression equation is obtained:

$$\hat{Y}(X_1) = -51.602 + .8997X_1$$

The scatter of data about this regression *line* is summarized by the standard error of the estimate

$$s_{Y \cdot X_1} = 1.8968 \text{ mpg}$$

The standard error for the regression plane is smaller, which indicates that a multiple regression analysis gives the better interpretation of the data. This conclusion follows from the amounts of variation in Y left unexplained by the respective regression analyses. The corresponding standard error of the estimate quantifies that variation. The fact that $S_{Y \cdot 12}$ is smaller than $s_{Y \cdot X_1}$ therefore indicates that the regression plane based on *two* independent variables as predictors (octane and speed) leaves less variation in mileage Y unexplained than does the regression line based on a *single* independent variable (octane only). Thus, including X_2 (speed) as the second variable actually increases the explained variation in Y and permits more accurate predictions of gasoline mileage than would be possible using octane as the only predictor.

Pitfalls in Multiple Regression Analysis

Multiple regression analysis is subject to all the potential complications discussed in Section 10-5 for simple regression. These include nonlinear relationships between variables, nonconstant variance in the error terms, outliers, and nonnormality in error terms. As in Chapter 10, an analysis of residuals may be helpful in identifying some of these problem areas, and many of the same remedies apply in multiple regression. Greater challenges are created by the existence of several variables and their interrelationships.

Multicollinearity Unique to multiple regression is a complication arising when two independent variables are highly correlated. For example, consider a regression evaluation for predicting a product's cost Y using assembly time X_1 and number of operations required X_2 as independent variables. Since assembly time may have a high correlation with number of operations, the three-variable regression relationship may provide sample data that are scattered tightly about a *single line* lying inside the estimated regression plane.

The problem is not in finding an estimated regression equation. A high correlation between X_1 and X_2 does not prevent that. But a second set of sample observations from the same underlying population would likely result in *very different* estimated regression coefficients, even though the linear scatter might be nearly identical for the two data sets. (Imagine two panes of glass suspended in space—each representing a different sample and regression equation—intersecting at a wide angle on the common line of scatter.) Because different regression planes can obtain the same line, such a situation exhibits **multicollinearity**.

Data exhibiting potential for multicollinearity reflect an inherent instability in the underlying regression relationship. An obvious remedy is to eliminate either X_1 or X_2 from the regression evaluation. That would make it impossible to draw conclusions about the removed variable. Another possible remedy may be to increase the number of sample observations so that the pattern of multicollinearity is broken.

Problems

11-1 In preparing a hydrological survey, a civil engineer has recorded the following sample data:

Seasonal Mean River Flow (cfps)	Watershed Rainfall (inches)	Upstream Dam Releases (cfps)	Average Daily High Temperature (°F)
240,000	25	90,000	62
210,000	19	100,000	62
220,000	23	65,000	61
175,000	16	120,000	64
200,000	20	135,000	64
215,000	17	100,000	60
150,000	14	120,000	62
175,000	15	125,000	64
220,000	21	75,000	61
230,000	16	95,000	60
235,000	24	65,000	60
190,000	19	95,000	64
210,000	21	75,000	63
210,000	20	80,000	62
230,000	23	65,000	62

(a) Using watershed rainfall and upstream dam releases as the independent variables and seasonal mean river flow as the dependent variable, determine the equation for the estimated regression plane.
(b) Complete a table showing actual, predicted, and residual values for seasonal mean river flow.
(c) Compute the standard error of the estimate for Y.

11-2 Repeat Problem 11-1 using instead watershed rainfall and average daily high temperature as the independent variables.

11-3 The following data provide the final product yield from 15 pilot batches tested at a chemical plant:

Final Product Yield (pounds)	Settling Time (hours)	Solid Catalyst (pounds)	Liquid Catalyst (gallons)
400	20	150	210
420	25	230	220
380	15	140	210
440	20	310	170
340	8	50	90
420	15	200	230
330	20	110	150
290	9	70	120
340	16	150	110
330	11	120	150
310	14	90	100
260	8	80	70
290	9	70	110
250	5	40	60
400	12	170	220

(a) Using settling time and weight of solid catalyst as the independent variables and final product yield as the dependent variable, determine the equation for the estimated regression plane.
(b) Complete a table showing actual, predicted, and residual values for final product yield.
(c) Compute the standard error of the estimate for Y.

11-4 Refer to the chemical processing data in Problem 11-3.
(a) Find the estimated regression equation using settling time and volume of liquid catalyst as the independent variables used to predict final product yield. Then compute the standard error of the estimate.
(b) Find the estimated regression equation using weight of solid catalyst and volume of liquid catalyst as the independent variables used to predict final product yield. Then compute the standard error of the estimate.
(c) Which of the above multiple regressions would provide the better predictions of the final product yield? Explain.

11-5 A software engineer wishes to predict processing times for thermodynamic evaluations made by an experimental program run on a particular computer. The following data resulted from 20 test runs:

Processing Time (minutes)	Required RAM (k bytes)	Amount of Input (k bytes)	Amount of Output (k lines)
5.2	19	5	1
17.3	105	10	2
15.5	70	15	5
23.4	80	20	8
15.4	24	12	10
9.5	15	2	5
6.2	22	3	4
10.0	35	10	3
7.7	42	5	2
6.3	15	2	2
7.2	8	4	5
8.5	7	5	6
8.9	12	10	3
5.6	15	7	2
4.1	17	4	1
9.7	18	3	6
13.4	24	8	5
11.7	25	8	4
8.4	32	10	3
12.1	79	12	2

(a) Using required RAM (random access memory) and amount of input as the independent variables and processing time as the dependent variable, determine the equation for the estimated regression plane.

(b) Complete a table showing actual, predicted, and residual values for processing time.

(c) Compute the standard error of the estimate for Y.

11-6 Refer to the software test data in Problem 11-5.

(a) Find the estimated regression equation using required RAM and output as the independent variables used to predict processing time. Then compute the standard error of the estimate.

(b) Find the estimated regression equation using amount of input and output as the independent variables used to predict processing time. Then compute the standard error of the estimate.

(c) Which of the foregoing multiple regressions would provide the better predictions of final processing time? Explain.

11-3 Regression with Many Variables: Computer Applications

Multiple regression may involve more than two predictor variables. For example, in addition to rainfall X_1 and dam releases X_2, the mean daily temperature X_3 in its watershed area might be used in predicting a river's flow Y. With three independent variables, the estimated regression equation may be expressed as

$$\hat{Y} = a + b_1X_1 + b_2X_2 + b_3X_3$$

This expression describes a four-dimensional linear surface referred to as a regression **hyperplane**. This surface may be fitted to the observed data points in the same way as in the case of two independent variables. In establishing values for the estimated regression coefficients, additional intermediate calculations must be made to find the constants in *four* normal equations.

Ordinarily, higher-dimensional regression analyses will be performed with the assistance of a digital computer. It is therefore not necessary to give a detailed discussion of the algebraic expressions that would be needed for solving such a problem by hand. Instead, a general model that is based on matrix algebra will be described in Section 11-7.

All the earlier multiple regression concepts extend to the case of three independent variables. The amount of variation in Y explained by the regression may be gauged by the standard error of the estimate, denoted by $S_{Y\cdot 123}$. This quantity is calculated in the same way as before, except that an additional term, $b_3 \sum X_3Y$, must be subtracted in the numerator. The new divisor is $n - 4$, which represents the number of degrees of freedom remaining after 4 regression coefficients have been estimated from the sample results.

Additional independent variables X_4, X_5, and so on, might be included in a multiple regression analysis. In general, we may denote the dimension of a regression by m, which includes one dependent variable and $m - 1$ independent variables. There is no theoretical limit to the number of independent variables that might be included in a multiple regression analysis. (However, ridiculous results may be expected when m approaches the sample size n; whenever $m = n$ the regression hyperplane will always provide a perfect fit to any sample data.) For purposes of constructing prediction intervals, the degrees of freedom will equal $n - m$.

We will now extend our gasoline mileage example one more dimension and illustrate computer applications of multiple regression with three independent variables.

Using the Computer in Multiple Regression Analysis

By now you should appreciate the advantages of using digital computers in multiple regression analysis, especially when the number of sample observations is large. A wide variety of existing programs may be tapped for this purpose. Individual computer systems and programs will vary in terms of I/O requirements and results provided. Generally, all multiple regression programs will provide values for a, b_1, b_2, and so on. Most will provide statistics for the individual variables and also give standard errors and simple correlations. Beyond these rudimentary results, individual programs will vary considerably in the quantity and quality of output.

To demonstrate both computer applications and higher-dimensional regression analysis at the same time, we will expand the gasoline mileage illustration. Table 11-4 shows the data for an augmented regression analysis that now includes the load X_3 carried during each test run. Figure 11-3 shows a printout for a computer run using the SAS package to perform a multiple regression with those expanded gasoline mileage data. The top portion of the printout shows the SAS commands and a portion of the raw data. Next is the SAS report, broken down into three parts. Shown first is the analysis of variance (which will be described in Section 11-5). That is followed by the regression results and the predicted values—with residuals—computed for all observations.

Although computer packages will vary in presentation, most provide the same information as shown in Figure 11-3, where the more important values are identified with annotated symbols added to the printout. (SAS uses a synonym "root MSE" for the standard error of the estimate $S_{Y\cdot123}$.)

Table 11-4 *Sample Gasoline Mileage Data from 20 Automobile Test Runs, Using Octane, Speed, and Load as Independent Variables*

Run	Gasoline Mileage (mpg) Y	Octane X_1	Average Speed (mph) X_2	Load (pounds) X_3
1	24.8	88	52	646
2	30.6	93	60	465
3	31.1	91	58	359
4	28.2	90	52	665
5	31.6	90	55	214
6	29.9	89	46	606
7	31.5	92	58	458
8	27.2	87	46	557
9	33.3	94	55	605
10	32.6	95	62	407
11	30.6	88	47	259
12	28.1	89	58	423
13	25.2	90	63	596
14	35.0	93	54	286
15	29.2	91	53	612
16	31.9	92	52	399
17	27.7	89	52	444
18	31.7	94	53	697
19	34.2	93	54	376
20	30.1	91	58	363

Figure 11-3 *SAS printout for multiple regression with gasoline mileage data using m = 4 variables.*

```
TITLE 'GASOLINE MILEAGE DATA';
DATA MILEAGE;
  INPUT MILEAGE @@ OCTANE @@ SPEED @@ LOAD @@;
  CARDS;
24.8 88 52 646
30.6 93 60 465
31.1 91 58 359
      :
30.1 91 58 363
;
PROC REG;
MODEL MILEAGE = OCTANE SPEED LOAD /P R;
QUIT;
```

GASOLINE MILEAGE DATA

Model: MODEL1
Dep Variable: MILEAGE

Analysis of Variance

Source	DF	Sum of Squares	Mean Square	F Value	Prob>F
Model	3	134.59603 *SSR*	44.86534 *MSR*	83.070	0.0001
Error	16	8.64147 *SSE*	0.54009 *MSE*		
C Total	19	143.23750 *SSTO*			

Root MSE	0.73491 $S_{Y\cdot 123}$	R-Square	0.9397 $R^2_{Y\cdot 123}$
Dep Mean	$\bar{Y}$ 30.22500	Adj R-Sq	0.9284
C.V.	2.43146		

Parameter Estimates

Variable	DF	Parameter Estimate	Standard Error	T for H0: Parameter=0	Prob > |T|
INTERCEP	1	-59.004933 a	7.09881835 s_a	-8.312	0.0001
X_1 OCTANE	1	1.209731 b_1	0.08812242 s_{b_1}	13.728	0.0001
X_2 SPEED	1	-0.296972 b_2	0.04181721 s_{b_2}	-7.102	0.0001
X_3 LOAD	1	-0.009833 b_3	0.00119102 s_{b_3}	-8.256	0.0001

Obs	MILEAGE Y	Predict Value $\hat{Y}$	Std Err Predict	Residual $e = Y - \hat{Y}$	Std Err Residual	Student Residual
1	24.8000	25.6566	0.344	-0.8566	0.649	-1.319
2	30.6000	31.1093	0.263	-0.5093	0.686	-0.742
3	31.1000	30.3261	0.248	0.7739	0.692	1.118
4	28.2000	27.8892	0.288	0.3108	0.676	0.460
5	31.6000	31.4331	0.356	0.1669	0.643	0.260
6	29.9000	29.0415	0.357	0.8585	0.642	1.337
7	31.5000	30.5623	0.208	0.9377	0.705	1.330
8	27.2000	27.1038	0.381	0.0962	0.629	0.153
9	33.3000	32.4272	0.340	0.8728	0.652	1.339
10	32.6000	33.5051	0.369	-0.9051	0.635	-1.424
11	30.6000	30.9469	0.431	-0.3469	0.595	-0.583
12	28.1000	27.2773	0.327	0.8227	0.658	1.250
13	25.2000	25.3010	0.483	-0.1010	0.554	-0.182
14	35.0000	34.6512	0.341	0.3488	0.651	0.536
15	29.2000	29.3231	0.236	-0.1231	0.696	-0.177
16	31.9000	32.9243	0.257	-1.0243	0.688	-1.488
17	27.7000	28.8526	0.223	-1.1526	0.700	-1.646
18	31.7000	32.1165	0.425	-0.4165	0.600	-0.695
19	34.2000	33.7663	0.280	0.4337	0.679	0.638
20	30.1000	30.2867	0.245	-0.1867	0.693	-0.270

Sum of Residuals	-2.59348E-13
Sum of Squared Residuals	8.6415
Predicted Resid SS (Press)	12.7560

The results indicate that gasoline mileage may be predicted from the estimated regression equation

$$\hat{Y} = -59.005 + 1.2097X_1 - .2970X_2 - .0098X_3$$

The estimated partial regression coefficient $b_1 = 1.2097$ indicates the estimated increase in gasoline mileage that might be achieved for each additional point in fuel octane rating. This is very close to the counterpart coefficient computed for the earlier two-independent-variable regression. There has also been modest change in the coefficient for speed, $b_2 = -.2970$, which gives .2970 as the estimated drop in mileage attributed to a 1-mph increase in speed. Both of these apply regardless of the levels for the other independent variables. The partial regression coefficient is $b_3 = -.0098$ for the load carried. This indicates an estimated decrease of almost .01 mile per gallon for each pound of load—again regardless of octane or speed. Stated differently, each 100 pounds of load is estimated to decrease gasoline mileage by about 1 mpg. The Y-intercept for the new regression hyperplane is $a = -59.005$, about 4.4 mpg higher than before.

We can see that including X_3 as an additional variable will improve predictions made using the resulting estimated regression equation. This is because the standard error, $S_{Y \cdot 123} = .7349$, is smaller than when X_1 and X_2 were the only predictors (and $S_{Y \cdot 12} = 1.6352$, as we found in the last section). Including X_3 in the regression brings about a substantial reduction in standard error. This indicates that the latest estimated multiple regression equation should considerably improve the precision of prediction intervals for Y.

Adding more dimensions to the regression analysis by including further independent predictor variables might lead to even more precision. But adding more variables could just as well raise the standard error and actually cloud the multiple regression results. Consider the observations in Table 11-5 for the stops per mile observed for each car during its test run. These data were incorporated into a five-dimensional regression computer run, the SAS output for which is shown in Figure 11-4. The estimated multiple regression equation is

$$\hat{Y} = -56.526 + 1.1763X_1 - .2814X_2 - .0093X_3 - .9555X_4$$

Notice that the previous regression coefficients change very little. The value $b_4 = -.9555$ signifies that gasoline mileage is estimated to drop by nearly 1 mpg for each stop per mile.

The standard error of the estimate for the 5-variable regression is $S_{Y \cdot 1234} = .6281$, which is smaller than before. This indicates that inclusion of X_4 (stops per mile) should further improve the quality of multiple regression predictions, although the improvement is much less substantial than that achieved by prior inclusion of X_3 (load).

Stepwise Multiple Regression

In some applications there may be a multitude of candidates to include as independent variables. An interesting side issue is choosing which of them to include in the final regression model. Sample observations of the dependent variable made under a range of settings for each candidate predictor may be used to make this decision.

Table 11-5
Observations of Stops per Mile as a Fourth Independent Variable in Gasoline Mileage Investigation

Run	Mileage (mpg) Y	Stops Per Mile X_4
1	24.8	1.2
2	30.6	.7
3	31.1	.3
4	28.2	.6
5	31.6	1.2
6	29.9	.1
7	31.5	.2
8	27.2	.1
9	33.3	.3
10	32.6	.4
11	30.6	.2
12	28.1	.7
13	25.2	.6
14	35.0	.3
15	29.2	.8
16	31.9	.7
17	27.7	1.4
18	31.7	1.1
19	34.2	.2
20	30.1	.1

For instance, a variety of additional predictors might be desired in a gasoline mileage investigation. These might include tire inflation pressure, degree of test-track roughness, humidity, driver skill, ambient temperature, average engine rpm, plus many others. A regression model may be fitted to the sample data by performing a **stepwise multiple regression**. Such an investigation extends the concepts presented, with a series of separate regressions being performed in stages. Each stage culminates in the selection of one predictor to be included in the chosen set of independent variables—possibly replacing a variable selected earlier. Of course with even a modest number of candidate predictors, the number of possible combinations for such a set can be huge; for c candidates, there are 2^c possibilities. Stepwise multiple regression is an efficient search procedure, since not all possible regression models need be considered—just the more promising ones.

A variety of criteria exist for finding which variables to include in the final regression model. A **forward procedure** is perhaps the most common. It begins by performing a simple regression of Y with each possible X. The candidate predictor that explains the greatest amount of variation (yielding the lowest standard error) is saved. Then a three-dimensional multiple regression is performed using the remaining candidates as the second independent variable. That candidate providing the greatest reduction in unexplained Y variation is saved for the next-higher dimensional regression. The process continues until no further reductions in unexplained Y-variation are found. Sometimes a **backward procedure** is used instead. This begins with all candidates serving as independent variables in a single

Figure 11-4 *SAS printout for multiple regression with gasoline mileage data using $m = 5$ variables.*

```
TITLE 'GASOLINE MILEAGE DATA';
DATA MILEAGE;
  INPUT MILEAGE @@ OCTANE @@ SPEED @@ LOAD @@ STOPS @@;
  CARDS;
24.8 88 52 646 1.2
30.6 93 60 465  .7
31.1 91 58 359  .3
      :
30.1 91 58 363  .1
;
PROC REG;
MODEL MILEAGE = OCTANE SPEED LOAD STOPS /P R;
QUIT;
```

GASOLINE MILEAGE DATA

Model: MODEL1
Dep Variable: MILEAGE

Analysis of Variance

Source	DF	Sum of Squares	Mean Square	F Value	Prob>F
Model	4	137.32040	34.33010	87.028	0.0001
Error	15	5.91710	0.39447		
C Total	19	143.23750			

Root MSE	0.62807	R-Square	0.9587
Dep Mean	30.22500	Adj R-Sq	0.9477
C.V.	2.07799		

Parameter Estimates

	Variable	DF	Parameter Estimate	Standard Error	T for H0: Parameter=0	Prob > |T|
	INTERCEP	1	-56.525667	6.13973412	-9.207	0.0001
X_1	OCTANE	1	1.176293	0.07637879	15.401	0.0001
X_2	SPEED	1	-0.281352	0.03622883	-7.766	0.0001
X_3	LOAD	1	-0.009309	0.00103723	-8.975	0.0001
X_4	STOPS	1	-0.955497	0.36358473	-2.628	0.0190

Obs	MILEAGE	Predict Value	Std Err Predict	Residual	Std Err Residual	Student Residual
1	24.8000	25.1975	0.342	-0.3975	0.527	-0.755
2	30.6000	30.9908	0.229	-0.3908	0.585	-0.668
3	31.1000	30.5699	0.231	0.5301	0.584	0.908
4	28.2000	27.9465	0.247	0.2535	0.577	0.439
5	31.6000	30.7276	0.406	0.8724	0.479	1.820
6	29.9000	29.4853	0.349	0.4147	0.522	0.794
7	31.5000	30.9202	0.224	0.5798	0.587	0.988
8	27.2000	27.5889	0.374	-0.3889	0.505	-0.771
9	33.3000	32.6528	0.303	0.6472	0.550	1.176
10	32.6000	33.6073	0.318	-1.0073	0.542	-1.860
11	30.6000	31.1624	0.378	-0.5624	0.502	-1.121
12	28.1000	27.2393	0.279	0.8607	0.562	1.530
13	25.2000	25.4939	0.419	-0.2939	0.468	-0.629
14	35.0000	34.7275	0.293	0.2725	0.555	0.491
15	29.2000	29.1437	0.213	0.0563	0.591	0.095
16	31.9000	32.6798	0.239	-0.7798	0.581	-1.342
17	27.7000	28.0631	0.356	-0.3631	0.517	-0.702
18	31.7000	31.5947	0.414	0.1053	0.472	0.223
19	34.2000	33.9852	0.253	0.2148	0.575	0.374
20	30.1000	30.7238	0.268	-0.6238	0.568	-1.098

Sum of Residuals	-2.70006E-13
Sum of Squared Residuals	5.9171
Predicted Resid SS (Press)	11.0738

multiple regression run. That candidate found to contribute least in explaining Y-variation is dropped, and another run is made. The process terminates when dropping additional variables worsens the latest model's predictive powers.

Caution should be taken when making inferences from a model fitted by stepwise multiple regression. When the estimated regression coefficients are based on the same sample observations used in selecting the predictors there will be some **prediction bias**, because the same data are used to select which X's to include and to fit their coefficients. This bias may be avoided by using different sample observations for estimating the regression coefficients once the X's have been established in a preliminary investigation.

Problems

11-7 Using the hydrological survey data in Problem 11-1, find the estimated multiple regression equation for the hyperplane, incorporating all three predictors to forecast seasonal mean river flow.

11-8 Using the chemical processing data in Problem 11-3, find the estimated multiple regression equation for the hyperplane, incorporating all three predictors to forecast final product yield.

11-9 Using the computer test run data in Problem 11-5, find the estimated multiple regression equation for the hyperplane, incorporating all three predictors to forecast processing time.

11-10 The following readings may be incorporated into the hydrological survey results given in Problem 11-1. Find the multiple regression equation for the hyperplane, incorporating irrigation withdrawals as the fourth independent variable.

Seasonal Mean River Flow (cfps)	Irrigation Withdrawals (cfps)	Seasonal Mean River Flow (cfps)	Irrigation Withdrawals (cfps)
240,000	78,000	220,000	141,000
210,000	81,000	230,000	100,000
220,000	152,000	235,000	139,000
175,000	103,000	190,000	95,000
200,000	82,000	210,000	81,000
215,000	88,000	210,000	101,000
150,000	275,000	230,000	64,000

11-11 The readings at the top of page 459 may be incorporated into the chemical processing data given in Problem 11-3:

(a) Find the multiple regression equation for the original variables plus chamber temperature as the fourth independent variable.

(b) Find the multiple regression equation for the hyperplane based on the original variables and the incorporation of chamber pressure as the fourth independent variable.

(c) Find the multiple regression equation for the hyperplane based on all five independent variables.

Final Product Yield (pounds)	Chamber Temperature (°F)	Chamber Pressure (psi)
400	280	150
420	320	175
380	310	155
440	290	160
340	265	190
420	295	145
330	325	160
290	340	155
340	290	145
330	340	170
310	285	140
260	355	170
290	315	130
250	325	135
400	305	150

11-12 The following readings may be incorporated into the computer test run data given in Problem 11-5:

Processing Time (minutes)	Number of Nonzero Parameters	Required Disk Storage (m bytes)
5.2	50	1.4
17.3	60	2.5
15.5	40	1.8
23.4	58	2.2
15.4	43	1.7
9.5	57	2.2
6.2	31	3.1
10.0	52	1.9
7.7	45	1.3
6.3	59	1.7
7.2	38	2.1
8.5	33	4.4
8.9	49	3.7
5.6	27	4.8
4.1	35	3.0
9.7	48	1.8
13.4	65	4.5
11.7	63	2.8
8.4	39	1.5
12.1	27	2.5

(a) Find the multiple regression equation for the original variables plus the number of nonzero parameters as the fourth independent variable.
(b) Find the multiple regression equation for the hyperplane based on the original variables and the incorporation of required disk storage as the fourth independent variable.
(c) Find the multiple regression equation for the hyperplane based on all five independent variables.

11-13 The following data apply to a random sample of 25 engineers:

Earnings Y	Professional Experience (years) X_1	Engineering Education (years) X_2	Positions Held X_3	Number of Employees X_4	Patents Held X_5
$36,000	10	4.0	1	0	0
58,700	4	5.0	2	50	0
52,400	7	5.5	4	30	0
153,200	25	4.0	3	500	2
53,100	9	4.5	4	2	15
38,200	9	4.0	5	0	0
41,400	8	4.0	3	0	5
40,800	12	5.0	5	0	1
43,700	11	6.0	4	12	0
42,500	3	5.5	2	6	0
47,300	5	4.5	1	5	8
56,300	6	8.5	2	0	23
34,500	5	4.0	3	0	0
35,900	5	7.0	3	2	0
42,200	8	9.0	5	3	0
63,400	13	6.5	7	0	15
56,800	21	10.5	12	4	8
45,700	12	4.5	5	8	4
61,300	8	5.0	12	25	3
55,900	23	5.0	10	47	0
97,500	14	4.0	5	123	0
42,400	15	6.0	4	0	0
43,800	8	4.5	3	5	1
37,900	3	5.0	2	2	2
43,500	2	5.0	1	10	3

(a) Determine the estimated multiple regression equation for income as a function of the remaining five independent variables.
(b) Holding all other variables constant, determine the estimated worth (increase in earnings) from: (1) one year of experience, (2) one year of engineering education, (3) one more position held, (4) an additional employee being supervised, (5) one patent.
(c) Calculate the standard error of the estimate.

Table 11-6 Results of Cerium (IV) Contact with Stainless Steel

Temp. X_2, °C	Contact Time X_3, h	Loss of Weight Y, mg for Cerium Concentration X_1, M				
		0.115 M	0.057 M	0.028 M	0.01 M	0.0025 M
25	1	3.85	1.89	0.5	—	—
	3	13.75	7.1	3.0	—	—
	6	34.6	15.4	6.4	—	—
	12	88.4	36.8	15.9	—	—
45	1	19.2	11.7	4.4	2.6	1.5
	3	76.9	39.7	16.9	10.8	3.2
	6	152.2	77.1	36.4	14.6	5.2
	12	220.2	100.2	49.3	19.3	5.1
65	1	101.5	41.0	15.4	7.1	2.9
	3	139.8	113.0	44.0	9.4	4.0
	6	228.0	100.0	54.7	17.0	4.6
	12	276.7	100.0	54.1	26.0	4.7
78	1	112.6	49.7	27.5	—	—
	3	211.1	86.9	43.9	—	—
	6	265.9	112.9	48.0	—	—
	12	280.0	122.0	65.8	—	—

Source: Bray and Thomas. *Transactions of American Nuclear Society*, Vol 55, November 15–19, 1987, pp. 230–231.

11-14 Table 11-6 shows the results obtained by two researchers studying methods for decontaminating stainless steel cannisters used in disposing of radioactive waste. A corrosive concentration of varying strength was applied to sample cannisters and the weight loss measured.

(a) Using loss of weight as the dependent variable Y, with cerium concentration X_1, temperature X_2, and contact time X_3 as the independent variables, determine the estimated regression equation.

(b) Find point estimates for each of the following cases:
(1) 0.01 M Ce, 25°C, 3 hours (3) 0.10 M Ce, 50°C, 1 hour
(2) 0.05 M Ce, 45°C, 6 hours
Do you notice anything strange about the estimated values?

(c) Holding temperature fixed at 65°C and time at 6 hours, plot a scatter diagram of weight loss versus cerium concentration. What does the graph suggest regarding using a *linear* regression model in evaluating these data?

11-4 A Generalized Multiple Regression Model

We have so far presented the more important concepts and procedures of multiple regression analysis. A general representation of the multiple regression model takes a matrix form. This representation provides a more convenient basis for making inferences regarding the multiple regression coefficients.

Matrix Representation of Multiple Regression

The general multiple regression model is based on the following relationship for the ith observed value of the dependent variable:

$$\hat{Y}_i = A + B_1 X_{i1} + B_2 X_{i2} + \cdots + B_{m-1} X_{i,m-1} + \epsilon_i$$

where X_{ik} denotes the fixed level for the kth independent variable used in making the ith sample observation. The parameters $A, B_1, B_2, \ldots$ are the true regression coefficients. These are estimated by their sample counterparts, $a, b_1, b_2, \ldots$. Altogether, $m - 1$ independent variables are used in predicting levels for the single dependent variable.

Each of the n observed sample values of the dependent variable may be represented by the $n \times 1$ column vector

$$\mathbf{Y} = \begin{bmatrix} Y_1 \\ Y_2 \\ \vdots \\ Y_n \end{bmatrix}$$

The entire collection of settings for each combination of observation and independent variable may be expressed by the following $n \times m$ matrix:

$$\mathbf{X} = \begin{bmatrix} 1 & X_{11} & X_{12} & \cdots & X_{1,m-1} \\ 1 & X_{21} & X_{22} & \cdots & X_{2,m-1} \\ & & \vdots & & \\ 1 & X_{n1} & X_{n2} & \cdots & X_{n,m-1} \end{bmatrix}$$

Each row represents a particular observed data point, and there is a column for each independent variable. The first column consists of 1s, reflecting the fact that each observation shares the same Y-intercept or constant term.

We may illustrate this matrix using the gasoline mileage data from Table 11-4. There are $n = 20$ sample observations and $m = 4$ variables. Thus, we have

$$\mathbf{Y} = \begin{bmatrix} 24.8 \\ 30.6 \\ \vdots \\ 30.1 \end{bmatrix} \qquad \mathbf{X} = \begin{bmatrix} 1 & 88 & 52 & 646 \\ 1 & 93 & 60 & 465 \\ & \vdots & & \\ 1 & 91 & 58 & 363 \end{bmatrix}$$

When the true regression parameters are represented as an $m \times 1$ column vector

$$\mathbf{B} = \begin{bmatrix} A \\ B_1 \\ B_2 \\ \vdots \\ B_{m-1} \end{bmatrix}$$

and the error terms as an $n \times 1$ column vector

$$\epsilon = \begin{bmatrix} \epsilon_1 \\ \epsilon_2 \\ \vdots \\ \epsilon_n \end{bmatrix}$$

then the set of n sample observations may be represented algebraically by

$$\underset{n \times 1}{\mathbf{Y}} = \underset{n \times m}{\mathbf{X}} \; \underset{m \times 1}{\mathbf{B}} + \underset{n \times 1}{\epsilon}$$

In an analogous fashion, the estimated regression equation evaluated at the ith set of independent variable values,

$$\hat{Y}_i = a + b_1 X_{i1} + b_2 X_{i2} + \cdots + b_{m-1} X_{i,m-1}$$

may be represented algebraically by

$$\hat{\mathbf{Y}} = \mathbf{Xb}$$

where $\hat{\mathbf{Y}}$ is an $n \times 1$ column vector of computed values and $\mathbf{b}$ is an $m \times 1$ column vector of estimated regression coefficients.

The normal equations for the least squares model may be represented by

$$\underset{m \times n}{\mathbf{X}'} \; \underset{n \times 1}{\mathbf{Y}} = \underset{m \times m}{(\mathbf{X}'\mathbf{X})} \; \underset{m \times 1}{\mathbf{b}}$$

Solving this for $\mathbf{b}$, we get the following expression, which provides the estimated regression coefficients

$$\underset{m \times 1}{\mathbf{b}} = \underset{m \times m}{(\mathbf{X}'\mathbf{X})^{-1}} \; \underset{m \times 1}{\mathbf{X}'\mathbf{Y}}$$

Returning to the gasoline mileage illustration, we have

$$\mathbf{X}'\mathbf{Y} = \begin{bmatrix} \sum Y \\ \sum X_{i1} Y \\ \sum X_{i2} Y \\ \sum X_{i3} Y \end{bmatrix} = \begin{bmatrix} 604.5 \\ 55{,}066.5 \\ 32{,}905.2 \\ 281{,}660.2 \end{bmatrix}$$

and also

$$\mathbf{X}'\mathbf{X} = \begin{bmatrix} n & \sum X_{i1} & \sum X_{i2} & \sum X_{i3} \\ \sum X_{i1} & \sum X_{i1}^2 & \sum X_{i1} X_{i2} & \sum X_{i1} X_{i3} \\ \sum X_{i2} & \sum X_{i1} X_{i2} & \sum X_{i2}^2 & \sum X_{i2} X_{i3} \\ \sum X_{i3} & \sum X_{i1} X_{i3} & \sum X_{i2} X_{i3} & \sum X_{i3}^2 \end{bmatrix}$$

$$= \begin{bmatrix} 20 & 1{,}819 & 1{,}088 & 9{,}437 \\ 1{,}819 & 165{,}535 & 99{,}063 & 858{,}048 \\ 1{,}088 & 99{,}063 & 59{,}626 & 511{,}505 \\ 9{,}437 & 858{,}048 & 511{,}505 & 4{,}842{,}223 \end{bmatrix}$$

The inverse of this matrix is

$$(\mathbf{X}'\mathbf{X})^{-1} = \begin{bmatrix} -.052661284 & .000624732 & -.000082878 & .000000683 \\ .000624732 & .001239470 & -.001876786 & -.000022600 \\ -.000082878 & -.001876786 & .003006517 & .000015139 \\ .000000683 & -.000022600 & .000015139 & .000002611 \end{bmatrix}$$

Thus, we may express the estimated regression coefficients in terms of the following:

$$\mathbf{b} = \begin{bmatrix} -.052661284 & .000624732 & -.000082878 & .000000683 \\ .000624732 & .001239470 & -.001876786 & -.000022600 \\ -.000082878 & -.001876786 & .003006517 & .000015139 \\ .000000683 & -.000022600 & .000015139 & .000002611 \end{bmatrix} \times \begin{bmatrix} 604.5 \\ 55{,}066.5 \\ 32{,}905.2 \\ 281{,}660.2 \end{bmatrix}$$

$$= \begin{bmatrix} -59.005 \\ 1.2097 \\ -\ .2970 \\ -\ .0098 \end{bmatrix}$$

We may use the algebraic model to express the standard error, $S_{Y\cdot 12\ldots}$. The square of this quantity is calculated by averaging the squared deviations, or residuals values, about the regression surface. In terms of the preceding vectors, each residual is an element in the column vector

$$\mathbf{Y} - \hat{\mathbf{Y}} = \mathbf{e}$$

and the sum of squared residuals may be expressed by the scalar

$$(\mathbf{Y} - \hat{\mathbf{Y}})'(\mathbf{Y} - \hat{\mathbf{Y}}) = \mathbf{e}'\mathbf{e}$$

This is mathematically equivalent to

$$\mathbf{Y}'\mathbf{Y} - \mathbf{b}'\mathbf{X}'\mathbf{Y}$$

so that the following result is obtained:

$$S^2_{Y\cdot 12\ldots} = \frac{1}{n-m}[\mathbf{Y}'\mathbf{Y} - \mathbf{b}'\mathbf{X}'\mathbf{Y}] = \frac{\mathbf{e}'\mathbf{e}}{n-m}$$

Problems

11-15 Consider the following sample data:

Y	X_1	X_2
10	5	4
15	6	5
24	10	8
30	9	12
18	4	10

(a) Express the following vectors and matrices:
   (1) $\mathbf{Y}$ (2) $\mathbf{X}$ (3) $\mathbf{X'X}$ (4) $\mathbf{X'Y}$
(b) Determine the following vectors and matrices:
   (1) $(\mathbf{X'X})^{-1}$ (2) $\mathbf{b}$ (3) $\hat{\mathbf{Y}}$ (4) $\mathbf{Y} - \hat{\mathbf{Y}}$
(c) Express the estimated multiple regression equation.
(d) Determine the following vectors and matrices:
   (1) $\mathbf{Y'Y}$ (2) $\mathbf{b'X'Y}$ (3) $S^2_{Y\cdot 12}(\mathbf{X'X})^{-1}$

11-5 Inferences in Multiple Regression Analysis

As with any statistical sampling procedure, any conclusions made from the estimated regression equation are subject to sampling error. As in the simple regression analysis of Chapter 10, a variety of inferences may be drawn. The existence of multiple independent variables makes for a richer mix of procedures, including confidence intervals for the conditional mean, prediction intervals for individual Y's, tests for nonzero regression coefficients, and estimates of regression coefficients.

Confidence and Prediction Intervals

Although they are conceptually similar to the intervals found for simple regression applications in Chapter 10, confidence and prediction intervals in multiple regression evaluations are much more challenging to compute. This arises from the covariability among the several variables involved.

Estimating the Conditional Mean The conditional mean $\mu_{Y\cdot 12\ldots}$ is estimated by the value of Y computed for the given levels of the independent variables. These given values may be denoted by the $1 \times m$ vector

$$\mathbf{X}_g = \begin{bmatrix} 1 \\ X_1 \\ X_2 \\ \vdots \\ X_{m-1} \end{bmatrix}$$

The standard error of $\hat{Y}$, denoted as $s_{\hat{Y}}$, depends on the given levels of the X_k's. It may be computed from

$$s_{\hat{Y}} = S_{Y\cdot 12\ldots}\sqrt{\mathbf{X}_g'(\mathbf{X'X})^{-1}\mathbf{X}_g}$$

where $(\mathbf{X'X})^{-1}$ is one of the matrices found from the original data (see page 463). Based on the above, the following expression may be used to construct the $100(1 - \alpha)\%$ confidence interval.

CONFIDENCE INTERVAL ESTIMATE OF THE CONDITIONAL MEAN

$$\mu_{Y \cdot 12 \cdots} = \hat{Y} \pm t_{\alpha/2} s_{\hat{Y}}$$

where $\hat{Y}$ is the computed value from the estimated regression equation for the given levels for the X_k's. The critical value $t_{\alpha/2}$ is read from Appendix Table G using $n - m$ degrees of freedom.

To illustrate, consider the second regression evaluation with the gasoline mileage data involving $m = 4$ variables altogether (one Y and three X's). In Section 11-3 we found the estimated regression equation

$$\hat{Y} = -59.005 + 1.2097X_1 - .2970X_2 - .0098X_3$$

We will estimate $\mu_{Y \cdot 123}$ when the fuel octane rating is $X_1 = 90$, the average speed is $X_2 = 50$ mph, and the load is $X_3 = 500$ pounds. The point estimate for mean gasoline mileage given these levels is

$$\hat{Y} = -59.005 + 1.2097(90) - .2970(50) - .0098(500) = 30.118 \text{ mpg}$$

The following vector is used in computing the standard error for $\hat{Y}$:

$$\mathbf{X_g} = \begin{bmatrix} 1 \\ 90 \\ 50 \\ 500 \end{bmatrix}$$

In Section 11-4 we obtained the matrix

$$(\boldsymbol{X'X})^{-1} = \begin{bmatrix} -.052661284 & .000624732 & -.000082878 & .000000683 \\ .000624732 & .001239470 & -.001876786 & -.000022600 \\ -.000082878 & -.001876786 & .003006517 & .000015139 \\ .000000683 & -.000022600 & .000015139 & .000002611 \end{bmatrix}$$

The vector products provide

$$\mathbf{X_g'(X'X)^{-1}X_g} = .09449$$

so that using the value found earlier, $S_{Y \cdot 123} = .7349$, we compute

$$s_{\hat{Y}} = .7349\sqrt{.09449} = .2259$$

A 95% confidence interval estimate of $\mu_{Y \cdot 123}$ with the above $\mathbf{X_g}$ may be found. With $n - m = 20 - 4 = 16$ degrees of freedom, Appendix Table G provides $t_{.025} = 2.120$, so that

$$\begin{aligned} \mu_{Y \cdot 123} &= 30.118 \pm 2.120(.2259) \\ &= 30.118 \pm .479 \end{aligned}$$

or

$$29.639 \le \mu_{Y \cdot 123} \le 30.597$$

With 95% confidence we may conclude from this that test runs using 90-octane fuel, conducted at an average speed of 50 mph, and with a 500-pound load, will have a mean gasoline mileage falling somewhere between 29.639 and 30.597 mpg.

Predicting an Individual Y The above procedure may be slightly adapted to predictions of individual Y's. Denoting by Y_I a forecast value when the vector $\mathbf{X}_g$ applies, the following expression provides the $100(1 - \alpha)\%$ prediction interval estimate.

PREDICTION INTERVAL ESTIMATE OF AN INDIVIDUAL VALUE OF Y

$$Y_I = \hat{Y} \pm t_{\alpha/2}\sqrt{S^2_{Y \cdot 12 \ldots} + s^2_{\hat{Y}}}$$

with $s_{\hat{Y}}$ computed as before and $t_{\alpha/2}$ the critical value of t for $n - m$ degrees of freedom.

To illustrate, we continue with the same gasoline mileage regression and keep $\mathbf{X}_g$ the same as before. Then, at a 95% level of confidence, we predict the mileage for a particular run at those levels to fall inside the interval

$$\begin{aligned} Y_I &= 30.118 \pm 2.120\sqrt{(.7349)^2 + (.2259)^2} \\ &= 30.118 \pm 1.630 \end{aligned}$$

or

$$28.488 \leq Y_I \leq 31.748$$

Notice that this interval is considerably wider than the earlier one for the conditional mean. The main reason for the reduced precision is that *individual* values for Y will vary more than *means*.

Using the Computer For different given levels of the independent variables, not only will the *centers* of the confidence intervals for $\mu_{Y \cdot 12 \ldots}$ or Y_I shift to the corresponding $\hat{Y}$, but the *widths* will also change in accordance with the different $\mathbf{X}_g$ vector. Since the computations in finding $s_{\hat{Y}}$ can be particularly onerous and error-prone, these confidence intervals are best computed with the help of a computer.

Inferences Regarding Regression Coefficients

The estimated intercept a and partial regression coefficients $b_1, b_2, \ldots$ are statistical estimators of the true regression coefficients $A, B_1, B_2, \ldots$. Confidence intervals for the B's are not ordinarily constructed until the following null hypothesis:

$$H_0\colon B_1 = B_2 = \cdots = B_{m-1} = 0$$

has been tested. The regression results are statistically significant when this null hypothesis is rejected. The procedure for making this determination is based on the *analysis-of-variance* approach (described in more detail in Chapters 14 and 15), which focuses on sums of squares.

Analogous to simple regression, the least-squares procedure minimizes the sum of the vertical deviations about the regression plane. Each deviation may be partitioned into the explained and unexplained component. Summing over all observations the squares of the individual components, the following generalized relationship applies:

$$\begin{array}{ccccc} \text{Total variation} & = & \text{Explained variation} & + & \text{Unexplained variation} \\ \sum_{i=1}^{n} [Y_i - \bar{Y}]^2 & = & \sum_{i=1}^{n} [\hat{Y}_i - \bar{Y}]^2 & + & \sum_{i=1}^{n} [Y_i - \hat{Y}_i]^2 \\ SSTO & = & SSR & + & SSE \end{array}$$

The **total sum of squares** $SSTO$ expresses the variation in observed Y's about their mean $\bar{Y}$ without regard to the regression relationship. It may be computed from

$$SSTO = \sum_{i=1}^{n} [Y_i - \bar{Y}]^2 = \mathbf{Y'Y} - n\bar{Y}^2$$

The **regression sum of squares** SSR summarizes the explained variation in estimated Y's about the sample mean. The following expression applies:

$$SSR = \sum_{i=1}^{n} [\hat{Y}_i - \bar{Y}]^2 = \mathbf{b'X'Y} - n\bar{Y}^2$$

The **error sum of squares** SSE expresses the unexplained variation in observed Y's about the estimated regression plane. It may be computed in a variety of ways:

$$SSE = \sum_{i=1}^{n} [Y_i - \hat{Y}_i]^2 = \sum_{i=1}^{n} e_i^2 = \mathbf{e'e} = (n - m)S^2_{Y \cdot 12 \ldots}$$

or

$$SSE = \mathbf{Y'Y} - \mathbf{b'X'Y}$$

The sums of squares in a multiple regression analysis are not ordinarily computed by hand, and they are standard output from the popular computer software packages.

The Analysis of Variance Approach To test the null hypothesis that all partial regression coefficients are zero, we must compare the explained and unexplained variations in Y. This is accomplished by a statistical test using a composite test statistic. This test statistic is based on the *means* of the squared deviations for the explained and unexplained variations. The first of these is the following.

REGRESSION MEAN SQUARE

$$MSR = \frac{SSR}{m - 1}$$

where m is the total number of variables in the analysis. The denominator $m - 1$ is referred to as the number of degrees of freedom for SSR. The second mean is the following.

ERROR MEAN SQUARE

$$MSE = \frac{SSE}{n - m} = S^2_{Y \cdot 12 \cdots}$$

where $n - m$ is the number of degrees of freedom for the unexplained variation.

Recall that the sample variance is found by averaging squared deviations about the central value. MSR and MSE are both computed that way, and thus both mean squares are actually sample variances. Since these form the basis for the test, the procedure is called analysis of *variance* (even though the null hypothesis itself says nothing about variances).

The following expression is the regression test statistic.

REGRESSION TEST STATISTIC

$$F = \frac{\text{Variance explained by regression}}{\text{Unexplained variance}} = \frac{MSR}{MSE}$$

This quantity has the F distribution (introduced in Chapter 7). Large values for F tend to refute the null hypothesis. Critical values may be read from Appendix Table J. There are two sets of entries: the lightface ones apply when the upper-tail area is $\alpha = .05$, and the boldface values correspond to $\alpha = .01$. Two parameters must be specified. These are the divisors used in computing the respective mean squares; for this test $m - 1$ applies for the numerator and $n - m$ for the denominator.

For the gasoline mileage illustration, the SAS computer printouts in Figures 11-3 ($m = 4$) and 11-4 ($m = 5$) give the respective sums of squares, mean squares, degrees of freedom, and F.

With the $m = 4$ variable regression, Figure 11-3 provides $SSR = 134.596$ and $SSE = 8.641$. Dividing by the respective degrees of freedom, the mean squares are

$$MSR = \frac{134.596}{4 - 1} = 44.865 \qquad MSE = \frac{8.641}{20 - 4} = .540$$

These indicate that the variation in gasoline mileage Y that is explained by that regression is quite high in relation to that left unexplained. This results in the computed value for the test statistic of

$$F = \frac{44.865}{.540} = 83.070$$

Using degrees of freedom of $m - 1 = 3$ for the numerator and $n - m = 16$ for the denominator, Appendix Table J provides the critical value $F_{.01} = 5.29$. Since the computed value is greater, the null hypothesis of zero-valued partial regression coefficients must be *rejected* at the $\alpha = .01$ significance level. We must conclude that B_1, B_2, and B_3 are not all equal to zero. Stated equivalently, the regression results are significant at the $\alpha = .01$ level. (Since $F = 83.070$ is so much larger than the tabled critical value, this conclusion would apply at a far lower significance level than .01. Figure 11-3 gives the rounded probability as .0001.)

Estimating Regression Coefficients The matrix algebraic model provides a basis for extending statistical inferences to the multiple regression coefficients. The sample variance-covariance values for the coefficient pairs are provided by the following $m \times m$ matrix:

$$S^2_{Y \cdot 12 \cdots}(\mathbf{X'X})^{-1} = MSE(\mathbf{X'X})^{-1} = \frac{1}{n - m}[\mathbf{Y'Y} - \mathbf{b'X'Y}](\mathbf{X'X})^{-1}$$

The respective standard error for b_k, denoted by s_{b_k}, is the square root of the kth diagonal element in this matrix, with $k = 0, 1, 2, \ldots, m - 1$, so that the intercept a corresponds to b_0. These are not ordinarily determined by hand and will be provided automatically by the popular computer software packages. (For the gasoline mileage illustration, these may be read directly from the computer printouts in Figures 11-3 or 11-4.)

Once the null hypothesis of zero B_j's has been rejected, investigators will generally estimate these with a set of intervals applicable
confidence. The following expression is used.

BONFERRONI INTERVAL ESTIMATES FOR PARTIAL REGRESSION COEFFICIENTS AT THE 100(1 − α)% COLLECTIVE CONFIDENCE LEVEL

$$B_j = b_j \pm t_{\alpha/2(m-1)} s_{b_j}$$

where $n - m$ degrees of freedom apply for the Student t statistic, and s_{b_j} is the standard error of b_j ordinarily provided by computer printouts of multiple regression results. (If a single estimate were to be made, the upper-tail area in determining the critical value would be $\alpha/2$. Since altogether $m - 1$ estimates are made *jointly*, we must divide $\alpha/2$ by $m - 1$ to establish the upper-tail area for the common critical value.)

The computer printout in Figure 11-3 gives the estimated regression coefficients and standard errors for the $m = 4$-variable multiple regression with the gasoline mileage data. The three partial regression coefficients will be estimated using a 94% collective confidence level. The corresponding upper-tail area is

$$\frac{\alpha}{2(m - 1)} = \frac{.06}{2(4 - 1)} = .01$$

so that using $n - m = 20 - 4 = 16$ degrees of freedom, Appendix Table G provides $t_{.01} = 2.583$.

Octane X_1 yields the following interval estimate of the partial regression coefficient:

$$B_1 = 1.2097 \pm 2.583(.0881)$$
$$= 1.2097 \pm .2276$$

or

$$.9821 \leq B_1 \leq 1.4373 \qquad \text{(octane)}$$

The following estimated regression coefficients are obtained analogously for speed X_2 and load X_3:

$$-.4050 \le B_2 \le -.1890 \qquad \text{(speed)}$$

$$-.0129 \le B_3 \le -.0067 \qquad \text{(load)}$$

Notice that the intervals are fairly wide, so that the regression analysis only roughly estimates the B_j's. All of them are significantly nonzero (at the collective .06 level), since no interval overlaps zero.

Problems

11-16 Refer to Problem 11-1 and your answer to that exercise. Using $s_{\hat{Y}} = 16{,}412$, construct a 95% confidence interval for the mean river flow when rainfall is 20 inches and 150,000 cfps are released from upstream dams.

11-17 Refer to Problem 11-2 and your answer to that exercise. Using $s_{\hat{Y}} = 5{,}795$, construct a 95% confidence interval for the mean river flow when rainfall is 20 inches and average daily temperature is 60°F.

11-18 Refer to Problem 11-3 and your answer to that exercise.

(a) Construct a 95% confidence interval for the mean final product yield when settling time is 10 hours and 100 pounds of solid catalyst are used. Use $s_{\hat{Y}} = 9.51$.

(b) Repeat (a), predicting instead the final product yield of a single batch.

11-19 Refer to Problem 11-5 and your answer to that exercise.

(a) Construct a 99% confidence interval for the mean processing time for all jobs requiring 50 k bytes of RAM and 10 k bytes of input. Use $s_{\hat{Y}} = .699$.

(b) Repeat (a), predicting instead the processing time for a single job.

11-20 Refer to Problem 11-7 and your answer to that exercise.

(a) Construct a 95% confidence interval for the mean river flow when rainfall is 20 inches, 150,000 cfps are released from upstream dams, and the average daily temperature is 60°F. Use $s_{\hat{Y}} = 20{,}380$.

(b) Compare your interval from (a) to the counterparts found in Problems 11-16 and 11-17. What do you notice? Explain.

11-21 Refer to Problem 11-15 and your answers to that exercise.

(a) Using given levels $X_1 = 8$ and $X_2 = 9$, determine the level for $s_{\hat{Y}}$.

(b) Using the above given levels, construct 95% interval estimates for (1) the conditional mean and (2) an individual prediction.

(c) Do the partial regression coefficients differ significantly from zero?

(d) Find the values for s_{b_1} and s_{b_2}. Then construct 90% collective confidence interval estimates for B_1 and B_2.

11-22 The following estimated regression equation and standard errors were obtained from a computer multiple regression run based on $n = 25$ observations:

$$\hat{Y} = 11.4 + .553X_1 + .002X_2 - .143X_3$$

$$s_{b_1} = .072 \qquad s_{b_2} = .053 \qquad s_{b_3} = .031$$

(a) Construct an 85% confidence interval estimate for each partial regression coefficient.

(b) Indicate for each case whether the null hypothesis that $B_k = 0$ should be accepted or rejected at the $\alpha = .15$ significance level.

11-23 The multiple regression results for the expanded gasoline mileage illustration are provided in the computer printout in Figure 11-4. The standard errors for each partial regression coefficient are provided there.

(a) Construct an interval estimate for each partial regression coefficient at a 92% collective confidence level.

(b) At the 8% significance level, for which independent variables must the null hypothesis of zero be accepted?

11-24 Refer to Problem 11-7 and your answer to that exercise.

(a) Are the regression results significant?

(b) Construct 90% collective confidence interval estimates for the partial regression coefficients. Which ones differ significantly from zero?

11-25 Referring to Problem 11-8 and your answer to that exercise, repeat (a) and (b) as above using instead a 97% collective confidence level.

11-26 Referring to Problem 11-9 and your answer to that exercise, repeat (a) and (b) as in Problem 11-24, using instead a 99.7% collective confidence level.

11-27 Refer to Problem 11-13 and your answer to that exercise.

(a) Construct 95% confidence intervals for the mean income of all engineers falling into each of the following groups:

Professional Experience	Engineering Education	Positions Held	Number of Employees	Patents Held
(1) 5 yr	5 yr	2	0	2
(2) 10	4	5	10	0
(3) 10	6	2	0	10
(4) 15	4	3	25	1

(b) Using the same groups as above, construct 95% prediction intervals for the income of a single engineer having those characteristics.

(c) Are the regression results significant?

(d) Construct 95% confidence interval estimates of the partial regression coefficients. Which ones differ significantly from zero?

11-6 Multiple Correlation

In Chapter 10 we saw that correlation analysis is sometimes useful in explaining the relationship between two variables. We saw that correlation might be used independently of regression analysis, and that it also may be used to help explain the regression relationship itself. In multivariate evaluations, correlation analysis is primarily an adjunct to multiple regression analysis.

The Coefficient of Multiple Determination

We have seen how the sample coefficient of determination, r^2, expresses the proportion of the total variation in Y that can be explained by the sample regression line. In the same way, when two independent variables X_1 and X_2 are used in predicting Y, a similar index indicates the relative amount of variation explained by the estimated multiple regression plane. This is called the **sample coefficient of multiple determination** and is denoted by the symbol $R^2_{Y \cdot 12}$. We define this quantity as follows:

$$R^2_{Y \cdot 12} = \frac{\text{Explained variation}}{\text{Total variation}} = \frac{SSTO - SSE}{SSTO} = 1 - \frac{\sum (Y - \hat{Y})^2}{\sum (Y - \bar{Y})^2}$$

Taking the positive square root of this, we obtain the sample multiple correlation coefficient, $R_{Y \cdot 12} = \sqrt{R^2_{Y \cdot 12}}$. The sample coefficient of determination $R^2_{Y \cdot 12 \cdots}$ may be found for any multiple regression evaluation involving m variables.

Computation using the above requires a complete set of residual values. The following mathematically equivalent expression is easier to use in calculation.

SAMPLE COEFFICIENT OF
MULTIPLE DETERMINATION

$$R^2_{Y \cdot 12 \cdots} = 1 - \frac{S^2_{Y \cdot 12 \cdots}}{s^2_Y}\left(\frac{n - m}{n - 1}\right)$$

As a practical matter, the sample coefficient of multiple determination is provided by most multiple regression computer programs.

Our earlier evaluation of the gasoline mileage data provides an illustration of the above. Using the data in Table 11-2, we compute the sample standard deviation for gasoline mileage Y:

$$s_Y = \sqrt{\frac{\sum Y^2 - \frac{1}{n}(\sum Y)^2}{n - 1}} = \sqrt{\frac{18{,}414.25 - (\frac{1}{20})(604.5)^2}{20 - 1}} = 2.7457$$

When octane rating X_1 and average speed X_2 are the only independent variables in predicting gasoline mileage Y, then there are $m = 3$ variables altogether, and using $S^2_{Y \cdot 12} = 1.6352$ from before,

$$R^2_{Y \cdot 12} = 1 - \frac{(1.6352)^2}{(2.7457)^2}\left(\frac{20 - 3}{20 - 1}\right) = .683$$

This value indicates the proportion of explained variation in Y. Thus, 68.3% of the variation in gasoline mileage is explained by the estimated regression plane based on octane and speed as the independent predictor variables. Only 31.7% of the variation in Y may be attributable to chance or any other factors not included in the regression evaluation, such as number of stops, load, and so forth.

When the data set was expanded to include load X_3 as the third independent variable, the estimated regression equation was recomputed to be $\hat{Y} = -59.005 + 1.2097X_1 - .2970X_2 - .0098X_3$. We noted that the standard error of the estimate about this estimated regression hyperplane was considerably smaller than that when X_1 and X_2 were the only predictors. This indicates that more of the total variation in Y may be explained by including X_3 in the analysis. The corresponding coefficient of multiple determination reflects this. Using the values from the computer printout in Figure 11-3, where $m = 4$, we have

$$R^2_{Y \cdot 123} = \frac{\text{Explained variation}}{\text{Total variation}} = 1 - \frac{(.7349)^2}{(2.7457)^2}\left(\frac{20-4}{20-1}\right) = .940$$

(This calculation was actually unnecessary here, since the sample coefficient of multiple determination was provided directly in the computer printout.) Thus, the percentage of total variation in gasoline mileage that is explained by regression increases from 68.3% to 94.0% when X_3 is incorporated into the analysis as a third independent variable.

These findings reflect that expanding a multiple regression to include more predictors might improve the reliability of predictions to be made. When a fourth independent variable X_4 (number of stops per mile) was included in a further-expanded multiple regression, the value $R^2_{Y \cdot 1234} = .959$ was determined for the corresponding sample coefficient of multiple determination.

Partial Correlation

Earlier in this chapter we indicated that simple correlation between Y and an independent variable may poorly reflect the worth of that predictor when used in a multiple regression analysis. This is because simple correlation measures the strength of association between only two variables at a time. But much of the effect of an independent variable on Y might be attributed to its relationship with one or more other variables. This indirect effect must be measured another way.

A good measure for assessing the strength of association between an X and Y in multiple regression is one that summarizes the impact of including that X in the analysis. One such quantity is the **coefficient of partial determination**, which expresses the proportional reduction in previously unexplained variation in Y that can be attributed to that predictor's inclusion. This measure of correlation is established by considering a succession of regressions, each followed by an expanded higher-dimensional evaluation. To illustrate, we will continue with the gasoline mileage investigation.

It was established earlier that a simple regression analysis based on octane rating X_1 as the only predictor of gasoline mileage Y provides the following estimated regression line

$$\hat{Y}(X_1) = -51.602 + .8997X_1$$

You may verify that the standard error for Y about this regression line is $s_{Y \cdot X_1} = 1.8968$. Using this, and the standard deviation in gasoline mileage $s_Y = 2.7457$, the sample coefficient of determination is computed:

$$r^2_{Y1} = 1 - \frac{s^2_{Y \cdot X_1}}{s^2_Y}\left(\frac{n-2}{n-1}\right) = 1 - \frac{(1.8968)^2}{(2.7457)^2}\left(\frac{20-2}{20-1}\right) = .548$$

In keeping with our notation, the subscript $Y1$ indicates that the relationship between Y and X_1 is being reported. As we have seen, this quantity expresses the proportion of variation in Y explained by the estimated regression line. When average speed X_2 is incorporated as a second independent variable, we have seen that predictions of mileage will improve. This we know, because the coefficient of multiple determination is $R^2_{Y \cdot 12} = .683$, which indicates that the resulting regression plane explains a somewhat higher proportion of the total variation in Y.

The following expression can be used.

COEFFICIENT OF PARTIAL DETERMINATION

$$r^2_{Y2 \cdot 1} = \frac{\textbf{Reduction in unexplained variation}}{\textbf{Previously unexplained variation}} = \frac{R^2_{Y \cdot 12} - r^2_{Y1}}{1 - r^2_{Y1}}$$

Here the subscript $Y2 \cdot 1$ indicates that the correlation between X_2 and Y is being measured, when the other independent variable X_1 is still considered but held constant. Analogously, by reversing the independent variables, the coefficient of partial determination for Y and X_1 (denoted with subscript $Y1 \cdot 2$) may be computed.

For the gasoline mileage investigation we have

$$r^2_{Y2 \cdot 1} = \frac{.683 - .548}{1 - .548} = .299$$

The difference $R^2_{Y \cdot 12} - r^2_{Y1}$ represents the reduction in unexplained variation (and also the increase in explained variation) in gasoline mileage achieved by incorporating average speed into the analysis. The denominator is the proportion of variation previously left unexplained by the regression line when octane is the only predictor. The coefficient of partial determination for X_2 and Y is .299. The positive square root of this quantity,

$$r_{Y2 \cdot 1} = \sqrt{.299} = .547$$

is the **partial correlation coefficient**. Either quantity expresses the net association between gasoline mileage Y and average speed X_2 when octane rating X_1 is held constant but still accounted for.

To appreciate the improvement in measuring association by the new measures, we may compare the partial and simple correlation coefficients. For mileage Y and speed X_2 the simple correlation coefficient $r_{Y2} = .081$, which in a lower-dimensional analysis would indicate practically no correlation between these variables. The much higher partial correlation coefficient $r_{Y2 \cdot 1} = .547$ more properly reflects the true extent to which speed and mileage are associated.

The coefficient of partial determination extends to higher dimensions. When a third independent variable X_3 is incorporated into the multiple regression analysis, this quantity may be calculated from

$$r^2_{Y3 \cdot 12} = \frac{\text{Reduction in unexplained variation}}{\text{Previously unexplained variation}} = \frac{R^2_{Y \cdot 123} - R^2_{Y \cdot 12}}{1 - R^2_{Y \cdot 12}}$$

Again employing the gasoline mileage data, we may use this expression to gauge the effect of incorporating load X_3 into the multiple regression analysis. With $R^2_{Y \cdot 123} = .940$, we have

$$r^2_{Y3 \cdot 12} = \frac{.940 - .683}{1 - .683} = .811$$

This indicates that including load in the regression analysis brings about a further reduction of 81.1% in gasoline mileage variation left unexplained by octane and speed. This signifies that a substantial benefit will be gained in predicting Y when X_3 is included as one of the predictors. The partial correlation coefficient for X_3 and Y, $r_{Y3 \cdot 12} = \sqrt{.811} = .90$, provides a better summary of the association between these variables than the one provided by the smaller simple correlation coefficient, $r_{Y3} = .478$.

Table 11-7 summarizes for the gasoline mileage investigation how successive higher-dimensional regressions have reduced the variation in Y. These reductions are reflected by the progressively smaller standard errors in column (3) and by the progression of increasing coefficients of determination in column (4). The latter values provide the proportion of total variation in Y that is explained by the respective estimated regression surface. Further computations in column (5) provide the proportional reduction in previously unexplained variation from including the newest independent variable. These quantities are the coefficients of partial determination, which most meaningfully measure correlation between Y and the new predictor variable.

Table 11-7 *Summary of How Successive Higher-Dimensional Regression Analyses Increase the Explained Variation in Y as Additional Independent Variables Are Included in the Gasoline Mileage Illustration*

(1) Independent Variables Included	(2) New Variable	(3) Standard Error of Estimate (mpg)	(4) Proportion of Variation Explained	(5) Proportional Reduction in Previously Unexplained Variation
none	—	$s_Y = 2.7457$	0	—
X_1	X_1	$s_{Y \cdot X_1} = 1.8968$	$r^2_{Y1} = .548$	$\frac{.548 - 0}{1 - 0} = .548 = r^2_{Y1}$
X_1, X_2	X_2	$S_{Y \cdot 12} = 1.6352$	$R^2_{Y \cdot 12} = .683$	$\frac{.683 - .548}{1 - .548} = .299 = r^2_{Y2 \cdot 1}$
X_1, X_2, X_3	X_3	$S_{Y \cdot 123} = .7349$	$R^2_{Y \cdot 123} = .940$	$\frac{.940 - .683}{1 - .683} = .811 = r^2_{Y3 \cdot 12}$
X_1, X_2, X_3, X_4	X_4	$S_{Y \cdot 1234} = .6281$	$R^2_{Y \cdot 1234} = .959$	$\frac{.959 - .940}{1 - .940} = .317 = r^2_{Y4 \cdot 123}$

Notice that the inclusion of X_4 does provide an estimated regression equation that explains more variation in Y than before, although the improvement is a modest one. In some cases, inclusion of a new predictor variable will actually provide an estimated multiple regression equation that explains less, rather than more, of the variation in Y. Although partial correlation analysis may help in selecting the final set of predictors, judgment plays the key role in establishing candidate independent variables that have some logical connection to Y.

Problems

11-28 From the data in Problem 11-1, the following results have been obtained for predicting seasonal mean river flow Y from rainfall X_1 and dam releases X_2:

$$r_{Y1} = .772 \qquad r_{Y2} = -.712 \qquad R^2_{Y\cdot12} = .647$$

(a) Compute the coefficient of partial determination $r^2_{Y2\cdot1}$ for river flow and dam releases, holding rainfall constant.
(b) Compute the coefficient of partial determination $r^2_{Y1\cdot2}$ for river flow and rainfall, holding dam releases constant.

11-29 From the data in Problem 11-3, the following results have been obtained for predicting yield Y from settling time X_1 and weight X_2 of solid catalyst:

$$r_{Y1} = .751 \qquad r_{Y2} = .879 \qquad R^2_{Y\cdot12} = .788$$

(a) Compute the coefficient of partial determination $r^2_{Y2\cdot1}$ for yield and weight of solid catalyst, holding settling time constant.
(b) Compute the coefficient of partial determination $r^2_{Y1\cdot2}$ for yield and settling time, holding weight of solid catalyst constant.

11-30 The same predictors used in the gasoline mileage investigation might be considered in a different sequence. Different values will then be achieved for the estimated multiple regression coefficients, standard errors, and coefficients of multiple and partial determination. Answer the following, assuming that load X_3 is the first independent variable, followed by octane X_1, then by number of stops X_4, and finally by average speed X_2. For your convenience, the following succession of computer-generated results apply:

$$\hat{Y}(X_3) = 34.555 - .0092X_3 \qquad s_{Y\cdot X_3} = 2.4771$$

$$\hat{Y} = -45.536 + .8777X_1 - .0086X_3 \qquad S_{Y\cdot13} = 1.4528$$

$$\hat{Y} = -42.907 + .8540X_1 - .0079X_3 - 1.4187X_4 \qquad S_{Y\cdot134} = 1.3626$$

$$\hat{Y} = -56.526 + 1.1763X_1 - .2814X_2 - .0093X_3 - .9555X_4 \qquad S_{Y\cdot1234} = .6281$$

(a) Compute the following coefficients of multiple determination:
(1) $R^2_{Y\cdot13}$ (2) $R^2_{Y\cdot134}$ (3) $R^2_{Y\cdot1234}$
(b) Compute the following coefficients of partial determination:
(1) $r^2_{Y1\cdot3}$ (2) $r^2_{Y4\cdot13}$ (3) $r^2_{Y2\cdot134}$
(c) How do you explain the difference (if any) between your value for $r^2_{Y4\cdot13}$ and the value $r^2_{Y4\cdot123} = .317$ computed in Table 11-7?

11-31 Refer to Problem 11-30. Assume instead the following sequence of variable incorporation: (1) X_2 (average speed), (2) X_3 (load), (3) X_4 (number of stops per mile), and (4) X_1 (octane rating). The following regression results apply:

$\hat{Y}(X_2) = 27.696 + .046X_2$ $\qquad s_{Y \cdot X_2} = 2.8116$

$\hat{Y} = 34.125 + .0076X_2 - .00914X_3$ $\qquad S_{Y \cdot 23} = 2.5486$

$\hat{Y} = 33.9377 + .0218X_2 - .0081X_3 - 1.888X_4$ $\qquad S_{Y \cdot 234} = 2.4935$

$\hat{Y} = -56.526 + 1.1763X_1 - .2814X_2 - .0093X_3 - .9555X_4$ $\qquad S_{Y \cdot 1234} = .6281$

(a) Compute the following coefficients of multiple determination:
(1) $R^2_{Y \cdot 23}$ (2) $R^2_{Y \cdot 234}$ (3) $R^2_{Y \cdot 1234}$

(b) Compute the following coefficients of partial determination:
(1) $r^2_{Y3 \cdot 2}$ (2) $r^2_{Y4 \cdot 23}$ (3) $r^2_{Y1 \cdot 234}$

11-32 Using the data in Problem 11-13, we apply the following progression of regressions:

New Independent Variable	Estimated Regression Equation	Standard Deviation of Error
none	—	24,668
(1) X_1	$\hat{Y}(X_1) = 28{,}992 + 2{,}437X_1$	20,131
(2) X_2	$\hat{Y} = 43{,}681 + 2{,}527X_1 - 2{,}851X_2$	19,963
(3) X_3	$\hat{Y} = 41{,}137 + 3{,}082X_1 - 1{,}733X_2 - 2{,}088X_3$	19,746
(4) X_4	$\hat{Y} = 37{,}301 - 170X_1 + 607X_2 + 1{,}410X_3 + 238X_4$	8,534
(5) X_5	$\hat{Y} = 39{,}647 - 203X_1 - 481X_2 + 1{,}600X_3 + 241X_4 + 833X_5$	7,044

(a) Determine the coefficient of (multiple) determination for each of the regressions in the table.

(b) Determine the coefficient of partial determination for the newly added variable, holding earlier variables constant, in each of the multiple regressions in the table.

11-33 A *linear* multiple regression equation might provide a poor fit to the data. Consider the experimental data in Table 11-6, where stainless steel weight losses Y are given for cerium concentration X_1, temperature X_2, and contact time X_3.

(a) Using the original raw data, find the estimated multiple regression equation and the sample coefficient of multiple determination. What percentage of the variation in Y is explained by the regression hyperplane?

(b) Transform each variable into its natural logarithm:

$$Y' = \ln(Y) \qquad X_i' = \ln(X_i)$$

The transformed raw data adhere fairly well to the linear model. Using those data, find the estimated multiple regression equation and the sample coefficient of multiple determination. What percentage of the variation in $\ln(Y)$ is explained by the resulting regression hyperplane?

(c) Make point forecasts for the following cases using regression equation (1) from (a) and (2) from (b). In the latter case, make the inverse transformation:

$$Y = e^{Y'}$$

A. 0.01 M Ce, 25°C, 3 hours
B. 0.05 M Ce, 45°C, 6 hours
C. 0.10 M Ce, 50°C, 1 hour

11-7 Polynomial Regression

Chapter 10 introduced curvilinear regression, in which various nonlinear relationships between X and Y could be treated as essentially linear. This was made possible by transforming one or both variables, using an inverse or a logarithm to a convenient base. But such a procedure is limited to a narrow class of functional shapes. More generally, it is possible to fit Y to a polynomial in the independent variable X. As long as the polynomial involves high enough powers of X, virtually any curve form can be selected to fit the data. This may be accomplished by extending the least-squares multiple regression procedure.

Parabolic Regression

The simplest such polynomial is an estimated regression equation of the form

$$\hat{Y} = a + b_1X + b_2X^2$$

which gives a parabola. This equation is equivalent to that for the estimated regression plane when $X_1 = X$ and $X_2 = X^2$. Making these variable substitutions, we can perform a multiple regression to find the values of a, b_1, and b_2.

Illustration: *Finding a Stress-Strain Curve*

Consider the stress-strain data in Table 11-8 obtained from testing an experimental alloy. The above least-squares procedure provides the stress-strain curve shown in Figure 11-5. A parabola was determined to provide the best fit over the middle-strength range.

Table 11-8 *Test Data for Experimental Alloy Tension Member*

Test	Stress (ksi) Y	Strain (in./in.) X
1	91	.001
2	97	.002
3	108	.003
4	111	.005
5	114	.006
6	110	.006
7	112	.009
8	105	.011
9	98	.016
10	91	.017

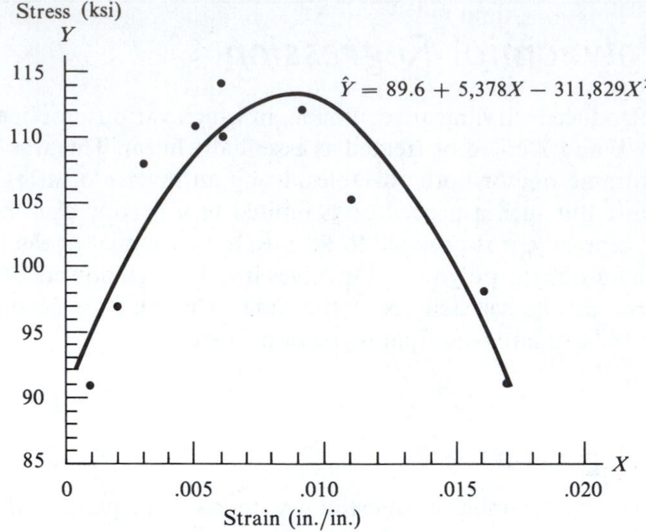

Figure 11-5 *Scatter diagram and parabolic stress-strain regression curve for experimental alloy tension member.*

The intermediate calculations are given in Table 11-9. Substituting the appropriate values into the expression given above, we obtain the following normal equations:

$$1{,}037 = 10a + .076b_1 + 858 \times 10^{-6}b_2$$

$$7.786 = .076a + 858 \times 10^{-6}b_1 + 11{,}662 \times 10^{-9}b_2$$

$$.085454 = 858 \times 10^{-6}a + 11{,}662 \times 10^{-9}b_1 + 173{,}574 \times 10^{-12}b_2$$

Table 11-9 *Intermediate Calculations in Support of Parabolic Regression to Find Stress-Strain Curve*

Y	X_1 (X)	X_1Y	X_2 (X^2)	X_2Y	X_1X_2	X_2^2
91	.001	.091	1×10^{-6}	.000091	1×10^{-9}	1×10^{-12}
97	.002	.194	4×10^{-6}	.000388	8×10^{-9}	16×10^{-12}
108	.003	.324	9×10^{-6}	.000972	27×10^{-9}	81×10^{-12}
111	.005	.555	25×10^{-6}	.002775	125×10^{-9}	625×10^{-12}
114	.006	.684	36×10^{-6}	.004104	216×10^{-9}	$1{,}296 \times 10^{-12}$
110	.006	.660	36×10^{-6}	.003960	216×10^{-9}	$1{,}296 \times 10^{-12}$
112	.009	1.008	81×10^{-6}	.009072	729×10^{-9}	$6{,}561 \times 10^{-12}$
105	.011	1.155	121×10^{-6}	.012705	$1{,}331 \times 10^{-9}$	$14{,}641 \times 10^{-12}$
98	.016	1.568	256×10^{-6}	.025088	$4{,}096 \times 10^{-9}$	$65{,}536 \times 10^{-12}$
91	.017	1.547	289×10^{-6}	.026299	$4{,}913 \times 10^{-9}$	$83{,}521 \times 10^{-12}$
1,037 $= \sum Y$	.076 $= \sum X_1$	7.786 $= \sum X_1Y$	858×10^{-6} $= \sum X_2$	.085454 $= \sum X_2Y$	$11{,}662 \times 10^{-9}$ $= \sum X_1X_2$	$173{,}574 \times 10^{-12}$ $= \sum X_2^2$

Solving these equations simultaneously, we obtain the following estimated regression coefficients:

$$a = 89.6 \qquad b_1 = 5{,}378 \qquad b_2 = -311{,}829$$

The estimated parabolic regression equation for the stress-strain curve is

$$\hat{Y} = 89.6 + 5{,}378X_1 - 311{,}829X_2$$

or,

$$\hat{Y} = 89.6 + 5{,}378X - 311{,}829X^2$$

Regression with Higher-Power Polynomials

The parabolic regression procedure may be extended by incorporating higher powers of X as additional independent variables. In the case of a third-degree polynomial, the estimated regression equation will be of the form

$$\hat{Y} = a + b_1X + b_2X^2 + b_3X^3$$

A perfect fit can be achieved for any data when the highest power of X matches the number of data points. A close fit to almost any data might be achieved using a polynomial involving modest powers. However, unless this is carefully done, predictions of Y made from a higher-power regression equation may substantially miss subsequent actual levels. The greatest challenge in such a curve-fitting exercise is providing a rationale that explains the regression coefficients. For this reason, polynomials involving powers higher than X^3 are rarely employed in regression analysis. Figure 11-6 shows a variety of curve shapes that might be encountered in third-power polynomial regression.

The least-squares procedure for higher-degree polynomial regression is really a simple extension of multiple regression, using X^2, X^3, and so on as additional independent variables. As we have seen, such regressions will usually be accomplished with the assistance of a digital computer.

Polynomial Multiple Regression

The linear multiple regression procedures described so far are based on a linear relationship between Y and two or more independent variables. In some applications the relationship between the several variables may be a nonlinear one. In such cases, a model expressing Y as a higher-degree polynomial in the several independent variables might be employed. When there are two independent variables, the following second-degree polynomial might be fitted to the sample data:

$$\hat{Y} = a + b_1X_1 + b_2X_2 + b_{11}X_1^2 + b_{22}X_2^2 + b_{12}X_1X_2$$

In solving this equation, a multiple regression may be performed using X_1^2, X_2^2, and X_1X_2 as additional independent predictors in the linear model. The coefficient b_{12} provides a measure of the *interaction* between X_1 and X_2.

Polynomial multiple regression is a valuable tool for an empirical investigation seeking a model to explain relationships not yet firmly established.

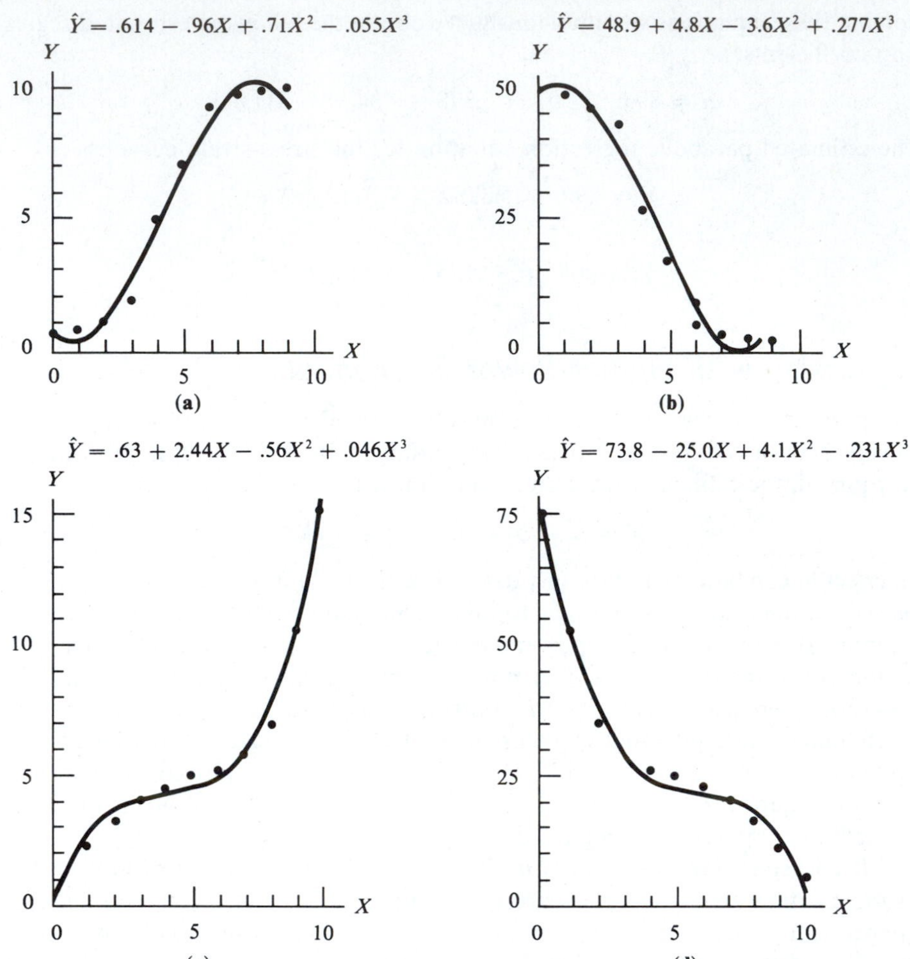

Figure 11-6 Scatter diagrams and regression curves for third-power polynomials.

Problems

11-34 Plot a scatter diagram for the aquifer contamination data in Problem 10-33. Then fit a regression parabola and sketch the curve on your graph.

11-35 Plot a scatter diagram for the bitumen viscosity data in Problem 10-37. Then fit a third-power regression polynomial and sketch the curve on your graph.

11-36 Consider the following sample observations:

Y	X
50	10
61	18
72	25
69	29
70	37

Y	X
55	38
41	41
21	43
15	49
18	55

(a) Plot these data on a scatter diagram.
(b) Fit a parabola to the data and sketch the resulting regression curve on your graph.
(c) Fit a third-power polynomial to the data and sketch the resulting regression curve on your graph.
(d) Which curve do you think provides the better fit?

11-37 The following stress-strain data were collected from testing a particular material:

Conventional Stress (ksi)	Strain (in./in.)	Conventional Stress (ksi)	Strain (in./in.)
18	.0015	25	.0091
23	.0022	22	.0109
29	.0029	18	.0118
28	.0054	19	.0119
33	.0069	15	.0125

Plot these data on a scatter diagram using conventional stress as the dependent variable. Then fit a regression parabola and sketch the resulting curve on your graph.

11-38 The following data apply to the material in Problem 11-37, where, after the onset of plastic instability, the smallest cross-sectional area of the metal bar is used in computing true stress.

True Stress (ksi)	Strain (in./in.)	True Stress (ksi)	Strain (in./in.)
18	.0015	39	.0091
23	.0022	40	.0109
30	.0029	49	.0118
32	.0054	58	.0119
35	.0069	69	.0125

Plot these data on a scatter diagram using true stress as the dependent variable. Then fit a third-power regression polynomial and sketch the resulting curve on your graph.

11-39 Consider the following data:

Y	X_1	X_2	Y	X_1	X_2
103	5	10	251	50	50
105	10	5	267	50	60
131	20	30	308	60	50
142	25	40	415	70	70
175	40	20	474	80	70

Extend the linear multiple regression procedure to fit a second-degree polynomial (including X_1^2, X_2^2, and X_1X_2 terms) to these data.

11-8 Multiple Regression with Indicator Variables

Our discussions of multivariate relationships have so far been limited to quantitative variables such as time, weight, and volume. Regression analysis is concerned mainly with finding an appropriate functional relationship between such values. An interesting extension of multiple regression analysis considers a wider class of applications involving *qualitative* variables as well as quantitative ones.

A variety of qualitative variables might be important in engineering evaluations. A chemical engineer might use regression to predict chemical yield in response to different temperature or pressure settings. But slightly different raw feedstocks from two suppliers might provide a different response, so it may be helpful to further categorize each production run by *type* of feedstock. Or consider a computer designer simulating a proposed microcircuit to predict how processing time will respond to the number of inputs and volume of computations. Different results may be achieved for test runs involving main memory only as opposed to those needing peripheral access as well; thus, *memory classification* may be a crucial qualitative variable. As a further illustration, a tire-tread evaluation might be based on test runs involving cars from one or more manufacturers, so an important variable might be *car model.*

Qualitative factors may be incorporated into regression evaluations by using **indicator variables**. Such a quantity assumes a value of 0 or 1, depending upon which category applies in a two-way classification.

The Basic Multiple Regression Model

A typical application involves estimating the relationship between Y and a single quantitative independent variable X_1. But the sample observations fall into two categories, so an indicator variable X_2 is incorporated as the second independent variable. The estimated regression equation takes the form

$$\hat{Y} = a + b_1X_1 + b_2X_2$$

where $X_2 = 0$ if one category applies and $X_2 = 1$ if the second is indicated. Two estimated regression lines will apply:

$$\hat{Y} = a + b_1X_1 + b_2(0) = a + b_1X_1 \qquad \text{(when } X_2 = 0\text{)}$$

and

$$\hat{Y} = a + b_1X_1 + b_2(1) = (a + b_2) + b_1X_1 \qquad \text{(when } X_2 = 1\text{)}$$

Illustration:* *Predicting Chemical Yields

Consider the chemical processing data in Table 11-10, where the dependent variable Y is the final yield and the independent variable is the amount X_1 of active ingredient. An expensive catalyst is used on some of the sample runs. The following

Table 11-10 *Chemical Yield and Ingredient Data from Processing Test Batches Where Catalyst Is Present or Absent*

Yield (kg) Y	Amount of Active Ingredient (kg) X_1	Catalyst	Indicator Variable X_2
21	50	Absent	0
33	100	Absent	0
41	90	Present	1
35	80	Present	1
24	60	Absent	0
20	60	Absent	0
27	70	Absent	0
23	50	Absent	0
24	80	Absent	0
26	90	Absent	0
39	90	Present	1
43	110	Present	1
47	130	Present	1
45	130	Present	1
53	140	Present	1
55	150	Present	1

indicator variable is defined:

$$X_2 = \begin{cases} 0 & \text{if no catalyst is present} \\ 1 & \text{if catalyst is present} \end{cases}$$

The following estimated multiple regression equation applies:

$$\hat{Y} = 9.000 + .225X_1 + 9.875X_2$$

The partial regression coefficient $b_1 = .225$ indicates that .225 kg of yield is estimated for every 1 kg of active ingredient used in processing a batch of chemical. The second coefficient, $b_2 = 9.875$, signifies that batches run with the catalyst present will provide yields that are estimated to be 9.875 kg greater than those run with an identical quantity of active ingredient, but with the catalyst absent. This is presumed true regardless of the amount of active ingredient used.

Figure 11-7 should help demonstrate the advantage of performing the multiple regression with the indicator variable. The crosses represent those sample observations resulting from test runs where the catalyst was present. The circles apply to the outcomes where the catalyst is absent. The bottom yield regression line

$$\hat{Y} = 9.000 + .225X_1 \qquad \text{(catalyst absent)}$$

applies when the catalyst is absent. The higher parallel line

$$\begin{aligned} \hat{Y} &= (9.000 + 9.875) + .225X_1 \\ &= 18.875 + .225X_1 \qquad \text{(catalyst present)} \end{aligned}$$

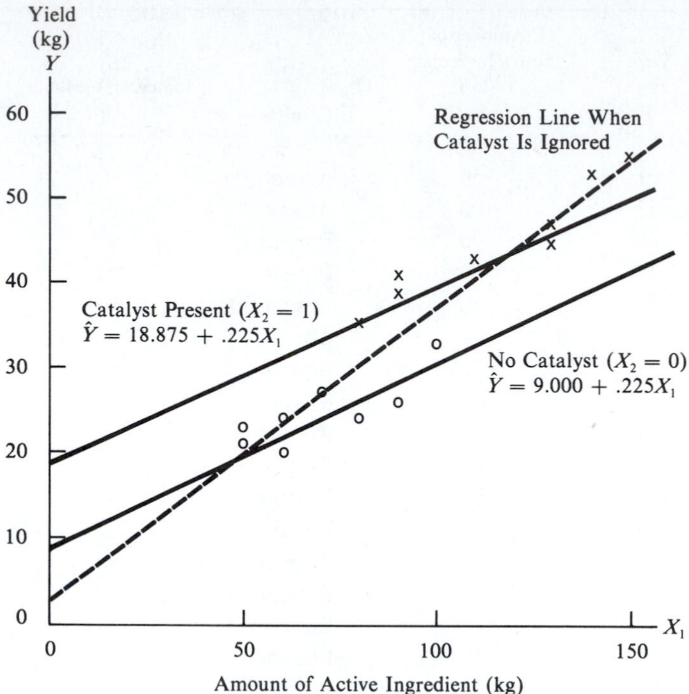

Figure 11-7 *Scatter diagram for chemical processing data showing linear projections from multiple regression with indicator variable for catalyst (no interaction assumed).*

is applicable when the catalyst is used. Notice that each line falls near the middle of the scatter for the respective data points for the category it represents. They were not, however, chosen so that the respective sums of squared deviations about them were minimized. These lines are instead the X_1-Y *projections* of perpendicular slices made through the estimated regression plane at $X_2 = 0$ and $X_2 = 1$, respectively. The least-squares criterion applies to that plane, not to the linear projections.

Advantages of Using Indicator Variables

The multiple regression using X_2 as a catalyst indicator variable is superior to the result that would be achieved with a single variable regression evaluation employing amount of active ingredient as the only predictor. Ignoring the catalyst effect, the dashed line in Figure 11-7 shows that a single line fits the sample data poorly.

But there is no reason why the effect of the catalyst should be ignored. Why not perform separate simple regression evaluations, the first based only on those points where no catalyst was used and the second for those involving the catalyst?

Upon reflection, the answer should be obvious. In any scientific sampling investigation, greater precision and reliability accrue as the number of observations increases. The standard error obtained from the multiple regression will be very close in value to what would be obtained by averaging the pooled squared deviations about the two separate regression lines. But the multiple regression represents a greater amount of information per observation. It should therefore provide more precise confidence intervals for the conditional mean Y than what would be possible from either simple regression analysis alone. In effect, the multiple regression takes greater advantage of existing information.

Of course, the underlying model should have a meaningful rationale. It seems perfectly logical that the Y-intercept should be greater for the catalyst case than it is for no catalyst since the greater yields are expected when the catalyst is included. But the same slope applies to both linear projections. This can be true only if it is assumed that yield responds to the active-ingredient levels at a constant rate—regardless of the presence or absence of the catalyst. That assumption may be inappropriate, and a greater slope might apply when the catalyst is present.

We next consider a procedure even better suited to the chemical processing illustration because it allows both for different *slopes* and for different Y-intercepts.

Interactive Multiple Regression with Indicator Variables

As we have seen, a linear model may not be appropriate for characterizing the relationship between two or more variables. In the previous section, we encountered a second-degree polynomial in two independent variables involving X_1^2, X_2^2, and X_1X_2 terms. As a special case, we may drop the squared terms, leaving X_1X_2 as the only additional term. This product characterizes the *interaction* between X_1 and X_2. When X_2 is a zero-one indicator variable, X_1X_2 becomes 0 when $X_2 = 0$; X_1X_2 becomes X_1 when $X_2 = 1$. If we apply a linear multiple regression using X_1X_2 as the third independent variable, the following estimated multiple regression equation applies:

$$\hat{Y} = a + b_1X_1 + b_2X_2 + b_{12}X_1X_2$$

This regression surface gives two linear projections:

$$\hat{Y} = a + b_1X_1 \qquad \text{(when } X_2 = 0\text{)}$$

and

$$\hat{Y} = (a + b_2) + (b_1 + b_{12})X_1 \qquad \text{(when } X_2 = 1\text{)}$$

A computer run with the data in Table 11-10 gives the following estimated multiple regression equation:

$$\hat{Y} = 12.208 + .179X_1 + 4.031X_2 + .069X_1X_2$$

Setting X_2 at the respective indicator values, we find that this equation gives

$$\hat{Y} = 12.208 + .179X_1 \qquad \text{(catalyst absent)}$$

and

$$\begin{aligned}\hat{Y} &= (12.208 + 4.031) + (.179 + .069)X_1 \\ &= 16.239 + .248X_1 \qquad \text{(catalyst present)}\end{aligned}$$

Figure 11-8 shows these linear projections.

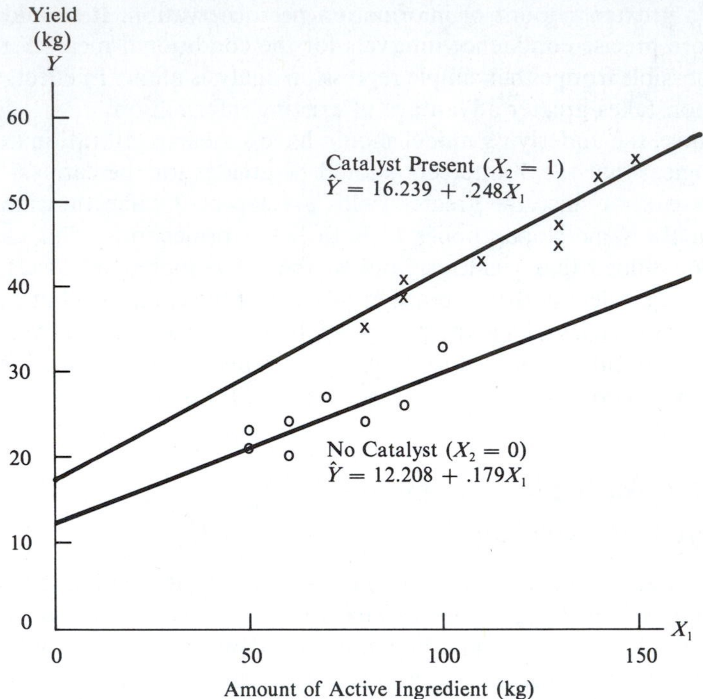

Figure 11-8 *Scatter diagram for chemical processing data showing linear projections from multiple regression with catalyst indicator variable. Interaction between independent variables is assumed.*

Illustration: *Polymer Production for Oil-Recovery Enhancement*

Researchers conducted an experiment with xanthan biopolymer broth, a chemical used to thicken water for oil-recovery enhancement. The final chemical is produced from fermentation using a beginning cell concentration and feed sugar. Table 11-11 shows partial results from 22 test runs. The main variable of interest is the viscosity Y of the final product. The level is affected by four variables: percentage weight of feed sugar X_1, starting cell concentration X_2, presence or absence of logarithmic growth inoculum X_3 (indicator variable), and whether or not sugar is sterilized separately X_4 (indicator variable).

Figure 11-9 shows the results of a multiple regression analysis run with SAS. Included is a fifth indicator interaction variable X_3X_4. The following estimated multiple regression equation was obtained:

$$\hat{Y} = -4{,}232.4 + 5{,}043.9X_1 - 20.29X_2 + 408.6X_3 + 2{,}500.7X_4 - 935.0X_3X_4$$

This equation provides an estimated increase in viscosity of 408.6 when $X_3 = 1$ (logarithmic growth inoculum is present) or an increase of 2,500.7 when $X_4 = 1$ (sugar is sterilized separately). The coefficient of the indicator-variable product is −935.0, which requires that the estimated viscosity be reduced by that amount when X_3 and X_4 each equal 1 (yes for both factors).

Table 11-11 *Results of Fermentation to Produce Polymer Used in Oil-Recovery Enhancement*

Test Run	Yield Viscosity (cp) Y	Feed Sugar (%) X_1	Beg. Cell Concent. ($\times 10^{-8}$/ml) X_2	Growth Inoculum (no = 0, yes = 1) X_3	Separate Sugar Steril. (no = 0, yes = 1) X_4	Interaction X_3X_4
1	6,300	2.00	5.2	0	0	0
2	3,900	1.50	2.2	0	0	0
3	4,900	2.00	2.7	0	0	0
4	3,000	1.50	3.6	0	0	0
5	8,000	2.00	4.6	1	1	1
6	5,200	1.50	0.9	1	1	1
7	3,700	1.25	3.6	1	1	1
8	2,400	1.25	4.0	1	0	0
9	3,900	1.25	2.8	1	1	1
10	4,200	1.25	1.9	1	1	1
11	4,400	1.25	1.5	1	1	1
12	4,000	1.25	1.5	1	1	1
13	2,700	1.25	2.0	1	1	1
14	4,100	1.25	1.9	1	1	1
15	4,300	1.25	12.0	0	1	0
16	4,400	1.25	10.0	0	1	0
17	4,200	1.25	1.2	1	1	1
18	4,600	1.25	1.8	1	1	1
19	4,500	1.25	0.2	1	1	1
20	3,500	1.25	0.2	1	1	1
21	3,300	1.25	1.8	1	1	1
22	4,900	1.25	1.0	1	1	1

Source: Norton et al., *Society of Petroleum Engineers Journal*, April 1981, pp. 205–217.

Problems

11-40 The following data have been obtained for the return signal achieved from radar bursts aimed at satellite targets:

Mean Signal-to-Noise Ratio	Target Altitude (miles)	Target in Earth's Shadow
43	90	Yes
42	45	Yes
17	250	No
11	285	No
25	155	No
26	258	Yes

Mean Signal-to-Noise Ratio	Target Altitude (miles)	Target in Earth's Shadow
27	212	Yes
39	78	Yes
34	85	No
24	135	No
30	215	Yes

Figure 11-9 *SAS multiple regression report for oil-recovery enhancement data.*

```
TITLE 'OIL-RECOVERY ENHANCEMENT';
DATA OIL;
  INPUT VISCOS @@ SUGAR @@ CELL @@ GROWTH @@ SEPARATE @@ INTERACT @@;
  CARDS;
6300 2.00 5.2 0 0 0
3900 1.50 2.2 0 0 0
4900 2.00 2.7 0 0 0
        •
        •
        •
4900 1.25 1.0 1 1 1
;
PROC REG;
MODEL VISCOS = SUGAR CELL GROWTH SEPARATE INTERACT;
QUIT;
```

```
                        OIL-RECOVERY ENHANCEMENT

Model: MODEL1
Dep Variable: VISCOS

                          Analysis of Variance

                          Sum of          Mean
       Source      DF     Squares        Square        F Value       Prob>F

       Model        5 23931512.261 4786302.4523        13.280        0.0001
       Error       16 5766669.5569 360416.84730
       C Total     21 29698181.818

           Root MSE      600.34727     R-Square        0.8058
           Dep Mean     4290.90909     Adj R-Sq        0.7451
           C.V.           13.99114

                          Parameter Estimates

                        Parameter        Standard       T for H0:
Variable      DF         Estimate           Error     Parameter=0      Prob > |T|

INTERCEP       1     -4232.362222    1230.9139387          -3.438          0.0034
SUGAR          1      5043.922101    781.97732144           6.450          0.0001
CELL           1       -20.292395    136.68487861          -0.148          0.8838
GROWTH         1       408.629178    800.97083972           0.510          0.6169
SEPARATE       1      2500.675946    1384.4019803           1.806          0.0897
INTERACT       1      -935.049305    1804.2575445          -0.518          0.6114
```

(a) Using an indicator variable (with value 0 if the target is not in Earth's shadow), determine the estimated multiple regression equation expressing mean signal-to-noise ratio as the dependent variable. Assume no interaction between the indicator and target altitude.

(b) Give the linear equation for predicting the mean signal-to-noise ratio when satellite targets are in the shadow.

(c) Give the linear equation for predicting the mean signal-to-noise ratio when satellite targets are not shadowed by Earth.

11-41 Refer to the reaction time and temperature data in Problem 10-40 for refining a chemical feedstock. A stabilizer was present for all test batches run below 250°, but because that additive breaks down at the higher temperatures, it could not be used for the 5 test batches having the hottest retort temperatures.

(a) Using reaction time as the dependent variable, find the estimated multiple regression equation when an indicator variable X_2 is used for the stabilizer (having value 1 when the stabilizer is present). Ignore any interaction between the indicator and temperature.

(b) Determine the linear equations for predicting reaction time from retort temperature when (1) the stabilizer is not present and (2) the stabilizer is present.

11-42 Refer to the mileage and speed data in Problem 10-43 for automobile test-track runs. The first four runs were made in the rain.

(a) Using mileage as the dependent variable, find the estimated multiple regression equation when an indicator variable X_2 is used for rain (having value 1 for test runs made in the rain). Ignore any interaction between the indicator and speed.

(b) Determine the linear equations for predicting mileage from speed when it is (1) not raining and (2) raining.

11-43 The following data apply to the total processing time of various sizes of computer runs with a magneto-hydrodynamics software package. Only some runs required peripheral memory:

Processing Time (sec)	Amount of Input (k bytes)	Peripheral Memory Needed
550	3.5	No
915	6.1	Yes
748	4.4	Yes
910	5.6	Yes
707	7.2	No
589	4.5	Yes
786	10.2	No
1,004	12.6	No
648	8.4	No
1,025	5.9	Yes
1,284	11.4	Yes
627	7.7	No
849	6.1	No
753	8.3	No
1,806	12.5	Yes
2,253	14.2	Yes

(a) Treating processing time as the dependent variable Y and amount of input X_1 as the independent variable, plot these data on a scatter diagram. Use a cross for those points involving no peripheral memory and circles for the rest.

(b) Determine the estimated regression equation using X_1 as the only independent variable. Plot this line on your scatter diagram.

(c) Let X_2 be an indicator variable having value $X_2 = 0$ if peripheral memory is not needed and $X_2 = 1$ if it is needed. Determine the estimated multiple regression equation when no interaction is assumed between X_1 and X_2.

(d) Determine the estimated multiple regression equation assuming interaction between amount of input and need for peripheral memory. Then determine the expression for the projected lines when $X_2 = 0$ and $X_2 = 1$. Plot these on your scatter diagram.

11-44 Refer to Problem 11-43. Separate the data into two sets: (1) the case of no peripheral memory and (2) computer runs made when peripheral memory is needed.

(a) Perform a simple regression analysis on the two data sets and determine the expressions for each estimated regression line. Determine for each the standard error of the estimate.

(b) Using the appropriate results from (a), construct a 95% confidence interval for the mean processing time for jobs involving 10 k bytes of input when: (1) no peripheral memory is needed and (2) peripheral memory is needed.

(c) Construct 95% confidence intervals for the same jobs using your interactive estimated multiple regression results from Problem 11-43.

11-45 Refer to the polymer production illustration on page 488. The researchers were interested in a second dependent variable, fermentation time (hours). The following data apply:

Run	Time	Run	Time	Run	Time
1	42	9	32	16	33
2	32	10	34	17	40
3	33	11	30	18	36
4	26	12	31	19	36
5	48	13	47	20	34
6	35	14	31	21	27
7	31	15	33	22	42
8	23				

Find the estimated multiple regression equation using fermentation time, instead of viscosity, as Y.

Comprehensive Problems

11-46 Consider the following data:

Y	X_1	X_2	X_3	Y	X_1	X_2	X_3
40	5	2	8	82	5	10	15
39	6	5	5	73	7	8	12
42	10	10	2	52	10	5	5
55	4	6	8	50	5	5	10
42	5	2	10	69	4	8	12
41	1	9	3	112	12	16	20
36	3	1	9	61	15	10	5
29	4	4	4				

Determine the estimated multiple regression equations for each of the following cases:

Independent Variables
(a) X_1 and X_2
(b) X_1 and X_3
(c) X_2 and X_3
(d) X_1, X_2, and X_3

11-47 The following data apply to a random sample of warehouses constructed under the supervision of a particular architectural and engineering firm:

Heating and Cooling Cost Y	Floorspace (k sq. ft.) X_1	Winter/Spring Mean Temp. (°F) X_2	Summer/Fall Mean Temp. (°F) X_3	Energy Cost Index X_4
$8,500	45	50	70	82
11,120	40	38	65	95
11,400	45	42	72	87
8,010	25	45	68	100
8,760	30	25	65	105
15,500	60	34	71	111
17,950	75	37	72	110
18,700	80	48	75	89
13,950	55	51	81	97
8,540	25	52	69	124
10,410	35	45	68	110
12,260	45	22	71	98
13,200	50	24	70	112
22,730	75	25	65	129
12,090	45	35	72	105

(a) Perform a simple regression analysis and find the estimated regression equation for the regression line expressing heating and cooling cost in terms of floorspace. Then calculate the standard error of the estimate using that regression line.

(b) Compute the estimated coefficient of determination applicable to the two variables being considered. What percentage of the variation in Y is explained by the regression line?

(c) Using your results from (a), predict the cost of a 50-thousand-square-foot warehouse. Construct a 95% confidence interval for the mean cost for all such warehouses.

11-48 Refer to the data in Problem 11-47.

(a) Perform a multiple regression analysis and determine the equation for the estimated regression plane using both floorspace and winter/spring mean temperature as predictors of heating and cooling cost. Then calculate the standard error of the estimate.

(b) Compute the estimated coefficient of multiple determination applicable to your results from (a). What percentage of the variation in Y is explained by the regression plane?

(c) Compute the coefficient of partial determination for Y and X_2. What is the proportional reduction in previously unexplained variation in heating and cooling cost attributable to the inclusion of winter/spring mean temperature?

(d) Using your results from (a), predict the cost of a 50-thousand-square-foot warehouse when the mean winter/spring temperature is 40 degrees. Construct a 95% confidence interval for the mean cost for all such warehouses.

11-49 Refer to the data in Problem 11-47 and your answers to Problem 11-48.

(a) Perform a multiple regression analysis and determine the equation for the estimated regression hyperplane incorporating summer/fall mean temperature as the third predictor of heating and cooling cost. Then calculate the standard error of the estimate.

(b) Compute the estimated coefficient of multiple determination applicable to your results from (a). What percentage of the variation in Y is explained by the regression hyperplane?

(c) Compute the coefficient of partial determination for Y and X_3. What is the proportional reduction in previously unexplained variation in heating and cooling cost attributable to the inclusion of summer/fall mean temperature?

(d) Using your results from (a), predict the cost of a 50-thousand-square-foot warehouse when the mean temperature is 40° for winter/spring and 70° for summer/fall. Construct a 95% confidence interval for the mean cost for all such warehouses.

11-50 Refer to the data in Problem 11-47 and your answers to Problem 11-49.

(a) Perform a multiple regression analysis and determine the equation for the estimated regression hyperplane incorporating energy cost index as the fourth predictor of heating and cooling cost. Then calculate the standard error of the estimate.

(b) Compute the estimated coefficient of multiple determination applicable to your results from (a). What percentage of the variation in Y is explained by the regression hyperplane?

(c) Compute the coefficient of partial determination for Y and X_4. What is the proportional reduction in previously unexplained variation in heating and cooling cost attributable to the inclusion of the energy cost index?

(d) Using your results from (a), predict the cost of a 50-thousand-square-foot warehouse when the mean temperature is 40° for winter/spring and 70° for summer/fall and the energy cost index is 100. Construct a 95% confidence interval for the mean cost for all such warehouses.

11-51 The following data were obtained from a strength of materials test, where the neighborhood of 50° is considered a critical threshold temperature:

Maximum Elastic Stress (ksi)	Temperature (°C)
43	35
40	38
40	42
38	45
37	48
32	52
26	58
20	63
16	65
10	72

Using temperature as the independent variable and maximum elastic stress as the dependent variable, complete the following:

(a) Plot the raw data on a scatter diagram.

(b) Fit a regression parabola to the data. Then sketch the curve on your graph.

(c) Each sample observation falls into one category: below critical temperature or above critical temperature. For the data points in each range perform a separate, simple linear least-squares regression. Then plot each line on your graph.

(d) Which relationship do you think better fits the observed data, the parabola from (b) or the kinked linear function found in (c)?

11-52 An industrial engineering student has collected the following sample data for 10 skilled workers. Productivity is a dependent variable to be predicted from years of experience.

Productivity (units/hr)	Years of Experience	Sex
60	2	Male
51	1	Female
53	2	Female
61	3	Female
72	5	Male
74	6	Male
73	6	Male
78	4	Female
75	7	Male
91	7	Female

(a) Using different symbols for men and women, plot these data on a scatter diagram. Do your data suggest that sex might be a helpful predictor?

(b) Using all 10 data points and ignoring sex of worker, find a simple regression line and plot it on your graph.

(c) Consider using an indicator variable for sex. Assuming interaction between that variable and experience, find the expression for the estimated regression equation. Then determine separate equations by worker sex for the linear projections. Plot each of these on your graph.

11-53 Refer to the strength of materials data in Problem 11-51. Include an indicator variable for whether the observed temperature falls below or above the critical level of 50°. Assuming interaction between the independent variables, determine the estimated multiple regression equation. Then express equations for the respective linear projections.

CHAPTER TWELVE

Controlled Experiments: Two-Sample Inferences

ONE OF THE most important applications of statistics involves comparisons of two populations. Often such comparisons are made to help select the alternative that gives the better results. Many statistical procedures were originally developed for biological and medical research, where experimentation is used to determine the better treatment. These same techniques apply to a wide variety of engineering applications where comparisons must be made between methods, materials, or designs.

12-1 The Nature of Two-Sample Inferences

The procedures described in this chapter all involve two samples. One is selected randomly from population A, the other from population B. Inferences are made regarding these populations. The letters A and B are used symbolically as subscripts to separate the two-sample groups, so that the sizes of the respective samples are denoted as n_A and n_B.

Controlled Experiments

In this chapter we consider appropriate statistical procedures for evaluating **controlled experiments**, in which separate samples are obtained from two populations whose characteristics are uncertain. On the basis of the sample evidence, inferences may be drawn about the two populations. Generally, one population is the traditional one, while the second is what would result if a substantial change were made. For example, a new material might be considered as a substitute for Portland cement in certain types of construction. A series of tests might be performed to establish the superior strength of the experimental substance. Identical test conditions would be used to obtain data points for both materials.

Table 12-1 *Control and Experimental Group Designations in Engineering Investigations*

Investigation	Control Group	Experimental Group
Microcircuit etching	Photographic	Laser gun
Computer language	FORTRAN	Pascal
Chemical reaction	Low pressure, high temperature	High pressure, low temperature
Laboratory course emphasis	Empirical	Theoretical
Programming skill	Electrical engineering major	Physics major
Secondary recovery process for oil	Ground water injection	Hot steam treatment

The test units fabricated from the new material belong to the **experimental group**. Those made from regular cement constitute the **control group**. Although these terms originate from biological studies having living subjects for sample units, they now encompass inanimate (and sometimes imaginary) units as well. Table 12-1 gives examples of how these designations might apply in engineering investigations. The sample results constitute a controlled experiment in the sense that any deviations between groups, other than chance-caused, should be attributable to inherent population differences rather than to the testing procedure itself.

Independent Samples and Matched Pairs

There are two basic approaches in conducting such a sampling investigation. One treats the two groups as independent collections of data. The other involves matching units in each group into observation pairs. Independent sampling is the less complex procedure and allows for different sample sizes. Matched-pairs sampling, on the other hand, must be more carefully monitored; this approach has the advantage of minimizing any differences between groups that might be explained by extraneous factors. Different methodology applies in the two cases.

The Sample Size Dichotomy

Two parallel methodologies apply for making inferences regarding two means. The choice of method depends on the sample sizes used.

Recall from Chapter 7 that when a *single* sample is randomly selected from a population with parameters μ and σ, the central limit theorem states that $\bar{X}$ is approximately normally distributed, with mean μ and standard deviation $\sigma/\sqrt{n}$. But the value for σ must be *known* in order for the normal curve to be used in making any probability statements regarding levels for $\bar{X}$. If σ is unknown, the Student t distribution is used as the basis for making inferences about μ. Recall also that the Student t distribution approaches the normal distribution as the sample size becomes larger.

With two independent samples, one having mean $\bar{X}_A$ and the other $\bar{X}_B$, these same issues apply. For large sample sizes, the normal curve is used as the sampling distribution for the difference $\bar{X}_A - \bar{X}_B$. That is done even though the population standard deviations σ_A and σ_B remain unknown. When the standard deviations are unknown, the normal distribution is strictly an *approximation*, which improves with larger n's. One important advantage of using the normal distribution in this way is that σ_A and σ_B are then allowed to have *different* (even if unknown) levels.

When the n's are small, the normal approximation is poor. In those cases, the Student t distribution is used in making inferences regarding the population means. But using this distribution requires that populations A and B have *identical standard deviations* (of some unknown level).

Further Considerations

The familiar dichotomy applies to the type of inference to be drawn—estimation versus hypothesis test. Both types of inferences will be described for the key population parameters, the mean and proportion. Often the underlying theoretical assumptions necessary for these procedures are overrestrictive or unrealistic. In those cases, the respective quantitative populations are better compared without consideration of parameters (μ's). This chapter describes **nonparametric statistics**, alternative procedures that involve fewer restrictive assumptions than traditional methods.

12-2 Confidence Intervals for the Difference Between Means

The most common type of estimate made in two-population investigations is the magnitude of *difference* between the respective means. As noted earlier, the common designation refers to one group as population A and the second as population B. Using the respective letter subscripts, the following designations apply:

$$\mu_A = \text{Mean of population A}$$

$$\mu_B = \text{Mean of population B}$$

The designation of A and B is completely arbitrary, although the A population is often believed to have the larger values.

The following example will be used to illustrate several techniques for drawing inferences about the means of two populations.

Illustration:* *The Effectiveness of Personal Computers in Engineering Education

During the early 1980s, personal computers came into widespread use as the hardware capabilities improved and the prices fell substantially below that of a good used car. One engineering dean, faced with the prospect of upgrading the central time-sharing system at his school, came to the conclusion that funds would be better spent by providing every student with a personal computer. Not only would the student have considerable flexibility, but every computer could still be connected via telephone line to the mainframe; in effect, the computers could also serve as "smart terminals" for time-sharing. The dean believed that students would benefit from personal computers in several ways. From their enhanced "hands-on" computer experience, they would certainly learn more about the various aspects of computer science. They would become familiar with the present state of the art in software, and all their course work could be enriched by their easy access to a digital computer. The lucky students would be more capable than otherwise.

The engineering school could actually save funds that would otherwise go to acquiring hardware and housing for an upgraded time-sharing system. Fewer data-processing personnel would be required, and less expensive disk storage would be required for the main computer system. Students would be allowed to purchase their computers upon graduation, so that funds would be generated for replacement. The manufacturer of one of the popular personal computers agreed to provide its machines to the proposed program at a substantial discount.

Two random samples of students were chosen to participate in a two-year experiment to measure the effectiveness of personal computers. The control group consisted of 40 new sophomores who were not given personal computers and who were left to fend for themselves on computer assignments through the existing campus computer center. The experimental group consisted of a like number of students with the same background but who were permitted to check out a personal computer on long-term loan. Each student in that group received a modest budget for procuring software and supplies. The effectiveness of personal computers in engineering education was to be gauged by the respective grade-point averages achieved by the populations represented by the two groups over the two-year period.

The sample results will be described shortly.

Independent Samples

The simplest procedure for estimating the difference between means involves two independent sample groups. We will denote the sample statistics computed for the respective groups by the subscripts A and B. In making inferences regarding the difference $\mu_A - \mu_B$, we use the following composite statistic.

DIFFERENCE BETWEEN SAMPLE MEANS

$$D = \bar{X}_A - \bar{X}_B$$

This equation serves as an unbiased estimator of the difference in population means. The respective sample sizes for the two groups are denoted as n_A and n_B. A different sample size may apply to each group.

The two sample means $\bar{X}_A$ and $\bar{X}_B$ are independent random variables and, by the central limit theorem, they are approximately normally distributed. Denoting by σ_A and σ_B the respective population standard deviations, the standard errors for the respective sample means are $\sigma_A/\sqrt{n_A}$ and $\sigma_B/\sqrt{n_B}$. The difference D will also be approximately normally distributed, with mean $\mu_A - \mu_B$ and variance equal to the sum of the individual variances,

$$\sigma_D^2 = \frac{\sigma_A^2}{n_A} + \frac{\sigma_B^2}{n_B}$$

The values of σ_A and σ_B are ordinarily unknown. Two parallel procedures apply for estimating $\mu_A - \mu_B$, depending on the sample sizes.

Large Sample Sizes When n_A and n_B are both greater than 30, the following estimate is used for the standard error for D:

$$s_D = \sqrt{\frac{s_A^2}{n_A} + \frac{s_B^2}{n_B}}$$

The $100(1 - \alpha)\%$ confidence interval estimate for the difference in population means may be determined in the usual manner:

$$\mu_A - \mu_B = D \pm z_{\alpha/2} s_D$$

This may be stated more conveniently by the following equivalent expression.

CONFIDENCE INTERVAL FOR THE DIFFERENCE BETWEEN MEANS USING INDEPENDENT SAMPLES (LARGE n's)

$$\mu_A - \mu_B = (\bar{X}_A - \bar{X}_B) \pm z_{\alpha/2} \sqrt{\frac{s_A^2}{n_A} + \frac{s_B^2}{n_B}}$$

As an illustration, consider the sample results in Table 12-2 for the personal computer investigation. We denote the experimental group by the letter A and the control group by B. These data will be used to estimate the difference between sophomore and junior grade-point averages (GPAs), $\mu_A - \mu_B$. Although 40 students were originally included in each group, each has suffered a reduction by 5 students, so that $n_A = n_B = 35$. The following statistics were computed:

$$\bar{X}_A = 3.38 \qquad \bar{X}_B = 3.26 \qquad s_A = .47 \qquad s_B = .46$$

A 95% confidence interval estimate of the difference in population means, using $z_{.025} = 1.96$, is

$$\mu_A - \mu_B = (3.38 - 3.26) \pm 1.96 \sqrt{\frac{(.47)^2}{35} + \frac{(.46)^2}{35}}$$

$$= .12 \pm .22$$

so that

$$-.10 \le \mu_A - \mu_B \le .34$$

This indicates that the mean GPA for all students who receive personal computers exceeds the corresponding figure for all students not having computers by some

Table 12-2
Sample Results for Personal Computer Evaluation

Experimental Group (A) Personal Computers		Control Group (B) No Computers	
Student	Sophomore-Junior GPA X_A	Student	Sophomore-Junior GPA X_B
R. A.	3.05	M. A.	3.10
W. A.	3.05	N. A.	3.45
B. B.	3.80	T. A.	3.10
A. C.	3.60	E. B.	3.50
T. C.	3.70	F. D.	3.80
A. D.	3.10	K. D.	2.30
P. D.	2.45	S. E.	3.05
E. E.	3.75	K. F.	3.60
H. E.	2.80	M. F.	3.55
Q. E.	3.60	N. J.	3.10
R. F.	4.00	B. K.	3.75
D. G.	2.45	A. K.	3.70
B. H.	3.30	J. K.	2.85
G. H.	3.60	D. L.	2.40
D. J.	3.35	L. L.	3.70
H. J.	3.45	R. L.	3.35
R. J.	4.00	C. M.	3.90
L. K.	2.55	E. M.	3.40
J. L.	3.80	J. M.	3.75
R. L.	4.00	W. M.	3.90
L. M.	4.00	F. P.	3.20
M. M.	3.60	D. Q.	2.60
M. N.	2.75	M. R.	3.30
L. P.	2.55	L. S.	2.65
O. P.	3.25	T. S.	3.50
M. R.	3.40	T. T.	2.85
S. R.	3.20	C. U.	3.65
C. S.	3.25	D. U.	2.90
E. S.	2.95	T. V.	2.25
D. T.	3.95	V. V.	3.20
S. T.	3.80	T. W.	3.75
R. U.	3.50	W. W.	3.45
S. V.	3.25	R. Y.	3.30
A. Z.	3.85	W. Y.	3.45
W. Z.	3.60	D. Z.	2.70

amount between $-.10$ (a disadvantage of a tenth of a grade point) and $+.34$ (a more than three-tenths advantage).

A proper interpretation of this interval is that if the sampling experiment were repeated several times, about 95% of all intervals computed in a similar fashion would contain the true $\mu_A - \mu_B$ difference. The engineering dean would probably not find these results particularly convincing, as the interval end points span a considerable range. However, there is a better procedure, described next, that is considerably more discerning.

Small Sample Sizes When either $n_A \leq 30$ or $n_B \leq 30$, the procedure employs the Student t distribution. The key theoretical assumption is that a common variance σ^2 is assumed to apply for the two populations, so that $\sigma_A^2 = \sigma_B^2 = \sigma^2$. The standard error of D may then be expressed by the following:

$$\sigma_D = \sqrt{\frac{\sigma_A^2}{n_A} + \frac{\sigma_B^2}{n_B}} = \sigma\sqrt{\frac{1}{n_A} + \frac{1}{n_B}}$$

The sample variances s_A^2 and s_B^2 each serve as unbiased estimators of σ^2. By pooling the sample results, the estimate of σ^2 may be improved. Thus, we may use a weighted average of the sample variances as an unbiased estimator of σ^2. That will also improve the estimator of σ_D, which may be determined from the following expression:

$$s_D = \sqrt{\frac{(n_A - 1)s_A^2 + (n_B - 1)s_B^2}{n_A + n_B - 2}}\sqrt{\frac{1}{n_A} + \frac{1}{n_B}}$$

Two sample means are used in computing the sample variances, and one degree of freedom is lost for each. In order for the weighted average of the sample variances to be an unbiased estimator of σ^2, the weights must each be one less than the respective sample size. Their sum serves as the denominator in the first fraction and equals the combined sample size, reduced by two.

The following expression is used to compute the $100(1 - \alpha)\%$ confidence interval.

CONFIDENCE INTERVAL FOR THE DIFFERENCE BETWEEN MEANS USING INDEPENDENT SAMPLES (SMALL n's)

$$\mu_A - \mu_B = (\bar{X}_A - \bar{X}_B) \pm t_{\alpha/2}\sqrt{\frac{(n_A - 1)s_A^2 + (n_B - 1)s_B^2}{n_A + n_B - 2}}\sqrt{\frac{1}{n_A} + \frac{1}{n_B}}$$

The critical value $t_{\alpha/2}$ is read from Appendix Table G for $n_A + n_B - 2$ degrees of freedom.

Example: Comparing Maintenance Policies

As part of a summer project, an industrial engineering student is investigating maintenance policies that might be adopted by a large trucking firm. One alternative (A) involves a regular schedule of preventive maintenance after every 25,000 miles of operation, when belts, seals, gaskets, and many bearings are automatically replaced, whether or not the item is actually unsatisfactory. This policy will be compared with a radically different one (B), where there is no fixed interval for preventive maintenance and nothing is repaired until there is a breakdown or some obvious problem. Then all of the above items are checked and replaced only if necessary.

Two independent sample groups of trucks were selected to be run under one of the proposed policies for one year, after which the total maintenance cost for that period would be determined. The following data were obtained:

Policy A		
\$12,366	\$12,575	\$13,589
11,950	13,820	12,276
11,786	12,479	13,125
12,659		

Policy B		
\$7,024	\$11,115	\$10,443
18,203	6,450	4,255
12,357	19,204	4,158
23,425	3,718	8,295
6,225	4,870	9,146

The sample sizes are $n_A = 10$ and $n_B = 15$. The following sample statistics were computed:

$$\bar{X}_A = \$12{,}663 \qquad \bar{X}_B = \$9{,}926$$

$$s_A = \$664 \qquad s_B = \$6{,}044$$

A 95% confidence interval estimate may be constructed for the difference in mean maintenance costs under the two policies. For $10 + 15 - 2 = 23$ degrees of freedom, Appendix Table G provides the critical value $t_{.025} = 2.069$. Thus, we have

$$\begin{aligned}\mu_A - \mu_B &= (\$12{,}663 - \$9{,}926)\\ &\quad \pm 2.069\sqrt{\frac{(10-1)(664)^2 + (15-1)(6{,}044)^2}{10+15-2}}\sqrt{\frac{1}{10}+\frac{1}{15}}\\ &= \$2{,}737 \pm \$3{,}998\end{aligned}$$

or

$$-\$1{,}261 \leq \mu_A - \mu_B \leq \$6{,}735$$

The mean cost advantage per truck of fix-only-when-broken (policy B) over scheduled preventive maintenance (policy A) is therefore estimated to fall somewhere between $-\$1{,}261$ and $+\$6{,}735$ per year.

This result is valid only if the population variances are identical. There is strong sample evidence that this is not the case. Although of little help in estimation, the nonparametric procedures described later in the chapter avoid this type of difficulty.

Matched-Pairs Samples

A matched-pairs experiment is achieved by matching each sample unit in the experimental group to a counterpart in the control group. Each resulting pair is then evaluated as a single entity. The matching should ordinarily be done before the sample observations are taken. The objective of matching is to obtain units in each pair which are alike in many of the factors that might contribute to the level of the variable being considered. Ideal matching in medical research would be achieved by a sample involving identical twins, with one sibling in the control group and the other in the experimental group. Any differences in recovery achieved by the twin pairs could then be explained by differences in treatments rather than by other causes. Of course experimentation with identical twins is only an ideal, and units must be paired so that "near twins" result.

In the personal computer evaluation, a matched pair should comprise students who are expected to perform equally well. Thus, the students in each pair should be of nearly the same demonstrated ability and achievement level. Table 12-3 shows how the students in the study were actually paired. That is, they were paired first in terms of their freshman GPA, deemed to be the best predictor of future academic achievement. In the case of ties in GPA, quantitative SAT scores were used to refine the final pairing choice.

Keep in mind that either the matched-pairs procedure or independent samples, but *not both*, will be used in a controlled experiment involving two populations. We have applied both procedures to the same data only to facilitate a comparison of their relative strengths and weaknesses.

Letting i denote the ith pair, with X_{A_i} and X_{B_i} representing the sample observations for that pair from the respective groups, we express the difference as follows.

MATCHED-PAIR DIFFERENCE

$$\boldsymbol{d_i = X_{A_i} - X_{B_i}}$$

Altogether there are n sample pairs. (We use a lowercase d to distinguish *pairwise* differences from the uppercase D used to represent the difference between two sample *means.*)

We denote the sample mean for the matched-pair differences by $\bar{d}$. The sample standard deviation is denoted by s_d. Both of these statistics are computed in the usual manner. The sample results for the personal computer illustration are shown in Table 12-4. From these, we obtain

$$\bar{d} = \frac{\sum d_i}{n} = \frac{4.30}{35} = .123$$

and

$$s_d = \sqrt{\frac{\sum (d_i - \bar{d})^2}{n - 1}} = .130$$

The mean matched-pair difference $\bar{d}$ is an unbiased estimator of $\mu_A - \mu_B$.

Although two populations, A and B, generate the respective samples, we consider the single *population of paired differences*. Theoretically, inferences regarding that population may be made in exactly the same way as for any single population, as we saw in Chapters 8 and 9. When the standard deviation of the paired-difference population is unknown, the Student t distribution may be applied in the same way as in those chapters.

The following expression is used in computing the $100(1 - \alpha)\%$ confidence interval.

Table 12-3 *Matched Pairs Used in the Personal Computer Experiment*

	Experimental Group			Control Group		
Pair	Student	Freshman GPA	Quantitative SAT	Student	Freshman GPA	Quantitative SAT
1	R. J.	4.00	740	C. M.	3.95	780
2	A. Z.	3.90	760	B. K.	3.90	750
3	S. T.	3.90	720	T. W.	3.90	690
4	L. M.	3.80	790	A. K.	3.85	720
5	B. B.	3.80	760	C. U.	3.80	710
6	R. F.	3.80	690	F. D.	3.75	730
7	R. L.	3.75	720	W. M.	3.75	670
8	Q. E.	3.60	685	L. L.	3.70	650
9	T. C.	3.60	585	J. M.	3.65	620
10	G. H.	3.50	640	W. W.	3.50	590
11	E. E.	3.40	600	K. F.	3.40	650
12	M. M.	3.40	560	E. B.	3.40	610
13	J. L.	3.25	635	T. S.	3.35	595
14	D. T.	3.25	625	M. F.	3.30	660
15	R. U.	3.25	615	R. Y.	3.25	640
16	H. J.	3.25	575	E. M.	3.25	605
17	W. Z.	3.20	670	W. Y.	3.20	640
18	A. C.	3.20	610	M. R.	3.20	585
19	O. P.	3.20	590	T. A.	3.20	580
20	S. R.	3.10	710	S. E.	3.20	570
21	M. R.	3.10	690	F. P.	3.10	730
22	B. H.	3.10	660	M. A.	3.10	710
23	R. A.	3.00	685	J. K.	3.10	610
24	D. J.	2.90	675	N. A.	3.00	575
25	S. V.	2.90	650	R. L.	2.90	650
26	A. D.	2.90	620	V. V.	2.90	640
27	C. S.	2.90	580	N. J.	2.85	595
28	W. A.	2.85	610	T. T.	2.80	550
29	H. E.	2.75	590	D. U.	2.70	580
30	E. S.	2.65	610	D. Z.	2.60	725
31	M. N.	2.50	575	L. S.	2.55	640
32	L. K.	2.40	585	D. Q.	2.45	655
33	P. D.	2.30	625	D. L.	2.25	610
34	L. P.	2.20	580	K. D.	2.25	575
35	D. G.	2.10	560	T. V.	2.20	640

CONFIDENCE INTERVAL FOR THE DIFFERENCE BETWEEN MEANS USING MATCHED PAIRS

$$\mu_A - \mu_B = \bar{d} \pm t_{\alpha/2} \frac{s_d}{\sqrt{n}}$$

Table 12-4 Matched-Pairs Differences for the Personal Computer Experiment

	Student Pair		Sophomore-Junior GPA		
i	Group A Student	Group B Student	Group A X_{A_i}	Group B X_{B_i}	Difference d_i
1	R. J.	C. M.	4.00	3.90	.10
2	A. Z.	B. K.	3.85	3.75	.10
3	S. T.	T. W.	3.80	3.75	.05
4	L. M.	A. K.	4.00	3.70	.30
5	B. B.	C. U.	3.80	3.65	.15
6	R. F.	F. D.	4.00	3.80	.20
7	R. L.	W. M.	4.00	3.90	.10
8	Q. E.	L. L.	3.60	3.70	−.10
9	T. C.	J. M.	3.70	3.75	−.05
10	G. H.	W. W.	3.60	3.45	.15
11	E. E.	K. F.	3.75	3.60	.15
12	M. M.	E. B.	3.60	3.50	.10
13	J. L.	T. S.	3.80	3.50	.30
14	D. T.	M. F.	3.95	3.55	.40
15	R. U.	R. Y.	3.50	3.30	.20
16	H. J.	E. M.	3.45	3.40	.05
17	W. Z.	W. Y.	3.60	3.45	.15
18	A. C.	M. R.	3.60	3.30	.30
19	O. P.	T. A.	3.25	3.10	.15
20	S. R.	S. E.	3.20	3.05	.15
21	M. R.	F. P.	3.40	3.20	.20
22	B. H.	M. A.	3.30	3.10	.20
23	R. A.	J. K.	3.05	2.85	.20
24	D. J.	N. A.	3.35	3.45	−.10
25	S. V.	R. L.	3.25	3.35	−.10
26	A. D.	V. V.	3.10	3.20	−.10
27	C. S.	N. J.	3.25	3.10	.15
28	W. A.	T. T.	3.05	2.85	.20
29	H. E.	D. U.	2.80	2.90	−.10
30	E. S.	D. Z.	2.95	2.70	.25
31	M. N.	L. S.	2.75	2.65	.10
32	L. K.	D. Q.	2.55	2.60	−.05
33	P. D.	D. L.	2.45	2.40	.05
34	L. P.	K. D.	2.55	2.30	.25
35	D. G.	T. V.	2.45	2.25	.20

where the critical value $t_{\alpha/2}$ is found from Appendix Table G for $n - 1$ degrees of freedom.

The above may be used to find a 95% confidence interval for the difference in GPAs between the experimental and control groups in the personal computer investigation. From Appendix Table G, for $n - 1 = 34$ degrees of freedom, we find $t_{.025} = 2.034$ (using linear interpolation). From this we have

$$\mu_A - \mu_B = .123 \pm 2.034\left(\frac{.130}{\sqrt{35}}\right)$$
$$= .123 \pm .045$$

so that

$$.078 \leq \mu_A - \mu_B \leq .168$$

This confidence interval is considerably tighter than the one found for the same data when the independent-sample procedure was applied. We may now compare the two procedures.

Example: *Evaluation of Computer-Aided Design*

The chief engineer in a machine parts manufacturing company is comparing CAD (computer-aided design) (A) to the traditional method (B). The two procedures are compared in terms of the mean time from start until production drawings and specifications are ready. A random sample of 20 parts has been selected, each to be designed twice, once by an engineer using borrowed time on the CAD system at a nearby facility and again by an engineer in-house working in the traditional manner. The two engineers designing each sample part have been matched in terms of the quality of their past performance. The engineers in the CAD group have each just completed an after-hours training program and have been judged proficient in the new system.

The following completion times (days) have been obtained:

	Design Time	
Part	**CAD**	**Tradit.**
1	8.2	11.4
2	15.4	25.2
3	4.6	4.0
4	5.5	6.0
5	8.0	11.5
6	10.6	9.4
7	5.3	8.1
8	5.0	6.4
9	19.3	20.1
10	7.7	9.3

	Design Time	
Part	**CAD**	**Tradit.**
11	4.3	6.1
12	2.7	3.5
13	5.2	4.8
14	13.8	16.9
15	22.5	30.0
16	15.3	18.4
17	7.5	9.0
18	12.7	14.4
19	14.5	13.5
20	6.3	8.1

From these data, the following sample results have been obtained:

$$\bar{d} = -2.085 \qquad s_d = 2.67$$

A 95% confidence interval estimate for the difference in mean time may be constructed from the above. Appendix Table G provides that for $20 - 1 = 19$

degrees of freedom the critical value $t_{.025} = 2.093$. The difference in mean design times is thus estimated to be

$$\mu_A - \mu_B = -2.085 \pm 2.093 \frac{2.67}{\sqrt{20}}$$
$$= -2.085 \pm 1.250$$

or

$$-3.335 \leq \mu_A - \mu_B \leq -.835 \text{ days}$$

This interval indicates that there is a mean time advantage in using computer-assisted design of about 1 to 3 days per part.

Matched Pairs Compared to Independent Samples

Suppose that the two groups have the same sample sizes so that $n_A = n_B$, and the two populations have nearly equal variances. It may be proved that, for any given reliability or confidence level and degree of precision, matched-pairs sampling will require a smaller sample size than independent sampling. Furthermore, the reduction in sample size is proportional to the correlation coefficient for the two populations. Thus, if A and B involve a correlation of 90%, a 90% saving in sample size can be achieved by using an ideal criterion to select matched pairs over the number of observations that would be required with independent samples.

It is natural to wonder at this point why independent samples are used at all. Keep in mind that matched-pairs sampling is the more complicated procedure, and each sample observation using such a design can be an order of magnitude more costly to make. Also, there may be no suitable matching criterion to apply, and even if there is, there may be no useful data base for exercising it. In addition, separate samples may be obtained from the two populations by different investigators or at different times, so the entire experiment would have to be repeated if matched pairs were to be used.

Problems

12-1 An engineering society wishes to determine how much, if at all, the mean income of practicing electrical engineers exceeds that of mechanical engineers. Two independent random samples provided the following data:

Electrical Engineers	Mechanical Engineers
$n_A = 76$	$n_B = 58$
$\bar{X}_A = \$37{,}246$	$\bar{X}_B = \$36{,}412$
$s_A = \$8{,}371$	$s_B = \$8{,}856$

Construct a 95% confidence interval estimate for the difference in population means.

12-2 Construct a 95% confidence interval estimate of the difference between the MTBFs of two types of power cells, where the following results have been obtained for independent samples:

Cell A	Cell B
$n_A = 12$	$n_B = 12$
$\bar{X}_A = 527$ days	$\bar{X}_B = 488$ days
$s_A = 123$	$s_B = 102$

12-3 An electrical engineer wishes to find the difference between the mean time between failures for transformers obtained from two different vendors. The following data were obtained from independent high-temperature testing:

Vendor A	Vendor B
$n_A = 76$	$n_B = 225$
$\bar{X}_A = 1{,}246$ hr	$\bar{X}_B = 1{,}347$ hr
$s_A = 157$	$s_B = 217$

Construct a 99% confidence interval for the difference between MTBF for the two vendors.

12-4 Suppose that the observations in Problem 12-2 have been matched according to type of application, environmental conditions, and power demands. The mean and standard deviation of the differences found by subtracting cell B's time to failure from that of cell A are

$$\bar{d} = 39 \text{ days} \qquad s_d = 17.3 \text{ days}$$

Construct a 95% confidence interval estimate for the difference in MTBF between the two types of power cell.

12-5 A chief engineer for construction wishes to determine how electrical subcontractors differ in their ability to complete jobs quickly. From past records he has found $n = 43$ jobs done by each firm that may be considered to be representative random samples. These have been paired in terms of size, overall difficulty, general environment, time of year, and several other factors. For each pair the logged completion duration for subcontractor B was subtracted from the corresponding time taken by subcontractor A. The following data were determined:

$$\bar{d} = .5 \text{ days} \qquad s_d = 1.4 \text{ days}$$

Construct a 95% confidence interval estimate for the difference between mean completion times for the two contractors.

12-6 A heavy-equipment foreman for a regional construction company wants to compare maintenance costs for two makes of diesel engines. He has selected a random sample of 50 vehicles equipped with brand A engines. Each of these has been paired with a counterpart brand B engine, also randomly selected. Each match was based on age and type of vehicle and on anticipated usage. The following sample results summarize the difference in annual maintenance cost found for each engine pair:

$$\bar{d} = -\$113.67 \qquad s_d = \$58.42$$

Construct a 99% confidence interval estimate for the difference in mean annual maintenance cost for the two makes of engine.

12-7 The following sample data have been collected independently for the drying times (in hours) of two brands of paint:

Paint A			
12.7	13.4	14.5	11.7
10.6	11.4	12.2	13.7
14.1	13.3	12.6	12.2
11.3	12.5	12.3	13.7
15.1	13.7	12.5	14.4
12.5	13.3	13.3	14.5
12.5	13.3	13.5	12.5
10.7	10.5	12.4	11.9
12.0	13.5	14.1	

Paint B			
9.8	10.4	12.6	13.7
12.3	11.7	12.1	10.8
12.6	11.9	10.1	9.9
12.3	12.1	11.6	10.8
13.1	11.5	10.9	11.4
10.2	10.4	12.7	12.6
11.2	11.7	12.4	13.1
11.0	12.0	13.1	12.0

Construct a 95% confidence interval estimate for the difference in mean drying times for the two paints.

12-8 The following grade-point data have been obtained from matched pairs taken randomly from two populations of engineering students. Group A consists of cooperative students who regularly rotate between school and industry. Group B consists of regular full-time students.

Coop.	Reg.	Coop.	Reg.	Coop.	Reg.	Coop.	Reg.
3.65	3.71	2.84	2.75	3.22	3.15	3.61	3.40
3.25	2.96	2.86	2.91	3.11	2.95	3.55	3.35
3.33	3.17	3.55	3.51	4.00	3.86	3.74	3.55
2.54	2.27	2.85	2.62	3.45	3.28	3.61	3.48
3.44	3.42	3.65	3.51	3.12	2.88	3.05	3.12
3.14	3.02	3.67	3.50	3.24	3.18	3.15	2.91
3.29	3.17	3.64	3.52	3.78	3.42	3.20	2.95
2.65	2.90	2.85	2.69	3.53	3.37	3.15	3.11
3.22	2.84	3.55	3.65	3.52	3.44		

Construct a 99% confidence interval for the difference in mean GPA for all students in the respective populations.

12-3 Hypothesis Tests for Comparing Two Means

We are now ready to establish the framework for testing hypotheses in controlled experiments where the decision is based on the relative levels of two population means. A typical engineering application of such a test would be a decision whether or not to adopt a new method, material, procedure, or vendor. The final action is based on data collected from two populations, one sample representing the status quo, the other reflecting the proposed population. The decision rule determines whether or not the sample results are significant in signaling that a difference exists between true levels of the population means. An appropriate null hypothesis must then be accepted or rejected.

That null hypothesis will take one of the following forms:

$$H_0\colon \mu_A \geq \mu_B \quad \text{or} \quad H_0\colon \mu_A - \mu_B \geq 0 \qquad \text{(lower-tailed test)}$$

$$H_0\colon \mu_A \leq \mu_B \quad \text{or} \quad H_0\colon \mu_A - \mu_B \leq 0 \qquad \text{(upper-tailed test)}$$

$$H_0\colon \mu_A = \mu_B \quad \text{or} \quad H_0\colon \mu_A - \mu_B = 0 \qquad \text{(two-sided test)}$$

These may be tested using either independent samples or matched pairs.

Independent Samples

The null hypothesis (in whatever form) is usually evaluated from independent samples using the difference in sample means, $D = \bar{X}_A - \bar{X}_B$, as the test statistic. As with estimating $\mu_A - \mu_B$, there are two methods for testing H_0, depending on whether the sample sizes are large or small.

Large Sample Sizes When $n_A > 30$ and $n_B > 30$, the normal deviate serves as the test statistic. The following expression is used.

NORMAL DEVIATE USING INDEPENDENT SAMPLES (LARGE *n*'s)

$$z = \frac{D}{s_D} = \frac{\bar{X}_A - \bar{X}_B}{\sqrt{\dfrac{s_A^2}{n_A} + \dfrac{s_B^2}{n_B}}}$$

Illustration:* *Evaluation of Test Stands

Consider a decision by a quality-control department regarding testing methods.

The quality-assurance director in an electronics fabricating facility wants to determine if new test stands might be effective in increasing the flow of production units. A sampling investigation on a borrowed unit will determine the final choice whether to acquire the new stands or to retain the present manual checking procedure. Sample data have already been obtained on the present procedure's testing times, so a further independent sample of times will be obtained using a borrowed test stand. The latter sample constitutes the experimental group, designated by the letter A (an arbitrary choice). A total of $n_A = 50$ observations will be made. The control group data involves $n_B = 100$ observations. (Fewer group A observations were made because the borrowed stand was available only for a short time.)

The same basic five hypothesis-testing steps introduced in Chapter 9 apply here. To facilitate discussions, a complementary null hypothesis is assumed but not stated.

Step 1. *Formulate the null hypothesis.* The director chooses as her null hypothesis

$$H_0: \mu_A \geq \mu_B \qquad \text{(Test stands are at least as slow as manual testing.)}$$

This designation reflects her strong desire to avoid adopting the test stands when they actually take longer than manual testing.

Step 2. *Select the test procedure and test statistic.* Although the difference D between sample means is the basis for the test, it will be transformed into the normal deviate expressed earlier, so that z will serve as the test statistic.

Step 3. *Establish the significance level and the acceptance and rejection regions for the decision rule.* Because the director regards so seriously the type I error of rejecting H_0 when it is true, she chooses a 1% significance level, so that $\alpha = .01$. From Appendix Table E this corresponds to a critical normal deviate of $z_{.01} = 2.33$. Extreme negative levels for z will refute H_0, and the test is lower-tailed. The decision rule is shown in Figure 12-1.

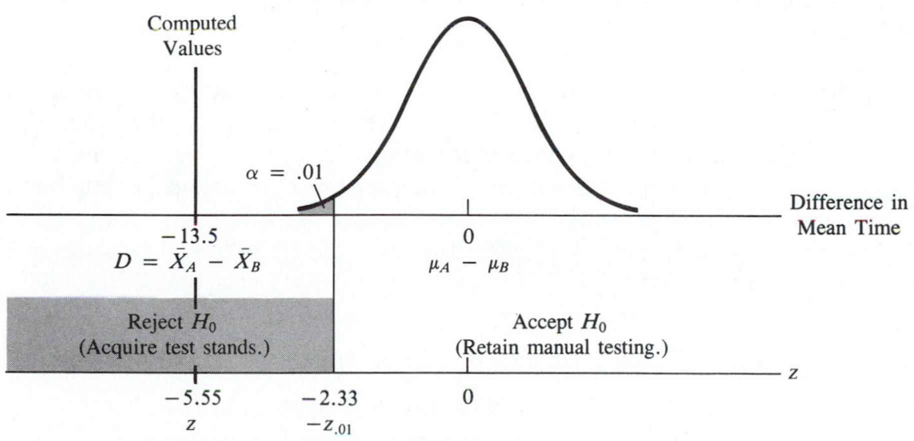

Figure 12-1

Step 4. *Collect the sample data and compute the value of the test statistic.* The sample results for the testing procedure evaluation are

Test Stand	Manual Testing
$\bar{X}_A = 45.6$ min	$\bar{X}_B = 59.1$ min
$s_A = 13.3$	$s_B = 15.4$

The computed value of the test statistic is

$$z = \frac{45.6 - 59.1}{\sqrt{\dfrac{(13.3)^2}{50} + \dfrac{(15.4)^2}{100}}} = -5.55$$

Step 5. *Make the decision.* The computed normal deviate falls inside the rejection region. The director must therefore reject H_0 and acquire the new test stands.

This illustration involves a lower-tailed test. An upper-tailed test with $H_0\colon \mu_A \leq \mu_B$ will have a rejection region encompassing large positive values for z. A two-sided test, when $H_0\colon \mu_A = \mu_B$, will involve a double rejection region.

Small Sample Sizes When either $n_A \leq 30$ or $n_B \leq 30$, the testing procedure is based on the Student t distribution. The following expression is used.

t STATISTIC FOR SMALL INDEPENDENT SAMPLES

$$t = \frac{D}{s_D} = \frac{\bar{X}_A - \bar{X}_B}{\sqrt{\dfrac{(n_A - 1)s_A^2 + (n_B - 1)s_B^2}{n_A + n_B - 2}}\sqrt{\dfrac{1}{n_A} + \dfrac{1}{n_B}}}$$

The critical value for t is read from Appendix Table G. The number of degrees of freedom is $n_A + n_B - 2$.

Illustration:* *Alloys for Armor Plating

Consider the following test performed by an engineer evaluating two candidate alloys for armor plating. One candidate is an all-metal alloy (A); the other alloy (B) is a high-quality steel with imbedded grains of Teflon. The test procedure involves shooting uranium darts at sample plates from a gun at successively higher muzzle velocities until penetration. The basis for comparison is the mean velocity at penetration. A total of $n_A = 15$ all-metal and $n_B = 10$ Teflon alloy plates were used in the experiment.

Step 1. *Formulate the null hypothesis.* The engineer chooses as his null hypothesis

$H_0\colon \mu_A \geq \mu_B$ (The all-metal alloy offers at least as much resistance as the Teflon alloy.)

This designation reflects his strong desire to avoid recommending Teflon when this unconventional material actually provides inferior resistance.

Step 2. *Select the test procedure and test statistic.* The Student t statistic computed for the difference in sample means is the basis for the test.

Step 3. *Establish the significance level and the acceptance and rejection regions for the decision rule.* The engineer chooses a type I error probability of .01. From

Appendix Table G this corresponds to a critical value of $t_{.01} = 2.500$, with $15 + 10 - 2 = 23$ degrees of freedom. Extreme negative levels for t will refute H_0, and the test is lower-tailed. The decision rule is shown in Figure 12-2.

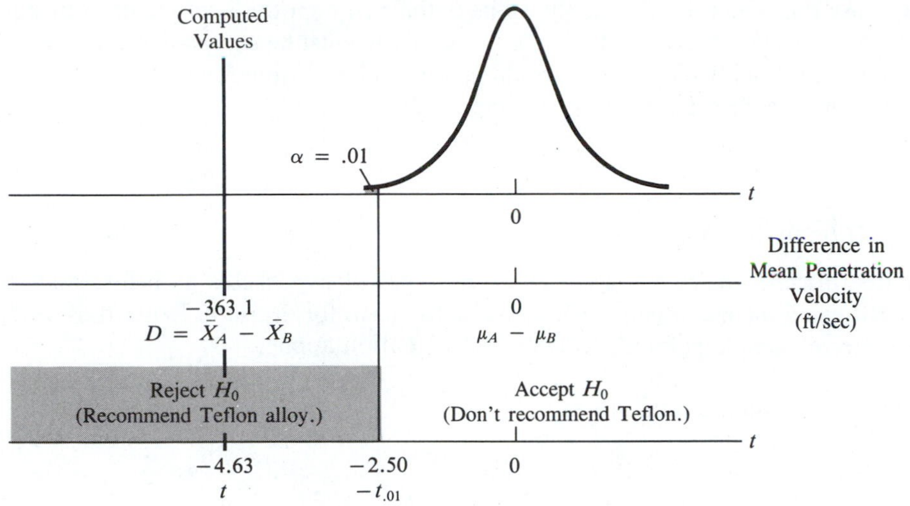

Figure 12-2

Step 4. *Collect the sample data and compute the value of the test statistic.* The sample results for the experiment are

All-Metal Alloy	Teflon Alloy
$\bar{X}_A = 4{,}245.0$ ft/sec	$\bar{X}_B = 4{,}608.1$ ft/sec
$s_A = 210.5$	$s_B = 159.4$

The computed value of the test statistic is

$$t = \frac{4{,}245.0 - 4{,}608.1}{\sqrt{\dfrac{14(210.5)^2 + 9(159.4)^2}{23}}\sqrt{\dfrac{1}{15} + \dfrac{1}{10}}} = -4.63$$

Step 5. *Make the decision.* The computed value for t falls inside the rejection region. The engineer must therefore reject H_0 and recommend the Teflon alloy as armor plating.

Two-Sided Tests Using Confidence Intervals A two-sided test may be conducted in conjunction with the confidence interval for $\mu_A - \mu_B$ described in the previous section. For example, in the personal computer evaluation described earlier we found the following 95% confidence interval estimate for the difference in mean sophomore-junior GPA:

$$-.10 \leq \mu_A - \mu_B \leq .34$$

Because this interval contains the value 0, the null hypothesis of identical means, H_0: $\mu_A = \mu_B$ (or, equivalently, H_0: $\mu_A - \mu_B = 0$), must be *accepted* at the $\alpha = .05$ significance level (which, again, would be a valid conclusion for that investigation only if no matched pairs had been employed).

Matched-Pairs Samples

When matched pairs are used, the same types of hypotheses as before may be tested using the mean pair difference $\bar{d}$ as the basis for decision. Regardless of the number of sample pairs, the Student t distribution applies.

t STATISTIC USING MATCHED PAIRS

$$t = \frac{\bar{d}}{s_d/\sqrt{n}}$$

The critical value for t corresponds to the desired significance level when the number of degrees of freedom is $n - 1$.

Illustration: *Cost Effectiveness of Electronic Mail*

Consider an experiment by a government agency to evaluate the effectiveness of electronic mail to replace interoffice memos. Although rapid and increased communication is the major benefit of the new technology, significant operating-cost reductions are needed to substantiate procurement of the new hardware. With equipment supplied by one manufacturer, a three-month test will be performed using electronic mail in a sample of 10 district offices. Each office in this experimental group is matched with a counterpart where all interoffice memos are generated on hard copy. The monthly cost for the office using electronic mail (B) includes equipment rental, less any salary recovery from a smaller clerical staff. That amount will be subtracted from the clerical communications cost of the matching office where the standard stenographic procedure (A) is used.

The following steps apply,

Step 1. *Formulate the null hypothesis.* The chosen null hypothesis is

$$H_0: \mu_A \leq \mu_B \qquad \text{(Electronic mail is not cheaper.)}$$

This designation was chosen to control the error of recommending electronic mail when it will not in fact reduce clerical costs.

Step 2. *Select the test procedure and test statistic.* The Student t statistic computed from the matched-pair differences is used.

Step 3. *Establish the significance level and the acceptance and rejection regions for the decision rule.* The chosen significance level is $\alpha = .05$. Using $10 - 1 = 9$ degrees of freedom from Appendix Table G, this corresponds to a critical value of $t_{.05} = 1.833$. Large positive levels for t will refute H_0, and the test is upper-tailed. The decision rule is shown in Figure 12-3.

Figure 12-3

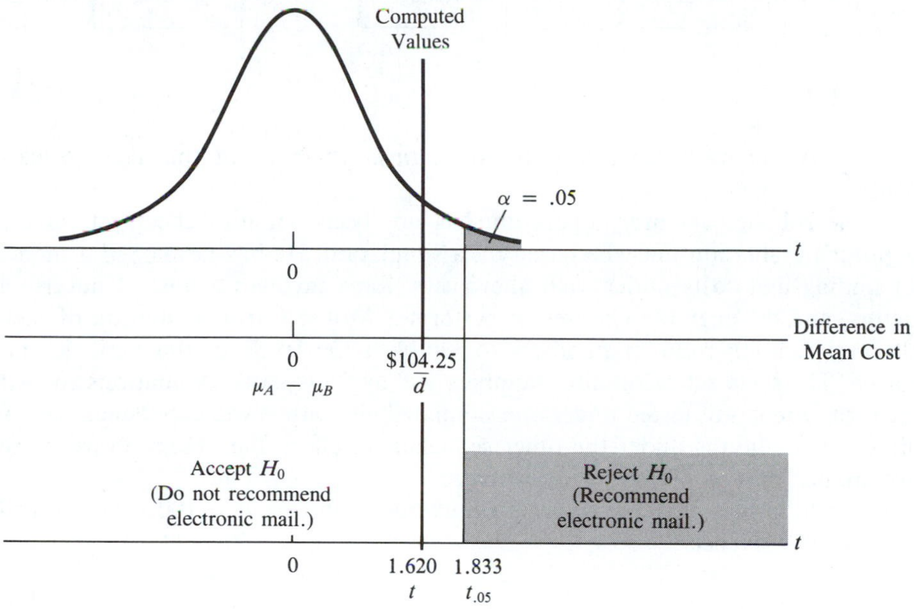

Computed Values

$\alpha = .05$

t

0

Difference in Mean Cost

0 $\mu_A - \mu_B$

\$104.25 $\bar{d}$

Accept H_0 (Do not recommend electronic mail.)

Reject H_0 (Recommend electronic mail.)

t

0

1.620 t

1.833 $t_{.05}$

Step 4. *Collect the sample data and compute the value of the test statistic.* The sample results for the experiment are

$$\bar{d} = \$104.25 \qquad s_d = \$203.50$$

The computed value of the test statistic is

$$t = \frac{\$104.25}{\$203.50/\sqrt{10}} = 1.620$$

Step 5. *Make the decision.* The computed value of the test statistic falls inside the acceptance region. There is not a significant enough cost advantage to warrant recommending adoption of electronic mail, and the null hypothesis is accepted.

Illustration: Monte Carlo Simulation to Evaluate Testing Sequences

Consider a second example based on Monte Carlo simulation.

An industrial engineer is faced with a different type of evaluation, also involving testing. His concern is with the sequencing of final testing operations for robot units. Two alternatives are shown in Figure 12-4. The engineer believes that the greater variability in mechanical testing time might result in a greater need for buffer stock under alternative A, increasing the overall unit-testing cost. But alternative B will cause difficulties with umbilical testing, increasing operator-time re-

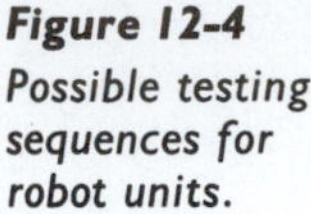
Figure 12-4 *Possible testing sequences for robot units.*

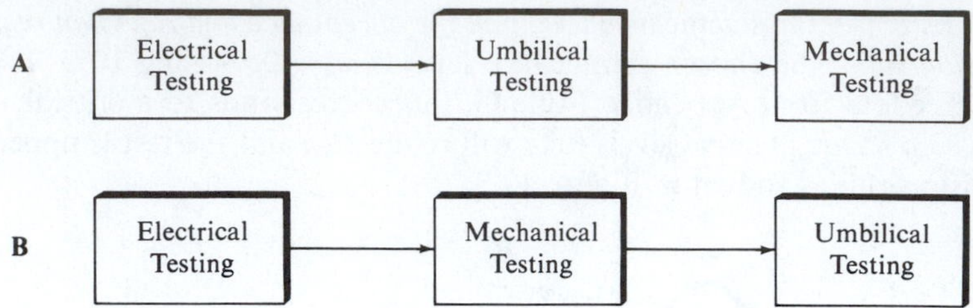

quirements. He wants to select the alternative which is on the average least costly.

The robotics assembly operation has not been put into effect yet, so the engineer's evaluation must be based on a simulation. He has developed a model for finding unit costs under each alternative. Each involves a host of uncertain factors, and the engineer chooses to perform a Monte Carlo simulation of each alternative, using random numbers to establish the levels of the variables involved. The same set of random numbers will apply to both evaluations so that each trial robot unit tested under one simulated alternative will experience exactly the same conditions under the other sequence. In effect, the observations in the two simulations, or samples, may be paired.

The following steps apply for a simulation involving $n = 100$ robots tested under each sequence.

Step 1. *Formulate the null hypothesis.* The engineer chooses the following null hypothesis:

$$H_0\colon \mu_A = \mu_B \qquad \text{(The two sequences have identical mean unit costs.)}$$

This designation results in a two-sided test and reflects the engineer's neutral stance toward the candidate sequences. He views the two ways of incorrectly rejecting H_0 to be equally serious.

Step 2. *Select the test procedure and test statistic.* The Student t statistic computed from the matched-pair differences is used.

Step 3. *Establish the significance level and the acceptance and rejection regions for the decision rule.* The engineer chooses a 5% significance level. This must be split evenly between the two ways of incorrectly rejecting H_0. From Appendix Table G the critical value for $\alpha/2 = .025$ is $t_{.025} = 1.987$. The decision rule, shown in Figure 12-5, indicates that the testing sequence chosen must be consistent with the sample results obtained. If the null hypothesis is accepted, then no significant difference between the population means has been found. The engineer might in that case pick one sequence arbitrarily. The following, however, instead indicates that further simulation trials will be taken after accepting the null hypothesis.

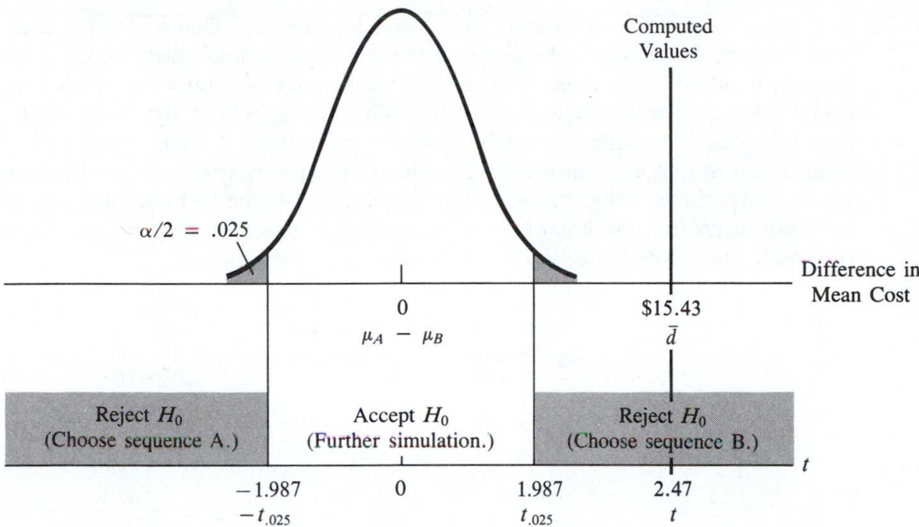

Figure 12-5

Step 4. *Collect the sample data and compute the value of the test statistic.* The following sample results were obtained for the matched-pairs differences

$$\bar{d} = \$15.43$$

$$s_d = \$62.47$$

so the t statistic is computed to be

$$t = \frac{\$15.43}{\$62.47/\sqrt{100}} = 2.47$$

Step 5. *Make the decision.* The computed value for t falls into the upper rejection region. The engineer must therefore reject H_0 and adopt sequence B.

Two-Sided Tests Using Confidence Intervals The same conclusion could have been reached instead by first constructing a 95% confidence interval estimate for the difference in population means. The mean matched-pair difference $\bar{d} = \$15.43$ is a point estimate of the true difference in population means. This is the central value in the following confidence interval,

$$\mu_A - \mu_B = \bar{d} \pm t_{.025}\frac{s_d}{\sqrt{n}} = \$15.43 \pm 1.987\frac{\$62.47}{\sqrt{100}}$$
$$= \$15.43 \pm \$12.41$$

so that

$$\$3.02 \leq \mu_A - \mu_B \leq \$27.84$$

This interval could have been used in testing the original null hypothesis. Since the entire confidence interval lies above zero, again we see that H_0 should be rejected at the $\alpha = .05$ significance level.

Problems

12-9 A chemical engineer is comparing two chemical processes. One involves a high-pressure purification stage. The second achieves purification at lower pressures but involves a catalyst. The levels of impurities (percentage of total volume) resulting from the two procedures are to be compared. The more expensive high-pressure process will be adopted only if it yields the lower mean impurity level. The chemical engineer wants only a 1% chance of adopting it when in fact that process is worse. His null hypothesis is that the mean impurity level from the high-pressure process is at least as great as that from the low-pressure procedure. Pilot batches have been run under the two methods, and the following data were obtained:

Low Pressure	High Pressure
$n_A = 50$	$n_B = 100$
$\bar{X}_A = 2.3$	$\bar{X}_B = 1.5$
$s_A = 1.0$	$s_B = 0.8$

Which process should the chemical engineer select?

12-10 A project engineer is comparing operating modes for a satellite tracking and identification radar. Her study is based on radar cross sections obtained from "chirp" pulses (short continuous transmissions of increasing frequency) and "burst" pulses (a series of very brief transmissions of constant frequency, punctuated by silent gaps). The cross sections are digitized and matched to templates stored in the memory of the system computer. The computer then identifies the satellite target. Although both procedures eventually provide correct identifications, the required time from target acquisition to final confirmation varies considerably and is uncertain. Since satellite position and atmospheric conditions are so crucial, and since only one acquisition radar and computer system was available for testing, independent sampling had to be used. Samples of 100 targets were evaluated under the two modes. The chirp mode's mean time was 35.3 seconds until identification, with a standard deviation of 8.1 seconds. Under the burst mode, the mean time was 33.1, with a standard deviation of 6.3.

At the 5% significance level, must the engineer accept or reject the null hypothesis that the chirp mode provides identification at least as fast as the burst mode?

12-11 Refer to the data for the maintenance policy example on page 503. The fix-only-when-broken policy will be adopted if it is found to be significantly cheaper. The null hypothesis is that this policy is at least as expensive as regular preventive maintenance. Using a 5% significance level, what action should be taken?

12-12 Consider the computer-aided design example on page 508. Only if the sample evidence indicates that CAD is significantly faster will it be adopted. To make the two sets of times comparable, the log-in procedure at the drafting department must be eliminated for the traditional design observations. Subtract one day from each of those times. Then test the null hypothesis that CAD will on the average take at least as long as the traditional method. Use a 1% significance level. What action should be taken?

12-13 An engineering curriculum committee wishes to decide about high school mathematics requirements for entering freshmen. Although it is widely believed that high school calculus does not improve college calculus performance (making it a waste of time for fulfilling mathematics requirements), the committee has been told that high school calculus students outperform the others in the first *physics* course. A sampling investigation was performed involving 50 pairs of students, who were matched by quantitative SAT scores and first-term GPA. The control group (A) consisted of students who did not have high school calculus. The experimental group (B) contained those who did. The physics final examination score of the experimental student in each pair was subtracted from that of the control student, and the mean difference was found to be -2.4 points with a standard deviation of 4.6 points.

At the 1% significance level, should the committee accept or reject the null hypothesis that mean physics final examination scores are at least as high for those who did not have high school calculus as for those who did?

12-14 An automotive engineer is evaluating an experimental fuel-mixture control. Independent test runs have been made using random samples of cars where the new device has been installed and where standard carburetion was used. The following mileage data have been obtained:

Experimental	Control
$n_A = 15$	$n_B = 17$
$\bar{X}_A = 33.4$ mpg	$\bar{X}_B = 32.1$ mpg
$s_A = 1.3$	$s_B = 1.5$

Should the engineer accept or reject the null hypothesis that the standard carburetion provides at least as great mileage as that of the experimental method? Use $\alpha = .05$.

12-15 A statistics instructor wishes to evaluate two methods for teaching statistics. One approach emphasizes computer applications, with all homework assignments requiring the use of "canned" programs in the central time-sharing library. One class of 20 students, forming the experimental group (A), was instructed under this format. The control group (B) was another statistics class of 25 students taught without any encouragement to use computers. The same final examination was given to the two groups. The instructor tested the null hypothesis that examination scores achieved by all students instructed under the two methods have identical means. The experimental group achieved a mean score of 78.5 with a standard deviation of 10.6; the controls achieved a mean of 81.2 and a standard deviation of 9.6. Using a 5% significance level, did the instructor accept or reject the null hypothesis?

12-16 A chemical engineer is evaluating a new catalyst to determine if it speeds up a chemical procedure. A sample of 10 pilot batches was processed with the new catalyst (group A). Each test run was carefully matched with a second batch processed with the present catalyst (group B). Each pair of runs was conducted under nearly identical control settings, used the same raw-material stock, and involved almost the same time-phasing. The processing time for batch B was subtracted from its batch A partner, and the mean difference was found to be .53 hour with a standard deviation of .16 hour.

(a) Construct a 95% confidence interval estimate of the difference between mean processing times for all batches run with the two catalysts.

(b) Should the null hypothesis that the mean processing times are identical be accepted or rejected at the 5% significance level?

12-17 A test similar to the one in Problem 12-16 was performed independently in two different plants, one using the new catalyst (A) and the other using the present one (B). A total of 8 sample batches were processed using the new catalyst, while 15 runs were made with the present one. The following processing time results were obtained:

$$\bar{X}_A = 12.37 \text{ hours} \qquad \bar{X}_B = 12.93 \text{ hours}$$

$$s_A = 2.14 \qquad s_B = 1.86$$

(a) Construct a 95% confidence interval estimate of the difference between mean processing times.

(b) Should the null hypothesis that the mean processing times are identical be accepted or rejected at the 5% significance level?

12-18 Refer to the confidence interval you found in Problem 12-1 for the difference in engineers' mean incomes. Should the society accept or reject the null hypothesis that the mean incomes are the same? Use $\alpha = .05$.

12-19 Refer to the confidence interval you found in Problem 12-3 for the difference in MTBFs for two different transformers. Should the electrical engineer accept or reject the null hypothesis that the mean times between failures are the same? Use $\alpha = .01$.

12-20 Refer to the confidence interval you found in Problem 12-5 for the difference in mean completion times. Should the chief engineer accept or reject the null hypothesis that the mean times are the same for both electrical subcontractors? Use $\alpha = .05$.

12-21 Refer to the confidence interval you found in Problem 12-6 for the difference in mean annual engine maintenance cost. Should the foreman accept or reject the null hypothesis that the mean costs are identical for the two makes? Use $\alpha = .01$.

12-22 Refer to the paint drying-time data in Problem 12-7. Test the null hypothesis that paint A involves drying times that are on the average less than or equal to those of paint B. At the 1% significance level, must H_0 be accepted or rejected?

12-23 Refer to the student GPA data in Problem 12-8. Test the null hypothesis that regular engineering students achieve grade-point averages at least as high as those of co-operative students. At the 5% significance level, must H_0 be accepted or rejected?

12-4 Hypothesis Tests for Comparing Proportions

Recall that many populations evaluated by statistical procedures are *qualitative* in nature and involve observations that place units into two or more categories. The *proportion* of observations falling into a desired category often provides a suitable basis for decision making, with hypothesis tests used with single samples to decide whether the population proportion falls above or below a particular level. That testing procedure may be extended to controlled experiments where two qualitative populations are compared. Many important engineering applications, such as quality control, mineral exploration, production, and reliability, involve such comparisons. The basic approach for testing two population proportions parallels the methods for comparing two means.

If we denote the two population proportions by π_A and π_B and the respective sample proportions by P_A and P_B, we can estimate the difference between population proportions.

DIFFERENCE IN SAMPLE PROPORTIONS

$$D = P_A - P_B$$

The forms of the hypotheses tested match the cases for testing two means. We have

$$\begin{array}{lll} H_0\colon \pi_A \geq \pi_B \quad \text{or} \quad H_0\colon \pi_A - \pi_B \geq 0 & & \text{(lower-tailed test)} \\ H_0\colon \pi_A \leq \pi_B \quad \text{or} \quad H_0\colon \pi_A - \pi_B \leq 0 & & \text{(upper-tailed test)} \\ H_0\colon \pi_A = \pi_B \quad \text{or} \quad H_0\colon \pi_A - \pi_B = 0 & & \text{(two-sided test)} \end{array}$$

Although the sampling distributions for P_A and P_B are represented by the binomial or hypergeometric distributions, we may ordinarily apply the normal approximation when sample sizes are sufficiently large. Unlike controlled experiments with the mean, matching does not apply in testing proportions, so only the case of two independent samples need be considered. The expected value of the difference in sample proportions, $D = P_A - P_B$, is $\pi_A - \pi_B$. Since the variance of the difference in two independent random variables is equal to the sum of the individual variances, it follows that

$$\sigma_D = \sqrt{\frac{\pi_A(1 - \pi_A)}{n_A} + \frac{\pi_B(1 - \pi_B)}{n_B}}$$

All of the preceding null hypotheses involve an extreme case of $\pi_A - \pi_B = 0$, so the sample results may be treated as if they were obtained from the same population. Under H_0, therefore, we may use the *pooled* sample results to estimate either π_A or π_B. Thus, the pooled data give the following.

COMBINED SAMPLE PROPORTION

$$P_C = \frac{n_A P_A + n_B P_B}{n_A + n_B}$$

This is used in estimating the standard error of D,

$$s_D = \sqrt{P_C(1 - P_C)\left(\frac{1}{n_A} + \frac{1}{n_B}\right)}$$

The following expression gives the normal deviate for comparing two proportions.

NORMAL DEVIATE FOR COMPARING
TWO PROPORTIONS

$$z = \frac{D}{s_D} = \frac{P_A - P_B}{\sqrt{P_C(1 - P_C)\left(\dfrac{1}{n_A} + \dfrac{1}{n_B}\right)}}$$

Illustration: Probability of Getting a Busy Signal

To illustrate this procedure we will consider a decision by the chief engineer for a long-distance telephone service supplier whether to augment the company's present capacity by acquiring communications satellite channels. The alternative is to add a series of microwave relay stations into the ground-based system. Since both approaches are equally costly, that alternative will be chosen which results in the lower probability of getting a busy signal for peak-period calls placed between the Pacific and Atlantic coasts. Those probabilities are equivalent to the population proportion of all attempted calls that cannot be completed because of busy circuits.

Under either scheme, the number of combinations of possible system states is astronomical, so the desired probabilities must be represented by the sample proportions obtained from Monte Carlo simulations of the two proposed linkages. Two simulations, each involving 10,000 trial observations, were to be performed.

The following hypothesis-testing steps were taken:

Step 1. *Formulate the null hypothesis.* The satellite linkage is designated by A and the ground relays by B. The engineer chose the following null hypothesis:

$$H_0: \pi_A = \pi_B \qquad \text{(The probability of busy is identical under the two linkages.)}$$

This designation reflects the engineer's neutrality about the alternative changes to the communications network.

Step 2. *Select the test procedure and test statistic.* The normal deviate based on the difference between sample proportions was selected for the test statistic.

Step 3. *Establish the significance level and the acceptance and rejection regions for the decision rule.* A 5% significance level was chosen for the simulation experiment. This corresponds to a critical normal deviate of $z_{.025} = 1.96$. The procedure is two-sided with the decision rule shown in Figure 12-6.

Figure 12-6

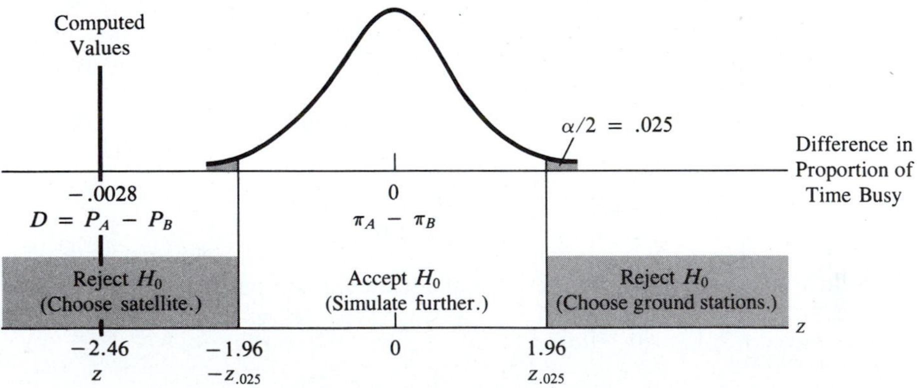

Step 4. *Collect the sample data and compute the value of the test statistic.* The sample results for the experiment are

$$P_A = .0051 \qquad P_B = .0079$$

These results provide the following difference in sample proportions:

$$D = .0051 - .0079 = -.0028$$

The combined sample proportion of simulation trials found busy is

$$P_C = \frac{10{,}000(.0051) + 10{,}000(.0079)}{10{,}000 + 10{,}000} = .0065$$

The computed value of the normal deviate is

$$z = \frac{.0051 - .0079}{\sqrt{(.0065)(.9935)\left(\frac{1}{10{,}000} + \frac{1}{10{,}000}\right)}} = -2.46$$

Step 5. *Make the decision.* The computed normal deviate falls inside the lower rejection region, which indicates that the satellite linkage provides the lower busy-signal probability. The engineer must therefore reject H_0 and recommend that system configuration.

Problems

12-24 A government agency studying science and engineering education is testing the null hypothesis that the proportion of graduates who will continue on to graduate school is the same for electrical and mechanical engineers. In a random sample of 50 electrical engineers, 13 planned to go to graduate school. A sample of 100 mechanical engineers provided 21 respondents desiring graduate work. At the 5% significance level, should the agency's null hypothesis be accepted or rejected?

12-25 A panel of 50 experts independently evaluated two alternative designs. Design A was found acceptable by 27 of the panel members. Design B was approved by 31 members. Assuming the panel is representative of all persons sharing their expertise, can the null hypothesis be rejected that design A will have a higher acceptance proportion? Use $\alpha = .01$.

12-26 An industrial engineer is auditing quality-control standards for the fabrication of a part at two different plants. A random sample of 100 parts has been selected independently from each plant, and the respective proportions defective have been computed. The engineer wishes to test the null hypothesis that defectives occur at the same rate in plant A as in plant B. The sample results provide $P_A = .08$ and $P_B = .04$. Using a 5% significance level, what conclusion should the engineer reach?

12-27 An engineer wishes to compare two alternative designs for a satellite electronics system. Design A is a maximum-redundancy design involving many inexpensive, low-reliability components in parallel. Design B is a high-quality design involving a minimum number of highly reliable but expensive components. He will select the second design unless the first is found to have a significantly smaller failure rate.

A 1,000-mission Monte Carlo simulation was conducted under each design. The results provided 47 failures with design A and 62 with design B. At the 5% significance level, which design should the engineer select?

12-28 A computer scientist wishes to compare an improved modem interface (A) as a possible replacement for the existing hardware (B) when the present microcomputer network is updated. She wishes to test the null hypothesis that the new equipment will provide an encoding/decoding error rate at least as high as the present modems. The new devices will replace the present modems if they yield a significantly lower error rate, as determined by statistical testing. This was done using an experimental network. First a test run was made with 10 million bits transmitted and received with the new modem interfaces; 3 errors were determined. Then a similar 10 million-bit test was made on the same network after the modems were replaced by the versions presently in use; 5 errors were recorded. At the 1% significance level, which modems will be selected in the updated network?

12-5 Nonparametric Statistics

The hypothesis-testing procedures described so far in this book are all concerned with drawing conclusions about population *parameters*—the mean, proportion, or variance. Reflecting their purpose, such procedures may be referred to as **parametric statistics**. Many of the parametric procedures are founded on assumptions that are necessary in establishing the sampling distributions for the test statistics employed. For instance, in testing means the Student t distribution requires that the underlying population itself be normally distributed. Although the Student t distribution may safely be applied to a wide variety of applications where the population is nonnormal, this might detract from the validity and credibility of the results.

Fortunately, there are statistical procedures that are based on less stringent assumptions about the underlying population. In the remainder of this chapter we describe some important tests that involve no assumptions regarding population parameters. For this reason, such procedures are categorized as **nonparametric statistics**.

Advantages of Nonparametric Statistics

The main advantage of nonparametric statistics is that these procedures usually involve few restrictive assumptions, which makes them valid for a wide variety of applications. As added benefits, many of the nonparametric methods are based on very simple probability concepts, which makes them easy to understand and use. They are particularly convenient when sample sizes are small—an important consideration when the cost of data is great or when human subjects are involved.

Nonparametric procedures are alternatives to traditional methods, which sometimes make assumptions regarding population parameters that are not even being tested. For instance, in Section 12-3 we introduced a procedure for comparing means using independent samples. That test requires that the two parent populations have the same variance.

An implicit assumption of many parametric procedures is that the underlying quantitative populations involve *continuous* values. This is appropriate for many engineering applications where physical quantities, such as time, weight, or volume, fall naturally on a continuous spectrum. But a host of variables—examination scores, grade-point averages, or the number of defectives, errors, or failures, for example—are inherently discrete quantities. A wide class of quantitative values cannot even be combined arithmetically; for example, subjective ratings in which a subject is asked to rate a product design on an arbitrary scale such as from 0 (poor) to 10 (superlative). The same type of information could be attained using limits of 0 and 100, −10 and 10, or 1 and 5.

Subjective ratings expressing preference, ranking, or priority are numerical values that mainly convey order and belong to an **ordinal scale**. Basic arithmetic operations such as differences, sums, products, or ratios cannot be consistently applied to such numbers. Distance between values has no inherent meaning. Thus, statistical tests involving computed means or variances are not ordinarily valid when population values are ordinal. Nonparametric tests are usually the only valid tests for drawing inferences regarding such populations.

Disadvantages of Nonparametric Statistics

There is another side to the ledger. Avoiding possibly overrestrictive assumptions tends to make nonparametric procedures less keen in discerning differences between populations. Thus, applied to the same data, a parametric test will usually be more powerful than its nonparametric counterpart. That is, it will allow the null hypothesis to be rejected with a lower type I error probability. Of course, the nonparametric test will be the correct procedure for a wider variety of applications. When samples are large, nonparametric tests can be computationally cumbersome (although this handicap is not too important when good computer software is available). Another drawback of nonparametric tests is that special tables must often be used to establish critical values for the test statistic. Paradoxically, the sampling distributions of many of those test statistics may be closely approximated by the normal distribution.

12-6 Comparisons Using Independent Samples: The Wilcoxon Rank-Sum Test

As an alternative to the parametric procedures described earlier for testing two population means using independent samples, we consider the **Wilcoxon rank-sum test**. This popular nonparametric test is named in honor of the statistician who proposed it in 1945. Unlike the t test, which requires that the populations be nearly normally distributed, this procedure makes no assumption regarding the population frequency distribution. For this reason, the test falls into the category of **distribution-free statistics**. This particular Wilcoxon test is equivalent to another popular nonparametric procedure referred to as the Mann-Whitney test.

Description of the Test

This test compares two samples, the control and experimental groups, each drawn from a different parent population. The null hypotheses tested under this procedure all assume that the two sample groups actually derive from the same common population. In that sense, the test is somewhat more stringent than the earlier ones, where the null hypotheses considered equality between means only (although the Student t procedure requires equal variances, too).

The procedure is based on a ranking by size of the pooled sample outcomes. The smallest observation receives a rank of 1, the next smallest is ranked as 2, and so on. After ranking, the observations are split into their original groups and the sum of ranks for group A is determined. The test statistic is based on that sum.

Illustration: *Comparing Design Ratings by Two Panels*

We illustrate the procedure with an evaluation of two types of architectural review panels. A large architectural and engineering firm presently relies on a panel of laypersons selected by its client to evaluate its work by rating various alternative designs. One partner proposes a faster and less expensive evaluation using an in-house panel of engineers and architects. The efficacy of that short-cut depends on the ability of the in-house professional panel (A) to exercise the same collective taste as the client lay group (B). An experiment was conducted using a hypothetical panel of each type. Each individual was asked to rate ten designs esthetically on a scale from 1 to 10; the sum of the ratings for each person was obtained. The hypotheses being tested are

H_0: Population A values = Population B values
(Professionals and laypersons have identical taste.)

H_1: Population A values $\neq$ Population B values
(Professionals and laypersons have different taste.)

The ratings from the two sample groups are provided in Table 12-5. Those data have been pooled and then ranked from lowest to highest rating. If the populations are identical, then the A observations should be ranked low and high about as many times as the B's. The sum of the ranks obtained for either group will indicate whether the results tend to support or refute that null hypothesis. For this purpose, we use

$$W = \text{Sum of ranks obtained by group A}$$

Under the null hypothesis of identical taste (identical rating populations), there is an equal chance that any particular rank will correspond to group A or group B. Here we have groups of different size, with $n_A = 8$ and $n_B = 10$. Under the null hypothesis, all combinations of 8 group-A ranks from 18 are equally likely elementary events.

To establish the probability distribution for possible levels of the group-A rank-sum W, it would be necessary to catalogue all of those combinations. We could then apply the basic concepts of Chapter 2 to establish the probability for

Table 12-5 Architectural Review Panel Ratings for Two Groups with Ranks Obtained from Pooled Results

Professionals (A)		Laypersons (B)	
Rating	Rank	Rating	Rank
60	3	77	12
67	7	82	14
75	11	65	5
83	15	85	16
64	4	72	9
53	1	69	8
58	2	79	13
66	6	74	10
	$W = 49$	87	17
		92	18
$n_A = 8$		$n_B = 10$	

the events $W = 36$, $W = 37$, and so on. (In the present illustration we need not consider sums less than 36. Why not?) This could be quite a chore. Although tables have been constructed for the smaller sample sizes, it is more convenient to make use of the fact that W has a sampling distribution that can be approximated for most n's by the normal curve. This allows us to use the following expression.

NORMAL DEVIATE FOR THE WILCOXON RANK-SUM TEST

$$z = \frac{W - \dfrac{n_A(n_A + n_B + 1)}{2}}{\sqrt{\dfrac{n_A n_B(n_A + n_B + 1)}{12}}}$$

The present illustration involves a two-sided test, since the null hypothesis of identical populations is refuted if W turns out to be untypically large or small. Suppose that the partner selects a 5% significance level. Appendix Table E provides $z_{.025} = 1.96$. The acceptance and rejection regions are provided in Figure 12-7. Notice that only if H_0 is accepted will the in-house professional design review panel be substituted for the laypersons now provided by clients.

In Table 12-5 the computed value for the group-A rank-sum is $W = 49$. The corresponding value of the test statistic z is

$$z = \frac{49 - \dfrac{8(8 + 10 + 1)}{2}}{\sqrt{\dfrac{8(10)(8 + 10 + 1)}{12}}} = -2.40$$

Figure 12-7

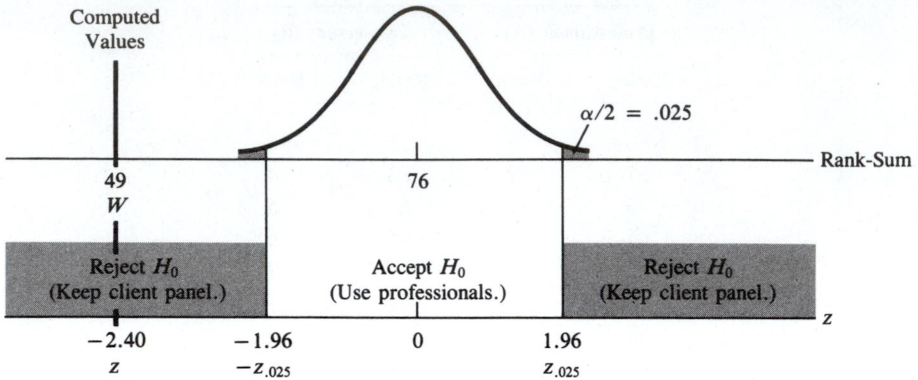

Since this falls in the lower rejection region, the null hypothesis must be rejected. It may then be concluded that laypersons and professionals have different tastes, and the use of client panels must be continued.

Application to One-Sided Tests

The procedure applies to one-sided tests as well.

Illustration: Computer Recognition of Shapes

As another illustration, consider an artificial intelligence experiment where the software will be chosen to control robots performing simple assembly tasks. The most difficult task is recognizing components and then fitting them together properly. This requires an optical controller, which is a program employing a central processing unit connected to the robot. Two basic approaches are being evaluated. Method A is based on comparing the optical image to shape templates in memory. Method B is a heuristic procedure in which the robot software is the consequence of a learning experience. Although it requires an extensive data base that must be updated each time there is a product change, the template method is easiest to program and will be chosen unless it can be established that the heuristic software will be faster in actual application.

Two independent sample runs were made under each optical control method. The resulting assembly times are provided in Table 12-6.

The following hypothesis-testing steps were taken:

Step 1. *Formulate the null hypothesis.* The following null hypothesis is being tested:

H_0: Population A times $\leq$ Population B times
(The template controller operates faster.)

Step 2. *Select the test statistic.* The normal deviate based on the assembly-time ranks serves as the test statistic.

Table 12-6 *Sample Results for Robot Assembly Test Runs Using Template and Heuristic Controllers*

Template Controller (A)		Heuristic Controller (B)	
Time (seconds)	Rank	Time (seconds)	Rank
37	14	43	19
*35	12.5	22	5
25	6	*35	12.5
38	15	18	3
5	1	15	2
47	20	30	8
27	7	32	9
40	16	32	10
41	17	20	4
42	18	34	11
55	21		
82	23		
65	22		
115	24		
150	25		
	$W = 241.5$		

* Tying observations.

Step 3. *Establish the significance level and the acceptance and rejection regions for the decision rule.* A 1% significance level was chosen for the simulation experiment. This corresponds to a critical normal deviate of $z_{.01} = 2.33$. The null hypothesis will be refuted if sample times for the template procedure tend to exceed those of the heuristic method. Should that be the outcome, then group A will receive preponderantly high ranks, and W will be large. The resulting normal deviate will be large and positive, so the test is an upper-tailed one. The procedure is therefore one-sided with the decision rule shown below.

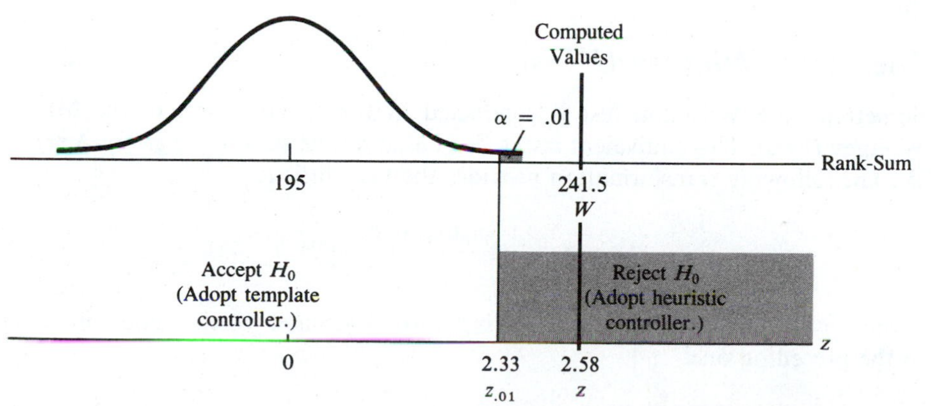

Figure 12-8

Step 4. *Collect the sample data and compute the value of the test statistic.* The ranks for the pooled observations are given in Table 12-6. The sum of the group-A ranks is $W = 241.5$. Using $n_A = 15$ and $n_B = 10$, we compute the following normal deviate:

$$z = \frac{241.5 - \dfrac{15(15 + 10 + 1)}{2}}{\sqrt{\dfrac{15(10)(15 + 10 + 1)}{12}}} = 2.58$$

Step 5. *Make the decision.* The computed normal deviate falls in the rejection region, which indicates that the heuristic controller will operate faster than the template controller. H_0 must therefore be rejected, and the heuristic controller should be adopted.

In this illustration it would not have been appropriate to use the t test, even if the underlying populations were normally distributed. The evidence is rather strong that groups A and B represent populations with dramatically different variances, as evidenced by the respective sample standard deviations of $s_A = 37.1$ and $s_B = 8.9$ seconds.

Treatment of Ties

Several ties were encountered in the pooled sample data. As long as the tying observations fall in the same sample group, there is no problem, and successive ranks may be assigned arbitrarily to the identical values. This happened in Table 12-6, where the seventh and eighth group-B observations tied at 32 seconds each; those times received the next larger ranks of 9 and 10 in the regular ascending sequence. But when ties occur between observed values in opposite groups, the standard procedure is to assign ranks equal to the average of the run of ranks that would have been assigned if there had been no ties at all. For example, there was a 35-second assembly time in both sample groups; the successive ranks of 12 and 13 were averaged, and each observation was assigned a rank of 12.5.

The Mann-Whitney U Test

Sometimes the Wilcoxon test is conducted in the variant form of the Mann-Whitney U test. This equivalent test is also based on the sum of the group-A ranks W. The following transformation provides the test statistic:

$$U = n_A n_B + \frac{n_A(n_B + 1)}{2} - W$$

From this, we can calculate the following normal deviate, which is equal in value to the preceding one:

$$z = \frac{U - \dfrac{n_A n_B}{2}}{\sqrt{\dfrac{n_A n_B(n_A + n_B + 1)}{12}}}$$

Identical results to those of the Wilcoxon will be achieved using this test.

Problems

12-29 The Wilcoxon rank-sum test will be used to determine the appropriate action to be taken in the following situations. Find the acceptance and rejection regions and indicate whether H_0 should be accepted or rejected at the stated significance level for each result provided.

(a)	(b)	(c)	(d)
H_0: A's $\geq$ B's	H_0: A's $\leq$ B's	H_0: A's $\geq$ B's	H_0: A's = B's
$n_A = 25$	$n_A = 16$	$n_A = 18$	$n_A = 35$
$n_B = 20$	$n_B = 12$	$n_B = 24$	$n_B = 50$
$\alpha = .05$	$\alpha = .01$	$\alpha = .05$	$\alpha = .01$
$W = 364$	$W = 346$	$W = 253$	$W = 2{,}170$

12-30 A test laboratory intends to change from vendor A to vendor B if the latter's batteries last longer. The null hypothesis is that vendor A's batteries last at least as long as vendor B's. A random sample of 8 brand A batteries was discharged until dead, while a similar test was conducted on 10 brand B batteries. The following times were obtained:

Battery A: 11.5, 13.3, 15.4, 14.6, 12.7, 13.8, 12.7, 13.6
Battery B: 13.5, 14.2, 15.1, 12.9, 13.5, 15.8, 15.2, 14.8, 15.2, 16.3

Which vendor will be selected at the 5% significance level?

12-31 Suppose that the laboratory in Problem 12-30 placed 50 batteries of each type into operation, and on the basis of rankings of the normal lifetimes the rank-sum $W = 1{,}350$ was achieved. Based on that experiment, should the null hypothesis that vendor A's batteries last at least as long as vendor B's be accepted or rejected? Use $\alpha = .05$.

12-32 Suppose that the chemical engineer in Problem 12-9 applies the Wilcoxon rank-sum test to a sample of 7 batches run under the low-pressure process and 9 batches under the high-pressure process. The following yields (grams/liter) of active ingredient were obtained:

(A) *Low-Pressure:* 25, 27, 28, 29, 30, 30, 31
(B) *High-Pressure:* 28, 31, 32, 33, 33, 34, 35, 35, 36

The engineer is testing the null hypothesis that both procedures provide identical yields. At the 5% significance level, can it be concluded that the high-pressure process provides greater yields?

12-33 An operations researcher is comparing two methods for shipping small packages of supplies to customers. The null hypothesis is that the U.S. Postal Service (A) is at least as fast as a private mailer (B) in completing delivery. The following numbers of days were taken for delivery of two independent random sample packages between equidistant points:

(A) *Postal Service:* 12, 13, 12, 14, 15, 15, 13, 15, 17, 20, 14, 13, 15, 16, 18, 19
(B) *Private Mailer:* 8, 9, 10, 12, 7, 11, 10, 12, 11, 10, 9

Can the researcher reject the null hypothesis, using $\alpha = .01$?

12-34 An engineering society wishes to establish whether there is any difference in consulting fees between firms on the East Coast and on the West Coast. The following average daily billings were obtained from independent random samples of firms:

East Coast: \$346, \$465, \$512, \$488, \$475, \$527, \$621
West Coast: \$372, \$395, \$415, \$452, \$389, \$364, \$417, \$434

At the 5% significance level, should the null hypothesis of equal fees be accepted or rejected?

12-35 Consider the *true* sampling distribution of W when $n_A = 7$ and $n_B = 8$.
(a) Compute the exact probabilities for
(1) $W = 28$ (2) $W = 29$ (3) $W = 30$ (4) $W = 31$ (5) $W = 32$
(b) The null hypothesis is H_0: A's $\geq$ B's. Using your answer to (a), determine the critical value for W that would apply when the type I error probability cannot exceed $\alpha = .0015$.

12-36 Consider the *true* sampling distribution of W when $n_A = 10$ and $n_B = 10$.
(a) Compute the exact probabilities for
(1) $W = 155$ (3) $W = 153$ (5) $W = 151$
(2) $W = 154$ (4) $W = 152$
(b) The null hypothesis is H_0: A's $\leq$ B's. Using your answer to (a), determine the critical value for W that would apply when the type I error probability cannot exceed $\alpha = .00005$.

12-7 Comparisons Using Matched Pairs: The Wilcoxon Signed-Rank Test

A second Wilcoxon test is popular in matched-pairs testing. This nonparametric procedure may be used instead of its counterparts described in Sections 12-2 and 12-3. It is based on the difference between the observed values for each matched pair. Unlike the parametric tests, it makes no stipulations regarding the nature of the population distributions, other than that they be *identical.*

Description of the Test: Using Subjective Ratings to Choose a Design

The procedure is based on the rank of absolute values of pair differences. The ranks for the positive differences are then summed. The test statistic V is then obtained.

We may summarize this procedure by the following three steps:

1. Calculate the differences $d_i = X_{A_i} - X_{B_i}$ for each sample pair.
2. Rank the absolute values, $|d_i|$. Do not include 0 differences.
3. Calculate V, the sum of ranks for positive d_i's.

This test may be one-sided or two-sided, and under all circumstances H_0 includes the possibility that the A and B populations are identical. In that case, there is a .5 chance for any pair difference of any absolute size to be positive. Under H_0, therefore, half the ranks are expected to correspond to positive differences, and the sum of these ranks V should be close in value to half of the total rank-sum value.

As with the preceding Wilcoxon procedure, an exact probability accounting can be made to establish the sampling distribution of V. But as before, it is more convenient to make the normal approximation and use the following as the test statistic.

NORMAL DEVIATE FOR THE WILCOXON SIGNED-RANK TEST

$$z = \frac{V - \dfrac{n(n+1)}{4}}{\sqrt{\dfrac{n(n+1)(2n+1)}{24}}}$$

The value for n represents the number of observed pairs, less those where the values tie.

Illustration: Establishing the Preferred Design

As an illustration, consider an evaluation by a panel of targeted users of two designs for a proposed product. Table 12-7 gives the sample results of these subjective evaluations, where each pair constitutes the rating made of the two designs by the same person.

The following null hypothesis is being tested:

H_0: Population A values = Population B values
(The two designs are regarded equally well.)

A 5% significance level was selected for this test. The critical normal deviate is thus $z_{.025} = 1.96$. Values of V that are very large or very small will refute the null hypothesis, and the two-sided rejection region shown in Figure 12-9 applies.

The ranks of the absolute differences are found in Table 12-7. These numbers were computed in exactly the same way as the earlier Wilcoxon test. Should ties occur between positive and negative differences, the successive ranks are averaged. Thus, the -9 and $+9$ differences each received ranks of $(9 + 10)/2 = 9.5$. Ties within the positive or within the negative groups do not matter.

The sum of the ranks for the 14 positive differences is

$$V = 8 + 3 + 7 + 6 + 18 + 14 + 15 + 13 + 9.5 + 2 + 16 + 17 + 12 + 5$$
$$= 145.5$$

Table 12-7 *Matched-Pairs Investigation of Two Alternative Designs for a New Product, with Assignment of Ranks*

Pair i	Design A X_{A_i}	Design B X_{B_i}	Difference $d_i = X_{A_i} - X_{B_i}$	Sign	Rank of Absolute Value of Difference
1	73	65	+8	+	8
2	44	47	−3	−	4
3	53	51	+2	+	3
4	65	57	+8	+	7
5	83	76	+7	+	6
6	55	64	−9	−	9.5
7	49	49	0	tie	—
8	78	57	+21	+	18
9	74	61	+13	+	14
10	88	74	+14	+	15
11	53	41	+12	+	13
12	37	47	−10	−	11
13	70	70	0	tie	—
14	67	58	+9	+	9.5
15	63	61	+2	+	2
16	57	42	+15	+	16
17	66	50	+16	+	17
18	85	74	+11	+	12
19	64	65	−1	−	1
20	60	54	+6	+	5

For $n = 18$ nonzero pair differences, the normal deviate is

$$z = \frac{145.5 - \dfrac{18(18+1)}{4}}{\sqrt{\dfrac{18(18+1)(36+1)}{24}}} = 2.61$$

Figure 12-9

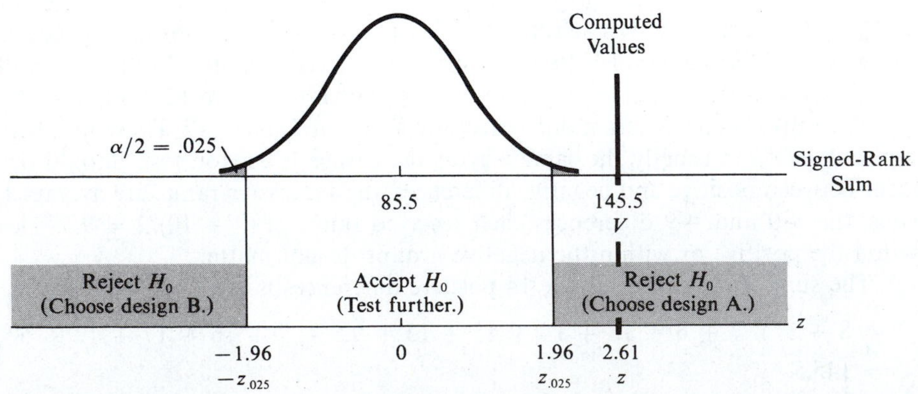

Since this falls inside the upper rejection region, H_0 must be rejected and design A should be chosen.

The Sign Test

Another nonparametric procedure used in testing with matched pairs is the **sign test**. This test is concerned only with the sign of the pair differences and not their relative sizes. It therefore makes less use of the available information than the Wilcoxon test does. The Wilcoxon is the more powerful procedure. However, the sign test is somewhat simpler to use and is also less restrictive, since under H_0 the A and B populations are not assumed to be identical. The sign test also works better when there are many ties. But on balance, the Wilcoxon signed-rank test should be preferred in most engineering applications where a nonparametric test is applied with matched pairs. A detailed discussion of the sign test may be found in the references in the Bibliography.

Problems

12-37 The Wilcoxon signed-rank test will be used to determine the appropriate action to be taken in the following situations. Find the acceptance and rejection regions and indicate whether H_0 should be accepted or rejected at the stated significance level for each result provided.

(a)	(b)	(c)	(d)
H_0: A's $\geq$ B's	H_0: A's $\leq$ B's	H_0: A's $\geq$ B's	H_0: A's $=$ B's
$n = 20$	$n = 15$	$n = 25$	$n = 40$
$\alpha = .05$	$\alpha = .01$	$\alpha = .05$	$\alpha = .01$
$V = 35$	$V = 95$	$V = 52$	$V = 765$

12-38 A consulting engineer for a large company must determine whether or not to recommend replacing the unleaded gasoline now being used in the car fleet with methyl alcohol. That action will be taken if it can be shown that the operating costs are significantly lower using the alcohol. The null hypothesis is that methyl alcohol would not be cheaper. The following sample cost-per-mile data were obtained from a test using two sets of automobiles matched into pairs by car type, driving conditions, and terrain.

Pair	Methyl Alcohol (A)	Unleaded Gas (B)	Pair	Methyl Alcohol (A)	Unleaded Gas (B)
1	\$.18	\$.20	6	\$.23	\$.27
2	.17	.18	7	.19	.21
3	.23	.23	8	.22	.22
4	.19	.18	9	.19	.24
5	.21	.24	10	.20	.26

Using a significance level of .05, what action should the engineer take?

12-39 A government synthetic-fuels study performed an experiment similar to the one in Problem 12-38 to compare the operating costs of ethyl alcohol (A) to methyl alcohol (B). In testing the null hypothesis that the two fuels are equally costly, a sample of cars was driven using both fuels. After eliminating the nonzero differences, 95 test pairs remained. The sum of the ranks for the positive differences was $V = 150$. At the 1% significance level what conclusion should be reached?

12-40 Apply the Wilcoxon signed-rank procedure to the design-time data on page 508 and, using $\alpha = .05$, test the null hypothesis that CAD and traditional methods require the same amount of time.

12-41 The admissions committee at an engineering school is investigating replacing the high school mechanical-drawing requirement with a computer programming elective, so that freshmen can begin their first term with a course in computer graphics. But the committee does not want this to reduce the applicants' scores in the visual perception test. Two random samples of high school students have been obtained. Group A students have had the equivalent of the computer programming course, but no mechanical drawing. Group B students have had mechanical drawing. The students have been paired in terms of general interest, artistic abilities, relevant hobbies, and academic background. Each has been given the standard visual perception test, and the following scores were obtained:

Pair	A Score	B Score	Pair	A Score	B Score
1	87	85	11	62	77
2	55	73	12	67	43
3	62	62	13	77	75
4	51	68	14	71	67
5	48	53	15	58	42
6	66	39	16	58	81
7	53	56	17	61	66
8	24	38	18	51	33
9	91	83	19	80	75
10	63	60	20	44	55

The committee has selected the null hypothesis that the computer background will yield visual perception scores no higher than those achieved by students who have had mechanical drawing. Mechanical drawing will continue to be required unless that hypothesis is rejected. What conclusion will be reached at the 1% significance level?

12-42 A chemical engineer is evaluating a new compound to serve as the stabilizer for powdered chlorine. Ten sample batches have been prepared under that formulation and ten batches with the old stabilizer. Batches from each group have been paired, with each pair then stored under a variety of environments. The following are the observed times until serious decomposition:

Pair	Decomposition Time (months) Old Stabilizer	New Stabilizer	Pair	Decomposition Time (months) Old Stabilizer	New Stabilizer
1	15.3	16.4	6	9.2	11.5
2	17.4	18.0	7	6.3	7.3
3	11.5	13.0	8	10.0	12.6
4	12.0	11.7	9	12.2	12.3
5	14.8	15.1	10	11.5	11.5

Determine whether the engineer should accept or reject the null hypothesis that the old formulation lasts at least as long as the new one. Use $\alpha = .05$.

12-43 Consider the *true* sampling distribution of V when the number of nonzero pair differences is $n = 10$.

(a) Compute the exact probabilities for

(1) $V = 0$ (3) $V = 2$ (5) $V = 4$

(2) $V = 1$ (4) $V = 3$ (6) $V = 5$

(b) The null hypothesis is H_0: A's $\geq$ B's. Using your answer to (a), determine the critical value for V that would apply when the type I error probability cannot exceed $\alpha = .005$.

12-44 Consider the *true* sampling distribution of V when the number of nonzero pair differences is $n = 15$.

(a) Compute the exact probabilities for

(1) $V = 120$ (3) $V = 118$ (5) $V = 116$ (7) $V = 114$

(2) $V = 119$ (4) $V = 117$ (6) $V = 115$ (8) $V = 113$

(b) The null hypothesis is H_0: A's $\leq$ B's. Using your answer to (a), determine the critical value for V that would apply when the type I error probability cannot exceed $\alpha = .0005$.

12-8 Hypothesis Tests for Comparing Variances

Although far less common than tests for means, there are engineering applications where comparisons must be made of two population variances. Such evaluations can be crucial to an engineer whose final product must be fabricated from components having consistent strength. By comparing variances in the strength properties of materials, the designer could eliminate those that might function erratically, even though their mean levels might be satisfactory. Testing for equality of variance can be important in justifying statistical procedures that require equal population variances, such as using the Student t for testing two means.

The testing procedure involves two population variances, σ_A^2 and σ_B^2. The null hypotheses take the usual forms:

$$H_0: \sigma_A^2 \leq \sigma_B^2$$
$$H_0: \sigma_A^2 \geq \sigma_B^2$$
$$H_0: \sigma_A^2 = \sigma_B^2$$

In testing these, the test statistic is based on computed levels of the sample variances, s_A^2 and s_B^2. A ratio of these quantities provides the test statistic, which is computed differently for each H_0 form. The following summarizes these test procedures.

TEST PROCEDURES FOR
COMPARING TWO VARIANCES

	$H_0: \sigma_A^2 \leq \sigma_B^2$	$H_0: \sigma_A^2 \geq \sigma_B^2$	$H_0: \sigma_A^2 = \sigma_B^2$
	$F = \dfrac{s_A^2}{s_B^2}$	$F = \dfrac{s_B^2}{s_A^2}$	$F = \dfrac{\text{Max}(s_A^2, s_B^2)}{\text{Min}(s_A^2, s_B^2)} = \dfrac{s_1^2}{s_2^2}$
Numerator df:	$n_A - 1$	$n_B - 1$	$n_1 - 1$
Denominator df:	$n_B - 1$	$n_A - 1$	$n_2 - 1$
Reject H_0 when:	$F > F_\alpha$	$F > F_\alpha$	$F > F_{\alpha/2}$

When the underlying populations are normally distributed with common variance, then the test statistic has the F distribution with degrees of freedom as indicated. That distribution was introduced in Chapter 7. The critical values for the F distribution may be read from Appendix Table J.

The following example illustrates a test with null hypothesis of the form $\sigma_A^2 \leq \sigma_B^2$.

Example:* *Comparing Two Arrival-Time Distributions

An industrial engineer tests the null hypothesis that the variance in times between arrivals at an assembly station are at least as great under layout B as they are under layout A. He collects the following data:

$$s_A^2 = 29.3 \qquad s_B^2 = 10.4 \text{ seconds}^2$$
$$n_A = 10 \qquad n_B = 15$$

What must he conclude?

He computes

$$F = \frac{29.3}{10.4} = 2.82$$

Using degrees of freedom $n_A - 1 = 9$ for the numerator and $n_B - 1 = 14$ for the denominator, we find, from Appendix Table J, the critical values $F_{.05} = 2.65$ (in lightface) and $F_{.01} = 4.03$ (in boldface). At the $\alpha = .05$ significance level he may *reject* H_0 and conclude that layout A provides greater variability in arrival times than layout B does. But at the $\alpha = .01$ significance level he must reach the opposite conclusion.

The next shows a test with null hypothesis of the form $\sigma_A^2 \geq \sigma_B^2$.

Example: *Comparing Two Chemical Processes*

A chemical engineer compares the variability in impurity levels for a low-pressure (A) and a high-pressure (B) separation process. He will use the more expensive low-pressure process if it gives lower variability in impurities. His null hypothesis is that the opposite actually applies. This is tested at the 5% significance level.

The engineer computes:

$$s_A^2 = 98.4 \qquad s_B^2 = 115.8 \text{ ppm}^2$$

$$n_A = 20 \qquad n_B = 15$$

The computed level of the test statistic is

$$F = \frac{115.8}{98.4} = 1.18$$

Using degrees of freedom $n_B - 1 = 14$ for the numerator and $n_A - 1 = 19$ for the denominator, we find, from Appendix Table J, the critical value $F_{.05} = 2.26$. Since the computed value falls below the critical value, he must *accept* H_0. The test results are not significant enough to warrant using the more expensive procedure.

The final example is concerned with testing H_0: $\sigma_A^2 = \sigma_B^2$.

Example: *Comparing Two Assembly-Time Distributions*

An industrial engineer collects sample data for the times taken by two teams to assemble a mechanical device. Each team-A worker repeatedly performs a single, well-defined task. Team-B workers interact, and nobody has a specific job. The engineer cares only whether the completion-time variabilities differ. She wishes to test the null hypothesis of identical variances at the $\alpha = .10$ significance level.

The following sample results were obtained:

$$s_A^2 = 19.8 \qquad s_B^2 = 55.6 \text{ hours}^2$$

$$n_A = 15 \qquad n_B = 10$$

The computed level of the test statistic is

$$F = \frac{\text{Max}\,(19.8, 55.6)}{\text{Min}\,(19.8, 55.6)} = \frac{55.6}{19.8} = 2.81$$

Using degrees of freedom $n_1 - 1 = 10 - 1 = 9$ for the numerator and $n_2 - 1 = 15 - 1 = 14$ for the denominator, Appendix Table J provides the critical value $F_{\alpha/2} = F_{.05} = 2.65$. Since the computed value exceeds this, she must *reject* the null hypothesis.

Problems

12-45 A metallurgist compares alloys A and B. She performs test-stand evaluations and collects observations of strength indexes. The following data apply:

$$s_A^2 = 203.5 \qquad s_B^2 = 77.4$$

$$n_A = 10 \qquad n_B = 10$$

If she wishes to protect herself at the 5% significance level against concluding that alloy A exhibits greater strength variability when the opposite is true, what should she conclude?

12-46 Consider the maintenance policy example on page 503. Can you conclude that the two populations of maintenance costs have the same variance? Justify your finding.

12-47 Refer to Problem 12-7 and your answers to that exercise. At the 10% significance level, can we conclude that the variabilities in drying times differ?

12-48 Refer to Problem 12-8 and your answers to that exercise. At the 5% significance level, must you accept or reject the null hypothesis that the variance in GPAs for cooperative students is at least as great as that for regular students?

Comprehensive Problems

12-49 An engineer is interested in the heat-exchanging properties of liquids carrying the products of chemical reactions. The output of either stage A or stage B will be used to warm the water feeding one of the plant boilers. The following data represent independent random observations of the effluent from the two stages:

Temperature (°F)			
Stage A		**Stage B**	
185	233	197	248
206	250	235	234
217	206	245	224
251	215	211	231
190	224	225	216
		245	258

(a) Construct a 95% confidence interval estimate for the difference in mean temperatures of the effluents from the two stages.
(b) Should the null hypothesis of identical mean temperatures be accepted or rejected at the 5% significance level?
(c) Should the null hypothesis of identical temperature variances be accepted or rejected at the 10% significance level?

12-50 The following assembly-time data have been obtained by an industrial engineer evaluating two methods. Each observation pair represents the number of seconds taken by a single operator to complete the required task using the respective method.

Operator	A Time	B Time	Operator	A Time	B Time
1	45	39	11	39	29
2	88	71	12	36	36
3	40	42	13	52	50
4	32	27	14	43	34
5	29	28	15	73	51
6	34	30	16	48	42
7	59	50	17	61	61
8	55	60	18	48	45
9	62	51	19	44	32
10	50	48	20	51	43

(a) Construct a 99% confidence interval estimate for the difference in mean assembly times under the two methods.

(b) Should the null hypothesis of identical means be accepted or rejected at the 1% significance level?

12-51 In conducting a Monte Carlo simulation of a nuclear power plant over a 30-year lifespan, two design alternatives were evaluated. Each simulation involved 10,000 independent trial reactor lifetimes. Under alternative A, the number of serious cooling interruptions was 37. Under alternative B only 18 such incidents occurred.

The null hypothesis being tested is that design A is at least as safe as design B. At the 1% significance level, should that assumption be accepted or rejected?

12-52 Consider the evaluation in Problem 12-49. The engineer is concerned with the proportion of time the effluent exceeds the boiling point of water. He has posed the null hypothesis that this proportion is identical from the two stages. At the 5% significance level, should that null hypothesis be accepted or rejected?

12-53 Perform the Wilcoxon rank-sum test with the data in Problem 12-49 to evaluate the null hypothesis that the stage-A effluent is at least as hot as that of stage B. At the 5% significance level should H_0 be accepted or rejected?

12-54 Perform the Wilcoxon signed-rank test with the data in Problem 12-50 to evaluate the null hypothesis that method A yields assembly times that are no longer than those of method B. At the 5% significance level should H_0 be accepted or rejected?

12-55 The following data have been collected by a student society comparing the grades achieved by full-time students to those of cooperative students:

Full-Time Students (A)				Cooperative Students (B)			
GPA	SAT	GPA	SAT	GPA	SAT	GPA	SAT
3.3	655	3.2	712	3.1	588	2.7	578
3.9	722	3.5	690	3.8	716	3.7	695
2.9	623	2.8	575	3.1	627	3.2	634
3.3	710	3.8	690	3.5	713	3.9	730

Answer the following, assuming that the samples were selected independently:

(a) Construct a 95% confidence interval estimate for the difference in population mean GPAs for full-time and cooperative students.

(b) Apply the t test with the null hypothesis that full-timers earn grades at least as high as cooperative students. At the 5% significance level, should that hypothesis be accepted or rejected?

(c) Test the same null hypothesis using the Wilcoxon rank-sum test. What conclusion should be reached at the 5% significance level?

(d) Should the null hypothesis of equal GPA variances for full-time and cooperative students be accepted or rejected at the 10% significance level?

12-56 Consider the GPA data in Problem 12-55. Match each full-time student with a partner in the cooperative group, using SAT score as the basis for assignment. Assuming that these were the matched pairs from the outset of the investigation, complete the following:

(a) Construct a 95% confidence interval estimate for the difference in population mean GPAs for full-time and cooperative students.

(b) Apply the t test with the null hypothesis that full-timers earn grades at least as high as cooperative students. At the 5% significance level, should that hypothesis be accepted or rejected?

(c) Test the same null hypothesis using the Wilcoxon signed-rank test. What conclusion should be reached at the 5% significance level?

12-57 A computer scientist determined the processing times of 50 jobs run under operating system A. Another 50 jobs were run under operating system B. The following sample results were obtained for the two independent investigations:

$$\bar{X}_A = 1.53 \text{ min} \qquad \bar{X}_B = 1.34 \text{ min}$$

$$s_A = .53 \qquad s_B = .49$$

(a) Construct a 95% confidence interval estimate for the difference in mean processing times under the respective operating systems.

(b) Using your results for (a), determine whether the null hypothesis of identical means should be accepted or rejected at the 5% significance level?

(c) The processing-time data were ranked and the sum of the ranks for the times under system A was computed to be $W = 2{,}713$. Does that result indicate that the null hypothesis of identical means should be accepted or rejected at the 5% level?

12-58 A statistician friend of the computer scientist in Problem 12-57 used the same data to illustrate the greater power of matched-pairs evaluations. Each of the 50 jobs in group A was matched with a job from group B, in terms of amount of input and amount of output. Subtracting the group B time from that of the group A partner, the standard deviation in matched-pair differences was computed to be $s_d = .08$ minute.

(a) Construct a 95% confidence interval estimate for the difference in mean processing times under the respective operating systems. Is this narrower than the one found in Problem 12-57?

(b) Using your results for (a), should the null hypothesis of identical means be accepted or rejected at the 5% significance level? Is this the same conclusion reached in Problem 12-57?

(c) The processing-time differences were ranked in absolute value, with 45 nonzero differences found. The sum of the ranks for the positive differences was computed to be $V = 739$. Does that result indicate that the null hypothesis of identical means should be accepted or rejected at the 5% level?

CHAPTER THIRTEEN

Analysis of Variance

THIS CHAPTER DESCRIBES procedures for evaluating several quantitative populations in terms of their respective means. To begin, we introduce a test that determines whether or not those parameters take on a common value. In Chapter 12, we encountered such a test in which just two populations are considered. Our present concern is with any number of populations greater than two, and our central focus is placed on a statistical application called **analysis of variance**. Although the terminology can be confusing, this procedure is actually concerned with levels of *means*.

Such a tool can be very useful in engineering evaluations, where there is need for a variety of multiple comparisons. For instance, software engineers must select the best of several candidate routines by evaluating them in terms of their mean job-completion times. The processing times required under a particular routine establish a target population that may be sampled by running test jobs. The collective results from all such samples can be used as the basis for a final choice. Likewise, a chemical engineer can use sample batches run under various temperature settings to establish the optimal level for future production runs to synthesize a compound.

Careful planning is especially important when several samples are used in making inferences about their counterpart populations. Because a large number of expensive or hard-to-get observations must usually be made, the efficiency of the statistical procedure is especially important. The analysis of variance procedure itself encompasses *experimental design* considerations that guide the investigator in setting up the details of a statistical experiment. This chapter provides some insight into the design of such experiments.

13-1 Framework for a Single-Factor Analysis

The procedure for a single-factor analysis compares several populations. Each of these groups represents a different **treatment**. This term originated with pioneering medical and biological applications. The methods developed for such investigations now have universal application in most experimental fields, and, regardless of application, any factor that the investigator wishes to evaluate is referred to as a treatment. Thus, a software engineer might refer to each candidate algorithm as a separate "treatment" to be administered to the data being processed. Similarly, a chemical engineer may view any particular retort temperature setting as a different treatment of the input compound. In all cases, sample data are generated by applying the respective treatments to randomly chosen units. Analysis of the data so obtained can indicate if the treatments differ.

To help our presentation of the analysis-of-variance procedure we will use a detailed illustration that extends the bacterial leaching investigation introduced in Chapter 9.

Testing for Equality of Means

Illustration: *Comparing Bacteria Strains for Leaching Copper Ore*

The data in Table 13-1 have been obtained for three different strains of *thiobacillus*, each representing a different bacterial leaching treatment. A mining engineer wishes to evaluate the strains in terms of their respective performance in the recovery of copper from low-grade ores.

Each observation represents the copper yield (lb/ton) obtained from a several-ton batch of ore removed from a random location in the deposit area. The yields obtained may be treated as random samples from the respective target populations of all batches that might be processed by the indicated bacteria strain.

Table 13-1 *Experimental Layout with Sample Copper Yields (lb/ton) from Leaching with Three Strains of Bacteria*

Observation i	*Thiobacillus* Treatment Strain 1	Strain 2	Strain 3
1	31	33	25
2	35	41	27
3	33	37	24
4	29	32	26
5	35	34	37
6	41	43	39
7	53	45	33
8	—	43	36
9	—	41	22
10	—	27	—
Totals	257	376	269
Means	$\bar{X}_{\cdot 1} = 36.71$	$\bar{X}_{\cdot 2} = 37.60$	$\bar{X}_{\cdot 3} = 29.89$

$$\bar{\bar{X}}_{\cdot\cdot} = \frac{257 + 376 + 269}{7 + 10 + 9} = 34.69$$

Two variables are involved in this statistical evaluation. Type of bacteria is referred to as the **treatment variable**. Although usually qualitative, treatment variables might also represent discrete quantities or numerical categories. The treatment plays a role analogous to the independent variable in regression analysis. Copper yield is the **response variable**, which must be quantitative. It has an interpretation similar to the dependent variable in a regression analysis.

The mining engineer must determine if the alternative strains of bacteria differ in their efficacy at leaching recoverable copper. He will begin by testing the null hypothesis that the population mean yield per ton of ore is the same for each treatment. Denoting the jth treatment population mean by μ_j, we have the following.

NULL HYPOTHESIS

$$H_0: \mu_1 = \mu_2 = \mu_3$$

This implies a complementary alternative: In at least one pair, the μ's differ.

The procedures described in this book actually test a more stringent hypothesis—the assumption that all treatment populations have a common frequency distribution. This implies that they have a common variance as well.

Figure 13-1 should help illustrate the underlying concepts. When the sample data are pooled, they look like observations from a single population with high variability, as in (a). When viewed separately, however, the same copper-yield data look like three separate populations with smaller variances, as shown in (b). Diagram (c) illustrates the single copper-yield frequency curve assumed by the null hypothesis.

Although we have become accustomed to using means to gauge differences between groups of data, we can also use the respective variabilities for this purpose. When several groups, each of which clusters at different levels, are *pooled*, they will generally exhibit greater dispersion than when viewed singly. This happens to be the case with the sample data from the bacterial leaching investigation. Notice that the frequency curve in Figure 13-1 (a) is flatter and has longer tails than do any of those in (b). The greater spread in the top curve can be attributed mainly to differences in sample treatment means. If the individual distributions have similar means and variances, the pooled sample will be indistinguishable from them.

In organizing evaluations, it is convenient to arrange sample data in the columnar format of Table 13-1. Such an arrangement is referred to as the **experimental layout**. In the usual layout, observed data are arranged in a matrix, with each observed value of the response variable denoted by X_{ij}. The subscript i represents the observation, or row, and the subscript j refers to the treatment, or column. Thus $X_{53} = 37$ lb/ton indicates the copper yield from the 5th observation (row 5) with treatment by strain 3 (column 3).

The Single-Factor Model

The models underlying analysis of variance treat each sample observation in terms of components. Here we describe the single-factor model.

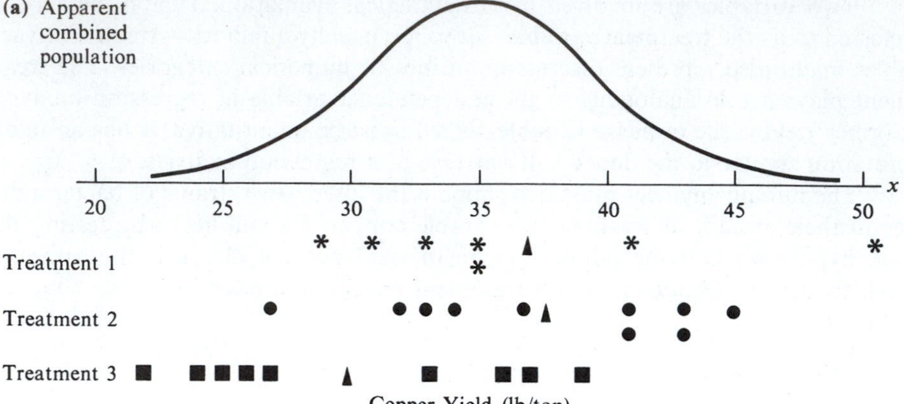

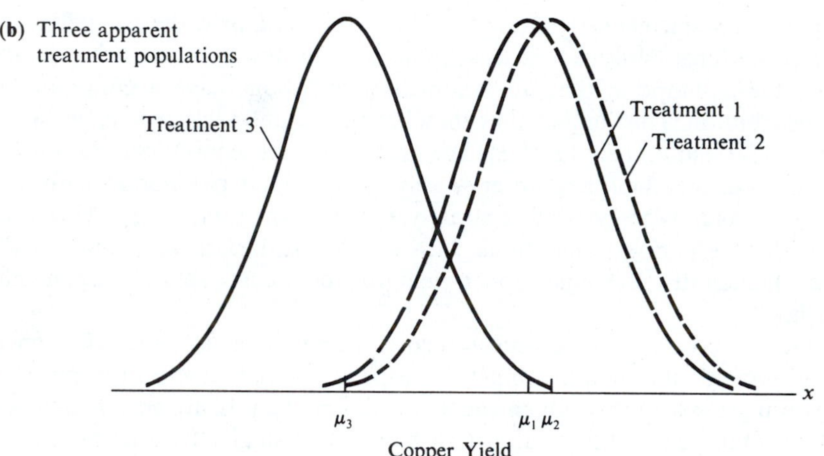

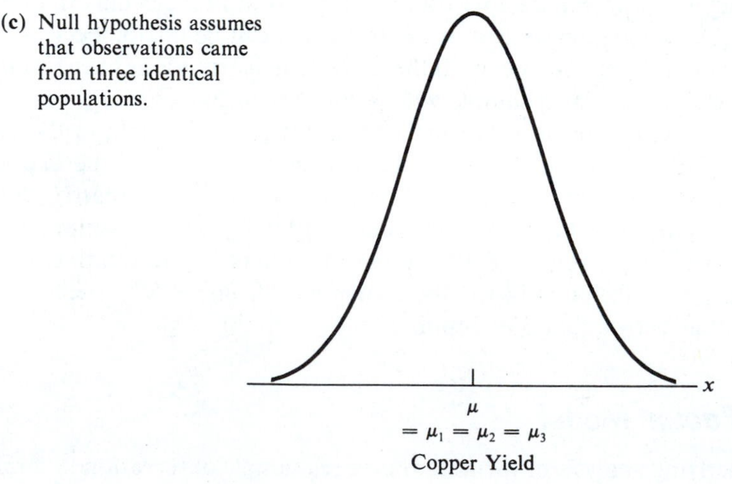

Figure 13-1 *Underlying concepts of analysis of variance.*

SINGLE-FACTOR MODEL

$$X_{ij} = \mu_j + \epsilon_{ij}$$

This expresses each sample observation in terms of two components. X_{ij} denotes the ith observation under treatment j. Altogether there are c treatments (one for each column) and $j = 1, 2, \ldots, c$. There are n_j observations made under the jth treatment, so $i = 1, 2, \ldots, n_j$.

In the bacterial leaching illustration, there are $c = 3$ treatments and $n_1 = 7$, $n_2 = 10$, and $n_3 = 9$. It is convenient to work with the combined sample.

COMBINED SAMPLE SIZE

$$n_T = \sum_{j=1}^{c} n_j$$

For the present illustration,

$$n_T = n_1 + n_2 + n_3 = 7 + 10 + 9 = 26$$

The **treatment population mean** μ_j is a constant of unknown value. The **error term** for each observation is denoted as ϵ_{ij}. The error terms are assumed to be independent, normally distributed random variables with mean 0 and standard deviation σ. Under this model, each sample observation is expected to be equal to the respective treatment population mean and to have a common variance σ^2, regardless of treatment j.

In a general sense, each treatment may be viewed as the level of a factor. The jth treatment population mean μ_j may also be referred to as the **mean for factor-level *j***. Each population mean may then be expressed as the sum of two components:

$$\mu_j = \mu. + B_j$$

The first term is the **overall population mean** $\mu.$. The dot in the subscript indicates that the overall mean is an average. The model assumes that the overall population mean is the average of the individual treatment population means:

$$\mu. = \frac{\sum_{j=1}^{c} \mu_j}{c}$$

The term B_j is defined as follows.

EFFECT FOR FACTOR-LEVEL j

$$B_j = \mu_j - \mu.$$

The null hypothesis of equal factor-level (treatment population) means may be stated equivalently as

$$H_0: B_1 = B_2 = \cdots = B_c$$

Sample Means

Treating each column as a separate sample, we calculate the following.

SAMPLE MEAN FOR FACTOR-LEVEL (TREATMENT) *j*

$$\bar{X}_{.j} = \frac{\sum_{i=1}^{n_j} X_{ij}}{n_j}$$

This indicates that the observations are averaged within each column. The subscript dot in the first position notation indicates that the averaging is done over the rows (i's).

In the bacterial leaching investigation, the sample mean for the first treatment is calculated using the yields in column 1 of Table 13-1:

$$\bar{X}_{.1} = \frac{31 + 35 + 33 + 29 + 35 + 41 + 53}{7} = 36.71 \text{ lb/ton}$$

The other treatment sample means are $\bar{X}_{.2} = 37.60$ and $\bar{X}_{.3} = 29.89$.

Since the expected value of each treatment sample mean is

$$E(\bar{X}_{.j}) = \mu_j$$

the above quantities are unbiased estimators of the respective treatment population means. The $X_{.j}$'s are also least-squares and maximum-likelihood estimators of the respective μ_j's.

The mean of the pooled sample data is also needed in the evaluation. The following expression is used.

GRAND MEAN FOR POOLED SAMPLE RESULTS

$$\bar{\bar{X}}_{..} = \frac{\sum_{j=1}^{c} \sum_{i=1}^{n_j} X_{ij}}{n_T}$$

We use two overbars to denote the grand mean ("called X-double-bar"). The two dots in the subscript indicate that the averaging is done over both rows and columns. The grand mean for the bacterial leaching illustration is $\bar{\bar{X}}_{..} =$ 34.69 lb/ton.

The grand mean $\bar{\bar{X}}_{..}$ is also an unbiased, least-squares, and maximum-likelihood estimator of the overall population mean $\mu_{..}$.

Problems

13-1 Show that $\bar{X}_{.j}$ is (a) a maximum-likelihood estimator of μ_j and (b) a least-squares estimator of μ_j.

13-2 Express $\bar{\bar{X}}_{..}$ equivalently as a weighted average of the $\bar{X}_{.j}$'s.

13-3 Show that the B_j's must sum to zero.

13-4 Assuming for the bacterial leaching illustration that $\mu_1 = 36$, $\mu_2 = 38$, and $\mu_3 = 29$, complete the following:

(a) Compute $\mu_{..}$.

(b) Compute for each treatment the value for B_j.

(c) Construct a table giving for each observation the associated error term.

13-2 Single-Factor Analysis of Variance

The analysis of variance procedure can help translate the sample data into action. First, the null hypothesis of identical treatment population means is accepted or rejected. Then, if significant differences are found in the μ_j's, a variety of comparisons may be made. Additional conceptual framework must be erected before we are ready to perform the basic test.

Deviations About Sample Means

There are three types of deviations, as shown in Figure 13-2 for the bacterial leaching illustration. The total deviations for each individual observation are shown in (a) as bars connecting the data point to the grand mean. To illustrate, if we subtract from the grand mean the first observation under treatment by bacteria strain 1, we obtain the

$$\text{Total deviation: } X_{11} - \bar{\bar{X}}_{..} = 31 - 34.69 = -3.69$$

which is negative since X_{11} lies below $\bar{\bar{X}}_{..}$.

Figure 13-2(b) plots the **treatment deviations** obtained for each factor level. These are found by subtracting the grand mean from the corresponding treatment (column) sample mean. For bacteria strain 1, the following difference provides the

$$\text{Treatment deviation: } \bar{X}_{.1} - \bar{\bar{X}}_{..} = 36.71 - 34.69 = 2.02$$

Figure 13-2(c) shows the **error deviations**. These are obtained by subtracting from each individual observation the respective computed value for the factor-level (treatment) sample mean. Again, using the first observation under treatment with strain 1, we obtain the

$$\text{Error deviation: } X_{11} - \bar{X}_{.1} = 31 - 36.71 = -5.71$$

Each total deviation may be *partitioned* as follows:

$$\begin{aligned} \text{Total deviation} &= \text{Treatment deviation} + \text{Error deviation} \\ (X_{11} - \bar{\bar{X}}_{..}) &= \quad (\bar{X}_{.1} - \bar{\bar{X}}_{..}) \quad + \quad (X_{11} - \bar{X}_{.1}) \end{aligned}$$

This expression justifies to some extent the following procedure for summarizing variation in sample data.

Figure 13-2
Deviation plots for bacterial leaching illustration.

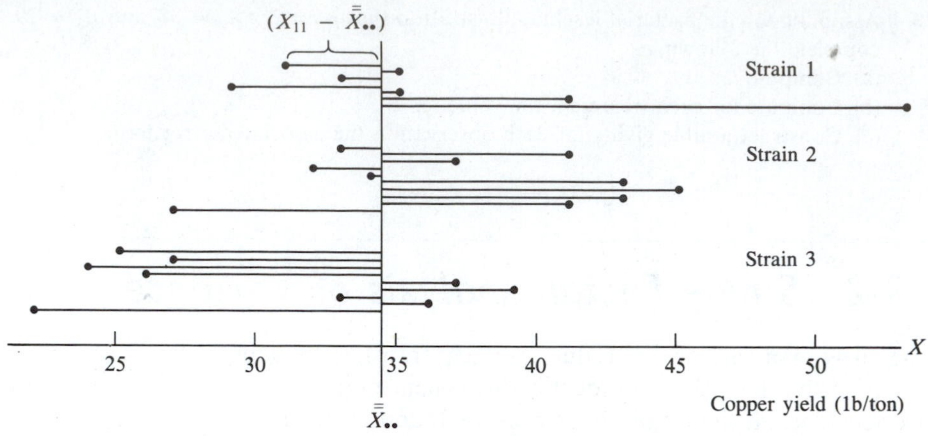

(a) Total deviations about grand mean

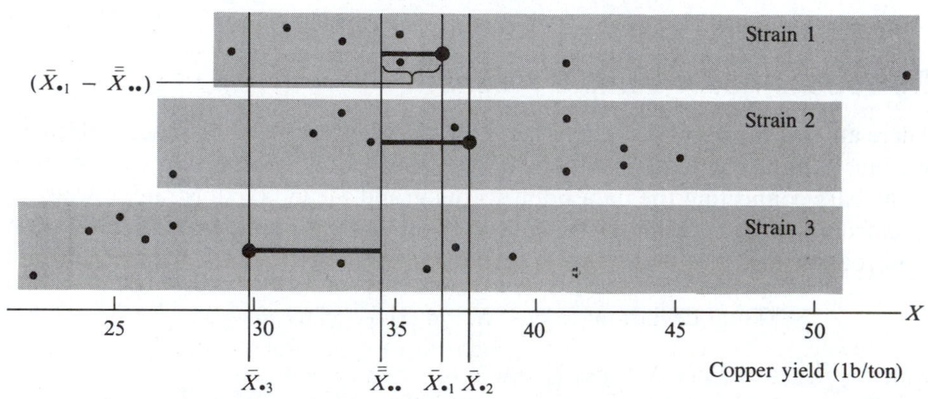

(b) Treatment deviations of factor-level means about grand mean

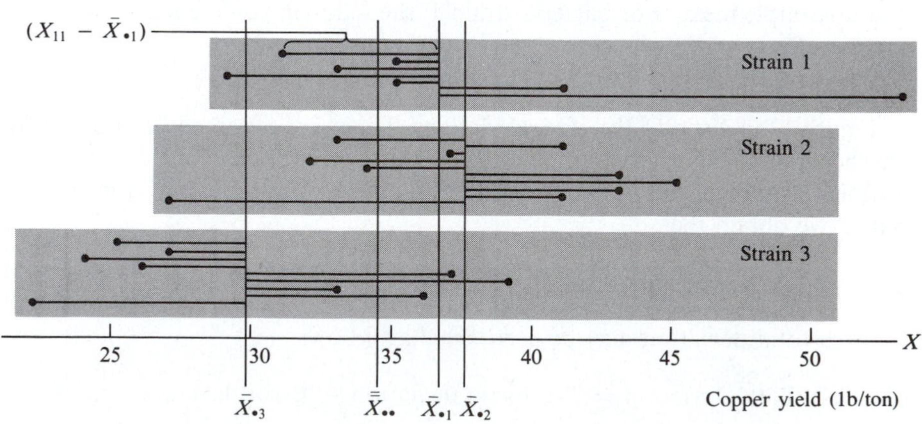

(c) Error deviations about factor-level (treatment) means

Summarizing Variation in Sample Data

The analysis of variance procedure considers the collective sums of individual *squared* deviations. These express the amount of *variation* about the respective means that is exhibited by the sample values. The next expression summarizes the relationships obtained by this:

$$\text{Total variation} = \text{Treatment variation} + \text{Error variation}$$

$$\sum_{j=1}^{c}\sum_{i=1}^{n_j}(X_{ij} - \bar{\bar{X}}_{..})^2 = \sum_{j=1}^{c} n_j(\bar{X}_{.j} - \bar{\bar{X}}_{..})^2 + \sum_{j=1}^{c}\sum_{i=1}^{n_j}(X_{ij} - \bar{X}_{.j})^2$$

Using the *sum of squares* notation encountered in regression analysis, we can write this relationship as

$$\text{Total variation} = \text{Treatment variation} + \text{Error variation}$$

$$SSTO = SSTR + SSE$$

Each sum of squares may be computed from the expressions described below.

The variability *between* sample groups or treatment columns is summarized by the following equation.

TREATMENTS SUM OF SQUARES

$$\boldsymbol{SSTR = \sum_{j=1}^{c} n_j(\bar{X}_{.j} - \bar{\bar{X}}_{..})^2}$$

There is one term in this summation for each of the c treatments (columns). In order that all observations be represented once, each squared deviation is multiplied by the number of observations n_j for that column. Since any differences between groups might be explained by different responses to treatment, rather than to chance alone, $SSTR$ is sometimes referred to as **explained variation**.

Using the results from the bacterial leaching illustration, we have

$$\begin{aligned} SSTR &= 7(36.71 - 34.69)^2 + 10(37.60 - 34.69)^2 + 9(29.89 - 34.69)^2 \\ &= 320.60 \end{aligned}$$

The variability *within* sample groups or treatment columns is summarized by the squared deviations of individual observations from their respective sample means. It is determined by the following expression.

ERROR SUM OF SQUARES

$$\boldsymbol{SSE = \sum_{j=1}^{c}\sum_{i=1}^{n_j}(X_{ij} - \bar{X}_{.j})^2 = SSTO - SSTR}$$

Since this quantity summarizes the differences between individual sample values that are due to chance and have no identifiable cause, SSE is sometimes referred to as **unexplained variation**.

Table 13-2 illustrates the computation of the error sum of squares for the bacterial leaching investigation. The value $SSE = 1{,}034.67$ is obtained. As with $SSTR$, all of the n_T sample observations are reflected in the final sum of squares.

Table 13-2 Calculation of the Error Sum of Squares for the Bacterial Leaching Investigation

i	$(X_{i1} - \bar{X}_{.1})^2$	$(X_{i2} - \bar{X}_{.2})^2$	$(X_{i3} - \bar{X}_{.3})^2$
1	$(31 - 36.71)^2 = 32.60$	$(33 - 37.60)^2 = 21.16$	$(25 - 29.89)^2 = 23.91$
2	$(35 - 36.71)^2 = 2.92$	$(41 - 37.60)^2 = 11.56$	$(27 - 29.89)^2 = 8.35$
3	$(33 - 36.71)^2 = 13.76$	$(37 - 37.60)^2 = 0.36$	$(24 - 29.89)^2 = 34.69$
4	$(29 - 36.71)^2 = 59.44$	$(32 - 37.60)^2 = 31.36$	$(26 - 29.89)^2 = 15.13$
5	$(35 - 36.71)^2 = 2.92$	$(34 - 37.60)^2 = 12.96$	$(37 - 29.89)^2 = 50.55$
6	$(41 - 36.71)^2 = 18.40$	$(43 - 37.60)^2 = 29.16$	$(39 - 29.89)^2 = 82.99$
7	$(53 - 36.71)^2 = 265.36$	$(45 - 37.60)^2 = 54.76$	$(33 - 29.89)^2 = 9.67$
8	—	$(43 - 37.60)^2 = 29.16$	$(36 - 29.89)^2 = 37.33$
9	—	$(41 - 37.60)^2 = 11.56$	$(22 - 29.89)^2 = 62.25$
10	—	$(27 - 37.60)^2 = 112.36$	—
	395.40	314.40	324.87

$$SSE = \sum\sum (X_{ij} - \bar{X}_{.j})^2 = 395.40 + 314.40 + 324.87 = 1{,}034.67$$

The final sum of squares expresses the variability resulting when all observations are treated as a combined sample coming from a common population. The following expression is used to calculate this quantity, the total sum of squares.

TOTAL SUM OF SQUARES

$$SSTO = \sum_{j=1}^{c} \sum_{i=1}^{n_j} (X_{ij} - \bar{\bar{X}}_{..})^2 = SSTR + SSE$$

For the bacterial leaching investigation the total sum of squares is computed to be

$$SSTO = (31 - 34.69)^2 + (35 - 34.69)^2 + \cdots + (22 - 34.69)^2 = 1{,}355.27$$

Note that if we add the treatments and error sums of squares, the same value is obtained:

$$SSTR + SSE = 320.60 + 1{,}034.67 = 1{,}355.27$$

A Basis for Comparison: Mean Squares

Total variation has two components: explained variation ($SSTR$) and unexplained variation (SSE). These may be used to determine if the explained variation is great enough to significantly refute the null hypothesis of identical treatment population means. That question will be settled by the test statistic (to be described shortly), which reflects the relative magnitudes of those two variations.

In any sampling experiment, unexplained variation within a sample is a natural result of chance sampling error. Under H_0, however, the amount of variation between sample treatment groups should be small, and the respective sample means should be close in value. If the responses to the several treatments are identical, then explained and unexplained variations should be about the same magnitude.

The sums of squares cannot be directly compared because $SSTR$ and SSE are each computed by totaling a different number of squared deviations. The former involves c terms, and the latter n_T terms. To make them comparable, each sum of squares must first be converted to an average. These sample variances, referred to as **mean squares**, are estimators of the respective population variances. To make these estimators unbiased, proper divisors are chosen, which are defined as follows.

TREATMENTS MEAN SQUARE,
ERROR MEAN SQUARE

$$MSTR = \frac{SSTR}{c - 1} \qquad MSE = \frac{SSE}{n_T - c}$$

Under the assumption of common frequency distributions, both $MSTR$ and MSE are equal to a common value, σ^2.

For the bacterial leaching investigation, the following mean squares apply:

$$MSTR = \frac{320.60}{3 - 1} = 160.30 \qquad \text{and} \qquad MSE = \frac{1{,}034.67}{26 - 3} = 44.99$$

The treatments mean square is almost four times as large as the error mean square. If H_0 is true and the treatment populations have identical means, these quantities ought to be close in value. Such a large discrepancy would be unlikely if the response of copper yield to treatment were identical for the three strains of bacteria.

Before we can determine how unlikely this result is, we must develop additional concepts.

The ANOVA Table and F Statistic

A convenient summary of the sample results is given in Table 13-3. Such an arrangement is referred to as an ANOVA table. It organizes the analysis-of-variance calculations, providing the sum of squares and mean square for each source of variation. Each source has a number of **degrees of freedom**, the divisors used in calculating the mean squares. The test statistic, which will now be described, is the quantity in the column labeled F.

Table 13-3 *ANOVA Table for Bacterial Leaching Investigation*

Variation	Degrees of Freedom	Sum of Squares	Mean Square	F
Explained by Treatments (between columns)	$c - 1 = 2$	$SSTR = 320.60$	$MSTR = \frac{320.60}{2} = 160.30$	$\frac{MSTR}{MSE} = \frac{160.30}{44.99} = 3.56$
Error or Unexplained (within columns)	$n_T - c = 23$	$SSE = 1{,}034.67$	$MSE = \frac{1{,}034.67}{23} = 44.99$	
Total	$n_T - 1 = 25$	$SSTO = 1{,}355.27$		

The treatments mean square $MSTR$ may be referred to as the variance explained by treatments. Likewise, the error mean square MSE can be called the unexplained variance. The *ratio* of these two quantities will be used to test the null hypothesis of equal population treatment means.

TEST STATISTIC FOR ANALYSIS OF VARIANCE

$$F = \frac{\textbf{Variance explained by treatments}}{\textbf{Unexplained variance}} = \frac{MSTR}{MSE}$$

Applying this to the bacterial leaching investigation, we have

$$F = \frac{160.30}{44.99} = 3.56$$

Under the null hypothesis the values of F are expected to be close to 1 since $MSTR$ and MSE are each unbiased estimators of the common population variance σ^2 and thus have a common expected value. This statistic has a sampling distribution referred to as the F distribution. It has two parameters, each of which expresses a number of degrees of freedom.

Degrees of Freedom The divisor $c - 1$ used in computing $MSTR$ is the number of degrees of freedom associated with using that statistic to estimate σ^2. Recall that this particular sum of squares involves calculating the grand mean $\bar{\bar{X}}..$, which itself can be expressed in terms of the individual treatment sample means, the $\bar{X}._j$'s. For a fixed level of $\bar{\bar{X}}..$, all but one of the c $\bar{X}._j$'s is free to vary.

Analogously, the divisor $n_T - c$ used in calculating MSE is the number of degrees of freedom associated with using MSE to estimate σ^2. This is easy to see since each $\bar{X}._j$ represents the mean of a column involving n_j quantities. Thus, with $\bar{X}._j$ at a fixed level, only $n_j - 1$ of the X_{ij}'s are free to assume any value. Since there are c columns, the number of free variables is

$$\sum_{j=1}^{c} n_j - 1 = n_T - c$$

A pair of degrees of freedom is associated with the F statistic. These sum to $n_T - 1$, the combined sample size minus 1. In the bacterial leaching investigation, we have

Degrees of Freedom	
Numerator, $c - 1$	$3 - 1 = 2$
Denominator, $n_T - c$	$26 - 3 = 23$
Total, $n_T - 1$	$26 - 1 = 25$

The theoretical sampling distribution for this test statistic is the F distribution, introduced in Chapter 7. Appendix Table J lists critical values F_α for either $\alpha = .05$ or $\alpha = .01$.

The Hypothesis-Testing Steps

Using analysis of variance, the mining engineer in the bacterial leaching investigation performed the following steps:

Step 1. *Formulate the hypothesis.* The null hypothesis is that the treatments by the three strains of bacteria yield identical mean copper-yield responses

$$H_0: \mu_1 = \mu_2 = \mu_3$$

Step 2. *Select the test procedure and test statistic.* The F statistic is used.

Step 3. *Establish the significance level and the acceptance and rejection regions for the decision rule.* The mining engineer selected a type I error probability of $\alpha = .05$ for incorrectly concluding that the bacterial leaching treatment means are not equal. For this investigation the degrees of freedom are 2 for the numerator and 23 for the denominator. Appendix Table J gives $F_{.05} = 3.42$. The acceptance and rejection regions for this test, which is *upper-tailed*, are shown in Figure 13-3.

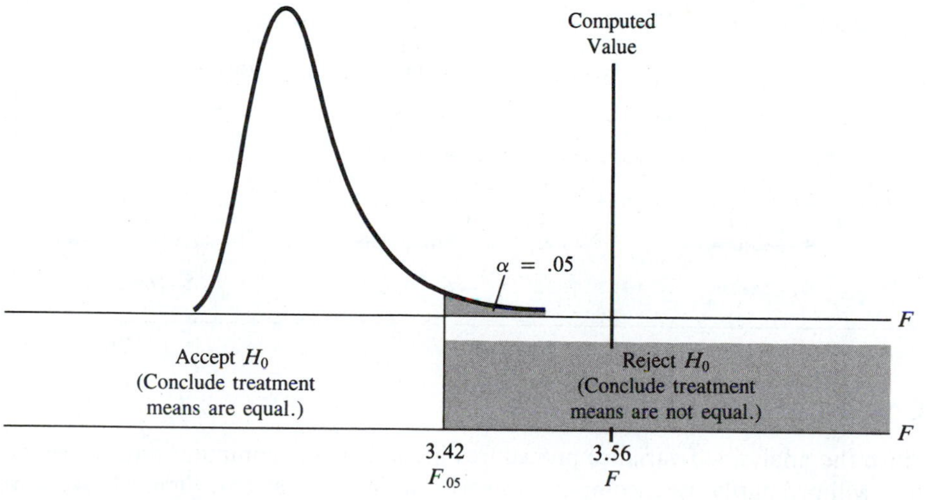

Figure 13-3

Step 4. *Collect the sample data and compute the value of the statistic.* Included in this step are the preliminary calculations and construction of the ANOVA table. Earlier, the computed value of $F = 3.56$ was found.

Step 5. *Make the decision.* The computed test statistic falls in the rejection region. The engineer must therefore reject H_0 and conclude that the treatment means are not equal. (The opposite conclusion would have been reached with $\alpha = .01$, in which case the critical value would have been $F_{.01} = 5.66$.)

Figure 13-4 *SAS printout for analysis of variance with bacterial leaching data.*

```
TITLE 'BACTERIAL LEACHING';
DATA BACT:
  INPUT STRAIN $ YIELD @@;
  CARDS:
S1 31 S1 35 S1 33 S1 29 S1 35 S1 41 S1 53
S2 33 S2 41 S2 37 S2 32 S2 34 S2 43 S2 45 S2 43 S2 41 S2 27
S3 25 S3 27 S3 24 S3 26 S3 37 S3 39 S3 33 S3 36 S3 22
;
PROC GLM;
CLASS STRAIN;
MODEL YIELD = STRAIN;
QUIT;
```

```
                         BACTERIAL LEACHING

                   General Linear Models Procedure

                       Class Level Information

                    Class      Levels     Values

                    STRAIN          3     S1 S2 S3


              Number of observations in data set = 26


Dependent Variable: YIELD
                                   Sum of           Mean
Source                  DF        Squares         Square     F Value     Pr > F

Model                    2   320.82108616   160.41054378        3.57       0.05

Error                   23  1034.82335962    44.99231998

Corrected Total         25  1355.64444578

                 R-Square           C.V.       Root MSE              YIELD Mean

                 0.236656      19.334635       6.707631               34.692307
```

Computer-Assisted Analysis

Since the analysis-of-variance procedures involve messy computations, investigators will ordinarily use a computer to perform the calculations. Figure 13-4 shows the output from a run with SAS for the data in the bacterial leaching illustration. The top portion of the printout lists the needed program commands and the input data. The lower portion shows the ANOVA table, which is nearly identical in format to the one in Table 13-3.

Deciding What to Do: Comparing Treatment Means

By rejecting the null hypothesis of equal population mean copper yield from ore treated under each bacteria strain, the mining engineer knows that at least one strain can be eliminated from further consideration. To identify the inferior strain or strains, additional comparisons must be made. This step will involve a pair-by-pair comparison. The basic method for doing this has already been encountered in Chapter 12. However, these earlier procedures must be amended when population pairs are compared after performing an analysis of variance. Section 13-3 describes those multiple comparisons.

More Discriminating Testing Procedures

The procedure just described uses only one treatment variable to explain differences between populations. It is therefore classified as a **one-way** analysis of variance. However, additional factors may explain those differences even more precisely. Chapter 14 describes **two factor** analysis-of-variance procedures. There, a second factor—grade of ore deposit—is used in addition to bacteria strain to explain copper yields. Analysis-of-variance techniques exist even for **three-factor** and higher experimental layouts.

The following example shows that the single-factor analysis-of-variance procedure might uncover no significant differences in population means. This may be the conclusion even when it seems obvious that the opposite should be concluded.

The sample data in Table 13-4 show that the column values all increase with subsequent observations, and that within each row the values are nearly the same. That is the case because each observation row represents a changing set of circumstances that have a similar effect on the response under all treatments. We will see in Chapter 14 how that fact may be incorporated to help reduce the unexplained variation by utilizing a second factor, or *blocking variable.*

Table 13-4 *Sample Results for Evaluation of Methods for Solving Quadratic Assignment Problems*

	Average Best Cost with Procedure (Treatment)			
Observation i	(1) Simulated Annealing	(2) CRAFT	(3) CRAFT Biased Sampling	(4) Revised Hillier
1	\$25	\$28.2	\$27.6	\$25
2	43	44.2	43.6	43
3	74	79.6	74.8	73.2
4	107	110	108.2	107.4
5	291	296.2	294.8	298.4
6	578.2	606	581	582.2
7	1,308.0	1,339	1,321	1,324.6
8	3,099.8	3,197.8	3,124	3,114.2
	$\bar{X}_{\cdot 1}$ = \$690.75	$\bar{X}_{\cdot 2}$ = \$712.63	$\bar{X}_{\cdot 3}$ = \$696.88	$\bar{X}_{\cdot 4}$ = \$696.00
		$\bar{\bar{X}}_{\cdot\cdot}$ = \$699.06		

Source: Wilhelm, Mickey R. and Thomas L. Ward, "Solving Quadratic Assignment Problems by 'Simulated Annealing,'" *IEE Transactions.* March 1987, pp. 107–118.

Table 13-5 *ANOVA Table for Test of Solution Procedures*

Variation	Degrees of Freedom	Sum of Squares	Mean Square	F
Explained by Treatments: Procedure (between cols.)	$c - 1 = 3$	$SSTR =$ 2,138	$MSTR = \dfrac{SSTR}{3}$ $= 712.5$	$\dfrac{MSTR}{MSE}$ $= .0006$
Error or Unexplained (within cols.)	$n_T - c = 28$	$SSE = 32{,}461{,}878$	$MSE = \dfrac{SSE}{28}$ $= 1{,}159{,}353$	
Total	$n_T - 1 = 31$	$SSTO = 32{,}464{,}016$		

***Example:* Evaluating Methods for Solving Quadratic Assignment Problems**

Two researchers have developed an algorithm that uses Monte Carlo simulation to pick intermediate solutions in solving quadratic assignment problems. Although it is not necessarily faster than traditional heuristic methods, it is believed that the proposed procedure will provide better, less costly solutions.

The researchers tested their model under a complex experimental design that provides for a range of settings for four parameters. Table 13-4 shows the results from one test. Each observation represents a set of computer runs made with problems of the same size, with bigger problems used for successive rows.

Table 13-5 is the ANOVA table for these results. Notice that the unexplained variation dwarfs that explained by treatments. The computed value for F is therefore too small to warrant rejection of the null hypothesis of identical means, and in accordance with the single-factor model, we should *accept* the null hypothesis of identical population means, concluding that average best costs do not differ under the four solution procedures.

The Type II Error

As with all hypothesis-testing procedures, whenever H_0 is accepted there is the chance that this is an incorrect action. Although generally judged to be the less serious error, this potential for type II error cannot be ignored. As with the chi-square tests, there can be many ways for H_0 not to be true, and the F distribution is not very helpful in quantifying type II errors in terms of β probabilities. As with earlier testing procedures, the chance of such errors can be kept within acceptable bounds by using large sample sizes.

Violations of the Underlying Model

The F test is predicated on the assumptions regarding error terms. As with regression analysis, violations of the model may be serious and could invalidate the conclusions. Two violations can be especially serious.

First, the error terms, and hence the X_{ij}'s and $\bar{X}_{.j}$'s, are assumed to be normally distributed. As with procedures based on the Student t distribution, this condition is usually of little consequence when the population frequency curves are not highly skewed. But if they are, or if response is discrete or measured on an ordinal scale, then the F test is not the proper one to use. In such cases a nonparametric procedure, the Kruskal-Wallis test, is a viable alternative. A detailed discussion of this test is given in Section 13-5.

A second violation arises when the variance in error terms is nonconstant. To remedy this, we could transform the response variable, perhaps using $X' = \sqrt{X}$ or $X' = \log(X)$. But as a practical matter, the nonconstant variance is most serious in the more complex models involving random effects (not described in this book). When factor effects are fixed (as in this book), the standard procedures still perform well and are therefore *robust* against this violation.

Problems

13-5 The superintendent at a chemical processing facility wishes to compare temperature settings at various stages of processing. Her response variable is the percentage impurity of the output at each stage, each of which is to be tested independently. In each test the null hypothesis is that the mean percentage of impurities is the same under all four temperature settings. A total of 10 batches has been run under each setting for each stage. For the following results, find the applicable critical value and indicate whether H_0 should be accepted or rejected when a 1% significance level applies:

(a) Stage 13, $F = 4.28$ (b) Stage 4, $F = 4.62$ (c) Stage 11, $F = 5.43$

13-6 A chemical engineer wishes to assess the effect of different pressure settings on the mean yield of a final product from a given level of raw ingredients. The following sample data (in grams/liter) have been obtained:

	Pressure			
Sample Batch	**(1) Low**	**(2) Moderate**	**(3) Strong**	**(4) High**
1	28	30	31	29
2	26	29	29	27
3	29	30	33	30
4	30	30	33	31
5	28	28	29	27
6	31	32	33	32
7	26	29	28	27
8	32	32	32	32
9	25	28	27	27
10	29	30	32	30

(a) Find the critical value of the test statistic and identify the acceptance and rejection regions. Use $\alpha = .01$.

Figure 13-5 *SPSS analysis-of-variance printout.*

```
    VARIABLE   TIME          PURIFICATION TIME IN HOURS
 BY VARIABLE   TYPE          FILTERING SYSTEM

                                      ANALYSIS OF VARIANCE

              SOURCE            D.F.     SUM OF SQUARES         MEAN SQUARES        F RATIO

        BETWEEN GROUPS             3         150.0808               50.0269           4.349

        WITHIN GROUPS            129        1491.0981               11.5589

        TOTAL                    132        1641.1789
```

(b) Construct the ANOVA table.

(c) Should the null hypothesis of identical mean yields under the various pressure settings be accepted or rejected?

13-7 Figure 13-5 shows the results of a computer run using SPSS. The sampling experiment considers the mean purification time (hours) of four filtering systems.

What can you conclude regarding mean filtering times of the respective systems?

13-8 The following data have been obtained by a systems analyst:

	Disk Drive		
Test Run	**(1) Flip-Flop**	**(2) Hard-Core**	**(3) Hy-Discus**
1	32%	17%	−5%
2	−11	23	8
3	14	15	12
4	9	7	10
5	16	13	—
6	8	—	—

The response expresses the percentage reduction in processing time over that taken when the identical job was run using existing disk hardware. The analyst wishes to test the null hypothesis that the three test drives yield identical mean responses.

(a) Find the critical value of the test statistic and identify the acceptance and rejection regions. Use $\alpha = .05$.

(b) Construct the ANOVA table.

(c) Should the null hypothesis of identical mean percentage reductions in processing times with the different disk drives be accepted or rejected?

13-9 An industrial engineering student helped an entomological research team evaluate various location strategies for gypsy moth scent-lure traps. The following data were obtained:

	Trap Location Strategy				
Observation	**(1) Scattered**	**(2) Concentrated**	**(3) Host Plant**	**(4) Aerial**	**(5) Ground**
1	90%	99%	95%	98%	87%
2	92	97	96	98	93
3	94	98	97	99	90
4	93	98	97	99	91
5	—	99	96	—	89

The response variable is the estimated percentage of the native male population trapped.

(a) Find the critical value of the test statistic and identify the acceptance and rejection regions. Use $\alpha = .01$.

(b) Construct the ANOVA table.

(c) Should the null hypothesis of identical mean percentage native males trapped under the various strategies be accepted or rejected?

13-10 A curriculum committee at an engineering school is evaluating basic laboratory policies for freshmen. They requested an experiment where laboratory sections were taught under three different syllabi. The following mean scores on a standard examination were achieved by each sample class in this experiment:

	Laboratory Syllabus		
Class	**(1) Theoretical**	**(2) Empirical**	**(3) Mixed Mode**
1	78	77	83
2	85	86	91
3	64	71	75
4	77	75	78
5	81	80	82
6	75	77	80

What should the committee conclude regarding the mean scores of all laboratory classes taught under the respective syllabi? Use $\alpha = .05$.

13-11 A petroleum engineer for a large oil company is evaluating secondary recovery techniques for oil wells. The following sample data provide the daily increase (barrels) in pumped crude obtained from sample wells:

	Secondary Treatment			
Well	(1) Explosive Fracture	(2) Water Injection	(3) Steam Injection	(4) Controlled Pumping
1	−5	11	15	8
2	11	14	10	10
3	13	22	25	18
4	−5	0	2	1
5	25	35	40	28
6	21	24	28	27
7	11	15	19	17
8	33	30	28	32
9	105	224	328	276

(a) Can the engineer conclude that some secondary treatments might be better than others? Use $\alpha = .01$.

(b) What factors other than secondary treatment type might explain variation in increased daily production?

13-3 Comparative Analysis of Treatments

Continuing with the bacterial leaching investigation of the previous section, we now consider what the investigator does next after rejecting the null hypothesis of identical treatment means. The mining engineer must find the best bacteria strain or at least narrow his search to fewer alternatives. How do we translate the analysis of variance into such tangible action? This involves either estimating individual population means or comparing pairs of population means in terms of their estimated differences.

Since several pairs of means can be compared, we encounter a new complication arising from the *collective* nature of the inferences made. Every confidence interval is a separate statement that affects the credibility of the other estimates, and all must be accounted for through construction of interval estimates somewhat wider than when only single estimates are made.

Inferences need not be restricted to individual population means, nor to pairwise differences between them. Interesting comparisons may also be made between *groupings* of population means. Multivariate analyses sometimes focus on linear combinations of the μ_j's called *constrasts.* These allow confidence intervals to be constructed for differences in average levels of grouped population means.

Single Inferences for a Mean or a Pairwise Difference

To set the stage, we first adapt the procedures of the previous chapters for making *single inferences.*

Confidence Interval for a Treatment Population Mean Recall that for a single sample and population, the unknown μ may be estimated from a $100(1 - \alpha)\%$ confidence interval of the following form

$$\mu = \bar{X} \pm t_{\alpha/2} \frac{s}{\sqrt{n}}$$

Slight modification must be made when several samples and populations are involved. Recall that a common variance σ^2 is assumed for all treatment populations (even when H_0 of identical means has been rejected). Rather than isolating the observations in a single sample group and using the sample standard variance s^2 to estimate σ^2, a more refined estimator is provided by the unexplained variance *MSE*, which reflects information from the pooled samples. The following expression is used.

SINGLE CONFIDENCE INTERVAL ESTIMATE OF μ_j

$$\boldsymbol{\mu_j = \bar{X}_{\cdot j} \pm t_{\alpha/2} \sqrt{\frac{MSE}{n_j}}}$$

The sample mean $\bar{X}_{\cdot j}$ for the jth treatment is used as the estimator of μ_j. The sample size is the number of observations n_j for that group. The number of degrees of freedom for $t_{\alpha/2}$ is $n_T - c$, the same quantity used as the divisor in computing *MSE*.

The above may be applied to the results from the bacterial leaching investigation. The 95% confidence interval estimate of the mean copper yield from *thiobacillus* strain 1 is constructed using $t_{.025} = 2.069$, taken from Appendix Table G for $n_T - c = 26 - 3 = 23$ degrees of freedom. Substituting $\bar{X}_{\cdot 1} = 36.71$ and $MSE = 44.99$, we have

$$\begin{aligned} \mu_1 &= \bar{X}_{\cdot 1} \pm t_{.025} \sqrt{\frac{MSE}{n_1}} \\ &= 36.71 \pm 2.069 \sqrt{\frac{44.99}{7}} \\ &= 36.71 \pm 5.25 \text{ lb/ton} \end{aligned}$$

or

$$31.46 \leq \mu_1 \leq 41.96 \text{ lb/ton}$$

Estimating a Pairwise Difference in Means In Chapter 12 we saw how to compare two population means using a confidence interval for their difference,

$$\mu_A - \mu_B = (\bar{X}_A - \bar{X}_B) \pm t_{\alpha/2} s_D$$

where s_D is the standard error for the difference $D = \bar{X}_A - \bar{X}_B$ and is the estimator of

$$\sigma_D = \sigma\sqrt{\frac{1}{n_A} + \frac{1}{n_B}}$$

where σ is the common population standard deviation. Using MSE as an unbiased estimator of σ^2 and using number subscripts instead of letters, the following equivalent expression is used in establishing the $100(1-\alpha)\%$ single confidence interval.

SINGLE CONFIDENCE INTERVAL FOR THE DIFFERENCE $\mu_j - \mu_k$

$$\mu_j - \mu_k = (\bar{X}_{.j} - \bar{X}_{.k}) \pm t_{\alpha/2}\sqrt{MSE\left(\frac{1}{n_j} + \frac{1}{n_k}\right)}$$

where, again, the number of degrees of freedom is $n_T - c$.

Continuing with the bacterial leaching investigation and using 95% confidence, we have the following estimated difference in mean copper yield between strains 2 and 3:

$$\mu_2 - \mu_3 = (\bar{X}_{.2} - \bar{X}_{.3}) \pm t_{.025}\sqrt{MSE\left(\frac{1}{n_2} + \frac{1}{n_3}\right)}$$

$$= (37.60 - 29.89) \pm 2.069\sqrt{44.99\left(\frac{1}{10} + \frac{1}{9}\right)}$$

$$= 7.71 \pm 6.38$$

or

$$1.33 \leq \mu_2 - \mu_3 \leq 14.09 \qquad \text{(significant)}$$

This interval suggests that the true difference in treatment population mean yield responses between strains 2 and 3 lies somewhere between 1.33 and 14.09 lb/ton. Since this confidence interval lies totally above zero, we may conclude that μ_2 and μ_3 differ significantly at the $\alpha = .05$ level.

Multiple Comparisons and Collective Inferences

The procedures just described have a glaring deficiency: *Each inference only applies individually*. Only a single estimate can be made with $100(1-\alpha)\%$ confidence, and only one test can be performed at the significance level. Those methods cannot be used in drawing a *family* of inferences involving several μ's and differences in means.

This inadequacy can be alleviated by one of two remedies: (1) Construct *wider* confidence intervals yielding less precise estimates or (2) *reduce* the confidence levels. This must be done systematically so that all the inferences are tied together. A key issue is deciding which quantities should be estimated and which pairwise comparisons should be made. The procedures of this chapter require that this issue be settled in advance, *before looking at the sample data.*

Table 13-6 *Results of Metallurgical Evaluation*

Observation i	Alloy (Treatment)			
	(1) Carbonized	(2) Titanium	(3) Composite	(4) Secret
1	13,550	13,440	16,430	17,130
2	14,240	13,750	15,880	17,770
3	11,970	12,880	16,930	16,550
4	12,560	14,110	17,330	16,230
5	13,220	12,330	15,350	18,030
$\bar{\bar{X}}_{..} = 14{,}984$	$\bar{X}_{.1} = 13{,}108$	$\bar{X}_{.2} = 13{,}302$	$\bar{X}_{.3} = 16{,}384$	$\bar{X}_{.4} = 17{,}142$

In a single-factor analysis, there is little difficulty in deciding which inferences will be made. The most important issue is determining which factor levels (treatments) provide the better mean response and which are not significantly different. This can usually be determined by making *every* pairwise comparison of the population means.

The following illustration will be helpful in describing the multiple comparison procedures.

***Illustration:** Metallurgical Evaluation of Strength of Materials*

Consider the results in Table 13-6 obtained for a metallurgical evaluation of the strength of materials. The sampling experiment involves $r = 5$ observations where test rods were stressed until permanent distortion. Rods of the same size were made of four different steel alloys. With each alloy as a separate treatment, $c = 4$. The response variable is the maximum stress level (psi) achieved. The ANOVA table in Table 13-7 shows that $F = 34.70$. Since the computed value exceeds $F_{.01} = 5.29$, at least one pair of the treatments provides significantly different population means.

Table 13-7 *ANOVA Table for Metallurgical Evaluation*

Variation	Degrees of Freedom	Sum of Squares	Mean Square	F
Explained by Treatments: Alloy	$c - 1 = 3$	$SSTR = 64{,}827{,}320$	$MSTR = 21{,}609{,}106.67$	34.70
Error or Unexplained	$n_T - c = (r - 1)c = 16$	$SSE = 9{,}963{,}760$	$MSE = 622{,}735.00$	
Total	$n_T - 1 = rc - 1 = 19$	$SSTO = 74{,}791{,}080$		

Figure 13-6
Plot of results from metallurgical evaluation.

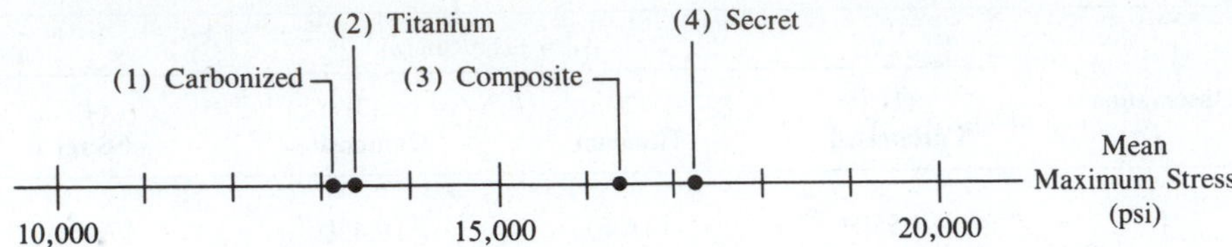

Figure 13-6 shows a plot of the sample data which indicates that alloys (1) and (2) both have substantially lower response than either of alloys (3) and (4). The first pair provides very close sample means, as does the latter. The procedures described next will be used to establish whether or not the differences are significant.

Multiple Comparisons of Treatment Population Means

There are two popular methods for estimating pairwise differences. The Tukey method usually provides the narrowest intervals. It presumes that every pair of μ_j's will be compared and requires equal sample sizes under all treatments. When fewer pairwise differences are of interest, the Bonferroni method can give more precise estimates. This latter accepts unequal sample sizes and can be generalized.

Tukey Method The Tukey method is based on the studentized range distribution, with critical values q listed in Appendix Table L. Like the F distribution, two parameters specify which particular distribution applies. Each treatment must have the same sample size, so that $n_1 = n_2 = \cdots = n_c = r$ (the number of rows in the experimental layout). The following expression is used to compute the Tukey interval estimates.

TUKEY INTERVAL ESTIMATES FOR DIFFERENCES IN FACTOR-LEVEL MEANS AT THE 100(1 − α)% COLLECTIVE CONFIDENCE LEVEL

$$\mu_j - \mu_k = (\bar{X}_{.j} - \bar{X}_{.k}) \pm q_{1-\alpha}\sqrt{\frac{MSE}{r}}$$

Parameters for q: numerator c, denominator $(r - 1)c$

Any pairwise difference having confidence intervals lying totally above or below zero corresponds to factor-level means that are significantly different at collective level α.

Let's use the metallurgical evaluation to illustrate how the Tukey intervals are constructed. Altogether there are $C_2^4 = 6$ pairs among the μ_j's. Using a 90% collective level of confidence, interval estimates will be constructed for each of these. All involve the same cirtical value $q_{.90}$, which will have parameters $c = 4$ for the numerator and $(r - 1)c = 16$ for the denominator. Appendix Table L shows that $q_{.90} = 3.52$.

Using $MSE = 622{,}735$, the following interval is constructed for the difference in mean maximum stress between the (3) composite and (1) carbonized alloys:

$$\mu_3 - \mu_1 = (\bar{X}_{.3} - \bar{X}_{.1}) \pm q_{.90}\sqrt{\frac{MSE}{5}}$$
$$= (16{,}384 - 13{,}108) \pm 3.52\sqrt{\frac{622{,}735}{5}}$$
$$= 3{,}276 \pm 1{,}242.3$$

or

Composite − carbonized: $2{,}033.7 \leq \mu_3 - \mu_1 \leq 4{,}518.3$ (significant)

This shows that the difference in mean maximum stress (psi) between the composite and carbonized alloys is significant at the collective $\alpha = .10$ level.

The remaining pairwise differences will have confidence intervals of the same width, as follows:

Titanium − carbonized:	$-1{,}048.3 \leq \mu_2 - \mu_1 \leq 1{,}436.3$	
Secret − carbonized:	$2{,}791.7 \leq \mu_4 - \mu_1 \leq 5{,}276.3$	(significant)
Composite − titanium:	$1{,}839.7 \leq \mu_3 - \mu_2 \leq 4{,}324.6$	(significant)
Secret − titanium:	$2{,}597.7 \leq \mu_4 - \mu_2 \leq 5{,}082.3$	(significant)
Secret − composite:	$-484.3 \leq \mu_4 - \mu_3 \leq 2{,}000.3$	

These intervals show significant difference between the composite alloy and either the carbonized or titanium alloys and also between the secret alloy and those two. Neither the titanium vs. carbonized nor the secret vs. composite pairings result in significant differences.

Bonferroni Method The Bonferroni method is based on the Student t distribution. Unlike the Tukey method, not every pairwise comparison need be included, but the total number of required estimates must be specified in advance.

Denoting by s the number of estimates to be made, the following expression is used to compute the Bonferroni interval estimates.

BONFERRONI INTERVAL ESTIMATES FOR DIFFERENCES IN FACTOR-LEVEL MEANS AT THE 100(1 − α)% COLLECTIVE CONFIDENCE LEVEL

$$\mu_j - \mu_k = (\bar{X}_{.j} - \bar{X}_{.k}) \pm t_{\alpha/2s}\sqrt{MSE\left(\frac{1}{n_j} + \frac{1}{n_k}\right)}$$

where

$$s = \text{number of intervals constructed}$$
$$n_T - c = \text{number of degrees of freedom}$$

and where $n_T = n_1 + n_2 + \cdots + n_c$. When all treatments have equal sample sizes, $n_T = rc$, and the number of degrees of freedom is $(r - 1)c$.

To illustrate, consider again the metallurgical evaluation results in Tables 13-6 and 13-7. With the same 90% overall level of confidence as before, there are altogether $s = 6$ pairwise differences for which interval estimates may be constructed. The critical value for t will correspond to an upper-tail area of

$$\frac{\alpha}{2s} = \frac{.10}{2(6)} = .0083$$

Here the number of degrees of freedom is $n_T - c = 16$. Appendix Table G has the critical value $t_{.0083} = 2.698$ (using linear interpolation).

The following interval estimate is obtained for the difference in mean maximum stress between the (3) composite and (1) carbonized alloys:

$$\begin{aligned}\mu_3 - \mu_1 &= (\bar{X}_{.3} - \bar{X}_{.1}) \pm t_{.0083}\sqrt{MSE\left(\frac{1}{n_3} + \frac{1}{n_1}\right)} \\ &= (16{,}384 - 13{,}108) \pm 2.698\sqrt{622{,}735\left(\frac{1}{5} + \frac{1}{5}\right)} \\ &= 3{,}276 \pm 1{,}346.6\end{aligned}$$

or

$$\text{Composite} - \text{carbonized:}\quad 1{,}929.4 \le \mu_3 - \mu_1 \le 4{,}622.6 \qquad \text{(significant)}$$

Like the earlier analysis, this interval lies totally above zero, so that the two alloys are again shown to have significantly different mean responses. When we apply the Bonferroni method, all of the pairwise differences in population means will have interval estimates with varying widths depending on the n's, and the same conclusions will ordinarily be reached as before.

Comparing Tukey and Bonferroni Methods The Bonferroni intervals will be wider than those obtained by the Tukey method (as the preceding illustration shows). Thus, when all pairwise comparisons are to be made, the Tukey intervals are the preferred ones. When not all pairs of means are included, the Bonferroni method is preferred since it can provide more precise estimates. (However, the investigator must then identify *ahead of time* which particular paired differences will be estimated.) A further advantage of the Bonferroni method is that treatments may have unequal sample sizes, making it more widely applicable than the Tukey method.

To illustrate, only Bonferroni confidence intervals can be constructed for the bacterial leaching investigation, which has unequal sample sizes. Suppose that a 95% collective confidence is used and that all $s = 3$ pairwise differences are estimated. The critical value will involve the upper-tail area

$$\frac{\alpha}{2s} = \frac{.05}{2(3)} = .0083$$

With $n_T - c = 23$ degrees of freedom, Appendix Table G provides the critical value $t_{.0083} = 2.604$ (using linear interpolation).

Consider the difference $\mu_2 - \mu_3$. The following interval estimate is obtained:

$$\mu_2 - \mu_3 = (\bar{X}_{.2} - \bar{X}_{.3}) \pm t_{.0083}\sqrt{MSE\left(\frac{1}{n_2} + \frac{1}{n_3}\right)}$$

$$= (37.60 - 29.89) \pm 2.604\sqrt{44.99\left(\frac{1}{10} + \frac{1}{9}\right)}$$

$$= 7.71 \pm 8.03$$

or

$$-.32 \leq \mu_2 - \mu_3 \leq 15.74$$

The remaining differences are estimated by the following intervals:

$$-7.72 \leq \mu_2 - \mu_1 \leq 9.50$$

$$-1.98 \leq \mu_1 - \mu_3 \leq 15.62$$

All intervals contain zero, so none of the pairs of treatment means differ significantly at the collective $\alpha = .05$ level.

Notice that the interval for $\mu_2 - \mu_3$ is wider than the *single* interval found on page 566. That is because all three pairwise differences are reflected by the latest set of intervals. Collective confidence intervals will generally be wider than single intervals since they provide a comprehensive set of inferences.

Computer-Assisted Evaluations

As noted in Section 13-2, computer-assisted evaluation can be very helpful in conducting an analysis of variance. Some software packages will construct collective confidence interval estimates for all pairwise differences in treatment population means. Consider the following example.

Example: ***Selecting the Best Fabrication Procedure***

A mechanical engineer is preparing her recommendations for production of a servomechanism assembly. She has done a test on four fabrication methods. The estimated cost of production was obtained from $c = 4$ samples, each of size $r = 8$, of assemblies fabricated under the respective methods. Figure 13-7 shows the ANOVA table from a SAS computer run. The engineer can reject the null hypothesis of identical means at better than $\alpha = .01$ level of significance.

The following sample means were obtained:

$$\bar{X}_{.1} = \$346 \qquad \bar{X}_{.2} = \$275 \qquad \bar{X}_{.3} = \$489 \qquad \bar{X}_{.4} = \$514 \qquad \bar{\bar{X}}_{..} = \$406$$

She chose to use the Tukey method to estimate all pairwise differences at a collective confidence level of 95%. Rather than compute these by hand, she made

Figure 13-7 SAS computer printout giving ANOVA table for fabrication procedure evaluation.

```
                     Analysis of Variance Procedure

Dependent Variable: EST. PROD. COST

                                     Sum of            Mean
Source                   DF         Squares          Square        F Value

Model                     3    117942.01003    39314.003345          46.42

Error                    28     23715.98732      846.999587

Corrected Total          31    140657.99736

                 R-Square            C.V.        Root MSE    EST. PROD. COST Mean

                 0.838508          7.1681        29.10326               406.01146
```

another SAS run, obtaining the printout in Figure 13-8. All the Tukey intervals are computed automatically, and significant differences are identified for pairs. All mean pairs except μ_4 with μ_3 are found to be significantly different at the collective $\alpha = .05$ level.

Figure 13-8 SAS computer printout giving Tukey intervals for fabrication procedure evaluation.

```
                     ANALYSIS OF VARIANCE PROCEDURE

TUKEY'S STUDENTIZED RANGE (HSD) TEST FOR VARIABLE: EST. PROD. COST
NOTE: THIS TEST CONTROLS THE TYPE I EXPERIMENTWISE ERROR RATE

    ALPHA=.05  CONFIDENCE=0.95  DF=28  MSE=846.9996
    CRITICAL VALUE OF STUDENTIZED RANGE=3.87
    MINIMUM SIGNIFICANT DIFFERENCE=71.000

COMPARISONS SIGNIFICANT AT THE 0.05 LEVEL ARE INDICATED BY '***'

                         SIMULTANEOUS                SIMULTANEOUS
                            LOWER      DIFFERENCE       UPPER
   EST. PROD. COST        CONFIDENCE     BETWEEN      CONFIDENCE
     COMPARISON             LIMIT         MEANS         LIMIT

   METH1  -  METH2         31.179        71.000       110.821    ***
   METH1  -  METH3       -182.821      -143.000      -103.179    ***
   METH1  -  METH4       -207.821      -168.000      -128.179    ***
   METH2  -  METH3       -253.821      -214.000      -174.179    ***
   METH2  -  METH4       -278.821      -239.000      -199.179    ***
   METH3  -  METH4        -64.821       -25.000        14.821
```

Estimating Individual Treatment Population Means

Bonferroni intervals may be constructed for individual treatment population means from the following expression.

BONFERRONI INTERVAL ESTIMATES FOR INDIVIDUAL FACTOR-LEVEL MEANS AT THE 100(1 − α)% COLLECTIVE CONFIDENCE LEVEL

$$\mu_j = \bar{X}_{\cdot j} \pm t_{\alpha/2s}\sqrt{\frac{MSE}{n_j}}$$

where

$$s = \text{number of intervals constructed}$$

$$n_T - c = \text{number of degrees of freedom}$$

(Except for the s divisor in establishing the critical value, this expression is the same as the first one in this section.)

To illustrate, consider the bacterial leaching investigation one more time. Suppose that a 95% collective confidence is used and that all $s = 3$ treatment population means are estimated. We make these suppositions because the sample sizes are unequal, which affects both the denominator in the fractional part and the number of degrees of freedom for t. Thus, slightly different interval widths will apply for each μ_j. Each critical value will involve the upper-tail area

$$\frac{\alpha}{2s} = \frac{.05}{2(3)} = .0083$$

Using linear interpolation, Appendix Table G provides, for $n_T - c = 16$, the critical value $t_{.0083} = 2.698$.

The following interval estimate is obtained for μ_j.

$$\begin{aligned}\mu_1 &= \bar{X}_{\cdot 1} \pm t_{.0083}\sqrt{\frac{MSE}{n_1}} \\ &= 36.71 \pm 2.698\sqrt{\frac{44.99}{7}} \\ &= 36.71 \pm 6.84\end{aligned}$$

or

$$29.87 \leq \mu_1 \leq 43.55$$

Notice that this interval is wider than the single interval constructed on page 565. Again, the above interval reflects that *three* estimates are being made, so that all intervals apply jointly. The interval estimates for the other population means are computed in like fashion:

$$31.88 \leq \mu_2 \leq 43.32$$

$$23.86 \leq \mu_3 \leq 35.92$$

These estimates can be combined with the earlier ones for the pairwise differences, which belong to a different family of inferences. However, when the two sets of inferences are combined, the overall level of confidence must be reduced.

Combining Two Families of Collective Estimates

Two separate groupings or families of inferences may be combined. Doing so only affects the overall confidence (significance) level. Consider two families of estimates, each constructed with its own confidence level:

Family	Confidence Level
1	$100(1 - \alpha_1)\%$
2	$100(1 - \alpha_2)\%$

The overall confidence level for the two families combined is *at least*

$$100(1 - \alpha_1 - \alpha_2)\%$$

This property can be illustrated for the bacterial leaching investigation by combining the interval estimates made earlier for the pairwise differences in μ's with the set of estimates just found for the individual population means. Both sets have separate 95% collective confidence levels, so that $\alpha_1 = \alpha_2 = .05$. The combined grouping of six confidence intervals will have collective confidence level of at least

$$100(1 - .05 - .05) = 90\%$$

It is possible to mix interval types in joining two confidence sets derived from a common sample. For instance, a set of 99% ($\alpha_1 = .01$) Tukey intervals for pairwise differences may be combined with a set of 98% ($\alpha_2 = .02$) Bonferroni estimates of individual means to provide a combined confidence set at the $100(1 - .01 - .02) = 97\%$ confidence level (with composite $\alpha = .03$).

Problems

13-12 The following ANOVA table and sample means were obtained by a builders' trade group. This organization wishes to determine if the mean amount of time taken by local governments to approve housing projects differs regionally. The regions are (1) East, (2) South, (3) Midwest, and (4) West. A sample of 10 projects was collected from each.

Variation	Degrees of Freedom	Sum of Squares	Mean Square	*F*
Treatments	3	5.30	1.77	4.12
Error	36	15.48	.43	
Total	39	20.78		

$\bar{X}_{\cdot 1} = 2.3$ yr $\quad \bar{X}_{\cdot 2} = 1.8$ yr $\quad \bar{X}_{\cdot 3} = 1.9$ yr $\quad \bar{X}_{\cdot 4} = 1.6$ yr $\quad \bar{\bar{X}}_{\cdot\cdot} = 1.9$ yr

(a) Construct a 95% confidence interval for the mean of treatment population (1).
(b) Construct a 95% confidence interval for the difference $\mu_1 - \mu_2$.
(c) Does your answer to (b) indicate that the mean approval times differ significantly, at the 5% level, between East and South?

13-13 Referring to Problem 13-8:
(a) Construct a 95% confidence interval for the mean percentage reduction in processing time using flip-flop disk drives.
(b) Construct a 95% confidence interval for the difference in mean percentage reduction using the flip-flop disk drive versus the hard-core disk drive.

13-14 Referring to Problem 13-9, determine whether the scattered and concentrated trap location strategies result in significantly different mean percentages of trapped native male gypsy moths. Use $\alpha = .05$.

13-15 The computer printout in Figure 13-9 was obtained, using SPSS, by an industrial engineer studying queue "discouragement." The percentage of arriving customers who left before receiving service was determined for five sample time periods, each under three different queue disciplines.
(a) With a 99% collective confidence level, use the Tukey method to construct interval estimates for the pairwise differences for all combinations of treatment population means.
(b) For which pairs, if any, do the mean percentages of discouraged customers differ significantly at the 1% level?

13-16 Refer to the data in Problem 13-15 and your answers to that exercise.
(a) With a 99% collective confidence level, use the Bonferroni method to construct interval estimates for all individual treatment population means.
(b) When the interval estimates from (a) are joined with those from Problem 13-15(a), what is the new collective level of confidence? Any means found different in Problem 13-15(b) now differ significantly at what level?

Figure 13-9 *SPSS computer printouts for investigation of queue discouragement.*

GROUP	COUNT	MEAN	STANDARD DEVIATION
1 FIFO	5	25.0000	8.7500
2 SIRO	5	35.0000	8.4167
3 PRIORITY	5	39.0000	7.7235
TOTAL	15	33.0000	5.4772

```
  VARIABLE   LEAVE PER.    PERCENTAGE LEAVING QUEUE
BY VARIABLE  DISCIPLINE    QUEUE DISCIPLINE
```

ANALYSIS OF VARIANCE

SOURCE	D.F.	SUM OF SQUARES	MEAN SQUARES	F RATIO
BETWEEN GROUPS	2	520.0000	260.0000	8.667
WITHIN GROUPS	12	360.0000	30.0000	
TOTAL	14	880.0000		

13-17 Refer to Problem 13-6.

(a) Using a 99% collective confidence level, apply the Tukey method to construct interval estimates for all pairwise differences in treatment population means.

(b) For which pairs, if any, do the mean yields differ significantly at the 1% level?

(c) Are your findings in (b) consistent with the original action regarding the null hypothesis of identical means? Comment on the need here to construct any intervals for pairwise differences.

13-18 Refer to Problem 13-9.

(a) Using a 99% collective confidence level, apply the Bonferroni method to construct interval estimates for all the pairwise differences in treatment population means.

(b) For which pairs, if any, do mean trapping percentages significantly differ at the 1% level?

13-4 Designing the Experiment

The analysis-of-variance procedure may be extended to a variety of experimental evaluations. Each may be classified in several major categories, or **experimental designs**. Selecting an appropriate experimental design is crucial when planning multivariate analyses.

Although this chapter is concerned largely with a single-factor experimental design, more elaborate structures do exist. These incorporate two or more factors, and there is a separate treatment for each combination in levels of these factors.

Example: *Evaluating Sequences for Etching and Layering Microcircuit Chips*

An industrial engineer is concerned with finding the best sequence for etching and layering microcircuit chips. Although even for a simple chip the number of possible combinations would be astronomical, she has identified four basic sequences, E1, E2, E3, and E4, for etching chips and five, L1, L2, L3, L4, and L5, for layering them. She proposes a series of test batch runs for each combination of both factors: E1-L1, E1-L2, . . . , E2-L1, E2-L2, . . . , and so on.

This experiment involves *two* factors: etching sequence and layering sequence. Altogether there are $4 \times 5 = 20$ distinct treatments. As her response variable, the engineer proposes using the percentage of acceptable chips per batch.

Another type of two-factor design involves treatments from only one of the factors. The second factor is a **blocking variable** that serves primarily to reduce the amount of unexplained variation in the response variable.

Example: *Methods for Training Assemblers*

A plant superintendent evaluates three training methods for assemblers: (1) on-the-job only, (2) classroom—then on-line, and (3) on-line—then classroom. A random sample of trainees is selected to undergo each method, or treatment. Because of trainer idiosyncrasies, each trainer indoctrinates one sample group under each treatment. Performance rating scores are obtained as the response variable.

Trainer is the blocking variable. Differences in trainer effectiveness are thereby effectively eliminated as a source of unexplained variation in scores. Trainer is not a treatment per se since it is the *method* of training that is being evaluated.

In Chapter 14 an expanded bacterial leaching experiment will again be evaluated using *site* as the blocking variable. Such an experiment is called a **randomized block design**. The sample ore batches from each site are randomly assigned bacterial treatments, nullifying differences between batches from different ore-bearing sites. Thus, no site systematically influences the measured bacterial efficacy. The assignment of ore batches to bacteria type is done randomly, a procedure called **randomization**.

In all of the illustrations described in this chapter the categories for each variable have been set in advance. This is an ideal arrangement, where the investigator has full control of the study—often referred to as a **fixed-effects experiment**. Often, however, the collection of data is not totally under the investigator's control, and levels for the treatment or blocking variables are really themselves random. Such an investigation is called a **random-effects experiment**. An extreme example arises in evaluating aircraft landing accidents, where crash distance from the airport (the response variable) might serve as the basis for comparing these events in terms of primary cause (the treatment variable). Using aircraft type as the blocking variable, an air safety investigator must take whatever data are available and obviously cannot create the conditions under which an observation is made.

The interaction between two variables might be considered. But the procedures for such evaluations are complicated and are not described here. One reason is that each combination of treatment and blocking levels must then be sampled more than once—a process referred to as **replication**. We encountered replication in the one-way procedure, where each treatment is evaluated several times.

The bacterial leaching experiment described earlier involved replication, with each treatment administered to from seven to ten batches. Two-factor evaluations may be made with or without replication. The primary advantage of having more than one observation per treatment combination is that conclusions can then be made regarding *interactions* between levels of the factors. Some combined effect could have greater influence than either factor individually.

When an observation is made of each combination of factor levels, the experiment has a **complete factorial design**. Since it may be impossible to observe all combinations of categories, some investigations leave out certain groupings. These involve **fractional factorial designs**. Such a design might arise in a chemical processing investigation, similar to that of the preceding section, where the apparatus is not strong enough to withstand high temperatures and high pressures simultaneously; thus, certain cells in the rectangular layout might be impossible to observe.

Economy considerations often limit the number of sample observations, in which case some treatments may not be evaluated under every combination of blocking variables. Such an **incomplete block design** arises when two blocking variables are used. A common incomplete design is the **Latin-square design**, which will be described in Chapter 14.

Problems

13-19 Establish design guidelines for a sampling experiment in each of the following situations. For each, suggest (1) appropriate factor(s) and possible levels that define the treatments, (2) a meaningful response variable, and (3) a possible blocking variable.

(a) A field-support manager is concerned with on-site time spent by customer engineers on calls to home–office management.

(b) A personnel director evaluates the placement of ads seeking engineering job applicants.

(c) A research engineer for a tire manufacturer evaluates tires in terms of conditions of use.

(d) A thermodynamics instructor compares how various strategies of assigning homework influence student achievement.

13-20 Give an example of a two-factor experiment where interactions might be an important consideration.

13-21 Refer to Problem 13-6. Suggest a blocking variable that the chemical engineer might use to reduce the level of unexplained variation.

13-22 Refer to Problem 13-11. Suggest a blocking variable that the petroleum engineer might use to reduce the level of unexplained variation.

13-5 Nonparametric Procedure: The Kruskal-Wallis Test

Sometimes the assumptions of the analysis-of-variance model discussed in the preceding sections may not apply. Two assumptions in particular—normal distributions in response (error terms) and constant variance under the various treatments—are important, and our inability to make these assumptions may present difficulties. Although the F test is to some degree robust against nonnormality, such a situation can cause difficulties if the underlying populations are highly skewed. Also, the fixed-effects procedures are robust against nonconstant variance.

There may be more fundamental difficulties as well. The data themselves might be ordinal, for example, so basic arithmetic operations cannot be consistently applied. As noted in Chapter 12, we might avoid these various difficulties by employing alternative procedures involving **nonparametric statistics**.

The main focus of analysis of variance is in making inferences about several treatment population means, with the F test primarily employed to accept or reject the null hypothesis of equality in these means. The nonparametric counterpart to the F test is the **Kruskal-Wallis test**. This procedure is based on a comparison of treatment in terms of the *ranks* of the observations rather than their nominal values.

This procedure will be described in conjunction with the following illustration.

Illustration: Student Ratings of Fundamental Courses

Samples of students from each foundation course at an engineering college were asked to rate the course, on a scale from 1 (very poor) to 10 (outstanding), in terms of five categories: usefulness, future value, support of advanced coursework, interest, and appropriate level of difficulty. The aggregate ratings in Table 13-8 were obtained for (1) surveying, (2) drawing, (3) statistics, and (4) materials, which here are separate treatments. No student rated more than one course. The ratings have been ranked from lowest to highest. Tied ratings have been assigned the average of the successive ranks for the sequence positions that would apply if they were slightly different.

The following null hypothesis is to be tested at the $\alpha = .05$ significance level.

H_0: There are no aggregate rating differences between courses.

All sample groups were selected from a common population. Thus, under H_0 each of the ratings in Table 13-8 would have the same probability of receiving any rank between 1 and 19.

***Table 13-8** Aggregate Ratings and Ranks for Foundation Course Evaluations*

(1) Surveying		(2) Drawing		(3) Statistics		(4) Materials	
Rating	Rank	Rating	Rank	Rating	Rank	Rating	Rank
38	13.5	44	19	35	11	17	1
27	5	42	17	33	10	21	2
41	16	38	13.5	31	9	29	7.5
36	12	39	15	28	6	26	4
	46.5	43	18	25	3		14.5
			82.5	29	7.5		
					46.5		
$T_1 = 46.5$		$T_2 = 82.5$		$T_3 = 46.5$		$T_4 = 14.5$	
$T_1^2 = 2{,}162.25$		$T_2^2 = 6{,}806.25$		$T_3^2 = 2{,}162.25$		$T_4^2 = 210.25$	
$n_1 = 4$		$n_2 = 5$		$n_3 = 6$		$n_4 = 4$	
$\frac{T_1^2}{n_1} = 540.56$		$\frac{T_2^2}{n_2} = 1{,}361.25$		$\frac{T_3^2}{n_3} = 360.38$		$\frac{T_4^2}{n_4} = 52.56$	

Rating:	17	21	25	26	27	28	29	29	31	33	35	36	38	38	39	41	42	43	44
Rank:	1	2	3	4	5	6	7.5	7.5	9	10	11	12	13.5	13.5	15	16	17	18	19

The test statistic is based on a comparison of variabilities in the ranks within each treatment column. The sum of ranks T_j for each has been computed in Table 13-8. The squares of these sums divided by the respective sample size are then obtained. The following expression is used.

KRUSKAL-WALLIS TEST STATISTIC

$$K = \frac{12}{n_T(n_T + 1)} \sum_{j=1}^{c} \left(\frac{T_j^2}{n_j}\right) - 3(n_T + 1)$$

where n_T is the sum of the sample sizes.

The sampling distribution of K is approximately chi-square with $c - 1$ degrees of freedom. Using $\alpha = .05$ and $c - 1 = 3$ degrees of freedom, Appendix Table H provides the critical value $\chi^2_{.05} = 7.815$.

The computed value for the test statistic is

$$\begin{aligned} K &= \frac{12}{19(19 + 1)}(540.56 + 1{,}361.25 + 360.38 + 52.56) - 3(19 + 1) \\ &= 13.097 \end{aligned}$$

Since $K = 13.097$ is greater than the critical value of 7.815, the null hypothesis of identical ratings populations must be rejected, and it must be concluded that the students rate the four foundation courses differently.

Problems

13-23 The following subjective ratings of activities were obtained from a sample of membership in a Tau Beta Pi chapter.

(1) Field Trips	(2) Speakers	(3) Discussions	(4) Dances	(5) Parties
88	73	78	37	65
82	76	81	44	52
93	86	83	55	54
94	85	77		70
92	90			

What can you conclude regarding member attitudes toward activities?

13-24 Refer to Problem 13-8. Apply the Kruskal-Wallis test. What conclusion should be made at the 5% significance level?

13-25 Refer to Problem 13-10. Apply the Kruskal-Wallis test. What conclusion should be made at the 5% significance level?

Comprehensive Problems

13-26 An engineer for a bridge authority wishes to evaluate several types of paint in terms of the time to onset of potentially damaging levels of oxidation. The following data (in months) have been obtained:

	Type of Paint			
Test Stand	**(1)**	**(2)**	**(3)**	**(4)**
1	5	6	5	7
2	6	8	7	6
3	5	9	7	5
4	5	9	7	5
5	6	8	7	6
6	5	9	8	7
7	7	8	—	6
8	—	9	—	7

(a) Construct the ANOVA table.
(b) Should the engineer conclude that the mean times are different for the paint types? Use $\alpha = .01$.
(c) What conclusion do you reach by using the Kruskal-Wallis test instead?

13-27 Refer to Problem 13-26 and your answer to that exercise.
(a) Using a collective confidence level of 94%, construct Bonferroni interval estimates for all six pairwise differences in mean paint lifetimes.
(b) Which paint, if any, provides a significantly longer lifetime for each pair?
(c) Which paint provides a significantly longer lifetime than all others?

13-28 A petroleum engineer is evaluating drilling bit shapes to be used in penetrating a deep stratum covering an oil field. On randomly selected days, each of the four shapes was used in boring ten portions of a particular shaft. The following data represent the observed drilling rate, or increased shaft depth (in feet) per 8-hour drilling day:

	Drill Bit Shape			
Observations	**(1)**	**(2)**	**(3)**	**(4)**
1	3	5	9	5
2	10	8	5	12
3	8	11	13	15
4	7	8	6	10
5	12	9	8	11
6	14	13	10	12
7	8	8	4	13
8	7	6	5	14
9	3	4	2	11
10	10	11	7	9

(a) Construct the ANOVA table.
(b) At the 1% significance level, can the engineer conclude that the mean drilling rate differs between bit shapes?

13-29 Refer to Problem 13-28 and your answer to that exercise.

(a) Using a collective confidence level of 95%, construct Tukey interval estimates for the difference in mean daily depths between all six pairs of bits.

(b) Which bit, if any, bores significantly deeper for each pair?

(c) Which bit bores significantly deeper than all others?

13-30 A software engineer conducted an experiment where four C language compilers were evaluated. In his experiment he first compiled a randomly selected program with the present compiler; he then recompiled the same program with the experimental routine. For each set of sample runs, the percentage reduction in compilation time was computed. The following results were obtained:

	Percentage Reduction in Compilation Time			
Observation i	**(1) Super C (U.S./cust.)**	**(2) C Squared (U.S./gen.)**	**(3) C Plus (imp./cust.)**	**(4) Triple C (imp./gen.)**
1	38.6	19.7	1.3	8.6
2	25.5	15.3	5.2	13.5
3	49.3	26.2	−6.7	12.6
4	42.2	21.9	2.0	7.3
5	36.0	22.2	0.5	11.5

Two of the compilers were written in the United States. Each compiler is either customizable or general-purpose.

(a) In this evaluation, a different set of five source programs were used with each version of C. Can you suggest how the engineer might have increased the efficiency of his test by using a slightly different experimental procedure?

(b) The researcher wants to estimate (1) all individual population means and (2) all pairwise differences in means. How many estimates will be made altogether?

(c) Construct Bonferroni interval estimates for (1) – (2) above. Give each subgroup the same overall level of confidence, so that the resulting family confidence level will be 94%.

(d) Which compiler provides the greatest percentage reduction in compilation time?

CHAPTER FOURTEEN

Multifactor Analysis of Variance

CHAPTER 13 LAYS the groundwork for and explains the concepts of analysis of variance. The procedures described there are based on experiments with a *single* treatment or factor. This chapter extends those procedures to experiments involving two factors, so that each combination of both factors defines a separate treatment. For example, a chemical process might be evaluated under different levels for both temperature and pressure. Such an experiment could also provide for *interactions* between the levels of the two factors. For example, chemical yield may vary not only as the temperature or the pressure change, but also with various combinations of temperature and pressure.

Single-treatment experiments can sometimes be improved by using a second factor as a *blocking variable* to reduce the amount of unexplained variation, thereby making the testing procedure more discriminating. Such experiments follow a *randomized block design*, described later in this chapter. For example, a computer algorithm may operate more efficiently on large problems than small ones, so that size of problem could serve as a blocking variable in evaluating various solution methods.

Third factors may also be considered. This chapter concludes with a procedure involving two blocking variables and one treatment factor. That method is based on the *Latin-square design*, which allows for a high degree of sampling efficiency.

14-1 The Two-Factor Analysis of Variance

The analysis-of-variance procedure involves two factors, denoted as A and B. Each factor involves several levels, so that the experimental layout takes the form of a matrix as shown in Table 14-1. There are r levels (rows) for factor A and c levels (columns) for factor B. A separate treatment (cell) applies for each combination of factor levels.

Table 14-1
Experimental Layout for Analysis of Variance with Two Factors

Levels for Factor A	Treatment Population Means Factor B Levels (1)	(2)	(3)	(4)	(5)	Factor A–Level Population Means
(1)	μ_{11}	μ_{12}	μ_{13}	μ_{14}	μ_{15}	$\mu_{1\cdot}$
(2)	μ_{21}	μ_{22}	μ_{23}	μ_{24}	μ_{25}	$\mu_{2\cdot}$
(3)	μ_{31}	μ_{32}	μ_{33}	μ_{34}	μ_{35}	$\mu_{3\cdot}$
(4)	μ_{41}	μ_{42}	μ_{43}	μ_{44}	μ_{45}	$\mu_{4\cdot}$
Factor B–Level Population Means	$\mu_{\cdot 1}$	$\mu_{\cdot 2}$	$\mu_{\cdot 3}$	$\mu_{\cdot 4}$	$\mu_{\cdot 5}$	$\mu_{\cdot\cdot}$

The Populations and Means

Under this layout, there are $r \times c$ treatment populations, each having the following mean.

TREATMENT (CELL) POPULATION MEAN

$$\mu_{ij} = \textbf{Population mean with factor } A \textbf{ at level } i \textbf{ and factor } B \textbf{ at level } j$$

There is a separate population for each factor level as well. The respective means are represented by the margins of Table 14-1, where the listed value is as follows.

FACTOR-LEVEL (ROW) POPULATION MEAN

$$\mu_{i\cdot} = \textbf{Factor } A \textbf{ population mean at level } i$$

FACTOR-LEVEL (COLUMN) POPULATION MEAN

$$\mu_{\cdot j} = \textbf{Factor } B \textbf{ population mean at level } j$$

Included in the two-factor layout is the **overall population mean** $\mu_{\cdot\cdot}$, which applies regardless of the particular factor level. The factor level (marginal) means $\mu_{i\cdot}$ and $\mu_{\cdot j}$ are equal to the mean of the treatment (cell) means μ_{ij} in their respective rows and columns:

$$\mu_{i\cdot} = \frac{\sum_{j=1}^{c} \mu_{ij}}{c} \qquad \text{and} \qquad \mu_{\cdot j} = \frac{\sum_{i=1}^{r} \mu_{ij}}{r}$$

The overall mean $\mu_{\cdot\cdot}$ is equal to the mean of the factor-level (marginal) means for both the rows and the columns:

$$\mu_{\cdot\cdot} = \frac{\sum_{i=1}^{r} \mu_{i\cdot}}{r} \qquad \text{and} \qquad \mu_{\cdot\cdot} = \frac{\sum_{j=1}^{c} \mu_{\cdot j}}{c}$$

The overall population mean is also equal to the mean of all treatment (cell) means:

$$\mu_{..} = \frac{\sum_{i=1}^{r} \sum_{j=1}^{c} \mu_{ij}}{rc}$$

The Underlying Model

The additive model for two-factor analysis of variance expresses the cell means as the sum of four components:

$$\mu_{ij} = \mu_{..} + A_i + B_j + (AB)_{ij}$$

The first component is the overall mean. The following components are the effects parameters. There are three types of parameters.

TWO-FACTOR EFFECT PARAMETERS

$$\mathbf{A_i = \text{main effect for factor } A \text{ at level } i}$$

$$\mathbf{B_j = \text{main effect for factor } B \text{ at level } j}$$

$$\mathbf{(AB)_{ij} = \text{interaction effect when factor } A \text{ is at level } i \text{ and factor } B \text{ is at level } j}$$

The effects parameters A_i, B_j, and $(AB)_{ij}$ are related to the population means as follows:

$$A_i = \mu_{i.} - \mu_{..}$$

$$B_j = \mu_{.j} - \mu_{..}$$

$$(AB)_{ij} = \mu_{ij} - \mu_{i.} - \mu_{.j} + \mu_{..}$$

These parameters are in the same units as the response variable. The following relationships apply:

$$\sum_i A_i = \sum_j B_j = \sum_i (AB)_{ij} = \sum_j (AB)_{ij} = 0$$

The effects parameters are helpful in explaining how the factor levels influence the response variable and any interactions between factors. The following examples illustrate the various cases.

Figure 14-1 shows the hypothetical means that might exist for an experiment that has paint drying time as the response variable. Factor A is temperature, with three levels, and factor B is humidity, also with three levels. The plot of the means shows that factor B has *nonzero* effects, since the mean drying times differ for each level of humidity. The horizontal lines indicate that factor A has *zero* effects, reflecting that mean drying times for this paint do not differ with temperature.

Figure 14-2 illustrates a similar example involving an industrial engineering evaluation of how assembly rates are affected by line configuration A (two levels) and locomotion B (three levels). The plots show increasing assembly rates for factor A from (1) the series configuration to (2) the parallel arrangement and also for factor B from (1) carts to (2) roller surfaces and then to (3) conveyor belts.

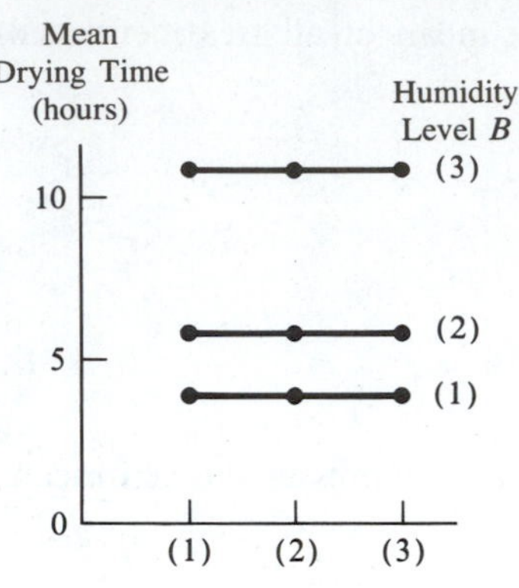

Figure 14-1 *Paint drying experiment.*

Temperature A	Humidity B (1) Dry	(2) Medium	(3) Damp	
(1) Cool	$\mu_{11} = 4$	$\mu_{12} = 6$	$\mu_{13} = 11$	$\mu_{1\bullet} = 7$
(2) Moderate	$\mu_{21} = 4$	$\mu_{22} = 6$	$\mu_{23} = 11$	$\mu_{2\bullet} = 7$
(3) Hot	$\mu_{31} = 4$	$\mu_{32} = 6$	$\mu_{33} = 11$	$\mu_{3\bullet} = 7$
	$\mu_{\bullet 1} = 4$	$\mu_{\bullet 2} = 6$	$\mu_{\bullet 3} = 11$	$\mu_{\bullet\bullet} = 7$

Zero Factor A Effects:

$$A_1 = \mu_{1\bullet} - \mu_{\bullet\bullet} = 7 - 7 = 0$$
$$A_2 = \mu_{2\bullet} - \mu_{\bullet\bullet} = 7 - 7 = 0$$
$$A_3 = \mu_{3\bullet} - \mu_{\bullet\bullet} = 7 - 7 = \underline{0}$$
$$0$$

Nonzero Factor B Effects:

$$B_1 = \mu_{\bullet 1} - \mu_{\bullet\bullet} = 4 - 7 = -3$$
$$B_2 = \mu_{\bullet 2} - \mu_{\bullet\bullet} = 6 - 7 = -1$$
$$B_3 = \mu_{\bullet 3} - \mu_{\bullet\bullet} = 11 - 7 = \underline{4}$$
$$0$$

No Interactions:

$$(AB)_{11} = \mu_{11} - \mu_{1\bullet} - \mu_{\bullet 1} + \mu_{\bullet\bullet} = 4 - 7 - 4 + 7 = 0$$
all $(AB)_{ij}$'s $= 0$

Thus, both factors have nonzero effects. The parallel plot lines indicate that the effects of one factor are equal under any level of the other factor. This indicates that there are *no interactions* among the factors.

The presence of interactions is shown in Figure 14-3, where a hypothetical experiment assesses how the performance of a digital computer, in millions of floating-point operations per second (FLOPS), relates to program complexity A (four levels) and CPU chip architecture B. The latter factor has two levels, (1) regular and (2) reduced instruction set computations (RISC) designs. Both factors tend to provide higher mean responses for successive levels. Moreover, the performance gap between levels of factor B, (1) regular and (2) RISC architectures, narrows as the level for factor A (program complexity) increases. Thus, there are underlying *interactions* between the factors. The presence of the interactions is reflected by the nonparallel plots.

Of course, for all these examples, the levels for the population means and the effects parameters will ordinarily by unknown. Inferences regarding these unknown quantities will be made using sample data. We next consider testing hypotheses regarding their levels. In Section 14-2 we will see how a variety of confidence intervals may be constructed for differences and contrasts in means.

The Null Hypotheses

Recall from Chapter 13 that the analysis-of-variance procedure is actually used to test for equality of the respective population means. In a two-factor evaluation,

Figure 14-2
Assembly layout experiment.

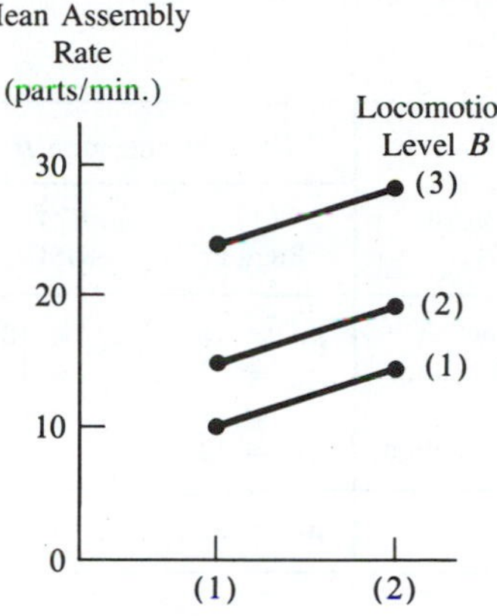

Configuration A	Locomotion B (1) Carts	(2) Rollers	(3) Belts	
(1) Series	$\mu_{11} = 10$	$\mu_{12} = 15$	$\mu_{13} = 26$	$\mu_{1\bullet} = 17$
(2) Parallel	$\mu_{21} = 14$	$\mu_{22} = 19$	$\mu_{23} = 30$	$\mu_{2\bullet} = 21$
	$\mu_{\bullet 1} = 12$	$\mu_{\bullet 2} = 17$	$\mu_{\bullet 3} = 28$	$\mu_{\bullet\bullet} = 19$

Nonzero Factor A Effects:

$$\begin{aligned} A_1 &= 17 - 19 = -2 \\ A_2 &= 21 - 19 = \underline{2} \\ && 0 \end{aligned}$$

Nonzero Factor B Effects:

$$\begin{aligned} B_1 &= 12 - 19 = -7 \\ B_2 &= 17 - 19 = -2 \\ B_3 &= 28 - 19 = \underline{9} \\ && 0 \end{aligned}$$

No Interactions:

$(AB)_{11} = 10 - 17 - 12 + 19 = 0$
all $(AB)_{ij}$'s $= 0$

three null hypotheses may be tested. The first two involve the factor-level population means, each hypothesis being that there is no main effect:

$$H_0\colon \mu_{1\cdot} = \mu_{2\cdot} = \cdots = \mu_{r\cdot} \qquad \text{(factor } A\text{)}$$
$$[\text{all } A_i\text{'s} = 0]$$

$$H_0\colon \mu_{\cdot 1} = \mu_{\cdot 2} = \cdots = \mu_{\cdot c} \qquad \text{(factor } B\text{)}$$
$$[\text{all } B_j\text{'s} = 0]$$

And, the third hypothesis is that there are no interaction effects:

$$H_0\colon \mu_{ij} = \mu_{\cdot\cdot} + A_i + B_j \text{ for all cells } (i, j) \qquad \text{(interactions)}$$
$$[\text{all } (AB)_{ij}\text{'s} = 0]$$

Sample Data in Two-Factor Experiments

The procedure requires that there be exactly n observations per treatment cell and that the same sample size apply to each combination of factor levels. Denoting by X_{ijk} the kth sample observation when factor A is at level i and B is at level j, the additive model provides

$$X_{ijk} = \mu_{ij} + \epsilon_{ijk}$$

or

$$X_{ijk} = \mu_{\cdot\cdot} + A_i + B_j + (AB)_{ij} + \epsilon_{ijk}$$

Figure 14-3 *Computer performance evaluation.*

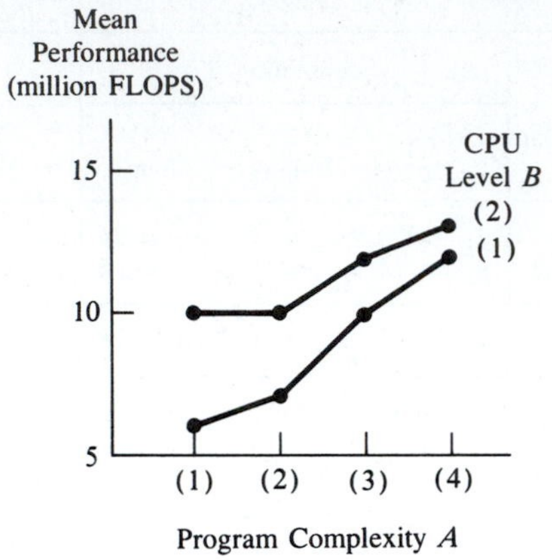

Program Complexity A	CPU Architecture B (1) Regular	(2) RISC	
(1) Low	$\mu_{11} = 6$	$\mu_{12} = 10$	$\mu_{1\bullet} = 8.0$
(2) Moderate	$\mu_{21} = 7$	$\mu_{22} = 10$	$\mu_{2\bullet} = 8.5$
(3) High	$\mu_{31} = 10$	$\mu_{32} = 12$	$\mu_{3\bullet} = 11.0$
(4) Very High	$\mu_{41} = 12$	$\mu_{42} = 13$	$\mu_{4\bullet} = 12.5$
	$\mu_{\bullet 1} = 8.75$	$\mu_{\bullet 2} = 11.25$	$\mu_{\bullet\bullet} = 10.0$

Nonzero Factor A Effects:

$$
\begin{aligned}
A_1 &= 8.0 - 10.0 = -2.0\\
A_2 &= 8.5 - 10.0 = -1.5\\
A_3 &= 11.0 - 10.0 = 1.0\\
A_4 &= 12.5 - 10.0 = \underline{2.5}\\
& 0.0
\end{aligned}
$$

Nonzero Factor B Effects:

$$
\begin{aligned}
B_1 &= 8.75 - 10.0 = -1.25\\
B_2 &= 11.25 - 10.0 = \underline{1.25}\\
& 0.00
\end{aligned}
$$

Interactions:

$$
\begin{aligned}
(AB)_{11} &= 6 - 8.0 - 8.75 + 10 = -.75\\
(AB)_{12} &= 10 - 8.0 - 11.25 + 10 = .75\\
(AB)_{21} &= 7 - 8.5 - 8.75 + 10 = -.25\\
(AB)_{22} &= 10 - 8.5 - 11.25 + 10 = .25\\
(AB)_{31} &= 10 - 11.0 - 8.75 + 10 = .25\\
(AB)_{32} &= 12 - 11.0 - 11.25 + 10 = -.25\\
(AB)_{41} &= 12 - 12.5 - 8.75 + 10 = .75\\
(AB)_{42} &= 13 - 12.5 - 11.25 + 10 = \underline{-.75}\\
& 0.00
\end{aligned}
$$

The last quantity ϵ_{ijk} expresses the residual value or **error term**. As with the model in Chapter 13, the analytical framework presented here assumes that the ϵ's are independent and normally distributed, with mean zero and common standard deviation σ.

The interaction effects cannot be assessed unless $n > 1$. In that case, sampling is done with **replication**.

Illustration: *Effectiveness of Timing and Positioning of Engineering Employment Newspaper Ads*

The employment director for a large aerospace company wishes to assess the effectiveness of newspaper display advertising, with day of week as factor A ($r = 7$ levels—one for each day), and section as factor B ($c = 3$ levels—news, sports, and business). He conducted a sampling experiment wherein $n = 3$ ads were run under each day-section combination. As his response variable the director chose the number of resumes received. The sample data from this experiment are shown in Table 14-2.

Table 14-2 *Sample Data for Experiment on Timing and Positioning of Engineering Employment Newspaper Ads*

Day of Week A	Newspaper Section B			Factor A Sample Mean
	(1) News	(2) Sports	(3) Business	
(1) Sunday	X_{111} (17) X_{112} (18) X_{113} (20)	X_{121} (21) X_{122} (18) X_{123} (15)	X_{131} (6) X_{132} (6) X_{133} (8)	
Cell Mean	$\bar{X}_{11\cdot} = 18.33$	$\bar{X}_{12\cdot} = 18.00$	$\bar{X}_{13\cdot} = 6.67$	$\bar{X}_{1\cdot\cdot} = 14.33$
(2) Monday	X_{211} (10) X_{212} (9) X_{213} (7)	X_{221} (4) X_{222} (2) X_{223} (6)	X_{231} (9) X_{232} (11) X_{233} (13)	
Cell Mean	$\bar{X}_{21\cdot} = 8.67$	$\bar{X}_{22\cdot} = 4.00$	$\bar{X}_{23\cdot} = 11.00$	$\bar{X}_{2\cdot\cdot} = 7.89$
(3) Tuesday	X_{311} (9) X_{312} (9) X_{313} (10)	X_{321} (5) X_{322} (4) X_{323} (6)	X_{331} (7) X_{332} (8) X_{333} (10)	
Cell Mean	$\bar{X}_{31\cdot} = 9.33$	$\bar{X}_{32\cdot} = 5.00$	$\bar{X}_{33\cdot} = 8.33$	$\bar{X}_{3\cdot\cdot} = 7.56$
(4) Wednesday	X_{411} (8) X_{412} (7) X_{413} (7)	X_{421} (5) X_{422} (9) X_{423} (8)	X_{431} (7) X_{432} (8) X_{433} (8)	
Cell Mean	$\bar{X}_{41\cdot} = 7.33$	$\bar{X}_{42\cdot} = 7.33$	$\bar{X}_{43\cdot} = 7.67$	$\bar{X}_{4\cdot\cdot} = 7.44$
(5) Thursday	X_{511} (4) X_{512} (4) X_{513} (3)	X_{521} (7) X_{522} (5) X_{523} (5)	X_{531} (9) X_{532} (7) X_{533} (5)	
Cell Mean	$\bar{X}_{51\cdot} = 3.67$	$\bar{X}_{52\cdot} = 5.67$	$\bar{X}_{53\cdot} = 7.00$	$\bar{X}_{5\cdot\cdot} = 5.44$
(6) Friday	X_{611} (12) X_{612} (12) X_{613} (11)	X_{621} (11) X_{622} (11) X_{623} (12)	X_{631} (10) X_{632} (9) X_{633} (8)	
Cell Mean	$\bar{X}_{61\cdot} = 11.67$	$\bar{X}_{62\cdot} = 11.33$	$\bar{X}_{63\cdot} = 9.00$	$\bar{X}_{6\cdot\cdot} = 10.67$
(7) Saturday	X_{711} (12) X_{712} (17) X_{713} (17)	X_{721} (14) X_{722} (13) X_{723} (14)	X_{731} (13) X_{732} (11) X_{733} (9)	
Cell Mean	$\bar{X}_{71\cdot} = 15.33$	$\bar{X}_{72\cdot} = 13.67$	$\bar{X}_{73\cdot} = 11.00$	$\bar{X}_{7\cdot\cdot} = 13.33$
Factor B Sample Mean	$\bar{X}_{\cdot 1\cdot} = 10.62$	$\bar{X}_{\cdot 2\cdot} = 9.29$	$\bar{X}_{\cdot 3\cdot} = 8.67$	$\bar{\bar{X}}_{\cdot\cdot\cdot} = 9.52$

The sample results in Table 14-2 are arranged in the same basic layout introduced earlier for the population means. The counterpart *sample* means are computed as follows.

CELL SAMPLE MEAN FOR TREATMENT COMBINATION (i, j)

$$\bar{X}_{ij\cdot} = \frac{\sum_{k=1}^{n} X_{ijk}}{n}$$

ROW SAMPLE MEAN FOR FACTOR A AT LEVEL i

$$\bar{X}_{i\cdot\cdot} = \frac{\sum_{j=1}^{c} \bar{X}_{ij\cdot}}{c}$$

COLUMN SAMPLE MEAN FOR FACTOR B AT LEVEL j

$$\bar{X}_{\cdot j\cdot} = \frac{\sum_{i=1}^{r} \bar{X}_{ij\cdot}}{r}$$

GRAND OVERALL SAMPLE MEAN

$$\bar{\bar{X}}_{\dots} = \frac{\sum_{i=1}^{r}\sum_{j=1}^{c} \bar{X}_{ij\cdot}}{rc} = \frac{\sum_{i=1}^{r} \bar{X}_{i\cdot\cdot}}{r} = \frac{\sum_{j=1}^{c} \bar{X}_{\cdot j\cdot}}{c}$$

In accordance with the underlying model, these sample means ($\bar{X}$'s) are unbiased least-squares estimators of the respective unknown treatment population means (μ's).

Analytical Framework

As with the single-factor procedures, the analytical framework may be explained in terms of the following three types of deviations:

$$\begin{array}{ccccc} \text{Total deviation} & = & \text{Treatments deviation} & + & \text{Error deviation} \\ (X_{ijk} - \bar{\bar{X}}_{\dots}) & = & (\bar{X}_{ij\cdot} - \bar{\bar{X}}_{\dots}) & + & (X_{ijk} - \bar{X}_{ij\cdot}) \end{array}$$

This formula *partitions* the total deviation about the grand mean into the treatments and error deviation components. The two-factor analysis further partitions the treatments deviation as follows:

$$\begin{array}{ccccccc} \begin{array}{c}\text{Treatments}\\\text{deviation}\end{array} & = & \begin{array}{c}A\text{: Main effect}\\\text{deviation}\end{array} & + & \begin{array}{c}B\text{: Main effect}\\\text{deviation}\end{array} & + & \begin{array}{c}AB\text{: Interaction effect}\\\text{deviation}\end{array} \\ (\bar{X}_{ij\cdot} - \bar{\bar{X}}_{\dots}) & = & (\bar{X}_{i\cdot\cdot} - \bar{\bar{X}}_{\dots}) & + & (\bar{X}_{\cdot j\cdot} - \bar{\bar{X}}_{\dots}) & + & (\bar{X}_{ij\cdot} - \bar{X}_{i\cdot\cdot} - \bar{X}_{\cdot j\cdot} + \bar{\bar{X}}_{\dots}) \end{array}$$

As with the single-factor experiments, the collective sums of individual squared deviations are used to express the amount of variation by groups. The first expression is generalized in terms of sums of squares as

$$\text{Total variation} = \text{Treatments variation} + \text{Error variation}$$
$$SSTO = SSTR + SSE$$

And, the treatments sum of squares is partitioned as follows:

$$\begin{matrix}\text{Treatments} \\ \text{variation}\end{matrix} = \begin{matrix}A\text{: Main effect} \\ \text{variation}\end{matrix} + \begin{matrix}B\text{: Main effect} \\ \text{variation}\end{matrix} + \begin{matrix}AB\text{: Interaction effect} \\ \text{variation}\end{matrix}$$
$$SSTR = SSA + SSB + SSAB$$

The respective sums of squares form the basis for evaluation. This is done through the ANOVA table.

The Two-Factor ANOVA Table

The general organization of the two-factor ANOVA table is shown in Table 14-3. The expressions for computing the respective sums of squares are also shown. Recall that the mean squares are obtained by dividing these by the respective degrees of freedom. The latter are obtained from the dimensions—reduced by 1—of the experimental layout matrix.

Notice that there are three test statistics, or values, for F—one for testing each of the above null hypotheses.

***Table 14-3** Expressions for Elements in Two-Factor Analysis and General Form of the ANOVA Table*

Variation	Degrees of Freedom	Sum of Squares	Mean Square	F
Explained by Factor A (between rows)	$r-1$	$SSA = nc\sum_{i=1}^{r}(\bar{X}_{i..} - \bar{\bar{X}}_{...})^2$	$MSA = \dfrac{SSA}{r-1}$	$\dfrac{MSA}{MSE}$
Explained by Factor B (between columns)	$c-1$	$SSB = nr\sum_{j=1}^{c}(\bar{X}_{.j.} - \bar{\bar{X}}_{...})^2$	$MSB = \dfrac{SSB}{c-1}$	$\dfrac{MSB}{MSE}$
Explained by Interactions (between cells)	$(r-1)(c-1)$	$SSAB = n\sum_{i=1}^{r}\sum_{j=1}^{c}(\bar{X}_{ij.} - \bar{X}_{i..} - \bar{X}_{.j.} + \bar{\bar{X}}_{...})^2$	$MSAB = \dfrac{SSAB}{(r-1)(c-1)}$	$\dfrac{MSAB}{MSE}$
Error or Unexplained (residual)	$rc(n-1)$	$SSE = \sum_{i=1}^{r}\sum_{j=1}^{c}\sum_{k=1}^{n}(X_{ijk} - \bar{X}_{ij.})^2$	$MSE = \dfrac{SSE}{rc(n-1)}$	
Total	$nrc-1$	$SSTO = \sum_{i=1}^{r}\sum_{j=1}^{c}\sum_{k=1}^{n}(X_{ijk} - \bar{\bar{X}}_{...})^2$		

Calculating Sums of Squares A set of streamlined expressions, equivalent to those in Table 14-3, would ordinarily be used when performing an analysis of variance by making *hand* computations with the assistance of a calculator. Since using these expressions does not significantly ease the computational burden, they are not given here. Rather, it is recommended that a *computer* be used for this task. The sums of squares will then be automatically computed and will appear in the ANOVA table, which is the standard output for all popular statistical software packages.

Continuing with the advertising experiment, the following sums of squares are computed from the sample data:

$$\begin{aligned} SSA &= 3(3)[(14.33 - 9.52)^2 + (7.89 - 9.52)^2 + \cdots + (13.33 - 9.52)^2] \\ &= 598.1587 \end{aligned}$$

$$\begin{aligned} SSB &= 3(7)[(10.62 - 9.52)^2 + (9.29 - 9.52)^2 + (8.67 - 9.52)^2] \\ &= 41.8095 \end{aligned}$$

$$\begin{aligned} SSAB &= 3[(18.33 - 14.33 - 10.62 + 9.52)^2 \\ &\qquad + \cdots + (11.00 - 13.33 - 8.67 + 9.52)^2] \\ &= 388.4127 \end{aligned}$$

$$\begin{aligned} SSE &= (17 - 18.33)^2 + (18 - 18.33)^2 + (20 - 18.33)^2 + (21 - 18.00)^2 \\ &\qquad + \cdots + (9 - 11.00)^2 \\ &= 103.3333 \end{aligned}$$

$$\begin{aligned} SSTO &= (17 - 9.52)^2 + (18 - 9.52)^2 + (20 - 9.52)^2 + (21 - 9.52)^2 \\ &\qquad + \cdots + (9 - 9.52)^2 \\ &= 1{,}131.7143 \end{aligned}$$

The ANOVA table is shown in Table 14-4.

Testing the null hypothesis that the mean numbers of resumes received are the same for all levels of factor A (day of week), the computed value of the test statistic is

$$F = \frac{MSA}{MSE} = 40.520$$

Reading Appendix Table J at 6 degrees of freedom for the numerator and 42 degrees of freedom for the denominator, and using $\alpha = .01$, we find that the critical value is $F_{.01} = 3.26$. Since the computed value exceeds this amount, the null hypothesis of no main factor A effects from day-of-week advertising placement must be *rejected*.

Also at the $\alpha = .01$ significance level, the computed value of $F = 8.497$ leads to *rejection* of the null hypothesis of identical mean resumes received for all levels of factor B, newspaper section. (The critical value using 2 and 42 degrees of freedom is $F_{.01} = 5.15$.) We must conclude that at least one of the factor B main effects is nonzero.

We may also test the null hypothesis that there are *no interactions*. Again using $\alpha = .01$, the computed value of $F = 13.156$, we see that the null hypothesis must also be *rejected*. (The critical value using 12 and 42 degrees of freedom is

Table 14-4 *Two-Factor ANOVA Table for Advertising Experiment*

Variation	Degrees of Freedom	Sum of Squares	Mean Square	F
Explained by Factor A: Day of Week (between rows)	$r - 1 = 6$	$SSA = 598.1587$	$MSA = \frac{SSA}{6} = 99.6931$	$\frac{MSA}{MSE} = 40.520$
Explained by Factor B: Section (between columns)	$c - 1 = 2$	$SSB = 41.8095$	$MSB = \frac{SSB}{2} = 20.9048$	$\frac{MSB}{MSE} = 8.497$
Explained by Interactions (between cells)	$(r - 1)(c - 1) = 12$	$SSAB = 388.4127$	$MSAB = \frac{SSAB}{12} = 32.367$	$\frac{MSAB}{MSE} = 13.156$
Error or Unexplained (residual)	$rc(n - 1) = 42$	$SSE = 103.3333$	$MSE = \frac{SSE}{42} = 2.4603$	
Total	$nrc - 1 = 62$	$SSTO = 1{,}131.7143$		

$F_{.01} = 2.64$.) We must conclude that a nonzero interaction effect exists for at least one treatment combination (cell).

Deciding What to Do

The preceding data indicate that the mean number of resumes received differs both by day of week and by newspaper section. Furthermore, the presence of interactions means that on some days, ads in a particular section might provide a higher mean response than on other days.

Multiple Comparison Procedures Depending on which of the null hypotheses is rejected, the next phase of the analysis involves comparing specific factor-level means, individual treatment (cell) means, or various groupings of treatment means. Section 14-2 describes the procedures for doing this. Depending on whether or not the sample data indicate the presence of interactions, one of the following approaches will be taken:

1. *Significant interactions are found.* Search the sample data for significant differences among treatment population (cell) means. The sample results may suggest further interesting comparisons.
2. *No interactions are significant.* Compare all factor-level means in each set for which the F test results in rejection of the null hypothesis of identical means. Isolate those pairs of means that differ significantly.

The placement director in the present illustration would take approach 1.

Computer-Assisted Analysis

A large number of calculations must be done in performing a two-factor analysis, and it is very desirable to use a computer to do these. Figure 14-4 shows a computer printout for an analysis of variance run with the advertising experiment

Figure 14-4 *SAS printout for analysis of variance in advertising experiment*

```
TITLE 'EMPLOYMENT NEWSPAPER ADS';
DATA ADS;
  INPUT DAY $ SECTION $ RESUMES @@;
  CARDS;
SUNDAY NEWS 17
SUNDAY NEWS 18
SUNDAY NEWS 20
SUNDAY SPORTS 21
      •
      •
      •
SAT BUSINESS 9
;
PROC ANOVA;
CLASS DAY SECTION;
MODEL RESUMES = DAY SECTION DAY*SECTION;
MEANS DAY SECTION DAY*SECTION;
QUIT;
```

```
                         EMPLOYMENT NEWSPAPER ADS

                      Analysis of Variance Procedure

                         Class Level Information

       Class     Levels    Values

       DAY            7    FRIDAY MONDAY SAT SUNDAY THURSDAY TUESDAY WED

       SECTION        3    BUSINESS NEWS SPORTS


                   Number of observations in data set = 63

Dependent Variable: RESUMES
                                   Sum of           Mean
Source                   DF       Squares         Square      F Value      Pr > F

Model                    20  1028.3809524     51.4190476        20.10      0.0001

Error                    42   103.3333333      2.4603175

Corrected Total          62  1131.7142857

                  R-Square           C.V.       Root MSE             RESUMES Mean

                  0.908693      16.469668      1.5685399               9.52380952


Source                   DF       Anova SS    Mean Square     F Value      Pr > F

DAY                       6    598.1587302     99.6931217       40.52      0.0001
SECTION                   2     41.8095238     20.9047619        8.50      0.0001
DAY*SECTION              12    388.4126984     32.3677249       13.16      0.0001
```

Level of DAY	N	RESUMES Mean	RESUMES SD
FRIDAY	9	10.6666667	1.41421356
MONDAY	9	7.8888889	3.48010217
SAT	9	13.3333333	2.59807621
SUNDAY	9	14.3333333	6.02079729
THURSDAY	9	5.4444444	1.87823794
TUESDAY	9	7.5555556	2.18581284
WED	9	7.4444444	1.13038833

Level of SECTION	N	RESUMES Mean	RESUMES SD
BUSINESS	21	8.6666667	2.12916259
NEWS	21	10.6190476	4.85258890
SPORTS	21	9.2857143	5.13948302

Level of DAY	Level of SECTION	N	RESUMES Mean	RESUMES SD
FRIDAY	BUSINESS	3	9.0000000	1.00000000
FRIDAY	NEWS	3	11.6666667	0.57735027
FRIDAY	SPORTS	3	11.3333333	0.57735027
MONDAY	BUSINESS	3	11.0000000	2.00000000
MONDAY	NEWS	3	8.6666667	1.52752523
MONDAY	SPORTS	3	4.0000000	2.00000000
SAT	BUSINESS	3	11.0000000	2.00000000
SAT	NEWS	3	15.3333333	2.88675135
SAT	SPORTS	3	13.6666667	0.57735027
SUNDAY	BUSINESS	3	6.6666667	1.15470054
SUNDAY	NEWS	3	18.3333333	1.52752523
SUNDAY	SPORTS	3	18.0000000	3.00000000
THURSDAY	BUSINESS	3	7.0000000	2.00000000
THURSDAY	NEWS	3	3.6666667	0.57735027
THURSDAY	SPORTS	3	5.6666667	1.15470054
TUESDAY	BUSINESS	3	8.3333333	1.52752523
TUESDAY	NEWS	3	9.3333333	0.57735027
TUESDAY	SPORTS	3	5.0000000	1.00000000
WED	BUSINESS	3	7.6666667	0.57735027
WED	NEWS	3	7.3333333	0.57735027
WED	SPORTS	3	7.3333333	2.08166600

Figure 14-4 *(Continued)*

data using SAS. The top portion shows both the commands needed by the program and the input data. This is followed by the ANOVA table. The last portion of the output lists in alphabetical order the row means, column means, and cell means.

Special Case—One Observation per Cell

A two-factor experiment may be conducted with a single observation per cell. Such an investigation provides no replication, and there is only $n = 1$ sample value per treatment combination. The experiment provides no special difficulty, but it will be less discriminating than one having replication. Perhaps the greatest drawback is that it does not allow us to make inferences regarding interactions when only one observation is made from each cell. (However, an alternative procedure, beyond the scope of this text, does exist for doing this.)

Table 14-5 Special Case—$n = 1$: Expressions and General Form of the ANOVA Table

Variation	Degrees of Freedom	Sum of Squares	Mean Square	F
Explained by Factor A (between rows)	$r - 1$	$SSA = c \sum_{i=1}^{r} (\bar{X}_{i\cdot} - \bar{\bar{X}}_{\cdot\cdot})^2$	$MSA = \frac{SSA}{r - 1}$	$\frac{MSA}{MSE}$
Explained by Factor B (between columns)	$c - 1$	$SSB = r \sum_{j=1}^{c} (\bar{X}_{\cdot j} - \bar{\bar{X}}_{\cdot\cdot})^2$	$MSB = \frac{SSB}{c - 1}$	$\frac{MSB}{MSE}$
Error or Unexplained (residual)	$(r - 1)(c - 1)$	$SSE = \sum_{i=1}^{r} \sum_{j=1}^{c} (X_{ij} - \bar{X}_{i\cdot} - \bar{X}_{\cdot j} + \bar{\bar{X}}_{\cdot\cdot})^2$	$MSE = \frac{SSA}{(r - 1)(c - 1)}$	
Total	$rc - 1$	$SSTO = \sum_{i=1}^{r} \sum_{j=1}^{c} (X_{ij} - \bar{\bar{X}}_{\cdot\cdot})^2$		

The earlier notation is somewhat simplified, since when $n = 1$, $\bar{X}_{ij\cdot}$, is equal to the single observation in the cell. We replace it here by X_{ij} (with no overbar). The third dot is also dropped from all subscripts in symbols representing the various sample means. The sums of squares SSA and SSB are computed as before. There is no $SSAB$ term. There are also some necessary modifications in computing $SSTO$ and SSE, and the number of degrees of freedom changes for the respective sources of variation. Table 14-5 shows the modified elements for constructing the ANOVA table.

Table 14-6 Sample Results for Chemical Processing-Parameter Evaluation (Yields in g/liter).

Factor A Temperature (°F)	Factor B Pressure (psi)				Factor A Sample Mean
	(1) 150	(2) 200	(3) 250	(4) 300	
(1) 300	30	32	33	33	$\bar{X}_{1\cdot} = 32.00$
(2) 325	28	30	31	31	$\bar{X}_{2\cdot} = 30.00$
(3) 350	28	28	30	30	$\bar{X}_{3\cdot} = 29.00$
(4) 375	27	28	29	30	$\bar{X}_{4\cdot} = 28.50$
(5) 400	27	26	27	27	$\bar{X}_{5\cdot} = 26.75$
Factor B Sample Mean	$\bar{X}_{\cdot 1} = 28.00$	$\bar{X}_{\cdot 2} = 28.80$	$\bar{X}_{\cdot 3} = 30.00$	$\bar{X}_{\cdot 4} = 30.20$	$\bar{\bar{X}}_{\cdot\cdot} = 29.25$

Example: Chemical Processing-Parameter Evaluation

The data in Table 14-6 represent the sample results of an investigation to establish operating parameters for processing a chemical. Two factors are considered. Factor A is temperature setting (degrees Fahrenheit, $r = 5$ levels) and factor B is pressure (psi, $c = 4$ levels). The response variable is yield (grams per liter of raw material). Two null hypotheses may be tested:

1. H_0: The mean yield (response) in grams per liter of raw material is the same under all temperature settings.
2. H_0: The mean yield (response) in grams per liter of raw material is the same under all pressure settings.

The sums of squares are calculated as follows:

$$\begin{aligned} SSA &= 4[(32.00 - 29.25)^2 + (30.00 - 29.25)^2 + \cdots + (26.75 - 29.25)^2] \\ &= 60.00 \end{aligned}$$

$$\begin{aligned} SSB &= 5[(28.00 - 29.25)^2 + (28.80 - 29.25)^2 + \cdots + (30.20 - 29.25)^2] \\ &= 16.15 \end{aligned}$$

$$\begin{aligned} SSTO &= (30 - 29.25)^2 + (32 - 29.25)^2 + \cdots + (27 - 29.25)^2 \\ &= 81.75 \end{aligned}$$

$$\begin{aligned} SSE &= (30 - 32.00 - 28.00 + 29.25)^2 + (32 - 32.00 - 28.80 + 29.25)^2 \\ &\quad + \cdots + (27 - 26.75 - 30.20 + 29.25)^2 \\ &= 5.60 \end{aligned}$$

The ANOVA table is given in Table 14-7. With significance levels of $\alpha = .01$, Appendix Table J shows critical values of $F_{.01} = 5.41$ (degrees of freedom 4 and 12) for factor A (temperature) and $F_{.01} = 5.95$ (degrees of freedom 3 and

Table 14-7 *Two-Way ANOVA Table for Chemical Processing-Parameter Investigation*

Variation	Degrees of Freedom	Sum of Squares	Mean Square	F
Explained by Factor A (between rows)	$r - 1 = 4$	$SSA = 60.00$	$MSA = \dfrac{60.00}{4} = 15.00$	$\dfrac{MSA}{MSE} = \dfrac{15.00}{.467} = 32.12$
Explained by Factor B (between columns)	$c - 1 = 3$	$SSB = 16.15$	$MSB = \dfrac{16.15}{3} = 5.383$	$\dfrac{MSB}{MSE} = \dfrac{5.383}{.467} = 11.53$
Error or Unexplained (within columns)	$(r - 1)(c - 1) = 12$	$SSE = 5.60$	$MSE = \dfrac{5.60}{12} = .467$	
Total	$rc - 1 = 19$	$SSTO = 81.75$		

12) for factor B (pressure). Since the respective computed values of F exceed these critical levels, both null hypotheses of identical means must be *rejected.*

Problems

14-1 Two researchers evaluating an algorithm for solving a dynamic lot-sizing problem formulated the original problem with a mathematical model allowing up to two sources and two price breaks. Their solution procedure is based on the branch-and-bound method. Table 14-8 shows the results obtained from several computer runs, each made using two similarly configured problems. The response variable used to assess the algorithm's efficiency is the average amount of CPU time.

(a) Construct the ANOVA table.

(b) At the $\alpha = .05$ levels of significance, indicate whether the following null hypotheses should be accepted or rejected:

(1) H_0: There are no interactions between problem duration and complexity.

(2) H_0: Means for all problem durations are identical.

(3) H_0: Means for all levels of complexity are identical.

Table 14-8 *Sample Data (Average CPU time in seconds) for Computer Evaluation of Algorithm Tested on Problems with Various Durations and Levels of Complexity*

	Complexity *B*			
Duration *A*	**(1) Simple**	**(2) Low**	**(3) Medium**	**(4) High**
(1) Short	.09 .13	.09 .16	.20 .39	.30 .30
(2) Middle	.11 .20	.15 .32	.39 2.46	2.48 7.12
(3) Long	.31 .49	.31 .33	.64 1.40	.75 1.99

Source: Erenguc, S. Selcuk, and Suleyman Tufekci, "A Branch-and-Bound Algorithm for a Single-Item Multi-Source Dynamic Lot-Sizing Problem with Capacity Constraints," *IIE Transactions*, March 1987, pp. 73–80.

14-2 A project engineer for a large construction firm conducted a sampling study to determine if the mean job completion delay of electrical jobs differs between subcontractors and also by type of job. The following sample data (days late) were obtained for jobs completed by four major subcontractors.

(a) Construct the ANOVA table.

(b) At the $\alpha = .05$ levels of significance, indicate whether the following null hypotheses should be accepted or rejected:

(1) H_0: There are no interactions between job type and subcontractor.

(2) H_0: Means for all job types are identical.

(3) H_0: Means for all subcontractors are identical.

Job Type	Subcontractor (1) Big M	(2) H-V Eng.	(3) CC&B	(4) Jones
(1) Remodeling	8.5	−3.0	7.5	13.0
	11.0	−7.5	9.0	11.5
	10.5	−4.5	7.5	20.5
(2) New Building	22.6	−2.0	34.0	45.0
	27.0	−8.5	32.0	39.5
	11.0	−4.5	24.0	35.5
(3) Groundwork	3.0	2.0	4.0	5.5
	−2.0	−3.0	1.5	2.5
	−1.0	−2.0	.5	4.0

14-3 An engineering admissions committee wishes to establish better guidelines for high schools regarding preparation for entering freshmen. A random sample of high school seniors has been selected and their scores on a freshman placement test have been determined. The following average test scores apply to students grouped by science courses taken and highest mathematics level.

Mathematics Preparation	Science Preparation (1) Chemistry Only	(2) Physics Only	(3) Chemistry and Physics	(4) Advanced Placement
(1) Trigonometry	57	60	73	74
	63	68	79	85
	66	67	65	67
	54	61	67	62
(2) Analysis	59	70	82	75
	64	62	77	79
	63	66	79	70
	62	62	66	74
(3) Calculus	67	69	83	76
	70	72	79	81
	72	75	77	88
	65	60	73	75

(a) Construct the ANOVA table.

(b) At the $\alpha = .05$ level of significance, indicate whether the following null hypotheses should be accepted or rejected:

(1) H_0: There are no interactions between mathematics and science preparations.

(2) H_0: Means for all mathematics preparations are identical.
(3) H_0: Means for all science preparations are identical.

14-4 The project engineer in Problem 14-2, also tested the two factors with average percentage cost overrun as the response variable. Only one observation per treatment combination was available. The following data apply:

	Subcontractor			
Job Type	**(1) Big M**	**(2) H-V Eng.**	**(3) CC&B**	**(4) Jones**
(1) Remodeling	15%	2%	13%	8%
(2) New Building	13	5	22	15
(3) Groundwork	5	−2	4	2

(a) Construct the ANOVA table.
(b) At the $\alpha = .05$ level of significance, indicate whether the following null hypotheses should be accepted or rejected:
(1) H_0: Means for all job types are identical.
(2) H_0: Means for all subcontractors are identical.

14-5 An engineering society wishes to assess the effects of member's specialty (A, with six levels) and employment sector (B, with four levels) on professional income. The following data represent observations of one randomly chosen person from each category combination.

$$SSA = 640{,}000 \qquad SSB = 710{,}000 \qquad SSTO = 1{,}500{,}000$$

(a) Construct a single-factor ANOVA table using specialty as the only treatment. At the $\alpha = .01$ significance level, can you conclude that the treatment means differ?
(b) Construct a single-factor ANOVA table using employment sector as the only treatment. At the $\alpha = .01$ significance level, can you conclude that the treatment means differ?
(c) Construct a two-factor ANOVA table ($n = 1$) using both specialty and employment sector as treatments. What can you conclude regarding the respective null hypotheses of identical mean incomes for specialties and for employment sectors at the $\alpha = .01$ level?
(d) Do any of your conclusions change in the two-factor evaluation from what they are in the respective single-factor evaluation?

14-6 The researchers in Problem 14-1 obtained a second response variable, average length of candidate list, in evaluating their algorithm for solving a dynamic lot-sizing problem. Figure 14-5 shows a partial SAS printout of the results. What conclusion should you reach?

14-7 Two researchers experimented with parameter settings for their "simulated annealing" algorithm for solving quadratic assignment problems. A portion of their results are shown in Table 14-9, where the response variable is the percentage by which their algorithm's solution falls above the best known solution for problems having the same parameters.

Figure 14-5 *SAS printout of analysis of variance for computer algorithm evaluation using list length as the response variable.*

```
                         Analysis of Variance Procedure

Dependent Variable: TIME
                                      Sum of          Mean
Source                    DF         Squares        Square      F Value

Model                     11     14.32500000    1.30200000        22.33

Error                     12      0.70000000    0.05833333

Corrected Total           23     15.02500000

                    R-Square            C.V.      Root MSE                TIME Mean

                    0.953411       778.39377      0.241523               1.88000000

Source                    DF         Anova SS   Mean Square     F Value

DURATION                   2      2.19000000    1.09500000        32.53
COMPLEX                    3      7.29800000    2.43266667        18.77
DURATION*COMPLEX           6      4.38700000    0.73116667        13.82

                  Level of               -------------TIME-------------
                  DURATION       N           Mean

                  LONG           8        1.70000000
                  MIDDLE         8        2.30000000
                  SHORT          8        1.63000000

                  Level of               -------------TIME-------------
                  COMPLEX        N           Mean

                  HIGH           6        2.70000000
                  LOW            6        1.50000000
                  MEDIUM         6        2.03000000
                  SIMPLE         6        1.27000000

        Level of      Level of                -------------TIME-------------
        DURATION      COMPLEX        N            Mean

        LONG          HIGH           2         2.30000000
        LONG          LOW            2         1.10000000
        LONG          MEDIUM         2         2.10000000
        LONG          SIMPLE         2         1.30000000
        MIDDLE        HIGH           2         4.10000000
        MIDDLE        LOW            2         1.80000000
        MIDDLE        MEDIUM         2         2.20000000
        MIDDLE        SIMPLE         2         1.10000000
        SHORT         HIGH           2         1.70000000
        SHORT         LOW            2         1.60000000
        SHORT         MEDIUM         2         1.80000000
        SHORT         SIMPLE         2         1.40000000
```

From these sample data, test at the 1% significance level null hypotheses of identical responses for all levels of each of two parameters—interchange complexity (factor A) and epoch interval (factor B)–plus the null hypothesis of no interactions from those levels. What conclusions do you reach?

Table 14-9 *Sample Data (Percentage Above Best Cost) for Simulated Annealing Solution of Quadratic Assignment Problems (with 30 Departments and .25 Error Constant)*

Interchange Complexity *A*	Epoch Interval *B*			
	(1) $e = 5$	(2) $e = 15$	(3) $e = 25$	(4) $e = 50$
(1) $N' = 10$	4.05	4.35	4.59	6.67
	3.97	4.55	4.78	6.44
	4.04	4.46	4.92	6.40
	4.02	4.49	5.02	6.37
	3.92	4.55	4.99	6.52
(2) $N' = 50$	2.81	2.01	1.82	2.33
	2.74	2.13	1.87	2.17
	2.74	2.19	1.99	2.18
	2.68	2.16	1.96	2.08
	2.78	2.26	2.11	2.24
(3) $N' = 100$	2.62	1.91	1.55	1.99
	2.65	1.87	1.59	2.18
	2.74	1.78	1.70	1.95
	2.70	1.75	1.64	1.73
	2.79	1.84	1.57	2.15

Source: Wilhelm, Mickey R., and Thomas L. Ward, "Solving Quadratic Assignment Problems by 'Simulated Annealing,'" *IEE Transactions*, March 1987, pp. 107–118.

Note: Only the means are actual values. The cell entries are for purposes of illustration only.

14-2 Analysis of Factor Effects

Constructing the ANOVA table and testing the null hypotheses are the easiest parts of a two-factor evaluation. The next phase is concerned with analyzing factor effects, which is accomplished largely by comparing various means.

In the absence of any significant interactions, pairwise comparisons will be made of the factor-level means, for the rows ($\mu_{i.}$'s), for the columns ($\mu_{.j}$'s), or for both, depending on which of these groups was found to have significantly different means by the respective F test. Such comparisons are done by constructing a set of collective confidence intervals, and identifying from these those pairs of means that differ significantly.

When the F test indicates the presence of interactions between factors, the analysis instead involves pairwise comparisons of the treatment (cell) population means (μ_{ij}'s), again through construction of collective confidence intervals and subsequent identification of μ's that differ significantly.

Recall from Chapter 13 that multiple comparisons must be accomplished under the umbrella of a collective confidence level. Two procedures are described here for doing this: the Tukey and Scheffé methods. The Tukey method ordinarily applies when no significant interactions have been found in the analysis-of-variance phase. The Scheffé method is used where there are interactions. Both methods maximize the comparisons that might be made (at some small price in reduced precision over alternative procedures). One important advantage of the Tukey and Scheffé methods is that they free the investigator from the concern of having to identify in advance which comparisons to make.

No Interactions Case: Multiple Pairwise Comparisons of Factor-Level Means

Two procedures, the Tukey and Bonferroni, were discussed in Chapter 13 for comparing factor-level means. Although both methods may be extended to multifactor analyses, the Tukey method tends to provide the more precise set of interval estimates when all mean pairs are to be examined.

Tukey Method The Tukey method is based on the studentized range distribution (with critical values listed in Appendix Table L). A separate set of intervals must be constructed independently for each factor—but only for those factors for which the respective H_0 of equal means (no main effects) has already been rejected. When the H_0's for both have been rejected, the overall level of confidence must be split for the two interval groups. Consider the following examples.

Group Collective Confidence Level		**Family Collective Confidence Level**
Factor *A* $100(1-\alpha_1)\%$	**Factor *B*** $100(1-\alpha_2)\%$	$100(1-\alpha_1-\alpha_2)\%$
(1) $100(1-.05)=95\%$	$100(1-.05)=95\%$	$100(1-.05-.05)=90\%$
(2) $100(1-.01)=99\%$	$100(1-.01)=99\%$	$100(1-.01-.01)=98\%$
(3) $100(1-.05)=95\%$	$100(1-.01)=99\%$	$100(1-.05-.01)=94\%$
(4) $100(1-.05)=95\%$	$100(1-.10)=90\%$	$100(1-.05-.10)=85\%$

The following expressions are used to compute the Tukey interval estimates.

TUKEY INTERVAL ESTIMATES FOR DIFFERENCES IN FACTOR-LEVEL MEANS AT THE $100(1-\alpha_1-\alpha_2)\%$ COLLECTIVE CONFIDENCE LEVEL

Factor *A*:

$$\mu_{i\cdot} - \mu_{m\cdot} = (\bar{X}_{i\cdot\cdot} - \bar{X}_{m\cdot\cdot}) \pm q_{1-\alpha_1}\sqrt{\frac{MSE}{cn}}$$

Parameters for q: numerator r, denominator $rc(n-1)$

Factor B:

$$\mu_{.j} - \mu_{.k} = (\bar{X}_{.j.} - \bar{X}_{.k.}) \pm q_{1-\alpha_2}\sqrt{\frac{MSE}{rn}}$$

Parameters for q: numerator c, denominator $rc(n-1)$

Any pairwise difference having confidence intervals lying totally above or below zero corresponds to factor-level means that are significantly different at the collective level $\alpha_1 + \alpha_2$.

Illustration: *Industrial Engineering Evaluation of an Assembly Operation*

To illustrate, consider the results in Table 14-10 obtained for an industrial engineering evaluation of an assembly operation. The sampling experiment involves $n = 5$ observations per treatment, one for each combination of $r = 4$ factor A levels (line configuration) and $c = 3$ factor B levels (method of locomotion). The response variable is the assembly rate (parts per minute).

Table 14-10 *Summary Sample Results for Assembly Operation Evaluation*

Sample Mean Assembly Rate (parts/minute)

Line Configuration A	Locomotion B (1) Carts	(2) Rollers	(3) Belts	Factor B Mean
(1) Staggered	$\bar{X}_{11.} = 10.6$	$\bar{X}_{12.} = 13.8$	$\bar{X}_{13.} = 25.8$	$\bar{X}_{1..} = 16.73$
(2) Straight	$\bar{X}_{21.} = 23.2$	$\bar{X}_{22.} = 24.6$	$\bar{X}_{23.} = 34.2$	$\bar{X}_{2..} = 27.33$
(3) U-Shaped	$\bar{X}_{31.} = 11.6$	$\bar{X}_{32.} = 17.2$	$\bar{X}_{33.} = 28.6$	$\bar{X}_{3..} = 19.13$
(4) Bent	$\bar{X}_{41.} = 11.0$	$\bar{X}_{42.} = 14.4$	$\bar{X}_{43.} = 26.8$	$\bar{X}_{4..} = 17.40$
Factor A Mean	$\bar{X}_{.1.} = 14.1$	$\bar{X}_{.2.} = 17.5$	$\bar{X}_{.3.} = 28.85$	$\bar{\bar{X}}_{...} = 20.15$

ANOVA Table

Variation	Degrees of Freedom	Sum of Squares	Mean Square	F
Explained by Factor A: Configuration	$r - 1 = 3$	$SSA = 1{,}078.05$	$MSA = 359.35$	32.746
Explained by Factor B: Locomotion	$c - 1 = 2$	$SSB = 2{,}386.30$	$MSB = 1{,}193.15$	108.715
Explained by Interactions	$(r-1)(c-1) = 6$	$SSAB = 56.50$	$MSAB = 9.42$	.858
Error or Unexplained	$rc(n-1) = 48$	$SSE = 526.80$	$MSE = 10.98$	
Total	$nrc - 1 = 59$	$SSTO = 4{,}047.65$		

The F test establishes that the factor-level means differ both for configuration and for locomotion, although no significant interactions are present.

Let's use 90% as the collective family confidence level, to be split evenly between both factors, so that $\alpha_1 = \alpha_2 = .05$. Then, using parameters $(4, 4[3][5-1]) = (4, 48)$ for factor A and $(3, 48)$ for factor B, we read the critical values for q from Appendix Table L. With linear interpolation, these values are $q_{.95} = 3.77$ for factor A and $q_{.95} = 3.42$ for factor B.

Consider first factor A, line configuration. The following interval is obtained for the difference in mean assembly rates between the (2) straight and (1) staggered configurations:

$$\begin{aligned}\mu_{2.} - \mu_{1.} &= (\bar{X}_{2..} - \bar{X}_{1..}) \pm q_{.95}\sqrt{\frac{MSE}{3(5)}}\\ &= (27.33 - 16.73) \pm 3.77\sqrt{\frac{10.98}{15}}\\ &= 10.60 \pm 3.23\end{aligned}$$

or

$$\text{Straight} - \text{staggered:} \quad 7.37 \leq \mu_{2.} - \mu_{1.} \leq 13.83 \qquad \text{(significant)}$$

All factor A pairwise differences will have confidence intervals of the same width, as follows:

U-Shaped − staggered:	$-.83 \leq \mu_{3.} - \mu_{1.} \leq 5.63$	
Bent − staggered:	$-2.56 \leq \mu_{4.} - \mu_{1.} \leq 3.90$	
U-Shaped − straight:	$-11.43 \leq \mu_{3.} - \mu_{2.} \leq -4.97$	(significant)
Bent − straight:	$-13.16 \leq \mu_{4.} - \mu_{2.} \leq -6.70$	(significant)
Bent − U-Shaped:	$-4.96 \leq \mu_{4.} - \mu_{3.} \leq 1.50$	

These intervals show a significant difference between $\mu_{2.}$ (straight) and each of the other configuration factor-level means. But since the remaining intervals all overlap zero, no significant differences exist between the mean assembly rates for (3) U-shaped and the (1) staggered or (4) bent configurations, nor between the latter two.

Because factor B, locomotion, has a different number of levels, confidence intervals for its pairwise differences will have a different width. The following interval is obtained for the difference in mean assembly rates between (2) rollers and (1) carts:

$$\begin{aligned}\mu_{.2} - \mu_{.1} &= (\bar{X}_{.2.} - \bar{X}_{.1.}) \pm q_{.95}\sqrt{\frac{MSE}{4(5)}}\\ &= (17.5 - 14.1) \pm 3.42\sqrt{\frac{10.98}{20}}\\ &= 3.4 \pm 2.53\end{aligned}$$

or

$$\text{Rollers} - \text{carts:} \quad 0.87 \leq \mu_{.2} - \mu_{.1} \leq 5.93 \qquad \text{(significant)}$$

All factor B pairwise differences will have confidence intervals of the above width, which is narrower than the intervals of the preceding set. These are as follows:

Belts − carts: $12.22 \leq \mu_{.3} - \mu_{.1} \leq 17.28$ (significant)

Belts − rollers: $8.82 \leq \mu_{.3} - \mu_{.2} \leq 13.88$ (significant)

The mean assembly rates differ significantly for all pairs of factor B levels for locomotion.

Special Case — One Observation per Cell When there is only $n = 1$ cell observation per factor-level combination, the above procedure still applies, but the q parameter must be modified by replacing the term $rc(n - 1)$ with $(r - 1)(c - 1)$.

Case of Significant Interactions: Multiple Contrasts Among Treatment Means

When there are interactions, the individual treatment population (cell) means can be meaningfully compared. Thus, collective confidence intervals involving the μ_{ij}'s may be constructed and will pinpoint specific significant differences. Various groupings of cell means — not just pairs — might be compared. Such comparisons may be made in terms of **contrasts**, denoted as follows.

CONTRAST IN TREATMENT (CELL) POPULATION MEANS

$$L = \sum_{i=1}^{r} \sum_{j=1}^{c} w_{ij}\mu_{ij}$$

where

$$\sum_{i=1}^{r} \sum_{j=1}^{c} w_{ij} = 0$$

All contrasts involve weighted sums of the population cell means, with the requirement that the weights be real numbers and sum to zero. (Pairwise differences in the μ_{ij}'s are contrasts where a single weight equals $+1$, a second weight is -1, and the remaining weights are zero.)

The Scheffé Method The Scheffé method allows collective confidence intervals to be constructed for *any number* of contrasts, without requiring that they be specified in advance. This allows the investigator to "snoop," sifting and sorting the data to find meaningful comparisons to make.

Being able to look at the data first is a very important advantage of the Scheffé method — especially in multifactor investigations where the number of pairwise differences can be huge. (For example, in a 7-by-3 layout there are $rc = 7(3) = 21$ cells, and thus 210 possible pairings of μ_{ij}'s.) And, for any investigation there is an unlimited number of contrast possibilities.

The following expression is used in constructing the Scheffé estimates.

SCHEFFÉ ESTIMATES FOR MULTIPLE CONTRASTS AT A 100(1 − α)% COLLECTIVE CONFIDENCE LEVEL

$$L = \sum_{i=1}^{r} \sum_{j=1}^{c} w_{ij}\bar{X}_{ij\cdot} \pm \sqrt{(rc - 1)F_{\alpha} \frac{MSE}{n} \sum_{i=1}^{r} \sum_{j=1}^{c} w_{ij}^2}$$

Degrees of freedom: numerator $rc - 1$, denominator $rc(n - 1)$

To illustrate the Scheffé method we will continue with the engineering-placement advertising example from Section 14-1. Before expressing any particular contrasts and constructing the corresponding confidence intervals, however, it is helpful to examine the sample data in detail to see which particular comparisons would be meaningful.

Finding Meaningful Comparisons: Plotting the Data When the number of treatments (cells) is large, it is helpful to plot the sample results, as in Figure 14-6, which illustrates the advertisement evaluation. Each data point shown was taken from Table 14-2. The mean responses are plotted for each level of both factor *A* and factor *B*.

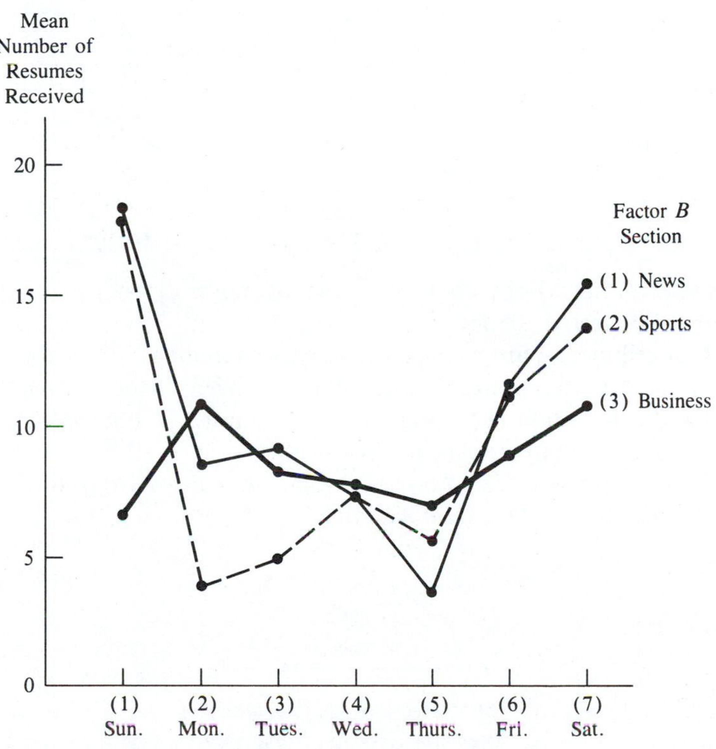

Figure 14-6 *Sample treatment (cell) means for engineering-placement advertisement evaluation.*

The presence of significant interactions between day-of-week A and newspaper section B is reflected by lines that oscillate and even cross. (The factor B lines would tend to be parallel if no interactions existed.)

Much of the fluctuation in plot lines is to be expected and may be attributed to chance sampling error. The Scheffé method will help us sort things out by isolating differences and contrasts that differ significantly from zero. In identifying the significant pairwise differences, it is not necessary to construct confidence intervals for each. All such intervals will have identical widths, so the essential information is given by the first confidence interval constructed. The cell pairs having significantly different means will be those with $\bar{X}_{ij\cdot}$'s separated by more than half the width of the first interval found. For the present illustration, a collective confidence level of 95% will be used.

Constructing the Intervals We can begin with the biggest gap, that between Sunday news (1, 1) and Sunday business (1, 3), which provides the contrast

$$L_1 = \mu_{11} - \mu_{13} \qquad \text{(with } w_{11} = 1,\ w_{13} = -1 \text{ and all other } w_{ij}\text{'s} = 0)$$

With degrees of freedom $rc - 1 = 20$ for the numerator and $rc(n - 1) = 42$ for the denominator, Appendix Table J provides the critical value $F_{.05} = 1.82$. Using $MSE = 2.4603$ and the sample means from Table 15-2, we find that the confidence interval is

$$L_1\!: \mu_{11} - \mu_{13} = [(1)\bar{X}_{11\cdot} + (-1)\bar{X}_{13\cdot}] \pm \sqrt{(7[3] - 1)(1.82)\frac{2.4603}{3}([1]^2 + [-1]^2)}$$
$$= [18.33 - 6.67] \pm 7.73$$
$$= 11.66 \pm 7.73$$

or

$$\text{Sunday news} - \text{Sunday business:} \quad 3.93 \le \mu_{11} - \mu_{13} \le 19.39 \qquad \text{(significant)}$$

Since this interval lies wholly above zero, the difference $\mu_{11} - \mu_{13}$ is significantly nonzero at the collective .05 level.

All other cell mean pairs having a difference exceeding 7.73 in absolute value will share this property. Altogether, about three dozen of the cell mean pairs in the advertising illustration show significant differences. Of course, the personnel director would only be interested in a few of these.

Quick perusal of Figure 14-6 shows that, on a day-by-day basis, only the following means are significantly different:

(1) Sunday
(Sports) $\mu_{12} > \mu_{13}$ (Business)
(News) $\mu_{11} > \mu_{13}$ (Business)

A sectional breakdown shows that the business section has *no days* that differ significantly. For the news section, only Friday and Saturday yield means that are significantly smaller than those of Sunday, while Saturday is significantly better than Wednesday and Thursday only. For sports, the Sunday advantage is significant for all days but Friday and Saturday; Saturday significantly outperforms all remaining days but Wednesday and Friday.

In addition to pairwise differences, some *group comparisons* are suggested by the data plot. The weekends, for example, show the strongest responses, but only for the news and sports sections. The following comparison is suggested:

Weekend sports or news vs. Weekday sports or news

The difference in the average of the respective cell means provides the following contrast:

$$L_2 = \frac{\mu_{11} + \mu_{12} + \mu_{71} + \mu_{72}}{4} - \frac{\mu_{21} + \mu_{22} + \mu_{31} + \mu_{32} + \mu_{41} + \mu_{42} + \mu_{51} + \mu_{52} + \mu_{61} + \mu_{62}}{10}$$

The respective weights are +.25 for the four cells represented by the first term and −.10 for the ten cells in the last term. In this contrast, all remaining cells have weights of zero. The point estimate of the above expression is

$$\begin{aligned}
&.25(\bar{X}_{11\cdot} + \bar{X}_{12\cdot} + \bar{X}_{71\cdot} + \bar{X}_{72\cdot}) \\
&\quad - .10(\bar{X}_{21\cdot} + \bar{X}_{22\cdot} + \bar{X}_{31\cdot} + \bar{X}_{32\cdot} + \bar{X}_{41\cdot} + \bar{X}_{42\cdot} + \bar{X}_{51\cdot} + \bar{X}_{52\cdot} + \bar{X}_{61\cdot} + \bar{X}_{62\cdot}) \\
&= .25(18.33 + 18.00 + 15.33 + 13.67) \\
&\quad - .10(8.67 + 4.00 + 9.33 + 5.00 + 7.33 + 7.33 + 3.67 + 5.67 + 11.67 + 11.33) \\
&= 8.93
\end{aligned}$$

and the sum of the squared weights is

$$4(.25)^2 + 10(-.10)^2 = .35$$

Thus

$$\begin{aligned}
L_2 &= 8.93 \pm \sqrt{(7[3] - 1)(1.82)\frac{2.4603}{3}(.35)} \\
&= 8.93 \pm 3.23
\end{aligned}$$

or

Weekend − weekday
(news and sports): $5.70 \leq L_2 \leq 12.16$ (significant)

The data plot in Figure 14-6 suggests a third contrast for Monday–Tuesday news or business vs. sports:

$$L_3 = \frac{\mu_{21} + \mu_{23} + \mu_{31} + \mu_{33}}{4} - \frac{\mu_{22} + \mu_{32}}{2}$$

The first four cells listed have weight .25, the last two each have weight −.50, and the unlisted ones have zero weights. The point estimate of the contrast is

$$\begin{aligned}&.25(\bar{X}_{21}. + \bar{X}_{23}. + \bar{X}_{31}. + \bar{X}_{33}. - .50(\bar{X}_{22}. + \bar{X}_{32}.)\\&= .25(8.67 + 11.00 + 9.33 + 8.33) - .50(4.00 + 5.00)\\&= 4.83\end{aligned}$$

and the sum of the squared weights is

$$4(.25)^2 + 2(-.50)^2 = .75$$

Thus

$$\begin{aligned}L_3 &= 4.83 \pm \sqrt{(7[3] - 1)(1.82)\frac{2.4603}{3}(.75)}\\&= 4.83 \pm 4.73\end{aligned}$$

or

Monday or Tuesday
news and
business − sports: $.10 \leq L_3 \leq 9.56$ (significant)

Confidence Summary Statement At the overall collective level of confidence, the separate statements may be combined into an omnibus statement. The following 95% confidence summary statement applies to the present illustration.

Sunday appears to provide the greatest mean response (resumes received) for ads placed in either the news or sports sections. (However, those two sections exhibit no significant Sunday difference between themselves). Saturday is a close second for the same positions. Weekends outperform weekdays for these two sections. On no day does the business section by itself significantly outperform the others. When news is grouped with sports, however, the two sections on the average outperform business on Mondays and Tuesdays only.

Estimating Effects Parameters

The theoretical model of Section 14-1 expresses the main effects in terms of the factor-level means and overall population mean:

$$A_i = \mu_{i\cdot} - \mu_{\cdot\cdot} \qquad B_j = \mu_{\cdot j} - \mu_{\cdot\cdot}$$

The interaction effects are expressed analogously:

$$(AB)_{ij} = \mu_{ij} - \mu_{i\cdot} - \mu_{\cdot j} + \mu_{\cdot\cdot}$$

All of these effects parameters may be expressed as contrasts in μ_{ij}'s and estimated by the Scheffé method. But since the usual applications are not concerned with explicit levels of these quantities, finding such estimates will be left as an exercise.

Problems

14-8 Refer to Problem 14-3 and your answers to that exercise. Assuming that no significant interactions are found, use a collective confidence level of 90% to find Tukey interval estimates for each pairwise difference in factor-level means. Assume $\alpha_1 = \alpha_2$. What may you conclude?

14-9 Refer to Problem 14-4 and your answers to that exercise. Since $n = 1$, testing for interactions is impossible. Use a collective confidence level of 94% to find Tukey interval estimates for each pairwise difference in factor-level means. Use $\alpha_1 = .05$ and $\alpha_2 = .01$. What may you conclude?

14-10 Refer to the chemical processing example in Section 14-1. Since $n = 1$, testing for interactions is impossible. Use a collective confidence level of 98% to find Tukey interval estimates for each pairwise difference in factor-level means. Use $\alpha_1 = \alpha_2$. What may you conclude?

14-11 Refer to Problem 14-2 and your answers to that exercise. All conclusions will be made at an overall 95% collective confidence level.

(a) Using job type as the horizontal axis and mean days late as the vertical axis, plot on a common graph the sample cell means for each contractor.

(b) For which cell pairs do the means significantly differ?

(c) Contrasts for the following comparisons are to be estimated:

(1) Above ground vs. groundwork

(2) Big contractors (Big M, CC&B) vs. small contractors (H-V Eng., Jones)

(3) H-V above ground vs. Others above ground

Express the contrast for each comparison in terms of differences in average cell means for applicable groupings. Then construct interval estimates for each.

(d) Incorporating any further contrasts you deem important, give an overall summary statement at the 95% collective confidence level.

14-12 Refer to Problem 14-6 and your answers to that exercise. All conclusions will be made at an overall 95% collective confidence level.

(a) Plot the cell mean response against levels for factor A, with one set of data points for each level of factor B.

(b) For which cell pairs do the means differ significantly?

(c) Find other meaningful comparisons to be estimated. For each of these express the contrast in terms of differences in average cell means for applicable groupings. Then construct interval estimates for each.

(d) Give an overall summary statement at the 95% collective confidence level.

14-13 Refer to Problem 14-7 and your answers to that exercise. All conclusions will be made at an overall 95% collective confidence level. Repeat (a)–(d) as in the preceding problem.

14-3 The Randomized Block Design

An important class of two-factor studies uses a randomized block design. Here the second factor is used primarily to reduce the amount of unexplained variation in response, thereby increasing the efficacy of the analysis of variance over what would be achieved with a single factor (treatment variable).

Table 14-11 *Layout for Bacterial Leaching Investigation and Expanded Sample Results under the Randomized Block Design*

Block i	*Thiobaccilus* Treatment j			Mean Yield (lb/ton)
	Strain 1	Strain 2	Strain 3	
Site 1	31	33	25	$\bar{X}_{1\cdot} = 29.67$
Site 2	35	41	27	$\bar{X}_{2\cdot} = 34.33$
Site 3	33	37	24	$\bar{X}_{3\cdot} = 31.33$
Site 4	29	32	26	$\bar{X}_{4\cdot} = 29.00$
Site 5	35	34	29	$\bar{X}_{5\cdot} = 32.67$
Site 6	41	43	37	$\bar{X}_{6\cdot} = 40.33$
Site 7	53	45	39	$\bar{X}_{7\cdot} = 45.67$
Site 8	41 new	43	33	$\bar{X}_{8\cdot} = 39.00$
Site 9	40 new	41	36	$\bar{X}_{9\cdot} = 39.00$
Site 10	31 new	27	22 new	$\bar{X}_{10\cdot} = 26.67$
Mean Yield (lb/ton)	$\bar{X}_{\cdot 1} = 36.90$	$\bar{X}_{\cdot 2} = 37.60$	$\bar{X}_{\cdot 3} = 29.80$	$\bar{\bar{X}}_{\cdot\cdot} = 34.77$

Illustration: *Bacterial Leaching Investigation*

To introduce the procedure, we extend the bacterial leaching investigation of Chapter 13, so that *deposit site* of the ore is the second factor in explaining copper yield variation. The mining engineer used ore taken from ten different major sites within the deposit complex. These were selected for their distinctive material composition, and the ores provided represent the wide range of types found at the mine. Each of the sites is referred to as a **block**. A random sample was then taken from each site or block. Table 14-11 shows the original bacterial leaching data arranged by block and treatment.

The earlier bacterial leaching data involved an unequal number of observations in each treatment column since not all blocks (sites) were represented in the earlier experiment. To meet the design requirements, new sample observations have been added to the original data under strains 1 and 3.*

A sampling procedure involving this type of layout is called a **randomized block design**. Although the terminology evokes neighboring rectangles, much like city blocks, there is no reason why the deposit sites must be contiguous or even the same size and shape. A block can be any homogeneous grouping of sample units. For example, such a design might be used to evaluate different languages (FORTRAN, BASIC, Pascal) in computer instruction, where test students might

* Of course, neither the single-factor nor the randomized blocking procedures should ordinarily be used on the same sample data. The same data set is used here only for purposes of illustration.

be blocked by available computer mode (batch, time-share, minicomputer), a contributing factor to programming proficiency in any given language. Although of secondary concern, a block might influence response by a similar order of magnitude as that of a particular treatment, and this factor is sometimes referred to as a **blocking variable**.

Analytical Framework and Theoretical Model

The underlying model for a two-factor randomized block design treats each observation as the following sum:

$$X_{ij} = \mu_{..} + B_i + T_j + \epsilon_{ij}$$

where $\mu_{..}$ denotes the overall population mean, B_i is the blocking effect for block i, T_j is the treatment effect, and ϵ_{ij} denotes the error term. The ϵ's are assumed to be independent, normally distributed random variables with mean zero and standard deviation σ. There are r blocks (levels for the blocking variable), one for each row. In addition, there is a separate column for each treatment (level of the treatments variable), and there are c of these. There must be exactly one observation under each treatment and for each block, and every treatment is represented exactly once by each block. Thus, $n = 1$ observation per cell.

The two-factor notation is simplified, with X_{ij} (with no overbar) used in place of $\bar{X}_{ij.}$, and with no third dot needed in the subscripts in symbols representing the various sample means. The analytical framework may be explained in terms of the following partitioning of the deviations:

$$\begin{array}{ccccccc} \text{Total} & & \text{Treatments} & & \text{Blocks} & & \text{Error} \\ \text{deviation} & = & \text{deviation} & + & \text{deviation} & + & \text{deviation} \end{array}$$

$$(X_{ij} - \bar{\bar{X}}_{..}) = (\bar{X}_{.j} - \bar{\bar{X}}_{..}) + (\bar{X}_{i.} - \bar{\bar{X}}_{..}) + (X_{ij} - \bar{X}_{i.} - \bar{X}_{.j} + \bar{\bar{X}}_{..})$$

As with the earlier procedures, the collective sums of individual squared deviations are used to express the amount of variation by groups:

$$\begin{array}{ccccccc} \text{Total variation} & = & \text{Treatments variation} & + & \text{Blocks variation} & + & \text{Error variation} \\ SSTO & = & SSTR & + & SSBL & + & SSE \end{array}$$

The respective sums of squares form the basis for evaluation. This is done through the ANOVA table, which has the generalized form shown in Table 14-12, where the expressions are given for the sums of squares, mean squares, and number of degrees of freedom.

The null hypothesis of equal treatment population means is as follows:

$$H_0\colon \mu_{.1} = \mu_{.2} = \cdots = \mu_{.c}$$

This is tested in the usual manner, by computing the F statistic,

$$F = \frac{\text{Variation explained by treatments}}{\text{Unexplained variation}} = \frac{MSTR}{MSE}$$

Table 14-12 Randomized Block Design: Expressions and General Form of the ANOVA Table

Variation	Degrees of Freedom	Sum of Squares	Mean Square	F
Explained by Treatments (between columns)	$c - 1$	$SSTR = r \sum_{j=1}^{c} (\bar{X}_{.j} - \bar{\bar{X}}_{..})^2$	$MSTR = \dfrac{SSTR}{c-1}$	$\dfrac{MSTR}{MSE}$
Explained by Blocks (between rows)	$r - 1$	$SSBL = c \sum_{i=1}^{r} (\bar{X}_{i.} - \bar{\bar{X}}_{..})^2$	$MSBL = \dfrac{SSBL}{r-1}$	*
Error or Unexplained (residual)	$(r-1)(c-1)$	$SSE = \sum_{i=1}^{r} \sum_{j=1}^{c} (X_{ij} - \bar{X}_{i.} - \bar{X}_{.j} + \bar{\bar{X}}_{..})^2$	$MSE = \dfrac{SSE}{(r-1)(c-1)}$	
Total	$rc - 1$	$SSTO = \sum_{i=1}^{r} \sum_{j=1}^{c} (X_{ij} - \bar{\bar{X}}_{..})^2$		

* In the randomized block design F is not ordinarily computed for blocks.

Returning to the bacterial leaching illustration, we obtain the ANOVA table in Table 14-13. The computed value for the test statistic is $F = 28.83$.

Suppose that the engineer chooses a smaller significance level, $\alpha = .01$, than used before. The acceptance and rejection regions shown in Figure 14-7 now apply. The critical value $F_{.01} = 6.01$ was read from Appendix Table J using $c - 1 = 2$ degrees of freedom for the numerator and $(r - 1)(c - 1) = 18$ degrees of freedom for the denominator.

The computed value $F = 28.83$ falls substantially inside the rejection region, providing strong evidence that the mean copper yields are unequal for different

Table 14-13 Two-Way ANOVA Table for Bacterial Leaching Investigation

Variation	Degrees of Freedom	Sum of Squares	Mean Square	F
Explained by Treatments (between columns)	$c - 1 = 2$	$SSTR = 372.47$	$MSTR = \dfrac{372.47}{2} = 186.23$	$\dfrac{MSTR}{MSE} = \dfrac{186.23}{6.46} = 28.83$
Explained by Blocks (between rows)	$r - 1 = 9$	$SSBL = 980.70$	$MSBL = \dfrac{980.70}{9} = 108.97$	
Error or Unexplained (within columns)	$(r-1)(c-1) = 18$	$SSE = 116.20$	$MSE = \dfrac{116.20}{18} = 6.46$	
Total	$rc - 1 = 29$	$SSTO = 1{,}469.37$		

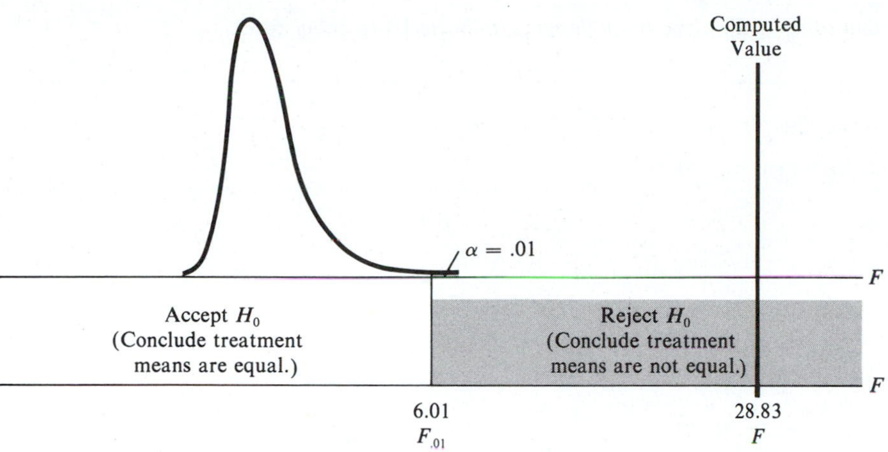

Figure 14-7

bacteria strains. The two-factor test provides a lower significance level than the single-factor procedure for nearly the same data. Although several degrees of freedom were lost by including the blocking variable, these were more than made up by the ensuing reduction in previously unexplained variation.

Computer-Assisted Analysis

It is advantageous to employ a computer in making the calculations for studies with randomized block designs. Figure 14-8 shows the printout from a SAS run with the bacterial leaching data. As with the earlier SAS printouts, the top portion gives the program commands and the input data. The essential ANOVA details are provided in the bottom two lines of the printout.

The following example further illustrates the advantage of using a randomized block design.

***Example:** Evaluating Methods for Solving Quadratic Assignment Problems (continued)*

Table 14-14 shows the sample results from an experiment testing four procedures for solving quadratic assignment problems. In Chapter 13 a single-factor analysis of those data resulted in a computed value of $F = .0006$; thus, that earlier statistical procedure leads to the conclusion that the mean response (average best cost) is identical under all methods.

The increased power of the randomized block design is illustrated now as we use the fact that successive sample observations involve problems of progressively larger size. That is, we now treat *problem size* as the blocking variable.

Table 14-15 is the ANOVA table. The computed value of $F = 3.219$ under the randomized block design has increased over 5,000-fold from the single-factor value found in Chapter 13. The results are now significant enough at the 5% level to *reject* the null hypothesis of identical mean costs.

Figure 14-8 *SAS printout using randomized block design with bacterial leaching data.*

```
TITLE 'BACTERIAL LEACHING';
DATA BACT;
  INPUT BLOCK TRTMENT $ YIELD;
  CARDS;
1 S1 31
1 S2 33
1 S3 25
2 S1 35
    •
    •
    •
10 S3 22
;
PROC ANOVA;
CLASS BLOCK TRTMENT;
MODEL YIELD = BLOCK TRTMENT;
QUIT;
```

```
                         BACTERIAL LEACHING

                   Analysis of Variance Procedure

                      Class Level Information

              Class       Levels    Values

              BLOCK           10    1 2 3 4 5 6 7 8 9 10

              TRTMENT          3    S1 S2 S3

              Number of observations in data set = 30

Dependent Variable: YIELD
                                    Sum of           Mean
Source                    DF       Squares         Square     F Value      Pr > F

Model                     11  1353.1666667    123.0151515       19.06      0.0001

Error                     18   116.2000000      6.4555556

Corrected Total           29  1469.3666667

                    R-Square          C.V.       Root MSE              YIELD Mean

                    0.920918     7.3080878      2.5407785             34.76666667

Source                    DF      Anova SS    Mean Square     F Value      Pr > F

BLOCK                      9    980.700000     108.966667       16.88      0.0001
TRTMENT                    2    372.466667     186.233333       28.85      0.0001
```

Table 14-14 Sample Results for Evaluation of Methods for Solving Quadratic Assignment Problems

Problem (Block)	Average Best Cost With Procedure (Treatment): (1) Simulated Annealing	(2) CRAFT	(3) CRAFT-Biased Sampling	(4) Revised Hillier	Mean Cost
1	\$25	\$28.2	\$27.6	\$25	$\bar{X}_{1\cdot}$ = \$26.45
2	43	44.2	43.6	43	$\bar{X}_{2\cdot}$ = 43.45
3	74	79.6	74.8	73.2	$\bar{X}_{3\cdot}$ = 75.40
4	107	110	108.2	107.4	$\bar{X}_{4\cdot}$ = 108.15
5	291	296.2	294.8	298.4	$\bar{X}_{5\cdot}$ = 295.10
6	578.2	606	581	582.2	$\bar{X}_{6\cdot}$ = 586.85
7	1,308.0	1,339	1,321	1,324.6	$\bar{X}_{7\cdot}$ = 1,323.15
8	3,099.8	3,197.8	3,124	3,114.2	$\bar{X}_{8\cdot}$ = 3,133.95
Mean Cost	\$690.75 = $\bar{X}_{\cdot 1}$	\$712.625 = $\bar{X}_{\cdot 2}$	\$696.875 = $\bar{X}_{\cdot 3}$	\$696.00 = $\bar{X}_{\cdot 4}$	$\bar{\bar{X}}_{\cdot\cdot}$ = \$699.0625

Source: Wilhelm, Mickey R., and Thomas L. Ward, "Solving Quadratic Assignment Problems by 'Simulated Annealing,'" *IEE Transactions*, March 1987, pp. 107–118.

Table 14-15 ANOVA Table for Evaluation of Methods for Solving Quadratic Assignment Problems

Variation	Degrees of Freedom	Sum of Squares	Mean Square	*F*
Explained by Treatments: Method (between columns)	$c - 1 = 3$	$SSTR =$ 2,138	$MSTR = \frac{SSTR}{3} = 712.5$	$\frac{MSTR}{MSE} = 3.219$
Explained by Blocks: Problem Size (between rows)	$r - 1 = 7$	$SSBL =$ 32,457,230	$MSBL = \frac{SSBL}{7} = 4{,}636{,}747$	
Error or Unexplained (residual)	$(r - 1)(c - 1) = 21$	$SSE =$ 4,648	$MSE = \frac{SSE}{21} = 221.3$	
Total	$rc - 1 = 31$	$SSTO =$ 32,464,015		

Multiple Comparisons of Factor-Level Means

The Tukey procedure for constructing several interval estimates at a collective level of confidence applies here. It is modified somewhat from the two-factor counterpart, since pairwise differences are estimated using *column* means ($\mu_{.j}$'s) only. (There is only one alpha, and the Scheffé method for estimating contrasts in the μ_{ij}'s does not apply.) Adapting the subscript notation, and changing the denominator parameter for q, we get the following expression.

TUKEY INTERVAL ESTIMATES FOR DIFFERENCES IN TREATMENT MEANS UNDER RANDOMIZED BLOCK DESIGN AT THE 100(1 − α)% COLLECTIVE CONFIDENCE LEVEL

$$\mu_{.j} - \mu_{.k} = (\bar{X}_{.j} - \bar{X}_{.k}) \pm q_{1-\alpha}\sqrt{\frac{MSE}{rc}}$$

Parameters for q: numerator c, denominator $(r - 1)(c - 1)$

The results of the preceding example will illustrate this procedure. Using the sample data from Tables 14-14 and 14-15, we make all pairwise comparisons of each procedure for solving quadratic assignment problems. At a collective confidence level of 95%, with q parameters of 4 for the numerator and $(8 - 1)(4 - 1) = 21$ for the denominator, Appendix Table L provides the critical value $q_{.95} = 3.95$ (with interpolation). The first estimate is the difference $\mu_{.2} - \mu_{.1}$:

$$\begin{aligned}\mu_{.2} - \mu_{.1} &= (\bar{X}_{.2} - \bar{X}_{.1}) \pm 3.95\sqrt{\frac{221.3}{8(4)}}\\ &= (\$712.625 - 690.75) \pm 10.39\\ &= \$21.875 \pm 10.39\end{aligned}$$

or

$$\$11.485 \le \mu_{.2} - \mu_{.1} \le \$32.265 \qquad \text{(significant)}$$

The five remaining pairwise differences are estimated by intervals with the same width:

$$\begin{aligned}-\$4.265 &\le \mu_{.3} - \mu_{.1} \le \$16.515\\ -\$5.140 &\le \mu_{.4} - \mu_{.1} \le \$15.640\\ -\$26.140 &\le \mu_{.3} - \mu_{.2} \le -\$5.360 \qquad \text{(significant)}\\ -\$27.015 &\le \mu_{.4} - \mu_{.2} \le -\$6.235 \qquad \text{(significant)}\\ -\$11.265 &\le \mu_{.4} - \mu_{.3} \le \$9.515\end{aligned}$$

We may conclude from this that mean cost under the CRAFT procedure is significantly higher than mean costs under the three other procedures, but those procedures do not differ significantly from each other in mean cost.

Problems

14-14 A randomized block design is being employed to test the null hypothesis that mean responses are identical under four treatments. Using six levels for the blocking variable, the following data are obtained:

$$SSTR = 315 \qquad SSBL = 723 \qquad SSTO = 1{,}457$$

(a) Construct the ANOVA table.
(b) Should the null hypothesis of identical treatment population means be accepted or rejected at the 1% significance level?

14-15 Suppose that the sample data in each row of Problem 13-6 represent observed yields using raw ingredients from a different supplier. Using supplier as a blocking variable, complete the following:
(a) Construct the ANOVA table.
(b) Compare the value of *SSE* to the one obtained in Problem 13-6. Does the blocking variable appear promising in reducing unexplained variation in yield?
(c) At the $\alpha = .01$ significance level, should the null hypothesis of identical mean yield under each pressure setting be accepted or rejected?

14-16 Refer to the bacterial leaching results on pages 612–615.
(a) With a collective confidence level of 95%, construct Tukey interval estimates for the difference in mean copper yields between (1) strains 1 and 2, (2) strains 1 and 3, and (3) strains 2 and 3.
(b) Which strain(s) induce the significantly higher yield for each pair?
(c) Which strain(s) do the data suggest ought to be used in order to maximize yield?

14-17 A quality-assurance engineer evaluates three different test stands to assess the characteristics of printed circuit boards. In conducting the test, he uses a different one of four circuit types for successive sample boards tested on the three stands. Prior to testing, each sample unit has been carefully prepared, with bugs and anomalies added. The engineer's response variable is the percentage of known defects detected. The following data apply:

	Percentage of Known Defects Detected		
Circuit	**(1) Stand *W***	**(2) Stand *Y***	**(3) Stand *Z***
1	92%	87%	78%
2	76	78	61
3	99	95	89
4	98	92	88

At the 5% significance level, what should the engineer conclude regarding the mean percentages of known defects detected?

14-18 Refer to the information in Problem 13-10. Suppose that each row represents a different instructor, with the three scores representing a separate class taught by that instructor using a different syllabus. Treating instructor as the blocking variable, complete the following:
(a) Construct the ANOVA table.
(b) Compare the value of *SSE* to the one obtained in Problem 13-10. Does the blocking variable appear promising in reducing unexplained variation in score?

(c) At the $\alpha = .01$ significance level, should the null hypothesis of identical mean examination score under each laboratory syllabus be accepted or rejected?

14-19 Refer to the information in Problem 13-11. Suppose that each row represents a different oil field, with the four increased daily yields representing a different well within the same field. Using oil field as the blocking variable, complete the following:

(a) Construct the ANOVA table.

(b) Compare the value of *SSE* to the one obtained in Problem 13-11. Does the blocking variable appear promising in reducing unexplained variation in yield increase?

(c) At the $\alpha = .01$ significance level, should the null hypothesis of identical mean daily increase in oil production under each method of secondary recovery be accepted or rejected?

14-20 Referring to Problem 14-17:

(a) With a collective confidence level of 95%, construct Tukey interval estimates for the difference in mean percentage of known defects detected between (1) stands *W* and *Y*, (2) stands *W* and *Z*, and (3) stands *Y* and *Z*.

(b) Which stand(s) provide the significantly higher mean detection percentage for each pair?

(c) Which stand(s) do the data suggest ought to be used in order to optimize detection of defects?

14-21 The partial SAS computer printout in Figure 14-9 was obtained by an industrial engineer studying the level of cross traffic (response, in item-feet) for three facility layouts (treatments). As her blocking variable she used product, four different types.

At the 5% significance level, can the engineer conclude that the mean cross-traffic levels differ?

Figure 14-9 *SAS printout for investigation of facility layout.*

```
                         FACILITY LAYOUT

                  Analysis of Variance Procedure

Dependent Variable: CROSS TRAFFIC

                                     Sum of             Mean
 Source                   DF        Squares           Square       F Value

Model                      5   234555.00000     46911.00000          5.76

Error                      6    48885.00000      8147.50000

Corrected Total           11   283440.00000

            R-Square             C.V.        Root MSE    CROSS TRAFFIC Mean

            0.827530          12.0031       0.2635000            752.000000


Source                    DF       Anova SS     Mean Square       F Value

BLOCK                      3     178444.000      59481.3333          7.30
TRTMENT                    2      56111.000      28055.5000          3.44
```

14-4 The Latin-Square Randomized Block Design

The analysis-of-variance procedure can extend to three or more factors. We now consider a three-factor analysis involving one treatment factor and two blocking variables. This popular and efficient procedure uses the **Latin-square design**.

Illustration: *Evaluating Computer Languages*

To illustrate the Latin-square design, we use a computer scientist's evaluation. He wishes to compare four versions of a language (the treatment variable) in terms of percentage improvement in running time (the response variable) over what has been achieved by the prototype language. Other factors than the language will affect running time. Two of these—job application and hardware configuration—were judged to be so significant that each is treated as a blocking variable.

Four categories are important for each. These are shown in Table 14-16, where there is a column for each application and a row for each hardware configuration. There are $4^2 = 16$ distinct application-hardware combinations. Each

Table 14-16 *Experimental Layout and Sample Results for Computer Language Evaluation Under Latin-Square Design*

Row Blocking Variable (Hardware)	Column Blocking Variable (Application)				Row Sample Mean
	(1) Accounting	(2) Graphical	(3) Simulation	(4) Technical	
(1) Mainframe Direct	*A* 18%	*B* 20%	*C* 24%	*D* 22%	$\bar{X}_{1.} = 21.0$
(2) Mainframe Remote	*B* 18	*A* 18	*D* 22	*C* 22	$\bar{X}_{2.} = 20.0$
(3) Minicomputer Alone	*C* 17	*D* 17	*A* 17	*B* 17	$\bar{X}_{3.} = 17.0$
(4) Minicomputer Network	*D* 19	*C* 19	*B* 17	*A* 17	$\bar{X}_{4.} = 18.0$
Column Sample Mean	$\bar{X}_{.1} = 18.0$	$\bar{X}_{.2} = 18.5$	$\bar{X}_{.3} = 20.0$	$\bar{X}_{.4} = 19.5$	$\bar{\bar{X}}_{..} = 19.0$
Treatment: Language Sample Mean	*A* $\bar{X}_A = 17.5$	*B* $\bar{X}_B = 18.0$	*C* $\bar{X}_C = 20.5$	*D* $\bar{X}_D = 20.0$	

of these is represented by a cell in the layout and can be considered as a block. A complete design would require every treatment language to be evaluated under each block combination. With no replication, this would involve $4^3 = 64$ sample test run observations.

Since it can be quite expensive or time-consuming to do a complete evaluation, the computer scientist elected to use an incomplete design. The Latin-square procedure limits the number of treatments to the number of blocks, and thus is an efficient method for minimizing sample size. Under this design, only 16 instead of 64 test runs would be required for the computer language investigation.

Each treatment is represented by a Latin letter. The letter in each cell indicates which treatment was applied for that block. Notice that each letter appears just once in each column and in each row. This means that each treatment is applied just once for each level of the two blocking variables.

The term *Latin-square* refers to any similar arrangement of letters. The following are examples:

$$\begin{matrix} B & A & D & C \\ D & C & B & A \\ A & B & C & D \\ C & D & A & B \end{matrix} \qquad \begin{matrix} A & B & C & D & E \\ B & A & E & C & D \\ C & D & A & E & B \\ D & E & B & A & C \\ E & C & D & B & A \end{matrix} \qquad \begin{matrix} A & B & C \\ B & C & A \\ C & A & B \end{matrix}$$

The sample column means apply to levels for the application blocking variable, and the sample row means for the hardware blocking variable. These are denoted as $\bar{X}_{.j}$ or $\bar{X}_{i.}$, as in the earlier designs. The treatment sample means are computed by summing the observed response in each cell with the same treatment (letter) and dividing by the number of observations in that category. These are denoted as $\bar{X}_A$, $\bar{X}_B$, and so on. Values for all these means are provided in Table 14-16.

As with the preceding analysis-of-variance procedures, the test statistic is determined first by finding the sums of squares for various groupings of sample data. Since three variables affect response, we must slightly modify the earlier notation. A sum of squares is computed for each blocking variable, denoted by *SSROW* (based on row means) and *SSCOL* (from column means). The treatment sum of squares is found, as before, using the deviations of individual treatment means from the grand mean.

The ANOVA table has the generalized form in Table 14-17, where the expressions are given for the sums of squares, mean squares, and the numbers of degrees of freedom. In Latin-square designs the number of rows and columns is the same, so that $r = c$, which is also equal to the number of treatments.

The ANOVA table for the computer scientist's investigation is shown in Table 14-18. The test statistic is computed in the usual manner. For the present illustration we have

$$F = \frac{\text{Variance explained by treatments}}{\text{Unexplained variance}} = 13.0$$

Table 14-17 *Three-Factor Randomized Block Latin-Square Design: Expressions and General Form of the ANOVA Table*

Variation	Degrees of Freedom	Sum of Squares	Mean Square	F
Explained by Treatments (between letters)	$r - 1$	$SSTR = r \sum_{k=A,B,\ldots} (\bar{X}_k - \bar{\bar{X}}_{..})^2$	$MSTR = \dfrac{SSTR}{r-1}$	$\dfrac{MSTR}{MSE}$
Explained by Row Blocks (between rows)	$r - 1$	$SSROW = r \sum_{i=1}^{r} (\bar{X}_{i\cdot} - \bar{\bar{X}}_{..})^2$	$MSROW = \dfrac{SSROW}{r-1}$	
Explained by Column Blocks (between columns)	$r - 1$	$SSCOL = r \sum_{j=1}^{r} (\bar{X}_{\cdot j} - \bar{\bar{X}}_{..})^2$	$MSCOL = \dfrac{SSCOL}{r-1}$	
Error or Unexplained (residual)	$(r-1)(r-2)$	$SSE = SSTO - SSROW - SSCOL - SSTR$	$MSE = \dfrac{SSE}{(r-1)(r-2)}$	
Total	$r^2 - 1$	$SSTO = \sum_{i=1}^{r} \sum_{j=1}^{r} (X_{ij} - \bar{\bar{X}}_{..})^2$		

For the computer language investigation the degrees of freedom are $r - 1 = 4 - 1 = 3$ for the numerator and $(r - 1)(r - 2) = 3(2) = 6$ for the denominator. The computer scientist chose a 1% significance level, so that from Appendix Table J the critical value is $F_{.01} = 9.78$. Since the computed value $F = 13.0$ exceeds 9.78, the computer scientist must *reject* his null hypothesis that the mean percentage reduction in processing time is the same under all four versions of the language.

Collective inferences may be made using differences of the form $\mu_B - \mu_A$. The Tukey or Bonferroni procedures of Section 14-2 may be adapted for this purpose.

The primary advantage of the Latin-square design is that it provides a three-factor analysis of variance evaluation using a minimum number of observations. But in doing so, it is less discriminating than a complete three-factor design (the layout for which would involve a separate matrix for each treatment). This is reflected by the low number for the degrees of freedom. The Latin-square procedure is also restricted to "square" problems, where the number of categories is the same for each blocking variable and the treatment variable. A more general model would incorporate a rectangular layout, so that $r \neq c$, and the number of treatments could be different from either the number of rows or the number of columns.

Table 14-18 *ANOVA Table for the Computer Language Investigation*

Variation	Degrees of Freedom	Sum of Squares	Mean Square	F
Explained by Treatments (between languages)	$r - 1 = 3$	$SSTR = 26$	$MSTR = \frac{26}{3} = 8.667$	$\frac{MSTR}{MSE} = \frac{8.667}{.667} = 13.0$
Explained by Row Blocks (between hardware)	$r - 1 = 3$	$SSROW = 40$	$MSROW = \frac{40}{3} = 13.33$	
Explained by Column Blocks (between applications)	$r - 1 = 3$	$SSCOL = 10$	$MSCOL = \frac{10}{3} = 3.33$	
Error or Unexplained (residual)	$(r - 1)(r - 2) = 6$	$SSE = 4$	$MSE = \frac{4}{6} = .667$	
Total	$r^2 - 1 = 15$	$SSTO = 80$		

It is also possible to expand the general two-factor procedure to three factors. In this case, the experimental layout would be three-dimensional, with each cell representing a combination of levels for three factors and having n observations per cell. (An application is left as an exercise.)

Problems

14-22 Indicate whether each of the following arrangements constitutes a Latin square.

(a)

A B D C
B A C D
C D A B

(b)

B D A E C
C B E A D
A C B D E
E A D C B
D E C B A

(c)

C B A D
D A B C
A C D B
C D B A

(d)

A B C D E F
B A D C F E
C D E F A B
D E F A B C
E F A B C D
F C B E D A

(e)

W X Y Z
Z W X Y
X Y W Z
Y Z X W

(f)

B A D F C E
A B C D E F
E F A B C D
F C B E D A
D E F A B C
C D E F A B

(g)

P Q R S T
Q R S T P
R S T P Q
S T P Q R

(h)

F E B C D A
D C F A B E
B A D E F C
E F A B C D
A B C D E F

14-23 The following sums of squares were obtained using a 5-by-5 Latin-square design:

$$SSTR = 107 \qquad SSCOL = 214 \qquad SSROW = 112 \qquad SSTO = 462$$

Construct the ANOVA table and test the null hypothesis of equal treatment means at the 5% significance level.

14-24 The following sums of squares were obtained using a 6-by-6 Latin-square design:

$$SSTR = 34 \qquad SSCOL = 105 \qquad SSROW = 43 \qquad SSTO = 215$$

Construct the ANOVA table and test the null hypothesis of equal treatment means at the 1% significance level.

14-25 A computer scientist is evaluating four compiler versions (A, B, C, and D) in terms of the compilation time per BASIC language instruction (milliseconds). Several test runs were made using application and source program length as the blocking variables. The following data were obtained:

Row Blocking Variable (Length)	Column Blocking Variable (Application)			
	(1) Accounting	(2) Graphical	(3) Simulation	(4) Technical
(1) Short	A 31	B 43	C 28	D 29
(2) Medium	B 29	A 44	D 29	C 28
(3) Long	C 27	D 38	A 26	B 27
(4) Very Long	D 22	C 37	B 27	A 25

Construct the ANOVA table. What can you conclude regarding the mean compilation time under the respective versions? Use $\alpha = .01$.

14-26 A telephone engineer is evaluating various materials for use in fabricating optical transmission fibers. Using five test fibers, each made from a different material, he conducted an experiment using frequency band and distance as blocking variables. The following data were obtained for the measured signal-to-noise ratios:

Row Blocking Variable (Distance)	Column Blocking Variable (Frequency Band)				
	(1) Low	(2) Low Medium	(3) Medium	(4) Moderately High	(5) High
(1) Tiny	A 41	B 44	C 45	D 45	E 44
(2) Short	B 45	A 46	E 52	C 45	D 46
(3) Middle	C 41	D 44	A 45	E 45	B 41
(4) Long	D 41	E 44	B 45	A 41	C 38
(5) Very Long	E 41	C 39	D 45	B 41	A 38

Construct the ANOVA table. What can you conclude regarding the mean signal-to-noise ratios for the different fiber materials? Use $\alpha = .01$.

14-27 Three-factor analysis of variance under the Latin-square design may be extended to three treatment factors. In that case one treatment variable is represented by the columns, another by the rows, and a third by the letters. Suppose that the following data were obtained for an experiment having a 5-by-5 Latin-square design (with no replication):

$$SSTR_1 = 3{,}924 \text{ (columns)}$$

$$SSTR_2 = 4{,}217 \text{ (rows)}$$

$$SSTR_3 = 8{,}652 \text{ (letters)}$$

$$SSTO = 19{,}384$$

(a) Construct the ANOVA table, including computation of the F statistic for each treatment.

(b) At the 5% significance level, what conclusion should be made regarding the population means under (1) factor 1, (2) factor 2, and (3) factor 3?

Comprehensive Problems

14-28 The researchers in Section 14-2 obtained the data in Table 14-19 for their experiment with the dynamic lot-size algorithm. Here, a different response variable, number of subproblems, applies.

(a) Construct the ANOVA table.

(b) At the $\alpha = .05$ level of significance, indicate whether the following null hypotheses should be accepted or rejected:

(1) H_0: There are no interactions between problem duration and complexity.

(2) H_0: Means for all problem durations are identical.

(3) H_0: Means for all levels of complexity are identical.

14-29 Refer to Problem 14-28 and your answers to that problem.

(a) Plot the cell mean response against levels for factor A, with one set of data points for each level of factor B.

Table 14-19 *Sample Data (Average Number of Subproblems) for Computer Evaluation of Algorithm Tested on Problems with Various Durations and Levels of Complexity*

	Complexity B			
Duration A	(1) Simple	(2) Low	(3) Medium	(4) High
(1) Short	4.2 6.6	3.8 8.2	7.8 14.6	11.0 11.0
(2) Middle	3.0 6.2	4.6 9.8	9.4 62.0	64.6 190.2
(3) Long	6.6 12.0	7.0 7.0	9.0 21.4	11.8 34.2

Source: Erenguc, S. Selcuk, and Suleyman Tufekci, "A Branch-and-Bound Algorithm for a Single-Item Multi-Source Dynamic Lot-Sizing Problem with Capacity Constraints," *IIE Transactions*, March 1987, p. 73–80.

(b) At the 95% collective confidence level, for which cell pairs do the means significantly differ? Do you find anything unusual? How do you explain this?
(c) Construct Tukey 90% confidence interval estimates for the differences in factor-B level means. Indicate which ones are significantly different.

14-30 An engineering consultant to a painting contractor is evaluating three brands of exterior paint in terms of drying time. Using drying condition as the blocking variable, he obtained the following data (in hours):

	Brand of Paint		
Drying Condition	(1) *A*	(2) *B*	(3) *C*
(1) Direct Sun	11.3	12.4	10.9
(2) Shade	14.2	14.3	13.7
(3) Humid	14.9	15.2	14.7
(4) Dry	12.2	11.3	12.5

At the 5% significance level, what should the engineer conclude regarding the mean drying times of the paints?

14-31 A petroleum refinery supervisor wishes to evaluate alternative procedures for removal of sulfur compounds from distillates in an intermediate stage of processing. The following sample data have been obtained for pilot runs under each procedure. The values represent the percentage reduction in sulfur content of the raw distillate processed at that stage. Each test run was made with crude from one of five major origins, which serve as levels for the blocking variable.

	Alternative Sulfur Removal Procedure			
Crude Source	(1) Evaporator	(2) Precipitator	(3) Slurry Reactor	(4) Tertiary Processing
(1) Alaska	80%	91%	88%	94%
(2) Arabia	30	28	10	40
(3) Louisiana	72	65	47	65
(4) Texas	65	57	62	72
(5) Venezuela	48	53	58	71

At the 5% significance level, what should the supervisor conclude regarding the mean percentage reductions in sulfur content?

14-32 A chemical engineer is evaluating alternative temperature settings for one stage of processing a particular product. The three alternatives are low temperature (A), middle temperature (B), and high temperature (C). As blocking variables, three levels each are used for catalyst and solvent. The following batch processing times (minutes) were obtained in pilot testing:

Solvent	Catalyst (1) Absent	(2) Powder	(3) Liquid
(1) Weak	A 30	B 27	C 25
(2) Standard	C 26	A 26	B 23
(3) Potent	B 25	C 22	A 22

At the $\alpha = .01$ significance level, what can the engineer conclude regarding the effect of temperature on mean processing time? If he chooses that temperature setting which minimizes mean processing time, which one would that be?

14-33 A university placement officer is evaluating the job market for engineers in terms of primary job functional area: sales (A), design (B), service (C), installation (D), and production (E). As blocking variables she uses specialization and industry. The following mean income data (in thousands of dollars per year) were obtained from compilation of a survey of recent placements:

Industry	Specialization (1) Mechanical	(2) Electrical	(3) Civil	(4) Industrial	(5) Others
(1) Defense	A $33	B $29	C $27	D $26	E $27
(2) Electronics	B 27	A 33	D 26	E 27	C 24
(3) Machine Tools	C 26	D 27	E 29	A 29	B 23
(4) Chemical	D 25	E 28	B 28	C 25	A 31
(5) Manufacturing	E 28	C 27	A 34	B 28	D 25

Do these data indicate that there are significant differences in mean incomes experienced by engineers in the various specialization areas? Use $\alpha = .01$.

14-34 Referring to Problem 14-33, complete the following:

(a) Construct a single-factor ANOVA table using functional area as the treatment and ignoring blocking variables. Can you reject the null hypothesis of identical population means at the $\alpha = .05$ significance level?

(b) Using industry as the only blocking variable, construct a two-factor ANOVA table for a randomized block design with functional area as the treatment. Can you reject the null hypothesis of identical population means at the $\alpha = .05$ significance level?

(c) Comparing your single-factor and two-factor ANOVA tables to each other and to the three-way table found in Problem 14-33, what conclusions can you draw regarding the advantages of using industry and specialization as the blocking variables?

14-35 Consider the main effects and interaction effects parameters for a two-factor evaluation.

(a) Determine for each type of effect a contrast that expresses the parameter as a linear combination of treatment population (cell) means. Specify exactly the weights that apply for each (i, j) combination.

(b) Using the data from the advertising evaluation in Table 14-2, construct 95% collective confidence interval estimates for (1) A_1, (2) B_1, and (3) $(AB)_{11}$.

(c) Identify which of the effects parameters in (b) differ significantly from zero.

CHAPTER FIFTEEN

Testing for Independence, Randomness, and Goodness-of-Fit

ALL THE STATISTICAL tests encountered so far in this book have involved comparisons. We began in Chapter 9 by comparing means, proportions, or variances to assumed constants. This process was extended in Chapter 12, where two populations were compared. Although this is ordinarily done in terms of population parameters such as means or proportions, the values of two populations might be compared using nonparametric procedures instead.

We now consider a different class of statistical tests that does not involve comparisons. These procedures are concerned with properties of populations that only indirectly involve variable size or the incidence of a particular characteristic. Three major properties are considered.

We first investigate a procedure for testing for *independence* between qualitative variables, or factors. As we have seen, independence is required in many statistical applications. Independence, or the lack of it, may be crucial in evaluations involving such dichotomies as design preference and group identity, engineering discipline and basic aptitudes, or types of mechanical failures and maintenance policies. But under some circumstances it may not be valid to assume independence. Inferential statistics can be extended to determine whether sample evidence supports or refutes the existence of independence.

The second important area involving noncomparative statistics is *randomness.* Along with independence, random selection of observed units is automatically assumed in most sampling experiments. But the sample evidence might be impossible to collect by means of a controlled experiment. This is often the case when the only available observations are those generated outside the investigator's purview. Evaluations of rare phenomena provide good examples. Consider aircraft accidents, structural failures, or the aftereffects of earthquakes. If supported by a statistical test of randomness, even the convenience samples arising from such events might be evaluated using models based on random selection.

The third testing category considered in this chapter involves the nature of the population frequency distribution. A *goodness-of-fit test* can establish whether or not the sample data support a particular distribution. Such a test can establish the normality of a particular class of populations. A finding like this would be especially crucial for applications involving Student t tests, which assume that the underlying population frequency distribution is a normal curve. Goodness-of-fit testing is helpful in validating other models. For example, queueing evaluations often assume exponential interarrival times.

15-1 The Chi-Square Test for Independence

When population variables are qualitative characteristics such as brand preference, sex, type of fabrication, method of data storage, or operating status, important conclusions may sometimes be reached depending on whether or not the variables are independent. An automobile designer who knows that a driver's preferred body style (sporty, conventional, or utilitarian) and driving traits (adventuresome, conservative, or abusive) are dependent can do a better job matching engine options to car models. Similarly, an equipment maintenance plan can consider operating environment (dusty, humid, extremes of temperature) when that variable is dependent with type of breakdown (mechanical, electrical). A highway engineer who knows that severity of automobile accidents (property damage, injury, fatality) depends on location (city streets, rural roads, or highways) can better evaluate where to place budget priorities.

Independence Between Qualitative Variables

Our first exposure to the concept of statistical independence concerned events and their probabilities. Recall that two events A and B are independent if the probability of one of them is unaffected by the occurrence of the other. For example, consider drawing a card at random from a deck of 52 ordinary playing cards. The event "queen" is statistically independent of the suit outcome "heart," which is easy to establish by the following:

$$\Pr[\text{queen} \mid \text{heart}] = \frac{1}{13} = \frac{4}{52} = \Pr[\text{queen}]$$

The concept of statistical independence extends to populations and samples. Consider a collection of units, each of which gives rise to two qualitative observations. These constitute two populations which, in the manner of our earlier designations, are identified as group A and group B. To illustrate, the units might be the students at a particular university. One population gives rise to qualitative variable A, which represents a student's level (undergraduate, graduate). The second population provides qualitative variable B, which we take to be the student's major (engineering, business, physics, music, and so on). If we assume that undergraduate students occur in the same proportion throughout the entire student body as they do among engineering majors, then we may conclude for a randomly chosen person that

$$\Pr[\text{undergraduate} \mid \text{engineer}] = \Pr[\text{undergraduate}]$$

so that the *events* "undergraduate" and "engineer" are statistically independent. And if this is so, then we know by the multiplication law that

$$\Pr[\text{undergraduate } \textit{and} \text{ engineer}] = \Pr[\text{undergraduate}] \times \Pr[\text{engineer}]$$

If a similar fact applies to all attribute combinations of level (variable A) and major (variable B), then the variables exhibit an important property.

Definition

Two qualitative population variables A and B are **independent** if the proportion of the total population having any particular attribute of A is the same as it is in the part of the population having a particular attribute of B, no matter which attributes are considered.

With students in most universities we know that level and major are *not* independent since there will be some programs that are exclusively for undergraduates and others that involve graduates only. And there is no reason to believe that any of the major areas will have undergraduates and graduates in even approximately the same proportion as the others or as the university as a whole.

But our concern is with variables for which we do not have such familiarity. We might believe that independence does or does not exist, but statistical evidence is needed before any firm conclusion can be made. This determination must ordinarily be made from sample evidence, which we know might deviate from the underlying populations. A hypothesis-testing procedure can translate the sample evidence into action while providing desired protection against sampling error.

Testing for Independence

Illustration: *A Malfunctioning Satellite*

Consider a decision by mission controllers regarding scheduling tasks for an orbiting astronomical platform. Because of its complexity and the hostile space environment, the satellite system experiences random malfunctions in its various subsystems. Controllers exert considerable effort to correcting these. One engineer has postulated that any greater or lesser incidence of malfunctions in some subsystems may be explained by differences in the satellite's environment when exposed directly to the sun as opposed to being shaded. If so, various tasks might be scheduled during particular phases of orbit, thereby greatly increasing the satellite's chances for meeting various mission goals.

Table 15-1 *Contingency Table for Actual Sample Results from Satellite Shakedown*

	Position		
Subsystem	**(1) Direct Sunlight**	**(2) Earth's Shadow**	**Total**
(1) Data Transmission	30	4	34
(2) Power	10	6	16
(3) Data Collection	43	11	54
(4) Reception	41	6	47
(5) Mechanical	12	10	22
Total	136	37	173

From the log of the shakedown orbits, during which system testing was done, the number of malfunctions was noted. These are summarized in Table 15-1, which involves two qualitative variables. The position variable is represented by a separate column for each attribute, one for when the platform is directly in line with the sun and the other for when it is in Earth's shadow. There is a different row for each attribute of the second variable, the subsystem that malfunctioned. Five subsystems have been deemed to be potentially affected by the sun: data-transmission, power, data collection, reception, and mechanical. The value within each cell is the number of observations sharing the respective attributes for the two populations. We refer to these cell numbers as **actual frequencies**. The actual frequency in the ith row and jth column is denoted symbolically by f_{ij}. The arrangement of data in Table 15-1 is referred to as a **contingency table** since it accounts for all combinations of factors under investigation.

The data in Table 15-1 are believed to be a representative random sample of the respective populations for all malfunctions that might arise during the lifetime of the satellite system. The engineer wishes to use these data to determine if solar radiation affects the incidence of subsystem malfunction. The null hypothesis being tested is that exposure to the sun has no effect on any subsystem's operating characteristics. In other words, *the null hypothesis is that subsystem and position when a malfunction occurs are two independent variables.*

In order to test this hypothesis it is necessary first to establish the nature of the sample results that would be *expected* if it were true. Those quantities can be compared to the actual ones to see if they differ significantly.

Table 15-2 is the contingency table for the expected results of this study. The total number of malfunctions experienced in each category is the same as actually observed, but the cell entries are different. These are referred to as **expected frequencies**. They are denoted symbolically by $\hat{f}_{ij}$ and appear in boldface in the table. Independence implies that the frequency of elementary units having any particular attribute pair may be found by multiplying the respective frequencies for the

Subsystem	Position		Total
	(1) Direct Sunlight	(2) Earth's Shadow	
(1) Data Transmission	$34 \times \frac{136}{173} = \mathbf{26.73}$	$34 \times \frac{37}{173} = \mathbf{7.27}$	34
(2) Power	$16 \times \frac{136}{173} = \mathbf{12.58}$	$16 \times \frac{37}{173} = \mathbf{3.42}$	16
(3) Data Collection	$54 \times \frac{136}{173} = \mathbf{42.45}$	$54 \times \frac{37}{173} = \mathbf{11.55}$	54
(4) Reception	$47 \times \frac{136}{173} = \mathbf{36.95}$	$47 \times \frac{37}{173} = \mathbf{10.05}$	47
(5) Mechanical	$22 \times \frac{136}{173} = \mathbf{17.29}$	$22 \times \frac{37}{173} = \mathbf{4.71}$	22
Total	136	37	173

Table 15-2 *Contingency Table for Expected Sample Results from Satellite Shakedown under the Null Hypothesis of Independence*

individual attributes and dividing by the total number of observations. The following is therefore used in computing the expected frequencies.

EXPECTED FREQUENCIES

$$\hat{f}_{ij} = \frac{\text{Row } i \text{ total} \times \text{Column } j \text{ total}}{n}$$

where $n = \sum \sum f_{ij}$ is the total number of sample observations.

Comparison of the actual and expected frequencies is accomplished by means of a test statistic. This quantity expresses a decision rule that provides a satisfactory balance between the type I error of rejecting independence when it actually exists and the type II error of accepting independence when that relationship does not actually apply. In actual practice when the sample size has been fixed, as in this satellite illustration, it is possible to control only the type I error probability α. For this purpose the chi-square distribution is used.

CHI-SQUARE STATISTIC FOR TESTING INDEPENDENCE

$$\chi^2 = \frac{1}{n} \sum_i \sum_j \frac{(f_{ij} - \hat{f}_{ij})^2}{\hat{f}_{ij}}$$

Table 15-3 shows the chi-square calculations for the satellite malfunction data. There we find that $\chi^2 = 14.015$.

The possible values of χ^2 range upward from zero. If the deviation between f_{ij} and $\hat{f}_{ij}$ is large for a particular cell, the squared deviation $(f_{ij} - \hat{f}_{ij})^2$ will also be large. To ensure that any differences are not exaggerated by a large number of observations in the respective categories, the squared deviations are divided by $\hat{f}_{ij}$ in computing χ^2. Large $(f_{ij} - \hat{f}_{ij})^2/\hat{f}_{ij}$ ratios occur when the actual and

Table 15-3 *Chi-Square Calculations for Satellite Malfunctions*

(i, j)	Actual Frequency f_{ij}	Expected Frequency $\hat{f}_{ij}$	$f_{ij} - \hat{f}_{ij}$	$(f_{ij} - \hat{f}_{ij})^2$	$\frac{(f_{ij} - \hat{f}_{ij})^2}{\hat{f}_{ij}}$
(1, 1)	30	26.73	3.27	10.6929	.400
(1, 2)	4	7.27	−3.27	10.6929	1.471
(2, 1)	10	12.58	−2.58	6.6564	.529
(2, 2)	6	3.42	2.58	6.6564	1.946
(3, 1)	43	42.45	.55	.3025	.007
(3, 2)	11	11.55	−.55	.3025	.026
(4, 1)	41	36.95	4.05	16.4025	.444
(4, 2)	6	10.05	−4.05	16.4025	1.632
(5, 1)	12	17.29	−5.29	27.9841	1.619
(5, 2)	10	4.71	5.29	27.9841	5.941
	173	173.00	0.00		$\chi^2 = 14.015$

expected results differ considerably, making the size of χ^2 large. The more the actual sample results deviate from what would be expected under the assumption of independence, the larger the value of χ^2 will be.

To determine whether or not the above result significantly refutes the null hypothesis of independence, the computed value for χ^2 must be compared to a critical value. This quantity will depend on the chosen significance level. Recall also that the chi-square distribution is parameterized in terms of the number of degrees of freedom. In testing for independence this quantity represents the minimum number of cells in a contingency table that are "free" to assume any particular level when the row and column totals are fixed. For this purpose we may use the following rule.

RULE FOR THE NUMBER OF DEGREES OF FREEDOM

(Number of rows − 1)(Number of columns − 1)

In the present illustration there are 5 rows and 2 columns, so the number of degrees of freedom is $(5 - 1)(2 - 1) = 4$. Referring to Table 15-2, we see that this means that with the totals fixed there is flexibility to make only 4 cell entries. These then dictate the values for the other cells. For instance, when the entry in cell (1, 1) is determined, the cell (1, 2) value is the difference between the row (1) total and that quantity. Similarly, once the first four entries are made in any column, the fifth cell's value is fixed.

The same basic five hypothesis-testing steps apply to the test for independence. The following pertain to the satellite malfunction illustration.

The Hypothesis-Testing Steps

Step 1. *Formulate the null hypothesis.* The following null hypothesis is being tested:

> H_0: The two qualitative population variables are independent. (Satellite position has no effect on which subsystem malfunctions.)

Step 2. *Select the test procedure and test statistic.* The chi-square statistic is used.

Step 3. *Establish the significance level and the acceptance and rejection regions for the decision rule.* A 1% significance level was chosen for the satellite investigation. This allows for a 1% chance of rejecting the null hypothesis of independence when in fact satellite position has no influence on the incidence of subsystem malfunctions. From Appendix Table H, using $\alpha = .01$ with 4 degrees of freedom, the critical value is $\chi^2_{.01} = 13.277$. The null hypothesis is refuted for large computed chi-square values, and the test is upper-tailed. (This is the case for all chi-square tests for independence.) The acceptance and rejection regions shown in Figure 15-1 apply.

Figure 15-1

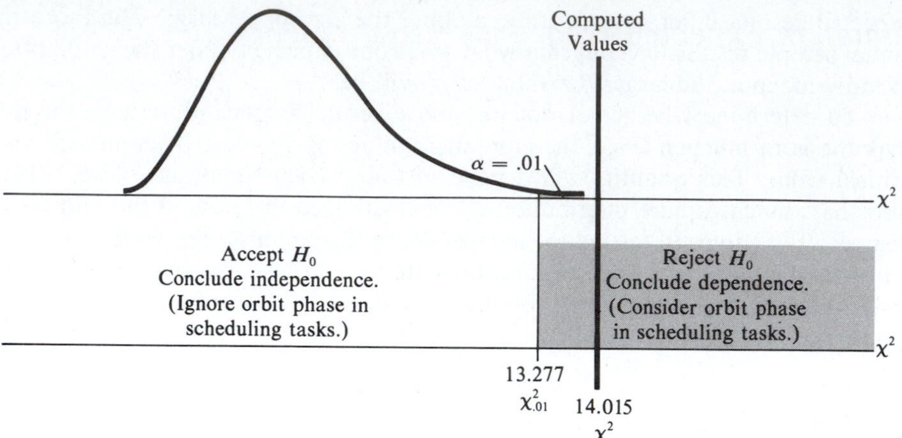

Owing to chance sampling error the value $\chi^2_{.01} = 13.277$ will be exceeded on the average by only 1% of the χ^2 values calculated from similar samples when in fact the population variables are independent.

Step 4. *Collect the sample data and compute the value of the test statistic.* This step has already been completed for the satellite malfunction investigation. This step includes finding the actual and expected frequencies and computing χ^2. We have already found $\chi^2 = 14.015$.

Step 5. *Make the decision.* The engineer's computed value of $\chi^2 = 14.015$ falls in the rejection region. The null hypothesis of independence must be rejected and the engineer must conclude that the two variables are dependent. Thus, the finding is that satellite position when a malfunction takes place does affect which subsystem will malfunction, and orbit phase should be considered in scheduling mission tasks.

Type II Error Considerations

Had the engineer in this satellite malfunction illustration chosen a smaller significance level, say .001, then the critical value from Appendix Table H would have been greater, $\chi^2_{.001} = 18.465$. In that case the reverse conclusion would have been reached, and H_0 would have been accepted. Whenever the null hypothesis is accepted, there is danger that the type II error of accepting a false hypothesis has been committed. (In our illustration, that would be concluding independence when satellite position at the time of malfunction and the particular subsystem involved are dependent variables.) There are many ways for H_0 to be untrue, so finding β probabilities is too difficult a task to undertake here. But, as with the earlier testing procedures, the type II error probability may be kept within reasonable bounds by making the sample size large.

Minimum Expected Frequency Requirement

The true sampling distribution of χ^2 calculated in testing independence is only *approximated* by the chi-square distribution. This approximation is similar to using the normal curve in place of the Student t or the binomial distribution. As with those approximations, satisfactory results may be achieved when the sample size is sufficiently large. The accepted rule of thumb for chi-square testing is that the expected frequency for each cell or category should be at least 5. Should an expected frequency fall below 5, separate categories may be combined into a single group so that two or more rows or columns would be combined before computing χ^2. This will necessitate a corresponding reduction in the number of degrees of freedom.

Computer Applications

Since hand calculation of χ^2 is tedious and prone to error, the chi-square procedure is often done with computer assistance. Many commercially available software packages have segments for reducing data and computing chi-square statistics. Figure 15-2 shows a SAS printout for the satellite shakedown evaluation.

Testing for Equality of Several Proportions

Chapter 12 describes a procedure for comparing two population proportions. When more than two populations are to be compared, the chi-square procedure just described may be adapted. The null hypothesis for such a test is that all the proportions are equal:

$$H_0: \pi_1 = \pi_2 = \cdots = \pi_k$$

To test this, we select independent samples from the respective populations, with the attribute of each observed unit noted. Under H_0, each sample proportion of observations having the desired characteristic is expected to be a common value.

Contingency tables may be constructed for the collective sample outcomes, with a row for each of the complementary attributes and a column for each population. The number of observations in each group provides the actual cell frequencies. The row totals and column totals may be used the same way as before to establish the respective expected frequencies. The chi-square statistic plays the same role as in testing for independence, and χ^2 may be computed in exactly the same way. In establishing the critical value, the number of degrees of freedom is determined from the same rule.

Example:* *Testing Suppliers' Quality

Three vendors have been chosen as the final candidates to be the main supplier for a part used in the assembly of disk drives. A random sample has been selected of the items to be furnished by each vendor, and the following results were

Figure 15-2
SAS printout for satellite shake-down evaluation.

```
DATA IND:
INPUT SUBSYS $ POSITION $ COUNT @@:
CARDS:
DATACOL SUNLIGHT  43           DATACOL SHADOW  11
RECEPTION SUNLIGHT  41         RECEPTION SHADOW  6
MECHANICAL SUNLIGHT  12        MECHANICAL SHADOW  10
TRANS SUNLIGHT  30             TRANS SHADOW  4
POWER SUNLIGHT  10             POWER SHADOW  6

PROC FREQ;
WEIGHT COUNT;
TABLES SUBSYS*POSITION / CHISQ;
```

```
                TABLE OF SUBSYSTEM BY POSITION

SUBSYSTEM    POSITION

Frequency|
 Percent |
 Row Pct |
 Col Pct |SHADOW  |SUNLIGHT|   TOTAL
---------+--------|--------+
DATACOL  |     11 |     43 |     54
         |   6.36 |  24.86 |  31.21
         |  20.37 |  79.63 |
         |  29.73 |  31.62 |
---------+--------+--------+
MECHANIC |     10 |     12 |     22
         |   5.78 |   6.94 |  12.72
         |  45.45 |  54.55 |
         |  27.03 |   8.82 |
---------+--------+--------+
POWER    |      6 |     10 |     16
         |   3.46 |   5.78 |   9.25
         |  37.50 |  62.50 |
         |  16.22 |   7.35 |
---------+--------+--------+
RECEPTIO |      6 |     41 |     47
         |   3.47 |  23.70 |  27.17
         |  12.77 |  87.23 |
         |  16.22 |  30.15 |
---------+--------+--------+
TRANS    |      4 |     30 |     34
         |   2.31 |  17.34 |  19.65
         |  11.76 |  88.24 |
         |  10.81 |  22.06 |
---------+--------+--------+
TOTAL          37      136      173
            21.39    78.61   100.00

      STATISTICS FOR TABLE OF SUBSYSTEM BY POSITION

STATISTIC                          DF      VALUE
-------------------------------------------------
CHI-SQUARE                          4     14.015

Sample Size = 173
```

obtained:

	Vendor 1	Vendor 2	Vendor 3
Number Satisfactory	145	78	94
Number Defective	23	11	7
Number of Parts	168	89	101
Proportion Defective	$P_1 = .137$	$P_2 = .124$	$P_3 = .069$

The total number of satisfactories is 317, the combined number of defectives is 41, and the total number of observations is 358. The contingency table for the expected frequencies is as follows:

	Supplier			
Quality	Vendor 1	Vendor 2	Vendor 3	Total
Satisfactory	148.76	78.81	89.43	317
Defective	19.24	10.19	11.57	41
Total	168	89	101	358

The computed value of the test statistic is

$$\chi^2 = \frac{(145 - 148.76)^2}{148.76} + \frac{(78 - 78.81)^2}{78.81} + \cdots + \frac{(7 - 11.57)^2}{11.57} = 2.941$$

The number of degrees of freedom is $(2 - 1)(3 - 1) = 2$. From Appendix Table H, $\chi^2_{.30} = 2.408$ is the largest critical chi-square value less than the computed value. This indicates that $\alpha = .30$ is the lowest significance level at which H_0 can be rejected. The sample evidence that the underlying population proportions differ is not very convincing.

Problems

15-1 A designer is testing to determine if a person's sex is independent of color preference. For each of the following situations the test statistic has been computed. Determine (1) the number of degrees of freedom, (2) the critical chi-square value, and (3) whether the null hypothesis of independence should be accepted or rejected.

(a) Pink, blue, and tan are considered. $\chi^2 = 8.133$ and $\alpha = .10$.
(b) Olive, rust, ochre, and amber are considered. $\chi^2 = 13.523$ and $\alpha = .05$.
(c) Chartreuse, aquamarine, charcoal, lime, and lavender are considered. $\chi^2 = 9.315$ and $\alpha = .01$.
(d) Burgundy, purple, plum, red, violet, and brown are considered. $\chi^2 = 15.892$ and $\alpha = .05$.

15-2 Several random samples have been collected. In each case independence between variables is to be tested at the indicated significance level using the computed chi-square statistics given below. Determine in each case (1) the number of degrees of freedom, (2) the critical chi-square value, and (3) whether H_0 should be accepted or rejected.

(a) Sex (male, female) versus college education (graduate, nongraduate), with $\alpha = .01$ and $\chi^2 = 5.371$.

(b) Engineering concentration (civil, electrical, industrial, mechanical, other) versus area of employment (government, education, industry), with $\alpha = .05$ and $\chi^2 = 10.835$.

(c) Patents held (0, 1–5, 6–10, >10) versus degrees earned (none, one, two, three or more), with $\alpha = .05$ and $\chi^2 = 12.164$.

(d) Build (small, medium, heavy) versus treadmill endurance (short, long), with $\alpha = .01$ and $\chi^2 = 15.144$.

15-3 An industrial engineer wants to determine if there are any significant differences between plants in terms of compliance with production schedules. A random sampling of orders has been traced, and the following results were obtained:

Schedule Compliance	Plant				Total
	Hong Kong	Seoul	Fresno	Athens	
Early	22	35	0	5	62
On Time	84	55	8	24	171
Late	25	17	22	12	76
Total	131	107	30	41	309

The null hypothesis is that schedule compliance on any given order and plant where placed are independent.

Assuming that the industrial engineer will tolerate only a 1% chance of incorrectly concluding that schedule compliance differs from plant to plant when this is not so, determine what conclusion the engineer should reach.

15-4 The industrial engineer in Problem 15-3 used the same sample orders to determine if there was any significant difference in the plants' fulfillment of reporting requirements. The following results were obtained:

Reporting Result	Plant				Total
	Hong Kong	Seoul	Fresno	Athens	
Incomplete	3	17	11	25	56
In Error	50	27	12	9	98
Late	44	35	3	7	89
Satisfactory	34	28	4	0	66
Total	131	107	30	41	309

The null hypothesis is that reporting fulfillment on any given order and plant where placed are independent.

Assuming that the industrial engineer will tolerate only a 5% chance of incorrectly concluding that reporting fulfillment differs from plant to plant when this is not so, determine what conclusion the engineer should reach. What would be your answer if the significance level were $\alpha = .01$?

15-5 In testing the null hypothesis that the region where an engineer went to school and that person's area of employment are independent, an investigator obtained the following sample results:

	Area of Employment			
Engineering School Region	**Government**	**Industry**	**Education**	**Total**
Northeast	24	76	15	115
Southeast	17	51	9	77
Southwest	32	38	5	75
Pacific	16	55	6	77
Total	89	220	35	344

At the 5% significance level, should the null hypothesis be accepted or rejected?

15-6 In testing the null hypothesis that level of heavy equipment usage and owner's maintenance policy are independent variables, a mechanical engineering student received replies to her questionnaire from a random sample of users. The following summary applies:

	Maintenance Policy			
Equipment Usage	**By Calendar**	**By Hours of Operation**	**As Required**	**Total**
Light	12	8	13	33
Moderate	7	15	22	44
Heavy	3	22	15	40
Total	22	45	50	117

At the 1% significance level, should the student accept or reject H_0?

15-7 An industrial engineer wants to know if the time of day influences the number of correctly assembled units. If she does find that time of assembly affects the error rate, she will investigate the possibility of rescheduling worker break times. The following sample data were obtained for randomly selected units:

	Time of Day			
	(1) Early A.M.	(2) Late A.M.	(3) Early P.M.	(4) Late P.M.
Number correct	52	48	51	47
Number incorrect	3	5	8	11

Using $\alpha = .05$, the engineer wants to test the null hypothesis that the proportions of incorrectly assembled units are identical. What conclusion should she reach?

15-8 A communications engineer wants to determine if the proportion of microwave transmissions encountering noise is affected by the frequency range. The following sample data have been obtained:

	Frequency Range		
	(1) Low	(2) Medium	(3) High
Number of transmissions with noise	42	28	29
Number of transmissions without noise	118	43	29

At the 1% significance level, should the engineer conclude that the noise rate is the same or different for all frequencies?

15-2 Testing for Randomness

This section describes another noncomparative testing procedure concerning the nature of the process by which events are generated. It focuses on finding whether the data obtained exhibit the characteristics of **randomness**. This is an important issue when statistical procedures are applied to data not obtained under the controlled conditions we usually associate with scientific random sampling. Such is often the case when sample observations are generated by single events over a period of time. Those data can be considered random only when the order in which certain attributes or values occur is unaffected by previous events—so that no serial dependencies exist. A test for randomness can establish whether or not the sequence of results obtained is similar to what would be expected in a random sample.

Testing for randomness can be helpful in many engineering applications of statistics. For instance, to control production quality it may be important to know not only the rate at which defectives have been occurring but also if bad items are being produced in clusters. A smoothly operating production system is typified by random occurrence of defectives over time. Should the poor items be batched into large groups, this could be evidence that equipment or operators are behaving erratically. A test for randomness could provide an early detection of such a production problem. A similar test might detect an analogous problem in a communications system, where under normal circumstances interruptions should occur randomly. A contrary finding from a sample investigation might indicate a malfunctioning subsystem. Similarly, a test for randomness could establish that the pattern of bad test samples from a chemical process has some explainable cause, such as contaminated feedstocks or equipment in need of maintenance.

Various approaches can be used to determine if a sequence of events exhibits randomness. These are based on a binary classification of events into complementary attributes, such as defective versus satisfactory, pass versus fail, or operable versus inoperable. The sequential pattern for the occurrence of these events may then be described in terms of **runs,** an uninterrupted string of events of the same type. A popular procedure for testing randomness is based on the number of runs obtained in the sample.

The Number-of-Runs Test

To illustrate the concept of number-of-runs testing, consider the sex of new engineering graduates hired by a large aerospace firm. Suppose that over a two-week period 20 persons were hired, half men and half women. Categorizing each hire by the employee's sex, the sample outcome could be expressed in terms of a string of letters arranged in the same chronological sequence as time of appointment. One hypothetical sample result is

Sample A: $\underbrace{a}\ \underbrace{b\,b\,b}\ \underbrace{a\,a\,a}\ \underbrace{b\,b\,b}\ \underbrace{a\,a}\ \underbrace{b}\ \underbrace{a\,a}\ \underbrace{b\,b\,b}\ \underbrace{a\,a\,a}$

$$a = \text{Men} \qquad b = \text{Women}$$

The braces indicate runs of a particular sex category. In this sequence there are five a runs (men) and four b runs (women).

The number of runs, either of a single category or altogether, may be used to express the degree of randomness. If the events do indeed occur randomly, then we would be unlikely to obtain a sample with too few runs or with too many runs. For example, the two-run result in

Sample B: $a\,a\,a\,a\,a\,a\,a\,a\,a\,a\,b\,b\,b\,b\,b\,b\,b\,b\,b\,b$

would be unlikely if persons were actually hired without consideration of their sex. It might be explained by a deliberate personnel policy of hiring enough women to comply with government sex discrimination statutes. That policy might explain the sample B result if after a large group of men were hired, a "catch-up" phase occurred, during which an equal number of women were appointed. Similarly, the 20-run result in

Sample C: $a\,b\,a\,b\,a\,b\,a\,b\,a\,b\,a\,b\,a\,b\,a\,b\,a\,b\,a\,b$

is also evidence of lack of randomness. A hiring policy of continuously balancing employee sex by alternate appointment could explain this sequence.

The number-of-runs test evaluates the null hypothesis that the sample was random. Therefore, each sequence position has the same prior chance of being assigned an a as any other. (However, there may be a higher proportion of one event type throughout the sample and the population. For instance, there could be 10% or 30% women, and the test can still support randomness.)

The test statistic employed here is based on the number of runs of type a:

$$R_a = \text{Number of type } a \text{ runs}$$

(Nearly identical results will be achieved by using instead the number of b runs or the number of runs of both types.) In this illustration we have $R_a = 5$ in sample A, $R_a = 1$ in sample B, and $R_a = 10$ in sample C.

The sampling distribution for R_a may be determined for any fixed number of events of the two types, denoted by n_a and n_b, respectively. R_a has the hypergeometric distribution, with parameters $n = n_a$, $N = n_a + n_b$, and $\pi = (n_b + 1)/N$. (It will be left as an exercise for the reader to show this.) Because hypergeometric probabilities are cumbersome, the normal approximation is employed instead. For this purpose, a normal deviate may then be used as the test statistic.

NORMAL DEVIATE FOR THE NUMBER-OF-RUNS TEST

$$z = \frac{R_a - \dfrac{n_a(n_b + 1)}{n_a + n_b}}{\sqrt{\dfrac{n_a(n_b + 1)(n_a - 1)}{(n_a + n_b)^2}\left(\dfrac{n_b}{n_a + n_b - 1}\right)}}$$

(For simplicity, the continuity correction has been ignored.)

Applying this to the hypothetical hiring samples, we have:

Sample A:

$$z = \frac{5 - \dfrac{10(10 + 1)}{10 + 10}}{\sqrt{\dfrac{10(10 + 1)(10 - 1)}{(10 + 10)^2}\left(\dfrac{10}{10 + 10 - 1}\right)}} = \frac{5 - 5.5}{1.1413} = -.44$$

Sample B: $z = \dfrac{1 - 5.5}{1.1413} = -3.94$

Sample C: $z = \dfrac{10 - 5.5}{1.1413} = 3.94$

These results may be tested at the 5% significance level. Since an unusually small or large number of runs would refute the null hypothesis of randomness, the procedure involves *two-sided* tests. Using $z_{.025} = 1.96$, the acceptance and rejection regions shown in Figure 15-3 apply.

The computed normal deviate of $z = -.44$ applies for sample A, indicating that the outcome for that sample is consistent with randomness and the null hy-

Figure 15-3

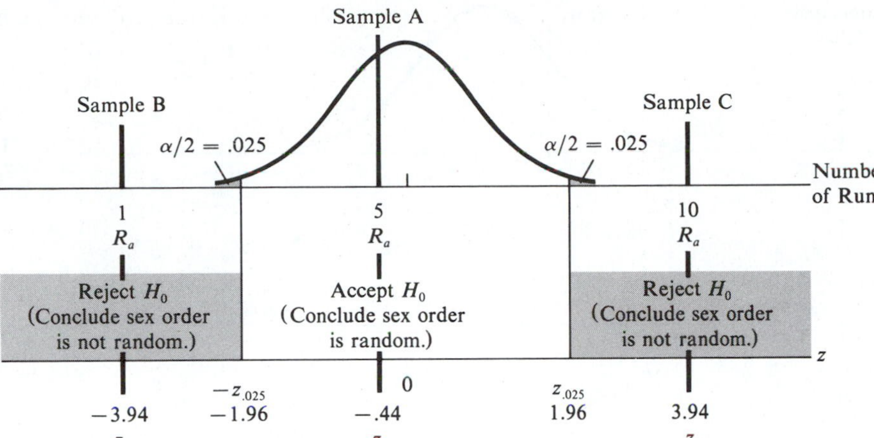

pothesis is accepted. But the null hypothesis of randomness must be rejected for sample B ($z = -3.94$), which has too few category-a runs, and for sample C ($z = 3.94$), which has too many.

***Illustration:* Randomness of Microwave Relay Noise**

To further illustrate the number-of-runs test, consider an investigation of noise on channels connecting two microwave relay stations. As part of his evaluation, an engineer recorded a test signal over a one-hour time span, and bursts of noise were identified. Breaking the time log into 360 10-second intervals, the engineer found noise in 38 of the periods, with no significant disturbances in the rest. The noise bursts were widely scattered, there being 25 runs of such disturbances.

The engineer believed that some systematic source of interference, such as air-control radar, might explain any nonrandom pattern of noise. He performed a number-of-runs test on the above results. The following hypothesis-testing steps apply.

Step 1. *Formulate the null hypothesis.* The following null hypothesis was tested:

H_0: The order in which noise bursts occur is random.

Step 2. *Select the test procedure and test statistic.* The test statistic is based on the number of runs of noisy time intervals, R_a. From this quantity, the normal deviate was determined and used as the basis for action.

Step 3. *Establish the significance level and the acceptance and rejection regions for the decision rule.* A 1% significance level was chosen by the investigator. This allowed for a 1% chance of rejecting the null hypothesis of randomness when in fact noise bursts do occur randomly. From Appendix Table E, with $\alpha = .01$, the critical normal deviate value is $z_{.005} = 2.57$. The acceptance and rejection regions in Figure 15-4 apply.

Figure 15-4

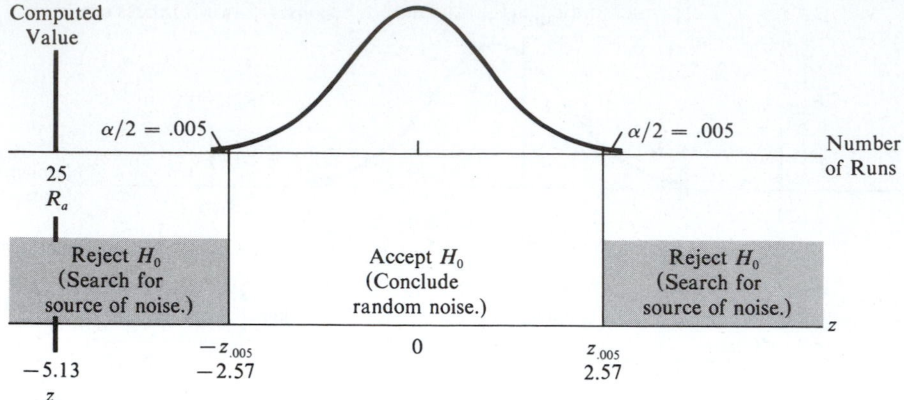

Step 4. *Collect the sample data and compute the value of the test statistic.* The sample results provide $n_a = 38$ 10-second intervals involving bursts of noise and $n_b = 360 - 38 = 322$ quiet intervals. The number of runs in noise bursts is $R_a = 25$. The computed value for the normal deviate is thus

$$z = \frac{25 - \dfrac{38(322 + 1)}{38 + 322}}{\sqrt{\dfrac{38(322 + 1)(38 - 1)}{(38 + 322)^2}\left(\dfrac{322}{38 + 322 - 1}\right)}} = -5.13$$

Step 5. *Make the decision.* The engineer's computed value of $z = -5.13$ falls in the lower rejection region. The null hypothesis of randomness was rejected and the engineer began a search for sources of noise. Once the source is located, shielding and other protective measures can be undertaken to minimize disturbances in future operations of the microwave relay.

Example:* *Testing 1970 Draft Lottery for Randomness*

The number-of-runs test was applied to the 1970 draft lottery to determine if it might support complaints that the procedure was not random. Lawsuits had been initiated on behalf of persons born in December, a month containing 26 high-priority birthdates. The plaintiffs complained that the capsules selected in the lottery had not been properly mixed, and that these were layered in the mixing barrel in reverse sequence, with December dates near the top and January dates on the bottom. (January received only 12 high-priority birthdates.)

With 366 dates altogether, each of which was assigned a priority number according to the sequence in which its capsule was drawn, the author placed each date into the vulnerable set (high-priority *a*'s) or the safe set (low-priority *b*'s). With groups of equal size, the respective sample sizes were $n_a = n_b = 183$. Proceeding chronologically through the calendar, the number of runs of vulnerable dates was determined to be $R_a = 86$. This provided a normal deviate value of $z = -1.25$, which indicates that the null hypothesis of randomness could be rejected at a significance level no lower than $\alpha = .21$.

Most statisticians would insist on a much lower level for α in order to reject any null hypothesis. They would accept the null hypothesis of randomness for the 1970 draft lottery based on the *results obtained*. The actual data suggest that a considerable element of randomness was still imparted by persons grabbing capsules from the barrel. Paradoxically, a majority of those same statisticians would criticize the *procedure used* as being highly nonrandom because capsules were layered by month.

* For a detailed discussion see Lawrence L. Lapin, *Statistics for Modern Business Decision*, 4th. ed., Harcourt Brace Jovanovich, Inc., 1987, pp. 538–542.

Other Tests for Randomness

We have considered only one type of test for randomness based on the *number* of runs. Similar tests may be performed using *length* of runs. Both procedures are based on the *sequence* of outcomes. Other tests might be used which consider the *frequency* of outcomes or their **serial correlation** (the tendency for one value or category to occur more or less frequently after another one than after some other category).

The "random" numbers used in Monte Carlo simulations conducted with digital computers nicely illustrate the limitations of runs testing. These numbers are called **pseudorandom numbers** since they are generated by an algorithm using one seed value. A number-of-runs test would accept as random the sequence of digits

2 7 7 2 2 2 2 7 7 7 2 2 7 7 2 2 2 7 7 7

even though such numbers would be inappropriate in generating cases where all digits between 0 and 9 are assumed to be equally likely. The number-of-runs test does not detect the frequency with which each digit occurs. A goodness-of-fit test (see Section 15-3) can establish whether or not the digits in a pseudorandom number generation scheme provide equal frequencies for all integers between 0 and 9. That test would reject the above sequence as refuting that property.

A test of serial correlation when applied to the sequence

1 2 3 7 5 4 4 0 1 2 0 0 3 5 6 7 6 5 9 8
9 1 2 4 9 8 7 7 6 3 4 2 1 2 6 4 9 7 8 2

might detect patterns that would not exist in truly random numbers, except by remote chance. In this sequence every 1 is followed by a 2. Both the number-of-runs and goodness-of-fit tests would ignore such unusual outcomes, and either procedure might accept the above as suitable for use as random numbers.

A detailed discussion of the various tests for randomness and applications to pseudorandom number evaluation may be found in some of the references in the Bibliography.

Problems

15-9 The following sequences represent the order in which 10 undergraduate (U) and 10 graduate (G) students might be hired by a university library:

(1) G G U G U G U U G G U G U G U U U G G U
(2) U G G U G U U G U G U G U G G U G U U G
(3) G G G G G G U U U U U G G G G U U U U U
(4) G U G U G U G U U G G U U G U G G U G U

(a) At the $\alpha = .05$ significance level, identify the appropriate acceptance and rejection regions for testing the null hypothesis that the class standing of the students in each sequence is randomly determined. Let R_a represent the number of runs of undergraduates.

(b) Applying your decision rule from (a), indicate whether a null hypothesis of randomness must be accepted or rejected for each sequence.

15-10 A die cube is rolled 18 times, and 10 even (E) faces and 8 odd (O) faces resulted. Letting R_a represent the number of even runs, determine whether the null hypothesis of random rolling should be accepted or rejected for each of the following sequences. Use $\alpha = .05$.

(a) E O O O E E O O E E O O E E O E E E
(b) E E E O O O O E E E E E E E O O O O
(c) E E O E E O E O O E E O E O E O E O
(d) E E E O O E E O O O E E E E O E O O

15-11 The engineer in the illustration who studied microwaves investigated another relay linkage. In a sequence of 50 contiguous 10-second transmission segments it was determined whether the signal was interrupted by noise (N) or was clear (C). The following data were obtained:

N C C C C N C C C N N C C C N C C C C C N N N C C
N N C C C C N C C C N N C C C C N N C C C C C N C

Letting R_a represent the number of runs of noisy segments, should the null hypothesis of randomness be accepted or rejected? Use $\alpha = .01$.

15-12 The successive items from a production process are categorized as satisfactory (a) or defective (b). The following results were obtained from a run of 100 items:

b b b b b a a a b b b b b b b b a a a a b a a a a
a a a a a a a b b b a a a a a a a a a a a a a a a
a a b b b b b a a a a a a a a a a a a a a a a a a
a a b b b b b a a a a a a a a a a a a a a a a a a

(a) At the 5% significance level, should the null hypothesis that defectives occur randomly be accepted or rejected?

(b) Do the above results indicate that some systematic influence might be causing defectives to occur in clusters? Explain.

15-13 The following numbers were generated in the listed sequence by an electromechanical device:

0 1 2 3 6 7 8 2 3 4 5 5 6 7 8 2 3 9 9 8 7 6 5 3 3
8 7 6 5 4 3 2 0 0 1 2 3 4 4 5 6 5 4 3 2 4 4 0 1 3
3 3 4 5 8 7 6 6 7 8 9 2 1 3 4 8 7 6 5 4 3 2 1 0 9
8 7 6 4 3 2 1 8 7 2 3 3 3 4 4 6 7 8 4 3 2 5 6 7 8

(a) Let R_a denote the number of runs of integers 0 through 4, inclusively. Should the null hypothesis of randomness with respect to magnitudes be accepted or rejected? Use $\alpha = .01$.

(b) Let R_a denote the number of runs of even values (with 0 an even integer). Should the null hypothesis of randomness with respect to evenness be accepted or rejected? Use $\alpha = .01$.

(c) The answers to (a) and (b) might lead to opposite conclusions regarding randomness. Explain why this might be so.

15-14 Toss a coin 50 times, recording whether a head (a) or tail (b) occurs for each. Then use the number-of-runs test to determine whether or not you have been tossing in a random fashion. Use $\alpha = .10$.

15-15 A computer science student has devised a pseudorandom number generator based on the hardware status of the computer system at the instant of request for a value. In a string of 100 random decimals, 56 fell at or above (a) .5 and 44 below (b). Testing the null hypothesis of the appearance of randomness, what conclusion should the student draw when $R_a = 27$? Use $\alpha = .05$.

15-16 In 1971 a second draft lottery was conducted in the United States. That time the priority numbers were determined by selection of two capsules, one from a barrel containing birthdates and the second from a bin where each capsule contained one integer from 1 to 365. Designating the lowest 183 dates as vulnerables (a) and the rest safe (b), the value $R_a = 95$ was determined.

(a) The expected value of R_a is $n_a(n_b + 1)/(n_a + n_b)$. Using the normal approximation, determine the probability that as many or more runs of vulnerable dates than the 95 actually found could be achieved if the lottery were repeated, assuming it was random. (Ignore the continuity correction.)

(b) Do the above findings suggest that the 1971 draft lottery randomly assigned priority numbers to birthdates? Explain.

15-3 Testing for Goodness-of-Fit

In this section we introduce the **goodness-of-fit test**, which is used in making inferences regarding the nature of the unknown population frequency distribution. This procedure determines if a particular *shape* for the population frequency curve is consistent with the sample results obtained. It can also establish whether or not sample data have been generated by a process characterized by a particular *probability distribution.* This is accomplished by comparing the sample frequency distribution to the hypothesized population counterpart.

The Importance of Knowing the Distribution

Many models used in statistical applications assume that events occur according to a particular probability distribution. For instance, important queueing results are predicated on specific probability distributions for arrival and service times.

Some reliability models rest on similar assumptions regarding system failure patterns. Several of the more general statistical procedures described in this book make implicit assumptions regarding the nature of the population distribution.

A common assumption of queueing models is that arrivals form a Poisson process over time. Earlier in the book we discussed two important distribution families associated with the Poisson process, the continuous *exponential distribution* and the discrete *Poisson distribution.* Recall that a Poisson process involves no memory, has a constant mean rate over time, and permits a low probability for more than one event occurring in a short span of time. There is no way except by actual observation to determine if these conditions are being met. The goodness-of-fit test can indicate whether or not observed data exhibit a pattern consistent with such generation.

We have seen that the Student t distribution is widely used in making inferences about means when the population standard deviation is unknown. But a theoretical condition is that the population itself have a normal frequency curve. It may therefore be desirable to use a goodness-of-fit test to establish whether or not the *normal distribution* closely fits the population. In the last section we indicated that pseudorandom number generation schemes can be exposed to a variety of tests for the "appearance" of randomness. But it is also important that all digits in the long run occur with equal frequency. A goodness-of-fit test can be employed to determine if a series of random digits are consistent with the *uniform distribution* implied by that condition.

Illustration:* *Arrivals at a Toll Plaza

The goodness-of-fit testing procedure is similar to the test for independence described in Section 15-1. As with that test, two sets of frequencies are compared for each category. In testing for fit, the categories represent either class intervals (when the variable is continuous) or integer values (for discrete variables). A chi-square statistic is used in the same way as the earlier test to determine whether or not the frequencies are significantly different.

To facilitate our discussion, we will use a queueing illustration. As part of a term project, an industrial engineering student is studying the toll-collecting system on one of the bridges over San Francisco Bay. She has collected a variety of sample information to determine parameters and distributions to use in her queueing evaluations. One set of data is shown in Table 15-4, which provides the sample frequency distribution for times between arrival at the toll plaza between 4 and 5 A.M. on weekday mornings.

The class frequencies are the *actual frequencies* and, as before, are denoted by f_i, where i denotes the category (class interval).

The above will be compared to the *expected frequencies*, which, as before, are denoted by $\hat{f}_i$. These correspond to the distribution assumed under the null hypothesis.

The student's queueing models assume that the times between arrivals are exponentially distributed. Under that null hypothesis, the probability that the time

Table 15-4 *Sample Frequency Distribution for Times Between Arriving Cars at a Toll Plaza*

(1) Interval i	(2) Time Between Arrivals (minutes)	(3) Actual Frequency f_i	(4) Interval Midpoint X_i	(5) f_iX_i
1	0–under .5	38	.25	9.50
2	.5–under 1.0	26	.75	19.50
3	1.0–under 1.5	12	1.25	15.00
4	1.5–under 2.0	10	1.75	17.50
5	2.0–under 2.5	8	2.25	18.00
6	2.5–under 3.0	3	2.75	8.25
7	3.0–under 3.5	2	3.25	6.50
8	3.5–under 4.0	1	3.75	3.75
9	≥ 4.0	0	—	—
		100		98.00

$$\bar{X} = 98.00/100 = .9800 = 1/\lambda$$

$$\lambda = 1.02$$

X between any two successive arrivals lies at or below x is

$$\Pr[X \leq x] = 1 - e^{-\lambda x}$$

where λ is the mean rate of arrivals. The true value for λ is unknown and may be estimated from the sample results. In Table 15-4 the sample mean time between arrivals was computed from the grouped data to be $\bar{X} = .98$ minute per car. The inverse of this provides $\lambda = 1.02$ cars per minute. Using that rate, the probability that under .5 minute transpires between two successive arrivals is

$$\Pr[X \leq .5] = 1 - e^{-1.02(.5)} = 1 - .6005 = .3995$$

This quantity may be used to determine the expected frequency for cars arriving in the interval 0–under .5 minute. Out of 100 arrivals, when the exponential distribution applies at the above rate, the expected number of interarrival times falling in that interval is

$$100 \times .3995 = 39.95$$

This quantity applies for the first class interval, so that the expected frequency for that interval is

$$\hat{f}_1 = 39.95$$

The expected frequencies for the remaining class intervals are shown in Table 15-5.

Under the null hypothesis, expected frequencies from the exponential distribution represent the underlying arrival time population. Figure 15-5 shows the corresponding frequency curve. Also shown is the sample histogram, representing the actual frequencies. A perfect fit between the histogram and curve would be

Table 15-5 *Expected Frequencies for Times Between Arrivals Assuming Exponential Distribution*

(1) Interval i	(2) Time Between Arrivals (minutes) x	(3) Exponential Cumulative Probability at Upper Limit $1 - e^{-\lambda x}$	(4) Interval Probability	(5) Expected Frequency [100 × (4)] $\hat{f}_i$
1	0–under .5	.3995	.3995	39.95
2	.5–under 1.0	.6394	.2399	23.99
3	1.0–under 1.5	.7835	.1441	14.41
4	1.5–under 2.0	.8700	.0865	8.65
5	2.0–under 2.5	.9219	.0519	5.19
6	2.5–under 3.0	.9531	.0312	3.12
7	3.0–under 3.5	.9718	.0187	1.87
8	3.5–under 4.0	.9831	.0113	1.13
9	≥ 4.0	1.0000	.0169	1.69
			1.0000	100.00

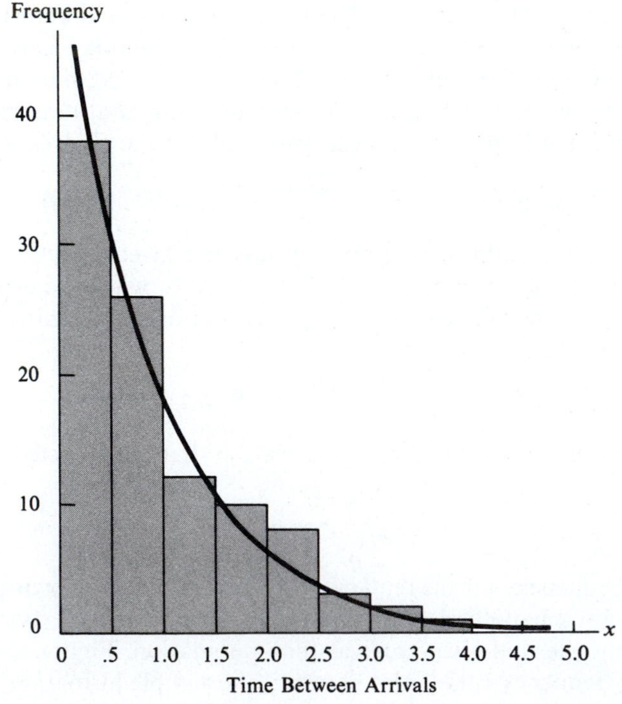

Figure 15-5 *Actual frequency histogram and expected frequency curve for times-between-arrivals at a bridge toll booth.*

Table 15-6 *Chi-Square Calculations for Times Between Arrivals of Cars at a Toll Plaza*

i	Actual Frequency f_i	Expected Frequency $\hat{f}_i$	$f_i - \hat{f}_i$	$(f_i - \hat{f}_i)^2$	$\frac{(f_i - \hat{f}_i)^2}{\hat{f}_i}$
1	38	39.95	−1.95	3.8025	.095
2	26	23.99	2.01	4.0401	.168
3	12	14.41	−2.41	5.8081	.403
4	10	8.65	1.35	1.8225	.211
5	8	5.19	2.81	7.8961	1.521
6	3 }	3.12 }			
7	2 } 6	1.87 } 7.81	−1.81	3.2761	.419
8	1 }	1.13 }			
9	0 }	1.69 }			
	100	100.00	0.00		$\chi^2 = 2.817$

highly unlikely even if the null hypothesis were true. The goodness-of-fit between the actual expected results is summarized in terms of a test statistic.

The test statistic is computed in exactly the same way as in testing for independence. Using single subscripts instead, we calculate the following.

CHI-SQUARE STATISTIC FOR A GOODNESS-OF-FIT TEST

$$\chi^2 = \sum_i \frac{(f_i - \hat{f}_i)^2}{\hat{f}_i}$$

Table 15-6 shows the chi-square calculations for the car arrival times. The value $\chi^2 = 2.817$ is obtained.

The chi-square distribution only approximates the true distribution for χ^2 as computed above. This approximation is close enough for testing goodness-of-fit whenever the expected frequency in any category is at least 5. Generally, the highest (or lowest) categories must be combined so that the resulting expected frequency for the upper (or lower) tail achieves this minimum. For this reason, the last four classes in Table 15-6 have been grouped into a single category, which provides a combined expected frequency of 7.81 and an actual frequency of 6.

Like the test for independence, the number of degrees of freedom is based on the number of frequencies that are free to assume any value, given that all of them must sum to the sample size used. An additional degree of freedom is lost for each distribution parameter estimated from the sample. The following summarizes the rule.

RULE FOR THE NUMBER OF DEGREES OF FREEDOM

Number of categories − Number of parameters estimated − 1

In the present illustration, there are only six categories left after grouping the frequencies, and since λ was estimated from the sample results, the number of degrees of freedom is $6 - 1 - 1 = 4$.

As with earlier testing procedures, the same basic five steps apply.

The Hypothesis-Testing Steps

Step 1. *Formulate the null hypothesis.* The following null hypothesis is being tested:

H_0: The sample data represent an exponential population distribution with mean rate $\lambda = 1.02$.

Step 2. *Select the test procedure and test statistic.* The chi-square statistic is used.

Step 3. *Establish the significance level and the acceptance and rejection regions for the decision rule.* A 5% significance level was chosen by the student investigator. This allows for a 5% chance of rejecting the null hypothesis when in fact the interarrival times have the stipulated exponential distribution. From Appendix Table H, with $\alpha = .05$ with 4 degrees of freedom, the critical value is $\chi^2_{.05} = 9.488$. The null hypothesis is refuted for large computed chi-square values, and the test is upper-tailed. (This is the case for all chi-square tests for goodness-of-fit.) The acceptance and rejection regions in Figure 15-6 apply.

Owing to chance sampling error, the value $\chi^2_{.05} = 9.488$ will be exceeded on the average by only 5% of the χ^2 values calculated from similar samples when in fact the true interarrival time distribution is exponential.

Figure 15-6

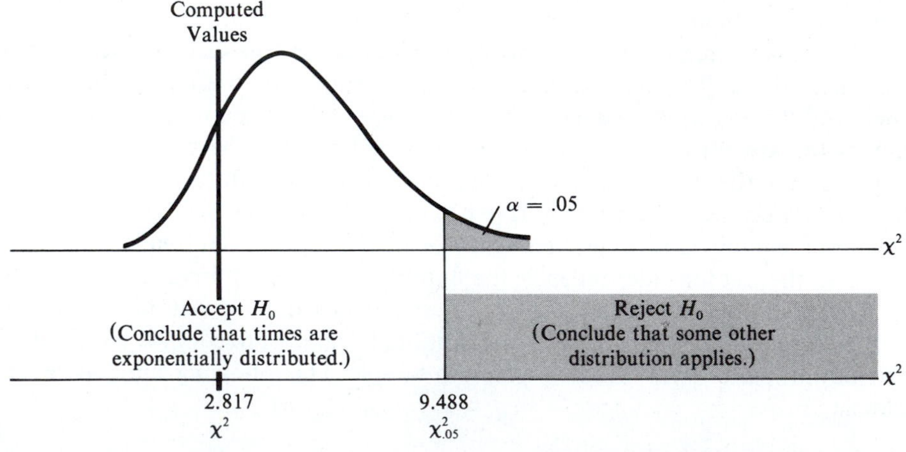

Step 4. *Collect the sample data and compute the value of the test statistic.* This step includes finding the actual and expected frequencies and computing χ^2.

Step 5. *Make the decision.* The student's computed value of $\chi^2 = 2.817$ falls in the acceptance region. The null hypothesis that the times between arrival are exponentially distributed must be accepted.

Testing with Given Population Parameters

The goodness-of-fit test also applies to situations where the population parameters are given. For instance, the student in the preceding example could test to see if the arrival-time distribution for a different time period fits the pattern just established, for which λ has already been estimated. Rather than reestimate λ, she could incorporate the value $\lambda = 1.02$ as part of the H_0 for the new sample investigation. This saves one more degree of freedom in establishing the critical value for χ^2. However, the resulting test might have a tendency to reject that null hypothesis should the sample exhibit a higher or lower mean rate of arrival than the one specified—even if the shape of the sample histogram is clearly exponential.

Large-scale testing programs such as those used in college or graduate school admissions are predicated upon a high degree of consistency between different versions of the same test. The normal distribution is assumed by test makers and users. In using the scores of a sample group to validate a new edition, however, a goodness-of-fit test must be applied not with just any normal curve but with that *particular* normal curve established over the history of that test. As part of the null hypothesis, historical figures might then be assumed for μ and σ. These parameters would not be estimated from the sample, and 2 degrees of freedom would thereby be saved in determining the critical chi-square value. Rejection of H_0 might require that changes be made to that edition's question set. Conversely, acceptance would then be sufficient grounds for using the new edition.

The following example illustrates an application where no parameters need to be estimated from the sample. It also illustrates how the goodness-of-fit test can be used with a discrete variable.

Example:* *Testing Pseudorandom Numbers for Uniform Frequency

The "random" numbers used in the typical computer simulation and often generated by special functions in languages such as FORTRAN, BASIC, Pascal, COBOL, FORTH, and APL, are not really random at all. Typically they are based on a simple algorithm that transforms a seed value into a string of numbers that have the appearance of randomly generated values. One important requirement of truly random numbers is that the possible values assumed by any particular digit be equally likely to be any integer between 0 and 9. This property is mimicked by pseudorandom numbers that exhibit uniform frequency, as if the generated values constituted a random sample from a discrete, uniformly distributed population, where each integer has an identical chance of occurring at each position in the string.

The frequency distribution of 150 integers in a string of values obtained with a particular pseudorandom number generator is as follows:

Digit i	Actual Frequency f_i	Expected Frequency $\hat{f}_i$	$\frac{(f_i - \hat{f}_i)^2}{\hat{f}_i}$
0	6	15	5.400
1	18	15	.600
2	17	15	.267
3	14	15	.067
4	13	15	.267
5	19	15	1.067
6	23	15	4.267
7	19	15	1.067
8	10	15	1.667
9	11	15	1.067
	150	150	$\chi^2 = 15.736$

These values may be treated as if they were a random sample of all digits that the algorithm eventually generates. The expected frequency of each digit of the ten possible digits is $\frac{150}{10} = 15$. The null hypothesis is

H_0: Digits are generated according to a discrete uniform distribution.

There are 10 categories, and no parameters are specified, so the number of degrees of freedom is $10 - 1 = 9$. Referring to Appendix Table H, we see that these results are significant enough to warrant rejecting H_0 at $\alpha = .10$ but not at $\alpha = .05$. Many scientists would demand a much smaller significance level, such as .005 or .001, before they would reject a null hypothesis.

Concluding Remarks

It should be emphasized once more that hypothesis-testing procedures focus on the type I error. The type II error in goodness-of-fit testing is to accept the null hypothesis when some other distribution applies. There is an infinitude of possibilities for such an outcome. If we could calculate it, each of these distributions would provide its own value for β. As with other tests, our best protection against the multitude of such type II errors is to use a large sample size.

The sample size must also be large in order for the chi-square procedure to work satisfactorily (with an expected frequency of at least 5 for each category). This may require combining less frequent categories so that some information is thereby lost. But some circumstances involve such tiny samples that the chi-square procedure would be unworkable anyway.

In Section 15-4 another test is described as an alternative. This is the Kolmogorov-Smirnov test, which is based on a test statistic that expresses the maximum deviation between cumulative levels of expected and actual frequencies. That procedure will tend to yield higher probabilities than the chi-square for rejecting a false null hypothesis. But it is basically a nonparametric procedure and does not allow for population parameters to be estimated from the sample.

Problems

15-17 The actual sample frequency distribution for the number of power failures occurring in a single day somewhere in a particular grid system is given in the following table. Assuming that the number of power failures has a Poisson distribution with a mean rate of 6 per day (a parameter not estimated from the sample), the expected frequencies in the last column were obtained.

Number of Power Failures	Actual Frequency (days)	Expected Frequency (days)
≤2	2	6.2
3	5	8.9
4	10	13.4
5	22	16.1
6	21	16.1
7	15	13.8
8	8	10.3
9	4	6.9
≥10	13	8.3
	100	100.0

Are the sample data consistent with the assumed Poisson distribution? Use $\alpha = .05$.

15-18 The student in the car arrival illustration obtained the following data for the time between arrivals during a different period of the day:

Time Between Arrivals (minutes)	Actual Frequency
0–under .5	61
.5–under 1.0	38
1.0–under 1.5	19
1.5–under 2.0	12
2.0–under 2.5	11
2.5–under 3.0	7
3.0–under 3.5	5
3.5–under 4.0	3
≥4.0	0
	156

Conduct a goodness-of-fit test to the exponential distribution having the same mean rate of arrivals as in the sample. Use $\alpha = .05$.

15-19 In the 1930s a group of British statisticians used the London telephone directory as a list of random numbers. An engineering student wants to determine if the digits in his local phone book occur with the uniform frequency that random decimals exhibit. Using only the last 4 digits of successive listed telephone numbers, converted to decimals by dividing by 10,000, the following sample frequency distribution was obtained for 1,000 entries:

Class Interval	Number of Values
.0001–.1000	163
.1001–.2000	137
.2001–.3000	119
.3001–.4000	122
.4001–.5000	113
.5001–.6000	105
.6001–.7000	93
.7001–.8000	72
.8001–.9000	54
.9001–1.0000	22
	1,000

The student's null hypothesis was that these numbers might have been generated by a uniform random process, with all values between 0 and 1 equally likely to occur each time.

At the $\alpha = .10$ significance level, can the student conclude that the phone book might be a suitable source for random decimals?

15-20 An engineer for a computer manufacturer wants to determine the population frequency distribution for the number of failures in a certain element during one year of mainframe operation. Over the past year records have been kept of the failures occurring in a sample of computer systems. The following frequency distribution applies:

Number of Failures	Frequency
0	2
1	5
2	14
3	10
4	7
5	5
6	6
7	3
8	2
9	1
10 or more	0
	55

(a) Estimate the mean number of failures per year.

(b) Construct a table of expected frequencies under the null hypothesis that the number of failures has a Poisson distribution with mean annual rate estimated in (a).

(c) Compute the chi-square statistic. At the 5% significance level should H_0 be accepted or rejected?

15-21 The following data represent the observed stress at the onset of a material's inelastic strain range:

Class Interval (ksi)	Frequency
45–under 46	1
46–under 47	3
47–under 48	7
48–under 49	12
49–under 50	23
50–under 51	16
51–under 52	13
52–under 53	10
53–under 54	8
54–under 55	4
55–under 56	2
56–under 57	1
	100

(a) From these sample data, estimate the population mean and standard deviation.

(b) For each class interval, determine the expected frequency under the null hypothesis that the population is normally distributed with parameter values determined in (a).

(c) Compute the chi-square statistic. Must the null hypothesis be accepted or rejected at the 5% significance level?

15-22 Refer to the stress data in Problem 15-21. Suppose that the population mean and standard deviation are assumed to be $\mu = 50$ ksi and $\sigma = 3$ ksi. At the 5% significance level, should the null hypothesis that the population is normally distributed be accepted or rejected?

15-23 In plotting a histogram for the telephone decimal data in Problem 15-19, one student commented that the observations seem to fit an exponential distribution. Letting λ denote the reciprocal of the mean decimal value, estimate this parameter from the sample. Then test the null hypothesis that the exponential distribution applies. At the $\alpha = .05$ significance level, should H_0 be accepted or rejected?

15-24 Refer to the transaction time regression results in Table 10-3.

(a) Construct for the residuals $e = Y(X) - \hat{Y}(X)$ a frequency distribution starting with "under -25.00" and continuing from there with class intervals of width 10.00 seconds, using "≥ 35.00" as the last interval.

(b) Test the null hypothesis that the residuals are samples from a normally distributed error population, with mean 0 and standard deviation estimated from the sample results. What must you conclude at the 5% significance level?

15-25 Refer to the gasoline mileage multiple regression results in Table 11-3. Test the null hypothesis that the residuals $e = Y - \hat{Y}$ are samples from a normally distributed error population, with mean 0 and standard deviation estimated from the sample results. What must you conclude at the 5% significance level?

15-4 The Kolmogorov-Smirnov Goodness-of-Fit Test

In the preceding section the chi-square goodness-of-fit test was introduced. Although the chi-square distribution only approximates the sampling distribution of the test statistic introduced there, that procedure works well as long as the sample size is large enough so that there are at least five observations per class interval grouping. Often, however, there may be too few sample data observations to effectively employ that procedure.

The **Kolmogorov-Smirnov test** for goodness-of-fit is an alternative procedure. This test compares the cumulative sample frequency distribution to the theoretical counterpart specified by the null hypothesis. The test statistic is the maximum deviation between matching cumulative frequencies of the respective distributions.

Illustration: *Service-Time Distribution in Queueing Investigations*

The most familiar queueing (waiting line) model assumes that arrivals occur randomly over time, so that a Poisson process applies and an *exponential distribution* generally serves well in characterizing the time between successive arrivals. Unfortunately, that same model assumes that completion times for service are also exponentially distributed (with a smaller mean). Recall from Chapter 4 that the exponential distribution assigns greatest density to zero, so that if service times were indeed exponential, then the most likely completion times would be those near zero! That is virtually impossible for most commonly encountered service, especially when human operators are involved.

As part of her term project involving a queueing investigation of information retrieval via time-sharing mode on a mainframe computer, an industrial engineering student obtained the sample data in Table 15-7. This table lists job completion times (in milliseconds) for 12 randomly selected queries; only central processor time is included.

The student believes that the mean job completion time is 200 milliseconds, so the mean rate of completions is $\frac{1}{200} = .005$ per millisecond. Before collecting her sample data she therefore chose as her null hypothesis:

H_0: Job processing times are exponentially distributed, with $\lambda = .005$.

The test statistic is based on the cumulative *relative* frequency distributions, shown in Table 15-8. Column (1) lists the actual sample data, sorted in increasing magnitude, and denoted by x. Each sample observation is given equal relative frequency of $\frac{1}{12} = .0833$. Column (2) lists the actual cumulative relative frequencies $F(x)$, starting with .0833 for the first value, and increasing by that amount for each successive row. Thus, the first entry in column (2) is $F(113) = .0833$, the second is $F(147) = F(113) + .0833 = .1666$, and so on.

Table 15-7 *Sample Results of Mainframe Information Retrieval CPU Times*

Job Completion CPU Time (milliseconds)		
148	157	147
113	179	163
247	306	198
178	345	237
$\bar{X} = 201.50$	$s = 69.44$	

Column (3) of Table 15-8 shows for each observed x the expected cumulative relative frequency under the null hypothesis. These are obtained from

$$\hat{F}(x) = 1 - e^{-\lambda x}$$

so that each expected frequency is equal to the cumulative exponential probability computed at the respective observed level for x. The first entry is

$$\hat{F}(113) = 1 - \exp[-.005(113)] = .4316$$

which indicates that under the null hypothesis, job completion times of 113 milliseconds or less should occur only with probability .4316, or about 43% of the time. This conclusion is not very consistent with the actual data since only one observation out of twelve fell within that range.

The deviation between the actual and expected cumulative relative frequencies when $x = 113$ shows a wide variance from what would be expected if H_0 were true:

$$F(113) - \hat{F}(113) = .0833 - .4316 = -.3483$$

An even greater deviation is obtained for the second data point:

$$F(147) - \hat{F}(147) = .1666 - .5205 = -.3539$$

Table 15-8 *Actual and Expected Cumulative Relative Frequencies for Job Completion CPU Times, with Calculation of the Test Statistic*

(1) Sample Job Completion Time (milliseconds) x	(2) Actual Cumulative Relative Frequency of Times $\leq x$ $F(x)$	(3) Expected Cumulative Relative Frequency of Times $\leq x$ $\hat{F}(x) = 1 - e^{-\lambda x}$	(4) Deviation $F(x) - \hat{F}(x)$
113	.0833	.4316	−.3483
147	.1666	.5205	−.3539
148	.2499	.5229	−.2730
157	.3332	.5439	−.2107
163	.4165	.5574	−.1409
178	.4998	.5893	−.0895
179	.5831	.5914	−.0083
198	.6664	.6284	.0380
237	.7497	.6943	.0554
247	.8330	.7092	.1238
306	.9163	.7835	.1328
345	.9996	.8218	.1778

$$D = \max|F(x) - \hat{F}(x)| = .3539$$

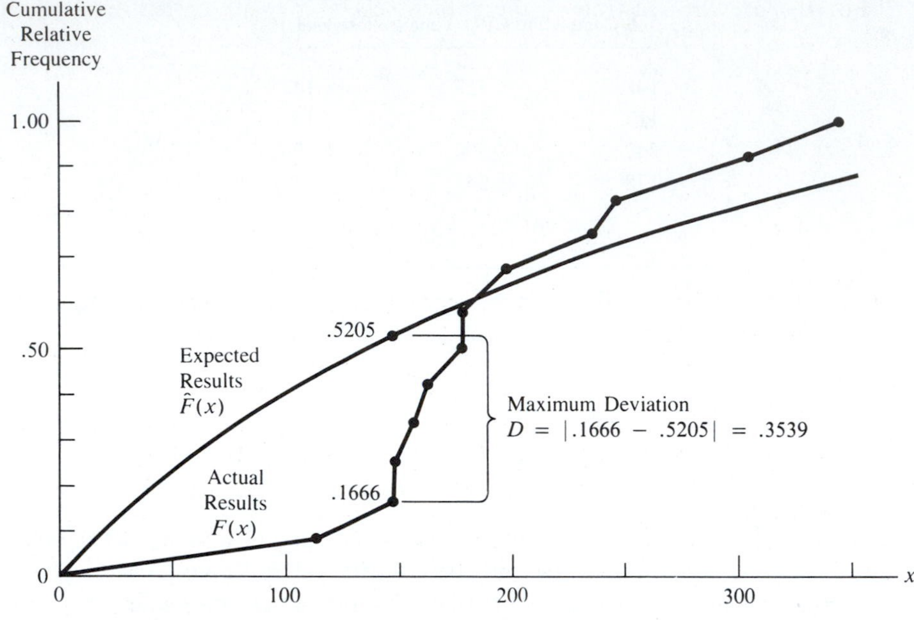

Figure 15-7 *Actual and expected cumulative frequency distributions for job completion CPU times.*

Column (4) of Table 15-8 lists all the deviations between actual and expected frequencies. Large deviations indicate a poor fit between actual data and what is expected. This type of comparison is illustrated in Figure 15-7, where the two cumulative relative frequency distributions are plotted. In absolute value, the greatest deviation occurs when $x = 147$.

The maximum absolute frequency deviation serves as the test statistic.

KOLMOGOROV-SMIRNOV TEST STATISTIC

$$D = \max|F(x) - \hat{F}(x)|$$

For the present illustration, the computed test statistic is $D = .3539$. We next determine if this is significantly large enough to warrant rejecting the null hypothesis of the exponential service time distribution.

Finding the Critical Value

Large levels for D will refute the null hypothesis. For the desired significance level α the critical value D_α may be obtained using the values listed in Appendix Table M. That table was constructed in accordance with the underlying sampling distribution for the random variable D. The key parameter is the sample size n, with D becoming smaller as n increases.

The following acceptance and rejection regions apply for the upper-tailed test:

Accept H_0 | Reject H_0

D_α D

For the present illustration, the student chose $\alpha = .05$ as her significance level. For $n = 12$, Table M provides $D_{.05} = .33815$. Since the computed value of $D = .3539$ is greater, she must reject H_0 and conclude that the exponential distribution does not fit her data.

Example: *Cycle Times in Switching Relays*

A quality-assurance engineer for a large city tests a sample of traffic-light switching relays to determine whether or not the supplier is meeting specifications. These include the requirement that actual relay cycle times for uncalibrated units be normally distributed with mean equal to the nominal setting and standard deviation of .5 second. The engineer tests units at various nominal settings and will compute for each observation the cycle-time error, clocked time minus nominal time. The following null hypothesis applies:

$$H_0\text{: Errors are normally distributed with } \mu = 0 \text{ and } \sigma = .5.$$

Using a sample of $n = 35$ relays, the engineer computes the cycle-time errors in column (1) of Table 15-9. Giving each observation frequency $\frac{1}{35} = .0286$, the actual cumulative relative frequencies $F(x)$ are computed in column (2). Column (3) shows the normal deviate applicable to each observation under H_0, and column (4) lists the expected frequencies $\hat{F}(x)$. The deviations between actual and expected frequencies are computed in column (5).

The computed Kolmogorov-Smirnov test statistic is $D = .3229$. Since this value is greater than the last tabled value, $D_{.005} = .26897$, the engineer rejected her null hypothesis at the $\alpha = .005$ significance level.

Comparison with the Chi-Square Test

When n is small, the Kolmogorov-Smirnov test is more discriminating than the chi-square. For larger sample sizes, the chi-square test performs better.

The above example demonstrates this feature of the Kolmogorov-Smirnov test. Table 15-10 shows the calculation of the chi-square statistic under the identical null hypothesis. There we find that $\chi^2 = 15.319$. With $6 - 1 = 5$ degrees of freedom, Appendix Table H shows $\chi^2_{.01} = 15.086$ as the smallest critical value exceeded by the computed figure. Thus, the chi-square procedure finds the sample results significant at the much higher $\alpha = .01$ level, compared to .005 found with the Kolmogorov-Smirnov test.

The main drawback of the Kolmogorov-Smirnov test is that it does not allow us to estimate any population parameters from the sample data. *All population parameters must be specified in advance of testing.* We could not use $1/\bar{X}$ to estimate λ in the job completion times investigation, nor could we use $\bar{X}$ and s to estimate μ and σ in a test involving the normal distribution.

When the sample sizes are great enough to allow for a choice in procedure, the chi-square test would be preferred by investigators primarily concerned with the *distribution family* rather than with *specific parameters*. The preceding example illustrates this point.

Table 15-9 *Actual Traffic Signal Cycle-Time Errors with Calculation of Kolmogorov-Smirnov Test Statistic*

(1) Cycle Time Error (seconds) x	(2) Actual Cumulative Relative Frequency $F(x)$	(3) $z = \dfrac{x - 0}{.5}$	(4) Expected Cumulative Relative Frequency $\hat{F}(x) = \Phi(z)$	(5) Deviation $F(x) - \hat{F}(x)$
−1.2	.0286	−2.40	.0082	.0204
−1.1	.0572	−2.20	.0139	.0433
−1.1	.0858	−2.20	.0139	.0719
−1.0	.1144	−2.00	.0228	.0916
−0.9	.1430	−1.80	.0359	.1071
−0.9	.1716	−1.80	.0359	.1357
−0.8	.2002	−1.60	.0548	.1454
−0.6	.2288	−1.20	.1151	.1137
−0.6	.2574	−1.20	.1151	.1423
−0.5	.2860	−1.00	.1587	.1273
−0.5	.3146	−1.00	.1587	.1559
−0.5	.3432	−1.00	.1587	.1845
−0.4	.3718	−0.80	.2119	.1599
−0.4	.4004	−0.80	.2119	.1885
−0.4	.4290	−0.80	.2119	.2171
−0.3	.4576	−0.60	.2743	.1833
−0.3	.4862	−0.60	.2743	.2119
−0.3	.5148	−0.60	.2743	.2405
−0.3	.5434	−0.60	.2743	.2691
−0.2	.5720	−0.40	.3446	.2274
−0.2	.6006	−0.40	.3446	.2560
−0.2	.6292	−0.40	.3446	.2846
−0.2	.6578	−0.40	.3446	.3132
−0.1	.6864	−0.20	.4207	.2657
−0.1	.7150	−0.20	.4207	.2943
−0.1	.7436	−0.20	.4207	.3229
0.0	.7722	0.00	.5000	.2722
0.0	.8008	0.00	.5000	.3008
0.1	.8294	0.20	.5793	.2501
0.2	.8580	0.40	.6554	.2026
0.3	.8866	0.60	.7257	.1609
0.4	.9152	0.80	.7881	.1271
0.6	.9438	1.20	.8849	.0589
0.7	.9724	1.40	.9192	.0532
0.9	1.0010	1.80	.9641	.0369

$\bar{X} = -.286 \quad s = .51 \quad D = .3229$

Table 15-10 Computation of Chi-Square Statistic for Traffic Signal Cycle-Time Errors—Prior Specification of Parameters

Error Upper Class Limit y	Actual Frequency f	$z = \frac{y - 0}{.5}$	$\Phi(z)$	Class Interval Area	Expected Frequency $\hat{f}$ = Area × 35	$\frac{(f - \hat{f})^2}{\hat{f}}$
≤ -0.75	7 } 12	−1.50	.0668	.0668	2.34 } 5.56	7.459
−0.50	5	−1.00	.1587	.0919	3.22	
−0.25	7	−0.50	.3085	.1498	5.24	.591
0.00	9	0.00	.5000	.1915	6.70	.790
0.25	2	0.50	.6915	.1915	6.70	3.297
0.50	2	1.00	.8413	.1498	5.24	2.003
0.75	2 } 3	1.50	.9332	.0919	3.22 } 5.56	1.179
>0.75	1		1.0000	.0668	2.34	
						$\chi^2 = 15.319$

Estimating the mean and standard deviation by their sample counterparts, we now test the following:

Modified H_0: Errors are normally distributed, with $\mu = -.286$ and $\sigma = .51$.

The chi-square test statistic is computed in Table 15-11, where $\chi^2 = 2.991$ is obtained. With the degrees of freedom now reduced to $6 - 2 - 1 = 3$, Table H

Table 15-11 Computation of Chi-Square Statistic for Traffic Signal Cycle-Time Errors—Parameters Estimated from Sample

Error Upper Class Limit y	Actual Frequency f	$\frac{y - (-.286)}{.51} = z$	$\Phi(z)$	Class Interval Area	Expected Frequency $\hat{f}$ = Area × 35	$\frac{(f - \hat{f})^2}{\hat{f}}$
≤ -0.75	7	−0.91	.1814	.1814	6.35	.067
−0.50	5	−0.42	.3372	.1558	5.45	.037
−0.25	7	−0.07	.4721	.1349	4.72	1.101
0.00	9	0.56	.7123	.2402	8.41	.041
0.25	2	1.05	.8531	.1408	4.93	1.741
0.50	2	1.54	.9382	.0851	2.98	
0.75	2 } 5	2.03	.9788	.0406	1.42 } 5.14	.004
>0.75	1		1.0000	.0212	.74	
						$\chi^2 = 2.991$

now shows $\chi^2_{.50} = 2.366$ as the smallest critical value at which the modified null hypothesis can be rejected. The effective significance level is $\alpha = .50$. The engineer would have to accept the null hypothesis of normality. (The Kolmogorov-Smirnov test could not be applied in testing the modified null hypothesis.)

Problems

15-26 A production foreman wishes to establish whether the times-between-arrivals of parts at a particular assembly station are exponentially distributed. From previous tests he has estimated the mean interarrival time to be 5 seconds. The following times (in seconds) were obtained for 10 randomly selected pairs of parts:

11	12	9	3	2
15	6	6	8	3

If he applies the Kolmogorov-Smirnov test, what should the foreman conclude?

15-27 Repeat Problem 15-26 using instead a mean interarrival time of 10 seconds.

15-28 Refer to Problem 15-20. Assuming that the mean number of failures per year is 4, apply the Kolmogorov-Smirnov test to determine whether or not a Poisson distribution applies.

15-29 Refer to the transaction time regression results in Table 10-3. Test the null hypothesis that the residuals $e = Y(X) - \hat{Y}(X)$ are samples from a normally distributed error population, with mean 0 and standard deviation 14.0 seconds. What must you conclude at the 5% significance level?

15-30 Refer to the gasoline mileage multiple regression results in Table 11-3. Test the null hypothesis that the residuals $e = Y - \hat{Y}$ are samples from a normally distributed error population, with mean 0 and standard deviation 1.5 miles per gallon. What must you conclude at the 5% significance level?

Comprehensive Problems

15-31 A statistician asked her students to toss a coin 50 times and perform a number-of-runs test. One student "dry-labbed" his results, providing the following list without actually tossing a coin:

H T H T H T H H H T T H T H T H H T T H T H H T T
H T H H H T H T H T H T T H H T H T H H T H T H T

Perform a number-of-runs test, letting *a* represent head. At the 5% significance level should the null hypothesis of randomness be accepted or rejected?

15-32 Statistics students have been asked to construct their own random decimals from scratch. The following sample results were obtained from one student as an illustration of how faulty this procedure can be:

1	3	0	6	7	9	3	0	5	0	4	6	7	8	3	3	4	2	9	9	2	3	4	8	7
2	2	3	4	5	7	8	4	5	6	9	1	2	8	5	4	2	4	2	5	7	9	3	7	9
3	4	8	9	7	9	5	7	4	6	3	5	4	6	2	4	7	9	5	6	7	3	4	3	5
2	8	3	7	2	8	1	8	3	7	2	9	2	5	3	8	2	4	1	6	8	4	9	0	1

(a) Letting R_u denote the number of runs of integers below 5, perform a number-of-runs test at the $\alpha = .10$ significance level. Do the values pass this test of randomness?

(b) Construct an actual frequency distribution for the integers in the above sample. Then apply the goodness-of-fit test to determine whether the values are consistent with a discrete uniform distribution. Use $\alpha = .10$. Do the values pass this test of randomness?

15-33 Shown below for a succession of power failures experienced by a large manufacturing complex are the actual times (in days) since the preceding failure. Also shown are the expected frequencies that would apply if the failure times were exponentially distributed with a mean rate of one every 20 days.

Time Between Failures (days)	Actual Frequency	Expected Frequency
0–under 10	30	39.3
10–under 20	28	23.9
20–under 30	16	14.5
30–under 40	10	8.8
40–under 50	6	5.3
50–under 60	4	3.2
≥ 60	6	5.0
	100	100.00

Are the sample data consistent with the indicated exponential distribution? Use the chi-square test with $\alpha = .10$.

15-34 The following data have been obtained by an automotive engineer interested in isolating different target categories for car owners:

	Engine Size			
Transmission	**Small**	**Medium**	**Large**	**Total**
4-Speed	34	19	12	65
5-Speed	24	28	5	57
Automatic	7	12	22	41
Total	65	59	39	163

He wishes to test the null hypothesis that transmission type and engine size chosen by the car-owning population are independent. Using a 5% significance level, determine if these data indicate that the hypothesis should be accepted or rejected.

15-35 An engineer is investigating the reliability of car brakes. He has collected the following data pertaining to accidents in which brake failure was the primary cause:

Type of Brake	Cause of Failure: Wheel Cylinder	Master Cylinder	Worn Lining	Power System	Total
Disk	35	15	0	10	60
Drum	40	10	10	15	75
Total	75	25	10	25	135

He wishes to test the null hypothesis that type of brake and cause of failure are independent variables. Using a 5% significance level, do the above data indicate that the hypothesis should be accepted or rejected?

15-36 Consider the labor cost data given in Problem 1-27.

(a) Construct a table for the sample frequency distribution using intervals of width \$.50, with \$18.50 serving as the lower limit of the first class.

(b) Determine from your grouped data in (a) the sample mean and sample standard deviation.

(c) Determine the expected frequencies for each class, assuming that the data were obtained from a normally distributed population having the mean and standard deviation found in (b).

(d) Compute the chi-square statistic. At the $\alpha = .05$ significance level should the null hypothesis of a normal population be accepted or rejected?

15-37 A computer-systems analyst has taken random samples of 100 terminal transactions for each of three types of jobs. The proportion of errors found were .07 (type 1), .13 (type 2), and .09 (type 3). At the 5% significance level, can the analyst conclude that the error rate differs among job types?

15-38 An industrial engineer periodically evaluates the equipment in a food-processing plant. The following sequences of results were obtained from different machines filling 16-oz bottles with vinegar. Each bottle was above (*a*) or below (*b*) the labeled content minimum.

Machine 1: *a a a a a b b b a a a a a a a a a a a a a a a a a*
a a a a a a a a a a b b b a a a a a a a a a a a a

Machine 2: *b a a a a a a b a a a a a a a a a a a b a a a a a*
a a a a a a a a a a a b a a a a a a a a a a a a a

Machine 3: *a b b a a a a a a a a a b a a a a a a a a a b b a*
a a a a a b b a a a a a a a a b a a a a a a a a a

Determine for each machine whether the sample data are consistent with the null hypothesis of randomness in the filling process. Let R_a be the number of runs above the minimum. Use $\alpha = .05$.

15-39 Refer to the information in Problem 15-38. Test the null hypothesis that the three machines yield a common proportion of below-minimum content volumes. Use $\alpha = .05$.

15-40 You want to determine whether or not a series of events are consistent with what would be expected from a Poisson process. The mean process rate λ is known. You may select one of the following procedures:

(a) Randomly select 10 time periods of the same duration and record the number of events that occur during each. Then test the null hypothesis that the Poisson distribution applies.

(b) Randomly select 10 event pairs, recording the time between the two occurrences. Then test the null hypothesis that the exponential distribution applies.

Which procedure would you use? Why?

PART FOUR

SPECIALIZED TOPICS IN STATISTICS

CHAPTER SIXTEEN

Statistical Quality Control

ONE OF THE most important statistical applications for engineers is **statistical quality control**. This importance is due to the nature of engineering itself, which is largely concerned with the creation of things, the operation of processes, or the design of public works and facilities. End products must adhere to the engineer's specifications, as should the components and materials from which they are fabricated. The design engineer depends on others to create the final product, and high quality is expected.

Quality control is ordinarily associated with manufacturing, and one of the main responsibilities of industrial engineers is to ensure that proper procedures are being employed to monitor production and to remedy any unusual problems that might prevent the final product from being satisfactory. Although the industrial engineer plays the terminal role, all engineers involved with the creation of any final product should be concerned with its quality.

There are two major groupings of statistical methods for quality control. **Acceptance sampling** is usually concerned with items obtained from vendors or suppliers, although manufacturers often apply the same methods to production batches. The objective of acceptance sampling is to determine whether items should be accepted or rejected. A manufacturer also wants to ensure that the production process is operating as intended. For this purpose **control charts** are used to monitor current production on a "real-time" basis.

The fundamental background for understanding statistical quality control is given in the earlier chapters of this text. Statistical quality control embodies many of the basic concepts and methods of descriptive and inferential statistics. As in earlier applications, parallel methods are employed, depending on whether the data are qualitative or quantitative. In evaluating qualitative data the *proportion* is used, whereas several measures are employed with quantitative data, including the *mean*, *standard deviation*, and *range*.

16-1 The Control Chart

Historically, statistical quality-control procedures have been developed mainly for the manufacturing environment, especially when production volumes are high. These methods permit early detection and speedy correction of trouble. Because uncorrected problems could result in the scrapping of thousands of finished units, the potential cost savings from preventive measures can be huge. In manufacturing applications of statistical quality control the primary focus is the production *process* itself. The disposition of those items already produced is an important but secondary consideration.

Problems are detected indirectly after a sample establishes at a given time that the process is either in **control** or is not. This is true as long as the observed units fall within reasonable bounds, which are predetermined in accordance with what would be expected from a process that is functioning as intended. They allow for the same kinds of chance variations that would arise in any sampling situation.

Successive samples falling within the limits are expected to differ from each other, and any variation is assumed to be caused by *chance*. Should a sample fall outside those limits, it is then presumed that there is some assignable cause. Such an occurrence is a signal for action, and a rigorous search should be made for the actual cause. It could be traced to a variety of sources, such as malfunctioning machinery, an inattentive operator, a poor batch of raw materials, or an environmental disturbance.

The mechanism for detecting quality problems is the **control chart**. To illustrate its construction and application, we will first describe investigations involving the proportion. Later we will describe control charts that are employed with quantitative data.

Establishing Control Limits for the Proportion

Illustration: *Unsatisfactory Memory Chips*

An industrial engineer at a microelectronics manufacturing plant is setting up statistical quality-control procedures for a new high-density computer memory chip. Production begins September 1 on a limited scale, with volume gradually increasing thereafter. The engineer wants to minimize the incidence of defective chips. There are many ways for a chip to be unsatisfactory—fuzzy lines, uneven layering, breaks, poor texturing, and so on. Rather than accounting separately for each possibility, the testers will simply classify each chip as satisfactory or defective.

The engineer selects a random sample of $n = 1{,}000$ chips daily for the entire month. Table 16-1 shows the results obtained. These data were used in constructing the control chart for the proportion of defective chips, shown in Figure 16-1.

The abscissa of the control chart represents successive samples. In the present illustration, there is one sample in each successive day. The ordinate shows possible levels for the sample proportion defective. The value of this statistic computed for each sample is plotted as a dot.

Notice that the dots fluctuate around a central value, the **combined proportion defective** $\bar{P}$. This value is computed from the data in Table 16-1, treating the 30

Table 16-1 *Sample Data from Production of Silicon Memory Chips*

Date	Number of Defectives	Proportion Defective P
September 1	41	.041
2	75	.075
3	77	.077
4	56	.056
5	55	.055
6	108	.108
7	127	.127
8	57	.057
9	46	.046
10	135	.135
11	47	.047
12	72	.072
13	76	.076
14	61	.061
15	59	.059
16	49	.049
17	66	.066
18	45	.045
19	54	.054
20	42	.042
21	74	.074
22	49	.049
23	70	.070
24	45	.045
25	44	.044
26	72	.072
27	51	.051
28	43	.043
29	57	.057
30	54	.054
	1,907	

days' observations as a single sample of size 30,000:

$$\bar{P} = \frac{\text{Combined number of defectives}}{\text{Combined sample size}} = \frac{1{,}907}{30{,}000} = .0636$$

Before the data are collected, the individual sample proportions P may be viewed as random variables, each having a binomial distribution with parameters n and π. The true population proportion defective π is unknown and may be estimated from the sample results. As long as the process is under control, a single value for π is assumed to apply as successive observations are made.

Figure 16-1 *Initial control chart for proportion of defective memory chips.*

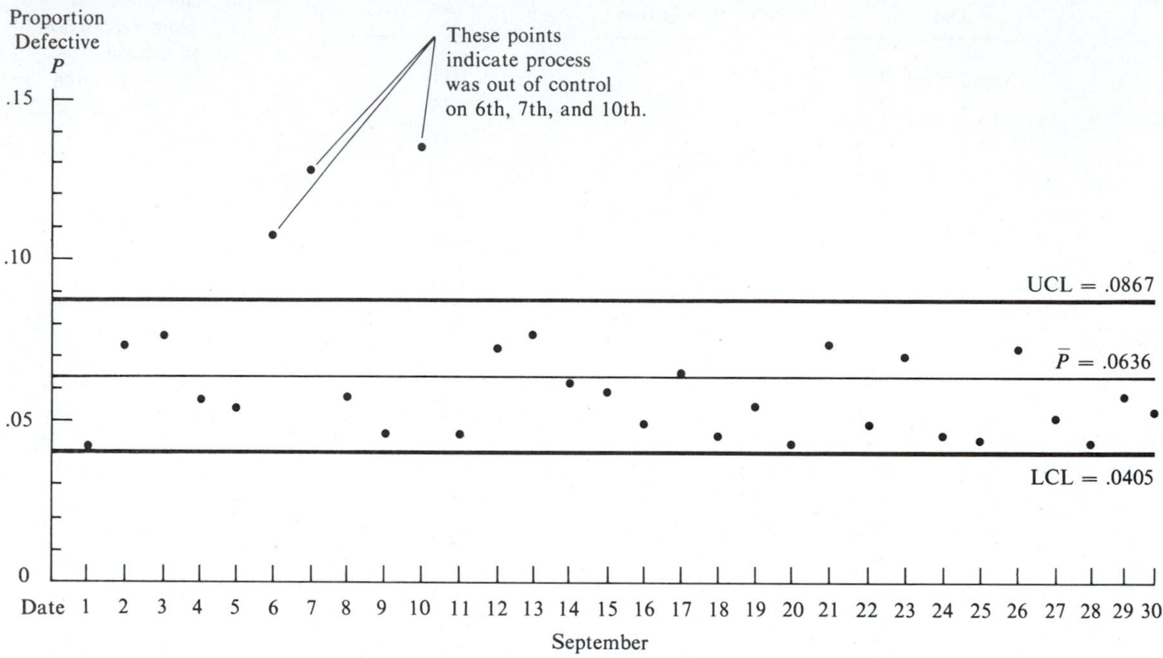

Notice from Figure 16-1 that the observed P's vary. This variability may be summarized by the standard error for P, which we have seen may be determined from

$$\sigma_P = \sqrt{\frac{\pi(1 - \pi)}{n}}$$

Using $\bar{P}$ as the point estimator, this quantity may be estimated by

$$\hat{\sigma}_P = \sqrt{\frac{\bar{P}(1 - \bar{P})}{n}}$$

The computed value for $\hat{\sigma}_P$ is used to establish the **control limits** that determine whether or not the process is in statistical control. For this purpose, *3-sigma limits* are traditionally computed from the sample data. Recall that about 99.7% of the values generated by a normal distribution will fall within $\mu \pm 3\sigma$, and a similar percentage is assumed to apply for values of P falling within $\pi \pm 3\sigma_P$ (since the binomial distribution is closely approximated by the normal distribution with parameters π and σ_P). For a computed sample proportion to fall outside those limits when in fact π is the true proportion of defectives would be extremely unlikely.

Of course π, and hence σ_P, are generally unknown. Furthermore, π may occasionally shift to some unusually high level. At those times the process is *out of control*, and these circumstances are what the control chart is designed to detect. The control limits are based on the observed data, with $\bar{P}$ replacing π and $\hat{\sigma}_P$ used instead of σ_P. The following expressions are used in establishing the corresponding levels.

CONTROL LIMITS FOR THE PROPORTION DEFECTIVE

$$\text{LCL} = \bar{P} - 3\sqrt{\frac{\bar{P}(1 - \bar{P})}{n}}$$

$$\text{UCL} = \bar{P} + 3\sqrt{\frac{\bar{P}(1 - \bar{P})}{n}}$$

The abbreviation LCL is used to denote the **lower control limit**, and UCL the **upper control limit**. The process is assumed out of control whenever a sample is obtained for which the proportion defective falls above or below these limits.

It may seem a bit odd that there is *any* lower control limit for the proportion defective. After all, it is desirable for P to be small. Sometimes zero is used as a lower limit so that the procedure is essentially "one sided." But there are obvious advantages to identifying samples that involve an inordinately low proportion defective. A low proportion can be symptomatic of a problem, such as too much expensive care being devoted by workers in certain operations. It can also be due to a breakdown in the inspection process itself, with too many defects going undetected. And of course, there is always the serendipitous possibility that the low-P sample will lead to the discovery of a factor whose presence or absence might actually improve the process.

For the silicon memory chip illustration, the following control limits apply:

$$\begin{aligned}\text{LCL} &= .0636 - 3\sqrt{\frac{.0636(1 - .0636)}{1{,}000}} \\ &= .0636 - 3(.0077) \\ &= .0405\end{aligned}$$

and

$$\begin{aligned}\text{UCL} &= .0636 + 3(.0077) \\ &= .0867\end{aligned}$$

These limits are represented by the boldface horizontal lines shown on the control chart in Figure 16-1.

Using the Control Chart

In our memory chips example, three points lie above the UCL—.108, .127, and .135—and it is assumed that the process was out of control on the corresponding dates—September 6, 7, and 10. Since no control limits had yet been established on those dates, the underlying production problems went undetected. (A later investigation showed that several key personnel were ill on the three days.) These three out-of-control data points must be eliminated from the data base before the control chart can be used in actual production decisions.

Figure 16-2 shows the revised control chart, which was then used to monitor October production. The combined proportion defective is recomputed to be $\bar{P} = \frac{1{,}537}{27{,}000} = .0569$. This proportion necessitates revised control limits: LCL = .0349 and UCL = .0789.

Figure 16-2 *Final control chart for proportion of defective memory chips.*

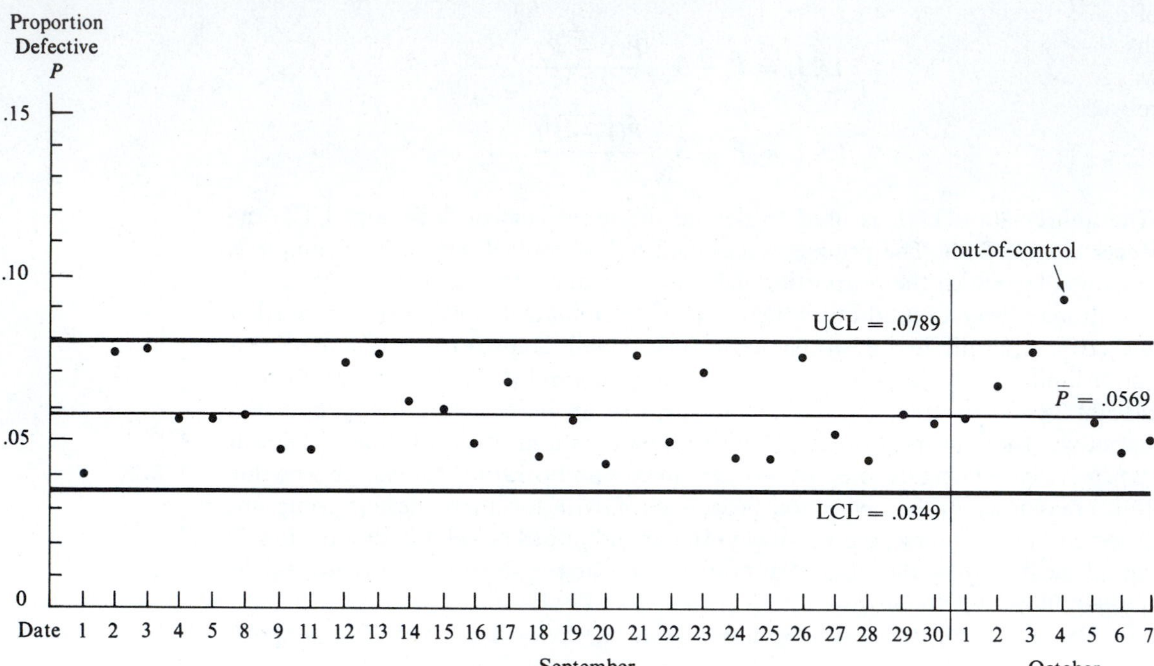

The 3-sigma limits are traditional and are supposed to provide an acceptable balance between triggering too many false alarms and letting quality problems go undetected too long. Of course, it is possible to use tighter limits, such as 2-sigma, or to set wider ones. This question is really the familiar issue of balancing the type I error (here, unnecessarily searching for a nonexistent cause when the sample fell outside the control limits just by chance) against the type II error (letting an out-of-control process continue uncorrected).

During the first week in October, the following observations were made:

Date	Proportion Defective P	Assumed Status
10/01	.055	in control
10/02	.067	in control
10/03	.075	in control
10/04	.093	out of control
10/05	.053	in control
10/06	.044	in control
10/07	.049	in control

The sample results for October 4 indicated that an inordinately high proportion of defective chips was being produced. A thorough investigation of the chips in the sample for that day showed a very high incidence of cracked layers. The cause was traced to a malfunctioning thermostat in one of the baking ovens, which was replaced.

Problems

16-1 Determine the UCL and LCL for P in each of the following situations:
(a) $\bar{P} = .10, n = 100$
(b) $\bar{P} = .05, n = 1{,}000$
(c) $\bar{P} = .07, n = 500$
(d) $\bar{P} = .08, n = 200$

16-2 The following numbers of defectives were found in successive samples of 1,000 items taken from a production process:

7/01	75	7/06	109	7/11	75	7/16	52
02	64	07	111	12	74	17	54
03	75	08	63	13	63	18	61
04	48	09	71	14	58	19	63
05	49	10	48	15	53	20	59

(a) Determine the control limits for the sample proportion. Construct the control chart and plot the points for these data.
(b) Identify those dates when the process was out of control.
(c) Determine the new control limits applicable after removing the out-of-control data points. Indicate for each of the following numbers of defectives found in samples of size 1,000 whether the process was or was not in control:
(1) 95 (2) 35 (3) 55 (4) 63 (5) 64

16-3 Determine the 2-sigma control limits for the sample proportion, using the data in Table 16-1. Then indicate on which dates the process was out of control.

16-4 The following numbers of defectives were found in successive samples of 100 items taken from a production process:

8/06	8	8/11	12	8/16	22	8/21	11
07	15	12	3	17	13	22	7
08	12	13	9	18	10	23	15
09	19	14	14	19	15	24	24
10	7	15	10	20	18	25	2

Answer parts (a)–(b) as in Problem 16-2.

16-5 Determine the 2-sigma control limits for the sample proportion, using the data in Problem 16-2. Then indicate on which dates the process was out of control.

16-6 Refer to the final set of control limits for the computer memory illustration. Use the normal approximation to find the probability that the process will be found (1) in control and (2) out of control when a particular sample of 1,000 items is taken and the true proportion defective is:
(a) .07 (b) .065 (c) .045 (d) .055

16-2 Control Charts for Quantitative Data

The quality of production processes is often measured in terms of some quantity, such as weight, dimension, strength, yield, or volume. Control charts using quantitative sample data may be constructed in a fashion analogous to the P-chart. Central tendency and variability are both important considerations in such evaluations, and it is possible to monitor quality using a separate control chart for each.

Control charts based on the *sample mean* are universal. Two types of control charts may be used in monitoring variability. One of these is based on the *sample standard deviation.* Although the standard deviation is very familiar to people with any formal statistical education, it presents conceptual problems to people with little or no statistical background, and it is cumbersome to compute. For those reasons, an alternative control chart that is based on the *sample range* is frequently used. Since this latter is more commonly encountered in practice, this is the type we describe in this book.

Quality management with two control charts involves more elaborate procedures than those encountered with one P-chart. Quantitative data also allow for far greater "resolution." Numerical measurements provide a great deal of information using much smaller sample sizes than is practicable with the P-chart. The smaller n's allow inspection resources to be stretched farther and make it possible to sample more often. The net result is a quicker response in detecting quality problems.

Illustration: *Controlling Thickness in Rolled Sheet Metal*

A detailed illustration will help us explain the concepts and procedures. An industrial engineer for a steel mill has implemented a procedure for controlling the quality of rolled sheet metal. A major product is sheets one-twentieth of an inch thick. Although individual thicknesses will vary within the same sheet and from sheet to sheet, no sheet should deviate much from the target dimension. The true population mean μ for individual sheet thickness is unknown and may be estimated from the sample data. The standard deviation σ of that population may also be estimated from the sample.

In initiating the quality-control procedure the baseline data in Table 16-2 were collected. Every hour, a sample of $n = 5$ sheets is selected and measured for thickness. The mean $\bar{X}$ and range R are computed for each sample. The rolling process is assumed to be in control throughout this initial phase.

The control charts in Figure 16-3 were constructed from the rolled sheet thickness data. Chart (a) applies to values for $\bar{X}$ and identifies shifts in the process mean μ. Chart (b) represents the levels of the sample range R. It is used to identify those times when the process output does not consistently meet the targeted dimensions, which would be the case when σ shifts to a higher than intended level owing to some assignable cause. Such a change might not be reflected in sample means, which tend to average out any fluctuations. The R-chart may thus uncover quality problems not readily detected by the $\bar{X}$-chart alone.

The procedure for constructing the control charts is based on the same principles as those used to construct the P-chart.

Table 16-2 *Thickness of Rolled Steel Sheets*

Sample Number	Average Thickness of Each Sheet in Sample (inches)					Mean $\bar{X}$	Range R
1	.0494	.0489	.0508	.0485	.0501	.0495	.0023
2	.0514	.0520	.0510	.0469	.0485	.0500	.0051
3	.0481	.0504	.0486	.0496	.0500	.0493	.0023
4	.0508	.0477	.0481	.0487	.0458	.0482	.0050
5	.0491	.0453	.0523	.0498	.0480	.0489	.0070
6	.0458	.0508	.0492	.0478	.0526	.0492	.0068
7	.0473	.0527	.0494	.0511	.0521	.0505	.0054
8	.0525	.0492	.0492	.0475	.0495	.0496	.0050
9	.0533	.0524	.0516	.0490	.0487	.0510	.0046
10	.0502	.0502	.0513	.0497	.0514	.0506	.0017
11	.0507	.0485	.0456	.0475	.0485	.0482	.0051
12	.0509	.0477	.0520	.0486	.0479	.0494	.0043
13	.0467	.0516	.0510	.0498	.0507	.0500	.0049
14	.0503	.0471	.0490	.0515	.0490	.0494	.0044
15	.0506	.0485	.0515	.0498	.0494	.0500	.0030
16	.0501	.0495	.0475	.0475	.0487	.0487	.0026
17	.0518	.0499	.0499	.0471	.0521	.0502	.0050
18	.0479	.0506	.0520	.0506	.0493	.0501	.0041
19	.0481	.0480	.0517	.0512	.0503	.0499	.0037
20	.0479	.0506	.0492	.0523	.0502	.0500	.0044
21	.0521	.0477	.0523	.0516	.0521	.0512	.0046
22	.0475	.0478	.0494	.0524	.0517	.0498	.0049
23	.0499	.0488	.0511	.0473	.0456	.0485	.0055
24	.0500	.0500	.0492	.0502	.0509	.0501	.0017
25	.0483	.0502	.0476	.0468	.0489	.0484	.0034
						1.2407	.1068

Control Chart for the Sample Mean

To estimate μ the sample results are pooled, and the **grand mean** $\bar{\bar{X}}$ is computed by averaging the individual sample means. Since the grand mean estimates the central value about which individual $\bar{X}$'s will cluster, $\bar{\bar{X}}$ provides the centerline of the $\bar{X}$-chart. For the rolled sheets we have

$$\bar{\bar{X}} = \frac{1.2407}{25} = .04963 \text{ inch}$$

Thus, the process mean μ is estimated to be .04963 inch, and the centerline for the $\bar{X}$-chart in Figure 16-3 is placed at a height of .04963.

Figure 16-3 *Control charts for thickness of rolled steel sheets.*

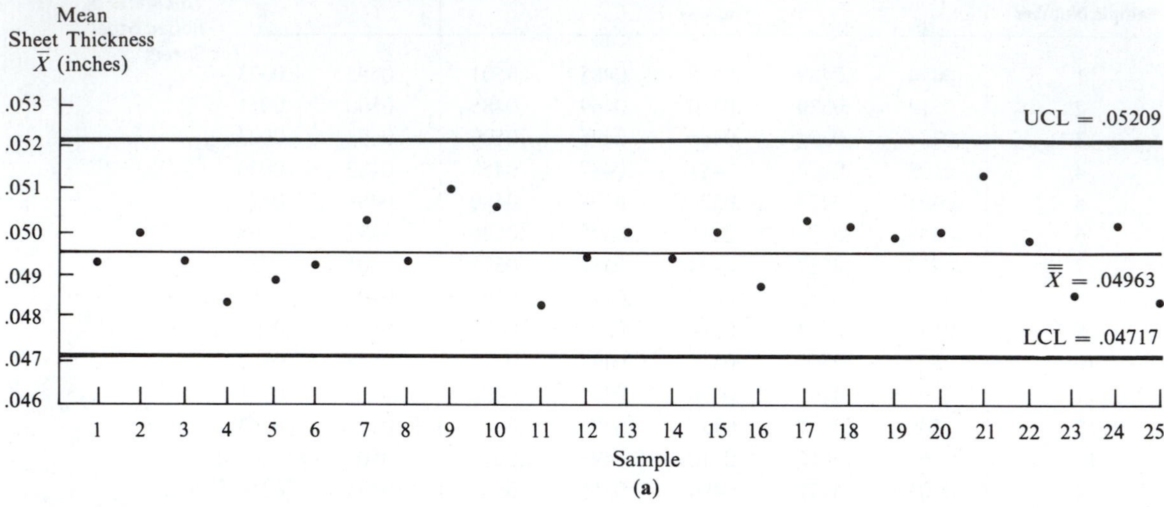

(a)

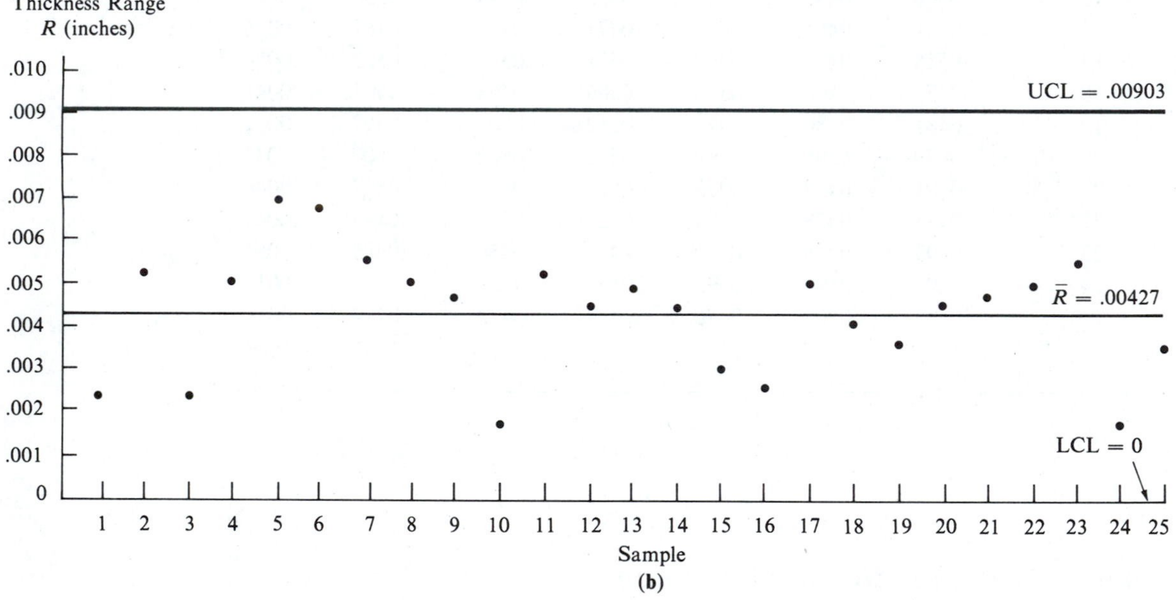

(b)

As before, 3-sigma limits must be found. The estimate of $\sigma_{\bar{X}}$ is more roundabout. Until now we have used the sample variance s^2 as the unbiased estimator of σ^2. The same result may be achieved using the sample range R.

Space does not permit a detailed explanation of the relation between R and σ. The essential property applies to normally distributed populations, for which it has been established that the ratio of the expected value of the sample range to the population standard deviation is equal to a constant. That quantity depends only on the sample size n:

$$d_2 = \frac{E(R)}{\sigma}$$

Using the mean of the sample ranges, $\bar{R}$, as the estimator of $E(R)$, the following estimator of the population standard deviation may be used:

$$\hat{\sigma} = \frac{\bar{R}}{d_2}$$

Values for d_2 are provided in Appendix Table K. For $n = 5$, $d_2 = 2.326$.

For the rolled sheet thickness data, we have

$$\bar{R} = \frac{.1068}{25} = .00427 \text{ inch}$$

so that

$$\hat{\sigma} = \frac{.00427}{2.326} = .001836 \text{ inch}$$

The standard error of $\bar{X}$ is estimated by $\hat{\sigma}/\sqrt{n}$.

In constructing the control chart for the mean, 3-sigma limits are generally used. With the above as estimators, the following expressions are used.

CONTROL LIMITS FOR THE SAMPLE MEAN

$$\mathbf{LCL} = \bar{\bar{X}} - \left(\frac{3}{d_2\sqrt{n}}\right)\bar{R} \qquad \mathbf{UCL} = \bar{\bar{X}} + \left(\frac{3}{d_2\sqrt{n}}\right)\bar{R}$$

For the rolled sheet thickness data, these provide

$$\text{LCL} = .04963 - \left(\frac{3}{2.326\sqrt{5}}\right)(.00427) = .04963 - .00246 = .04717$$

$$\text{UCL} = .04963 + .00246 = .05209$$

Control Chart for the Sample Range

The control chart for the sample range is found analogously. The centerline for individual sample ranges is provided by $\bar{R}$. The 3-sigma limits about this value are determined in the same general way as before. The standard error of the sample range, σ_R, is estimated by

$$\hat{\sigma}_R = \frac{\sigma_w}{d_2}\bar{R}$$

where σ_w is a constant that expresses the ratio of σ_R to the population standard deviation. Under the assumption of a normally distributed population, σ_w depends only on the sample size n. Along with d_2, σ_w may be read from Appendix Table K. With $n = 5$, $\sigma_w = .8641$.

The 3-sigma control limits are thus based on $\bar{R} \pm 3\hat{\sigma}_R$. Substituting the above for $\hat{\sigma}_R$ and simplifying, we obtain the following.

CONTROL LIMITS FOR THE SAMPLE RANGE

$$\textbf{LCL} = \left(1 - \frac{3\sigma_w}{d_2}\right)\bar{R} \qquad \textbf{UCL} = \left(1 + \frac{3\sigma_w}{d_2}\right)\bar{R}$$

The factors of R are sometimes separately tabled (usually denoted as D_3 for the LCL and D_4 for the UCL). For the rolled sheet thickness data, we have

$$\text{LCL} = \left(1 - \frac{3(.8641)}{2.326}\right)(.00427) = -.00049 \quad \text{or} \quad 0$$

and

$$\text{UCL} = \left(1 + \frac{3(.8641)}{2.326}\right)(.00427) = .00903$$

Note that the LCL *cannot be negative*, and thus the lower limit is taken to be *zero*.

Implementing Statistical Control

The industrial engineer kept a log of the quality-control findings during a 200-hour run with $\frac{1}{20}$-inch rolled steel sheets. In that time $\bar{X}$ exceeded the UCL twice and fell below the LCL once. The problems were promptly remedied by adjusting roller settings and replacing a malfunctioning optical scanner. In two separate samples R exceeded its UCL. In one case many of the starting ingots were too cool, a condition corrected when holding-oven thermometers were replaced. The other defect, an increase in thickness variability, was traced to a broken seal in a pneumatic press.

Without the quality management capability provided by the control charts, all these problems would have been detected many hours later, and several thousand square feet of rolled sheet would have had to be scrapped. The engineer estimated that using control charts saved the mill over $50,000 in costs in just this one production run. The costs of inspection involved only $500 in direct labor. Furthermore, future business was preserved with customers who might have found different suppliers had poor-quality sheets been shipped.

Problems

16-7 Determine the LCL and UCL for (1) $\bar{X}$ and (2) R in each of the following situations:

(a) $\bar{\bar{X}} = .10$, $\bar{R} = .005$, $n = 5$ (c) $\bar{\bar{X}} = 5.62$, $\bar{R} = .13$, $n = 4$

(b) $\bar{\bar{X}} = .005$, $\bar{R} = .001$, $n = 10$ (d) $\bar{\bar{X}} = 1.46$, $\bar{R} = .22$, $n = 7$

16-8 The following sample data apply to the thickness of $\frac{1}{20}$-inch rolled steel sheets. Compute the sample mean for each. Then, using the control limits found on page 683, indicate whether or not $\bar{X}$ places the process in control.

(a) .0505 .0475 .0516 .0452 .0555

(b) .0504 .0485 .0492 .0515 .0502

(c) .0523	.0515	.0531	.0528	.0541
(d) .0488	.0435	.0505	.0562	.0512
(e) .0453	.0479	.0462	.0474	.0458

16-9 Using the data in Problem 16-8, compute the sample range for each sample. Then, using the control limits found on page 684, indicate whether or not R places the process in control.

16-10 The following sample data apply to diameters of $\frac{1}{4}$-inch ball bearings:

(1) .254	.262	.245	.266	.259
(2) .241	.235	.241	.231	.242
(3) .255	.245	.253	.249	.252
(4) .255	.263	.265	.271	.255
(5) .241	.249	.250	.252	.253
(6) .255	.253	.247	.244	.248
(7) .235	.260	.253	.242	.259
(8) .254	.246	.246	.252	.260
(9) .243	.254	.256	.252	.246
(10) .259	.257	.253	.246	.249

(a) Compute the mean and range for each sample. Then find the control limits for $\bar{X}$ and construct the control chart, including the data point for each sample. Indicate for which samples (if any) the process is out of control.

(b) Find the control limits for R and construct the control chart, including the data point for each sample. Indicate for which samples (if any) the process is out of control.

(c) Eliminating the samples where the process was found to be out of control in parts (a) and (b), determine the revised control limits for (1) $\bar{X}$ and (2) R.

16-11 Assuming that the control limits on page 683 apply, determine the probabilities that the process for rolling steel sheet will for any particular sample of size $n = 5$ be found out of control in terms of $\bar{X}$ when the true average thickness is represented by a population with:

(a) $\mu = .052$ and $\sigma = .002$
(b) $\mu = .055$ and $\sigma = .003$
(c) $\mu = .047$ and $\sigma = .0015$
(d) $\mu = .048$ and $\sigma = .0025$

16-3 Acceptance Sampling

The second major area of statistical quality control is acceptance sampling. The objective is to determine how to dispose of items that have already been produced or have been provided by a supplier or vendor. The main consideration is the **lot**, which may be either a production batch or a shipment. This lot is treated as a statistical population with unknown characteristics. Two actions are generally considered: to *accept* the lot or to *reject* it, and these choices are predicated on the assumption that the lot is *good* or *bad*. Depending on circumstances, the items in a rejected production batch may be reworked, sold as seconds, or scrapped. A rejected shipment will usually be returned to the supplier.

Since it would be uneconomic to inspect every item in the lot, sample data are used to make the decision. We first encountered the underlying concepts of acceptance sampling in Chapter 9, where hypothesis testing was introduced. Because of the special terminology and notation used in acceptance sampling, it will be helpful to review some of those concepts here in the context of quality control.

The Underlying Procedure

Inspected sample items from a lot are ordinarily classified as *defective* or *satisfactory*. In the case of quantitative measurements for some variate such as volume or dimension, those items falling outside of tolerance specifications are classified as defectives. A good lot is one having a low proportion of defectives π. The level of this parameter is unknown, although various values are hypothesized. We have previously used the sample proportion P as the test statistic for decisions involving π (and have applied the normal approximation after transforming P to z). Acceptance sampling decision rules are often equivalently expressed in terms of the number of defectives, all of which may be summarized by an **acceptance number** C. When the number of sample defectives is $\leq C$, the lot is accepted, whereas it is rejected if the number of defectives is $>C$.

The decision structure is the same as that of more general hypothesis-testing procedures. The null hypothesis is H_0: $\pi \leq \pi_0$ (the lot is good), while the alternative is that $\pi > \pi_0$ (the lot is bad). The value π_0 is referred to as the **acceptable quality level**. This value is one of two extreme cases used in evaluating acceptance sampling procedures. The other is the **lot tolerance proportion defective**, denoted as π_1. In keeping with our earlier notation, we represent these quantities as decimal fractions. (In practice these are often stated as percentages.)

Accepting the shipment is the same as accepting H_0, while rejecting the shipment is to reject H_0. There are two correct decisions: accepting a good lot and rejecting a bad one. There are two incorrect ones: rejecting a good shipment and accepting a bad one.

Producer's and Consumer's Risk

The incorrect action of rejecting a good shipment is the type I error. This result is what suppliers or vendors want most to avoid for shipments received by their customers. The probability α of this error is often referred to as the **producer's risk**. The second incorrect action is to accept a bad shipment, which is to accept the null hypothesis when it is false. This is the type II error, for which the probability β is sometimes called the **consumer's risk**.

These designations of risks are historical and may be explained from the manufacturer's point of view. The producer inspects each production batch before accepting it for shipment to the consumer. In such a sampling investigation, the rejection of a good production batch is very expensive and is ordinarily judged to be the more serious error. The error of accepting a bad lot is less troublesome in the short run. The customer or consumer would then be stuck with an excessive number of defective items, which the producer will typically replace once identified.

But the consumer's risk must be kept small, too, for in the long run the customer may become discouraged and search for a new source of supply.

Selecting the Decision Rule

For any given levels for π_0, sample size n, and lot size N, the probability for the type I error may be determined. In Chapter 3 we saw that the number of sample defectives D is a random variable having the hypergeometric distribution with mass function $h(r; n, \pi, N)$ described in Section 3-5. Thus,

$$\begin{aligned}\alpha = \Pr[\text{reject } H_0 \mid \pi = \pi_0] &= \Pr[D > C] \\ &= 1 - \Pr[D \leq C] \\ &= 1 - \sum_{r=0}^{C} h(r; n, \pi_0, N)\end{aligned}$$

To illustrate, consider a lot of $N = 500$ items from which a random sample of $n = 25$ is taken and each unit is classified as defective or satisfactory. Assuming an acceptable quality level of $\pi_0 = .05$, then for an acceptance number of $C = 3$ defectives, the type I error probability is

$$\alpha = 1 - .9699 = .0301$$

Computation of hypergeometric probabilities is burdensome. Since there are too many combinations of parameter values to completely table hypergeometric probabilities, they are often approximated by the binomial (or the normal) distribution.

As we have seen, the usual approach in a statistical test is to fix the level of α and then establish the acceptance and rejection regions (which is the same as finding C). For a fixed n, it is always possible to find C, which will be the greatest level of r for which the computed type I error probability is $\leq$ the α target. But we have seen that the subsequent type II error probabilities (β's) can be huge unless n is kept large. The operating characteristic (OC) curve provides a useful summary of accept probabilities for levels of π lying above π_0.

Figure 16-4 shows the OC curve constructed using $C = 3$ with $n = 25$ and $N = 500$. Notice that the probability of accepting not-so-good lots can be substantial: .8700 when $\pi = .08$ and .4677 when $\pi = .15$. To provide a satisfactory balance between the type I error probability and the various type II error probabilities, targets are often set for both α and β. The latter applies when $\pi = \pi_1$, the level chosen as the lot tolerance proportion defective (an arbitrary quantity representing a barely satisfactory lot). When N, π_0, π_1, α, and β are prescribed, it is possible to find the combination of C and n that meets those targets. Procedures for finding (C, n) decision rules are elaborate and are described in many of the references listed in the Bibliography.

Sequential Sampling Plans

Acceptance sampling procedures may be based on more than one sample. Typical of such schemes is **sequential sampling**. This procedure involves a series of separate random samples. After each is collected, one of three actions is taken: (1) The lot

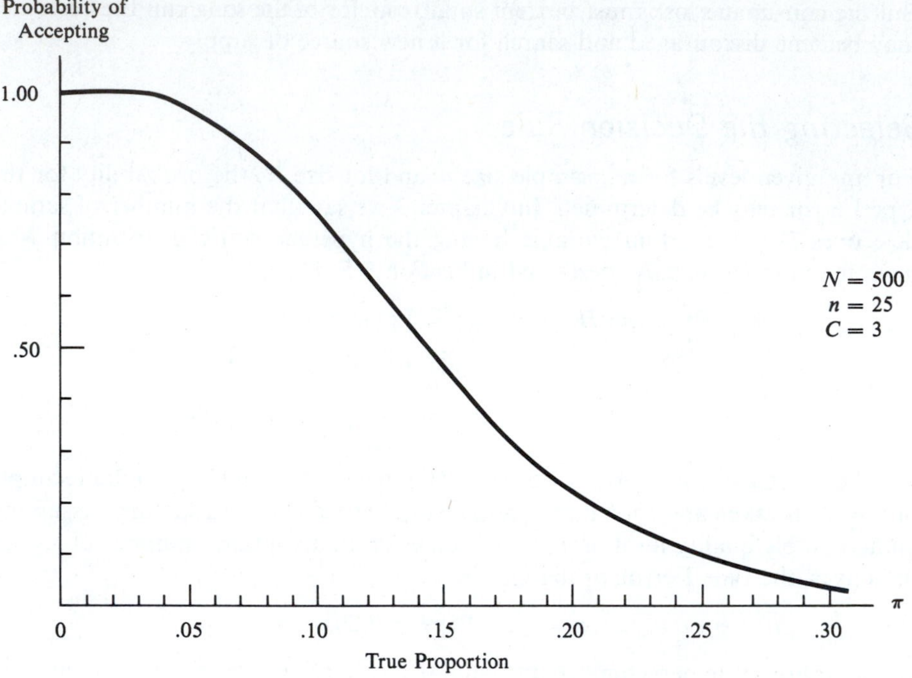

Figure 16-4 *Operating characteristic curve for acceptance sampling.*

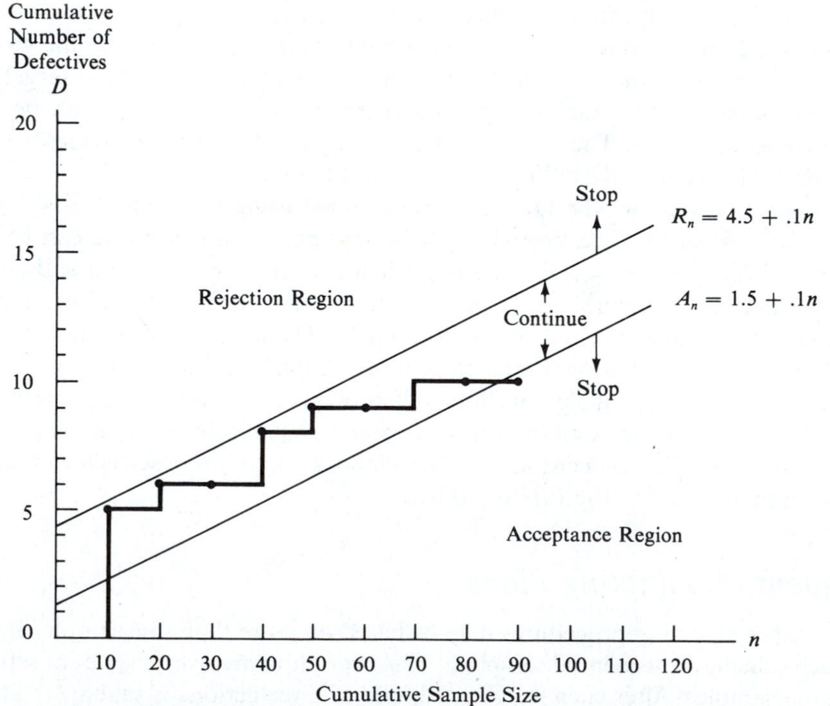

Figure 16-5 *Illustration of a sequential sampling scheme.*

is accepted, (2) the lot is rejected, or (3) ultimate lot disposition is deferred until the result of one more set of sample observations becomes known. The decision rule involves both an acceptance number A_n and a **rejection number** R_n. Each number may be expressed as a function of the cumulative sample size n. Sampling continues as long as $A_n \leq D \leq R_n$. The chart in Figure 16-5 illustrates a sequential sampling plan involving successive samples of size 10 taken from a lot having $N = 500$ items.

Sequential sampling schemes make more efficient use of the observations than can single samples. This efficiency is reflected in lower sample size requirements for the same α and β targets. The major drawback of sequential sampling schemes is that finding the decision rules is a more complicated process. They also demand a higher level of training for inspection personnel.

Space does not permit a detailed discussion here. Further information on sequential sampling may be found in many references listed in the Bibliography.

Problems

16-12 A sample of 100 items is taken from a lot containing 10,000. Assuming an acceptable quality level of .10, use the binomial approximation to find the acceptance number when the type I error probability can be no greater than

(a) $\alpha = .01$ (b) $\alpha = .025$ (c) $\alpha = .05$ (d) $\alpha = .10$

16-13 Repeat Problem 16-12 using a sample of size 20.

16-14 Repeat Problem 16-12 using an acceptable quality level of .05.

16-15 A sample of size 100 is taken from a lot containing 2,000. The acceptance number is 6 defectives. Using the binomial approximation, determine the probability of accepting a lot when the true proportion of defectives is

(a) $\pi = .01$ (b) $\pi = .05$ (c) $\pi = .10$ (d) $\pi = .20$

16-16 Consider the following numbers of defectives obtained in successive samples of 10 items:

(1) 1	(3) 2	(5) 1	(7) 1	(9) 0
(2) 1	(4) 2	(6) 3	(8) 1	(10) 1

Determine for each of the following sequential sampling plans which of these would be the last set of observations and whether the lot would have been accepted or rejected:

(a) $A_n = 1 + .1n$, $R_n = 3 + .1n$ (c) $A_n = .05n$, $R_n = 1 + .05n$

(b) $A_n = 2 + .05n$, $R_n = 4 + .05n$ (d) $A_n = .2n$, $R_n = 1 + .2n$

CHAPTER SEVENTEEN

Reliability Analysis

As modern society has become progressively more complex, there has been increasing dependency on devices and man–machine systems. The complexity of these has been growing exponentially over time, making it ever more important that malfunctions occur only infrequently. Since engineers are charged with responsibility for designing, fabricating, and often controlling these products and systems, they must also ensure high reliability. Reliability analysis is therefore one of the most important statistical applications for modern engineering.

As a statistical concept, the following definition is given to reliability.

> *Definition* An entity's **reliability** is the *probability* that it will survive fully functional throughout a particular time span.

The entity may be a part, such as a fuse, or a complex system, such as the communication capability of a surveillance satellite. The reliability of the system may sometimes be expressed as a function of the reliabilities of the components or subsystems. Reliability will often be an integral part of system design, so that a final configuration will be determined that meets overall reliability specifications.

Time to failure is the key variable in reliability analysis. A variety of probability distributions characterize how long an entity will survive. One aspect of reliability analysis is finding a distribution model that provides an appropriate fit to the item or system under investigation. Among the important probability distributions are the exponential, gamma, and normal (in truncated form)—encountered in earlier chapters. The exponential and gamma distributions have attractive mathematical properties, but they are based on assumptions that may not apply to the lifetimes of certain items and systems. A more flexible distribution that is quite useful in reliability analysis is the Weibull distribution, which is introduced in this chapter.

Reliability specifications typically stipulate a target for the **mean time between failures**, or MTBF. More generally, reliability engineers are concerned with identifying the entire appropriate failure-time distribution, which may be used to *predict* the reliability of a proposed product or system. Such predictions provide an interesting and useful application of probability concepts. Through experimentation, these investigators will also *assess* the reliability of existing designs, a process involving statistical inference, including estimation and hypothesis testing.

17-1 Underlying Concepts of Reliability Analysis

Failure time T is the key variable in reliability analysis. This is a continuous random variable with probability density function $f(x)$. The failure-time cumulative probability distribution function

$$F(t) = \Pr[T \leq t] = \int_0^t f(x)\,dx$$

establishes the reliability function.

RELIABILITY FUNCTION

$$\mathbf{\bar{F}(t) = Pr[T > t] = 1 - F(t)}$$

Since this gives the probability that the item or system survives longer than time t, $\bar{F}(t)$ is sometimes referred to as the *survival function.*

Exponential Failure-Time Distribution

The exponential distribution is one of the common failure-time distributions. In Chapter 4 it was established that this family of distributions has probability density function

$$f(x) = e^{-\lambda x}$$

The failure-time cumulative distribution function is

$$F(t) = 1 - e^{-\lambda t}$$

and the reliability function is therefore

$$\bar{F}(t) = \lambda e^{-\lambda t}$$

To illustrate, suppose that a particular type of printed circuit board has a mean lifetime of 10,000 hours, so that the mean failure rate is

$$\lambda = \frac{1}{10{,}000} = .0001 \text{ per hour}$$

The following reliabilities apply:

$$\bar{F}(1{,}000) = \exp[-.0001(1{,}000)] = .9048$$

$$\bar{F}(5{,}000) = \exp[-.0001(5{,}000)] = .6065$$

$$\bar{F}(10{,}000) = \exp[-.0001(10{,}000)] = .3679$$

$$\bar{F}(50{,}000) = \exp[-.0001(50{,}000)] = .0067$$

The exponential distribution is appropriate to a large class of lifetimes. All of these involve failures that occur randomly over time. Recall that the exponential distribution applies to a Poisson process and that failures in one time period are statistically independent of those in any other nonoverlapping period. In effect, the process has no memory, so the remaining lifetime expected for an old printed circuit board of the above type is the same as that for a new one. If the exponential distribution literally applies to an item's lifetime, then the tendency for it to fail remains unchanged as it ages.

Human mortality provides an analogy for explaining this critical assumption. If the exponential distribution were a valid model for representing time until death, then people would die only from random causes—such as an exotic disease or an accident—and the probability of death would be the same for a baby as a 100-year-old or 500-year-old. (There would be no such thing as "old age" as the cause of death.) The death rate would be the same for all ages, and everybody would sip from a perpetual fountain of youth.

The Failure Rate Function

Of course, human mortality is not constant. As people become older, they wear out and eventually meet their end because of old age. Many products, machines, and complex systems exhibit lifetimes similar to those of people, so their failure rates change over time. More generally, the failure rate is a function of an entity's lifetime so far.

FAILURE RATE FUNCTION

$$h(t) = \frac{f(t)}{1 - F(t)} = \frac{f(t)}{\bar{F}(t)}$$

This function is found by dividing the density function at time t by the reliability for that duration. It is sometimes referred to as the *hazard rate function.*

Figure 17-1 shows the representative shape for a failure rate function. Early in a system's life the failure rate is high due to the presence of defective original components. As these are replaced, the failure rate drops steeply. For electronic systems this initial stage of life systems is commonly referred to as the "burn-in" period. The next phase is a long period of stable or nearly constant failure rate. Eventually, components start to deteriorate physically, and the failure rate rises rapidly. The representative failure rate function creates a plot called the *bathtub curve*, as suggested by its shape. Human lifetimes follow this pattern, with the highest mortality rates occurring during early infancy, falling throughout childhood, and stabilizing during early adulthood through middle age, when causes of death are largely random. In later years, people experience increasing mortality as physical deterioration progresses.

The exponential distribution provides a *constant* failure rate over time, so that $h(t) = \lambda$. It empirically fits the actual experience of many systems only over the flat portions of their bathtub curves—that is, after they have survived the initial burn-in period and before the onset of physical deterioration. The applicability of the exponential distribution is limited (although it is favored because of

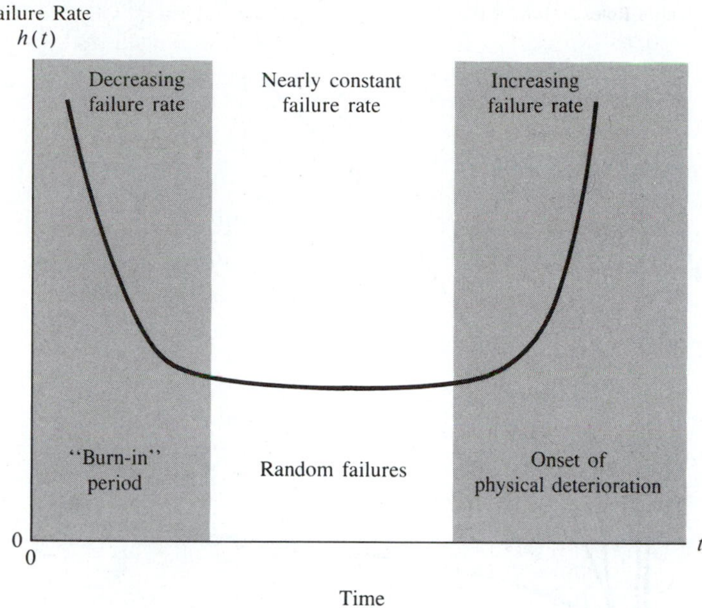

Figure 17-1 *Representative failure rate over time, or bathtub curve.*

its nice mathematical properties). Since a reliability analysis may involve items or systems with decreasing or increasing failure rates, a more general distribution is important to reliability analysis. The *Weibull distribution* fills this bill.

The Weibull Distribution

Along with the exponential, the Weibull is one of the most useful failure-time distributions in reliability analysis. The following expressions are used.

WEIBULL PROBABILITY DISTRIBUTION

$$f(t) = \lambda\beta(\lambda t)^{\beta-1}e^{-(\lambda t)^\beta}$$

$$F(t) = 1 - e^{-(\lambda t)^\beta}$$

where the parameters λ and β are nonnegative constants. The reliability function is

$$\bar{F}(t) = e^{-(\lambda t)^\beta}$$

The Weibull distribution has the following failure-rate function:

$$h(t) = \frac{\lambda\beta(\lambda t)^{\beta-1}e^{-(\lambda t)^\beta}}{e^{-(\lambda t)^\beta}} = \lambda\beta(\lambda t)^{\beta-1}$$

Figure 17-2 shows the failure rates for selected members of the Weibull family. Notice that depending on the level for the parameter β, as t becomes greater, $h(t)$ decreases (when $\beta < 1$), remains constant (when $\beta = 1$), or increases (when $\beta > 1$).

The Weibull density function may take a variety of forms. The parameter β is the *shape parameter*. Figure 17-3 shows Weibull distributions for selected levels

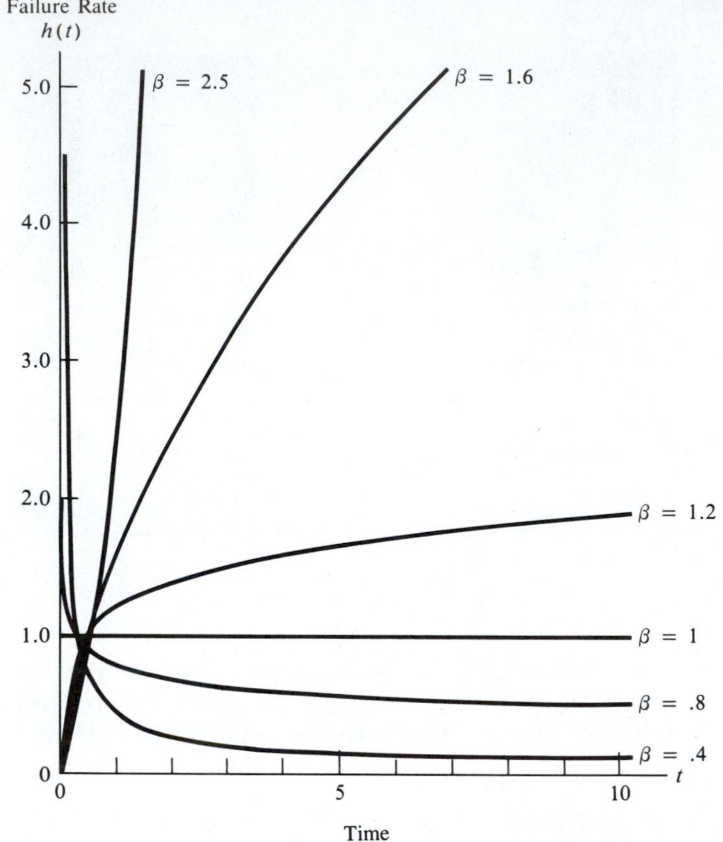

Figure 17-2 *Failure rate curves for Weibull distribution with $\lambda = 1$.*

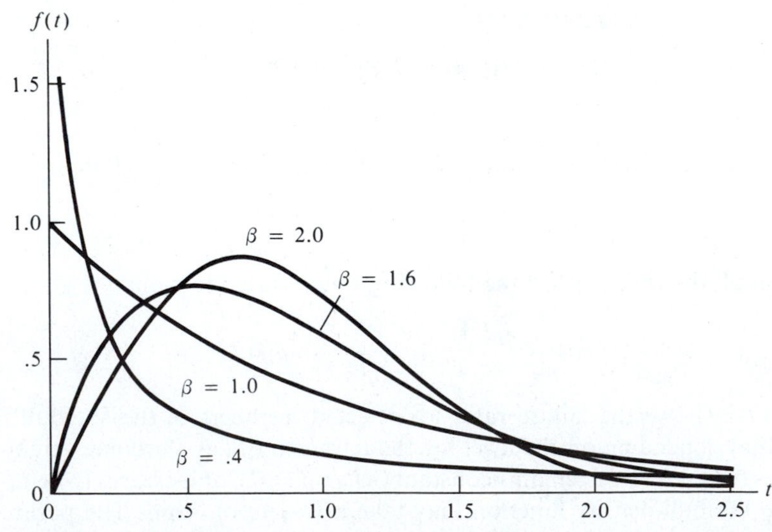

Figure 17-3 *Weibull distributions with selected values for shape parameter ($\lambda = 1$).*

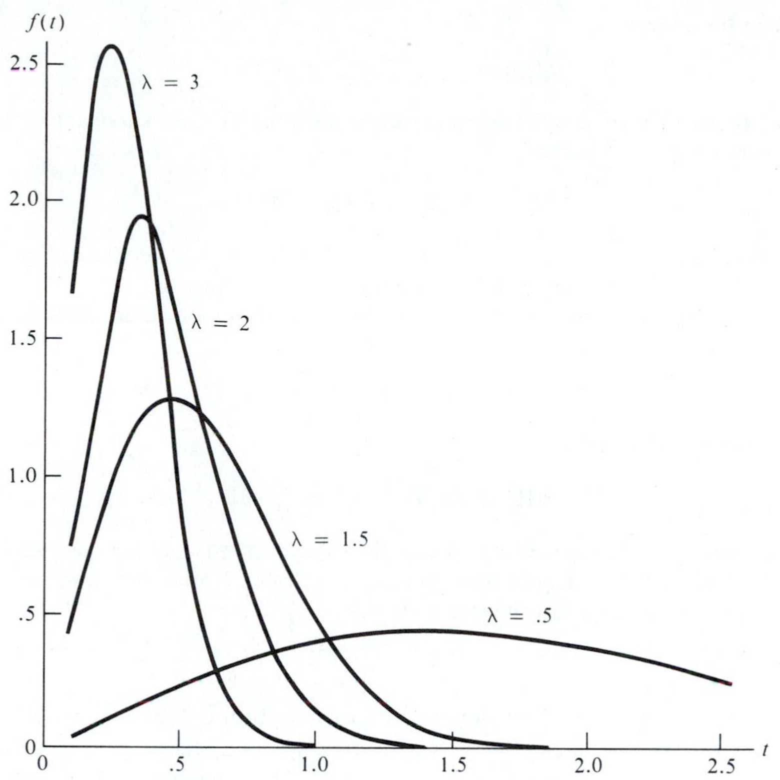

Figure 17-4 *Weibull distributions with selected values for scale parameter ($\beta = 2$).*

of β when $\lambda = 1$. Notice that the curves may be one-tailed or two-tailed with a positive skew. The second parameter λ is the *scale parameter*. Figure 17-4 shows Weibull distributions for several λ's when $\beta = 2$. Section 17-3 describes how Weibull parameters may be estimated from sample data.

Example: *Fuse Lifetimes*

A reliability engineer assumes that the lifetimes of a particular fuse will follow the Weibull distribution, with parameters $\lambda = .005$ per year and $\beta = .10$. He determines the probability that a randomly chosen fuse will function for more than $t = 10$ years.

The reliability function provides the answer:

$$\bar{F}(10) = \exp\{-[.005(10)]^{.10}\}$$

He first evaluates .005(10) = .05 to the .10 power by taking the natural logarithm

$$\ln[(.05)^{.10}] = .10 \ln(.05) = .10(-2.99573) = -.299573$$

so that

$$(.05)^{.10} = \exp(-.299573) = .74113$$

Using the above,

$$\bar{F}(10) = \exp(-.74113) = .4766$$

In a similar fashion the engineer obtained probabilities for survival beyond 5 years and past 15 years:

$$\bar{F}(5) = .5008 \qquad \bar{F}(15) = .4622$$

He was surprised to see that the probability that the fuse will survive beyond 5 years is so close to the probability for a longer-than-15-year life.

He found an explanation for this by first computing the failure rates when $t = 5$,

$$h(5) = .005(.10)[.005(5)]^{.10-1} = .0138$$

and when $t = 10$ and $t = 15$:

$$h(10) = .0074 \qquad h(15) = .0051$$

The failure rates are decreasing (always the case for $\beta < 1$) and are quite small, so that failure within a short time span is a rare event. Notice that from $t = 10$ years to $t = 15$ years, $h(t)$ changes very little.

An interesting property of the Weibull distribution is that it becomes the exponential distribution when $\beta = 1$. It also includes as a second special case—the *Rayleigh distribution*—when $\beta = 2$.

Other Failure-Time Distributions

Several less important distributions are encountered in reliability analysis. Some common ones are described here.

Gamma Distribution The gamma distribution described in Chapter 4 can have decreasing, increasing, or constant failure rates, depending on the parameter settings. This distribution is especially important in reliability analysis because of its relation to the exponential distribution. In a system with several independent components, each having identical exponential failure-time distributions, the gamma distribution provides probabilities for that time by which a specified number of these will fail. The following expression may be used to compute gamma survival probabilities.

GAMMA RELIABILITY FUNCTION (EXPONENTIAL COMPONENTS)

$$\bar{F}_r(t) = \sum_{k=0}^{r-1} \frac{(\lambda t)^k}{k!} e^{-\lambda t}$$

where $\lambda > 0$ and r is the specified number of failing components.

Example: Survival of Satellite Power System

A satellite is powered by 10 solar cells connected in a parallel system. There will be enough power for full operation if at least 3 cells are working. The cell lifetimes are independent random variables, each having exponential distribution, with failure rate $\lambda = 1$ per year. What is the probability that the satellite will have a useful life of at least 2 years?

The system lifetime has a gamma distribution with parameters $\lambda = 1$ and $r = 7$ failures. Using $\lambda t = 1(2) = 2$, the probability is

$$\begin{aligned}\Pr[T > 2] &= \bar{F}_7(2)\\ &= \left(\frac{2^0}{0!} + \frac{2^1}{1!} + \frac{2^2}{2!} + \frac{2^3}{3!} + \frac{2^4}{4!} + \frac{2^5}{5!} + \frac{2^6}{6!}\right)e^{-2}\\ &= (1 + 2 + 2 + 1.3333 + .6667 + .2667 + .0889)(.1353)\\ &= .9952\end{aligned}$$

Truncated Normal Distribution The normal distribution as used earlier in this book would be unsuitable as a failure-time distribution because it allows for negative values of t. That obstacle may be surmounted by using only that portion of the normal curve where t is nonnegative. This modification results in the **truncated normal distribution.** One property of this distribution is that failure rates are *increasing* functions of t, a feature limiting the truncated normal's usefulness for reliability analysis.

This distribution is not to be confused with the *lognormal distribution* introduced in Chapter 4, which also has been proposed as a failure-time distribution. The lognormal is unsuitable for reliability analysis because it has a failure rate that first increases and then decreases rapidly to zero.* It would, in effect, turn the bathtub curve upside down—a feature not supported by empirical failure-rate data.

Problems

17-1 The exponential distribution applies to lifetimes of a certain component. Its failure rate λ is unknown. Find the probability that the component will survive past 5 years assuming: (a) $\lambda = .5$/year, (b) $\lambda = 1$/year, and (c) $\lambda = 2$/year.

17-2 A particular capacitor has a mean time between failures of 5 years. Assuming that the exponential distribution applies, determine the probability that a capacitor placed in continuous operation will survive beyond (a) 2 years, (b) 5 years, (c) 10 years, and (d) 20 years.

17-3 Consider the reliability engineer's problem in the example on page 695. Change β to .20, then answer the following.

(a) Recompute the survival probabilities for (1) 5, (2) 10, and (3) 15 years.

(b) Recompute the failure rates at each of these times.

* Barlow, Richard E., and Frank Proschan, *Mathematical Theory of Reliability* (New York: Wiley, 1965), p.11.

17-4 Repeat Problem 17-3 using $\lambda = .01$ per year while keeping $\beta = .10$, the original level.

17-5 You are conducting a reliability analysis for a new product. Based on prior testing with a similar product, you believe the Weibull failure-time distribution, with parameter $\lambda = .20$ per year, applies. But you have no basis for establishing β. Compute at a time span of 5 years (1) the failure rate and (2) the survival probability assuming that (a) $\beta = .5$, (b) $\beta = .8$, (c) $\beta = 1.2$, and (d) $\beta = 2.0$.

17-6 A solar-powered foghorn is attached to a buoy moored off-shore from a remote coast. Five solar collectors independently feed the power supply system. As long as at least 2 collectors remain functional, the buoy will operate satisfactorily. Each collector has an exponential failure-time distribution with a mean failure rate of 1 per year. Find the probability that the buoy will continue to operate without interruption beyond 3 years.

17-7 The failure-time distribution for an item is represented by a normal distribution with mean $\mu = 2{,}000$ hours and standard deviation $\sigma = 500$ (so that negative values for t have negligible probabilities and there is no need to truncate). Find the probability that the item survives beyond (a) 1,800 hours, (b) 2,500 hours, (c) 3,000 hours, and (d) 3,500 hours.

17-8 The density function $f_X(x)$ for the normal distribution is given on page 168. Adapt this to establish the density function $f_T(t)$ for the truncated normal. This may be accomplished by replacing x by t and multiplying $f_X(t)$ by an appropriate constant so that the integral from 0 to infinity of the resulting density is equal to 1.

17-2 Predicting the Reliability of Systems

An important application of reliability analysis is the prediction of an overall system's reliability, using as building blocks the reliabilities of individual components. It is convenient to use the notation

$$R_i(t) = \Pr[\text{component } i \text{ survives beyond time } t]$$

The reliability of the system is then a function of the reliabilities of its components:

$$R_S(t) = f\{R_1(t), R_2(t), \ldots\}$$

When t is fixed at a specified level, a closed form may be obtained for $R_S(t)$ that coincides with the system logic. There are two primary system forms: one applies when the components are arranged in *parallel* and another when they are joined in a *series*. More complex cases involve modular subsystems as building blocks. We can portray the logic of systems schematically in similar fashion to electrical circuits.

Systems with Series Components

A **series system** is one that performs satisfactorily as long as *all* components are fully functional. Figure 17-5 shows the logic of a small system. The system is analogous to an electrical circuit that works as long as current flows through all com-

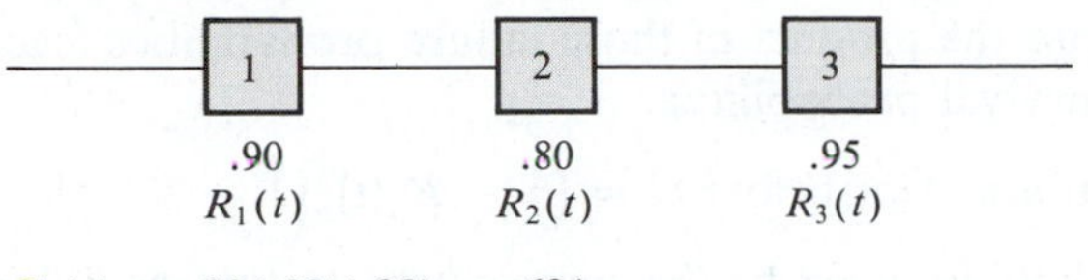

Figure 17-5 *Logic for system with series components.*

ponents. Failure of any component will cause the entire system to fail. Propulsion on the NASA space shuttle is an example of a series system since failure on any of the booster rockets will result in an aborted mission.

Under the series logic, the system will survive past time t only if its components do so. System reliability, therefore, is equal to the product of the reliabilities for the components:

$$R_S(t) = R_1(t)R_2(t)\ldots R_n(t)$$

The system in Figure 17-5 involves three components with reliabilities $R_1(t) = .90$, $R_2(t) = .80$, and $R_3(t) = .95$. The system reliability is

$$R_S(t) = .90(.80)(.95) = .684$$

Systems with Parallel Components

A **parallel system** is one that performs as long as *any one* of its components remains operational. Figure 17-6 is a schematic for a simple parallel system. The electrical analogy is a circuit through which current flows as long as any component works, providing an open path. All components must fail before the parallel system does. For example, batteries for a hand-held calculator might be arranged in parallel, so that the calculator may function as long as at least one battery provides sufficient power.

A parallel system will survive past time t if at least one component survives. Failure of the system occurs only if all components fail. The probability of system

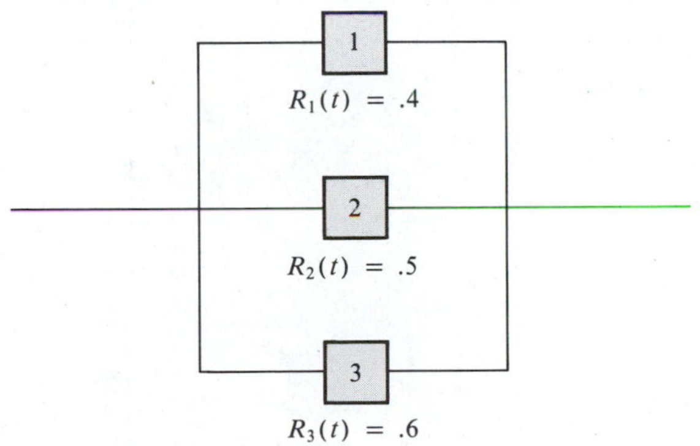

Figure 17-6 *Logic for system with parallel components.*

failure is therefore the product of those failure probabilities (each being 1 minus the respective survival probability):

$$\Pr[\text{system failure at or before } t] = [1 - R_1(t)][1 - R_2(t)]\ldots[1 - R_n(t)]$$

Thus, system reliability must be the complement to this, as follows:

$$R_S(t) = 1 - [1 - R_1(t)][1 - R_2(t)]\ldots[1 - R_n(t)]$$

The parallel system in Figure 17-6 has component reliabilities of $R_1(t) = .4$, $R_2(t) = .5$, and $R_3(t) = .6$. That system's reliability is

$$\begin{aligned} R_S(t) &= 1 - (1 - .4)(1 - .5)(1 - .6) \\ &= 1 - .12 \\ &= .88 \end{aligned}$$

Modular Systems

Individual components may be aggregated into subsystems according to their interrelationships. Figure 17-7 shows a modular system with four primary subsystems—A, B, C, and D—arranged in series. To find the system reliability,

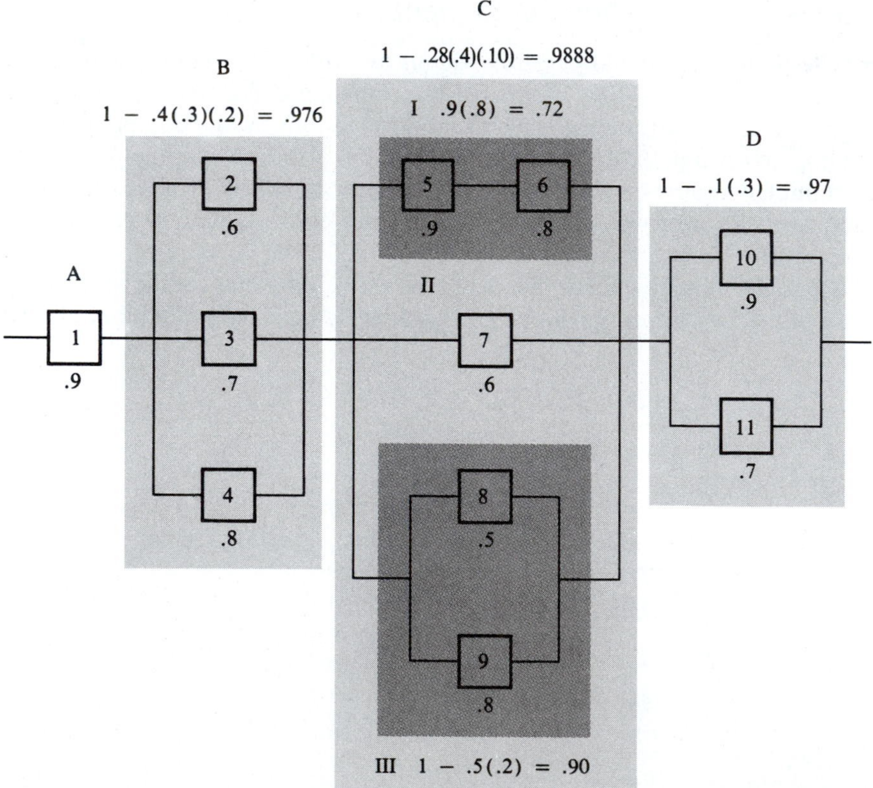

Figure 17-7 *Modular system with series subsystems.*

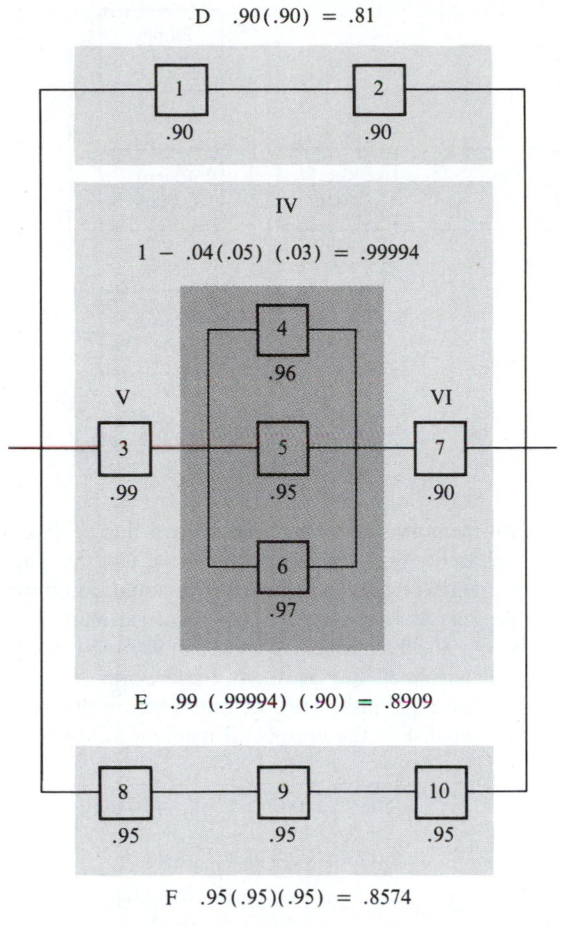

Figure 17-8 *Modular system with parallel subsystems.*

we must first establish separate reliabilities for each subsystem. The most complex module is C, with three parallel components or subsystems of its own—I, II, and III. Before we can get the reliability of C, these must each be evaluated.

Figure 17-8 shows another module with three subsystems—D, E, and F—arranged in parallel. Each primary subsystem involves a series structure, and their reliabilities must be established before we can compute overall system reliability. Subsystem E has three components or subsystems, including IV, itself a parallel subsystem.

Example: *Reliability of a Personal Computer System*

A software engineer's personal computer system is shown in Figure 17-9. The reliabilities given there represent the probability that the component performs satisfactorily throughout a specific workday. This setup involves 5 components: the power supply (reliability .995), PC unit (.999), keyboard (.9999), peripheral memory subsystem, and printer subsystem. These are shown in *series* since all must function in order for the system to be usable.

Figure 17-9
Logic for personal computer system.

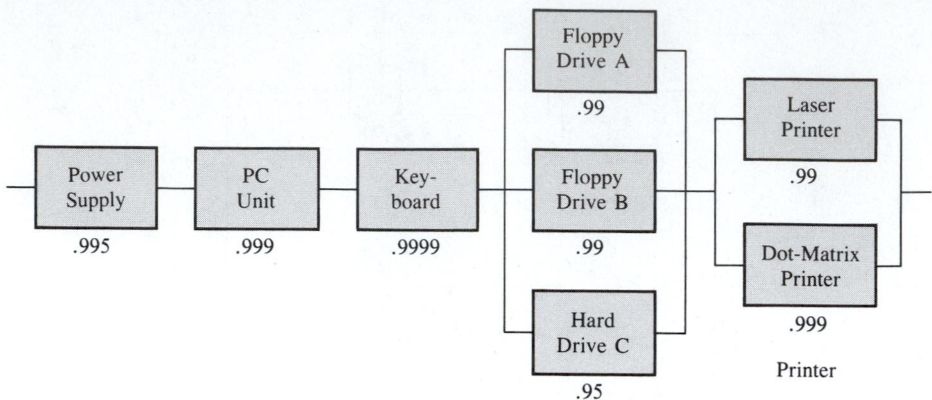

The peripheral memory subsystem includes two floppy disk drives (A and B, both with .99 reliability) and one hard disk drive C (.95). The system logic shows those components in *parallel*, since the personal computer system will be functional as long as at least one of them is operational. The printer subsystem also has two parallel components, one laser (.99) and one dot-marix (.999) printer; the system is usable as long as one of these works.

The two subsystem reliabilities must be found before the overall system reliability can be computed. For the peripheral memory subsystem, we have the reliability

$$1 - (1 - .99)(1 - .99)(1 - .95) = .999995$$

And, the reliability of the printer subsystem is

$$1 - (1 - .99)(1 - .999) = .99999$$

The overall system reliability is

$$R_S = .995(.999)(.9999)(.999995)(.99999) = .994$$

Notice that a system reliability of .994 is only slightly lower than the .995 reliability of the power supply by itself. Although minuscule improvements in system reliability might be achieved by substituting hardier components, no significant improvement in survival probability can be achieved without greater power reliability. Since power is supplied by a public utility, that might only be achieved by installing an auxillary power system.

Increasing System Reliability

A design issue in systems is how to increase system reliability. One way this might be accomplished is by raising the reliability of individual components (perhaps by substituting better materials, such as gold wiring instead of copper). To illustrate, consider the system in Figure 17-10(a). Suppose that the reliability of component 2 can be raised to .95, as in Figure 17-10(b), by substituting a more expensive item.

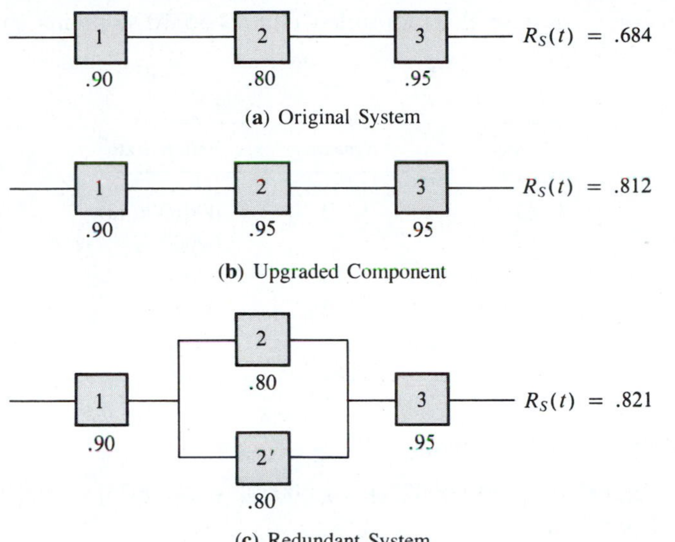

Figure 17-10 *Two approaches for increasing system reliability.*

The system reliability then becomes

$$R_S(t) = .90(.95)(.95) = .812$$

which is a substantial boost over the original level of .684.

A second approach to boosting system reliability is to duplicate certain components, replacing individual items with a parallel subsystem composed of two or more identical units. In effect, reliability is increased through system *redundancy*. Consider the system again, but now modified as in Figure 17-10(c). In this system, component 2 has an identical partner 2′, joined in parallel. The new subsystem has reliability

$$1 - [1 - R_2(t)][1 - R_{2'}(t)] = 1 - (1 - .8)(1 - .8) = .96$$

and the overall system reliability becomes

$$R_S(t) = .90(.96)(.95) = .821$$

Exponential Series Systems

One reason why the exponential distribution is so common in reliability analysis is its nice mathematical properties. We see this in a *series system* having N types of components. We assume that each has an exponential distribution, with the failure rate for each type i component denoted as λ_i. There may be several items of each type, with n_i denoting the quantity of each. We further assume that the lifetimes of all are *statistically independent*. Then *the system failure-time distribution will itself be exponentially distributed*, with mean failure rate

$$\lambda_S = \sum_{i=1}^{N} n_i \lambda_i$$

To illustrate, suppose that a printed circuit board contains the following components:

Item	Quantity	Failure Rate
1. Chip	5	.00010/hour
2. Resistor	20	.00002
3. Capacitor	10	.00050
4. Switch	2	.00025
5. Relay	5	.00015

The system's mean failure rate is

$$\lambda_S = 5(.00010) + 20(.00002) + 10(.00050) + 2(.00025) + 5(.00015)$$
$$= .00715$$

The mean time to failure (MTTF) is

$$\text{MTTF} = \frac{1}{\lambda_S} = \frac{1}{.00715} = 139.9 \text{ hours}$$

and the $t = 100$-hour reliability is

$$R_S(100) = \bar{F}(100) = \exp[-.00715(100)] = .489$$

Exponential Parallel Systems: Gamma Reliability

In Section 17-1 we encountered the gamma failure-time distribution. That distribution applies for the time until r failures occur among n independent items, each of which shares a common exponential failure-time distribution with failure rate λ. The gamma provides the reliability for a *parallel* system of such components.

For a system that survives as long as at least one out of n components does, the gamma reliability function gives

$$R_S(t) = \bar{F}_n(t) = \sum_{k=0}^{n-1} \frac{(\lambda t)^k}{k!} e^{-\lambda t}$$

To illustrate, suppose that a redundant system has $n = 4$ components, each with mean failure rate $\lambda = .01$ per hour. The $t = 200$-hour system reliability is, using $\lambda t = 2$,

$$R_S(200) = \bar{F}_4(2) = \left[\frac{2^0}{0!} + \frac{2^1}{1!} + \frac{2^2}{2!} + \frac{2^3}{3!}\right] e^{-2}$$
$$= (1 + 2 + 2 + 1.3333)(.1353)$$
$$= .857$$

Additional Comments

Of course, system components may have failure-time distributions other than the exponential. Closed-form expressions for the reliabilities of such systems can be very difficult, if not impossible, to achieve when t is not fixed. However, one approach is to find bounds on the system reliability by making simplifying assumptions. Such sophisticated approaches still assume that the components fail independently, though, which may not be the case. The more complex systems can sometimes be evaluated using Monte Carlo simulation techniques. More information on those procedures may be found in the references listed in the Bibliography in the back of this book.

Problems

17-9 Consider a system having four components with reliabilities through time t of (1) .9, (2) .7, (3) .8, and (4) .6. Find the system reliability assuming (a) series logic and (b) parallel logic.

17-10 Which has greater reliability: (a) a series system with 4 components, each having .99 reliability, or (b) a parallel system with 4 components, each having .90 reliability?

17-11 Consider a sound system with logic and single-hour listening reliability as given in Figure 17-11. Find the system reliability.

17-12 Consider the system in Figure 17-12. The reliabilities apply to the first 100 hours of operation.

(a) Determine the system reliability.
(b) Upgrade component 4 to .9 reliability. Recompute the system reliability.
(c) A redundant system may be created instead by substituting two type-4 items, each at the original reliability. Recompute the reliability for that system.

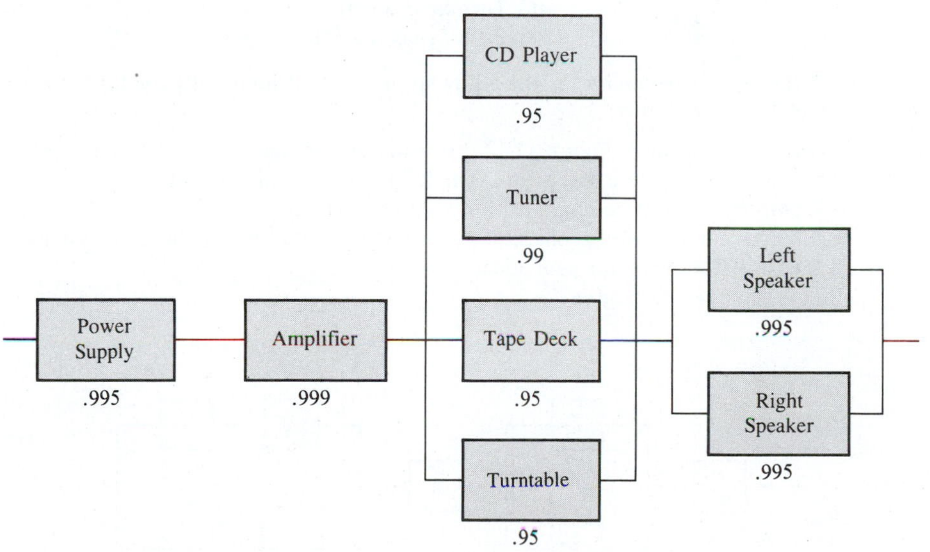

Figure 17-11
Logic for sound system.

Figure 17-12

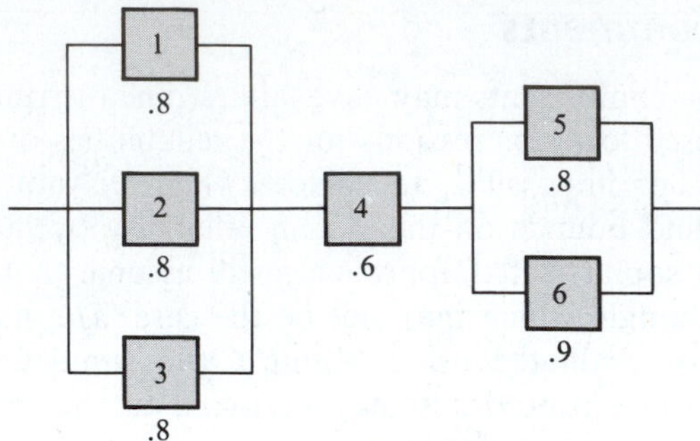

17-13 Consider the system in Figure 17-13. Suppose that each component has a Weibull failure-time distribution with $\beta = .5$ and $\lambda = .01$/hour. Find the system reliability for a span of (a) $t = 100$ hours, (b) $t = 50$, and (c) $t = 10$.

17-14 A series system consists of 100 independent units, each with exponential distribution with $\lambda = .005$. Find the system reliability over a span of (a) $t = 10$, (b) $t = 20$, and (c) $t = 50$.

17-15 A parallel system consists of five independent components, each having common exponential failure-time distribution with $\lambda = .10$. Find the system reliability over a span of (a) $t = 2$, (b) $t = 3$, and (c) $t = 5$.

17-16 Refer to the sound system in Figure 17-11. Suppose that all components have independent exponential failure-time distributions, with the following mean failure rates:

(1) Power: .0001/hour
(2) Amplifier: .0002
(3) CD player: .0050
(4) Tuner: .0010
(5) Tape Deck: .0020
(6) Turntable: .0090
(7,8) Each Speaker: .0010

Find the system reliability for the following number of hours of play: (a) 5, (b) 10, and (c) 100.

17-17 Consider the system in Figure 17-12. For each percentage point increase in system reliability there will be a \$10 payoff. The following possibilities apply:
(a) Component 4 can be upgraded for \$100 to reliability .9.
(b) Component 4 can be duplicated any number of times in a redundant system at a cost of \$50 per extra item added.
What is the optimal solution?

Figure 17-13

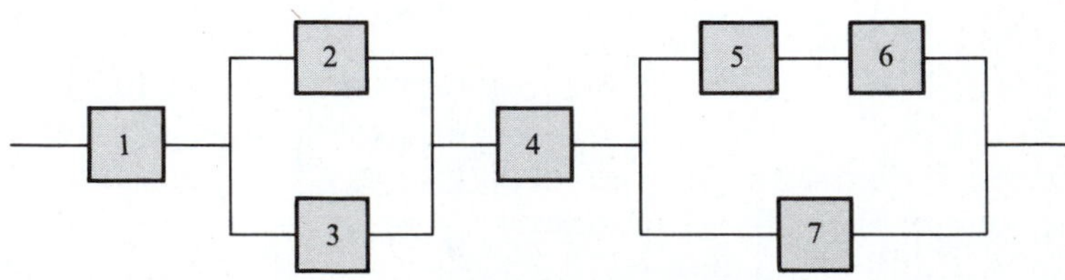

17-3 Life Testing and Reliability Assessment

The preceding two sections describe the framework for reliability analysis, focusing on predicting reliability using probability. *Reliability assessment* is concerned with the evaluation of existing components and systems. With reliability assessment sample information is used to draw inferences regarding the nature of the underlying reliability or the failure-time distributions.

Life Testing

Reliability assessments are made through **life testing**, an experiment in which test items are placed in actual or simulated operation until they fail. The failure times constitute the sample observations. Such experiments can be particularly challenging because they may involve products and systems with very long lifetimes. We generally cannot wait until all test items fail before beginning statistical analysis; thus, the conventional sampling approach of completely observing every sample unit is impossible to follow.

The resulting data may have to be *censored*, in the sense that some failure times cannot be measured. Censoring most commonly occurs when the test is terminated at the conclusion of a predetermined period, with some sample items surviving without failure to the end. Since the sample represents failure times for the nonsurvivors only, it consists of **right-censored data**. Sample data is *left*-censored when failed items are removed periodically and the exact failure times cannot be recorded.

Even allowing for censored data, lifetimes can be too long to assess within a reasonably short test period. A variety of procedures may then be used to create **accelerated life testing**. Electrical items may be tested under much higher than nominal voltages, for example. Sample results can then be mathematically transformed to the results that would be expected under testing at nominal levels. Such super-stressing can be applied to physical items as well, so testing may take place under higher than nominal temperatures, pressures, or loads.

Finding the Failure-Time Distribution

Often the form of the underlying failure-time distribution (exponential, Weibull, gamma, and so on) is known, so the test results provide estimated values for unknown parameter(s). The Weibull is often assumed to apply because it can fit items with failure rates that decrease ($\beta < 1$), increase ($\beta > 1$), or remain constant ($\beta = 1$). In the last case, the Weibull reduces to the exponential distribution.

The failure rate $h(t)$ is fundamental in establishing whether a particular distribution fits. The actual fitting uses the *cumulative* failure rate. If failure time t is treated as a variable, it may be expressed as follows.

CUMULATIVE FAILURE RATE FUNCTION

$$\boldsymbol{H(t) = \int_0^t h(x)\,dx}$$

A closed form for this expression may be found for many common failure-time distributions. For the Weibull distribution we have

$$H(t) = \int_0^t \lambda\beta(\lambda x)^{\beta-1}\,dx = (\lambda t)^\beta$$

The cumulative failure rate can be empirically measured over time based on life-testing data. Using those data, a specific function can be established by regression analysis, treating $H(t)$ as the dependent variable and failure time t as the independent variable.

Establishing Weibull Distribution Parameters

The Weibull distribution is often chosen as the failure-time distribution because of its versatility in representing a wide class of item types. The parameters of the particular Weibull distribution may be found by first transforming the data, taking the natural logarithms of the observed values for t and $H(t)$, and then finding the regression line. That line may be found either graphically through visual fit or by using the method of least squares. The estimated values for λ and β are then determined from the regression line by making the inverse transformation.

The simplest procedure involves simultaneously placing n items into the test at time zero. These are removed (but not replaced) as they fail, and the time of failure is recorded. After the testing period is over, some items will have ordinarily survived, so the data are usually right-censored. (More elaborate testing schemes can be employed. See the references listed in the Bibliography for details.)

The observed cumulative failure rate is based on the order of an item's removal and the time at which that occurs. For the ith item removed, failing at time t_i,

$$H(t_i) = \frac{1}{n} + H(t_{i-1})$$

where $t_0 = 0$ and $H(0) = 0$.

Illustration:* *Life Testing of Floppy Diskettes

Table 17-1 shows the results from a life test involving $n = 20$ floppy diskettes. In that experiment a variety of files were written onto the disks and then read back and compared to the original contents. This was done continuously until the diskette failed or the test was terminated (at 25.3 hours from the start).

At termination, there were 8 surviving floppy disks still in operation. The cumulative failure-time distribution for the first 12 failed items is shown in columns (2) and (3).

Table 17-1
Results of Life Testing with Floppy Disks

(1) Order of Removal i	(2) Removal or Failure Time t_i	(3) Cumulative Failure Rate $H(t_i)$
1	2.3	.05
2	5.9	.10
3	6.1	.15
4	9.0	.20
5	9.5	.25
6	17.9	.30
7	18.8	.35
8	19.7	.40
9	22.6	.45
10	23.0	.50
11	24.3	.55
12	24.8	.60
13	25.3	X
14	25.3	X
15	25.3	X
16	25.3	X
17	25.3	X
18	25.3	X
19	25.3	X
20	25.3	X

Regression Analysis with Cumulative Failure-Rate Function

The failure-time distribution of the floppy disk data is assumed to be a member of the Weibull family, so a regression equation of the form

$$\hat{H}(t) = (\lambda t)^{\beta}$$

will be fit to the data.

Method of Least Squares We find the regression line first by the method of least squares. Taking the natural logarithm of both sides, we find the equivalent equation:

$$\ln [\hat{H}(t)] = \beta \ln(\lambda) + \beta \ln(t)$$

The natural logarithms were taken of the data from columns (2) and (3). Then using those values with the sample size *reduced* to $n = 12$ (accounting only for failed items), a simple least-squares regression analysis was performed. This gives the following estimated regression line:

$$\ln [\hat{H}(t)] = -3.81612 + .983944 \ln(t)$$

The estimated Weibull shape parameter is therefore

$$\hat{\beta} = .983944$$

Using this, we find the estimated Weibull scale parameter:

$$\hat{\beta}\ln(\hat{\lambda}) = -3.81612$$

so that

$$\hat{\lambda} = \exp\left(\frac{-3.81612}{.983944}\right) = \exp(-3.878391)$$
$$= .021$$

Graphical Solution The above results may be achieved alternatively by plotting the original data on log-log paper and getting the scatter diagram shown in Figure 17-14. A visual fit can determine the regression line, which may then be drawn onto the graph with a straight edge. A special preprinted grid, *Weibull paper*, is available for making such plots. The horizontal axis represents the *dependent* variable $H(t)$. (The axes are reversed from the usual regression convention.)

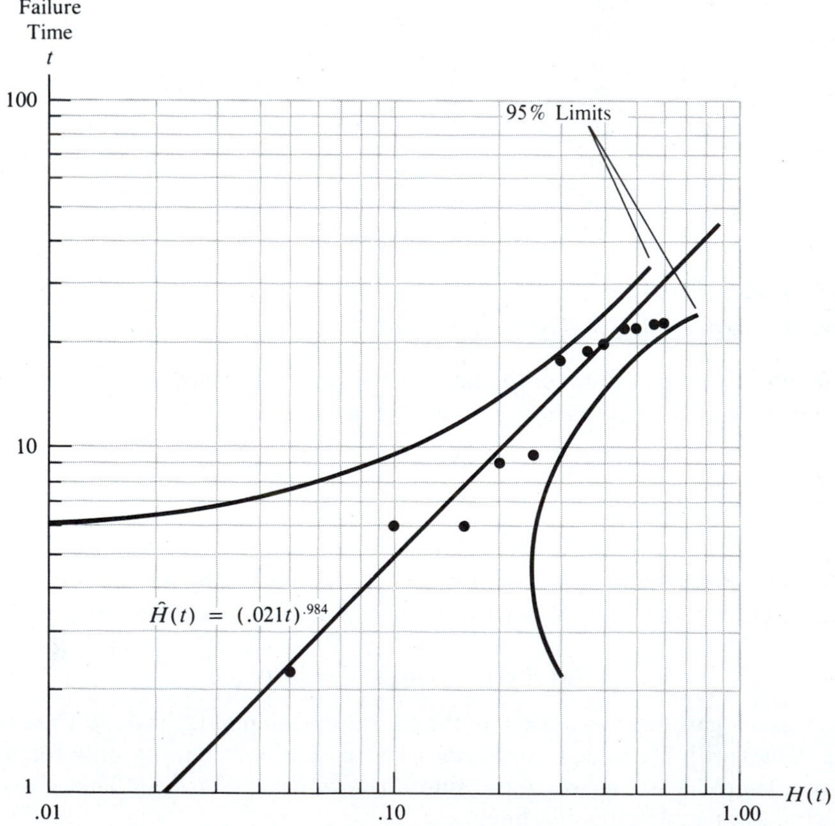

Figure 17-14 *Cumulative plot of floppy disk failures.*

The results indicate that the true value for λ may be close to 1, so the data are consistent with an underlying *exponential* failure-time distribution. When least-squares regression results are available, confidence intervals may be constructed for the regression coefficients, and from these, confidence intervals may be obtained for λ and β.

Since the regression results are subject to sampling error, the true regression line, and hence true expression for $H(t)$, remains unknown. Confidence intervals may be constructed for $H(t)$ for any given level of t. These may be plotted on the graph to form a confidence band, as shown in Figure 17-14 for a 95% level of confidence.

Concluding Remarks

The reliability field is a rich one that includes many assessment procedures for more elaborate life tests, for other types of distributions, and for estimating the reliability probability. Unfortunately, there is not space enough in this book to give further details. You may read more about reliability assessment in the references listed in the Bibliography at the back of this book.

Problems

17-18 The following results have been obtained from a life test with a particular fuse:

Item	Failure Time	Item	Failure Time
1	503 hours	6	1,928 hours
2	948	7	2,510
3	971	8	3,614
4	1,232	9	undetermined
5	1,414	10	undetermined

A Weibull distribution is assumed to apply.

(a) Find the estimated values for the Weibull parameters.

(b) Does the exponential distribution fit the sample data? Explain.

17-19 The following results have been obtained from a life test with a fan blade:

Item	Failure Time	Item	Failure Time
1	981 hours	6	11,145 hours
2	1,153	7	undetermined
3	2,055	8	undetermined
4	5,231	9	undetermined
5	10,533	10	undetermined

A Weibull distribution is assumed to apply.

(a) Find the estimated values for the Weibull parameters.

(b) Does the exponential distribution fit the sample data? Explain.

17-20 Refer to the data in Problem 17-18 and your answer to that exercise. Construct individual 95% confidence interval estimates for λ and β. (In estimating λ assume $\beta = 1$.)

17-21 Refer to the data in Problem 17-19 and your answer to that exercise. Construct a 95% confidence interval estimate for $H(t)$ when $t = 5{,}000$.

17-22 A student conducted a life test with Spotsylvania 100-watt light bulbs in her home. She removed all the present bulbs in two rooms, substituting new Spotsylvanias. She left the lights on continuously in those rooms. She determined the lifetimes for most of the test items to the nearest 10 hours. The following data were found:

Bulb	Failure Time	Bulb	Failure Time
1	150 hours	8	950 hours
2	170	9	1,150
3	310	10	undetermined
4	450	11	undetermined
5	510	12	undetermined
6	520	13	undetermined
7	810		

An exponential distribution is assumed to apply.

(a) Use regression analysis to find the estimated values for the Weibull parameters.

(b) Does the exponential distribution fit the sample data? Explain.

(c) Assuming $\beta = 1$, construct a 95% confidence interval estimate for λ.

(d) Using your answer from (c), construct a 95% confidence interval estimate for the MTTF.

(e) The manufacturer claims that Spotsylvania light bulbs should have a mean lifetime of 700 hours. Is their claim supported at the .05 level of significance?

Comprehensive Problems

17-23 The following results have been obtained from a life test with an electronic component:

Item	Failure Time	Item	Failure Time
1	10 hours	6	44 hours
2	11	7	51
3	17	8	undetermined
4	33	9	undetermined
5	38	10	undetermined

A Weibull distribution is assumed to apply. Find the estimated parameters.

17-24 Consider the system in Figure 17-7. Suppose that all components have the same failure-time distribution as the one tested in Problem 17-23. Assuming that the parameter values estimated in Problem 17-23 apply, find the system reliability when $t = 20$ hours.

17-25 The following results have been obtained from a life test with an electronic component:

Item	Failure Time	Item	Failure Time
1	100 hours	6	370 hours
2	150	7	430
3	200	8	undetermined
4	250	9	undetermined
5	350	10	undetermined

A Weibull distribution is assumed to apply. Find the estimated parameters.

17-26 Consider the system in Figure 17-8. Suppose that all components have the same failure-time distribution as the one tested in Problem 17-25. Assuming that the parameter values estimated in Problem 17-25 apply, find the system reliability when $t = 200$ hours.

17-27 You are designing a system that must have reliability at time t of .95. The logic requires that at least one item each of types 1, 2, and 3 remain functional for the system to work. The following reliabilities apply:

$$R_1(t) = .75 \qquad R_2(t) = .60 \qquad R_3(t) = .90$$

You may introduce redundancy by placing items of the same type in parallel subsystems.

Diagram the system logic that provides exactly nine items altogether of the various types.

CHAPTER EIGHTEEN

Bayesian Decision Making

UNTIL NOW WE have been primarily concerned with **traditional statistics**, the common thread of which is making inferences about populations using evidence obtained from samples. The procedures of statistical inference were originally developed for applications in agriculture, biology, and medicine. There the main focus was, and still is, performing hypothesis tests to arrive at decisions consistent with sample evidence. Indeed, the procedures developed by the pioneer statisticians in these fields have been updated and are now used throughout the physical, life, and social sciences, as well as in business and engineering.

But decision making in the business world, and to a lesser extent in engineering as well, must often be made under uncontrolled conditions where the formalism of traditional statistics is at best impractical—and sometimes even impossible to implement. That is, *there is often no population from which a sample can be drawn.* And yet uncertainties about the outcomes of all possible choices may still exist.

For example, backed only by the skimpiest of indirect evidence, an oil wildcatter must still decide whether or not to drill. Unless he commits to drilling, there is no way for him to sample the rocks under the site or to measure the porosity of the oil-bearing structure. Useful sample data can be obtained only after the fact and can be of no help whatsoever in evaluating the primary decision of whether or not to drill. Similar, though less dramatic, situations are faced by engineers who must "take the plunge" and propose designs to their clients. They must risk considerable time and effort on sketches, models, or prototypes—yet they could reach a bitter end similar to that experienced by a wildcatter who has bored a dry hole.

Because the usual sample data are not available in so many decisions that must be made under uncertainty, the approach taken must be different from the usual hypothesis test. This approach involves so-called **Bayesian statistics**, a body of procedures named in honor of the Reverend Thomas Bayes, a pioneer in the development of modern probability theory. Bayes originally proposed using subjective probabilities to quantify a person's judgment regarding uncertain events. Bayesian procedures incorporate measures for events arising from nonrepeatable random experiments or when sampling itself is often a moot consideration.

The heavy role played by subjective probability in Bayesian statistics is the source of some controversy. It is not unusual for traditional statisticians to reject subjective probabilities with statements like: "The probability of oil must be either one or zero because either there is oil or there isn't oil, so that the long-run frequency for striking oil must be one (it will be hit every time we drill) or zero (a dry hole will be found every time)." The *nonrepeatability* of the drilling experiment makes gibberish of such a statement, and traditional statistics will not help a petroleum engineer decide whether or not to drill. Nevertheless, engineers can establish "betting odds" that they feel comfortable with and which will help in making the evaluation. After all, the engineer will bear the ultimate consequences of the choice.

Although subjective probability can be valuable in such a private evaluation, it has no satisfactory role in public decision making. The safety of a new drug cannot be left to someone's judgmental hunch. Subjective evaluations cannot substitute for rigorous strength testing of materials. And no physical model should be applied unless the values of all constants have been empirically justified. Most of the controversy surrounding subjective probabilities can be defused if we limit their use to investigations where potential disagreements regarding their level are unimportant.

18-1 Structuring and Evaluating Decisions Involving Uncertainty

Our present concern is with decision making *under uncertainty*, where the decision-maker does not know the outcome. A convenient model of such a decision is two-dimensional. One dimension, that of *choice*, is under control of the decision-maker. The second dimension is that of *chance*, which we might say is under nature's control. Procedures for evaluating decisions under uncertainty fall into the general field of **statistical decision theory**.

The basic model for a decision under uncertainty involves three elements: **acts** (the decision-maker's available choices), uncertain **events**, and **outcomes**. A distinct outcome exists for each act–event combination. We will describe two structures that organize these elements. One is the **decision table**, a matrix having a row for each event and a column for each act; the outcomes serve as the elements of the array. The second structure is the **decision tree diagram**, a display similar to the probability trees encountered earlier in this book. Under both representations the decision is quantified by ranking the outcomes in terms of how closely each one brings the decision-maker to his goal. The selected yardstick, or measure of effectiveness, is referred to as the **payoff measure**.

Decision theory is partly concerned with selecting a suitable payoff measure. But its central focus is on finding an appropriate **decision criterion** or rule for making a choice. Making the decision would be a trivial matter except for the uncertainties involved. The uncertainties are quantified by attaching probabilities to all events in the decision structure. A variety of criteria might be used in making the final choice of act. We will focus on the most universal one, **maximizing expected payoff**.

The Decision Table

The simplest structural representation of a decision under uncertainty is provided by the decision table.

Illustration:* *Evaluating a Design

An example is shown in Table 18-1 for a design evaluation involving three acts: accept, request modification, and reject. Each of these is represented by a separate column. Three mutually exclusive and collectively exhaustive events also pertain to the present design: unworkable, inefficient, and satisfactory. Nine outcomes are listed in the body of the table, each representing the unique consequence resulting from choice of a particular act and occurrence of a specific event.

Although the decision table is a convenient structural representation, it does not provide enough information for evaluating the decision and ultimately selecting the act. Two additional features are required before a decision can be systematically evaluated. When these are incorporated, a more comprehensive display is obtained.

The Payoff Table

The first additional feature that's required is that a probability must be established for each event. As noted at the outset, such values must often be based on judgment. In the present illustration, there is only one design, and its quality events are the result of a unique random experiment. Like drilling for oil, this random experiment is not repeatable. The only probabilities available to the project engineer are subjective ones. Later in this chapter we will present methods for establishing such numbers. The following values apply:

$$\Pr[\text{unworkable}] = .30$$
$$\Pr[\text{inefficient}] = .20$$
$$\Pr[\text{satisfactory}] = .50$$

Table 18-1 *Decision Table for Design Review*

Events	Acts		
	Accept	**Request Modification**	**Reject**
Unworkable	Wasted Resources	Salvaged Design	Saved from Further Losses
Inefficient	Damaged Reputation	Improved Design	Reputation Preserved
Satisfactory	Opportunity Achieved	Delayed Opportunity	Missed Opportunity

Table 18-2
Payoff Table for Design Review

Events	Probability	Acts		
		Accept	Request Modification	Reject
Unworkable	.30	\$−50,000	\$−5,000	\$0
Inefficient	.20	−20,000	10,000	0
Satisfactory	.50	100,000	50,000	0
	1.00			

Our second requirement is finding a suitable payoff for each outcome. These values must be in units of a measure of effectiveness that ranks the outcomes in terms of how closely they bring the decision-maker to his goal. Depending on the nature of the decision, a variety of measures might apply, such as:

Mean Time Between Failures	Return on Investment
Mechanical Efficiency	Unit Cost Savings
Reliability Probability	Operational Life
Mean Operator Training Rate	Net Profit

For the present evaluation net profit was chosen as the payoff measure, and a dollar figure was reached for each outcome.

To ease our discussions, all payoffs in this book have been selected so that the *greater* number is the better one. It is natural to maximize profit, reliability, or lifetime. However, keep in mind that some measures should be minimized. In those cases, we will substitute an equivalent measure so that maximization of the new payoff value matches the underlying goal. Thus, in a decision where cost is to be minimized we can instead maximize cost *savings*. Similar adjustments are possible for nonmonetary measures. In a queueing evaluation, we would maximize waiting time *advantage* over some nominal level. In selecting a maintenance policy, downtime minimization would be equivalent to maximization of operating time. And to assess reliability, failure rates could be minimized by maximizing success rates.

It is convenient to arrange all information pertaining to a decision in a single display, the simplest of which is the **payoff table**. An example of a payoff table is given in Table 18-2 for the design evaluation illustration.

An analysis of the payoff table will indicate which act the decision-maker should select.

The Universal Criterion: Maximizing Expected Payoff

A common decision criterion for making choices under uncertainty is to choose that act which maximizes expected payoff. Recall that an expected value is the average value for a random variable computed using probabilities as weights. In the context of a decision structure, there is a separate random variable (in units of the payoff measure) for each act (column). From the payoff table for the design evaluation we compute the expected payoff for each act as follows:

Payoff × Probability		
Accept	**Request Modification**	**Reject**
.30($−50,000) = $−15,000	.30($−5,000) = $−1,500	.30($0) = $0
.20($−20,000) = $ −4,000	.20($10,000) = $2,000	.20($0) = $0
.50($100,000) = $50,000	.50($50,000) = $25,000	.50($0) = $0
$31,000	$25,500	$0

The maximum expected payoff is $31,000, computed for the first act. The indicated choice is to *accept* the design.

The interpretation of the expected values in the table must be differentiated from what we have encountered so far. If the random experiment were repeatable we would say that an expected payoff represents the long-run average dollar payoff. That would certainly be the case when placing $1 bets on red in roulette. In that casino game a wheel having 38 equal-sized slots is spun. A ball is then released, and whichever slot the ball falls into determines the outcome. There are 18 red, 18 black, and 2 green slots (for which the house always wins). The red-player's probabilities are therefore $\frac{18}{38}$ for winning and $\frac{20}{38}$ for losing. The expected winnings from placing a $1 red bet are thus

$$\left(\frac{18}{38}\right)(\$1) + \left(\frac{20}{38}\right)(\$-1) = \$-\frac{2}{38} = \$-.053$$

On the average, the red bettor will lose $.053 per wager when placing a long string of $1 bets.

But the design evaluation decision will be made just once. There is no long-run average value, and the probabilities themselves cannot be thought of as long-run frequencies (as in roulette). They are subjective values and should more properly be considered as indices of the evaluator's strength of conviction regarding whether or not the respective events will occur. In that case, the amount $31,000 may be interpreted as the decision-maker's average conviction as to what the payoff will be if he chooses to accept the design.

The Decision Tree Diagram

An alternative to the payoff table is a decision tree diagram, which can also be used to show underlying structure. Like probability trees, such a diagram represents each event by a branch. Each act is also represented by a branch. Figure 18-1 shows the decision tree for the design evaluation. There are two kinds of branching

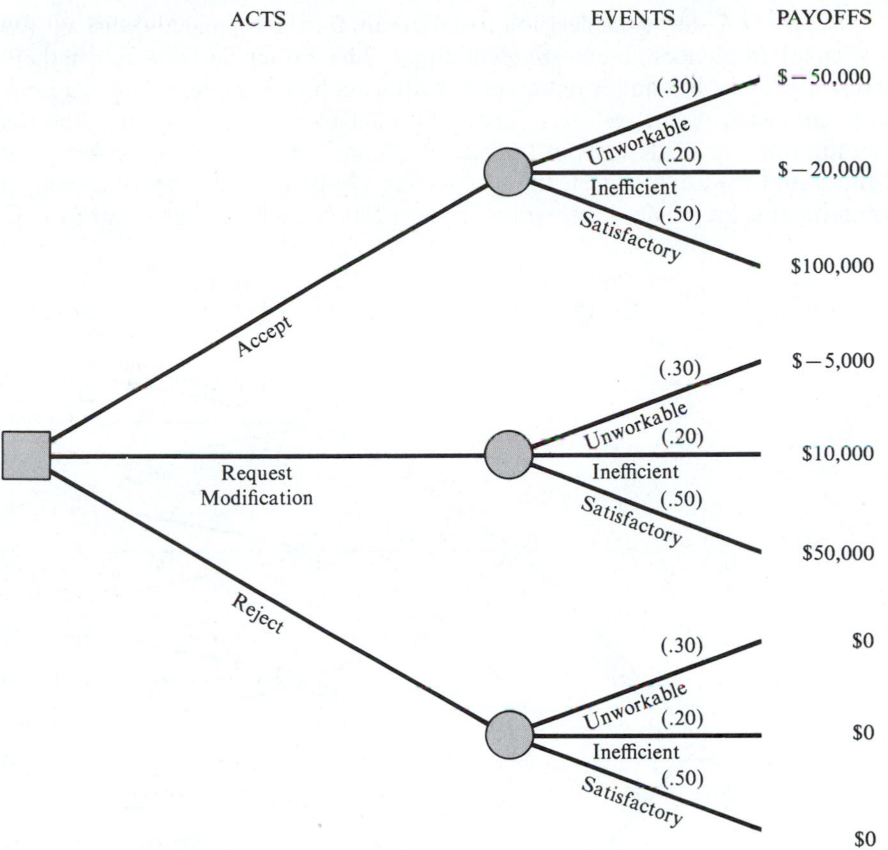

Figure 18-1 Decision tree diagram for design evaluation.

points, which are called forks. One type is a decision point, or **act fork**, the branches of which represent all choices possible at that point. The other branching point is the **event fork**, a point of uncertainty having one branch for each possible event. As a diagrammatical convenience, the nodes for act forks are shown as boxes, and those for event forks are portrayed as circles.

The decision tree represents the logical progression of acts and events that might arise. Usually, the sequence proceeds chronologically from left to right, with the earlier branching points appearing first. In Figure 18-1 the initial fork is a decision point, since the design action must be taken before its operating characteristics are determined. Notice that each act leads to a separate event fork, each of which involves the same three events. Altogether there are three acts and three events for each, so there are nine paths from the start to the end of the tree. Each of these paths culminates in an outcome. The corresponding payoffs are listed in the final column and agree with those given in Table 18-2.

There is no particular advantage in using a decision tree when there is a single decision point. But a decision tree is very helpful when there is more than one decision point or when there are two or more points of uncertainty. This would be the case if we expand the design evaluation to incorporate a possible preliminary test.

Figure 18-2 shows the decision tree diagram for the expanded decision, which now involves choices at two different times. The earlier decision is whether to perform a model test and is represented by the act fork at node *a*. There, a branch represents each of the test acts. If it is decided to have no test, then the main decision must be made without further ado, as reflected by the act fork at *e*. But if the initial choice is to test, the test results are determined before any further commitments are made. These are represented as branches in the event fork at *b*.

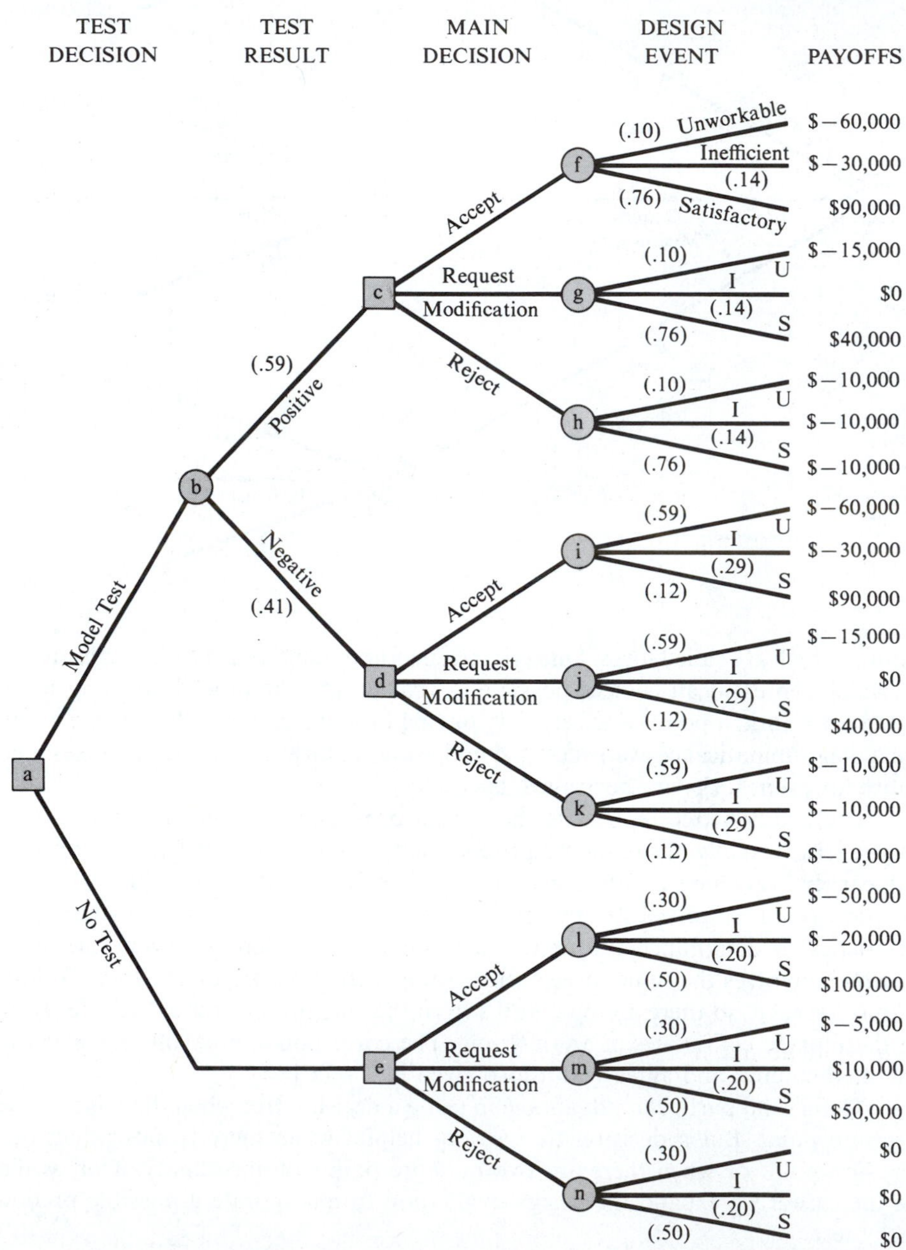

Figure 18-2 *Decision tree diagram for design evaluation with test decision.*

To ease our discussion, the test findings will be classified as either positive or negative, so that only two branches appear.

The probability values are .59 for getting a positive test result and .41 for a negative one. (Section 18-2 provides a detailed discussion of how such probabilities may be established.) Of course the operating characteristics of the design are events dependent on those results, so a different set of probability values applies in the final event forks, depending on the result obtained. For instance, the probability for unworkable is a low value of .10 in nodes *f*, *g*, and *h*, since these branching points all follow a positive test result; similarly, unworkable has a higher probability of .59 in *i*, *j*, and *k*, which follows a negative test result. A further set of probabilities applies when no test is used, as shown in nodes *l*, *m*, and *n*, where the unworkable probability is .30.

The model test will cost $10,000, so the payoff values for all outcomes in which it is involved are reduced from the original levels by that amount. For instance, from node *l* a payoff value is indicated of $-50,000 for accept followed by unworkable. That same outcome has a payoff of $-60,000 from nodes *f* and *i*, which are both preceded by the model test.

The decision tree diagram clearly shows the chronological sequence of acts and events, tying them together in a logical summary. It can be very useful simply as an informational organizing device. But it also serves as the analytical framework for maximizing expected payoff.

Backward Induction Analysis

The expanded design evaluation can no longer be resolved by picking one act. A complete decision must specify which act to select at the two decision stages under all contingencies for the intervening events. This may be accomplished by "pruning" the decision tree at each act fork, leaving only the branch that leads to the maximum expected payoff. But the expected payoff from the model test branch at node *a* cannot be found without knowing the later choices at nodes *c* and *d*. One way to reach the answer would be to consider every possible pruned version of the basic tree. The remnant having the greatest expected payoff would then indicate the best action.

A more efficient procedure is to prune the tree just once, working backward in time—a procedure called **backward induction analysis**. To begin, the futuremost branching points are evaluated. In the case of event forks, the expected payoff is calculated and that value is saved at that node. Within act forks the branch is preserved that leads to the greatest payoff, either directly to a number in the payoff column itself or to an adjacent node from which the expected payoff has already been determined; all other branches are pruned. (Branches for events are *never* pruned.) The maximum payoff value is brought back to the present node and saved. The process continues until just one branch remains at each act fork in the tree.

Figure 18-3 shows the backward induction result for the design evaluation. The first step was to compute the expected payoffs for the event forks at nodes *f* through *n*. These values were obtained in the usual manner by averaging the payoffs, using the respective probability weights. For instance, at node *f* the expected

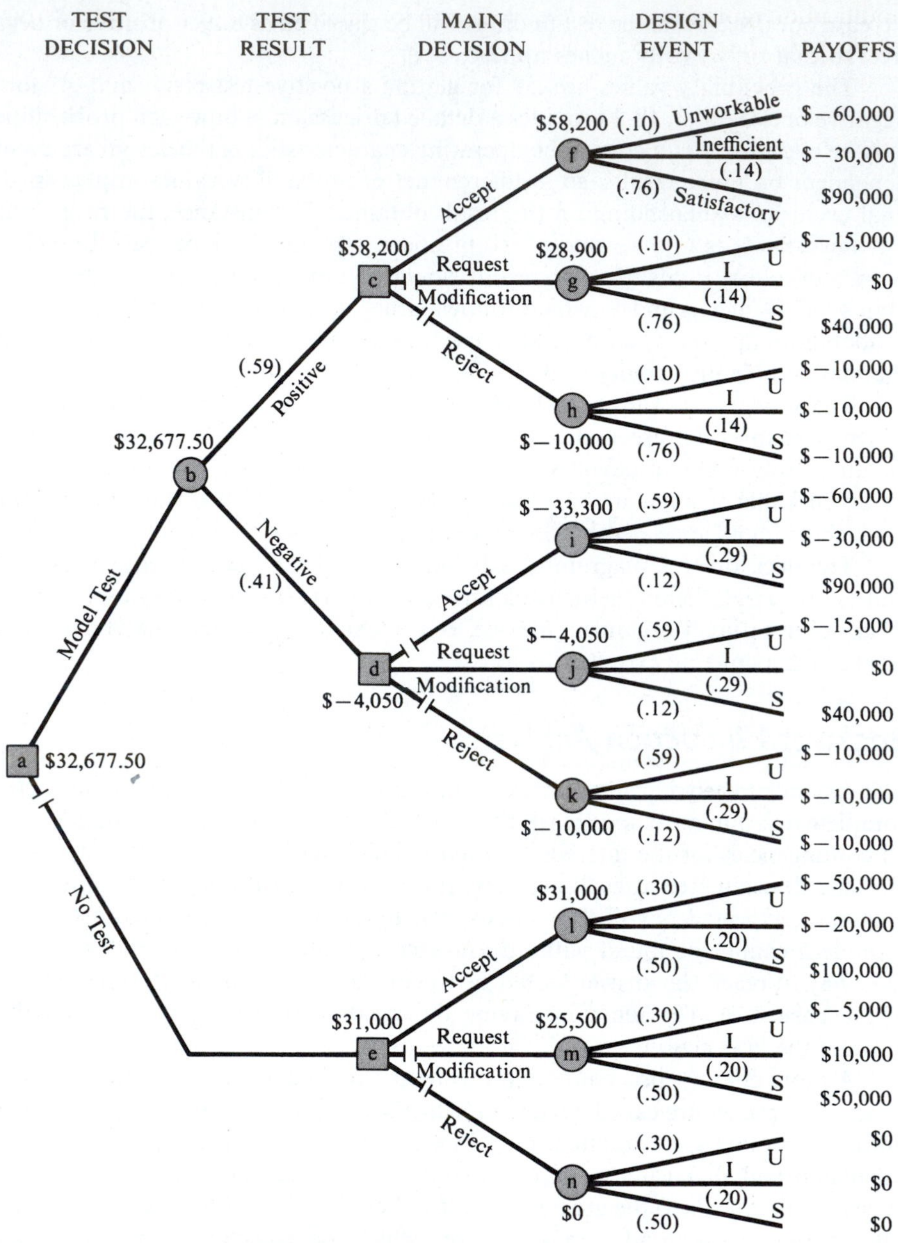

Figure 18-3 Decision tree diagram for design evaluation with backward induction analysis.

payoff is

$$\$-60{,}000(.10) + \$-30{,}000(.14) + \$90{,}000(.76) = \$58{,}200$$

This value is placed on the tree above node f. Next, the act forks at c, d, and e are evaluated and the branches that end at nodes having the lower expected payoffs are pruned. For instance from node c we preserve the branch for accept, since it leads to the greatest expected payoff of \$58,200 at f; the branches for request modification (going to g with \$28,900) and for reject (going to h with \$−10,000) are pruned. The amount \$58,200 is brought back to node c to be used in further evaluations. Similar actions are indicated at d, where request modification provides the greatest expected payoff of \$−4,050 and at e, where accept yields the greatest expected payoff of \$31,000.

The final expected payoff calculation is then made for the event fork at b,

$$\$58{,}200(.59) + (\$-4{,}050)(.41) = \$32{,}677.50$$

using the amounts saved at the adjacent nodes, c and d. The amount \$32,677.50 is saved at b. In making the final comparison of expected payoffs in the initial act fork at a, the branch for no test is pruned. The greatest expected payoff at that point is \$32,677.50, reached by conducting the model test.

Ignoring the eliminated branches, the pruned tree indicates the course of action that maximizes expected payoff:

> Conduct the model test. Then if the result is positive, accept the design. But if the test is negative, request modification of the design.

A Problem with Using Monetary Payoffs

As a decision-making criterion, maximization of expected payoff has many desirable features. One compelling argument in its favor, for instance, is that it makes the greatest use of available information.

But that does not make it a perfect criterion. It is easy to see that it often leads to the less desirable choice. Consider the decision to buy casualty insurance (fire, theft, health, and so on). Assuming that the individual and the insurance company use the same probabilities for size of claim, the purchaser must pay more for coverage than the expected loss. (That is, the insurance company must establish a premium charge high enough to cover both the claims and its operating expenses.) Under these conditions, buying casualty insurance would never maximize expected *monetary* payoff (treating expenditures and losses as negative payoff values). And yet people buy insurance and are happier with it than without it.

The problem lies with using monetary values as payoffs, not with maximizing expected payoff as a criterion. When outcomes are extreme (say, losing a \$100,000 home in a fire), the face amount of the monetary payoff is a poor representation of the relative worth of these outcomes. This difficulty can be rectified by using *utility* values as payoffs. Utility theory is a branch of decision theory. It is described in detail in many of the references listed in the Bibliography at the back of this book.

Problems

18-1 Suppose that the probabilities for the design evaluation in Table 18-2 are .25 for unworkable, .40 for inefficient, and .35 for satisfactory.

(a) Determine the expected payoffs for each act.

(b) Assuming the decision-maker will select the act that maximizes expected payoff, which act should be chosen?

18-2 The following data apply to the decision of a city regarding surfacing new roadways:

Method	Cost Savings (dollars)	Lifetime (years)	Index of Ride Smoothness
A	0	25	4
B	−300,000	30	2
C	−500,000	15	6

For each of the following decision-makers, determine the appropriate payoff measure and the best method if that person were to make the final choice:

(a) The city engineer wants to build long-lasting roads.

(b) The transportation director is concerned with vibration damage to city vehicles and wants smooth streets.

(c) The city controller must cut current costs.

18-3 Suppose that the design evaluation illustration includes the choice of model test, with probabilities .525 for positive and .475 for negative. The following probabilities now apply to the main events:

Event	No Test	Positive	Negative
Unworkable	.25	.095	.421
Inefficient	.40	.305	.505
Satisfactory	.35	.600	.074

All monetary amounts are the same as before, except that the model test now costs $6,000.

(a) Redraw the decision tree diagram, then perform the backward induction analysis.

(b) Which course of action maximizes expected payoff?

18-4 An industrial engineer must choose a method for manufacturing parts. The size of the production run will depend on the number of parts ordered, which is an uncertain quantity. The following quantities are believed to be equally likely: 2,000, 3,000, 4,000, and 5,000. The parts will be sold for $100 each, regardless of quantity ordered.

The following data apply to the three methods of production:

	Method A	Method B	Method C
Set-up Cost	$100,000	$80,000	$50,000
Unit Materials Cost	10	15	20
Unit Labor Cost	10	10	15

(a) Using total profit from the parts as the payoff measure, construct the engineer's payoff table.
(b) Determine the expected payoff for each act. Which method will maximize expected profit?

18-5 An industrial engineer is evaluating two machines for weaving tire cords. The following data apply:

	Machine A	Machine B
Capacity (tires/day)	100	70
Labor (hours/tire)	.70	1.00
Annual Lease Cost	$100,000	$70,000

Labor costs $20 per hour, and the value added to each tire by the operation is $30. The plant operates 200 days per year. If there is not enough capacity to meet the production demand, the weaving must be subcontracted at a cost of $30 per tire.
(a) Determine the break-even volumes for the two machines. At what volume does machine A contribute more net value than machine B?
(b) The following probability distribution applies for annual production demand:

Volume	Probability
4,000	.15
8,000	.20
12,000	.30
16,000	.25
20,000	.10

Which machine should the engineer choose if she wishes to maximize the expected difference between value added and cost?

18-6 A chemical engineer must decide whether to process an order or to contract it out at a cost of $10,000. The final product batch will be sold for $20,000. In-house processing involves direct costs for raw materials of $2,000. The first step is hydrogenation, for which there is a 90% chance of getting a satisfactory intermediate chemical base at a cost of $1,000. If the base is unsatisfactory, there will not be sufficient time to start a new batch, but there will still be a choice of turning down the order or contracting out the production. In the latter case, there is a 50-50 chance of being too late and having to dump the final product. The final stage of in-house processing may be a low-pressure process costing $5,000 or a high-pressure process costing $8,000. There is a 20% chance that the low-pressure process will fail, so that the resulting chemicals must be dumped; it would then be too late to go outside. The high-pressure procedure is certain to work.
(a) Using net cash flow (revenue minus costs) as the payoff measure, diagram the engineer's decision tree.
(b) What action maximizes expected payoff?

18-7 A traffic engineer for a port authority wishes to determine how many lanes and collection booths to build for a bridge toll plaza. Three options are available. For a cost of \$3 million, roadway and facilities can be built to support 10 lanes, which will be sufficient if demand is as planned. At a cost of \$500,000 more, a pad large enough for adding 5 more lanes can be made ready for future expansion. Alternatively, at a cost of \$1 million, the 5 lanes can be added into the original construction for a total cost of \$4 million.

The actual demand might turn out as planned, or it could be higher (probability .30) so that 15 lanes will be required. But the follow-on construction will cost \$.75 million if the pad is sufficiently large and \$1.5 million if the original pad has to be expanded.

(a) Construct the traffic engineer's decision tree diagram. Quantify each outcome in terms of total cost.

(b) The engineer wants to minimize expected total cost. What action should be taken?

18-8 Consider the following payoff table:

Events	Probability	Acts				
		A_1	A_2	A_3	A_4	A_5
E_1	.15	\$500	\$300	\$−100	\$200	\$100
E_2	.25	400	500	600	300	600
E_3	.40	300	400	200	300	500
E_4	.20	−200	−100	100	−200	0

Determine the expected payoff for each act. Which act maximizes expected payoff?

18-9 The **opportunity loss** for an outcome expresses the difference between the payoff that might have been achieved if the best act for the indicated event were chosen and the payoff of the outcome resulting from the act actually chosen. These are found by first finding the greatest payoff for each row. The opportunity losses for each row are then obtained by subtracting all payoffs for that row from the maximum. An opportunity loss is not a loss in the usual sense, but rather a measure of *regret*.

(a) Refer to the payoff table in Problem 18-8. Using those data, construct the opportunity loss table.

(b) Compute the expected opportunity loss for each act.

(c) Which act has the minimum expected opportunity loss? (This will always be the same act that maximizes expected payoff.)

18-10 An **inadmissible act** is a choice that is under no circumstance better than one other single act, and for at least one event is strictly worse. The inadmissible act is **dominated** by the other. An inadmissible act will never be chosen by a rational decision maker regardless of criterion used and no matter what the decision-maker's underlying attitudes happen to be. When identified, inadmissible acts may be removed from the decision structure prior to any further analysis.

Identify the inadmissible act(s), if any, in the payoff table in Problem 18-8.

18-2 Decision Making with Experimental Information

Decision-makers who must make choices under conditions of uncertainty will often seek additional information before finally selecting an act. Such information can be very elaborate, involving years of special study to obtain, or it can be as minor as a quick perusal of literature pertaining to the subject area. Empirical tests are helpful sources of information for decision making in engineering and science.

Facing a structure involving uncertain events, a decision-maker will value any information that might help predict which event will occur. Such data that serve to realign uncertainties are called **experimental information**. We will use a petroleum engineer's decision whether or not to drill a wildcat oil well to illustrate the role of experimental information. Based on an initial investigation of the surrounding geology and regional oil exploration history, the engineer can make a preliminary assignment of probabilities for geological events. As the engineer studies rock samples and correlates evidence, he will revise those probabilities. Each bit of experimental information can realign current judgments. Dramatic shifts in probabilities can be expected to follow a high-reliability experiment, such as a seismic survey.

Experimental information can realign uncertainties in many types and levels of engineering evaluations. Consider the development of a product. In the last section we saw how a model test could be used to sharpen the uncertainty regarding the operating characteristics of its design. Once the design concept is firmly established, a survey of user opinions might be helpful in choosing the final product configuration. And before production begins, samples of parts from potential vendors can be collected and evaluated to narrow the choice of suppliers.

In this section we formally incorporate experimental information into the decision structure. A preliminary evaluation provides the **expected value of perfect information**, which establishes an upper bound on the worth of any data that might be used to predict events. This quantity provides a crude test for evaluating sources of information in terms of expected payoff, making it easy to eliminate overexpensive candidates. Sources passing that test must be subjected to a more rigorous evaluation that includes revising probabilities in accordance with Bayes' theorem (encountered in Chapter 2).

The Expected Value of Perfect Information

Perfect information is that ideal predictive instrument which removes all uncertainty and indicates exactly which event will occur. No such data exist in the real world, except under artificial circumstances. For instance, perfect information in a gambling situation would only be available after the fact or if the game were rigged. Nevertheless, some knowledge of perfect information can be very helpful in guiding decision-makers as they cope with less-than-perfect information.

Table 18-3
Payoff Table for Oil-Wildcatting Decision

Events	Prior Probability	Acts	
		Drill	Abandon
Oil	.20	$500,000	$0
Dry	.80	−100,000	0
	1.00		

Illustration: *A Petroleum Engineer's Oil-Wildcatting Decision*

To illustrate the concept, we will consider in more detail a petroleum engineer's oil-wildcatting decision. The engineer wishes to determine whether to drill on a leased site or to abandon it. His payoff table is shown in Table 18-3. Based on personal judgment formed during a preliminary investigation, the engineer has established subjective probabilities for the two events. These values, .20 for oil and .80 for dry, are referred to as **prior probabilities**, since they apply prior to getting any experimental information.

It is easy to see that if the engineer must choose now, the expected payoffs are

$$\$500{,}000(.20) + (\$-100{,}000)(.80) = \$20{,}000$$

for drill and $0 for abandon. To maximize expected payoff, the engineer would choose to drill.

Although not in the table, another course of action would be to seek out some kind of experimental information, postponing the main decision until a result is achieved from that investigation. Before we consider any concrete source of predictive data however, we will consider the ideal case, when perfect information is available (realizing, of course, that such information may not actually exist). A perfect predictor will indicate, without error, that there is oil or that the site is dry. Should the decision-maker have that information, he would pick the better act for each case.

Table 18-4 helps illustrate the perfect information evaluation. Once the perfect information is in hand, the petroleum engineer would drill if it predicts oil,

Table 18-4
Perfect Information Evaluation for Oil-Wildcatting Decision

Events	Probability	Acts		With Perfect Information		
		Drill	Abandon	Maximum Payoff	Chosen Act	Payoff × Probability
Oil	.20	$500,000	$0	$500,000	Drill	$100,000
Dry	.80	−100,000	0	0	Abandon	0
					Expected Payoff with Perfect Information =	$100,000

since that act provides the maximum payoff of \$500,000 should the predicted oil event occur. Should the forecast instead be for a dry hole, he would abandon, achieving the best payoff of \$0 under that circumstance. In effect, perfect information allows the decision-maker to experience the maximum payoff in each row; these numbers are copied into the maximum payoff column.

Of course perfect information will eliminate any uncertainty. But since the engineer *does not yet have the information*, he is still uncertain about what its *prediction* will be. By applying the probability weights to the maximum payoffs, as shown in the last column of Table 18-4, the engineer obtains the **expected payoff with perfect information**, which he calculates to be \$100,000.

The \$100,000 amount expresses "on the average" how well off the decision-maker is with perfect information. The net worth of that information must be found by comparing the engineer's position when it becomes available to where he would be without it. In terms of expected payoff, we have already seen that the engineer will experience a **maximum expected payoff (with no information)** of \$20,000 by choosing to drill. The difference between these two amounts provides the expected value of perfect information.

EXPECTED VALUE OF PERFECT INFORMATION

EVPI = Expected payoff with perfect information
− Maximum expected payoff (with no information)

For the oil-wildcatting decision the expected value of perfect information is

$$\text{EVPI} = \$100{,}000 - \$20{,}000 = \$80{,}000$$

which expresses on the average how much better off the petroleum engineer would be with a perfect predictor than without any information beyond that used in establishing his prior probabilities.

The practical significance of the EVPI is its implication regarding less-than-perfect information. Any imperfectly reliable predictor must have a worth to the decision-maker of some amount less than the EVPI, which sets the threshold. Thus, if any such information were to cost more than the EVPI, it should be rejected out of hand—regardless of its purported quality. For instance, if the petroleum engineer were offered a highly reliable seismic survey for a fee of \$105,000, it could be turned down without a second thought.

The absolute level of the EVPI is itself a useful number. Imagine an engineering decision where the computed value is EVPI = \$50.00. Getting any kind of predictive information could hardly be worth considering when the maximum benefit (in terms of expected payoff) cannot exceed such a low amount. The main decision under such circumstances should be made immediately on the basis of present knowledge alone. On the other hand, a high level such as EVPI = \$1,000,000 suggests that a great deal of effort should be devoted to the search for potential sources of predictive information. A large EVPI indicates that substantial improvement might be achieved if the main decision is deferred until after an exhaustive search has been made.

Stages for Analyzing a Decision

It is helpful to identify the various stages involved in analyzing decisions under uncertainty. These are described as follows.

1. *Prior analysis.* This evaluation stage includes identification of the decision structure, selection of a payoff measure, and a determination of prior probabilities. The expected payoff for each act and the EVPI are then computed. If the latter is small, no further investigation is required, and the main decision should be made. Otherwise, the next stage begins.
2. *Preposterior analysis.* At this point in the evaluation, sources are sought for predictive information regarding the events in the decision structure. Only those that have low cost (in relation to the EVPI) and a history of reliable predictions need be evaluated. That evaluation (to be described shortly) includes both probability and decision tree analyses.
3. *Posterior action.* This action stage occurs after the predictive information has been procured and actual results are known. The analysis of what to do at this point will have already been performed in conjunction with stage 2. The main decision must be made in accordance with the action indicated in that analysis.
4. *Future analysis.* It is common for one evaluation to raise further questions, and later decisions involving some of the same uncertainties might have to be made. In some cases, posterior probability values for events in the present investigation will serve as prior probabilities for those or similar events in a future investigation. Data from ongoing evaluations should be collected in a data base to be used in future decisions.

Preposterior Analysis

We will describe preposterior analysis by continuing with the petroleum engineer's oil-wildcatting evaluation. For a fee of $25,000 a consulting firm will perform a complete seismic survey of the structure beneath the lease site. The outfit has a very good "batting average," confirming oil-bearing structure in 85% of all surveys performed in fields known to have oil. The firm has been even more reliable in denying oil under sites that are known to be dry, being correct 90% of the time.

Figure 18-4 shows the petroleum engineer's decision tree diagram, with an initial act fork for the decision as to whether to order the seismic survey or not. Notice that if the seismic survey is ordered, the main decision will not be made until after its results have become known. The payoff values for the applicable outcomes are the original ones less the $25,000 cost of the survey. Notice that there are no geological event forks following the abandon acts—reflecting the fact that oil and dry never become known unless drilling is done. (Although potentially misleading, it would not be improper to include those missing forks in the tree.)

The *prior probabilities* for oil and dry apply in the event fork at the bottom, since no seismic findings are applicable there. But we have yet to establish the probabilities for the remaining events on the decision tree.

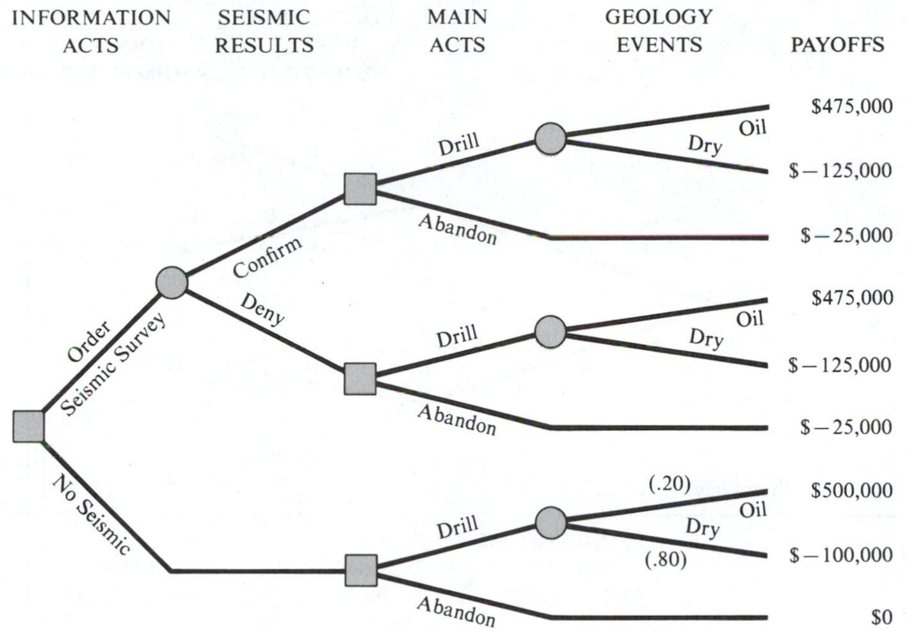

Figure 18-4 *Petroleum engineer's decision tree diagram for wildcat well.*

Values must be established for the test results, Pr [confirm] and Pr [deny], which are necessary for the branches in the leftmost event fork. Depending on the particular result obtained, a different set of probabilities will apply for oil and dry in the subsequent forks. Those values are the *posterior probabilities*, which can be computed from the given information using Bayes' theorem. Since the evaluation is already in tree form, it is usually convenient to use probability trees to establish the required values.

Figure 18-5 shows two applicable probability tree diagrams. At the top in (a) is the tree for the **actual chronology** of events. This tree represents the actual sequence of occurrences: (1) geology events oil and dry (determined millions of years ago), followed by (2) seismic events confirm and deny (to be established from the survey). The prior probabilities of .20 for oil and .80 for dry appear beside the respective branches. The given information on the reliability of the seismic firm provides the following probabilities:

$$\Pr[\text{confirm} \mid \text{oil}] = .85 \qquad \text{and} \qquad \Pr[\text{deny} \mid \text{dry}] = .90$$

from which the complementary values are obtained:

$$\Pr[\text{deny} \mid \text{oil}] = 1 - .85 = .15 \qquad \text{and} \qquad \Pr[\text{confirm} \mid \text{dry}] = 1 - .90 = .10$$

These values will be referred to as **conditional result probabilities**. Multiplying the probabilities on each two-branch sequence, the *joint probabilities* listed along the right in tree (a) are found.

The bottom tree in Figure 18-5(b) portrays the **informational chronology** of events. Events are presented in the sequence in which they become known to the decision-maker and coincide with the timing implicit in the decision structure itself.

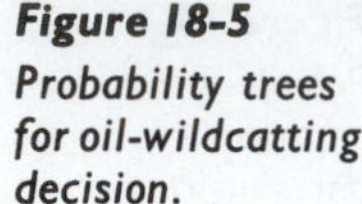

Figure 18-5 *Probability trees for oil-wildcatting decision.*

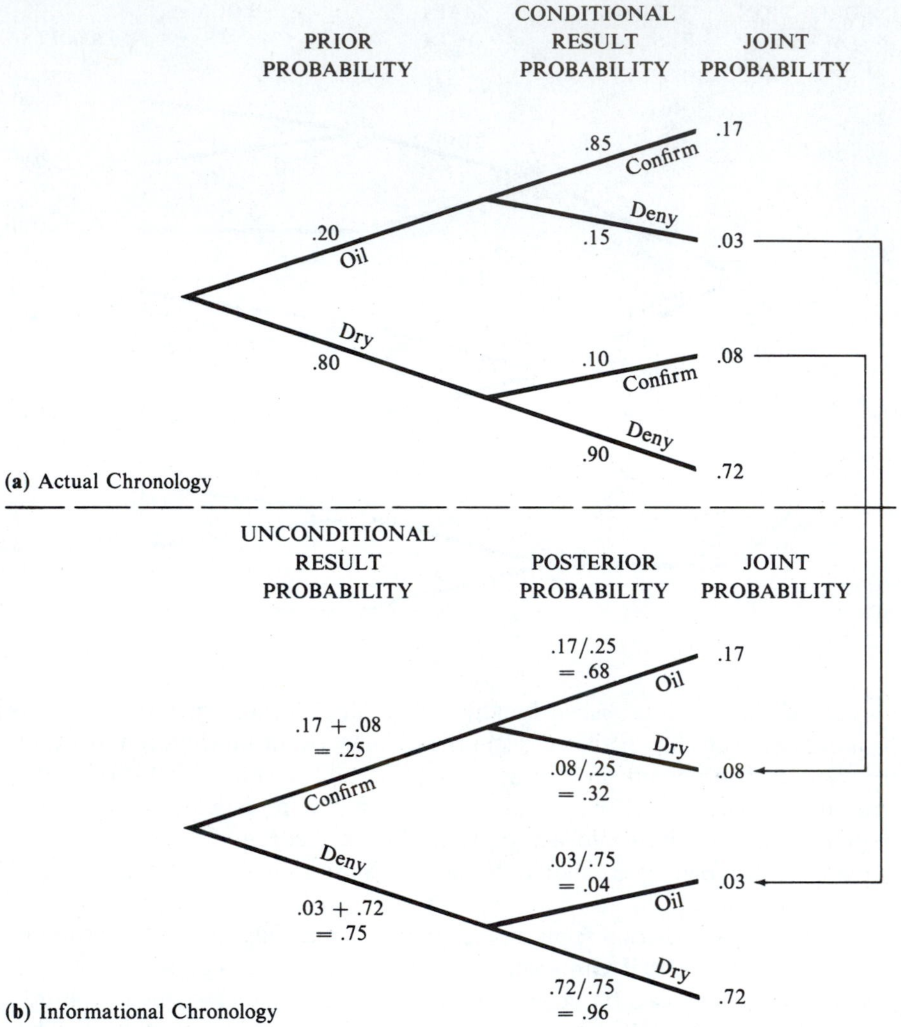

There are also three sets of probability values in that tree. The values are established in the following order. First, the joint probabilities are transferred down from the tree for the actual chronology. For instance, the joint probability .08 for "confirm *and* dry" in (b) is the same as that for "dry *and* confirm" found in (a). Second, the probability for each initial event is found by adding together the joint probabilities for the end positions that can be reached after that event occurs. We have

$$\Pr[\text{confirm}] = .17 + .08 = .25$$

$$\Pr[\text{deny}] = .03 + .72 = .75$$

These values are referred to as **unconditional result probabilities** since they pertain to the seismic results with no stipulations regarding the site geology.

The last step in completing the probability tree for the informational chronology is to calculate the *posterior probabilities*, which are actually conditional probabilities for the main events given an experimental result. Since the joint probability value at each end point must be the product of the probabilities on the branches leading to it, the posterior probability for a particular branch must be that quantity which fits. This is found by dividing the respective joint probability by the unconditional result probability on the preceding branch. We obtain

$$\Pr[\text{oil}\,|\,\text{confirm}] = \frac{.17}{.25} = .68 \qquad \Pr[\text{dry}\,|\,\text{confirm}] = \frac{.08}{.25} = .32$$

and

$$\Pr[\text{oil}\,|\,\text{deny}] = \frac{.03}{.75} = .04 \qquad \Pr[\text{dry}\,|\,\text{deny}] = \frac{.72}{.75} = .96$$

Every number on the tree for the informational chronology may be directly transferred onto the corresponding spot in the decision tree diagram. Figure 18-6 shows the complete tree for the petroleum engineer's oil-wildcatting decision. There the results of the backward induction analysis are shown. The following course of action will maximize the engineer's expected payoff:

> Order the seismic survey. Then if it confirms an oil-bearing structure, drill. But if it denies oil, abandon the site.

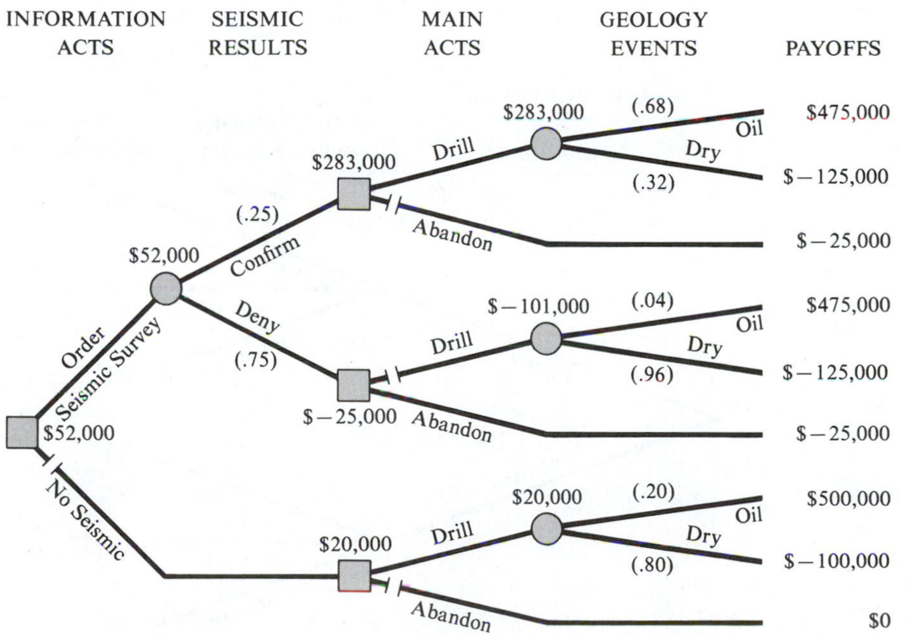

Figure 18-6 *Petroleum engineer's decision tree diagram for wildcat well with backward induction.*

Establishing Conditional Result Probabilities

Unlike the prior probabilities for the main events, which are often based on judgment and are therefore subjective values, conditional result probabilities are generally based on historical experience with the testing mechanism itself. These numbers are then more properly considered objective probabilities. Indeed, it was Thomas Bayes himself who advocated merging judgment with experience by melding subjective prior probabilities with objective conditional result probabilities.

Proper probabilities for test results such as seismic surveys are not easy to come by. For instance, Pr[deny | dry] cannot be estimated by any percentage

Figure 18-7 *Probability trees for design evaluation.*

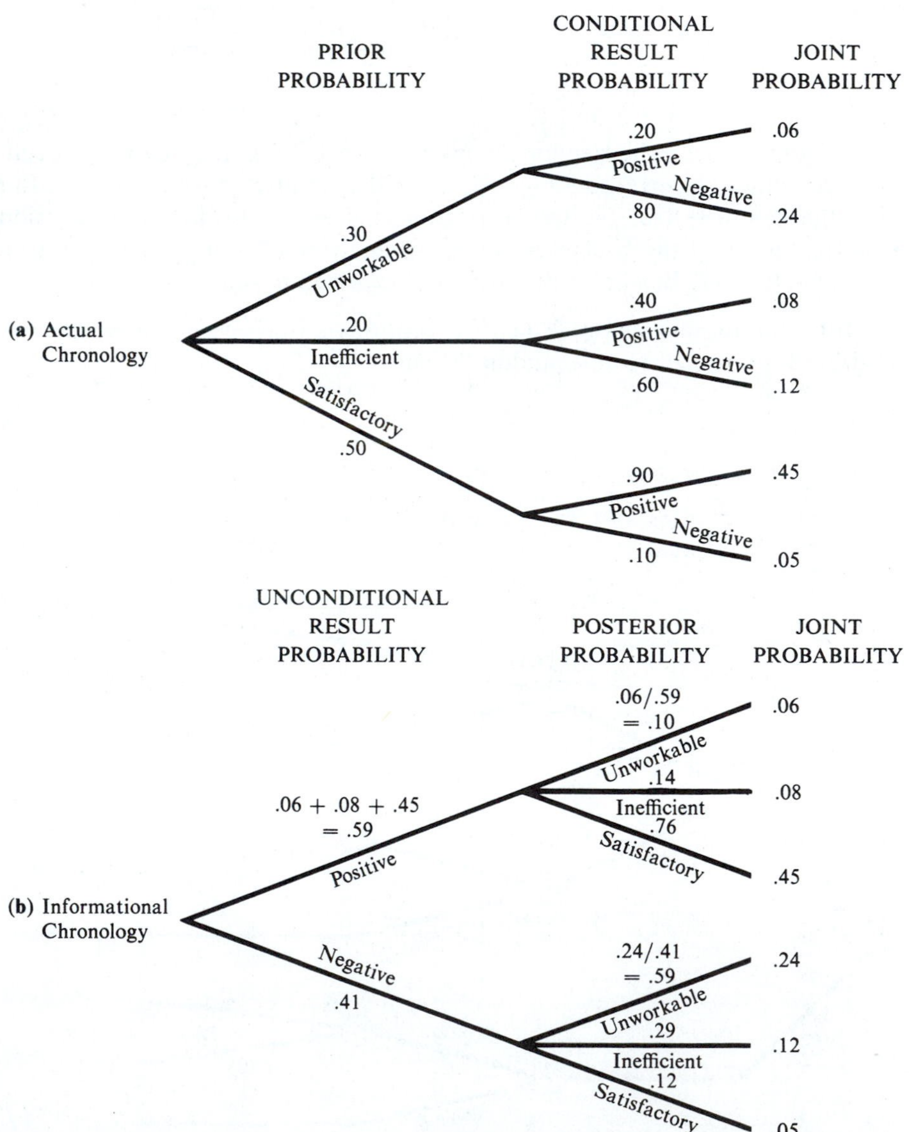

used in actual wildcatting, simply because most situations where the negative prediction is made are likely to involve sites abandoned and not drilled on. That those sites are dry cannot be firmly established. A proper way to establish Pr[deny|dry] is to "test the tester" by giving the seismologist data from sites known to be dry and determining the percentage of simulated predictions that are actually correct. Similarly, Pr[confirm|oil] can be established by simulated forecasts using seismic data from sites already known to be producing oil.

The design evaluation presented in Section 18-1 included a similar preposterior analysis regarding the model test. Figure 18-7 shows how the probabilities were obtained for that evaluation. The following conditional result probabilities were assumed to apply:

	Model Test Result		
Event	**Unworkable**	**Inefficient**	**Satisfactory**
Positive	.20	.40	.90
Negative	.80	.60	.10
	1.00	1.00	1.00

The values in the table were obtained through experience and by simulations with previous designs assuming use of a model test similar to the one proposed.

Some experimental information involves conditional result probabilities that are based on pure deduction, using probability concepts alone. This is the usual case when *sampling experiments* are the source of information. For example, we will see in Section 18-3 that the binomial distribution can be used to generate probabilities for the number of defectives to be obtained in a sample of size n when π is given to be some level and that parameter is itself treated as a random variable.

Problems

18-11 Suppose that the petroleum engineer in the illustration used .30 instead as the prior probability for oil.
(a) Construct the probability tree diagrams for the actual and informational chronologies.
(b) What is the posterior probability for oil when the seismic survey (1) confirms? (2) denies?

18-12 Consider the aeronautical engineer's evaluation in Problem 2-48.
(a) Construct the probability tree diagrams for the actual and informational chronologies.
(b) What is the posterior probability of an acceptable wing assuming that the wind-tunnel test is (1) favorable? (2) unfavorable?

18-13 Refer to the admissions committee selections in Problem 2-51.
(a) Construct the probability tree diagrams for the actual and informational chronologies.

(b) What is the posterior probability that a student will pass his first year given that (1) he scores 675 on the quantitative SAT? (2) he scores 580 on the quantitative SAT?

18-14 A project engineer must decide whether to request production of a new design or abandon the product altogether. If the product is a success (.40 prior probability), the net present value of future profits to the firm will be $500,000; if it is a failure, the profit will be $-200,000, owing largely to production and promotional expenses. Should the product be abandoned now, there will be no effect on future profits.

(a) Using net profit as the payoff measure, construct the engineer's payoff table and compute the expected payoff for each act.

(b) Compute the EVPI. What is the most the engineer would pay for information that would predict the success or failure of the design?

18-15 Refer to Problem 18-14. The engineer might conduct a field test before committing herself in the main decision. At a cost of $25,000, the test will provide a favorable or an unfavorable prediction. Previous testing has established the probabilities to be .90 for a favorable prediction given product success and .95 for an unfavorable prediction given a failed design.

(a) Construct the engineer's decision tree diagram incorporating all choices and events. Include all the net profit payoff values.

(b) Construct the probability trees incorporating the field test results.

(c) Enter onto your decision tree at the appropriate branches the probabilities from your informational tree from (b). Then perform a backward induction. Which course of action maximizes expected profit?

18-16 An electrical engineer responsible for new-product development for a computer manufacturer must decide whether to develop or drop a new modem interface. The prior probability for success is .60 if the device is developed, in which case company profits are expected to increase by $1,000,000 over the lifetime of the product. Development costs of $200,000 will not be recouped if the device fails to catch on. No direct costs have yet been incurred for the device.

(a) Using net profit as the payoff measure, construct the engineer's payoff table and compute the expected payoff for each act.

(b) Compute the EVPI. What is the most the engineer would pay for information that would predict the success or failure of the modem interface?

18-17 Consider the device in Problem 18-16. A test can be conducted involving a panel of potential users with a mock-up device and simulation. The panel consensus will be to dislike the device, have mixed opinions, or like it. Previous testing has established probabilities for these respective outcomes of .10, .30, and .60 given that the test item becomes successful and .80, .15, and .05 given that it fails. The cost of the test is $30,000.

(a) Construct the engineer's decision tree diagram incorporating all choices and events. Include all the net profit payoff values.

(b) Construct the probability trees incorporating the mock-up test results.

(c) Enter onto your decision tree at the appropriate branches the probabilities from your informational tree from (b). Then perform backward induction. Which course of action maximizes expected profit?

18-18 Suppose that the petroleum engineer described in the text ordered the seismic survey and that it confirmed oil. He may now use a further test involving a magnetic anomaly detector before deciding whether or not to drill. For a cost of $20,000 this test has forecast positively on 95% of fields already known to have oil and negatively on 90% of known dry tracts.

(a) Using the applicable posterior probabilities in Figure 18-5(b) as prior probabilities in the present evaluation, construct the engineer's payoff table, assuming that no further information will be used. (Include the cost of the seismic survey in the final payoffs.)
(b) Using your table from (a), compute the expected payoffs for each act. Then establish the applicable EVPI. What is the most that the engineer would now pay for additional information?
(c) Construct the engineer's decision tree diagram incorporating the possible use of the magnetic anomaly detector.
(d) Construct the probability tree diagrams for this situation.
(e) Transfer the probabilities from your informational chronology tree onto your decision tree. Then perform backward induction to determine what action maximizes expected payoff.

18-3 Decision Making with Sample Information

Bayesian analysis with experimental information extends to sampling investigations. The usual objective of such an evaluation is to find which of two complementary actions to take. This is the same end reached by hypothesis testing, first encountered in Chapter 9. We will ultimately compare traditional statistical testing with Bayesian analysis. To facilitate that comparison, we will again represent the decision-maker's acts as "accept" and "reject," although no hypotheses will be formulated.

As with traditional testing, the central focus of the Bayesian approach is a population parameter, ordinarily π in the case of qualitative populations and μ with quantitative ones. Another similarity is basing the final choice of act on the observed value of a test statistic. The sample proportion P or number R of successes (with qualitative populations) or the sample mean $\bar{X}$ (with quantitative populations) serves this purpose.

But the Bayesian approach is very much like the oil-wildcatting illustration of Section 18-2 and bears only superficial resemblance to traditional sampling procedures. It involves an underlying decision structure having the levels of π or μ as *events*. Prior probability values are established for each of these events. As with any decision under uncertainty, each combination of act and event results in an outcome, and each outcome is represented by a payoff value. The main decision is based on maximizing expected payoff. In this context, sampling is nothing more than a source of experimental information analogous to a seismic survey, and the decision whether or not to sample may itself be evaluated in terms of cost and payoff, just as the seismic survey was in Section 18-2. When viewed in this context, the sampling decision can be expanded to include choice of sample size itself, and we will see that an optimal sample size can actually be found.

Familiarity with Bayesian procedures leads to some surprising insights. One is that a good decision might be reached without any sampling—which may be too expensive in relation to the value of perfect information. Also, a great deal of information may be gleaned from very small sample sizes.

Decision Making with a Qualitative Population

To illustrate the procedure, we will use a decision involving a qualitative population for which the proportion π of units having a particular characteristic is unknown. Since this circumstance is so common in quality-control applications of engineering, we will use a detailed illustration from acceptance sampling.

Illustration: *Acceptance Sampling for Printed Circuit Boards*

An inspector for an electronics assembly division wishes to establish a procedure for determining whether to accept or reject shipments of printed circuit boards from a new supplier.

Each board in an accepted shipment will be directly installed into computer logic modules without being tested. Should such a board be defective, the entire logic unit must later be disassembled and the board replaced. That large expense might be avoided by testing each board first. An alternative is to reject the shipment, routing it to an inspection area where all boards are tested.

The primary consideration is the proportion π of defective boards in each shipment. Should π be small, it might be cheaper to install all boards without testing. But for a large π it might be considerably less expensive to test them all. Of course, the value of π for any given shipment is an uncertain quantity.

Table 18-5 shows the inspector's payoff table applicable to each shipment containing 1,000 printed circuit boards. Five possible levels are provided for π, with subjective prior probabilities based on previous experience with similar suppliers and on industry "grapevine."

The measure of effectiveness for this decision is based on a $100 value added to the final product from each installed board. This establishes a basic value of $100,000 for each shipment. Any relevant costs are subtracted from that amount, establishing a net payoff for each outcome. It is assumed that testing costs only $20 per board, whereas replacing a defective board in an assembled module costs $200, so the cost of replacing assembled boards from the entire shipment is

Table 18-5 *Payoff Table for Printed Circuit Board Decision*

Proportion Defective Events	Probability	Acts	
		Accept	Reject
$\pi = .05$	.05	$90,000	$62,500
$\pi = .10$	.15	80,000	62,500
$\pi = .15$	.20	70,000	62,500
$\pi = .20$	.25	60,000	62,500
$\pi = .25$	.35	50,000	62,500

$\$200\pi(1{,}000) = \$200{,}000\pi$. The cost of 100% testing of the boards in rejected shipments is \$20 per unit. However, rejecting involves delays and other increases in operating costs, amounting to an additional \$17,500 per shipment processed. Since the supplier gives full credit for returned boards, the total cost of rejecting a shipment is \$20(1,000) + \$17,500 = \$37,500.

The inspector's payoffs may be expressed as *linear functions.* We have:

$$\text{Payoff} = \begin{cases} \$100{,}000 - 200{,}000\pi & \text{for accept} \\ \$62{,}500 & \text{for reject} \end{cases}$$

The expected payoff for accept is \$63,000. This payoff is greater than the \$62,500 achieved by reject, indicating that without pursuing further any experimental information the inspector should accept each shipment, replacing only those boards found defective after already being installed.

However, the EVPI = \$5,000, which indicates that considerable savings might be achieved by sampling each incoming shipment and testing randomly selected boards.

Posterior Analysis Using Sample Information

Proper evaluation of a wildcat oil well when a seismic survey is employed is to defer the main decision to drill until after the seismic results become known. Analogously, if the inspector uses a sample, he won't decide what to do with a shipment until after obtaining the sample results. Should the sample contain many defectives, he would reject the shipment; if it contains few, he would accept. This is the same approach taken in traditional statistical testing, except that the point of demarcation between accepting and rejecting is here based on maximizing expected payoff.

The cost of small-scale sample testing is \$50 per board (two and one-half times the unit cost of testing all boards in a 1,000-unit shipment). This cost is offset somewhat by intercepting any defective boards in the sample before they are installed, at a saving of \$200 each. Denoting by n the sample size and by R the number of defectives found in the sample, the payoff functions for sampling are

$$\text{Payoff} = \begin{cases} \$100{,}000 - 200{,}000\pi + 200R - 50n & \text{if accept} \\ \$62{,}500 - 30n & \text{if reject} \end{cases}$$

(The coefficient of n for rejecting is -30, reflecting that \$20 will be saved by not having to retest any of the sample items in the 100% testing that will follow.)

The choice of sample size and the question of whether to sample at all will be considered shortly. For now, suppose that the inspector will use a sample of $n = 3$ items. The decision tree diagram in Figure 18-8 applies. Notice that there are 4 possible levels for the number of defectives R found in the sample. In each case, an act fork follows, involving accept and reject as the choices. The event forks in the final stage each have 5 branches, one for each π event. The final payoffs were determined from the above expression. For brevity, there is no event fork following reject, since the payoffs will be the same regardless of the level for π.

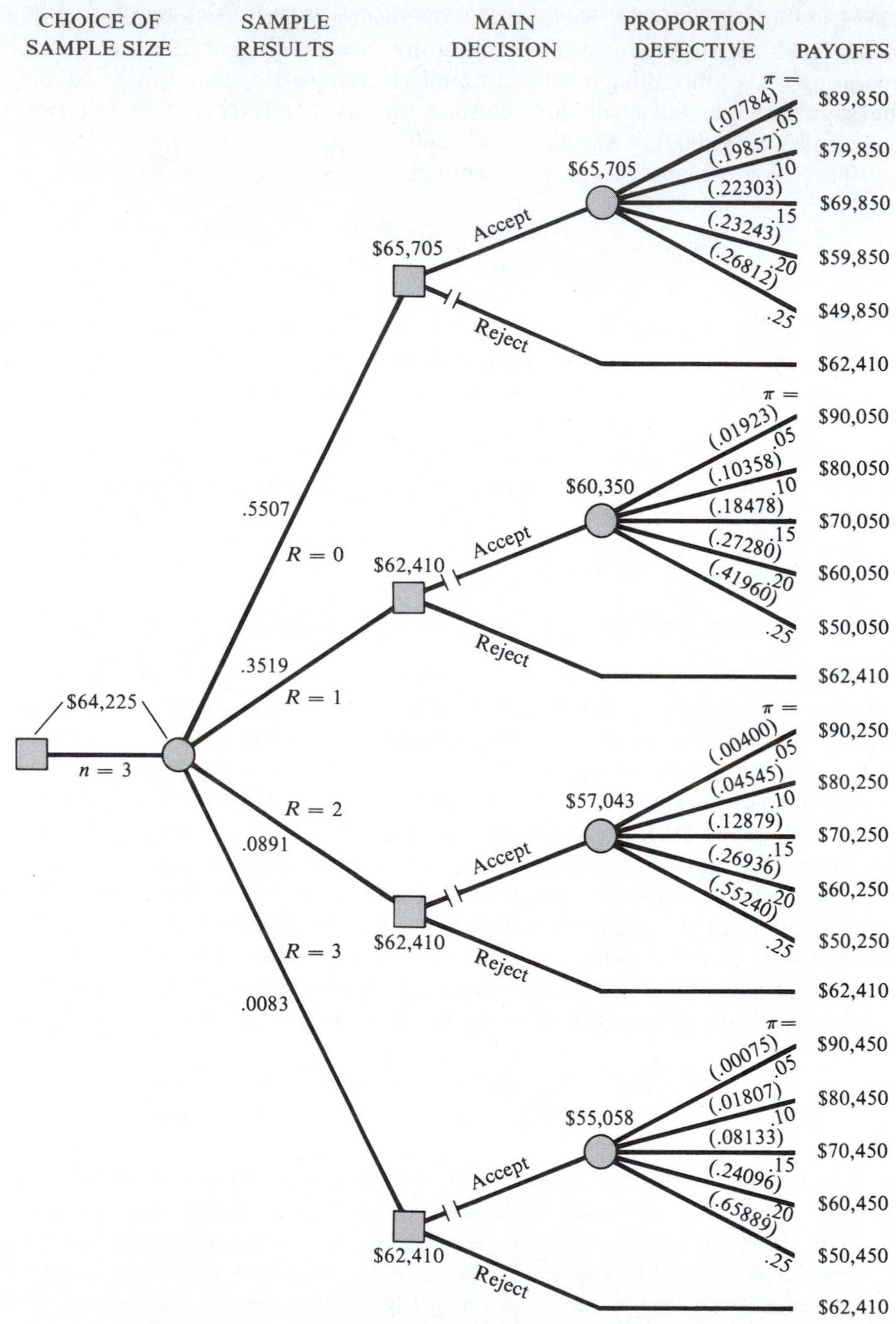

Figure 18-8 *One of several decision tree diagrams for acceptance sampling of printed circuit boards.*

The probability values were transferred from the tree for the informational chronology shown in Figure 18-9(b).

The revised set of probabilities is found in the same way as any other kind of experimental information. In the present illustration the conditional result prob-

Figure 18-9 *Probability trees for acceptance sampling of printed circuit boards.*

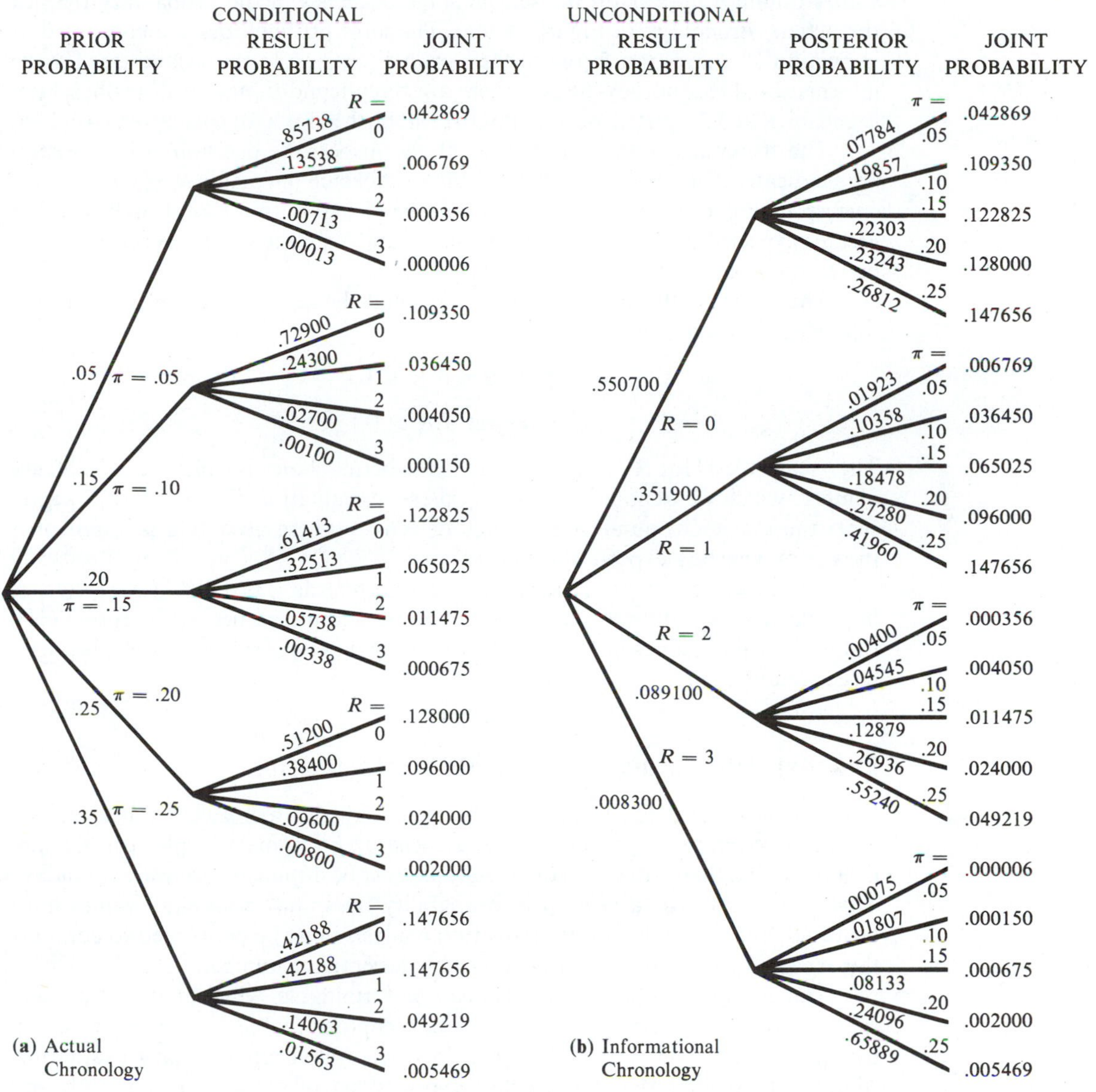

abilities were obtained using the binomial probability mass function,*

$$\Pr[R = r] = b(r; n, \pi) = \frac{n!}{r!(n-r)!}\,\pi^r(1-\pi)^{n-r}$$

* The binomial serves only as an approximation when sampling from a finite population. The true sampling distribution for R in those cases is the hypergeometric, which should be used instead when the population size N is small.

Thus, they follow logically from the given n and the assumed level for π. The values obtained appear on the second-stage branches in the probability tree for the actual chronology in Figure 18-9(a). The joint probabilities are computed as in Section 18-2 and transferred to their proper location in the second tree for the informational chronology. Shown there are the unconditional result probabilities found for R and the posterior probabilities established for π for each given level of R.

The backward induction analysis of the inspector's decision tree in Figure 18-8 indicates that when $n = 3$ the maximum expected payoff for $R = 0$ is \$65,705, corresponding to the act *accept* that would be picked in that case. For $R = 1$, the greater expected payoff is \$62,410 for *reject*, and that act would be selected. The better act at all the other act forks is *reject*.

The result of the backward induction may be summarized in terms of the following decision rule:

$$\text{Accept if } R \leq 0$$

$$\text{Reject if } R > 0$$

The greatest level for R at which accept is the better choice is called the **acceptance number**, which we denote by C. In the present evaluation $C = 0$. Here C represents the maximum number of sample defectives where accepting a shipment is the act maximizing expected payoff.

The expected payoff in using $n = 3$ tested boards is \$64,225, a considerable improvement over the expected payoff of \$63,000 found earlier for accepting each shipment without any sample testing. Obviously, *a sample should be used*. But how large should n be?

Preposterior Analysis of Sample Information

Since in the present illustration it is better to sample than not, the analysis can be continued into the preposterior stage, where an optimal sample size is established. To find that n, the preceding analysis must be duplicated for each candidate. That is, in each case, a new set of probability trees and probability values must be found, and a fresh backward induction analysis must be performed to compute the expected payoff and find the applicable acceptance number.

The ensuing computational task can be formidable, which makes it a chore that can be properly relegated to a digital computer. Table 18-6 shows a partial listing obtained from a computer run for the printed circuit board evaluation. Those data suggest that the greatest expected payoff occurs for a sample size somewhere between 20 and 50. Figure 18-10 shows the expected payoffs over this range for n. Notice that payoff function has cusps for each level for C; there is a *local* maximum expected payoff for the n's within each group. The *global* maximum is \$65,386, corresponding to $n = 32$. That is the optimal sample size. Sampling with the optimal n will provide an expected payoff improvement of \$2,386 per shipment over accepting without benefit of a sample.

Depending on the payoffs and prior probability distribution for π, the Bayesian procedure may indicate that no sample should be used. In some cases the optimal n is quite tiny, no larger than 5 or 10 observations. Such small sample sizes are unheard of in traditional statistics.

Sample Size n	Expected Payoff	Acceptance Number C
0	$63,000	—
1	63,616	0
2	64,003	0
3	64,225	0
4	64,328	0
7	64,587	1
10	64,869	1
20	65,253	3
30	65,346	5
40	65,350	6
50	65,293	8
60	65,178	10
100	64,458	46

Table 18-6 *Expected Payoffs and Acceptance Numbers for Various Sample Sizes in Printed Circuit Board Evaluation*

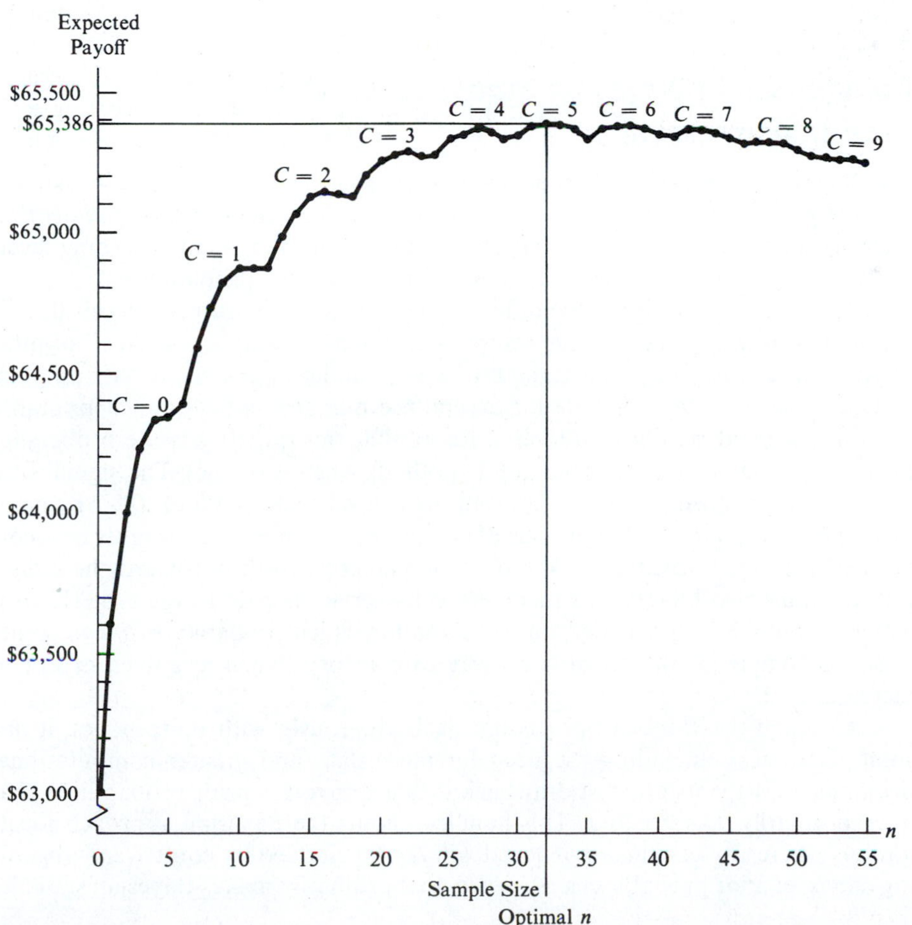

Figure 18-10 *Expected payoffs for possible sample sizes in printed circuit board acceptance sampling.*

Decision Making with the Mean

In this book we give only the details of sampling from qualitative populations, where payoffs may be expressed as functions of the proportion π. However, a parallel approach may be used in decision making with the mean, where quantitative observations are made. In this case, payoffs may be expressed as functions of the population mean μ. A prior probability distribution would then apply to μ, and if sampling is used, the main decision will be based on the computed level for $\bar{X}$. Conditional result probabilities for $\bar{X}$ may be obtained by using an appropriate normal distribution (permitted by the central limit theorem). Although that distribution is continuous, it might be approximated by a discrete distribution so that $\bar{X}$ values could be represented by a finite number of branches in the ensuing tree diagrams.

When the prior probability distribution for μ is itself a normal distribution, the continuous case may be evaluated directly without making discrete approximations. Rather than employing tree diagrams, the resulting procedure relies on a **loss function**, which is tabled in a fashion similar to normal curve areas. A detailed discussion of that approach is provided in some of the references listed in the Bibliography.

Comparison of Bayesian Statistics with Traditional Testing

The Bayesian approach embodies several elements. Its acts include choices for n and the possibility of *no sample at all.* This procedure incorporates payoffs that quantify all outcomes and also reflect sampling costs. Judgment regarding levels for the population parameter is reflected in a set of prior probabilities.

By contrast, traditional hypothesis testing seems very ad hoc. Recall that in the standard hypothesis test, the sample size has been established ahead of time, often to meet budgetary constraints. It is automatically assumed that sampling will be done. The decision rule (or acceptance number) is determined primarily from the targeted significance level α. Recall that this quantity is the probability for the type I error (rejecting the null hypothesis when it is true). The significance level is usually chosen at some "conventional" level such as .05 or .01. Any judgment concerning relative likelihoods of population parameter levels must be incorporated into the choice of α. The decision-maker's attitudes toward the consequences must also be reflected there. Some lip service is paid to the type II error, which might occur in a variety of ways and for which probabilities β can sometimes be computed. But there is no way to control β when n is fixed except to increase α.

Although the Bayesian procedure deals rigorously with more issues, it has some drawbacks, including the need for more data and greater computational demands. Perhaps its greatest drawback is that it involves prior probabilities that must ordinarily be subjective. This handicap limits the Bayesian approach to situations where any conclusion reached will not be clouded by controversy regarding choice of prior probabilities. As noted, that ordinarily makes Bayesian statistics unsuitable for evaluating decisions involving societal considerations or public debate. Besides, such situations often involve outcomes for which no natural payoff measure exists, so that maximizing expected payoff is itself a moot issue. Hypothesis testing was first employed in making decisions of that kind, and even today, that approach seems to be the only appropriate one. But when circumstances permit a choice of procedure, the Bayesian approach should be given serious consideration.

Problems

18-19 Consider the acceptance sampling illustration described in Section 18-3. For a sample of size $n = 2$ complete the following:

(a) Construct the probability tree diagrams for the actual and informational chronologies.

(b) Construct the decision tree diagram for this sample size. What is the acceptance number?

18-20 Suppose that the inspector in the circuit board illustration is faced with a similar evaluation regarding a second supplier for the same printed circuit boards. That supplier has a better reputation than the first, so the prior probabilities for the proportion of defectives in a 1,000-board shipment are .20 for $\pi = .10$ and .80 for $\pi = .20$. The payoffs for both suppliers are the same.

(a) Compute the expected payoffs for accepting and rejecting shipments without sampling. Which act maximizes expected payoff?

(b) Compute the inspector's EVPI. Does it appear that sampling might be beneficial in this case?

(c) Suppose that a sample size of $n = 10$ is used. Construct the probability tree diagrams for the actual and informational chronologies.

(d) For $n = 10$, find the acceptance number and the expected payoff. Would the inspector be better off using $n = 10$ or no sample at all?

18-21 An electrical engineer is uncertain about the proportion of major customers for an aircraft computer who will favor a new display. The following prior probabilities apply: .25 for $\pi = .4$, .45 for $\pi = .5$, and .30 for $\pi = .6$. Suppose that if the new design is used, the present value of future profits will decrease by \$10,000 when $\pi = .4$, will remain unchanged when $\pi = .5$, and will increase by \$10,000 when $\pi = .6$. Sampling is expensive, costing \$200 per observation, since each user opinion is based on in-flight evaluations.

(a) Which act provides the greatest expected payoff: use no sample at all or sample with $n = 1$?

(b) Which is better, to sample with $n = 1$ or $n = 2$?

18-22 A petroleum engineer's decision regarding bit shape depends on the proportion π of hard rock in the planned trajectory of the bore. The drilling time saved by using the hard-rock bit throughout drilling is $-10 + 20\pi$ days over what might be achieved by using a regular bit. Once drilling is under way, the bit will not be changed. The prior probabilities are .2 for $\pi = .4$, .3 for $\pi = .5$, .4 for $\pi = .6$, and .1 for $\pi = .7$.

(a) Using days saved as the payoff measure, determine which bit maximizes expected payoff. What is the EVPI?

(b) Suppose each day saved results in a cost saving of \$2,000. A summer employee suggested that the engineer use core samples from the shaft to revise his prior probabilities for π. The employee noted that a \$5 analysis of the drilling mud can establish the hardness of the present structure. The employee noted from the EVPI that sampling might have considerable value.

But the engineer informed the employee that even if each sample observation only cost \$.50, it still would not help. Explain the engineer's reasoning.

18-23 Consider again the acceptance sampling illustration. Suppose that a sample of size $n = 3$ was chosen and that no defective printed circuit boards were found. Instead of accepting the shipment, however, the decision-maker wants to know if further sampling in a *second stage* before making the main decision might be helpful.

(a) Using the posterior probabilities for π from the first sampling stage as prior probabilities, construct actual and informational chronology probability trees, again assuming the second sample will involve $n = 3$ observations.

(b) Construct a decision tree diagram similar to the one in Figure 18-8, revising all the payoff values to reflect the outcome from the first sampling stage. Is the expected payoff for the best action greater than that for accepting without a further sample?

18-24 A chemical engineer must decide whether to adjust or leave alone the settings in an automated process. The payoff is determined by the mean yield μ of active ingredient per liter of raw material.

$$\text{Payoff} = \begin{cases} \$10{,}000 - 200\mu & \text{if leave alone} \\ \$4{,}000 & \text{if adjust} \end{cases}$$

The prior probabilities are .20 for $\mu = 29$, .50 for $\mu = 30$, and .30 for $\mu = 31$.

(a) Which act maximizes expected payoff? Find the expected value of perfect information.

(b) By the central limit theorem, $\bar{X}$ is normally distributed with mean μ and standard deviation $\sigma/\sqrt{n}$. If we assume that the population standard deviation is $\sigma = 1$ gram/liter and σ is the same for all levels of μ, then for $n = 4$ the following approximate conditional result probabilities apply for the sample mean:

Sample Mean	Probability
$\bar{X} = \mu - 1.0$	.07
$\bar{X} = \mu - .5$	.24
$\bar{X} = \mu$	.38
$\bar{X} = \mu + .5$	.24
$\bar{X} = \mu + 1.0$	.07

Determine the probability trees for the actual and informational chronologies.

(c) Suppose that sampling costs \$5 per observation. Assuming that sampling will be done using $n = 4$, construct the engineer's decision tree and enter the probabilities from (b).

(d) Perform backward induction to determine the expected payoff. Give the decision rule for $\bar{X}$ that corresponds.

18-4 Finding Subjective Probabilities Using Judgment

At the heart of Bayesian analysis is the quantification of judgment regarding uncertainties. This process results in subjective probabilities, which are often the only available expressions of likelihood for nonrepeatable random experiments. Subjective probability has no role in traditional statistics, where all measures of probability are objective—either deduced by counting possibilities or estimatable from experimentation. But many interesting decisions involve nonrepeatable random experiments, for which subjective probabilities are the only available measures of uncertainty.

A subjective probability is best thought of as an expression of one person's *strength of conviction* that an event will occur. Since people may disagree about the likelihood of an event, a subjective probability is sometimes referred to as a **personal probability**. As applied to decision making, a succinct interpretation of subjective probabilities is that they represent the decision-maker's **betting odds** regarding uncertain events.

It is often helpful to use a lottery as a prop to assist in arriving at a subjective probability. Such gambles also provide some justification for using those values in the same way any other probability value might be used.

Subjective Probabilities and Hypothetical Lotteries

A decision-maker might be willing to exchange his or her decision tree diagram for an equivalent tree in which one or more of the event forks have been replaced by forks representing **hypothetical lotteries**. Each lottery has the same outcomes as the event fork it replaces, but the probabilities for the events in the lottery are established by the decision-maker. The chosen probabilities should reflect his or her *judgment* regarding the likelihoods of the events, and *not* the decision-maker's attitude toward the consequences.

The substitutability of the hypothetical lottery for the real uncertainty and the equivalence (in the decision-maker's mind) of the resulting decision tree to the original one can remove most objections to the use of subjective probabilities and to mixing them with objective probabilities. The resulting evaluation may then be carried out *on the substitute decision tree*, all events of which involve *objective* probabilities.

Subjective Probabilities for Discrete Events

To establish a subjective probability, you need only ask yourself what hypothetical lottery would make you indifferent between achieving the same outcomes by using it or by "letting nature take its course."

Figure 18-11

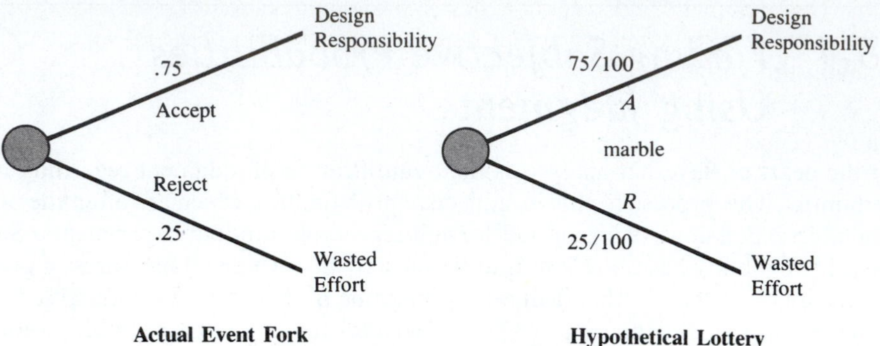

Consider an engineer wishing to establish a probability that the budget committee will accept his proposal to modify a product's design. To help find that number he will use a hypothetical lottery in which the events are generated by some random mechanism. It is simplest to think of a box containing 100 marbles, some stamped with the letter *A* for accept, the rest with *R* for reject. Suppose that the actual outcome were to be determined by which marble gets selected in that lottery, and that the engineer can choose any mixture of the two marble types that would make him indifferent between the actual event fork and the hypothetical lottery.

After some introspection, the engineer might decide he is indifferent between the actual event fork on the left and the hypothetical lottery on the right, shown in Figure 18-11. Since the lottery involves 75 *A* marbles and 25 *R* marbles, he has in effect established his subjective probability for accept to be $\frac{75}{100} = .75$.

Based on his judgment with new product designs, the chairman of the budget committee might similarly arrive at a 40% chance that the effort will result in an improvement to the product design. For him the equivalencies shown in Figure 18-12 apply. In this hypothetical lottery there are 40 *I* marbles, for improvement to design, and 60 not-*I* ones.

For persons unfamiliar with probability concepts the hypothetical lottery can be a helpful device for arriving at probabilities. This step can usually be skipped by investigators who feel comfortable with these concepts.

Figure 18-12

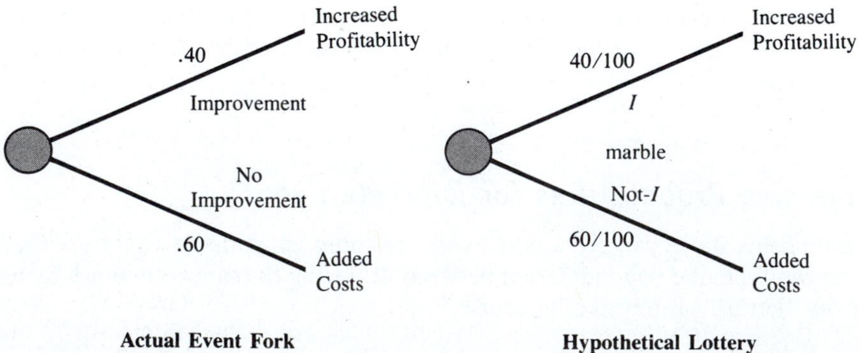

The hypothetical lottery can overcome some philosophical objections to subjective probability. Once we accept that equivalencies can be established and that substitutions are permitted, it is unnecessary actually to contort the decision structure to reflect all these props. Hereafter, we will assume that all of that could be done, and the same optimal course of action will be found in either case.

Example:* *Your Subjective Probability for Getting an A

Consider the event that you will get an A on your next statistics exam. Suppose that your instructor might allow you to elect to receive your grade by lottery. Pause for a moment and think what probability would make you indifferent between that lottery or letting your grade be determined by the usual procedure. (You will still have to study and will still have to sit for the exam. We are only referring to the grade you will receive.)

When faced with this prospect, there is a temptation to set an unrealistically high probability for getting that A. One student commented:

"What have I got to lose by using a high probability? This lottery can only improve my GPA."

One of his wiser classmates informed him that he should place the question in a decision-making perspective.

"Suppose you must decide how to allocate your studying time between statistics and your other courses. Your overall GPA will be affected by all the exams for the remainder of the term. Your statistics exam grade will be just one of many. Indeed, you should assign a higher probability for an A in statistics given that you study 10 hours than if you only study 5 hours. If you lie to yourself, you will only blow the decision evaluation and end up allocating too many study hours to some courses and not enough to others. If you are careful about how you pick your probabilities, you can actually perform a decision tree analysis and thereby find how to allocate your study time so as to maximize your expected GPA."

The use of hypothetical lotteries as props to help establish subjective probabilities extends to entire probability distributions.

Finding a Subjective Normal Curve

A variety of random variables have probability distributions that are closely approximated by the normal distribution. This is true of many uncertain dimensions, including those for anthropometric measurements and a variety of physical objects. Recall that a normal distribution is completely specified by its mean μ and standard deviation σ, so that these are the only necessary quantities to be found. Two 50-50 gambles can quickly establish judgmental levels for these parameters.

Because of the symmetry of the normal curve, the median and the mean are a common value. Thus, all that is necessary for finding μ is to establish the **median value**, that point above or below which the decision-maker believes it is equally likely that the uncertain quantity in question will fall. For example, if you believe that the height of a random male engineering student is just as likely to fall above 5′9″ as below, then your subjective probability distribution for the height of a randomly chosen man may be represented by a normal curve with mean $\mu = 5'9''$.

The standard deviation may be found analogously. This can be easily identified by establishing the **middle 50% limits** within which the random variable is believed to fall. Assuming those limits are equidistant above and below the mean, the area under the normal curve between those limits is .50. When the area between the mean and a point above it is .25, then a distance of .67 standard deviations separates that point from μ. Thus, the normal deviate is $z = .67$ and the middle 50% limits are equal to $\mu - .67\sigma$ and $\mu + .67\sigma$. This allows us to use the following expression.

STANDARD DEVIATION OF A SUBJECTIVE NORMAL DISTRIBUTION

$$\sigma = \frac{\textbf{Middle 50\% upper limit } - \mu}{\textbf{.67}}$$

Continuing again with the heights of male engineering students, you might establish the middle 50% limits as 5′7.5″ to 5′10.5″, so that

$$\sigma = \frac{5'10.5'' - 5'9''}{.67} = 2.24''$$

Example:* *Robot Service Times in an Automated Warehouse

An industrial engineering student is performing a queueing evaluation of alternative system configurations for an automated warehouse. She is presently concerned with the time taken by a robot retrieval mechanism to fetch a pallet and bring it to the loading dock. The probability distribution must be specified as one of the inputs to the queueing model. The student knows the overall specifications of the robot and has observed humans performing the same task, although no stopwatch observations are available for the robot. She believes that a normal distribution would be appropriate and judges that the time taken by the robot to fulfill a request is just as likely to fall at or below 5 minutes as above that duration. She furthermore feels that there is a 50-50 chance that any request will take between 4 and 6 minutes to fill versus taking some time shorter than 4 or longer than 6 minutes. The following parameters apply:

$$\mu = 5 \text{ minutes} \qquad \sigma = \frac{6 - 5}{.67} = 1.49 \text{ minutes}$$

Even under ideal circumstances, the normal distribution is only an approximate representation. It clearly should not be used if the random variable is believed to have a skewed distribution. But symmetry alone is not sufficient justification for using the normal curve. The underlying density must coincide with the bell shape and have tails that taper off rapidly. Other common distributions might be appropriate, such as the uniform or exponential. The latter applies when the underlying event-generating mechanism is a Poisson process (which would be the likely case for the *arrivals* of requests to the robot dispatcher in the preceding example).

But there is no reason why a random variable must fall into a particular distribution family. The essential characteristics may be established directly, and no formal background in probability is necessary. We now consider a procedure that extends the concept of 50-50 gambles to finding a general probability distribution.

Finding a Subjective Probability Distribution Using Cumulative Probabilities

An easy-to-understand method for establishing the subjective probability distribution involves finding a series of midpoints for the random variable. Each quantity divides successively smaller intervals into two equally likely halves, in the same way that μ for a normal distribution was established. The resulting values may be used to establish a graph for the underlying cumulative probability distribution.

Illustration: *Prior Probabilities for Proportion of Defectives in a Shipment*

To explain the procedure, we consider again the printed circuit boards used in the evaluation in Section 18-3. There the evaluation was predicated on a set of subjective probabilities for the proportion of defectives π. We now see how the inspector established those prior probabilities. We will use a different type of board supplied by a vendor for which the quality of shipments is judged to be considerably higher than before.

The inspector's first step is to consider the entire range of possibilities. Here the random variable is true proportion π of defectives in a shipment, values for which are represented by:

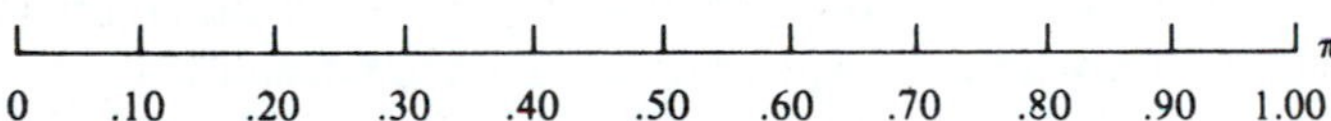

Believing that π is just as likely to fall above .09 as below, the inspector chose .09 as the median for π. This divides the possible π range into two equally likely halves:

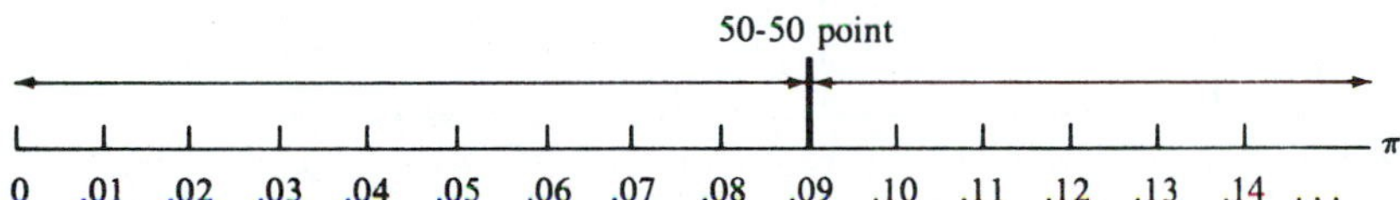

This determination establishes the cumulative probability to be .50 that π will fall at or below .09. We can refer to the point of demarcation as the **.50-fractile** for the random variable π. Other fractiles may be found by dividing the intervals (.0, .09) and (.09, 1.00).

The inspector established that *given* $\pi \leq .09$, it would be just as likely for that quantity to fall at or below .07 as above that level. Similarly, he determined

.12 as the 50-50 point for π *given* that $\pi > .09$. These findings are summarized by:

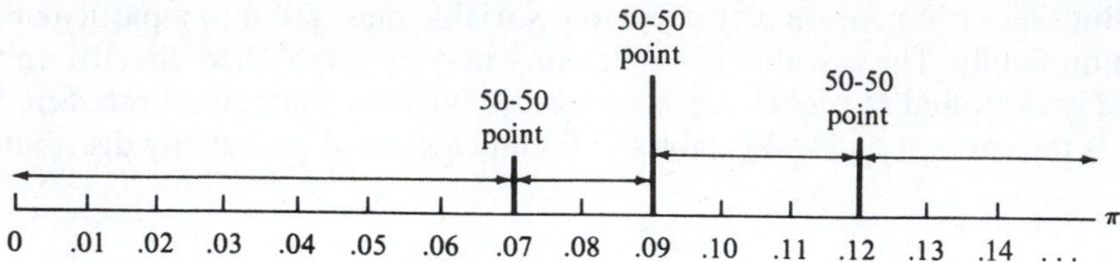

This implies that the cumulative probability for falling at or below .07 is .50(.50) = .25, and for falling at or below .12 is .50 + .50(.50) = .75. The value .07 is the **.25-fractile** for π, and .12 is the **.75-fractile**.

Working out from the middle, we can continue to divide the outside ranges into equally likely halves. Given that $\pi \leq .07$, the inspector decides that .055 is the 50-50 point. Given that $\pi \leq .055$, the median value is .045. This is summarized by:

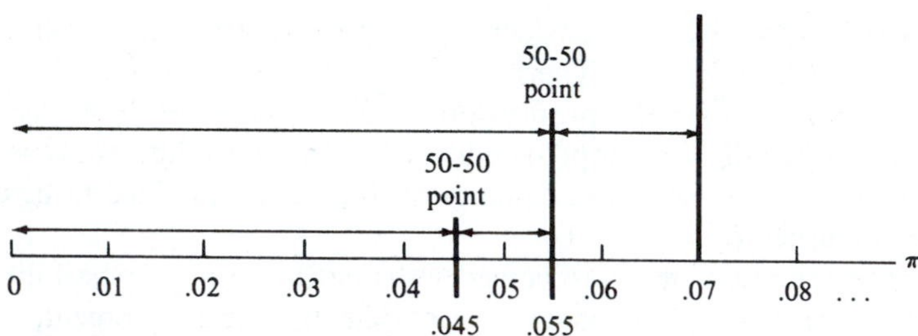

The cumulative probability for $\pi \leq .055$ is .5(.25) = .125, whereas for $\pi \leq .045$ it is .5(.125) = .0625. Thus, .055 is the **.125-fractile** for π (half the fractile of the next higher dividing point), whereas .045 serves as the **.0625-fractile.**

At the upper end of the scale, the inspector used .15 as the dividing point given $\pi > .12$ and .19 assuming $\pi > .15$. This is summarized by:

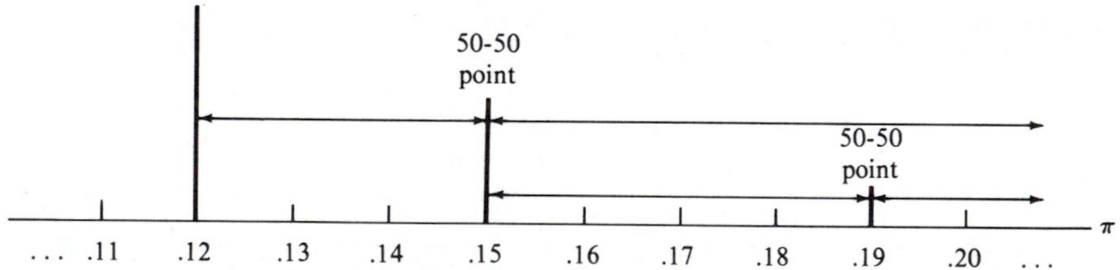

The cumulative probability for $\pi \leq .15$ is .75 + .5(1 − .75) = .875, whereas that for $\pi \leq .19$ is .875 + .5(1 − .875) = .9375. We may refer to the value .15 as the **.875-fractile** and .19 as the **.9375-fractile** for the proportion of defectives π.

The preceding establishes seven points on a cumulative probability graph. The points are sufficiently spread out that an accurate curve may be fitted to them once an underlying shape has been established. Unimodal distributions having both an upper and a lower tail will generally be represented by an S-shaped cumulative probability distribution. That is true for the present illustration, and the

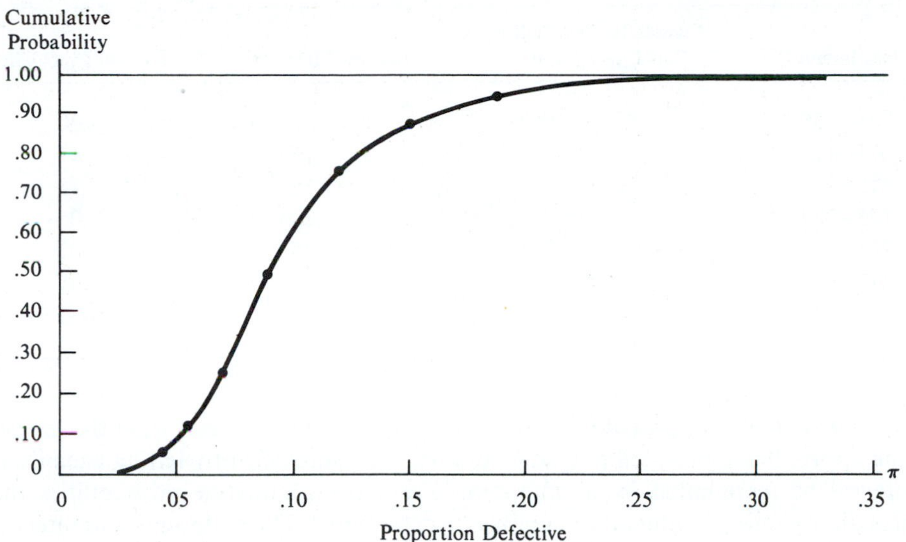

Figure 18-13 *Cumulative subjective probability distribution for proportion of defective printed circuit boards.*

cumulative probability distribution for π is shown in Figure 18-13. The curve touches the π axis at .025 and reaches a maximum height of 1 around $\pi = .30$. Those points can be determined fairly accurately in the process of sketching the curve.

The cumulative probability curve for the proportion of defectives has an S-shape that is elongated at the top. The corresponding density curve will be *positively skewed.* Another commonly encountered shape is an S elongated at the bottom, which indicates a *negatively skewed* density. Should the S be regular in shape, the density will be bell-shaped, resembling a normal curve. It is not unusual for a double-S shape to be obtained. This shape is indicative of a *bimodal* density, which reflects a nonhomogeneous influence. Such a finding necessitates splitting the variable into two cases, each of which must be evaluated separately.

A linear cumulative probability graph corresponds to a *uniform distribution,* and a "logarithmic" shape represents an *exponential distribution.* If the investigator has reason to believe that either of these applies, the easiest approach would be to establish the parameters directly by judgment rather than using the graphical approach. (Those would be the limits a and b for the uniform distribution and process rate λ for the exponential.)

Discrete Approximation to Continuous Subjective Distribution

A discrete approximation to the probability distribution for π can be established by fitting a series of steps to the cumulative probability curve. That result is accomplished by first dividing the range for the random variable into intervals. We approximate the curve in Figure 18-13 by dividing the range for π into six intervals of width .05, starting at .025 and ending at .325—a span virtually certain to occur. The resulting class intervals are listed in Table 18-7.

Table 18-7
Discrete Approximation to Continuous Subjective Probability Distribution for Proportion of Defectives

Class Interval	Cumulative Probability at Upper Limit	Interval Midpoint	Interval Probability
.025–.075	.30	$\pi = .05$	.30
.075–.125	.77	.10	.47
.125–.175	.92	.15	.15
.175–.225	.97	.20	.05
.225–.275	.99	.25	.02
.275–.325	1.00	.30	.01
			1.00

The cumulative probability for each class upper limit is read from the graph. The approximate probability distribution is then established by letting each class interval be represented by its midpoint. From the cumulative probabilities the probability for each interval can readily be determined. The midpoints and interval probabilities provide the approximate distribution. These are shown in the last two columns of Table 18-7.

The advantage of the approximation is that it allows us to represent the probability distribution for a variable by an event fork with a finite number of branches.

Problems

18-25 Establish the probability that you will receive an A grade on your next examination in statistics.

18-26 The mean repair time for the arms of robot automobile assemblers has been judged to be normally distributed. There is a 50-50 chance that any repair will take 2 hours or less. It is an equal chance that any repair time will fall within ± 20 minutes of that time versus falling either above or below that range.

(a) Find μ and σ.

(b) Find the subjective probability that any repair time will fall (1) below 2.5 hours, (2) above 3 hours, (3) between 1.5 and 2.4 hours, and (4) below 1.4 hours.

18-27 A petroleum engineer establishes the following fractiles for the size of an oil field's recoverable reserves:

Reserves (barrels)	50-50 Point
all	10,000,000
below 10,000,000	8,000,000
above 10,000,000	15,000,000
below 8,000,000	7,000,000
above 15,000,000	22,000,000
below 7,000,000	6,500,000
above 22,000,000	30,000,000

(a) Sketch the engineer's subjective cumulative probability distribution for the size of the field.
(b) From your graph, estimate the following fractiles:
(1) .30 (2) .70 (3) .90 (4) .95
(c) From your graph, determine the cumulative probability that the reserves fall below
(1) 6,000,000 (2) 14,000,000 (3) 18,000,000 (4) 25,000,000

18-28 Consider your earnings from engineering during the first full calendar year following graduation. For that uncertain quantity, use your judgment in completing the following:
(a) Establish your .50-, .25-, and .75-fractiles.
(b) Establish your .0625-, .125-, .875-, and .9375-fractiles.
(c) Sketch your curve on a graph.
(d) What is the probability that you will earn at least $25,000?

18-29 Refer to the subjective probability curve found in Problem 18-27 for the level of reserves in an oil field.
(a) Determine a discrete approximation, using class intervals of width 5,000,000 barrels with the lower limit of the first interval at 2,500,000 barrels. Represent each interval by its midpoint.
(b) Using your approximate distribution, compute the expected level of reserves.

18-30 Determine your subjective probability distribution for the number of hours it will take to write a 10-page A paper on the outlook for careers in engineering. Plot your cumulative probability curve.

Comprehensive Problems

18-31 A petroleum engineer assigns the following prior probabilities for the outcome of drilling on a leased site: .1 for oil only, .2 for gas only, .3 for both oil and gas, and .4 for dry. The rights to any gas found will be sold to a pipeline for $1,000,000, whereas those for oil will be transferred to a holding company for $2,000,000. Drilling costs are $500,000, plus an additional $100,000 lining cost if oil is found, with or without gas.
(a) Determine the engineer's payoff table, using total profit as the measure and assuming the two choices are to drill or to abandon.
(b) Compute expected payoffs. Which act provides the maximum?
(c) Compute the expected value of perfect information. What is the most the engineer would be willing to pay for additional information about the underground geology?

18-32 Suppose the engineer in Problem 18-31 can order a seismic survey costing $100,000. The test will predict favorably given oil or gas with probability .80, while given no oil or gas, it will predict unfavorably with probability .70. The test cannot give any indication of the type of fossil deposits.
(a) Construct probability trees for the actual and informational chronologies.
(b) Construct the engineer's decision tree, incorporating all choices and events. Then perform backward induction. What action maximizes expected payoff?

18-33 An engineer is faced with a decision regarding a new product design. Her choices are to abandon the design or to place it into production. In the latter case, the product will be a success with probability .6 or a failure with probability .4. A successful product will be acquired by the parent firm for $1,000,000, which will be credited to the engineer's research budget. A failure will result in a reduction of $350,000 in the engineer's budget.

(a) Using the net change to the engineer's budget as the payoff measure, construct her payoff table.
(b) Determine the expected payoff for each act. Which act has the maximum? Find the EVPI.

18-34 Refer to the data in Problem 18-33. The engineer may field-test prototype units before committing herself to the main decision. Previous testing has established that 80% of successful products receive positive results, while 90% of the failures get negative results. The test will cost $25,000.
(a) Construct the probability trees for the actual and informational chronologies.
(b) Construct the engineer's decision tree, reflecting all choices and events. Then perform backward induction to determine the course of action that maximizes expected payoff.

18-35 Again, refer to Problem 18-33. Suppose that production outcomes are represented instead by the number of units sold. The following judgment was expressed by the engineer:

Units X	50-50 Point
all	40,000
below 40,000	35,000
above 40,000	45,000
below 35,000	30,000
above 45,000	50,000
below 30,000	25,000
above 50,000	55,000

(a) Sketch the engineer's subjective cumulative probability curve.
(b) Determine a discrete approximation using intervals of width 10,000, starting at 10,000 as the lower limit of the first interval.
(c) Suppose that the engineer's budget will be increased by $\$-200{,}000 + 20X$. Determine the expected payoff for producing the item.

18-36 The proportion of missed welds in automated car assembly operations has a prior probability of .30 for $\pi = .10$, .40 for $\pi = .20$, and .30 for $\pi = .30$. There are two actions—overhaul the welder for a payoff of $\$-1{,}000$ or leave it alone for a payoff of $\$10{,}000 - 50{,}000\pi$.
(a) Construct the payoff table.
(b) Compute the expected payoffs. Which act has the greater payoff? Compute the EVPI.

18-37 Refer to the data in Problem 18-36. Sampling costs $2 per weld. Consider a sample of size $n = 3$.
(a) Construct the probability trees for the actual and informational chronologies.
(b) Construct the decision tree assuming that $n = 3$ observations will be made. Then perform backward induction to determine the acceptance number (for leaving alone) and the expected payoff for that sample size.

APPENDIX A

Tables

Table A
Cumulative Values for the Binomial Probability Distribution

$$B(r; n, \pi) = \Pr[R \leq r] = \sum_{k=0}^{r} b(k; n, \pi)$$

	r	π = .01	π = .05	π = .10	π = .20	π = .30	π = .40	π = .50
n = 1	0	0.9900	0.9500	0.9000	0.8000	0.7000	0.6000	0.5000
	1	1.0000	1.0000	1.0000	1.0000	1.0000	1.0000	1.0000
n = 2	0	0.9801	0.9025	0.8100	0.6400	0.4900	0.3600	0.2500
	1	0.9999	0.9975	0.9900	0.9600	0.9100	0.8400	0.7500
	2	1.0000	1.0000	1.0000	1.0000	1.0000	1.0000	1.0000
n = 3	0	0.9703	0.8574	0.7290	0.5120	0.3430	0.2160	0.1250
	1	0.9997	0.9927	0.9720	0.8960	0.7840	0.6480	0.5000
	2	1.0000	0.9999	0.9990	0.9920	0.9730	0.9360	0.8750
	3	1.0000	1.0000	1.0000	1.0000	1.0000	1.0000	1.0000
n = 4	0	0.9606	0.8145	0.6561	0.4096	0.2401	0.1296	0.0625
	1	0.9994	0.9860	0.9477	0.8192	0.6517	0.4752	0.3125
	2	1.0000	0.9995	0.9963	0.9728	0.9163	0.8208	0.6875
	3	1.0000	1.0000	0.9999	0.9984	0.9919	0.9744	0.9375
	4	1.0000	1.0000	1.0000	1.0000	1.0000	1.0000	1.0000
n = 5	0	0.9510	0.7738	0.5905	0.3277	0.1681	0.0778	0.0313
	1	0.9990	0.9774	0.9185	0.7373	0.5282	0.3370	0.1875
	2	1.0000	0.9988	0.9914	0.9421	0.8369	0.6826	0.5000
	3	1.0000	1.0000	0.9995	0.9933	0.9692	0.9130	0.8125
	4	1.0000	1.0000	1.0000	0.9997	0.9976	0.9898	0.9688
	5				1.0000	1.0000	1.0000	1.0000
n = 10	0	0.9044	0.5987	0.3487	0.1074	0.0282	0.0060	0.0010
	1	0.9957	0.9139	0.7361	0.3758	0.1493	0.0464	0.0107
	2	0.9999	0.9885	0.9298	0.6778	0.3828	0.1673	0.0547
	3	1.0000	0.9990	0.9872	0.8791	0.6496	0.3823	0.1719
	4	1.0000	0.9999	0.9984	0.9672	0.8497	0.6331	0.3770
	5	1.0000	1.0000	0.9999	0.9936	0.9526	0.8338	0.6230
	6	1.0000	1.0000	1.0000	0.9991	0.9894	0.9452	0.8281
	7				0.9999	0.9999	0.9877	0.9453
	8				1.0000	1.0000	0.9983	0.9893
	9						0.9999	0.9990
	10						1.0000	1.0000
n = 20	0	0.8179	0.3585	0.1216	0.0115	0.0008	0.0000	0.0000
	1	0.9831	0.7358	0.3917	0.0692	0.0076	0.0005	0.0000
	2	0.9990	0.9245	0.6769	0.2061	0.0355	0.0036	0.0002
	3	1.0000	0.9841	0.8670	0.4114	0.1071	0.0160	0.0013
	4	1.0000	0.9974	0.9568	0.6296	0.2375	0.0510	0.0059
	5	1.0000	0.9997	0.9887	0.8042	0.4164	0.1256	0.0207
	6	1.0000	1.0000	0.9976	0.9133	0.6080	0.2500	0.0577
	7	1.0000	1.0000	0.9996	0.9679	0.7723	0.4159	0.1316
	8	1.0000	1.0000	0.9999	0.9900	0.8867	0.5956	0.2517
	9	1.0000	1.0000	1.0000	0.9974	0.9520	0.7553	0.4119
	10				0.9994	0.9829	0.8725	0.5881
	11				0.9999	0.9949	0.9435	0.7483
	12				1.0000	0.9987	0.9790	0.8684
	13					0.9997	0.9935	0.9423
	14					1.0000	0.9984	0.9793
	15						0.9997	0.9941
	16						1.0000	0.9987
	17							0.9998
	18							1.0000

Table A (continued)

	r	π = .01	π = .05	π = .10	π = .20	π = .30	π = .40	π = .50
n = 50	0	0.6050	0.0769	0.0052	0.0000	0.0000	0.0000	0.0000
	1	0.9106	0.2794	0.0338	0.0002	0.0000	0.0000	0.0000
	2	0.9862	0.5405	0.1117	0.0013	0.0000	0.0000	0.0000
	3	0.9984	0.7604	0.2503	0.0057	0.0000	0.0000	0.0000
	4	0.9999	0.8964	0.4312	0.0185	0.0002	0.0000	0.0000
	5	1.0000	0.9622	0.6161	0.0480	0.0007	0.0000	0.0000
	6	1.0000	0.9882	0.7702	0.1034	0.0025	0.0000	0.0000
	7	1.0000	0.9968	0.8779	0.1904	0.0073	0.0001	0.0000
	8	1.0000	0.9992	0.9421	0.3073	0.0183	0.0002	0.0000
	9	1.0000	0.9998	0.9755	0.4437	0.0402	0.0008	0.0000
	10	1.0000	1.0000	0.9906	0.5836	0.0789	0.0022	0.0000
	11	1.0000	1.0000	0.9968	0.7107	0.1390	0.0057	0.0000
	12	1.0000	1.0000	0.9990	0.8139	0.2229	0.0133	0.0002
	13	1.0000	1.0000	0.9997	0.8894	0.3279	0.0280	0.0005
	14	1.0000	1.0000	0.9999	0.9393	0.4468	0.0540	0.0013
	15	1.0000	1.0000	1.0000	0.9692	0.5692	0.0955	0.0033
	16				0.9856	0.6839	0.1561	0.0077
	17				0.9937	0.7822	0.2369	0.0164
	18				0.9975	0.8594	0.3356	0.0325
	19				0.9991	0.9152	0.4465	0.0595
	20				0.9997	0.9522	0.5610	0.1013
	21				0.9999	0.9749	0.6701	0.1611
	22				1.0000	0.9877	0.7660	0.2399
	23					0.9944	0.8438	0.3359
	24					0.9976	0.9022	0.4439
	25					0.9991	0.9427	0.5561
	26					0.9997	0.9686	0.6641
	27					0.9999	0.9840	0.7601
	28					1.0000	0.9924	0.8389
	29						0.9966	0.8987
	30						0.9986	0.9405
	31						0.9995	0.9675
	32						0.9998	0.9836
	33						0.9999	0.9923
	34						1.0000	0.9967
	35							0.9987
	36							0.9995
	37							0.9998
	38							1.0000

Table A (continued)

	r	π = .01	π = .05	π = .10	π = .20	π = .30	π = .40	π = .50
n = 100	0	0.3660	0.0059	0.0000	0.0000	0.0000	0.0000	0.0000
	1	0.7358	0.0371	0.0003	0.0000	0.0000	0.0000	0.0000
	2	0.9206	0.1183	0.0019	0.0000	0.0000	0.0000	0.0000
	3	0.9816	0.2578	0.0078	0.0000	0.0000	0.0000	0.0000
	4	0.9966	0.4360	0.0237	0.0000	0.0000	0.0000	0.0000
	5	0.9995	0.6160	0.0576	0.0000	0.0000	0.0000	0.0000
	6	0.9999	0.7660	0.1172	0.0001	0.0000	0.0000	0.0000
	7	1.0000	0.8720	0.2061	0.0003	0.0000	0.0000	0.0000
	8	1.0000	0.9369	0.3209	0.0009	0.0000	0.0000	0.0000
	9	1.0000	0.9718	0.4513	0.0023	0.0000	0.0000	0.0000
	10	1.0000	0.9885	0.5832	0.0057	0.0000	0.0000	0.0000
	11	1.0000	0.9957	0.7030	0.0126	0.0000	0.0000	0.0000
	12	1.0000	0.9985	0.8018	0.0253	0.0000	0.0000	0.0000
	13	1.0000	0.9995	0.8761	0.0469	0.0001	0.0000	0.0000
	14	1.0000	0.9999	0.9274	0.0804	0.0002	0.0000	0.0000
	15	1.0000	1.0000	0.9601	0.1285	0.0004	0.0000	0.0000
	16	1.0000	1.0000	0.9794	0.1923	0.0010	0.0000	0.0000
	17	1.0000	1.0000	0.9900	0.2712	0.0022	0.0000	0.0000
	18	1.0000	1.0000	0.9954	0.3621	0.0045	0.0000	0.0000
	19	1.0000	1.0000	0.9980	0.4602	0.0089	0.0000	0.0000
	20	1.0000	1.0000	0.9992	0.5595	0.0165	0.0000	0.0000
	21	1.0000	1.0000	0.9997	0.6540	0.0288	0.0000	0.0000
	22	1.0000	1.0000	0.9999	0.7389	0.0479	0.0001	0.0000
	23	1.0000	1.0000	1.0000	0.8109	0.0755	0.0003	0.0000
	24				0.8686	0.1136	0.0006	0.0000
	25				0.9125	0.1631	0.0012	0.0000
	26				0.9442	0.2244	0.0024	0.0000
	27				0.9658	0.2964	0.0046	0.0000
	28				0.9800	0.3768	0.0084	0.0000
	29				0.9888	0.4623	0.0148	0.0000
	30				0.9939	0.5491	0.0248	0.0000
	31				0.9969	0.6331	0.0398	0.0001
	32				0.9984	0.7107	0.0615	0.0002
	33				0.9993	0.7793	0.0913	0.0004
	34				0.9997	0.8371	0.1303	0.0009
	35				0.9999	0.8839	0.1795	0.0018

Table A (continued)

	r	π = .01	π = .05	π = .10	π = .20	π = .30	π = .40	π = .50
n = 100	36				0.9999	0.9201	0.2386	0.0033
	37				1.0000	0.9470	0.3068	0.0060
	38					0.9660	0.3822	0.0105
	39					0.9790	0.4621	0.0176
	40					0.9875	0.5433	0.0284
	41					0.9928	0.6225	0.0443
	42					0.9960	0.6967	0.0666
	43					0.9979	0.7635	0.0967
	44					0.9989	0.8211	0.1356
	45					0.9995	0.8689	0.1841
	46					0.9997	0.9070	0.2421
	47					0.9999	0.9362	0.3086
	48					0.9999	0.9577	0.3822
	49					1.0000	0.9729	0.4602
	50						0.9832	0.5398
	51						0.9900	0.6178
	52						0.9942	0.6914
	53						0.9968	0.7579
	54						0.9983	0.8159
	55						0.9991	0.8644
	56						0.9996	0.9033
	57						0.9998	0.9334
	58						0.9999	0.9557
	59						1.0000	0.9716
	60							0.9824
	61							0.9895
	62							0.9940
	63							0.9967
	64							0.9982
	65							0.9991
	66							0.9996
	67							0.9998
	68							0.9999
	69							1.0000

Source: Lawrence L. Lapin, *Quantitative Methods for Business Decisions*, 4th ed. Copyright © 1987 Harcourt Brace Jovanovich, Inc., New York. Reproduced by permission of the publisher.

Table B
Exponential Functions

y	e^y	e^{-y}	y	e^y	e^{-y}
0.00	1.0000	1.000000	5.00	148.41	.006738
0.10	1.1052	.904837	5.10	164.02	.006097
0.20	1.2214	.818731	5.20	181.27	.005517
0.30	1.3499	.740818	5.30	200.34	.004992
0.40	1.4918	.670320	5.40	221.41	.004517
0.50	1.6487	.606531	5.50	244.69	.004087
0.60	1.8221	.548812	5.60	270.43	.003698
0.70	2.0138	.496585	5.70	298.87	.003346
0.80	2.2255	.449329	5.80	330.30	.003028
0.90	2.4596	.406570	5.90	365.04	.002739
1.00	2.7183	.367879	6.00	403.43	.002479
1.10	3.0042	.332871	6.10	445.86	.002243
1.20	3.3201	.301194	6.20	492.75	.002029
1.30	3.6693	.272532	6.30	544.57	.001836
1.40	4.0552	.246597	6.40	601.85	.001662
1.50	4.4817	.223130	6.50	665.14	.001503
1.60	4.9530	.201897	6.60	735.10	.001360
1.70	5.4739	.182684	6.70	812.41	.001231
1.80	6.0496	.165299	6.80	897.85	.001114
1.90	6.6859	.149569	6.90	992.27	.001008
2.00	7.3891	.135335	7.00	1096.6	.000912
2.10	8.1662	.122456	7.10	1212.0	.000825
2.20	9.0250	.110803	7.20	1339.4	.000747
2.30	9.9742	.100259	7.30	1480.3	.000676
2.40	11.023	.090718	7.40	1636.0	.000611
2.50	12.182	.082085	7.50	1808.0	.000553
2.60	13.464	.074274	7.60	1998.2	.000501
2.70	14.880	.067206	7.70	2208.3	.000453
2.80	16.445	.060810	7.80	2440.6	.000410
2.90	18.174	.055023	7.90	2697.3	.000371
3.00	20.086	.049787	8.00	2981.0	.000336
3.10	22.198	.045049	8.10	3294.5	.000304
3.20	24.533	.040762	8.20	3641.0	.000275
3.30	27.113	.036883	8.30	4023.9	.000249
3.40	29.964	.033373	8.40	4447.1	.000225
3.50	33.115	.030197	8.50	4914.8	.000204
3.60	36.598	.027324	8.60	5431.7	.000184
3.70	40.447	.024724	8.70	6002.9	.000167
3.80	44.701	.022371	8.80	6634.2	.000151
3.90	49.402	.020242	8.90	7332.0	.000136
4.00	54.598	.018316	9.00	8103.1	.000123
4.10	60.340	.016573	9.10	8955.3	.000112
4.20	66.686	.014996	9.20	9897.1	.000101
4.30	73.700	.013569	9.30	10938	.0000914
4.40	81.451	.012277	9.40	12088	.0000827
4.50	90.017	.011109	9.50	13360	.0000749
4.60	99.484	.010052	9.60	14765	.0000677
4.70	109.95	.009095	9.70	16318	.0000613
4.80	121.51	.008230	9.80	18034	.0000555
4.90	134.29	.007447	9.90	19930	.0000502
5.00	148.41	.006738	10.00	22026	.0000454

Table C
Cumulative Probability Values for the Poisson Distribution

$$P(x; \lambda, t) = \Pr[X \leq x] = \sum_{k=0}^{x} p(k; \lambda, t)$$

	λt									
x	**1.0**	**2.0**	**3.0**	**4.0**	**5.0**	**6.0**	**7.0**	**8.0**	**9.0**	**10.0**
0	0.3679	0.1353	0.0498	0.0183	0.0067	0.0025	0.0009	0.0003	0.0001	0.0000
1	0.7358	0.4060	0.1991	0.0916	0.0404	0.0174	0.0073	0.0030	0.0012	0.0005
2	0.9197	0.6767	0.4232	0.2381	0.1247	0.0620	0.0296	0.0138	0.0062	0.0028
3	0.9810	0.8571	0.6472	0.4335	0.2650	0.1512	0.0818	0.0424	0.0212	0.0103
4	0.9963	0.9473	0.8153	0.6288	0.4405	0.2851	0.1730	0.0996	0.0550	0.0293
5	0.9994	0.9834	0.9161	0.7851	0.6160	0.4457	0.3007	0.1912	0.1157	0.0671
6	0.9999	0.9955	0.9665	0.8893	0.7622	0.6063	0.4497	0.3134	0.2068	0.1301
7	1.0000	0.9989	0.9881	0.9489	0.8666	0.7440	0.5987	0.4530	0.3239	0.2202
8		0.9998	0.9962	0.9786	0.9319	0.8472	0.7291	0.5926	0.4557	0.3328
9		1.0000	0.9989	0.9919	0.9682	0.9161	0.8305	0.7166	0.5874	0.4579
10			0.9997	0.9972	0.9863	0.9574	0.9015	0.8159	0.7060	0.5830
11			0.9999	0.9991	0.9945	0.9799	0.9466	0.8881	0.8030	0.6968
12			1.0000	0.9997	0.9980	0.9912	0.9730	0.9362	0.8758	0.7916
13				0.9999	0.9993	0.9964	0.9872	0.9658	0.9262	0.8645
14				1.0000	0.9998	0.9986	0.9943	0.9827	0.9585	0.9165
15					0.9999	0.9995	0.9976	0.9918	0.9780	0.9513
16					1.0000	0.9998	0.9990	0.9963	0.9889	0.9730
17						0.9999	0.9996	0.9984	0.9947	0.9857
18						1.0000	0.9999	0.9993	0.9976	0.9928
19							0.9999	0.9997	0.9989	0.9965
20							1.0000	0.9999	0.9996	0.9984
21								1.0000	0.9998	0.9993
22									0.9999	0.9997
23									1.0000	0.9999
24										0.9999
25										1.0000

Source: Lawrence L. Lapin, *Quantitative Methods for Business Decisions,* 4th ed. Copyright © 1987 Harcourt Brace Jovanovich, Inc., New York. Reproduced by permission of the publisher.

Table C (continued)

	λt									
x	**11.0**	**12.0**	**13.0**	**14.0**	**15.0**	**16.0**	**17.0**	**18.0**	**19.0**	**20.0**
0	0.0000	0.0000	0.0000	0.0000	0.0000	0.0000	0.0	0.0	0.0	0.0
1	0.0002	0.0001	0.0000	0.0000	0.0000	0.0000	0.0000	0.0000	0.0000	0.0
2	0.0012	0.0005	0.0002	0.0001	0.0000	0.0000	0.0000	0.0000	0.0000	0.0000
3	0.0049	0.0023	0.0011	0.0005	0.0002	0.0001	0.0000	0.0000	0.0000	0.0000
4	0.0151	0.0076	0.0037	0.0018	0.0009	0.0004	0.0002	0.0001	0.0000	0.0000
5	0.0375	0.0203	0.0107	0.0055	0.0028	0.0014	0.0007	0.0003	0.0002	0.0001
6	0.0786	0.0458	0.0259	0.0142	0.0076	0.0040	0.0021	0.0010	0.0005	0.0003
7	0.1432	0.0895	0.0540	0.0316	0.0180	0.0100	0.0054	0.0029	0.0015	0.0008
8	0.2320	0.1550	0.0998	0.0621	0.0374	0.0220	0.0126	0.0071	0.0039	0.0021
9	0.3405	0.2424	0.1658	0.1094	0.0699	0.0433	0.0261	0.0154	0.0089	0.0050
10	0.4599	0.3472	0.2517	0.1757	0.1185	0.0774	0.0491	0.0304	0.0183	0.0108
11	0.5793	0.4616	0.3532	0.2600	0.1847	0.1270	0.0847	0.0549	0.0347	0.0214
12	0.6887	0.5760	0.4631	0.3585	0.2676	0.1931	0.1350	0.0917	0.0606	0.0390
13	0.7813	0.6815	0.5730	0.4644	0.3632	0.2745	0.2009	0.1426	0.0984	0.0661
14	0.8540	0.7720	0.6751	0.5704	0.4656	0.3675	0.2808	0.2081	0.1497	0.1049
15	0.9074	0.8444	0.7636	0.6694	0.5681	0.4667	0.3714	0.2866	0.2148	0.1565
16	0.9441	0.8987	0.8355	0.7559	0.6641	0.5660	0.4677	0.3750	0.2920	0.2211
17	0.9678	0.9370	0.8905	0.8272	0.7489	0.6593	0.5640	0.4686	0.3784	0.2970
18	0.9823	0.9626	0.9302	0.8826	0.8195	0.7423	0.6549	0.5622	0.4695	0.3814
19	0.9907	0.9787	0.9573	0.9235	0.8752	0.8122	0.7363	0.6509	0.5606	0.4703
20	0.9953	0.9884	0.9750	0.9521	0.9170	0.8682	0.8055	0.7307	0.6472	0.5591
21	0.9977	0.9939	0.9859	0.9711	0.9469	0.9108	0.8615	0.7991	0.7255	0.6437
22	0.9989	0.9969	0.9924	0.9833	0.9672	0.9418	0.9047	0.8551	0.7931	0.7206
23	0.9995	0.9985	0.9960	0.9907	0.9805	0.9633	0.9367	0.8989	0.8490	0.7875
24	0.9998	0.9993	0.9980	0.9950	0.9888	0.9777	0.9593	0.9317	0.8933	0.8432
25	0.9999	0.9997	0.9990	0.9974	0.9938	0.9869	0.9747	0.9554	0.9269	0.8878
26	1.0000	0.9999	0.9995	0.9987	0.9967	0.9925	0.9848	0.9718	0.9514	0.9221
27		0.9999	0.9998	0.9994	0.9983	0.9959	0.9912	0.9827	0.9687	0.9475
28		1.0000	0.9999	0.9997	0.9991	0.9978	0.9950	0.9897	0.9805	0.9657
29			1.0000	0.9999	0.9996	0.9989	0.9973	0.9940	0.9881	0.9782
30				0.9999	0.9998	0.9994	0.9985	0.9967	0.9930	0.9865
31				1.0000	0.9999	0.9997	0.9992	0.9982	0.9960	0.9919
32					0.9999	0.9999	0.9996	0.9990	0.9978	0.9953
33					1.0000	0.9999	0.9998	0.9995	0.9988	0.9973
34						1.0000	0.9999	0.9997	0.9994	0.9985
35							0.9999	0.9999	0.9997	0.9992
36							1.0000	0.9999	0.9998	0.9996
37								1.0000	0.9999	0.9998
38									1.0000	0.9999
39										0.9999
40										1.0000

Table D

Cumulative Probability Distribution Function for the Standard Normal Distribution

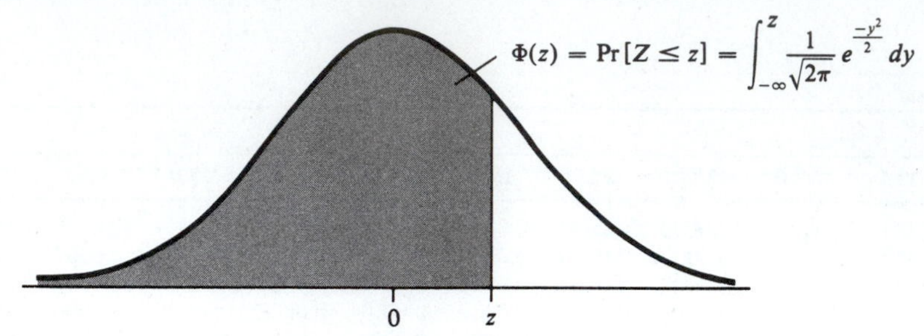

Normal Deviate z	.00	.01	.02	.03	.04	.05	.06	.07	.08	.09
−4.0	.00003									
−3.2	.0007	.0007	.0006	.0006	.0006	.0006	.0006	.0005	.0005	.0005
−3.1	.0010	.0009	.0009	.0009	.0008	.0008	.0008	.0008	.0007	.0007
−3.0	.0014	.0013	.0013	.0012	.0012	.0011	.0011	.0011	.0010	.0010
−2.9	.0019	.0018	.0018	.0017	.0016	.0016	.0015	.0015	.0014	.0014
−2.8	.0026	.0025	.0024	.0023	.0023	.0022	.0021	.0021	.0020	.0019
−2.7	.0035	.0034	.0033	.0032	.0031	.0030	.0029	.0028	.0027	.0026
−2.6	.0047	.0045	.0044	.0043	.0041	.0040	.0039	.0038	.0037	.0036
−2.5	.0062	.0060	.0059	.0057	.0055	.0054	.0052	.0051	.0049	.0048
−2.4	.0082	.0080	.0078	.0075	.0073	.0071	.0069	.0068	.0066	.0064
−2.3	.0107	.0104	.0102	.0099	.0096	.0094	.0091	.0089	.0087	.0084
−2.2	.0139	.0136	.0132	.0129	.0125	.0122	.0119	.0116	.0113	.0110
−2.1	.0179	.0174	.0170	.0166	.0162	.0158	.0154	.0150	.0146	.0143
−2.0	.0228	.0222	.0217	.0212	.0207	.0202	.0197	.0192	.0188	.0183
−1.9	.0287	.0281	.0274	.0268	.0262	.0256	.0250	.0244	.0239	.0233
−1.8	.0359	.0351	.0344	.0336	.0329	.0322	.0314	.0307	.0301	.0294
−1.7	.0446	.0436	.0427	.0418	.0409	.0401	.0392	.0384	.0375	.0367
−1.6	.0548	.0537	.0526	.0516	.0505	.0495	.0485	.0475	.0465	.0455
−1.5	.0668	.0655	.0643	.0630	.0618	.0606	.0594	.0582	.0571	.0559
−1.4	.0808	.0793	.0778	.0764	.0749	.0735	.0721	.0708	.0694	.0681
−1.3	.0968	.0951	.0934	.0918	.0901	.0885	.0869	.0853	.0838	.0823
−1.2	.1151	.1131	.1112	.1093	.1075	.1056	.1038	.1020	.1003	.0985
−1.1	.1357	.1335	.1314	.1292	.1271	.1251	.1230	.1210	.1190	.1170
−1.0	.1587	.1562	.1539	.1515	.1492	.1469	.1446	.1423	.1401	.1379
−0.9	.1841	.1814	.1788	.1762	.1736	.1711	.1685	.1660	.1635	.1611
−0.8	.2119	.2090	.2061	.2033	.2005	.1977	.1949	.1922	.1894	.1867
−0.7	.2420	.2388	.2358	.2327	.2296	.2266	.2236	.2206	.2177	.2148
−0.6	.2743	.2709	.2676	.2643	.2611	.2578	.2546	.2514	.2482	.2451
−0.5	.3085	.3050	.3015	.2981	.2946	.2912	.2877	.2843	.2810	.2776
−0.4	.3446	.3409	.3372	.3336	.3300	.3264	.3228	.3192	.3156	.3121
−0.3	.3821	.3783	.3745	.3707	.3669	.3632	.3594	.3557	.3520	.3483
−0.2	.4207	.4168	.4129	.4090	.4052	.4013	.3974	.3936	.3897	.3859
−0.1	.4602	.4562	.4522	.4483	.4443	.4404	.4364	.4325	.4286	.4247
−0.0	.5000	.4960	.4920	.4880	.4840	.4801	.4761	.4721	.4681	.4641

Table D (continued)

Normal Deviate z	.00	.01	.02	.03	.04	.05	.06	.07	.08	.09
0.0	.5000	.5040	.5080	.5120	.5160	.5199	.5239	.5279	.5319	.5359
0.1	.5398	.5438	.5478	.5517	.5557	.5596	.5636	.5675	.5714	.5753
0.2	.5793	.5832	.5871	.5910	.5948	.5987	.6026	.6064	.6103	.6141
0.3	.6179	.6217	.6255	.6293	.6331	.6368	.6406	.6443	.6480	.6517
0.4	.6554	.6591	.6628	.6664	.6700	.6736	.6772	.6808	.6844	.6879
0.5	.6915	.6950	.6985	.7019	.7054	.7088	.7123	.7157	.7190	.7224
0.6	.7257	.7291	.7324	.7357	.7389	.7422	.7454	.7486	.7518	.7549
0.7	.7580	.7612	.7642	.7673	.7704	.7734	.7764	.7794	.7823	.7852
0.8	.7881	.7910	.7939	.7967	.7995	.8023	.8051	.8078	.8106	.8133
0.9	.8159	.8186	.8212	.8238	.8264	.8289	.8315	.8340	.8365	.8389
1.0	.8413	.8438	.8461	.8485	.8508	.8531	.8554	.8577	.8599	.8621
1.1	.8643	.8665	.8686	.8708	.8729	.8749	.8770	.8790	.8810	.8830
1.2	.8849	.8869	.8888	.8907	.8925	.8944	.8962	.8980	.8997	.9015
1.3	.9032	.9049	.9066	.9082	.9099	.9115	.9131	.9147	.9162	.9177
1.4	.9192	.9207	.9222	.9236	.9251	.9265	.9279	.9292	.9306	.9319
1.5	.9332	.9345	.9357	.9370	.9382	.9394	.9406	.9418	.9429	.9441
1.6	.9452	.9463	.9474	.9484	.9495	.9505	.9515	.9525	.9535	.9545
1.7	.9554	.9564	.9573	.9582	.9591	.9599	.9608	.9616	.9625	.9633
1.8	.9641	.9649	.9656	.9664	.9671	.9678	.9686	.9693	.9699	.9706
1.9	.9713	.9719	.9726	.9732	.9738	.9744	.9750	.9756	.9761	.9767
2.0	.9772	.9778	.9783	.9788	.9793	.9798	.9803	.9808	.9812	.9817
2.1	.9821	.9826	.9830	.9834	.9838	.9842	.9846	.9850	.9854	.9857
2.2	.9861	.9864	.9868	.9871	.9875	.9878	.9881	.9884	.9887	.9890
2.3	.9893	.9896	.9898	.9901	.9904	.9906	.9909	.9911	.9913	.9916
2.4	.9918	.9920	.9922	.9925	.9927	.9929	.9931	.9932	.9934	.9936
2.5	.9938	.9940	.9941	.9943	.9945	.9946	.9948	.9949	.9951	.9952
2.6	.9953	.9955	.9956	.9957	.9959	.9960	.9961	.9962	.9963	.9964
2.7	.9965	.9966	.9967	.9968	.9969	.9970	.9971	.9972	.9973	.9974
2.8	.9974	.9975	.9976	.9977	.9977	.9978	.9979	.9979	.9980	.9981
2.9	.9981	.9982	.9982	.9983	.9984	.9984	.9985	.9985	.9986	.9986
3.0	.9986	.9987	.9987	.9988	.9988	.9989	.9989	.9989	.9990	.9990
3.1	.9990	.9991	.9991	.9991	.9992	.9992	.9992	.9992	.9993	.9993
3.2	.9993	.9993	.9994	.9994	.9994	.9994	.9994	.9995	.9995	.9995
3.3	.99952									
3.4	.99966									
3.5	.99977									
3.6	.99984									
3.7	.99989									
3.8	.99993									
3.9	.99995									
4.0	.99997									
4.5	1.00000									
5.0	1.00000									

Table E
Critical Normal Deviate Values

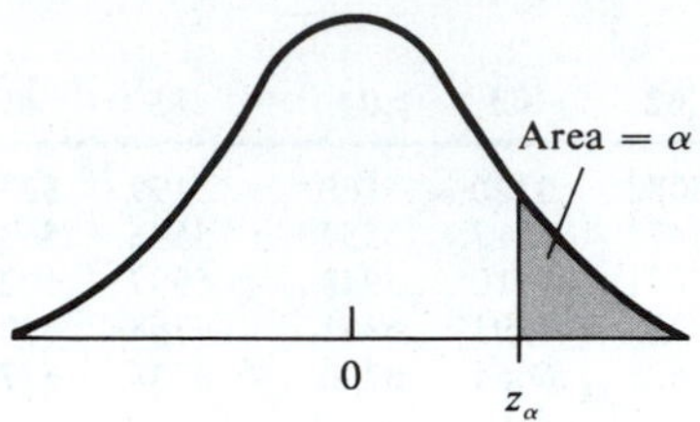

Upper-Tail Area α	Normal Deviate z_α
.10	1.28
.05	1.64
.025	1.96
.01	2.33
.005	2.57
.001	3.08
.0005	3.30

Table F
Random Numbers

12651	61646	11769	75109	86996	97669	25757	32535	07122	76763
81769	74436	02630	72310	45049	18029	07469	42341	98173	79260
36737	98863	77240	76251	00654	64688	09343	70278	67331	98729
82861	54371	76610	94934	72748	44124	05610	53750	95938	01485
21325	15732	24127	37431	09723	63529	73977	95218	96074	42138
74146	47887	62463	23045	41490	07954	22597	60012	98866	90959
90759	64410	54179	66075	61051	75385	51378	08360	95946	95547
55683	98078	02238	91540	21219	17720	87817	41705	95785	12563
79686	17969	76061	83748	55920	83612	41540	86492	06447	60568
70333	00201	86201	69716	78185	62154	77930	67663	29529	75116
14042	53536	07779	04157	41172	36473	42123	43929	50533	33437
59911	08256	06596	48416	69770	68797	56080	14223	59199	30162
62368	62623	62742	14891	39247	52242	98832	69533	91174	57979
57529	97751	54976	48957	74599	08759	78494	52785	68526	64618
15469	90574	78033	66885	13936	42117	71831	22961	94225	31816
18625	23674	53850	32827	81647	80820	00420	63555	74489	80141
74626	68394	88562	70745	23701	45630	65891	58220	35442	60414
11119	16519	27384	90199	79210	76965	99546	30323	31664	22845
41101	17336	48951	53674	17880	45260	08575	49321	36191	17095
32123	91576	84221	78902	82010	30847	62329	63898	23268	74283
26091	68409	69704	82267	14751	13151	93115	01437	56945	89661
67680	79790	48462	59278	44185	29616	76531	19589	83139	28454
15184	19260	14073	07026	25264	08388	27182	22557	61501	67481
58010	45039	57181	10238	36874	28546	37444	80824	63981	39942
56425	53996	86245	32623	78858	08143	60377	42925	42815	11159
82630	84066	13592	60642	17904	99718	63432	88642	37858	25431
14927	40909	23900	48761	44860	92467	31742	87142	03607	32059
23740	22505	07489	85986	74420	21744	97711	36648	35620	97949
32990	97446	03711	63824	07953	85965	87089	11687	92414	67257
05310	24058	91946	78437	34365	82469	12430	84754	19354	72745
21839	39937	27534	88913	49055	19218	47712	67677	51889	70926
08833	42549	93981	94051	28382	83725	72643	64233	97252	17133
58336	11139	47479	00931	91560	95372	97642	33856	54825	55680
62032	91144	75478	47431	52726	30289	42411	91886	51818	78292
45171	30557	53116	04118	58301	24375	65609	85810	18620	49198
91611	62656	60128	35609	63698	78356	50682	22505	01692	36291
55472	63819	86314	49174	93582	73604	78614	78849	23096	72825
18573	09729	74091	53994	10970	86557	65661	41854	26037	53296
60866	02955	90288	82136	83644	94455	06560	78029	98768	71296
45043	55608	82767	60890	74646	79485	13619	98868	40857	19415
17831	09737	79473	75945	28394	79334	70577	38048	03607	06932
40137	03981	07585	18128	11178	32601	27994	05641	22600	86064
77776	31343	14576	97706	16039	47517	43300	59080	80392	63189
69605	44104	40103	95635	05635	81673	68657	09559	23510	95875
19916	52934	26499	09821	87331	80993	61299	36979	73599	35055
02606	58552	07678	56619	65325	30705	99582	53390	46357	13244
65183	73160	87131	35530	47946	09854	18080	02321	05809	04898
10740	98914	44916	11322	89717	88189	30143	52687	19420	60061
98642	89822	71691	51573	83666	61642	46683	33761	47542	23551
60139	25601	93663	25547	02654	94829	48672	28736	84994	13071

Source: Reprinted from page 103 of *A Million Random Digits with 100,000 Normal Deviates,* by The Rand Corporation. New York: The Free Press, 1955.

Table G
Student t Distribution

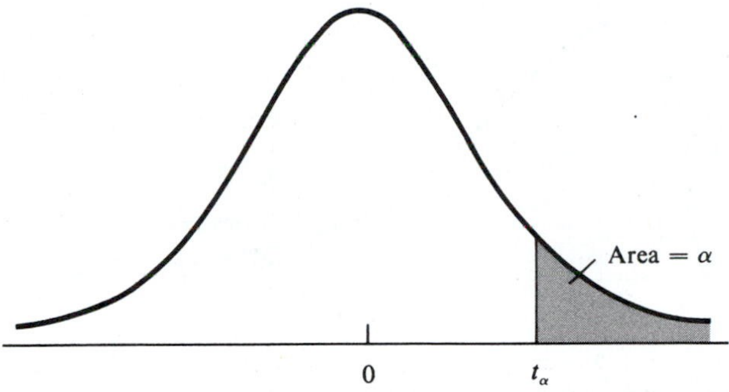

The following table provides the values of t_α that correspond to a given upper-tail area α and a specified number of degrees of freedom.

Degrees of Freedom	Upper-Tail Area α									
	.4	.25	.1	.05	.025	.01	.005	.0025	.001	.0005
1	.325	1.000	3.078	6.314	12.706	31.821	63.657	127.32	318.31	636.62
2	.289	.816	1.886	2.920	4.303	6.965	9.925	14.089	22.327	31.598
3	.277	.765	1.638	2.353	3.182	4.541	5.841	7.453	10.214	12.924
4	.271	.741	1.533	2.132	2.776	3.747	4.604	5.598	7.173	8.610
5	.267	.727	1.476	2.015	2.571	3.365	4.032	4.773	5.893	6.869
6	.265	.718	1.440	1.943	2.447	3.143	3.707	4.317	5.208	5.959
7	.263	.711	1.415	1.895	2.365	2.998	3.499	4.029	4.785	5.408
8	.262	.706	1.397	1.860	2.306	2.896	3.355	3.833	4.501	5.041
9	.261	.703	1.383	1.833	2.262	2.821	3.250	3.690	4.297	4.781
10	.260	.700	1.372	1.812	2.228	2.764	3.169	3.581	4.144	4.587
11	.260	.697	1.363	1.796	2.201	2.718	3.106	3.497	4.025	4.437
12	.259	.695	1.356	1.782	2.179	2.681	3.055	3.428	3.930	4.318
13	.259	.694	1.350	1.771	2.160	2.650	3.012	3.372	3.852	4.221
14	.258	.692	1.345	1.761	2.145	2.624	2.977	3.326	3.787	4.140
15	.258	.691	1.341	1.753	2.131	2.602	2.947	3.286	3.733	4.073
16	.258	.690	1.337	1.746	2.120	2.583	2.921	3.252	3.686	4.015
17	.257	.689	1.333	1.740	2.110	2.567	2.898	3.222	3.646	3.965
18	.257	.688	1.330	1.734	2.101	2.552	2.878	3.197	3.610	3.922
19	.257	.688	1.328	1.729	2.093	2.539	2.861	3.174	3.579	3.883
20	.257	.687	1.325	1.725	2.086	2.528	2.845	3.153	3.552	3.850
21	.257	.686	1.323	1.721	2.080	2.518	2.831	3.135	3.527	3.819
22	.256	.686	1.321	1.717	2.074	2.508	2.819	3.119	3.505	3.792
23	.256	.685	1.319	1.714	2.069	2.500	2.807	3.104	3.485	3.767
24	.256	.685	1.318	1.711	2.064	2.492	2.797	3.091	3.467	3.745
25	.256	.684	1.316	1.708	2.060	2.485	2.787	3.078	3.450	3.725
26	.256	.684	1.315	1.706	2.056	2.479	2.779	3.067	3.435	3.707
27	.256	.684	1.314	1.703	2.052	2.473	2.771	3.057	3.421	3.690
28	.256	.683	1.313	1.701	2.048	2.467	2.763	3.047	3.408	3.674
29	.256	.683	1.311	1.699	2.045	2.462	2.756	3.038	3.396	3.659
30	.256	.683	1.310	1.697	2.042	2.457	2.750	3.030	3.385	3.646
40	.255	.681	1.303	1.684	2.021	2.423	2.704	2.971	3.307	3.551
60	.254	.679	1.296	1.671	2.000	2.390	2.660	2.915	3.232	3.460
120	.254	.677	1.289	1.658	1.980	2.358	2.617	2.860	3.160	3.373
∞	.253	.674	1.282	1.645	1.960	2.326	2.576	2.807	3.090	3.291

Source: E. S. Pearson and H. O. Hartley, *Biometrika Tables for Statisticians,* Vol. I. London: Cambridge University Press, 1966. Partly derived from Table III of Fisher and Yates, *Statistical Tables for Biological, Agricultural and Medical Research,* published by Longman Group Ltd., London (previously published by Oliver & Boyd, Edinburgh, 1963). Reproduced with permission of the authors and publishers.

Table H
Chi-Square Distribution

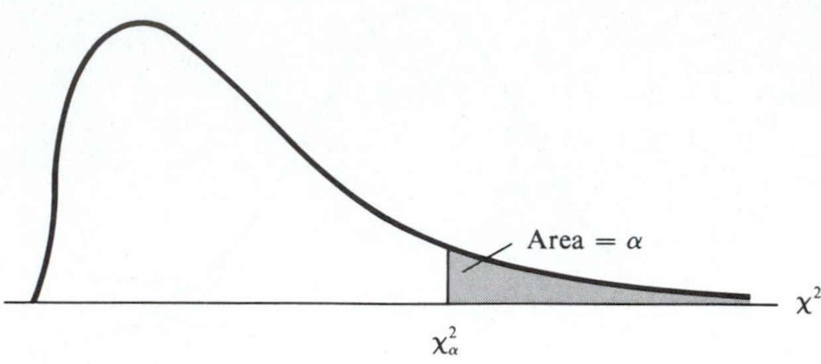

The following table provides the values of χ^2_α that correspond to a given upper-tail area α and a specified number of degrees of freedom.

Degrees of Freedom	Upper-Tail Area α						
	.99	**.98**	**.95**	**.90**	**.80**	**.70**	**.50**
1	.000157	.000628	.00393	.0158	.0642	.148	.455
2	.0201	.0404	.103	.211	.446	.713	1.386
3	.115	.185	.352	.584	1.005	1.424	2.366
4	.297	.429	.711	1.064	1.649	2.195	3.357
5	.554	.752	1.145	1.610	2.343	3.000	4.351
6	.872	1.134	1.635	2.204	3.070	3.828	5.348
7	1.239	1.564	2.167	2.833	3.822	4.671	6.346
8	1.646	2.032	2.733	3.490	4.594	5.527	7.344
9	2.088	2.532	3.325	4.168	5.380	6.393	8.343
10	2.558	3.059	3.940	4.865	6.179	7.267	9.342
11	3.053	3.609	4.575	5.578	6.989	8.148	10.341
12	3.571	4.178	5.226	6.304	7.807	9.034	11.340
13	4.107	4.765	5.892	7.042	8.634	9.926	12.340
14	4.660	5.368	6.571	7.790	9.467	10.821	13.339
15	5.229	5.985	7.261	8.547	10.307	11.721	14.339
16	5.812	6.614	7.962	9.312	11.152	12.624	15.338
17	6.408	7.255	8.672	10.085	12.002	13.531	16.338
18	7.015	7.906	9.390	10.865	12.857	14.440	17.338
19	7.633	8.567	10.117	11.651	13.716	15.352	18.338
20	8.260	9.237	10.851	12.443	14.578	16.266	19.337
21	8.897	9.915	11.591	13.240	15.445	17.182	20.337
22	9.542	10.600	12.338	14.041	16.314	18.101	21.337
23	10.196	11.293	13.091	14.848	17.187	19.021	22.337
24	10.856	11.992	13.848	15.659	18.062	19.943	23.337
25	11.524	12.697	14.611	16.473	18.940	20.867	24.337
26	12.198	13.409	15.379	17.292	19.820	21.792	25.336
27	12.879	14.125	16.151	18.114	20.703	22.719	26.336
28	13.565	14.847	16.928	18.939	21.588	23.647	27.336
29	14.256	15.574	17.708	19.768	22.475	24.577	28.336
30	14.953	16.306	18.493	20.599	23.364	25.508	29.336

Source: From Table IV of Fisher and Yates, *Statistical Tables for Biological, Agricultural and Medical Research*, published by Longman Group Ltd., London (previously published by Oliver & Boyd, Edinburgh, 1963). Reproduced with permission of the authors and publishers.

Table H (continued)

Degrees of Freedom	Upper-Tail Area α						
	.30	.20	.10	.05	.02	.01	.001
1	1.074	1.642	2.706	3.841	5.412	6.635	10.827
2	2.408	3.219	4.605	5.991	7.824	9.210	13.815
3	3.665	4.642	6.251	7.815	9.837	11.345	16.268
4	4.878	5.989	7.779	9.488	11.668	13.277	18.465
5	6.064	7.289	9.236	11.070	13.388	15.086	20.517
6	7.231	8.558	10.645	12.592	15.033	16.812	22.457
7	8.383	9.803	12.017	14.067	16.622	18.475	24.322
8	9.524	11.030	13.362	15.507	18.168	20.090	26.125
9	10.656	12.242	14.684	16.919	19.679	21.666	27.877
10	11.781	13.442	15.987	18.307	21.161	23.209	29.588
11	12.899	14.631	17.275	19.675	22.618	24.725	31.264
12	14.011	15.812	18.549	21.026	24.054	26.217	32.909
13	15.119	16.985	19.812	22.362	25.472	27.688	34.528
14	16.222	18.151	21.064	23.685	26.873	29.141	36.123
15	17.322	19.311	22.307	24.996	28.259	30.578	37.697
16	18.418	20.465	23.542	26.296	29.633	32.000	39.252
17	19.511	21.615	24.769	27.587	30.995	33.409	40.790
18	20.601	22.760	25.989	28.869	32.346	34.805	42.312
19	21.689	23.900	27.204	30.144	33.687	36.191	43.820
20	22.775	25.038	28.412	31.410	35.020	37.566	45.315
21	23.858	26.171	29.615	32.671	36.343	38.932	46.797
22	24.939	27.301	30.813	33.924	37.659	40.289	48.268
23	26.018	28.429	32.007	35.172	38.968	41.638	49.728
24	27.096	29.553	33.196	36.415	40.270	42.980	51.179
25	28.172	30.675	34.382	37.652	41.566	44.314	52.620
26	29.246	31.795	35.563	38.885	42.856	45.642	54.052
27	30.319	32.912	36.741	40.113	44.140	46.963	55.476
28	31.391	34.027	37.916	41.337	45.419	48.278	56.893
29	32.461	35.139	39.087	42.557	46.693	49.588	58.302
30	33.530	36.250	40.256	43.773	47.962	50.892	59.703

Table I
Conversion Table for Correlation Coefficient and Fisher's Z′

ρ	Z′	ρ	Z′	ρ	Z′	ρ	Z′
.00	.0000	.25	.2554	.50	.5493	.75	.973
.01	.0100	.26	.2661	.51	.5627	.76	.996
.02	.0200	.27	.2769	.52	.5763	.77	1.020
.03	.0300	.28	.2877	.53	.5901	.78	1.045
.04	.0400	.29	.2986	.54	.6042	.79	1.071
.05	.0500	.30	.3095	.55	.6184	.80	1.099
.06	.0601	.31	.3205	.56	.6328	.81	1.127
.07	.0701	.32	.3316	.57	.6475	.82	1.157
.08	.0802	.33	.3428	.58	.6625	.83	1.188
.09	.0902	.34	.3541	.59	.6777	.84	1.221
.10	.1003	.35	.3654	.60	.6931	.85	1.256
.11	.1104	.36	.3769	.61	.7089	.86	1.293
.12	.1206	.37	.3884	.62	.7250	.87	1.333
.13	.1307	.38	.4001	.63	.7414	.88	1.376
.14	.1409	.39	.4118	.64	.7582	.89	1.422
.15	.1511	.40	.4236	.65	.7753	.90	1.472
.16	.1614	.41	.4356	.66	.7928	.91	1.528
.17	.1717	.42	.4477	.67	.8107	.92	1.589
.18	.1820	.43	.4599	.68	.8291	.93	1.658
.19	.1923	.44	.4722	.69	.8480	.94	1.738
.20	.2027	.45	.4847	.70	.8673	.95	1.832
.21	.2132	.46	.4973	.71	.8872	.96	1.946
.22	.2237	.47	.5101	.72	.9076	.97	2.092
.23	.2342	.48	.5230	.73	.9287	.98	2.298
.24	.2448	.49	.5361	.74	.9505	.99	2.647

Source: Abridged from Table 14 in E. S. Pearson and H. O. Hartley, *Biometrika Tables for Statisticians,* Vol. 1. Cambridge, England: Cambridge University Press, on behalf of The Biometrika Society, 1966. By permission of the authors and publishers.

Table J
Critical Values for F Distribution

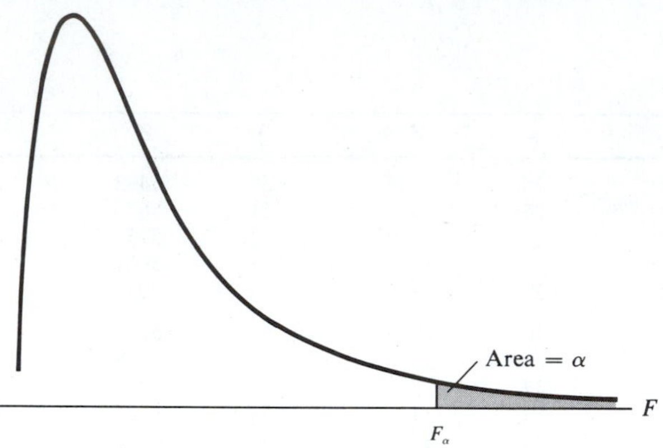

The following table provides the values of F_α that correspond to a given upper-tail area α and a specified degrees of freedom pair. The values of $F_{.05}$ are in lightface type, while those for $F_{.01}$ are given in boldface type.

Degrees of Freedom for Denominator	Degrees of Freedom for Numerator											
	1	2	3	4	5	6	7	8	9	10	11	12
1	161	200	216	225	230	234	237	239	241	242	243	244
	4,052	**4,999**	**5,403**	**5,625**	**5,764**	**5,859**	**5,928**	**5,981**	**6,022**	**6,056**	**6,082**	**6,106**
2	18.51	19.00	19.16	19.25	19.30	19.33	19.36	19.37	19.38	19.39	19.40	19.41
	98.49	**99.01**	**99.17**	**99.25**	**99.30**	**99.33**	**99.34**	**99.36**	**99.38**	**99.40**	**99.41**	**99.42**
3	10.13	9.55	9.28	9.12	9.01	8.94	8.88	8.84	8.81	8.78	8.76	8.74
	34.12	**30.81**	**29.46**	**28.71**	**28.24**	**27.91**	**27.67**	**27.49**	**27.34**	**27.23**	**27.13**	**27.05**
4	7.71	6.94	6.59	6.39	6.26	6.16	6.09	6.04	6.00	5.96	5.93	5.91
	21.20	**18.00**	**16.69**	**15.98**	**15.52**	**15.21**	**14.98**	**14.80**	**14.66**	**14.54**	**14.45**	**14.37**
5	6.61	5.79	5.41	5.19	5.05	4.95	4.88	4.82	4.78	4.74	4.70	4.68
	16.26	**13.27**	**12.06**	**11.39**	**10.97**	**10.67**	**10.45**	**10.27**	**10.15**	**10.05**	**9.96**	**9.89**
6	5.99	5.14	4.76	4.53	4.39	4.28	4.21	4.15	4.10	4.06	4.03	4.00
	13.74	**10.92**	**9.78**	**9.15**	**8.75**	**8.47**	**8.26**	**8.10**	**7.98**	**7.87**	**7.79**	**7.72**
7	5.59	4.74	4.35	4.12	3.97	3.87	3.79	3.73	3.68	3.63	3.60	3.57
	12.25	**9.55**	**8.45**	**7.85**	**7.46**	**7.19**	**7.00**	**6.84**	**6.71**	**6.62**	**6.54**	**6.47**
8	5.32	4.46	4.07	3.84	3.69	3.58	3.50	3.44	3.39	3.34	3.31	3.28
	11.26	**8.65**	**7.59**	**7.01**	**6.63**	**6.37**	**6.19**	**6.03**	**5.91**	**5.82**	**5.74**	**5.67**
9	5.12	4.26	3.86	3.63	3.48	3.37	3.29	3.23	3.18	3.13	3.10	3.07
	10.56	**8.02**	**6.99**	**6.42**	**6.06**	**5.80**	**5.62**	**5.47**	**5.35**	**5.26**	**5.18**	**5.11**
10	4.96	4.10	3.71	3.48	3.33	3.22	3.14	3.07	3.02	2.97	2.94	2.91
	10.04	**7.56**	**6.55**	**5.99**	**5.64**	**5.39**	**5.21**	**5.06**	**4.95**	**4.85**	**4.78**	**4.71**
11	4.84	3.98	3.59	3.36	3.20	3.09	3.01	2.95	2.90	2.86	2.82	2.79
	9.65	**7.20**	**6.22**	**5.67**	**5.32**	**5.07**	**4.88**	**4.74**	**4.63**	**4.54**	**4.46**	**4.40**
12	4.75	3.89	3.49	3.26	3.11	3.00	2.92	2.85	2.80	2.76	2.72	2.69
	9.33	**6.93**	**5.95**	**5.41**	**5.06**	**4.82**	**4.65**	**4.50**	**4.39**	**4.30**	**4.22**	**4.16**
13	4.67	3.80	3.41	3.18	3.02	2.92	2.84	2.77	2.72	2.67	2.63	2.60
	9.07	**6.70**	**5.74**	**5.20**	**4.86**	**4.62**	**4.44**	**4.30**	**4.19**	**4.10**	**4.02**	**3.96**

Source: Reprinted by permission from *Statistical Methods* by George W. Snedecor and William G. Cochran, 7th ed. © 1980 by Iowa State University Press, Ames, Iowa 50010.

Table J (continued)

Degrees of Freedom for Denominator	Degrees of Freedom for Numerator											
	14	16	20	24	30	40	50	75	100	200	500	∞
1	245	246	248	249	250	251	252	253	253	254	254	254
	6,142	**6,169**	**6,208**	**6,234**	**6,258**	**6,286**	**6,302**	**6,323**	**6,334**	**6,352**	**6,361**	**6,366**
2	19.42	19.43	19.44	19.45	19.46	19.47	19.47	19.48	19.49	19.49	19.50	19.50
	99.43	**99.44**	**99.45**	**99.46**	**99.47**	**99.48**	**99.48**	**99.49**	**99.49**	**99.49**	**99.50**	**99.50**
3	8.71	8.69	8.66	8.64	8.62	8.60	8.58	8.57	8.56	8.54	8.54	8.53
	26.92	**26.83**	**26.69**	**26.60**	**26.50**	**26.41**	**26.30**	**26.27**	**26.23**	**26.18**	**26.14**	**26.12**
4	5.87	5.84	5.80	5.77	5.74	5.71	5.70	5.68	5.66	5.65	5.64	5.63
	14.24	**14.15**	**14.02**	**13.93**	**13.83**	**13.74**	**13.69**	**13.61**	**13.57**	**13.52**	**13.48**	**13.46**
5	4.64	4.60	4.56	4.53	4.50	4.46	4.44	4.42	4.40	4.38	4.37	4.36
	9.77	**9.68**	**9.55**	**9.47**	**9.38**	**9.29**	**9.24**	**9.17**	**9.13**	**9.07**	**9.04**	**9.02**
6	3.96	3.92	3.87	3.84	3.81	3.77	3.75	3.72	3.71	3.69	3.68	3.67
	7.60	**7.52**	**7.39**	**7.31**	**7.23**	**7.14**	**7.09**	**7.02**	**6.99**	**6.94**	**6.90**	**6.88**
7	3.52	3.49	3.44	3.41	3.38	3.34	3.32	3.29	3.28	3.25	3.24	3.23
	6.35	**6.27**	**6.15**	**6.07**	**5.98**	**5.90**	**5.85**	**5.78**	**5.75**	**5.70**	**5.67**	**5.65**
8	3.23	3.20	3.15	3.12	3.08	3.05	3.03	3.00	2.98	2.96	2.94	2.93
	5.56	**5.48**	**5.36**	**5.28**	**5.20**	**5.11**	**5.06**	**5.00**	**4.96**	**4.91**	**4.88**	**4.86**
9	3.02	2.98	2.93	2.90	2.86	2.82	2.80	2.77	2.76	2.73	2.72	2.71
	5.00	**4.92**	**4.80**	**4.73**	**4.64**	**4.56**	**4.51**	**4.45**	**4.41**	**4.36**	**4.33**	**4.31**
10	2.86	2.82	2.77	2.74	2.70	2.67	2.64	2.61	2.59	2.56	2.55	2.54
	4.60	**4.52**	**4.41**	**4.33**	**4.25**	**4.17**	**4.12**	**4.05**	**4.01**	**3.96**	**3.93**	**3.91**
11	2.74	2.70	2.65	2.61	2.57	2.53	2.50	2.47	2.45	2.42	2.41	2.40
	4.29	**4.21**	**4.10**	**4.02**	**3.94**	**3.86**	**3.80**	**3.74**	**3.70**	**3.66**	**3.62**	**3.60**
12	2.64	2.60	2.54	2.50	2.46	2.42	2.40	2.36	2.35	2.32	2.31	2.30
	4.05	**3.98**	**3.86**	**3.78**	**3.70**	**3.61**	**3.56**	**3.49**	**3.46**	**3.41**	**3.38**	**3.36**
13	2.55	2.51	2.46	2.42	2.38	2.34	2.32	2.28	2.26	2.24	2.22	2.21
	3.85	**3.78**	**3.67**	**3.59**	**3.51**	**3.42**	**3.37**	**3.30**	**3.27**	**3.21**	**3.18**	**3.16**

Table J (continued)

Degrees of Freedom for Denominator	Degrees of Freedom for Numerator											
	1	2	3	4	5	6	7	8	9	10	11	12
14	4.60	3.74	3.34	3.11	2.96	2.85	2.77	2.70	2.65	2.60	2.56	2.53
	8.86	**6.51**	**5.56**	**5.03**	**4.69**	**4.46**	**4.28**	**4.14**	**4.03**	**3.94**	**3.86**	**3.80**
15	4.54	3.68	3.29	3.06	2.90	2.79	2.70	2.64	2.59	2.55	2.51	2.48
	8.68	**6.36**	**5.42**	**4.89**	**4.56**	**4.32**	**4.14**	**4.00**	**3.89**	**3.80**	**3.73**	**3.67**
16	4.49	3.63	3.24	3.01	2.85	2.74	2.66	2.59	2.54	2.49	2.45	2.42
	8.53	**6.23**	**5.29**	**4.77**	**4.44**	**4.20**	**4.03**	**3.89**	**3.78**	**3.69**	**3.61**	**3.55**
17	4.45	3.59	3.20	2.96	2.81	2.70	2.62	2.55	2.50	2.45	2.41	2.38
	8.40	**6.11**	**5.18**	**4.67**	**4.34**	**4.10**	**3.93**	**3.79**	**3.68**	**3.59**	**3.52**	**3.45**
18	4.41	3.55	3.16	2.93	2.77	2.66	2.58	2.51	2.46	2.41	2.37	2.34
	8.28	**6.01**	**5.09**	**4.58**	**4.25**	**4.01**	**3.85**	**3.71**	**3.60**	**3.51**	**3.44**	**3.37**
19	4.38	3.52	3.13	2.90	2.74	2.63	2.55	2.48	2.43	2.38	2.34	2.31
	8.18	**5.93**	**5.01**	**4.50**	**4.17**	**3.94**	**3.77**	**3.63**	**3.52**	**3.43**	**3.36**	**3.30**
20	4.35	3.49	3.10	2.87	2.71	2.60	2.52	2.45	2.40	2.35	2.31	2.28
	8.10	**5.85**	**4.94**	**4.43**	**4.10**	**3.87**	**3.71**	**3.56**	**3.45**	**3.37**	**3.30**	**3.23**
21	4.32	3.47	3.07	2.84	2.68	2.57	2.49	2.42	2.37	2.32	2.28	2.25
	8.02	**5.78**	**4.87**	**4.37**	**4.04**	**3.81**	**3.65**	**3.51**	**3.40**	**3.31**	**3.24**	**3.17**
22	4.30	3.44	3.05	2.82	2.66	2.55	2.47	2.40	2.35	2.30	2.26	2.23
	7.94	**5.72**	**4.82**	**4.41**	**3.99**	**3.76**	**3.59**	**3.45**	**3.35**	**3.26**	**3.18**	**3.12**
23	4.28	3.42	3.03	2.80	2.64	2.53	2.45	2.38	2.32	2.28	2.24	2.20
	7.88	**5.66**	**4.76**	**4.26**	**3.94**	**3.71**	**3.54**	**3.41**	**3.30**	**3.21**	**3.14**	**3.07**
24	4.26	3.40	3.01	2.78	2.62	2.51	2.43	2.36	2.30	2.26	2.22	2.18
	7.82	**5.61**	**4.72**	**4.22**	**3.90**	**3.67**	**3.50**	**3.36**	**3.25**	**3.17**	**3.09**	**3.03**
25	4.24	3.38	2.99	2.76	2.60	2.49	2.41	2.34	2.28	2.24	2.20	2.16
	7.77	**5.57**	**4.68**	**4.18**	**3.86**	**3.63**	**3.46**	**3.32**	**3.21**	**3.13**	**3.05**	**2.99**
26	4.22	3.37	2.89	2.74	2.59	2.47	2.39	2.32	2.27	2.22	2.18	2.15
	7.72	**5.53**	**4.64**	**4.14**	**3.82**	**3.59**	**3.42**	**3.29**	**3.17**	**3.09**	**3.02**	**2.96**
27	4.21	3.35	2.96	2.73	2.57	2.46	2.37	2.30	2.25	2.20	2.16	2.13
	7.68	**5.49**	**4.60**	**4.11**	**3.79**	**3.56**	**3.39**	**3.26**	**3.14**	**3.06**	**2.98**	**2.93**
28	4.20	3.34	2.95	2.71	2.56	2.44	2.36	2.29	2.24	2.19	2.15	2.12
	7.64	**5.45**	**4.57**	**4.07**	**3.76**	**3.53**	**3.36**	**3.23**	**3.11**	**3.03**	**2.95**	**2.90**
29	4.18	3.33	2.93	2.70	2.54	2.43	2.35	2.28	2.22	2.18	2.14	2.10
	7.60	**5.52**	**4.54**	**4.04**	**3.73**	**3.50**	**3.33**	**3.20**	**3.08**	**3.00**	**2.92**	**2.87**
30	4.17	3.32	2.92	2.69	2.53	2.43	2.34	2.27	2.21	2.16	2.12	2.09
	7.56	**5.39**	**4.51**	**4.02**	**3.70**	**3.47**	**3.30**	**3.17**	**3.06**	**2.98**	**2.90**	**2.84**
32	4.15	3.30	2.90	2.67	2.51	2.40	2.32	2.25	2.19	2.14	2.10	2.07
	7.50	**5.34**	**4.46**	**3.97**	**3.66**	**3.42**	**3.25**	**3.12**	**3.01**	**2.94**	**2.86**	**2.80**
34	4.13	3.28	2.88	2.65	2.49	2.38	2.30	2.23	2.17	2.12	2.08	2.05
	7.44	**5.29**	**4.42**	**3.93**	**3.61**	**3.38**	**3.21**	**3.08**	**2.97**	**2.89**	**2.82**	**2.76**
36	4.11	3.26	2.86	2.63	2.48	2.36	2.28	2.21	2.15	2.10	2.06	2.03
	7.39	**5.25**	**4.38**	**3.89**	**3.58**	**3.35**	**3.18**	**3.04**	**2.94**	**2.86**	**2.78**	**2.72**
38	4.10	3.25	2.85	2.62	2.46	2.35	2.26	2.19	2.14	2.09	2.05	2.02
	7.35	**5.21**	**4.34**	**3.86**	**3.54**	**3.32**	**3.15**	**3.02**	**2.91**	**2.82**	**2.75**	**2.69**
40	4.08	3.23	2.84	2.61	2.45	2.34	2.25	2.18	2.12	2.08	2.04	2.00
	7.31	**5.18**	**4.31**	**3.83**	**3.51**	**3.29**	**3.12**	**2.99**	**2.88**	**2.80**	**2.73**	**2.66**

Table J (continued)

Degrees of Freedom for Denominator	Degrees of Freedom for Numerator											
	14	16	20	24	30	40	50	75	100	200	500	∞
14	2.48	2.44	2.39	2.35	2.31	2.27	2.24	2.21	2.19	2.16	2.14	2.13
	3.70	**3.62**	**3.51**	**3.43**	**3.34**	**3.26**	**3.21**	**3.14**	**3.11**	**3.06**	**3.02**	**3.00**
15	2.43	2.39	2.33	2.29	2.25	2.21	2.18	2.15	2.12	2.10	2.08	2.07
	3.56	**3.48**	**3.36**	**3.29**	**3.20**	**3.12**	**3.07**	**3.00**	**2.97**	**2.92**	**2.89**	**2.87**
16	2.37	2.33	2.28	2.24	2.20	2.16	2.13	2.09	2.07	2.04	2.02	2.01
	3.45	**3.37**	**3.25**	**3.18**	**3.10**	**3.01**	**2.96**	**2.89**	**2.86**	**2.80**	**2.77**	**2.75**
17	2.33	2.29	2.23	2.19	2.15	2.11	2.08	2.04	2.02	1.99	1.97	1.96
	3.35	**3.27**	**3.16**	**3.08**	**3.00**	**2.92**	**2.86**	**2.79**	**2.76**	**2.70**	**2.67**	**2.65**
18	2.29	2.25	2.19	2.15	2.11	2.07	2.04	2.00	1.98	1.95	1.93	1.92
	3.27	**3.19**	**3.07**	**3.00**	**2.91**	**2.83**	**2.78**	**2.71**	**2.68**	**2.62**	**2.59**	**2.57**
19	2.26	2.21	2.15	2.11	2.07	2.02	2.00	1.96	1.94	1.91	1.90	1.88
	3.19	**3.12**	**3.00**	**2.92**	**2.84**	**2.76**	**2.70**	**2.63**	**2.60**	**2.54**	**2.51**	**2.49**
20	2.23	2.18	2.12	2.08	2.04	1.99	1.96	1.92	1.90	1.87	1.85	1.84
	3.13	**3.05**	**2.94**	**2.86**	**2.77**	**2.69**	**2.63**	**2.56**	**2.53**	**2.47**	**2.44**	**2.42**
21	2.20	2.15	2.09	2.05	2.00	1.96	1.93	1.89	1.87	1.84	1.82	1.81
	3.07	**2.99**	**2.88**	**2.80**	**2.72**	**2.63**	**2.58**	**2.51**	**2.47**	**2.42**	**2.38**	**2.36**
22	2.18	2.13	2.07	2.03	1.98	1.93	1.91	1.87	1.84	1.81	1.80	1.78
	3.02	**2.94**	**2.83**	**2.75**	**2.67**	**2.58**	**2.53**	**2.46**	**2.42**	**2.37**	**2.33**	**2.31**
23	2.14	2.10	2.04	2.00	1.96	1.91	1.88	1.84	1.82	1.79	1.77	1.76
	2.97	**2.89**	**2.78**	**2.70**	**2.62**	**2.53**	**2.48**	**2.41**	**2.37**	**2.32**	**2.28**	**2.26**
24	2.13	2.09	2.02	1.98	1.94	1.89	1.86	1.82	1.80	1.76	1.74	1.73
	2.93	**2.85**	**2.74**	**2.66**	**2.58**	**2.49**	**2.44**	**2.36**	**2.33**	**2.27**	**2.23**	**2.21**
25	2.11	2.06	2.00	1.96	1.92	1.87	1.84	1.80	1.77	1.74	1.72	1.71
	2.89	**2.81**	**2.70**	**2.62**	**2.54**	**2.45**	**2.40**	**2.32**	**2.29**	**2.23**	**2.19**	**2.17**
26	2.10	2.05	1.99	1.95	1.90	1.85	1.82	1.78	1.76	1.72	1.70	1.69
	2.86	**2.77**	**2.66**	**2.58**	**2.50**	**2.41**	**2.36**	**2.28**	**2.25**	**2.19**	**2.15**	**2.13**
27	2.08	2.03	1.97	1.93	1.88	1.84	1.80	1.76	1.74	1.71	1.68	1.67
	2.83	**2.74**	**2.63**	**2.55**	**2.47**	**2.38**	**2.33**	**2.25**	**2.21**	**2.16**	**2.12**	**2.10**
28	2.06	2.02	1.96	1.91	1.87	1.81	1.78	1.75	1.72	1.69	1.67	1.65
	2.80	**2.71**	**2.60**	**2.52**	**2.44**	**2.35**	**2.30**	**2.22**	**2.18**	**2.13**	**2.09**	**2.06**
29	2.05	2.00	1.94	1.90	1.85	1.80	1.77	1.73	1.71	1.68	1.65	1.64
	2.77	**2.68**	**2.57**	**2.49**	**2.41**	**2.32**	**2.27**	**2.19**	**2.15**	**2.10**	**2.06**	**2.03**
30	2.04	1.99	1.93	1.89	1.84	1.79	1.76	1.72	1.69	1.66	1.64	1.62
	2.74	**2.66**	**2.55**	**2.47**	**2.38**	**2.29**	**2.24**	**2.16**	**2.13**	**2.07**	**2.03**	**2.01**
32	2.02	1.97	1.91	1.86	1.82	1.76	1.74	1.69	1.67	1.64	1.61	1.59
	2.70	**2.62**	**2.51**	**2.42**	**2.34**	**2.25**	**2.20**	**2.12**	**2.08**	**2.02**	**1.98**	**1.96**
34	2.00	1.95	1.89	1.84	1.80	1.74	1.71	1.67	1.64	1.61	1.59	1.57
	2.66	**2.58**	**2.47**	**2.38**	**2.30**	**2.21**	**2.15**	**2.08**	**2.04**	**1.98**	**1.94**	**1.91**
36	1.89	1.93	1.87	1.82	1.78	1.72	1.69	1.65	1.62	1.59	1.56	1.55
	2.62	**2.54**	**2.43**	**2.35**	**2.26**	**2.17**	**2.12**	**2.04**	**2.00**	**1.94**	**1.90**	**1.87**
38	1.96	1.92	1.85	1.80	1.76	1.71	1.67	1.63	1.60	1.57	1.54	1.53
	2.59	**2.51**	**2.40**	**2.32**	**2.22**	**2.14**	**2.08**	**2.00**	**1.97**	**1.90**	**1.86**	**1.84**
40	1.95	1.90	1.84	1.79	1.74	1.69	1.66	1.61	1.59	1.55	1.53	1.51
	2.56	**2.49**	**2.37**	**2.29**	**2.20**	**2.11**	**2.05**	**1.97**	**1.94**	**1.88**	**1.84**	**1.81**

Table J (continued)

Degrees of Freedom for Denominator	Degrees of Freedom for Numerator											
	1	2	3	4	5	6	7	8	9	10	11	12
42	4.07	3.22	2.83	2.59	2.44	2.32	2.24	2.17	2.11	2.06	2.02	1.99
	7.27	**5.15**	**4.29**	**3.80**	**3.49**	**3.26**	**3.10**	**2.96**	**2.86**	**2.77**	**2.70**	**2.64**
44	4.06	3.21	2.82	2.58	2.43	2.31	2.23	2.16	2.10	2.05	2.01	1.98
	7.24	**5.12**	**4.26**	**3.78**	**3.46**	**3.24**	**3.07**	**2.94**	**2.84**	**2.75**	**2.68**	**2.62**
46	4.05	3.20	2.81	2.57	2.42	2.30	2.22	2.14	2.09	2.04	2.00	1.97
	7.21	**5.10**	**4.24**	**3.76**	**3.44**	**3.22**	**3.05**	**2.92**	**2.82**	**2.73**	**2.66**	**2.60**
48	4.04	3.19	2.80	2.56	2.41	2.30	2.21	2.14	2.08	2.03	1.99	1.96
	7.19	**5.08**	**4.22**	**3.74**	**3.42**	**3.20**	**3.04**	**2.90**	**2.80**	**2.71**	**2.64**	**2.58**
50	4.03	3.18	2.79	2.56	2.40	2.29	2.20	2.13	2.07	2.20	1.98	1.95
	7.17	**5.06**	**4.20**	**3.72**	**3.41**	**3.18**	**3.02**	**2.88**	**2.78**	**2.70**	**2.62**	**2.56**
55	4.02	3.17	2.78	2.54	2.38	2.27	2.18	2.11	2.05	2.00	1.97	1.93
	7.12	**5.01**	**4.16**	**3.68**	**3.37**	**3.15**	**2.98**	**2.85**	**2.75**	**2.66**	**2.59**	**2.53**
60	4.00	3.15	2.76	2.52	2.37	2.25	2.17	2.10	2.04	1.99	1.95	1.92
	7.08	**4.98**	**4.13**	**3.65**	**3.34**	**3.12**	**2.95**	**2.82**	**2.72**	**2.63**	**2.56**	**2.50**
65	3.99	3.14	2.75	2.51	2.36	2.24	2.15	2.08	2.02	1.98	1.94	1.90
	7.04	**4.95**	**4.10**	**3.62**	**3.31**	**3.09**	**2.93**	**2.79**	**2.70**	**2.61**	**2.54**	**2.47**
70	3.98	3.13	2.74	2.50	2.35	2.32	2.14	2.07	2.01	1.97	1.93	1.80
	7.01	**4.92**	**4.08**	**3.60**	**3.29**	**3.07**	**2.91**	**2.77**	**2.67**	**2.59**	**2.51**	**2.45**
80	3.96	3.11	2.72	2.48	2.33	2.21	2.12	2.05	1.99	1.95	1.91	1.88
	6.95	**4.88**	**4.04**	**3.56**	**3.25**	**3.04**	**2.87**	**2.74**	**2.64**	**2.55**	**2.48**	**2.41**
100	3.94	3.09	2.70	2.46	2.30	2.19	2.10	2.03	1.97	1.92	1.88	1.85
	6.90	**4.82**	**3.98**	**3.51**	**3.20**	**2.99**	**2.82**	**2.69**	**2.59**	**2.51**	**2.43**	**2.36**
125	3.92	3.07	2.68	2.44	2.29	2.17	2.08	2.01	1.95	1.90	1.86	1.83
	6.84	**4.78**	**3.94**	**3.47**	**3.17**	**2.95**	**2.79**	**2.65**	**2.56**	**2.47**	**2.40**	**2.33**
150	3.91	3.06	2.67	2.43	2.27	2.16	2.07	2.00	1.94	1.89	1.85	1.82
	6.81	**4.75**	**3.91**	**3.44**	**3.13**	**2.92**	**2.76**	**2.62**	**2.53**	**2.44**	**2.37**	**2.30**
200	3.89	3.04	2.65	2.41	2.26	2.14	2.05	1.98	1.92	1.87	1.83	1.80
	6.76	**4.71**	**3.88**	**3.41**	**3.11**	**2.90**	**2.73**	**2.60**	**2.50**	**2.41**	**2.34**	**2.28**
400	3.86	3.02	2.62	2.39	2.23	2.12	2.03	1.96	1.90	1.85	1.81	1.78
	6.70	**4.66**	**3.83**	**3.36**	**3.06**	**2.85**	**2.69**	**2.55**	**2.46**	**2.37**	**2.29**	**2.23**
1,000	3.85	3.00	2.61	2.38	2.22	2.10	2.02	1.95	1.89	1.84	1.80	1.76
	6.66	**4.62**	**3.80**	**3.34**	**3.04**	**2.82**	**2.66**	**2.53**	**2.43**	**2.34**	**2.26**	**2.20**
∞	3.84	2.99	2.60	2.37	2.21	2.09	2.01	1.94	1.88	1.83	1.79	1.75
	6.64	**4.60**	**3.78**	**3.32**	**3.02**	**2.80**	**2.64**	**2.51**	**2.41**	**2.32**	**2.24**	**2.18**

Table J (continued)

Degrees of Freedom for Denominator	Degrees of Freedom for Numerator											
	14	**16**	**20**	**24**	**30**	**40**	**50**	**75**	**100**	**200**	**500**	**∞**
42	1.94	1.89	1.82	1.78	1.73	1.68	1.64	1.60	1.57	1.54	1.51	1.49
	2.54	**2.46**	**2.35**	**2.26**	**2.17**	**2.08**	**2.02**	**1.94**	**1.91**	**1.85**	**1.80**	**1.78**
44	1.92	1.88	1.81	1.76	1.72	1.66	1.63	1.58	1.56	1.52	1.50	1.48
	2.52	**2.44**	**2.32**	**2.24**	**2.15**	**2.06**	**2.00**	**1.92**	**1.88**	**1.82**	**1.78**	**1.75**
46	1.91	1.87	1.80	1.75	1.71	1.65	1.62	1.57	1.54	1.51	1.48	1.46
	2.50	**2.42**	**2.30**	**2.22**	**2.13**	**2.04**	**1.98**	**1.90**	**1.86**	**1.80**	**1.76**	**1.72**
48	1.90	1.86	1.79	1.74	1.70	1.64	1.61	1.56	1.53	1.50	1.47	1.45
	2.43	**2.40**	**2.28**	**2.20**	**2.11**	**2.02**	**1.96**	**1.88**	**1.84**	**1.78**	**1.73**	**1.70**
50	1.90	1.85	1.78	1.74	1.69	1.63	1.60	1.55	1.52	1.48	1.46	1.44
	2.46	**2.39**	**2.26**	**2.18**	**2.10**	**2.00**	**1.94**	**1.86**	**1.82**	**1.76**	**1.71**	**1.68**
55	1.88	1.83	1.76	1.72	1.67	1.61	1.58	1.52	1.50	1.46	1.43	1.41
	2.43	**2.35**	**2.23**	**2.15**	**2.06**	**1.96**	**1.90**	**1.82**	**1.78**	**1.71**	**1.66**	**1.64**
60	1.86	1.81	1.75	1.70	1.65	1.59	1.56	1.50	1.48	1.44	1.41	1.39
	2.40	**2.32**	**2.20**	**2.12**	**2.03**	**1.93**	**1.87**	**1.79**	**1.74**	**1.68**	**1.63**	**1.60**
65	1.85	1.80	1.73	1.68	1.63	1.57	1.54	1.49	1.46	1.42	1.39	1.37
	2.37	**2.30**	**2.18**	**2.09**	**2.00**	**1.90**	**1.84**	**1.76**	**1.71**	**1.64**	**1.60**	**1.56**
70	1.84	1.79	1.72	1.67	1.62	1.56	1.53	1.47	1.45	1.40	1.37	1.35
	2.35	**2.28**	**2.15**	**2.07**	**1.98**	**1.88**	**1.82**	**1.74**	**1.69**	**1.63**	**1.56**	**1.53**
80	1.82	1.77	1.70	1.65	1.60	1.54	1.51	1.45	1.42	1.38	1.35	1.32
	2.32	**2.24**	**2.11**	**2.03**	**1.94**	**1.84**	**1.78**	**1.70**	**1.65**	**1.57**	**1.52**	**1.49**
100	1.79	1.75	1.68	1.63	1.57	1.51	1.48	1.42	1.39	1.34	1.30	1.28
	2.26	**2.19**	**2.06**	**1.98**	**1.89**	**1.79**	**1.73**	**1.64**	**1.59**	**1.51**	**1.46**	**1.43**
125	1.77	1.72	1.65	1.60	1.55	1.49	1.45	1.39	1.36	1.31	1.27	1.25
	2.23	**2.15**	**2.03**	**1.94**	**1.85**	**1.75**	**1.68**	**1.59**	**1.54**	**1.46**	**1.40**	**1.37**
150	1.76	1.71	1.64	1.59	1.54	1.47	1.44	1.37	1.34	1.29	1.25	1.22
	2.20	**2.12**	**2.00**	**1.91**	**1.83**	**1.72**	**1.66**	**1.56**	**1.51**	**1.43**	**1.37**	**1.33**
200	1.74	1.69	1.62	1.57	1.52	1.45	1.42	1.35	1.32	1.26	1.22	1.19
	2.17	**2.09**	**1.97**	**1.88**	**1.79**	**1.69**	**1.62**	**1.53**	**1.48**	**1.39**	**1.33**	**1.28**
400	1.72	1.67	1.60	1.54	1.49	1.42	1.38	1.32	1.28	1.22	1.16	1.13
	2.12	**2.04**	**1.92**	**1.84**	**1.74**	**1.64**	**1.57**	**1.47**	**1.42**	**1.32**	**1.24**	**1.19**
1,000	1.70	1.65	1.58	1.53	1.47	1.41	1.36	1.30	1.26	1.19	1.13	1.08
	2.09	**2.01**	**1.89**	**1.81**	**1.71**	**1.61**	**1.54**	**1.44**	**1.38**	**1.28**	**1.19**	**1.11**
∞	1.69	1.64	1.57	1.52	1.46	1.40	1.35	1.28	1.24	1.17	1.11	1.00
	2.07	**1.99**	**1.87**	**1.79**	**1.69**	**1.59**	**1.52**	**1.41**	**1.36**	**1.25**	**1.15**	**1.00**

Table K
Constants for Computing Control Chart Limits

Sample Size n	Conversion Constant for Range d_2	Standard Deviation Ratio $\sigma_w = \sigma_R/\sigma$
2	1.128	.8525
3	1.693	.8884
4	2.059	.8798
5	2.326	.8641
6	2.534	.8480
7	2.704	.8332
8	2.847	.8198
9	2.970	.8078
10	3.078	.7971
11	3.173	.7873
12	3.258	.7785

Source: Values for d_2 were obtained from American Society of Testing Materials, *A.S.T.M. Manual on Quality Control of Materials,* Philadelphia, 1951. Values for σ_w were obtained from E. S. Pearson and H. O. Hartley, *Biometrika Tables for Statisticians,* Volume I, Cambridge University Press, 1966. Reproduced with permission of the authors and publishers.

Table L

Critical Values for Studentized Range

$q_{.90}$

Parameter for Denominator	Parameter for Numerator								
	2	3	4	5	6	7	8	9	10
1	8.93	13.4	16.4	18.5	20.2	21.5	22.6	23.6	24.5
2	4.13	5.73	6.77	7.54	8.14	8.63	9.05	9.41	9.72
3	3.33	4.47	5.20	5.74	6.16	6.51	6.81	7.06	7.29
4	3.01	3.98	4.59	5.03	5.39	5.68	5.93	6.14	6.33
5	2.85	3.72	4.26	4.66	4.98	5.24	5.46	5.65	5.82
6	2.75	3.56	4.07	4.44	4.73	4.97	5.17	5.34	5.50
7	2.68	3.45	3.93	4.28	4.55	4.78	4.97	5.14	5.28
8	2.63	3.37	3.83	4.17	4.43	4.65	4.83	4.99	5.13
9	2.59	3.32	3.76	4.08	4.34	4.54	4.72	4.87	5.01
10	2.56	3.27	3.70	4.02	4.26	4.47	4.64	4.78	4.91
11	2.54	3.23	3.66	3.96	4.20	4.40	4.57	4.71	4.84
12	2.52	3.20	3.62	3.92	4.16	4.35	4.51	4.65	4.78
13	2.50	3.18	3.59	3.88	4.12	4.30	4.46	4.60	4.72
14	2.49	3.16	3.56	3.85	4.08	4.27	4.42	4.56	4.68
15	2.48	3.14	3.54	3.83	4.05	4.23	4.39	4.52	4.64
16	2.47	3.12	3.52	3.80	4.03	4.21	4.36	4.49	4.61
17	2.46	3.11	3.50	3.78	4.00	4.18	4.33	4.46	4.58
18	2.45	3.10	3.49	3.77	3.98	4.16	4.31	4.44	4.55
19	2.45	3.09	3.47	3.75	3.97	4.14	4.29	4.42	4.53
20	2.44	3.08	3.46	3.74	3.95	4.12	4.27	4.40	4.51
24	2.42	3.05	3.42	3.69	3.90	4.07	4.21	4.34	4.44
30	2.40	3.02	3.39	3.65	3.85	4.02	4.16	4.28	4.38
40	2.38	2.99	3.35	3.60	3.80	3.96	4.10	4.21	4.32
60	2.36	2.96	3.31	3.56	3.75	3.91	4.04	4.16	4.25
120	2.34	2.93	3.28	3.52	3.71	3.86	3.99	4.10	4.19
∞	2.33	2.90	3.24	3.48	3.66	3.81	3.93	4.04	4.13

$q_{.95}$

Parameter for Denominator	Parameter for Numerator								
	2	3	4	5	6	7	8	9	10
1	18.0	27.0	32.8	37.1	40.4	43.1	45.4	47.4	49.1
2	6.08	8.33	9.80	10.9	11.7	12.4	13.0	13.5	14.0
3	4.50	5.91	6.82	7.50	8.04	8.48	8.85	9.18	9.46
4	3.93	5.04	5.76	6.29	6.71	7.05	7.35	7.60	7.83
5	3.64	4.60	5.22	5.67	6.03	6.33	6.58	6.80	6.99
6	3.46	4.34	4.90	5.30	5.63	5.90	6.12	6.32	6.49
7	3.34	4.16	4.68	5.06	5.36	5.61	5.82	6.00	6.16
8	3.26	4.04	4.53	4.89	5.17	5.40	5.60	5.77	5.92
9	3.20	3.95	4.41	4.76	5.02	5.24	5.43	5.59	5.74
10	3.15	3.88	4.33	4.65	4.91	5.12	5.30	5.46	5.60
11	3.11	3.82	4.26	4.57	4.82	5.03	5.20	5.35	5.49
12	3.08	3.77	4.20	4.51	4.75	4.95	5.12	5.27	5.39
13	3.06	3.73	4.15	4.45	4.69	4.88	5.05	5.19	5.32
14	3.03	3.70	4.11	4.41	4.64	4.83	4.99	5.13	5.25
15	3.01	3.67	4.08	4.37	4.59	4.78	4.94	5.08	5.20
16	3.00	3.65	4.05	4.33	4.56	4.74	4.90	5.03	5.15
17	2.98	3.63	4.02	4.30	4.52	4.70	4.86	4.99	5.11
18	2.97	3.61	4.00	4.28	4.49	4.67	4.82	4.96	5.07
19	2.96	3.59	3.98	4.25	4.47	4.65	4.79	4.92	5.04
20	2.95	3.58	3.96	4.23	4.45	4.62	4.77	4.90	5.01
24	2.92	3.53	3.90	4.17	4.37	4.54	4.68	4.81	4.92
30	2.89	3.49	3.85	4.10	4.30	4.46	4.60	4.72	4.82
40	2.86	3.44	3.79	4.04	4.23	4.39	4.52	4.63	4.73
60	2.83	3.40	3.74	3.98	4.16	4.31	4.44	4.55	4.65
120	2.80	3.36	3.68	3.92	4.10	4.24	4.36	4.47	4.56
∞	2.77	3.31	3.63	3.86	4.03	4.17	4.29	4.39	4.47

Table L (continued)

$q_{.90}$

Parameter for Denominator	Parameter for Numerator 11	12	13	14	15	16	17	18	19	20
1	25.2	25.9	26.5	27.1	27.6	28.1	28.5	29.0	29.3	29.7
2	10.0	10.3	10.5	10.7	10.9	11.1	11.2	11.4	11.5	11.7
3	7.49	7.67	7.83	7.98	8.12	8.25	8.37	8.48	8.58	8.68
4	6.49	6.65	6.78	6.91	7.02	7.13	7.23	7.33	7.41	7.50
5	5.97	6.10	6.22	6.34	6.44	6.54	6.63	6.71	6.79	6.86
6	5.64	5.76	5.87	5.98	6.07	6.16	6.25	6.32	6.40	6.47
7	5.41	5.53	5.64	5.74	5.83	5.91	5.99	6.06	6.13	6.19
8	5.25	5.36	5.46	5.56	5.64	5.72	5.80	5.87	5.93	6.00
9	5.13	5.23	5.33	5.42	5.51	5.58	5.66	5.72	5.79	5.85
10	5.03	5.13	5.23	5.32	5.40	5.47	5.54	5.61	5.67	5.73
11	4.95	5.05	5.15	5.23	5.31	5.38	5.45	5.51	5.57	5.63
12	4.89	4.99	5.08	5.16	5.24	5.31	5.37	5.44	5.49	5.55
13	4.83	4.93	5.02	5.10	5.18	5.25	5.31	5.37	5.43	5.48
14	4.79	4.88	4.97	5.05	5.12	5.19	5.26	5.32	5.37	5.43
15	4.75	4.84	4.93	5.01	5.08	5.15	5.21	5.27	5.32	5.38
16	4.71	4.81	4.89	4.97	5.04	5.11	5.17	5.23	5.28	5.33
17	4.68	4.77	4.86	4.93	5.01	5.07	5.13	5.19	5.24	5.30
18	4.65	4.75	4.83	4.90	4.98	5.04	5.10	5.16	5.21	5.26
19	4.63	4.72	4.80	4.88	4.95	5.01	5.07	5.13	5.18	5.23
20	4.61	4.70	4.78	4.85	4.92	4.99	5.05	5.10	5.16	5.20
24	4.54	4.63	4.71	4.78	4.85	4.91	4.97	5.02	5.07	5.12
30	4.47	4.56	4.64	4.71	4.77	4.83	4.89	4.94	4.99	5.03
40	4.41	4.49	4.56	4.63	4.69	4.75	4.81	4.86	4.90	4.95
60	4.34	4.42	4.49	4.56	4.62	4.67	4.73	4.78	4.82	4.86
120	4.28	4.35	4.42	4.48	4.54	4.60	4.65	4.69	4.74	4.78
∞	4.21	4.28	4.35	4.41	4.47	4.52	4.57	4.61	4.65	4.69

$q_{.95}$

Parameter for Denominator	Parameter for Numerator 11	12	13	14	15	16	17	18	19	20
1	50.6	52.0	53.2	54.3	55.4	56.3	57.2	58.0	58.8	59.6
2	14.4	14.7	15.1	15.4	15.7	15.9	16.1	16.4	16.6	16.8
3	9.72	9.95	10.2	10.3	10.5	10.7	10.8	11.0	11.1	11.2
4	8.03	8.21	8.37	8.52	8.66	8.79	8.91	9.03	9.13	9.23
5	7.17	7.32	7.47	7.60	7.72	7.83	7.93	8.03	8.12	8.21
6	6.65	6.79	6.92	7.03	7.14	7.24	7.34	7.43	7.51	7.59
7	6.30	6.43	6.55	6.66	6.76	6.85	6.94	7.02	7.10	7.17
8	6.05	6.18	6.29	6.39	6.48	6.57	6.65	6.73	6.80	6.87
9	5.87	5.98	6.09	6.19	6.28	6.36	6.44	6.51	6.58	6.64
10	5.72	5.83	5.93	6.03	6.11	6.19	6.27	6.34	6.40	6.47
11	5.61	5.71	5.81	5.90	5.98	6.06	6.13	6.20	6.27	6.33
12	5.51	5.61	5.71	5.80	5.88	5.95	6.02	6.09	6.15	6.21
13	5.43	5.53	5.63	5.71	5.79	5.86	5.93	5.99	6.05	6.11
14	5.36	5.46	5.55	5.64	5.71	5.79	5.85	5.91	5.97	6.03
15	5.31	5.40	5.49	5.57	5.65	5.72	5.78	5.85	5.90	5.96
16	5.26	5.35	5.44	5.52	5.59	5.66	5.73	5.79	5.84	5.90
17	5.21	5.31	5.39	5.47	5.54	5.61	5.67	5.73	5.79	5.84
18	5.17	5.27	5.35	5.43	5.50	5.57	5.63	5.69	5.74	5.79
19	5.14	5.23	5.31	5.39	5.46	5.53	5.59	5.65	5.70	5.75
20	5.11	5.20	5.28	5.36	5.43	5.49	5.55	5.61	5.66	5.71
24	5.01	5.10	5.18	5.25	5.32	5.38	5.44	5.49	5.55	5.59
30	4.92	5.00	5.08	5.15	5.21	5.27	5.33	5.38	5.43	5.47
40	4.82	4.90	4.98	5.04	5.11	5.16	5.22	5.27	5.31	5.36
60	4.73	4.81	4.88	4.94	5.00	5.06	5.11	5.15	5.20	5.24
120	4.64	4.71	4.78	4.84	4.90	4.95	5.00	5.04	5.09	5.13
∞	4.55	4.62	4.68	4.74	4.80	4.85	4.89	4.93	4.97	5.01

Table L (continued)

$q_{.99}$

Parameter for Denominator	Parameter for Numerator								
	2	**3**	**4**	**5**	**6**	**7**	**8**	**9**	**10**
1	90.0	135	164	186	202	216	227	237	246
2	14.0	19.0	22.3	24.7	26.6	28.2	29.5	30.7	31.7
3	8.26	10.6	12.2	13.3	14.2	15.0	15.6	16.2	16.7
4	6.51	8.12	9.17	9.96	10.6	11.1	11.5	11.9	12.3
5	5.70	6.97	7.80	8.42	8.91	9.32	9.67	9.97	10.2
6	5.24	6.33	7.03	7.56	7.97	8.32	8.61	8.87	9.10
7	4.95	5.92	6.54	7.01	7.37	7.68	7.94	8.17	8.37
8	4.74	5.63	6.20	6.63	6.96	7.24	7.47	7.68	7.87
9	4.60	5.43	5.96	6.35	6.66	6.91	7.13	7.32	7.49
10	4.48	5.27	5.77	6.14	6.43	6.67	6.87	7.05	7.21
11	4.39	5.14	5.62	5.97	6.25	6.48	6.67	6.84	6.99
12	4.32	5.04	5.50	5.84	6.10	6.32	6.51	6.67	6.81
13	4.26	4.96	5.40	5.73	5.98	6.19	6.37	6.53	6.67
14	4.21	4.89	5.32	5.63	5.88	6.08	6.26	6.41	6.54
15	4.17	4.83	5.25	5.56	5.80	5.99	6.16	6.31	6.44
16	4.13	4.78	5.19	5.49	5.72	5.92	6.08	6.22	6.35
17	4.10	4.74	5.14	5.43	5.66	5.85	6.01	6.15	6.27
18	4.07	4.70	5.09	5.38	5.60	5.79	5.94	6.08	6.20
19	4.05	4.67	5.05	5.33	5.55	5.73	5.89	6.02	6.14
20	4.02	4.64	5.02	5.29	5.51	5.69	5.84	5.97	6.09
24	3.96	4.54	4.91	5.17	5.37	5.54	5.69	5.81	5.92
30	3.89	4.45	4.80	5.05	5.24	5.40	5.54	5.65	5.76
40	3.82	4.37	4.70	4.93	5.11	5.27	5.39	5.50	5.60
60	3.76	4.28	4.60	4.82	4.99	5.13	5.25	5.36	5.45
120	3.70	4.20	4.50	4.71	4.87	5.01	5.12	5.21	5.30
∞	3.64	4.12	4.40	4.60	4.76	4.88	4.99	5.08	5.16

Table L (continued)

$q_{.99}$

Parameter for Denominator	Parameter for Numerator									
	11	12	13	14	15	16	17	18	19	20
1	253	260	266	272	277	282	286	290	294	298
2	32.6	33.4	34.1	34.8	35.4	36.0	36.5	37.0	37.5	37.9
3	17.1	17.5	17.9	18.2	18.5	18.8	19.1	19.3	19.5	19.8
4	12.6	12.8	13.1	13.3	13.5	13.7	13.9	14.1	14.2	14.4
5	10.5	10.7	10.9	11.1	11.2	11.4	11.6	11.7	11.8	11.9
6	9.30	9.49	9.65	9.81	9.95	10.1	10.2	10.3	10.4	10.5
7	8.55	8.71	8.86	9.00	9.12	9.24	9.35	9.46	9.55	9.65
8	8.03	8.18	8.31	8.44	8.55	8.66	8.76	8.85	8.94	9.03
9	7.65	7.78	7.91	8.03	8.13	8.23	8.32	8.41	8.49	8.57
10	7.36	7.48	7.60	7.71	7.81	7.91	7.99	8.07	8.15	8.22
11	7.13	7.25	7.36	7.46	7.56	7.65	7.73	7.81	7.88	7.95
12	6.94	7.06	7.17	7.26	7.36	7.44	7.52	7.59	7.66	7.73
13	6.79	6.90	7.01	7.10	7.19	7.27	7.34	7.42	7.48	7.55
14	6.66	6.77	6.87	6.96	7.05	7.12	7.20	7.27	7.33	7.39
15	6.55	6.66	6.76	6.84	6.93	7.00	7.07	7.14	7.20	7.26
16	6.46	6.56	6.66	6.74	6.82	6.90	6.97	7.03	7.09	7.15
17	6.38	6.48	6.57	6.66	6.73	6.80	6.87	6.94	7.00	7.05
18	6.31	6.41	6.50	6.58	6.65	6.72	6.79	6.85	6.91	6.96
19	6.25	6.34	6.43	6.51	6.58	6.65	6.72	6.78	6.84	6.89
20	6.19	6.29	6.37	6.45	6.52	6.59	6.65	6.71	6.76	6.82
24	6.02	6.11	6.19	6.26	6.33	6.39	6.45	6.51	6.56	6.61
30	5.85	5.93	6.01	6.08	6.14	6.20	6.26	6.31	6.36	6.41
40	5.69	5.77	5.84	5.90	5.96	6.02	6.07	6.12	6.17	6.21
60	5.53	5.60	5.67	5.73	5.79	5.84	5.89	5.93	5.98	6.02
120	5.38	5.44	5.51	5.56	5.61	5.66	5.71	5.75	5.79	5.83
∞	5.23	5.29	5.35	5.40	5.45	5.49	5.54	5.57	5.61	5.65

Source: Portions reprinted by permission from *Biometrika Tables for Statisticians*, Vol I, by E. S. Pearson and H. O. Hartley, published by Biometrica Trustees, Cambridge University Press, Cambridge (1954). Also, portions by J. Pachera.

Table M
Critical Values of D for Kolmogorov-Smirnov Maximum Deviation Test for Goodness of Fit

The following table provides the critical values D_α corresponding to an upper-tail probability α of the test statistic D. The following relationship holds

$$\Pr[D_\alpha \leqslant D] = \alpha$$

n	$\alpha = .10$	$\alpha = .05$	$\alpha = .025$	$\alpha = .01$	$\alpha = .005$
1	.90000	.95000	.97500	.99000	.99500
2	.68377	.77639	.84189	.90000	.92929
3	.56481	.63604	.70760	.78456	.82900
4	.49265	.56522	.62394	.68887	.73424
5	.44698	.50945	.56328	.62718	.66853
6	.41037	.46799	.51926	.57741	.61661
7	.38148	.43607	.48342	.53844	.57581
8	.35831	.40962	.45427	.50654	.54179
9	.33910	.38746	.43001	.47960	.51332
10	.32260	.36866	.40925	.45662	.48893
11	.30829	.35242	.39122	.43670	.46770
12	.29577	.33815	.37543	.41918	.44905
13	.28470	.32549	.36143	.40362	.43247
14	.27481	.31417	.34890	.38970	.41762
15	.26588	.30397	.33760	.37713	.40420
16	.25778	.29472	.32733	.36571	.39201
17	.25039	.28627	.31796	.35528	.38086
18	.24360	.27851	.30936	.34569	.37062
19	.23735	.27136	.30143	.33685	.36117
20	.23156	.26473	.29408	.32866	.35241
21	.22617	.25858	.28724	.32104	.34427
22	.22115	.25283	.28087	.31394	.33666
23	.21645	.24746	.27490	.30728	.32954
24	.21205	.24242	.26931	.30104	.32286
25	.20790	.23768	.26404	.29516	.31657
26	.20399	.23320	.25907	.28962	.31064
27	.20030	.22898	.25438	.28438	.30502
28	.19680	.22497	.24993	.27942	.29971
29	.19348	.22117	.24571	.27471	.29466
30	.19032	.21756	.24170	.27023	.28987
31	.18732	.21412	.23788	.26596	.28530
30	.18445	.21085	.23424	.26189	.28094
33	.18171	.20771	.23076	.25801	.27677
34	.17909	.20472	.22743	.25429	.27279
35	.17659	.20185	.22425	.25073	.26897
36	.17418	.19910	.22119	.24732	.26532
37	.17188	.19646	.21826	.24404	.26180
38	.16966	.19392	.21544	.24089	.25843
39	.16753	.19148	.21273	.23786	.25518
40	.16547	.18913	.21012	.23494	.25205

Table M (continued)

n	$\alpha = .10$	$\alpha = .05$	$\alpha = .025$	$\alpha = .01$	$\alpha = .005$
41	.16349	.18687	.20760	.23213	.24904
42	.16158	.18468	.20517	.22941	.24613
43	.15974	.18257	.20283	.22679	.24332
44	.15796	.18053	.20056	.22426	.24060
45	.15623	.17856	.19837	.22181	.23798
46	.15457	.17665	.19625	.21944	.23544
47	.15295	.17481	.19420	.21715	.23298
48	.15139	.17302	.19221	.21493	.23059
49	.14987	.17128	.19028	.21277	.22828
50	.14840	.16959	.18841	.21068	.22604
51	.14697	.16796	.18659	.20864	.22386
52	.14558	.16637	.18482	.20667	.22174
53	.14423	.16483	.18311	.20475	.21968
54	.14292	.16332	.18144	.20289	.21768
55	.14164	.16186	.17981	.20107	.21574
56	.14040	.16044	.17823	.19930	.21384
57	.13919	.15906	.17669	.19758	.21199
58	.13801	.15771	.17519	.19590	.21019
59	.13686	.15639	.17373	.19427	.20844
60	.13573	.15511	.17231	.19267	.20673
61	.13464	.15385	.17091	.19112	.20506
62	.13357	.15263	.16956	.18960	.20343
63	.13253	.15144	.16823	.18812	.20184
64	.13151	.15027	.16693	.18667	.20029
65	.13052	.14913	.16567	.18525	.19877
66	.12954	.14802	.16443	.18387	.19729
67	.12859	.14693	.16322	.18252	.19584
68	.12766	.14587	.16204	.18119	.19442
69	.12675	.14483	.16088	.17990	.19303
70	.12586	.14381	.15975	.17863	.19167
71	.12499	.14281	.15864	.17739	.19034
72	.12413	.14183	.15755	.17618	.18903
73	.12329	.14087	.15649	.17498	.18776
74	.12247	.13993	.15544	.17382	.18650
75	.12167	.13901	.15442	.17268	.18528
76	.12088	.13811	.15342	.17155	.18408
77	.12011	.13723	.15244	.17045	.18290
78	.11935	.13636	.15147	.16938	.18174
79	.11860	.13551	.15052	.16832	.18060
80	.11787	.13467	.14960	.16728	.17949
81	.11716	.13385	.14868	.16626	.17840
82	.11645	.13305	.14779	.16526	.17732
83	.11576	.13226	.14691	.16428	.17627
84	.11508	.13148	.14605	.16331	.17523
85	.11442	.13072	.14520	.16236	.17421

Table M (continued)

n	$\alpha = .10$	$\alpha = .05$	$\alpha = .025$	$\alpha = .01$	$\alpha = .005$
86	.11376	.12997	.14437	.16143	.17321
87	.11311	.12923	.14355	.16051	.17223
88	.11248	.12850	.14274	.15961	.17126
89	.11186	.12779	.14195	.15873	.17031
90	.11125	.12709	.14117	.15786	.16938
91	.11064	.12640	.14040	.15700	.16846
92	.11005	.12572	.13965	.15616	.16755
93	.10947	.12506	.13891	.15533	.16666
94	.10889	.12440	.13818	.15451	.16579
95	.10833	.12375	.13746	.15371	.16493
96	.10777	.12312	.13675	.15291	.16408
97	.10722	.12249	.13606	.15214	.16324
98	.10668	.12187	.13537	.15137	.16242
99	.10615	.12126	.13469	.15061	.16161
100	.10563	.12067	.13403	.14987	.16081

Source: Reprinted by permission from L. H. Miller, "Table of Percentage Points of Kolmogorov Statistics," *Journal of the American Statistical Association*, 51 (1956), pages 111–121.

APPENDIX B

Bibliography

Probability

Feller, William. *An Introduction to Probability Theory and Its Applications,* Vol. 1, 3rd ed. New York: John Wiley & Sons, 1968.

Fisz, Marek. *Probability Theory and Mathematical Statistics.* New York: John Wiley & Sons, 1963.

Laplace, Pierre Simon, Marquis de. *A Philosophical Essay on Probabilities.* New York: Dover Publications, 1951.

Parzen, Emmanuel. *Modern Probability Theory and Its Applications.* New York: John Wiley & Sons, 1960.

Regression and Correlation Analysis

Draper, N. R., and H. Smith. *Applied Regression Analysis,* 2nd ed. New York: John Wiley & Sons, 1981.

Ezekial, Mordecai, and Karl A. Fox. *Methods of Correlation and Regression Analysis,* 3rd ed. New York: John Wiley & Sons, 1959.

Neter, John, William Wasserman, and Michael H. Kutner. *Applied Linear Regression Models.* Homewood, Ill.: Richard D. Irwin, 1983.

Analysis of Variance and Design of Experiments

Cochran, William G., and Gertrude M. Cox. *Experimental Designs,* 2nd ed. New York: John Wiley & Sons, 1957.

Cox, David R. *Planning of Experiments.* New York: John Wiley & Sons, 1958.

Guenther, W. C. *Analysis of Variance.* Englewood Cliffs, N.J.: Prentice-Hall, 1964.

Mendenhall, William. *An Introduction to Linear Models and the Design and Analysis of Experiments.* Belmont, Calif.: Wadsworth, 1968.

Neter, John, William Wasserman, and Michael H. Kutner. *Applied Linear Regression Models.* Homewood, Ill.: Richard D. Irwin, 1983.

Scheffe, Henry. *The Analysis of Variance.* New York: John Wiley & Sons, 1959.

Nonparametric Statistics

Bradley, James V. *Distribution-Free Statistical Tests.* Englewood Cliffs, N.J.: Prentice-Hall, 1968.

Conover, W. J. *Practical Nonparametric Statistics.* New York: John Wiley & Sons, 1971.

Gibbons, Jean D. *Nonparametric Statistical Inference.* New York: McGraw-Hill, 1970.

Hollander, M., and D. Wolfe. *Nonparametric Statistical Methods.* Boston: Houghton Mifflin Co., 1973.

Kraft, Charles H., and Constance van Eeden. *A Nonparametric Introduction to Statistics.* New York: Macmillan, 1968.

Siegel, Sidney. *Nonparametric Statistics for Behavioral Sciences.* New York: McGraw-Hill, 1956.

Decision Theory

Chernoff, Herman, and L. E. Moses. *Elementary Decision Theory.* New York: John Wiley & Sons, 1959.

Luce, R. D., and H. Raiffa. *Games and Decisions.* New York: John Wiley & Sons, 1957.

Pratt, J. W., H. Raiffa, and R. Schlaifer. *Introduction to Statistical Decision Theory.* New York: McGraw-Hill, 1965.

Raiffa, H. *Decision Analysis: Introductory Lectures on Choices Under Uncertainty.* Reading, Mass.: Addison-Wesley, 1968.

Schlaifer, R. *Analysis of Decisions Under Uncertainty.* New York: McGraw-Hill, 1969.

Schlaifer, R. *Introduction to Statistics for Business Decisions.* New York: McGraw-Hill, 1961.

Statistical Quality Control

American Society of Testing Materials. *A.S.T.M. Manual on Quality Control of Materials.* Philadelphia, 1951.

Cowden, Dudley J. *Statistical Methods in Quality Control.* Englewood Cliffs, N.J.: Prentice-Hall, 1957.

Deming, W. E. *Some Theory of Sampling.* New York: John Wiley & Sons, 1966.

Dodge, H. F., and H. G. Romig. *Sampling Inspection Tables—Single and Double Sampling*, 2nd ed. New York: John Wiley & Sons, 1959.

Enrick, N. L. *Quality Control and Reliability*, 5th ed. New York: The Industrial Press, 1966.

Grant, Eugene L., and Richard S. Leavenworth. *Statistical Quality Control*, 4th ed. New York: McGraw-Hill, 1972.

Juran, J. M. (ed.). *Quality Control Handbook*, 2nd ed. New York: McGraw-Hill, 1962.

Wadsworth, Harrison M., Kenneth S. Stephens, and B. Blanton Godfrey. *Modern Methods for Quality Control and Improvement*. New York: John Wiley & Sons, 1986.

Reliability Analysis

Amstadter, B. L. *Reliability Mathematics: Fundamentals, Practices, Procedures*. New York: McGraw-Hill, 1971.

Barlow, R. E., and F. Proschan. *Mathematical Theory of Reliability*. New York: John Wiley & Sons, 1965.

Barlow, R. E., and F. Proschan. *Statistical Theory of Reliability and Life Testing*. New York: Holt, Rinehart & Winston, 1975.

Goldberg, H. *Extending the Limits of Reliability Theory*. New York: John Wiley & Sons, 1981.

LLoyd, D. K., and M. Lipow. *Reliability: Management, Methods, and Mathematics*, 2nd ed. Redondo Beach, Calif.: the authors, 1977.

Mann, N. R., R. E. Schafer, and N. D. Singpurwalla. *Methods for Statistical Analysis of Reliability and Life Data*. New York: John Wiley & Sons, 1974.

Nelson, W. *Applied Life Data Analysis*. New York: John Wiley & Sons, 1982.

Wadsworth, Harrison M., Kenneth S. Stephens, and B. Blanton Godfrey. *Modern Methods for Quality Control and Improvement*. New York: John Wiley & Sons, 1986.

Mathematical Statistics

Fisz, Marek. *Probability Theory and Mathematical Statistics*. New York: John Wiley & Sons, 1963.

Hoel, P. G. *Introduction to Mathematical Statistics*, 3rd ed. New York: John Wiley & Sons, 1962.

Hogg, Robert V., and Allen T. Craig. *Introduction to Mathematical Statistics*, 4th ed. New York: Macmillan, 1978.

Statistical Tables

Beyer, William H. (ed.). *Handbook of Tables for Probability and Statistics*, 2nd ed. Cleveland, Ohio: The Chemical Rubber Co., 1968.

Burington, Richard S., and Donald C. May. *Handbook of Probability and Statistics with Tables*, 2nd ed. New York: McGraw-Hill, 1970.

Dodge, H. F., and H. G. Romig. *Sampling Inspection Tables—Single and Double Sampling*, 2nd ed. New York: John Wiley & Sons, 1959.

Fisher, Ronald A., and F. Yates. *Statistical Tables for Biological, Agricultural and Medical Research*, 6th ed. London: Longman Group, 1978.

Military Standard 105D. *Sampling Procedures and Tables for Inspection by Attributes.* Washington, D.C.: U.S. Government Printing Office, 1963.

Military Standard 414. *Sampling Procedures and Tables by Variables for Percent Defective.* Washington, D.C.: U.S. Government Printing Office, 1957.

Military Standard 1235 (Ord). *Single- and Multi-Level Continuous Sampling Procedures and Tables for Inspection by Attributes.* Washington, D.C.: U.S. Government Printing Office, 1962.

Pearson, E. S., and H. O. Hartley. *Biometrika Tables for Statisticians*, 3rd ed. Cambridge, England: Cambridge University Press, 1966.

The Rand Corportion. *A Million Random Digits with 100,000 Normal Deviates.* New York: The Free Press, 1955.

APPENDIX C

Answers to Selected Problems

1-1 (a)

Unit Cost (dollars/pound)	Frequency
10.00—under 11.00	1
11.00—under 12.00	8
12.00—under 13.00	16
13.00—under 14.00	9
14.00—under 15.00	6
15.00—under 16.00	4
16.00—under 17.00	3
17.00—under 18.00	2
18.00—under 19.00	1
	50

1-10 (a)

Gasoline Mileage (miles/gallon)	Frequency
10.0—under 15.0	1
15.0—under 20.0	2
20.0—under 25.0	4
25.0—under 30.0	12
30.0—under 35.0	17
35.0—under 40.0	14
	50

1-16

Depth of Well	(a) Relative Frequency	(b) Frequency	(c) Cumulative Frequency
0—under 1,000	.09	63	63
1,000—under 2,000	.26	182	245
2,000—under 3,000	.37	259	504
3,000—under 4,000	.16	112	616
4,000—under 5,000	.07	49	665
5,000—under 6,000	.03	21	686
6,000—under 7,000	.01	7	693
7,000—under 8,000	.01	7	700
	1.00	700	

1-18 (a) 3.1 (b) 3 (c) 2 (d) 2 (e) 4

1-21 (a) 8.109 (b) 6.25 (c) 2.5

1-25 (a) 29.827 (b) 29.98

1-30 (a) 1.30
(b) $\bar{X} = 3.043$ $s^2 = .1676$ $s = .409$
(c) (1) 2.695 (2) 3.110 (3) 3.338
(d) .643

1-39 (a) .47 (b) .85 (c) .26 (d) .05

1-44 (a) 3.4 (b) 3.5 (c) 4 (d) 10
(e) 6.57 (f) 2.56

1-49 (a) (1) 29.5 (2) 30.0 (3) 32.5
(b) (1) 43.75 (2) 6.61

2-1 (a) 21/38 (b) 19/38 (c) 10/38 (d) 16/38

2-4 (a) 27/55 (b) 28/55 (c) 24/55 (d) 31/55 (e) 3/55 (f) 21/55 (g) 24/55 (h) 7/55

2-6 (a) .091 (b) .60 (c) .25 (d) .143

2-11 (a) 7/10 (b) 6/9 (c) 7/9 (d) dependent

2-21 (a) (1) .81 (2) .01 (3) .99
(b) (1) .729 (2) .001 (3) .999
(c) (1) .6561 (2) .0001 (3) .9999
(d) (1) .59049 (2) .00001 (3) .99999

2-26 (a) .368 (b) .421 (c) .421 (d) .316 (e) .583 (f) .500 (g) .500 (h) .600

2-27 (a) (1) .27 (2) .49

(b)

	G	no *G*	
C	.27	.21	.48
D	.03	.49	.52
	.30	.70	1.00

(c) (1) .563 (2) .563

2-33 (b) (1) .855056 (2) .004998 (3) .139917 (4) .000029 (5) .144944 (6) .999971

2-37 (a) 15 (b) 252 (c) 1,287 (d) 6.3501×10^{11}

2-38 75,000

2-42 (a) 20! (b) 2(10!) (c) $2[10!\,(2^{10})]$ (d) 20(10!) (e) $2 \times (10!)^2$

2-45 (a) C_5^{48} (b) C_5^8/C_5^{48} (c) $4C_5^{12}/C_5^{48}$

2-46 (a) .927 (b) .20

2-50 (a) 3/10 (b) .719 (c) .885

2-8

Sample Space	Sum	Prob.
(1,1)	2	1/36
(1,2) (2,1)	3	2/36
(1,3) (3,1) (2,2)	4	3/36
(1,4) (4,1) (2,3) (3,2)	5	4/36
(1,5) (5,1) (2,4) (4,2) (3,3)	6	5/36
(1,6) (6,1) (2,5) (5,2) (3,4) (4,3)	7	6/36
(2,6) (6,2) (3,5) (5,3) (4,4)	8	5/36
(3,6) (6,3) (4,5) (5,4)	9	4/36
(4,6) (6,4) (5,5)	10	3/36
(5,6) (6,5)	11	2/36
(6,6)	12	1/36
		1

2-52

	M	*F*	
FT	25/71	14/71	39/71
PT	25/71	7/71	32/71
	50/71	21/71	1

2-58 (a) 2/3 (b) 7/12 (c) 6/7 (d) 1/6 (e) 1/12 (f) 1/8

2-61 (a) .24768 (b) .02505 (c) .27347

2-63 (a) .81 (b) .90 (c) .90 (d) .81 (e) .90 (f) .90

2-65 (a) .20 (b) .429 (c) .034

2-66 (a) 2,118,760 (b) 15,504; .0073 (c) 1,462,006; .6900

3-2

Amount	Probability
\$250,000	.05
300,000	.30
350,000	.45
400,000	.20
	1.00

3-4 (b) GPA

x	$\Pr[X = x]$
2.75	.010
3.00	.125
3.25	.355
3.50	.375
3.75	.135
	1.000

3-6 Points

x	$\Pr[X = x]$
0	.504
1	.182
2	.216
3–5	.000
6	.078
7	.014
8–13	.000
14	.006
	1.000

3-8 $E(X) = .99$ $\text{Var}(X) = .8699$ $SD(X) = .933$

3-11 $E(X) = 7.0$ $\text{Var}(X) = 5.833$ $SD(X) = 2.415$

3-12 (a) .0081 (b) .0729 (c) .5905 (d) .4095

3-17 (a) .0794 (b) .3233 (c) .5802 (d) .8817 (e) .9981

3-19 (a) .9829 (b) .1144 (c) .0051 (d) .3920 (e) .5823

3-21 (a) .1326 (b) .2707 (c) .2734 (d) .1823

3-25 (a) .0067 (b) .2240 (c) .9817 (d) .9596

3-28 (a) .9319 (b) .2378 (c) .5575 (d) .1044 (e) .0181

3-34 (a) .5421 (b) .3840

3-36 .0215

3-39 .64118

3-42 .0004952

3-44 (a) .31056 (b) .09798

3-47 (a) 1/8 (b) 1/4 (c) 1 (d) 7/16

3-49 (a) .00674 (b) .641514 (c) .148991 (d) .000480

3-50 (a) 10; 5 hours (b) 7; 3.5 hours

3-54 (a) binomial .0029 (b) Poisson .12234 (c) geometric .36973 (d) neg. binomial .00155

3-56 (a) 1 (b) 3.33 (c) 100; 69

3-59 (a) 20 (b) (1) .40126 (2) .64151 (3) .92305

3-63 (a) .0242 (b) .0019 (c) .0234

3-65 (a) .999975 (b) .99925

4-1 (a) (1) .3125 (2) .9844 (3) .1406 (4) 1
(b) $E(X) = 4/3$ $\text{Var}(X) = 2/9$

4-5 (a) (1) 6 (2) 12 (3) 3.4641 (4) .0833
(b) (1) .2 (2) .04 (3) .2 (4) .0815
(c) (1) 1.61 (2) .0742 (3) .272 (4) .4643
(d) (1) 1 (2) .1667 (3) .408 (4) .75

4-6 (a) 8 (b) 3 (c) 1 (d) 12

4-11 (a) −1.28 NM (b) −.67 NM (c) 1.04 NM (d) 1.64 NM (e) 2.33 NM

4-12 (a) (1) .0113 (2) .3001 (3) .7826 (4) between .00069 and .0008
(b) (1) 33.82″ (2) 34.29″ (3) 36.80″ (4) 38.06″

4-15 (a) .0062 (b) .2643 (c) .7888 (d) .2944

4-23 (a) .0485 (b) .2388

4-25 (a) .1292 (b) .0838 (c) .0215 (d) .6841 (e) .0043 (f) 0

4-27

r	(a)	(b)	(c)
0	.0565	.0565	.0668
1	.2569	.3134	.3085
2	.3854	.6988	.6915
3	.2372	.9360	.9332
4	.0593	.9953	.9938
5	.0047	1.0000	.99977

4-31 (a) .0067 (b) .0000 (c) .3679 (d) .0821

4-33 (a) (1) $200,000 (2) $100,000
(b) $100,000/year (c) 10 years

4-35 (a) .5934 (b) .0111 (c) .0001 (d) .0399

4-38 (a) (1) .6321 (2) .7769 (3) .9933
(b) (1) .00495 (2) .01233 (3) .02451
(c) 928

4-41 (a) .50 (b) .20 (c) .40 (d) .30

4-44 (a) 1 (b) .75 (c) .25 (d) 0

4-46 (a) ±48.82 degrees
(b) (1) .2712 (2) .5424

4-49 (a) .2642 (b) .5940 (c) .9084

4-51 (a) .7619 (b) .9862 (c) .2243

4-57 (a) 1 P.M. (b) 6:36 P.M. (c) .071

4-59 (a) (1) 34.5 (2) 26.565 (3) 5.1541
(b) .5491

4-61 (a) .2493 (b) .50 (c) .8882

4-65 (a) .0505 (b) .00255 (c) .7548

5-1 (a) 1.0 (b) 4.0

5-3 (a) (1) .50 (2) .01
(b) (1) 0 (2) 1.0000

5-6 (a) (1) 15.17 (2) 149.14
(b) (1) 27.5 (2) 72.925
(c) (1) 75.17 (2) 849.14

5-12 (a) 2/3 (2) 1/5 (3) 2/5
(b) (1) $f_X(x) = 1/5 \quad 0 \le x \le 5$
$= 0 \quad$ otherwise
(2) $f_Y(y) = 1/6 \quad 0 \le y \le 6$
$= 0 \quad$ otherwise

5-14 (a) .01128 (b) .000009 (c) .0293

5-17 (a)

x	$P_{X\mid Y}(x\mid 40)$
70	.057
75	.286
80	.086
85	.257
90	.114
95	.057
100	.000
105	.029
110	.029
115	.000
120	.086
125	.000
	1.001

5-19 (a) −6.43 (b) −.09

5-21 (a) 2.60 (b) .596

5-23 (a) (1) .0003 (2) 0 (3) .1003 (4) .0170
(b) (1) .9990 (2) 1 (3) .8485 (4) .9803

5-28 (a) (1) 3.5 (2) 2.917
(b) (1) 17.5 (2) 35 (3) 87.5
(c) (1) 14.585 (2) 29.17 (3) 72.925

5-30 (a) (1) .2643 (2) .0301 (3) .0994
(b) (1) .1056 (2) .2643 (3) .6301

5-32 (a) .7291 (b) .6993 (c) .4424

5-35 (a)

x	$P_X(x)$
1	.31
2	.30
3	.18
4	.14
5	.07
	1.00

(b)

y	$P_Y(y)$
1	.39
2	.28
3	.15
4	.11
5	.07
	1.00

(c)

x	$P_{X\|Y}(x\|1)$
1	0
2	20/39
3	10/39
4	6/39
5	3/39
	1

(d)

y	$P_{X\|Y}(y\|2)$
1	20/30
2	0
3	5/30
4	3/30
5	2/30
	1

5-36 (a) (1) 2.36 (2) 2.19 (3) 1.5504 (4) 1.5739
(5) 4.75
(b) (1) −.4184 (2) −.268

5-41 .000013

5-44 .8997

5-48 (a) .048 (b) 1.50 (c) .117 (d) .000720

5-50 $\dfrac{\pi e^u}{1 - (1 - \pi)e^u}$

6-4 (a) Evaluation (b) Planning
(c) Data Collection
(d) Evaluation and Planning (e) Evaluation
(f) Descriptive

6-5 (a) Deductive (b) Descriptive (c) Inferential
(d) Inferential (e) Inferential (f) Descriptive

6-10 (a) Nonsampling Error (b) Sampling Error
(c) Nonsampling Error
(d) Nonsampling Error

6-13 61 Bardeen, 74 Jensen, 98 Mottelson, 54 Walton, 15 Dalen, 47 Pauli, 64 Lee, 17 von Laue, 00 Richter, 53 Cockcroft

6-14 (a) Convenience (b) Judgment/Convenience
(c) Random (d) Judgment

7-1 (a) 5 lb (b) $2 (c) .004 m (d) .83 kHz

7-4

$\bar{x}$	$\Pr[\bar{X} = \bar{x}]$
65	2/15
70	3/15
75	5/15
80	3/15
85	2/15
	1

7-6 (a)

$\bar{x}$	$\Pr[\bar{X} = \bar{x}]$
2.0	4/25
2.5	8/25
3.0	8/25
3.5	4/25
4.0	1/25
	1

(b) (1) 2.8 (2) .529

7-9 (a) .4714 (b) .7888 (c) .9232 (d) .9948

7-11 (a) .05
(b) (1) .0456 (2) .0630 (3) .1587 (4) .0228

7-13 (a) .0475 (b) .0764 (c) .1587

7-15 (a) .9050 (b) .7888 (c) .6826

7-17 (a) .0833
(b) (1) .1151 (2) .0359 (3) .0359 (4) .9986

7-19 (a) .7689 (b) .0826

7-21 (a) 1.761 (b) 2.624 (c) 2.977 (d) 4.140

7-22 (a) .05 (b) .75 (c) .995 (d) .0025 (e) .3995

7-25 between .0005 and .001

7-26 (a) .8078 (b) .5987 (c) .4129

7-29 (a) .5319 (b) .2148

7-31 (a) .0778 (b) .1562 (c) .3300 (d) .5000

7-34 (a) 1.009 (b) 1.212 (c) 1.479 (d) 1.764
(e) 2.007

7-35 (a) .00000 (b) .9918 (c) .1112 (d) .0132

7-37 (a) (.10, .20) (b) (.50, .70)
(c) (.80, .90) (d) (.20, .30)

7-39 (a)

p	$\Pr[P = p]$
0	3/10
.50	6/10
1.00	1/10
	1

(b)

r	$\Pr[R = r]$
0	2/10
1	6/10
2	2/10
	1

7-41 (a) $2,007.33 (b) .7888

7-44 (a) .0668 (b) .00000

8-1 (a) No (b) No (c) Yes (d) No

8-4 (a) .71 min. (b) 1.065 min.; 21.3 miles

8-6 $SD(\bar{X}) = 4.714$ $SD(M) = 6.42$ $\bar{X}$ is more efficient.

8-10 (54.55, 60.25)

8-13 (a) ($95.51, $104.99)
(b) ($96.91, $102.59)
(c) ($99.48, $102.02)
(d) ($99.60, $101.40)

8-16 (69.39″, 71.01″)

8-17 (.206, .236)

8-21 (2.77, 11.85)

8-25 (a) (.087, .213) (b) (.023, .077)
(c) (.0604, .0896) (d) (.8262, .9238)

8-27 (a) (.002, .078) (b) .0184

8-34 (.0208, .0394)

8-36 (a) 384 (b) 68,295 (c) 753 (d) 96

8-41 (a) 250 psi (b) 24

8-43 10

8-44 (54.23, 56.37)

8-45 (a) $\bar{X} = \$104.10$ (b) $s^2 = 25.433$
(c) $P = .70$

8-47 (a) 238 (b) 138 (c) 59

8-49 (a) (11.69, 14.99) (b) (2.15, 4.82)

8-50 (.248, .523)

9-1 H_0: $\mu \leq 37$ lb/ton H_1: $\mu > 37$

9-3 (a) (1) type I error (2) type II error
(3) correct (4) correct
(b) (1) type I error (2) correct
(3) correct (4) type II error
(c) (1) type II error (2) correct
(3) type I error (4) correct

9-4 (a) $\alpha = .0764$ $\beta = .0162$

9-8 (a) Accept H_0 (continue search)
if $\bar{X} \leq 10{,}139{,}159$
Reject H_0 (use new) if $\bar{X} > 10{,}139{,}159$
(b) Accept H_0 and continue search.

9-12 (a) Accept H_0 if $\bar{X} \leq 5{,}031.3$
Reject H_0 if $\bar{X} > 5{,}031.3$
(b) Rejected

9-13 (a) H_0: $\mu = 1$ cm H_1: $\mu \neq 1$ cm
$\bar{X}_1^* = .99743$ cm $\bar{X}_2^* = 1.00257$ cm 57
(b) Reject H_0 and correct for undersized output.
(c) (.99973, 1.00487); yes;
accept H_0 and leave alone.

9-14 Reject

9-16 Yes

9-19 (a) H_0: $\mu = .55$ mHz H_1: $\mu \neq .55$ mHz
(b) (.550128, .550472); reject shipment

9-20 (b) Replace

9-24 Accept H_0 and not recommend new procedure.

9-28 (a) (.486, .624) (b) Accept

9-34 No

9-35 Accept

9-44 Yes

9-45 (a) Rejected (b) Rejected

9-47 (a) Yes (b) .001

9-49 .001

9-51 Conclude that springs do *not* meet specifications.

9-53 (a) 53 (b) 9 (c) 2,368 (d) 5

10-1 (b) $\hat{Y}(X) = 7.94 + .442X$ (c) 72.55
(d) 12.49 (e) (1) 96.34 (2) 140.54
(3) 184.74 (4) 228.94

10-4 (a) $\hat{Y}(X) = \$4{,}016.76 + 8.7116X$
(b) (1) $4,016.76 (2) $8.7116
(c) (1) $21,440 (2) $10.72

10-6 (a) $\sigma = -4{,}410 + 10{,}650{,}000\varepsilon$
(c) 10,650,000

10-9 (a) (11.30, 125.58) (b) (121.47, 177.29)

10-11 (a) (.405, .479) (b) Reject

10-14 (a) ($11,994.62, $13,462.10)
(b) ($7.9049, $9,5183)

10-16 (7,329,000, 13,972,000)

10-18 .986

10-19 (a) (.96, .99) (b) No; rejected

10-22 −.973

10-34 $\hat{Y}(X) = (3.73)X^{1.24}$

10-37 (a) $\hat{Y}(X) = .0006615(4{,}226)^{1.000/X}$

10-40 (a) $\hat{Y}(X) = 5.24 - .01007X$
(b) (1) .237 (2) .078
Yes

10-47 (a) .846 (b) .91 (c) Transformed

11-1 (a) $\hat{Y} = 161{,}813 + 4{,}024X_1 - .3532X_2$
(c) 16,275

11-2 $\hat{Y} = 581{,}468 + 4{,}982X_1 - 7{,}596X_2$
(c) 12,725

11-7 $\hat{Y} = 597{,}162 + 5{,}276X_1 + .0692X_2 - 8{,}046X_3$

11-10 $\hat{Y} = 801{,}017 + 3{,}922X_1 + .0538X_2 - 10{,}472X_3 - .2284X_4$

11-22 (a) (1) (.403, .703) (2) (−.108, .112)
(3) (−.207, −.079)
(b) Accept for B only.

11-28 (a) .126 (b) .284

11-31 (a) (1) .229 (2) .305 (3) .959
(b) (1) .224 (2) .099 (3) .941

11-35 $\hat{Y}(X) = 3{,}804 + 9.09X - .0415X^2 + .00003X^3$

11-38 $\hat{Y}(X) = -7.48 + 21{,}024.46X - 3{,}381{,}430X^2 + 173{,}234{,}902X^3$

11-39 $\hat{Y} = 103.2 + .199X_1 - .460X_2 + .020X_1^2 - .0072X_2^2 + .053X_1X_2$

11-40 (a) $\hat{Y} = 38.76 - .091X_1 + 9.36X_2$
(b) $\hat{Y} = 48.12 - .091X_1$
(c) $\hat{Y} = 38.76 - .091X_1$

11-43 (b) $\hat{Y} = 110.0 + 106.5X_1$
(c) $\hat{Y} = -104.72 + 105.65X_1 + 442.83X_2$
(d) $\hat{Y} = 427.22 + 39.16X_1 - 339.10X + 97.45X_1X_2$

11-47 (a) $\hat{Y} = 1{,}812.74 + 227.30X_1$; 1,580.01
(b) .873; 87.3% (c) ($12,293.85, $14,061.63)

11-48 (a) $\hat{Y} = 4{,}497.26 + 221.80X_1 - 63.27X_2$; 1,484.00
(b) .897; 89.7%
(c) .189 (d) ($12,221.54, $13,891.38)

11-52 (a) Yes (b) $\hat{Y} = 47.58 + 4.93X_1$
(c) $\hat{Y} = 42.20 + 7.24X_1 + 12.37X_2 - 4.11X_1X_2$

12-2 (−56.7, 134.7)

12-3 (−160.4, −41.6)

12-4 (28.0, 50.0)

12-6 ($−135.84, $−91.50)

12-7 (.60, 1.64)

12-8 (.068, .188)

12-10 Reject

12-11 Keep regular procedure.

12-12 Retain traditional process.

12-13 Reject

12-25 No

12-27 Design B

12-30 $W = 55$; Vendor B

12-32 $W = 33.0$; yes

12-33 $W = 310$; yes

12-35 (a) (1) 1/6, 435 (2) 1/6, 435 (3) 2/6, 435
(4) 3/6, 435 (5) 5/6, 435
(b) 31

12-38 $V = 1.5$; recommend alcohol.

12-40 $V = 17$; reject

12-41 $V = 93$; continue with mechanical drawing.

12-44 (a) (1) $1/2^{15}$ (2) $1/2^{15}$ (3) $1/2^{15}$ (4) $2/2^{15}$
(5) $2/2^{15}$ (6) $3/2^{15}$ (7) $4/2^{15}$ (8) $5/2^{15}$
(b) 113

12-45 (a) (−30, 74, 4.64) (b) Accepted

12-50 (a) (1.84, 10.16) (b) Rejected

12-53 $W = 93.5$; accepted

12-54 $V = 160.5$; rejected

12-56 (a) $(-2.779, .2029)$ (b) Accepted
(c) $V = 13$; accepted

13-4 (a) $\mu_{\cdot} = 34.33$
(b) $B_1 = 1.67$ $B_2 = 3.67$ $B_3 = -5.33$
10 27 $-38 = 11$

13-5 Using $4 - 1 = 3$ degrees of freedom for the numerator and $(40 - 4)4 = 36$ degrees of freedom for the denominator, the critical value is $F_{.01} = 4.38$.
(a) accepted (b) rejected (c) rejected

13-6 (a) $F_{.01} = 4.38$
(b)

Variation	Degrees of Freedom	Sum of Squares	Mean Square	F
Treatments	3	28.275	9.425	2.27
Error	36	149.700	4.158	
Total	39	177.975		

(c) Accept H_0

13-9 (a) $F_{.01} = 4.41$
(b)

Variation	Degrees of Freedom	Sum of Squares	Mean Square	F
Treatments	4	262.65	65.66	33.43
Error	18	35.35	1.96	
Total	22	298.00		

(c) Reject H_0

13-12 (a) $\mu_1 = 2.3 \pm .47$
(b) $\mu_1 - \mu_2 = -.5 \pm .67$
(c) No

13-15 (a) $\mu_1 - \mu_2 = -10.0 \pm 11.36\%$
$\mu_1 - \mu_3 = -14.0 \pm 11.36\%$
$\mu_2 - \mu_3 = -4.0 \pm 11.35\%$
(b) Only pair (1, 3), FIFO vs. priority, shows a significant difference in mean percentage of discouraged customers. FIFO yields a significantly lower mean.

13-18 (a) $\bar{X}_{\cdot 1} = 92.25\%$ $\bar{X}_{\cdot 2} = 98.20\%$ $\bar{X}_{\cdot 3} = 96.20\%$ $\bar{X}_{\cdot 4} = 90.00\%$ MSE = 1.96
$\mu_1 - \mu_2 = -5.95 \pm 3.68\%$
$\mu_1 - \mu_3 = -3.95 \pm 3.68\%$
$\mu_1 - \mu_4 = -6.25 \pm 3.88\%$
$\mu_1 - \mu_5 = 2.25 \pm 3.68\%$
$\mu_2 - \mu_3 = 2.00 \pm 3.47\%$
$\mu_2 - \mu_4 = -.30 \pm 3.68\%$
$\mu_2 - \mu_5 = 8.20 \pm 3.47\%$
$\mu_3 - \mu_4 = -2.30 \pm 3.68\%$
$\mu_3 - \mu_5 = 6.20 \pm 3.47\%$
$\mu_4 - \mu_5 = 8.50 \pm 3.68\%$

(b) For the following pairs the mean trapping percentages are significantly different:

Pair	Greater Mean	Smaller Mean
(1,2)	(2) concentrated	(1) scattered
(1,3)	(3) host plant	(1) scattered
(1,4)	(4) aerial	(1) scattered
(2,5)	(2) concentrated	(5) ground
(3,5)	(3) host plant	(5) ground
(4,5)	(4) aerial	(5) ground

13-23 $K = 16.202$ Reject H_0

14-2 (a)

Sample Means	(1) Big M	(2) H-V Eng.	(3) CC&B	(4) Jones	Factor *A* Mean
(1) Remodeling	10.0	−5.0	8.0	15.0	7.00
(2) New Bldg.	20.2	−5.0	30.0	40.0	21.30
(3) Groundwork	0.0	−1.0	2.0	4.0	1.25
Factor B Mean	10.067	−3.667	13.333	19.667	9.85

ANOVA Table

Variation	Degrees of Freedom	Sum of Squares	Mean Square	*F*
Explained by Factor *A*	2	2,558.22	1,279.11	85.123
Explained by Factor *B*	3	2,621.23	873.74	58.146
Explained by Interactions	6	1,431.86	238.64	15.881
Error or Unexplained	24	360.64	15.03	
Total	35	6,971.95		

(b) (1) $F_{.05} = 2.51$ rejected
(2) $F_{.05} = 3.40$ rejected
(3) $F_{.05} = 3.01$ rejected

14-8 For Factor *A* (mathematics preparation)
$\mu_{1\cdot} - \mu_{2\cdot} = -2.625 \pm 4.816$
$\mu_{1\cdot} - \mu_{3\cdot} = -7.125 \pm 4.816$
$\mu_{2\cdot} - \mu_{3\cdot} = -4.500 \pm 4.816$
For Factor *B* (science preparation)
$\mu_{\cdot 1} - \mu_{\cdot 2} = -2.5 \pm 6.13$
$\mu_{\cdot 1} - \mu_{\cdot 3} = -11.5 \pm 6.13$
$\mu_{\cdot 1} - \mu_{\cdot 4} = -12.0 \pm 6.13$
$\mu_{\cdot 2} - \mu_{\cdot 3} = -9.0 \pm 6.13$
$\mu_{\cdot 2} - \mu_{\cdot 4} = -10.5 \pm 6.13$
$\mu_{\cdot 3} - \mu_{\cdot 4} = -.5 \pm 6.13$

For mathematics preparation *A*, calculus (3) provides significantly higher mean scores than trigonometry (1). Analysis (2) is not significantly different from the other two. For science preparation *B*, both chemistry and physics (3) and advanced placement (4) result in higher mean scores than either chemistry only (1) or physics, but no other significant differences apply.

14-11 (b) Only the 27 pairs having $\bar{X}_{ij}$'s differing (in absolute value) by more than 15.64 are significantly different.
$F_{.05} = 2.22$
$-25.84 \le \mu_{11} - \mu_{21} \le 5.44$ days

(c) (1) $L_1 = (\mu_{11} + \mu_{12} + \mu_{13} + \mu_{14} + \mu_{21} + \mu_{22} + \mu_{23} + \mu_{24})/8 - (\mu_{31} + \mu_{32} + \mu_{33} + \mu_{34})/4$
$= 12.90 \pm 6.77$ days
(2) $L_2 = (\mu_{11} + \mu_{21} + \mu_{31} + \mu_{13} + \mu_{23} + \mu_{33})/6 - (\mu_{12} + \mu_{22} + \mu_{32} + \mu_{14} + \mu_{24} + \mu_{34})/6$
$= 3.70 \pm 6.39$ days
(3) $L_3 = (\mu_{12} + \mu_{22})/2 - (\mu_{11} + \mu_{21} + \mu_{13} + \mu_{23} + \mu_{14} + \mu_{24})/6$
$= -25.53 \pm 9.03$ days

14-25

Variation	Degrees of Freedom	Sum of Squares	Mean Square	F
Treatments (Compiler)	3	12.75	4.25	2.04
Columns (Application)	3	520.25	173.42	
Rows (Length)	3	70.25	23.42	
Error	6	12.50	2.08	
Total	15	615.75		

Accept H_0

15-2 (a) (1) 1 (2) 6.635 (3) accepted
(b) (1) 8 (2) 15.507 (3) accepted
(c) (1) 9 (2) 16.919 (3) accepted
(d) (1) 2 (2) 9.210 (3) rejected

15-5 $\chi^2 = 15.826$; rejected

15-6 $\chi^2 = 13.997$; reject

15-8 $\chi^2 = 11.782$; different

15-10 Accepted for (a, b) and rejected for (c, d).

15-12 (a) $R_a = 6$ $n_a = 27$ $n_b = 73$; rejected
(b) Yes

15-13 (a) $R_a = 11$ $n_a = 54$ $n_b = 46$; rejected
(b) $R_a = 38$ $n_a = 52$ $n_b = 48$; rejected

15-16 (a) .3192 (b) Yes

15-17 $\chi^2 = 13.568$; yes

15-21 (a) $\bar{X} = 50.54$ $s = 2.206$
(b)

Upper Limit for Interval	$\hat{f}$	
46	1.97	5.48
47	3.51	
48	7.03	
49	11.69	
50	15.93	
51	18.19	
52	16.22	
53	12.32	
54	7.32	
55	3.65	5.82
56	1.51	
57	.49	
∞	.17	
	100.00	

(c) 5.188; accepted

15-23 $\lambda = 2.558$; $\chi^2 = 241.628$; rejected

15-32 $R_a = 27$ $n_a = 50$ $n_b = 50$; accepted

15-34 $\chi^2 = 33.416$; rejected

15-37 $\chi^2 = 2.137$; accepted

15-38 Not random for machine 1 and random for machines 2 and 3.

16-2 (a) LCL = .042 UCL = .090
(b) 7/06 and 7/07
(c) LCL = .038 UCL = .084
(1) out (2) out (3) in
(4) in (5) in

16-6 (a) (1) .8770 (2) .1230
(b) (1) .96773 (2) .03227
(c) (1) .9306 (2) .0694
(d) (1) .99636 (2) .00364

16-10 (a) LCL = .24264 UCL = .25948
2 and 4 are out-of-control.
(b) LCL = 0 UCL = .031
All in control.
(c) (1) LCL = .24277 UCL = .25993
(2) LCL = 0 UCL = .0315

16-11 (a) .4602 (b) .9850 (c) .5987 (d) .22974

16-12 (a) 17 (b) 16 (c) 15 (d) 14

16-15 (a) .9999 (b) .7660 (c) .1172 (d) .0001

17-1 (a) .0821 (b) .0067 (c) .000045

17-4 (a) (1) .4776 (2) .4519 (3) .4373
(b) (1) .01482 (2) .00794 (3) .00551

17-9 (a) .3024 (b) .9976

17-13 (a) .0369 (b) .1114 (c) .4297

17-17 (a) The current reliability is .5833. $R_s(t) = .8749$, a 29.16% increase. The net payoff is \$191.60.
(b) The optimal choice is to add 2 units, with net payoff \$226.70.

17-20 $-11.30 \leq A \leq -6.28$
$.0000124 \leq \lambda \leq .00187$

17-25 $\hat{\beta} = 1.28$ $\hat{\lambda} = .0018135$

18-3 (b) Conduct the model test. Accept if it is positive and request modification if it is negative.

18-5 (a) 6,250 for A and 7,000 for B
(b) Machine A

18-6 (b) Contracting

18-10 A_4

18-11 (b) (1) .785 (2) .067

18-13 (b) (1) .766 (2) .583

18-16 (b) \$80,000

18-19 (b) 0

18-20 (a) \$64,000 for accept (the maximum) and \$62,500 for reject.
(b) \$2,000; yes
(d) 2; sampling would be better.

18-22 (a) .4 day
(b) Sampling will not help.

18-26 (a) $\mu = 2$ hours $\sigma = .49$ hour
(b) (1) .8461 (2) .0207 (3) .6400
(4) .1112

Index